Lecture Notes in Computer Science 7317

Commenced Publication in 1973
Founding and Former Series Editors:
Gerhard Goos, Juris Hartmanis, and Jan van Leeuwen

Alessandro Cimatti   Roberto Sebastiani (Eds.)

Theory and Applications of Satisfiability Testing – SAT 2012

15th International Conference
Trento, Italy, June 17-20, 2012
Proceedings

 Springer

Volume Editors

Alessandro Cimatti
Fondazione Bruno Kessler
Center for Information Technology
via Sommarive 18, Povo, 38123 Trento, Italy
E-mail: cimatti@fbk.eu

Roberto Sebastiani
University of Trento
Dept. of Information Engineering and Computer Science
via Sommarive 14, Povo, 38123 Trento, Italy
E-mail: roberto.sebastiani@disi.unitn.it

ISSN 0302-9743   e-ISSN 1611-3349
ISBN 978-3-642-31611-1   e-ISBN 978-3-642-31612-8
DOI 10.1007/978-3-642-31612-8
Springer Heidelberg Dordrecht London New York

Library of Congress Control Number: 2012941226

CR Subject Classification (1998): F.3.1, F.3, F.1, F.2, I.2, B.7

LNCS Sublibrary: SL 1 – Theoretical Computer Science and General Issues

Typesetting: Camera-ready by author, data conversion by Scientific Publishing Services, Chennai, India

Printed on acid-free paper

Springer is part of Springer Science+Business Media (www.springer.com)

Preface

This volume contains the papers presented at SAT 2012, the 15th International Conference on Theory and Applications of Satisfiability Testing, held during June 16–20 in Trento, Italy. SAT 2012 was co-organized and hosted by Fondazione Bruno Kessler (FBK) and the University of Trento (UniTN), Italy.

The SAT series originated in 1996 as a series of workshops, and later developed into the primary annual meeting for researchers studying the propositional satisfiability problem. Importantly, here SAT is interpreted in a rather broad sense: besides plain propositional satisfiability, it includes the domains of MaxSAT and Pseudo-Boolean (PB) constraints, Quantified Boolean Formulae (QBF), Satisfiability Modulo Theories (SMT), Constraints Programming (CSP) techniques for word-level problems and their propositional encoding. To this extent, many hard combinatorial problems can be encoded as SAT instances, in the broad sense mentioned above, including problems that arise in hardware and software verification, AI planning and scheduling, OR resource allocation, etc. The theoretical and practical advances in SAT research over the past 20 years have contributed to making SAT technology an indispensable tool in these domains. The topics of the conference span practical and theoretical research on SAT (in the broader sense above) and its applications, and include, but are not limited to, theoretical issues, solving and advanced functionalities, and applications.

SAT 2012 hosted two workshops: CSPSAT 2012 (Second International Workshop on the Cross-Fertilization Between CSP and SAT), and PoS 2012 (Third International Workshop on Pragmatics of SAT), and four competitive events: Max-SAT 2012 (7th Max-SAT Evaluation), PB12 (Pseudo-Boolean Competition 2012), QBFEVAL 2012 (QBF Competition 2012), and SAT Challenge 2012.

In SAT 2012 we introduced for the first time the possibility of submitting tool-presentation papers, and of directly submitting poster-presentation papers (2-page abstracts). Overall there were 112 submissions (88 full, 10 tool, and 14 poster papers). Each submission was reviewed by at least three Program Committee members; for the fist time for SAT, the review process also involved a rebuttal phase. The committee decided to accept 52 papers (29 full, 7 tool and 16 poster papers). Note that seven full papers were accepted as posters.

The program also included two remarkable invited talks:

- *Aaron Bradley* from the University of Colorado at Boulder, presented "Understanding IC3"
- *Donald Knuth* from Stanford University presented "Satisfiability and The Art of Computer Programming"

Given the interest of the scientific community outside SAT for the work of Donald Knuth, his talk was open to non-SAT 2012 attendees, and included a question-answering session on general topics in computer science.

SAT 2012 was co-located with the Second International SAT/SMT Summer School, with a program over four days that hosted 16 speakers. The school gave many students the opportunity to attend SAT 2012.

Our first thanks go to the Program Committee members and to the additional reviewers, who did a thorough and knowledgeable job and enabled the assembly of this body of high-quality work.

We thank the authors for their submissions, and for their collaboration in further improving their papers. A special thank goes to our invited speakers, Aaron Bradley and Donald Knuth, for accepting our invitation and for their very stimulating contributions.

We thank the organizers of the school, of the workshops and of the competitive events: Alberto Griggio and Stefano Tonetta for SAT/SMT School, Yael Ben Haim and Yehuda Naveh for CSPSAT 2012, Daniel Le Berre and Allen Van Gelder for PoS 2012, Josep Argelich, Chu Min Li, Felip Manya and Jordi Planes for Max-SAT 2012, Vasco Manquinho and Olivier Roussel for PB12, Massimo Narizzano for QBFEVAL 2012, Adrian Balint, Anton Belov, Matti Jarvisalo and Carsten Sinz for SAT Challenge 2012.

A special thank goes to Martina Lorenzi, Silvia Malesardi, Moira Osti, and to all the other members of the Ufficio Eventi of FBK and Ufficio Convegni of UniTN, who largely contributed to the success of this event.

We also thank the developers and maintainers of the EasyChair conference management system, which was of great help with the paper submission, reviewing, discussion, and with the assembly of the proceedings.

We gratefully acknowledge the generous contributions of our sponsors (in alphabetical order): IBM Research, Intel^{TM} Corporation, Jasper Technologies, Microsoft Research INRIA, Microsoft Research, NEC, plus the support of FBK, of UniTN and of the SAT Association. SAT 2012 was held also under the auspices of TrentoRise and of the European Association for Theoretical Computer Science, Italian Chapter.

May 2012   Alessandro Cimatti
                                                      Roberto Sebastiani

Organization

Program Committee

Dimitris Achlioptas	UC Santa Cruz, USA
Fahiem Bacchus	University of Toronto, Canada
Paul Beame	University of Washington, USA
Armin Biere	Johannes Kepler University, Austria
Randal Bryant	Carnegie Mellon University, USA
Uwe Bubeck	University of Paderborn, Germany
Alessandro Cimatti	FBK-Irst, Trento, Italy
Nadia Creignou	Université d'Aix-Marseille, France
Leonardo De Moura	Microsoft Research, USA
John Franco	University of Cincinnati, USA
Malay Ganai	NEC Labs America, USA
Enrico Giunchiglia	DIST - Univ. Genova, Italy
Youssef Hamadi	Microsoft Research, UK
Ziyad Hanna	University of Oxford, UK
Marijn Heule	TU Delft, The Netherlands
Holger Hoos	University of British Columbia, Canada
Kazuo Iwama	Kyoto University, Japan
Oliver Kullmann	Swansea University, UK
Daniel Le Berre	CRIL-CNRS UMR 8188 / Université d'Artois, France
Ines Lynce	IST/INESC-ID, Technical University of Lisbon, Portugal
Panagiotis Manolios	Northeastern University, USA
Joao Marques-Silva	University College Dublin, Ireland
David Mitchell	Simon Fraser University, Canada
Alexander Nadel	Intel, Israel
Jussi Rintanen	NICTA, Australia
Lakhdar Sais	CRIL - CNRS, Université Lille Nord, France
Karem A. Sakallah	University of Michigan, USA
Roberto Sebastiani	DISI, University of Trento, Italy
Bart Selman	Cornell University, USA
Laurent Simon	LRI - Univ. Paris Sud, France
Carsten Sinz	Karlsruhe Institute of Technology (KIT), Germany
Ofer Strichman	Technion, Israel
Stefan Szeider	Vienna University of Technology, Austria
Niklas Sörensson	Chalmer University of Technology, Sweden

Allen Van Gelder   University of California, Santa Cruz, USA
Toby Walsh   NICTA and UNSW, Australia
Xishun Zhao   Insitute of Logic and Cognition, Sun Yat-Sen
                                 University, China

Additional Reviewers

Aavani, Amir
Audemard, Gilles
Bailleux, Olivier
Barbanchon, Régis
Bayless, Sam
Belov, Anton
Bjork, Magnus
Bordeaux, Lucas
Bueno, Denis
Chauhan, Pankaj
Chen, Zhenyu
Ciardo, Gianfranco
Compton, Kevin
Davies, Jessica
Dequen, Gilles
Dilkina, Bistra
Een, Niklas
Egly, Uwe
Elenbogen, Dima
Erdem, Ozan
Ermon, Stefano
Fawcett, Chris
Fichte, Johannes Klaus
FranzÈn, Anders
Gao, Sicun
Goldberg, Eugene
Goultiaeva, Alexandra
Griggio, Alberto
Grundy, Jim
Gwynne, Matthew
Harrison, John
Hjort, HÂkan
Huang, Jinbo
Hugel, Thomas
Ivrii, Alexander
Jabbour, Said
Janota, Mikolas
Jha, Susmit
Jovanovic, Dejan
Järvisalo, Matti

Kleine Büning, Hans
Kroc, Lukas
Krstic, Sava
Lettmann, Theodor
Liedloff, Mathieu
Liffiton, Mark
Lundgren, Lars
Manquinho, Vasco
Martins, Ruben
Marx, Dániel
Matsliah, Arie
Mazure, Bertrand
Meier, Arne
Merz, Florian
Oliveras, Albert
Ordyniak, Sebastian
Ostrowski, Richard
Papavasileiou, Vasilis
Piette, Cedric
Ramanujan, Raghuram
Roussel, Olivier
Ryvchin, Vadim
Sabharwal, Ashish
Seger, Carl-Johan
Seidl, Martina
Semerjian, Guilhem
Shen, Yuping
Sinha, Nishant
Slivovsky, Friedrich
Soos, Mate
Styles, James
Tamaki, Sugur
Tamaki, Suguru
Tompkins, Dave
Urquhart, Alasdair
Veksler, Michael
Wintersteiger, Christoph
Wintersteiger, Christoph M.
Wolfovitz, Guy
Xu, Lin

Table of Contents

Complexity Analysis

Circuits and Encodings

Tool Presentations

Poster Presentations

Understanding IC3*

Aaron R. Bradley

ECEE Department, University of Colorado at Boulder
bradleya@colorado.edu

Abstract. The recently introduced model checking algorithm, IC3, has proved to be among the best SAT-based safety model checkers. Many implementations now exist. This paper provides the context from which IC3 was developed and explains how the originator of the algorithm understands it. Then it draws parallels between IC3 and the subsequently developed algorithms, FAIR and IICTL, which extend IC3's ideas to the analysis of ω-regular and CTL properties, respectively. Finally, it draws attention to certain challenges that these algorithms pose for the SAT and SMT community.

1 Motivation

In *Temporal Verification of Reactive Systems: Safety*, Zohar Manna and Amir Pnueli discuss two strategies for strengthening an invariant property to be inductive [13]: "(1) Use a stronger assertion, or (2) Conduct an incremental proof, using previously established invariants." They "strongly recommend" the use of the second approach "whenever applicable," its advantage being "modularity." Yet they note that it is not always applicable, as a conjunction of assertions can be inductive when none of its components, on its own, is inductive. In this paper, the first method is referred to as "monolithic"—all effort is focused on producing one strengthening formula—while the second method is called "incremental."

1.1 Monolithic and Incremental Proof Methods

A simple pair of transition systems clarifies the two strategies and the limitations of the second:

```
x, y := 1, 1                 1      x, y := 1, 1                 1
while *:                     2      while *:                     2
    x, y := x + 1, y + x     3          x, y := x + y, y + x     3
```

The star-notation indicates nondeterminism. Suppose that one wants to prove, for both systems, that $P : y \geq 1$ is invariant.

Consider the first system. To attempt to prove the invariant property P, one can apply induction:

---

* Work supported in part by the Semiconductor Research Corporation under contract GRC 2271.

A. Cimatti and R. Sebastiani (Eds.): SAT 2012, LNCS 7317, pp. 1–14, 2012.

– It holds initially because

$$\underbrace{x = 1 \wedge y = 1}_{\text{initial condition}} \Rightarrow \underbrace{y \geq 1}_{P} .$$

– But it does not hold at line 3 because

$$\underbrace{y \geq 1}_{P} \wedge \underbrace{x' = x + 1 \wedge y' = y + x}_{\text{transition relation}} \not\Rightarrow \underbrace{y' \geq 1}_{P'} .$$

The first step of an inductive proof of an invariant property is sometimes called *initiation*; the second, *consecution* [13]. In this case, consecution fails. Hence, an *inductive strengthening* of P must be found.

The first step in strengthening P is to identify why induction fails. Here, it's obvious enough: without knowing that x is nonnegative, one cannot know that y never decreases. The assertion $\varphi_1 : x \geq 0$ is inductive:

– it holds initially: $x = 1 \wedge y = 1 \Rightarrow x \geq 0$, and
– it continues to hold at line 3, where x is updated:

$$\underbrace{x \geq 0}_{\varphi_1} \wedge \underbrace{x' = x + 1 \wedge y' = y + x}_{\text{transition relation}} \Rightarrow \underbrace{x' \geq 0}_{\varphi_1'} .$$

Now $P : y \geq 1$ is inductive *relative to* φ_1 because consecution succeeds in the presence of φ_1:

$$\underbrace{x \geq 0}_{\varphi_1} \wedge \underbrace{y \geq 1}_{P} \wedge \underbrace{x' = x + 1 \wedge y' = y + x}_{\text{transition relation}} \Rightarrow \underbrace{y' \geq 1}_{P'} .$$

This use of "previously established invariants" makes for an "incremental proof": first establish φ_1; then establish P using φ_1. Here, each assertion is simple and discusses only one variable of the system. The inductive strengthening of P : $y \geq 1$ is thus $x \geq 0 \wedge y \geq 1$. Of course, the stronger assertion $x \geq 1 \wedge y \geq 1$ would work as well.

In the second transition system, neither $x \geq 0$ nor $y \geq 1$ is inductive on its own. For example, consecution fails for $x \geq 0$ because of the lack of knowledge about y:

$$x \geq 0 \wedge x' = x + y \wedge y' = y + x \not\Rightarrow x' \geq 0 .$$

Establishing $y \geq 1$ requires establishing the two assertions together:

– initiation: $x = 1 \wedge y = 1 \Rightarrow x \geq 0 \wedge y \geq 1$
– consecution: $x \geq 0 \wedge y \geq 1 \wedge x' = x + y \wedge y' = y + x \Rightarrow x' \geq 0 \wedge y' \geq 1$.

An incremental proof seems impossible in this case, as only the conjunction of the two assertions is inductive, not either on its own. Thus, for this system, one must invent the inductive strengthening of P all at once: $x \geq 0 \wedge y \geq 1$.

Notice that the assertion $x \geq 0 \wedge y \geq 1$ is inductive for the first transition system as well and so could have been proposed from the outset. However, especially in more realistic settings, an incremental proof is simpler than inventing a single inductive strengthening, when it is possible.

1.2   Initial Attempts at Incremental, Inductive Algorithms

IC3 is a result of asking the question: if the incremental method is often better for humans, might it be better for algorithms as well? The first attempt at addressing this question was in the context of linear inequality invariants. Previous work had established a constraint-based method of generating individual inductive linear inequalities [7]. Using duality in linear programming, the constraint-based method finds instantiations of the parameters $a_0, a_1, \ldots, a_n$ in the template

$$a_0 x_0 + a_1 x_1 + \cdots + a_{n-1} x_{n-1} + a_n \geq 0$$

that result in inductive assertions. A practical implementation uses previously established invariants when generating a new instance [17]. However, an enumerative algorithm generates the strongest possible over-approximation—for that domain—of the reachable state space, which may be far stronger than what is required to establish a given property.

A property-directed, rather than enumerative, approach is to guide the search for inductive instances with counterexamples to the inductiveness (CTIs) of the given property [5]. A CTI is a state (more generally, a set of states represented by a cube; that is, a conjunction of literals) that is a counterexample to consecution. In the first system above, consecution fails for $P : y \geq 1$:

$$y \geq 1 \wedge x' = x + 1 \wedge y' = y + x \not\Rightarrow y' \geq 1 \ .$$

A CTI, returned by an SMT solver, is $x = -1 \wedge y = 1$. Until this state is eliminated, P cannot be established. The constraint system for generating an inductive instance of the template $ax + by + c \geq 0$ is augmented by the constraint $a(-1) + b(1) + c < 0$. In other words, the generated inductive assertion should establish that the CTI $x = -1 \wedge y = 1$ is unreachable. If no such assertion exists, other CTIs are examined instead. The resulting lemmas may be strong enough that revisiting this CTI will reveal an assertion that is inductive relative to them, finally eliminating the CTI. But in this example, the instance $x \geq 0$ ($a = 1, b = 0, c = 0$) is inductive and eliminates the CTI.

In the context of hardware model checking, this approach was developed into a complete model checker, called FSIS, for invariance properties [4]. Rather than linear inequality assertions, it generates clauses over latches. While the algorithm for generating strong inductive clauses is not trivial, understanding it is not essential for understanding the overall model checking algorithm, which is simple. The reader is thus referred to previous papers to learn about the clause-generation algorithm [4, 3]. Consider finite-state system $S : (\bar{i}, \bar{x}, I(\bar{x}), T(\bar{x}, \bar{i}, \bar{x}'))$ with primary inputs $\bar{i}$, state variables (latches) $\bar{x}$, a propositional formula $I(\bar{x})$ describing the initial configurations of the system, and a propositional formula $T(\bar{x}, \bar{i}, \bar{x}')$ describing the transition relation, and suppose that one desires to establish the invariance of assertion P. First, the algorithm checks if P is inductive with two SAT queries, for initiation and consecution, respectively:

$$I \Rightarrow P \qquad \text{and} \qquad P \wedge T \Rightarrow P' \ .$$

If they hold, P is invariant. If the first query fails, P is falsified by an initial state, and so it does not hold. If consecution fails—the likely scenario—then there is a state s that can lead in one step to an error; s is a CTI.

The inductive clause generation algorithm then attempts to find a clause c that is inductive and that is falsified by s. If one is found, c becomes an incremental lemma, φ_1, relative to which consecution is subsequently checked:

$$\varphi_1 \wedge P \wedge T \Rightarrow P' .$$

If consecution still fails, another CTI t is discovered, and again the clause generation algorithm is applied. This time, however, the generated clause need only be inductive relative to φ_1, in line with Manna's and Pnueli's description of incremental proofs. In this manner, a list of assertions, $\varphi_1, \varphi_2, \ldots, \varphi_k$, is produced, each inductive relative to its predecessors, until $P \wedge \bigwedge_i \varphi_i$ is inductive.

But what is to be done if no clause exists that both eliminates s and is inductive? In this case, the target is expanded: the error states grow from $\neg P$ to $\neg P \vee s$; said otherwise, the property to establish becomes $P \wedge \neg s$. Every CTI is handled in this way: either a relatively inductive clause is generated to eliminate it, or it is added to the target. The algorithm is complete for finite-state systems.

One important, though subtle, point in applying the incremental method is that the invariance property, P, that is to be established can be assumed when generating new inductive assertions. That is, a generated assertion need only be inductive relative to P itself. For suppose that auxiliary information ψ is inductive relative to P, and P is inductive relative to ψ:

$$\psi \wedge P \wedge T \Rightarrow \psi' \qquad \text{and} \qquad \psi \wedge P \wedge T \Rightarrow P' .$$

Then clearly $\psi \wedge P$ itself is inductive.

In the second transition system of Section 1.1, this extra information makes a difference. Consider consecution again for P:

$$y \geq 1 \wedge x' = x + y \wedge y' = y + x \Rightarrow y' \geq 1 .$$

It fails with, for example, the CTI $x = -1 \wedge y = 1$. While $x \geq 0$ eliminates this CTI, it is not inductive on its own. However, it is inductive relative to P:

$$\underbrace{y \geq 1}_{P} \wedge \underbrace{x \geq 0}_{\varphi_1} \wedge \underbrace{x' = x + y \wedge y' = y + x}_{\text{transition relation}} \Rightarrow \underbrace{x' \geq 0}_{\varphi_1'} .$$

By assuming P, an incremental proof is now possible. Once sufficient strengthening information is found, this seemingly circular reasoning straightens into an inductive strengthening.

However, this trick does not fundamentally strengthen the incremental proof methodology. There are still many situations in which the purely incremental approach is impossible.[1] Experiments with FSIS made it clear that this weakness had to be addressed.

---

[1] Consider, for example, a similar transition relation with three variables updated according to **x, y, z := x + y, y + z, z + x**. Neither $x \geq 0$ nor $z \geq 0$ is inductive relative to $P : y \geq 1$.

1.3   Other SAT-Based Approaches

This section considers the strengths and weaknesses, which motivate IC3, of other SAT-based approaches.

At one extreme are solvers based on backward search. Exact SAT-based symbolic model checking computes the set of states that can reach an error, relying on cube reduction to accelerate the analysis [14]. Conceptually, it uses the SAT solver to find a predecessor, reduces the resulting cube, and then blocks the states of that cube from being explored again. At convergence, the blocking clauses describe the weakest possible inductive strengthening of the invariant. Sequential SAT similarly reduces predecessor cubes, but it also reduces state cubes lacking unexplored predecessors via the implication graph of the associated (unsatisfiable) SAT query [12]. This latter approach computes a convenient inductive strengthening—not necessarily the weakest or the strongest. FSIS is like this latter method, except that, when possible, it uses induction to reduce a predecessor state cube, which can allow the exploration of backward paths to end earlier than in sequential SAT, besides producing stronger clauses.

The strength of pure backward search is that it does not tax the SAT solver. Memory is not an issue. Its weakness is that the search is blind with respect to the initial states. In the case of FSIS, its selection of new proof obligations is also too undirected; some predecessors trigger more informative lemmas than others, but FSIS has no way of knowing which. Perhaps because of this lack of direction, successful modern SAT-based model checkers, other than IC3, derive from BMC [1]. BMC is based on unrolling the transition relation between the initial and error states. Thus, the SAT solver considers both ends in its search.

While BMC is strong at finding counterexamples, it is practically incomplete. Interpolation (ITP) [15] and k-induction [18] address this practical incompleteness. The latter combines BMC (which becomes initiation) with a consecution check in which the transition relation is unrolled k times and the property is asserted at each non-final level. When that check fails, k is increased; in a finite-state context, there is a k establishing P if P is invariant. In practice, the sufficient k is sometimes small, but it can also be prohibitively large. Like exact model checking, k-induction cannot find a convenient strengthening; rather, its strengthening is based on a characteristic of the transition system.

ITP goes further. Rather than unrolling from the initial states (BMC) or applying induction directly (k-induction), it unrolls from the current frontier F_i, which contains at least all states at most i steps from an initial state. If the associated SAT query is unsatisfiable, the algorithm extracts an interpolant between F_i and the k-unrolling leading to a violation of P, which serves as the $(i + 1)$-step over-approximation F_{i+1}. If the query is satisfiable, the algorithm increases k, yielding a finer over-approximating post-condition computation. The size of the unrolling that yields a proof can be smaller in practice than that of k-induction. Tuning the interpolant finder can allow it to find convenient assertions, potentially accelerating convergence to some inductive strengthening.

BMC-based approaches have the advantage of giving meaningful consideration to both initial and error states. However, they have the disadvantage of being monolithic. They search for a single, often complex, strengthening, which can require many unrollings in practice, overwhelming the SAT solver.

IC3 addresses the weaknesses of both types of solvers while maintaining their strengths. Like the backward search-based methods, it relies on many simple SAT queries (Section 2.1) and so requires relatively little memory in practice. Like the BMC-based methods, it gives due consideration to the initial and error states (Section 2.2). It can be run successfully for extended periods, and—for the same reasons—it is parallelizable. Compared to FSIS, it uses the core idea of incrementally applying relative induction but applies it in a context in which every state cube is inductively generalizable. Hence, induction becomes an even more powerful method for reducing cubes in IC3.

2   IC3

Manna's and Pnueli's discussion of incremental proofs is in the context of manual proof construction, where the ingenuity of the human is the only limitation to the discovery of intermediate lemmas. In algorithms, lemma generation is typically restricted to some abstract domain [8] such as linear inequalities [9] or a fixed set of predicates [10]. Thus, the case in which a CTI cannot be eliminated through the construction of a relatively inductive assertion arises all too frequently, making FSIS, in retrospect, a rather naive algorithm.

The goal in moving beyond FSIS was to preserve its incremental character while addressing the weakness of backward search and the weakness of the incremental proof method: the common occurrence of mutually inductive sets of assertions that cannot be linearized into incremental proofs. In other words, what was sought was an algorithm that would smoothly transition between Manna's and Pnueli's incremental methodology, when possible, and monolithic inductive strengthening, when necessary.

This section discusses IC3 from two points of view: IC3 as a prover and IC3 as a bug finder. It should be read in conjunction with the formal treatment provided in the original paper [3]. Readers who wish to see IC3 applied to a small transition system are referred to [19].

2.1   Perspective One: IC3 as a Prover

IC3 maintains a sequence of stepwise over-approximating sets, $F_0 = I, F_1, F_2, \ldots, F_k, F_{k+1}$, where each set F_i over-approximates the set of states reachable in at most i steps from an initial state. Every set except F_{k+1} is a subset of P: $F_i \Rightarrow P$. Once F_k is refined so that it excludes all states that can reach a $\neg P$-state in one transition, F_{k+1}, too, is strengthened to be a subset of P by conjoining P to it. F_k is considered the "frontier" of the analysis. A final characteristic of these sets is that $F_i \wedge T \Rightarrow F_{i+1}'$. That is, all successors of F_i-states are F_{i+1}-states.

This description so far should be relatively familiar. Forward BDD-based reachability [16], for example, computes exact i-step reachability sets, and if any such set ever includes a $\neg P$-state, the conclusion is that the property does not hold. ITP also computes i-step reachability sets, and like IC3's, they are over-approximating. However, when ITP encounters an over-approximating set that contains a $\neg P$-state, it refines its approximate post-image operator by further unrolling the transition relation, rather than addressing the weaknesses of the current stepwise sets directly. The crucial difference in the use of these sets between IC3 and ITP is that IC3 refines all of the sets throughout its execution.[2]

Putting these properties together reveals two characteristics of the reach sets. First, any state reachable in i steps is an F_i-state. Second, any F_i-state cannot reach a $\neg P$-state for at least $k - i + 1$ steps. For example, an F_{k+1}-state can actually be a $\neg P$-state, and an F_k-state may reach an error in one step. But an F_{k-1}-state definitely cannot transition to a $\neg P$-state (since $F_{k-1} \wedge T \Rightarrow F_k'$ and $F_k \Rightarrow P$).

Now, the property to check is whether P is inductive relative to F_k. Since $F_k \Rightarrow P$, the following query, corresponding to consecution for P relative to F_k, is executed:

$$F_k \wedge T \Rightarrow P' \, . \tag{1}$$

Suppose that the query succeeds and that F_k is itself inductive: $F_k \wedge T \Rightarrow F_k'$. Then F_k is an inductive strengthening of P that proves P's invariance.

Now suppose that the query succeeds but that F_k is not inductive. F_{k+1} can be strengthened to $F_{k+1} \wedge P$, since all successors of F_k-states are P-states. Additionally, a new frame F_{k+2} is introduced. IC3 brings in monolithic inductive strengthening by executing a phase of what can be seen as a simple predicate abstraction (propagateClauses [3]). Every clause that occurs in any F_i is treated as a predicate. A clause's occurrence in F_i means that it holds for at least i steps. This phase allows clauses to propagate forward from their current positions. Crucially, subsets of clauses can propagate forward together, allowing the discovery of mutually inductive clauses. For i ranging from 1 to k, IC3 computes the largest subset $C \subseteq F_i$ of clauses such that the following holds (consecution for C relative to F_i):

$$F_i \wedge T \Rightarrow C' \, .$$

These clauses C are then conjoined to F_{i+1}. Upon completion, F_{k+1} becomes the new frontier. Many of the stepwise sets may be improved as lemmas are propagated forward in time. If $F_k = F_{k+1}$, then F_k is inductive, which explains how F_k is determined to be inductive in the case above.

Finally, suppose that query (1) fails, revealing an F_k-state s (more generally, a cube of F_k-states) that can reach a $\neg P$-state in one transition; s is a CTI. In other words, the problem is not just that F_k is not inductive; the problem is that it is not even strong enough to rule out a $\neg P$-successor, and so more

---

[2] Of course, one might implement ITP to reuse previous over-approximating sets, so that it too could be seen to refine them throughout execution. Similarly, one might use transition unrolling in IC3. *But for completeness*, ITP relies on unrolling but not continual refinement of all stepwise sets, whereas IC3 relies on continual refinement of all stepwise sets but not unrolling.

reachability information must be discovered. IC3 follows the incremental proof methodology in this situation: it uses induction to find a lemma showing that s cannot be reached in k steps from an initial state. This lemma may take the form of a single clause or many clauses, the latter arising from analyzing transitive predecessors of s.

Ideally, the discovered lemma will prove that s cannot *ever* be reached. Less ideally, the lemma will be good enough to get propagated to future time frames once P becomes inductive relative to F_k. But at worst, the lemma will at least exclude s from frame F_k, and even F_{k+1}[3].

Specifically, IC3 first seeks a clause c whose literals are a subset of those of $\neg s$ and that is inductive relative to F_k; that is, it satisfies initiation and consecution:

$$I \Rightarrow c \qquad \text{and} \qquad F_k \wedge c \wedge T \Rightarrow c' \, .$$

Such a clause proves that s cannot be reached in $k + 1$ steps. However, there may not be any such clause. It may be the case that a predecessor t exists that is an F_k-state and that eliminates the possibility of a relatively inductive clause. In fact, t could even be an F_{k-1}-state.[4]

Here is where IC3 is vastly superior to FSIS, and where it sidesteps the fundamental weakness of the incremental proof method. A failure to eliminate s at F_k is not a problem. Suppose that a clause c is found relative to F_{k-2} rather than F_{k-1} or F_k (the worst case):

$$I \Rightarrow c \qquad \text{and} \qquad F_{k-2} \wedge c \wedge T \Rightarrow c' \, .$$

Because c is inductive relative to F_{k-2}, it is added to F_{k-1}: no successor of an $(F_{k-2} \wedge c)$-state is a $\neg c$-state. If even with this update to F_{k-1}, s still cannot be eliminated through inductive generalization (the process of generating a relatively inductive subclause of $\neg s$), then the failing query

$$F_{k-1} \wedge \neg s \wedge T \Rightarrow \neg s' \tag{2}$$

reveals a predecessor t that was *irrelevant for F_{k-2} but is a reason why inductive generalization fails relative to F_{k-1}*. This identification of a reason for failure of inductive generalization is one of IC3's insights. The predecessor is identified *after* the generation of c relative to F_{k-2} so that c focuses IC3 on predecessors of s that matter for F_{k-1}. The predecessor t is not just any predecessor of s: it is specifically one that prevents s's inductive generalization at F_{k-1}. IC3 thus has a meaningful criterion for choosing new proof obligations.

Now IC3 focuses on t until eventually a clause is produced that is inductive relative to frame F_{k-2} and that eliminates t as a predecessor of s through frame F_{k-1}. Focus can then return to s, although t is not forgotten. Inductive generalization of s relative to F_{k-1} may succeed this time; and if it does not, the newly discovered predecessor would again be a reason for its failure.

---

[3] The clause eventually generated for s relative to F_k strengthens F_{k+1} since it is inductive relative to F_k.

[4] However, it cannot be an F_{k-2}-state, for then s would be an F_{k-1}-state, and its successor $\neg P$-state an F_k-state. But it is known that $F_k \Rightarrow P$.

It is important that, during the recursion, all transitive predecessors of s be analyzed all the way through frame F_k. This analysis identifies mutually inductive (relative to F_k) sets of clauses. Only one of the clauses may actually eliminate s, but the clauses will have to be propagated forward together since they support each other. It may be the case, though, that some clauses are too specific, so that the mutual support breaks down during the clause propagation phase. This behavior is expected. As IC3 advances the frontier, it forces itself to consider ever more general situations, until it finally discovers the real reasons why s is unreachable. It is this balance between using stepwise-specific information and using induction to be as general as possible that allows IC3 to synthesize the monolithic and incremental proof strategies into one strategy.

2.2   Perspective Two: IC3 as a Bug Finder

Although IC3 is often inferior to BMC for finding bugs quickly, industry experience has shown that IC3 can find deep bugs that other formal techniques cannot. This section presents IC3 as a guided backward search. While heuristics for certain decision points may improve IC3's performance, the basic structure of the algorithm is already optimized for finding bugs. In particular, IC3 considers both initial and error states during its search. The following discussion develops a hypothetical, but typical, scenario for IC3, one which reveals the motivation behind IC3's order of handling proof obligations. Recall from the previous section that IC3 is also intelligent about choosing new proof obligations.

Suppose that query (1) revealed state s, which was inductively generalized relative to F_{k-2}; that query (2) revealed t as an F_{k-1}-state predecessor of s; and that t has been inductively generalized relative to F_{k-3}. At this point, IC3 has the proof obligations $\{(s,\ k-1),\ (t,\ k-2)\}$, indicating that s and t must next be inductively generalized relative to F_{k-1} and F_{k-2}, respectively. As indicated in the last section, neither state will be forgotten until it is generalized relative to F_k, even if s happens to be generalized relative to F_k first.

At this point, with the proof obligations $\{(s,\ k-1),\ (t,\ k-2)\}$, it is fairly obvious that until t is addressed, IC3 cannot return its focus to s; t would still cause problems for generalizing s relative to F_{k-1}. While focusing on t, suppose that u is discovered as a predecessor of t during an attempt to generalize t relative to F_{k-2}. Although t cannot be generalized relative to F_{k-2}, u may well be; it is, after all, a different state with different literals. Indeed, it may even be generalizable relative to F_k. In any case, suppose that it is generalizable relative to F_{k-2} but not higher, resulting in one more proof obligation: $(u,\ k-1)$. Overall, the obligations are now $\{(s,\ k-1),\ (t,\ k-2),\ (u,\ k-1)\}$.

The question IC3 faces is which proof obligation to consider next. It turns out that correctness requires considering the obligation $(t,\ k-2)$ first [3]. Suppose that t and u are mutual predecessors.[5] Were $(u,\ k-1)$ treated first, t could

---

[5] The current scenario allows it. For t's generalization relative to F_{k-3} produced a clause at F_{k-2} that excludes t, which means that the generalization of u relative to F_{k-2} ignores t. Therefore, it is certainly possible for u to be generalized at F_{k-2}, leaving obligation $(u,\ k-1)$.

be discovered as an F_{k-1}-state predecessor of u, resulting in a duplicate proof obligation and jeopardizing termination. But one might cast correctness aside and argue that u should be examined first anyway—perhaps it is "deeper" than s or t given that it is a predecessor of t.

Actually, the evidence is to the contrary. The obligations $(u, \ k - 1)$ and $(t, \ k - 2)$ show that u is at least k steps away from an initial state (recall that u has been eliminated from F_{k-1}), whereas t is at least only $k - 1$ steps away. That is, IC3's information predicts that t is "closer" to an initial state than is u and so is a better state to follow to find a counterexample trace.

Thus, there are two characteristics of a proof obligation $(a, \ i)$ to consider: (1) the length of the suffix of a, which leads to a property violation, and (2) the estimated proximity, $i + 1$, to an initial state. In the example above, s, t, and u have suffixes of length 0, 1, and 2, respectively; and their estimated proximities to initial states are k, $k - 1$, and k, respectively. Both correctness and intuition suggest that pursuing a state with the lowest proximity is the best bet. In the case that multiple states have the lowest proximity, one can heuristically choose from those states the state with the greatest suffix length (for "depth") or the shortest suffix length (for "short" counterexamples)—or apply some other heuristic.

From this perspective, IC3 employs inductive generalization as a method of dynamically updating the proximity estimates of states that lead to a violation of the property. Inductive generalization provides for not only the update of the proximities of explicitly observed CTI states but also of many other states. When $F_k \wedge T \Rightarrow P'$ holds, all proximity estimates of $\neg P$-predecessors are $k + 1$, and so another frame must be added to continue the guided search.

The bug-finding and proof-finding perspectives agree on a crucial point: even if the initial CTI s has been inductively generalized relative to F_k, its transitive predecessors should still be analyzed through F_k in order to update their and related states' proximity estimates. A consequence of this persistence is that IC3 can search deeply even when k is small.

3   Beyond IC3: Incremental, Inductive Verification

Since IC3, the term *incremental, inductive verification* (IIV) has been coined to describe algorithms that use induction to construct lemmas in response to property-driven hypotheses (e.g., the CTIs of FSIS and IC3). Two significant new incremental, inductive model checking algorithms have been introduced. One, called FAIR, addresses ω-regular (e.g., LTL) properties [6]. Another, called IICTL, addresses CTL properties [11]. While this section does not describe each in depth, it attempts to draw meaningful parallels among IC3, FAIR, and IICTL. Most superficially, FAIR uses IC3 to answer reachability queries, and IICTL uses IC3 and FAIR to address reachability and fair cycle queries, respectively.

An IIV algorithm can be characterized by (1) the form of its hypotheses, (2) the form of its lemmas, (3) how it uses induction, and (4) the basis of generalization. For IC3, these characterizations are as follows:

1. *Hypotheses*: Counterexamples to induction (CTIs). When consecution fails, the SAT solver returns a state explaining its failure, which IC3 then inductively generalizes, possibly after generating and addressing further proof obligations.
2. *Lemmas*: Clauses over state variables. A clause is generated in response to a CTI, using only the negation of the literals of the CTI.
3. *Induction*: Lemmas are inductive relative to stepwise information.
4. *Generalization*: Induction guides the production of minimal clauses—clauses that do not have any relatively inductive subclauses. The smaller the clause, the greater is the generalization; hence, induction is fundamental to generalization in IC3.

FAIR searches for reachable fair cycles, or "lasso" counterexamples. The fundamental insight of FAIR is that SCC-closed sets can be described by sequences of inductive assertions. In other words, an inductive assertion is a barrier across the state space which the system can cross in only one direction. A transition from one side to the other is a form of progress, since the system can never return to the other side. FAIR is characterized as an IIV algorithm as follows:

1. *Hypotheses*: Skeletons. A skeleton is a set of states that together satisfy all Büchi fairness conditions and that all appear on one side of every previously generated barrier. The goal is to connect the states into a "lasso" through reachability queries.
2. *Lemmas*: An inductive assertion. Each lemma provides one of two types of information: (1) global reachability information, which is generated when IC3 shows that a state of a skeleton cannot be reached; (2) SCC information, which is generated when IC3 shows that one state of the skeleton cannot reach another. In the latter case, all subsequent skeletons must be chosen from one "side" of the assertion.
3. *Induction*: SCC-closed sets are discovered via inductive assertions.
4. *Generalization*: Proofs constructed by IC3 can be refined to provide stronger global reachability information or smaller descriptions of one-way barriers. Furthermore, new barriers are generated relative to previous ones and transect the entire state space, not just the "arena" from which the skeleton was selected. Exact SCC computation is not required.

IICTL considers CTL properties hierarchically, as in BDD-based model checking [16], but rather than computing exact sets for each node, it incrementally refines under- and over-approximations of these sets. When a state is undecided for a node—that is, it is in the over-approximation but not in the under-approximation—its membership is decided via a set of SAT (for EX nodes), reachability (for EU nodes), or fair cycle (for EG nodes) queries. IICTL is characterized as an IIV algorithm as follows:

1. *Hypotheses*: A state is undecided for a node if it is included in the upper-bound but excluded from the lower-bound. If it comes up during the analysis, its status for the node must be decided.
2. *Lemmas*: Lemmas refine the over- and under-approximations of nodes, either introducing new states into under-approximations or removing states from over-approximations.

3. *Induction*: Induction is used to answer the queries for EU and EG nodes.
4. *Generalization*: Generalization takes two forms. For negatively answered queries, the returned proofs are refined to add (remove) as many states as possible to the under-approximations (from the over-approximations) of nodes, rather than just the motivating hypothesis state. For positively answered queries, the returned traces are generalized through a "forall-exists" generalization procedure, again to decide as many states as possible in addition to the hypothesis state.

All three algorithms are "lazy" in that they only respond to concrete hypotheses but are "eager" in that they generalize from specific instances to strong lemmas. Furthermore, all hypotheses are derived from the given property, so that the algorithms' searches are property-directed.

4   Challenges for SAT and SMT Solvers

The queries that IIV methods pose to SAT solvers differ significantly in character from those posed by BMC, k-induction, or ITP. There is thus an opportunity for SAT and SMT research to directly improve the performance of IIV algorithms.

IC3 is the first widespread verification method that requires highly efficient incremental solvers. An incremental interface for IC3 must allow single clauses to be pushed and popped; it must also allow literal assumptions. IIV algorithms pose many thousands to millions of queries in the course of an analysis, and so speed is crucial. FAIR requires even greater incrementality: the solver must allow sets of clauses to be pushed and popped.

IIV methods use variable orders to direct the generation of inductive clauses. An ideal solver would use these variable orders to direct the identification of the core assumptions or to direct the lifting of an assignment.

The inductive barriers produced in FAIR provide opportunities for generalization (in the *cycle queries* [6]) but are not required for completeness. Using all such barriers overwhelms the solver, yet using too few reduces opportunities for generalization. Therefore, currently, a heuristic external to the solver decides whether to use a new barrier or not. Ideally, a solver would provide feedback on whether a group of clauses has been used or not for subsequent queries. Those clause groups that remain unused for several iterations of FAIR would be removed. This functionality would allow direct identification of useful barriers.

IIV algorithms gradually learn information about a system in the form of lemmas. Thus, a core set of constraints, which includes the transition relation, grows and is used by every worker thread. On a multi-core machine, replicating this set of constraints in each thread's solver instance uses memory—and memory bandwidth—poorly, and this situation will grow worse as the number of available cores grows. An ideal solver for IIV algorithms would provide access to every thread to a growing core set of constraints. Each thread would then have a thread-specific view in which to push and pop additional information for incremental queries.

Finally, IC3 has shown itself to be highly sensitive to the various behaviors of SAT solvers. Swapping one solver for another, or even making a seemingly innocuous adjustment to the solver, can cause widely varying performance—even if the average time per SAT call remains about the same. For example, a SAT solver that is too deterministic can cause IC3 to dwell on one part of the state space by returning a sequence of similar CTIs, so that IC3 must generate more lemmas. Identifying the desirable characteristics of a solver intended for IC3 will be of great help.

5   What's Next?

Fundamentally, IC3 should not be seen as a clause-generating algorithm. Rather, the insight of IC3 is in how it can harness seemingly weak abstract domains to produce complex inductive strengthenings. At first glance, it seems that IC3's abstract domain is CNF over state variables. In fact, the abstract domain is *conjunctions of state variables* (over the inverse transition relation; see next paragraph), which is, practically speaking, the simplest possible domain.

If that perspective seems unclear, think about how IC3 works: to address CTI s, it performs what is essentially a simple predicate abstraction over the inverse of the transition relation, where the predicates are the literals of s. This process produces a cube $d \subseteq s$—that is, a conjunction of a subset of the predicates—that lacks $\neg d$-predecessors (within a stepwise context F_i); therefore, the clause $\neg d$ is inductive (relative to F_i). It is the incremental nature of IC3 that produces, over time, a conjunction of clauses.

Recalling the linear inequality domain [5] seems to confuse the issue, however. Where are the disjunctions in that context? To understand it, consider the polyhedral domain [9]. If it were to be used in the same way as the state variable domain of IC3, then each CTI would be analyzed with a full polyhedral analysis, each lemma would take the form of a disjunction of linear inequalities, and IC3 would produce proofs in the form of a CNF formula of linear inequalities. That approach would usually be unnecessarily expensive. Clearly, the polyhedral domain is not being used. Instead, the domain is much simpler: it is a domain of half-spaces.[6]

Therefore, the next step, in order to achieve word-level hardware or software model checking, is to introduce new abstract domains appropriate for IC3—domains so simple that they could not possibly work outside the context of IC3, yet sufficiently expressive that IC3 can weave together their simple lemmas into complex inductive strengthenings.

Acknowledgments. Thanks to Armin Biere, Zyad Hassan, Fabio Somenzi, and Niklas Een for insightful discussions that shaped my thinking on how better to explain IC3, and to the first three for reading drafts of this paper.

---

[6] A Boolean clause can be seen as a half-space over a Boolean hypercube. The author first pursued inductive clause generation (in FSIS) because of the parallel with linear inequalities.

References

[1] Biere, A., Cimatti, A., Clarke, E., Zhu, Y.: Symbolic Model Checking without BDDs. In: Cleaveland, W.R. (ed.) TACAS 1999. LNCS, vol. 1579, pp. 193–207. Springer, Heidelberg (1999)

[2] Bradley, A.R.: k-step relative inductive generalization. Technical report, CU Boulder (March 2010), http://arxiv.org/abs/1003.3649

[3] Bradley, A.R.: SAT-Based Model Checking without Unrolling. In: Jhala, R., Schmidt, D. (eds.) VMCAI 2011. LNCS, vol. 6538, pp. 70–87. Springer, Heidelberg (2011)

[4] Bradley, A.R., Manna, Z.: Checking safety by inductive generalization of counterexamples to induction. In: FMCAD (November 2007)

[5] Bradley, A.R., Manna, Z.: Verification Constraint Problems with Strengthening. In: Barkaoui, K., Cavalcanti, A., Cerone, A. (eds.) ICTAC 2006. LNCS, vol. 4281, pp. 35–49. Springer, Heidelberg (2006)

[6] Bradley, A.R., Somenzi, F., Hassan, Z., Zhang, Y.: An incremental approach to model checking progress properties. In: FMCAD (November 2011)

[7] Colón, M.A., Sankaranarayanan, S., Sipma, H.B.: Linear Invariant Generation Using Non-linear Constraint Solving. In: Hunt Jr., W.A., Somenzi, F. (eds.) CAV 2003. LNCS, vol. 2725, pp. 420–432. Springer, Heidelberg (2003)

[8] Cousot, P., Cousot, R.: Abstract interpretation: A unified lattice model for static analysis of programs by construction or approximation of fixpoints. In: POPL (1977)

[9] Cousot, P., Halbwachs, N.: Automatic discovery of linear restraints among variables of a program. In: POPL (January 1978)

[10] Graf, S., Saidi, H.: Construction of Abstract state Graphs with PVS. In: Grumberg, O. (ed.) CAV 1997. LNCS, vol. 1254, pp. 73–83. Springer, Heidelberg (1997)

[11] Hassan, Z., Bradley, A.R., Somenzi, F.: Incremental, inductive CTL model checking. In: CAV (July 2012)

[12] Lu, F., Iyer, M.K., Parthasarathy, G., Wang, L.-C., Cheng, K.-T., Chen, K.C.: An Efficient Sequential SAT Solver With Improved Search Strategies. In: DATE (2005)

[13] Manna, Z., Pnueli, A.: Temporal Verification of Reactive Systems: Safety. Springer, New York (1995)

[14] McMillan, K.L.: Applying SAT Methods in Unbounded Symbolic Model Checking. In: Brinksma, E., Larsen, K.G. (eds.) CAV 2002. LNCS, vol. 2404, pp. 250–264. Springer, Heidelberg (2002)

[15] McMillan, K.L.: Interpolation and SAT-Based Model Checking. In: Hunt Jr., W.A., Somenzi, F. (eds.) CAV 2003. LNCS, vol. 2725, pp. 1–13. Springer, Heidelberg (2003)

[16] McMillan, K.L.: Symbolic Model Checking. Kluwer, Boston (1994)

[17] Sankaranarayanan, S., Sipma, H.B., Manna, Z.: Scalable Analysis of Linear Systems Using Mathematical Programming. In: Cousot, R. (ed.) VMCAI 2005. LNCS, vol. 3385, pp. 25–41. Springer, Heidelberg (2005)

[18] Sheeran, M., Singh, S., Stålmarck, G.: Checking Safety Properties Using Induction and a SAT-Solver. In: Johnson, S.D., Hunt Jr., W.A. (eds.) FMCAD 2000. LNCS, vol. 1954, pp. 108–125. Springer, Heidelberg (2000)

[19] Somenzi, F., Bradley, A.R.: IC3: Where monolithic and incremental meet. In: FMCAD (November 2011)

Satisfiability and
The Art of Computer Programming

Donald Knuth

Stanford University, CA, USA

Abstract. The speaker will describe his adventures during the past eight months when he has been intensively studying as many aspects of SAT solvers as possible, in the course of preparing about 100 pages of new material for Volume 4B of *The Art of Computer Programming*.

A. Cimatti and R. Sebastiani (Eds.): SAT 2012, LNCS 7317, p. 15, 2012.

Choosing Probability Distributions for Stochastic Local Search and the Role of Make versus Break

Adrian Balint and Uwe Schöning

Ulm University
Institute of Theoretical Computer Science
89069 Ulm, Germany
{adrian.balint,uwe.schoening}@uni-ulm.de

Abstract. Stochastic local search solvers for SAT made a large progress with the introduction of probability distributions like the ones used by the SAT Competition 2011 winners Sparrow2010 and EagleUp. These solvers though used a relatively complex decision heuristic, where probability distributions played a marginal role.

In this paper we analyze a pure and simple probability distribution based solver probSAT, which is probably one of the simplest SLS solvers ever presented. We analyze different functions for the probability distribution for selecting the next flip variable with respect to the performance of the solver. Further we also analyze the role of *make* and *break* within the definition of these probability distributions and show that the general definition of the *score* improvement by flipping a variable, as *make* minus *break* is questionable. By empirical evaluations we show that the performance of our new algorithm exceeds that of the SAT Competition winners by orders of magnitude.

1 Introduction

The propositional satisfiability problem (SAT) is one of the most studied $\mathcal{NP}$-complete problems in computer science. One reason is the wide range of SAT's practical applications ranging from hardware verification to planning and scheduling. Given a propositional formula in conjunctive normal form (CNF) with variables $\{x_1, \ldots, x_N\}$ the SAT-problem consists in finding an assignment for the variables such that all clauses are satisfied.

Stochastic local search (SLS) solvers operate on complete assignments and try to find a solution by flipping variables according to a given heuristic. Most SLS-solvers are based on the following scheme: Initially, a random assignment is chosen. If the formula is satisfied by the assignment the solution is found. If not, a variable is chosen according to a (possibly probabilistic) variable selection heuristic, which we further call *pickVar*. The heuristics mostly depend on some score, which counts the number of satisfied/unsatisfied clauses, as well as other aspects like the "age" of variables, and others. It was believed that a good

A. Cimatti and R. Sebastiani (Eds.): SAT 2012, LNCS 7317, pp. 16–29, 2012.

flip heuristic should be designed in a very sophisticated way to obtain a really efficient solver. We show in the following that it is worth to "come back to the roots" since a very elementary and (as we think) elegant design principle for the *pickVar* heuristic just based on probability distributions will do the job extraordinary well.

It is especially popular (and successful) to pick the flip variable from an unsatisfied clause. This is called *focussed* local search in [11,14]. In each round, the selected variable is flipped and the process starts over again until a solution is eventually found. Depending on the heuristic used in *pickVar* SLS-solvers can be divided into several categories like GSAT, WalkSAT, and dynamic local search (DLS).

Most important for the flip heuristic seems to be the *score* of an assignment, i.e. the number of satisfied clauses. Considering the process of flipping one variable, we get the *relative score change* produced by a candidate variable for flipping as: (*score after flipping* minus *score before flipping*) which is equal to (*make* minus *break*). Here *make* means the number of newly satisfied clauses which come about by flipping the variable, and *break* means the number of clauses which become false by flipping the respective variable. To be more precise we will denote $make(x, \mathbf{a})$ and $break(x, \mathbf{a})$ as functions of the respective flip variable x and the actual assignment $\mathbf{a}$ (before flipping). Notice that in case of focussed flipping mentioned above the value of *make* is always at least 1.

Most of the SLS solvers so far, if not all, follow the strategy that whenever the score improves by flipping a certain variable from an unsatisfied clause, they will indeed flip this variable without referring to probabilistic decisions. Only if no improvement is possible as is the case in local minima, a probabilistic strategy is performed, which is often specified by some decision procedure. The winner of the SAT Competition 2011 category random SAT, Sparrow, mainly follows this strategy but when it comes to a probabilistic strategy it uses a probability distribution function instead of a decision procedure [2]. The probability distribution in Sparrow is defined as an exponential function of the *score*. In this paper we analyze several simple SLS solvers that use only probability distributions within their search.

2   The New Algorithm Paradigm

We propose a new class of solvers here, called probSAT, which base their probability distributions for selecting the next flip variable solely on the *make* and *break* values, but not necessarily on the value of (*make* minus *break*), as it was the case in Sparrow. Our experiments indicate that the influence of *make* should be kept rather weak – it is even reasonable to ignore *make* completely, like in implementations of WalkSAT. The role of make and break in these SLS-type algorithms should be seen in a new light. The new type of algorithm presented here can also be applied for general constraint satisfaction problems and works as follows.

Algorithm 1. ProbSAT

> **Input** : Formula F, $maxTries$, $maxFlips$
> **Output**: satisfying assignment **a** or UNKNOWN
> 1 **for** $i = 1$ *to* $maxTries$ **do**
> 2   **a** ← randomly generated assignment
> 3   **for** $j = 1$ *to* $maxFlips$ **do**
> 4   **if** *(**a** is model for F)* **then**
> 5   return **a**
> 6 C_u ← randomly selected unsat clause
> 7   **for** x *in* C_u **do**
> 8   compute $f(x, \mathbf{a})$
> 9 var ← random variable x according to probability $\frac{f(x,\mathbf{a})}{\sum_{z \in C_u} f(z,\mathbf{a})}$
> 10   flip(var)
> 11 return UNKNOWN;

The idea here is that the function f should give a high value to variable x if flipping x seems to be advantageous, and a low value otherwise. Using f the probability distribution for the potential flip variables is calculated. The flip probability for x is proportional to $f(x, \mathbf{a})$. Letting f be a constant function leads in the k-SAT case to the probabilities $(\frac{1}{k}, \ldots, \frac{1}{k})$ morphing the probSAT algorithm to the random walk algorithm that is theoretically analyzed in [12]. In all our experiments with various functions f we made f depend on $break(x, \mathbf{a})$ and possibly on $make(x, \mathbf{a})$, and no other properties of x and $\mathbf{a}$. In the following we analyze experimentally the effect of several functions to be plugged in for f.

2.1 An Exponential Function

First we considered an exponential decay, 2-parameter function:

$$f(x, \mathbf{a}) = \frac{(c_m)^{make(x,\mathbf{a})}}{(c_b)^{break(x,\mathbf{a})}}$$

The parameters are c_b and c_m. Because of the exponential functions used here (think of $c^x = e^{\frac{1}{T}x}$) this is reminiscence of the way Metropolis-like algorithms (see [14]) select a variable. We call this the *exp-algorithm*. Notice that we separate into the two base constants c_m and c_b which allow us to find out whether there is a different influence of the make and the *break* value – and there is, indeed.

It seems reasonable to try to maximize *make* and to minimize *break*. Therefore, we expect $c_m > 1$ and $c_b > 1$ to be good choices for these parameters. Actually, one might expect that c_m should be identical to c_b such that the above formula simplifies to $c^{make-break} = c^{scorechange}$ for an appropriate parameter c.

To get a picture on how the performance of the solver varies for different values of c_m and c_b, we have done a uniform sampling of $c_b \in [1.0, 4.0]$ and

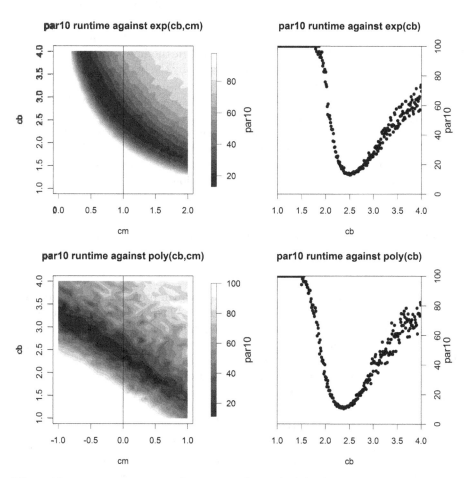

Fig. 1. Parameter space performance plot: The left plots show the performance of different combinations of c_b and c_m for the exponential (upper left corner) and the polynomial (lower left corner) functions. The darker the area the better the runtime of the solver with that parameter settings. The right plots show the performance variation if we ignore the make values (correspond to the cut in the left plots) by setting $c_m = 1$ for the exponential function and $c_m = 0$ for the polynomial function.

of $c_m \in [0.1, 2.0]$ for this exponential function and of $c_m \in [-1.0, 1.0]$ for the polynomial function (see below). We have then ran the solver with the different parameter settings on a set of randomly generated 3-SAT instances with 1000 variables at a clause to variable ratio of 4.26. The cutoff limit was set to 10 seconds. As a performance measure we use *par10*: penalized average runtime, where a timeout of the solver is penalized with 10·(cutoff limit). A parameter setting where the solver is not able to solve anything has a par10 value of 100.

In the case of 3-SAT a very good choice of the parameters is $c_b > 1$ (as expected) and $c_m < 1$ (totally unexpected), for example, $c_b = 3.6$ and $c_m = 0.5$ (see Figure 1 left upper diagram and the survey in Table 1) with a small variation depending on the considered set of benchmarks. In the interval $c_m \in [0.3, 1.8]$ the optimal choice of parameters can be described by the hyperbola-like function $(c_b - 1.3) \cdot c_m = 1.1$. Almost optimal results were also obtained if c_m is set to 1 (and c_b to 2.5), see Figure 1, both upper diagrams. In other words, the value of *make* is not taken into account in this case.

As mentioned, it turns out that the influence of make is rather weak, therefore it is reasonable, and still leads to very good algorithms – also because the implementation is simpler and has less overhead – if we ignore the make-value completely and consider the one-parameter function:

$$f(x, \mathbf{a}) = (c_b)^{-break(x,\mathbf{a})}$$

We call this the *break-only-exp-algorithm*.

2.2   A Polynomial Function

Our experiments showed that the exponential decay in probability with growing *break*-value might be too strong in the case of 3-SAT. The above formulas have an exponential decay in probability comparing different (say) *break*-values. The relative decay is the same when we compare $break = 0$ with $break = 1$, and when we compare, say, $break = 5$ with $break = 6$. A "smoother" function for high values would be a polynomial decay function. This led us to consider the following, 2-parameter function ($\epsilon = 1$ in all experiments):

$$f(x, \mathbf{a}) = \frac{(make(x, \mathbf{a}))^{c_m}}{(\epsilon + break(x, \mathbf{a}))^{c_b}}$$

We call this the *poly-algorithm*. The best parameters in case of 3-SAT turned out to be $c_m = -0.8$ (notice the minus sign!) and $c_b = 3.1$ (See Figure 1, lower part). In the interval $c_m \in [-1.0, 1.0]$ the optimal choice of parameters can be described by the linear function $c_b + 0.9c_m = 2.3$. Without harm one can set $c_m = 0$, and then take $c_b = 2.3$, and thus ignore the make-value completely.

Ignoring the make-value (i.e. setting $c_m = 0$) brings us to the function

$$f(x, \mathbf{a}) = (\epsilon + break(x, \mathbf{a}))^{-c_b}$$

We call this the *break-only-poly-algorithm*.

2.3   Some Remarks

As mentioned above, in both cases, the exp- and the poly-algorithm, it was a good choice to ignore the make-value completely (by setting $c_m = 1$ in the exp-algorithm, or by setting $c_m = 0$ in the poly-algorithm). This corresponds to the vertical lines in Figure 1, left diagrams. But nevertheless, the optimal choice in both cases, was to set $c_m = 0.5$ and $c_b = 3.6$ in the case of the exp-algorithm (and similarly for the poly-algorithm.) We have $\frac{0.5^{make}}{3.6^{break}} \approx 3.6^{-(break+make/2)}$. This can be interpreted as follows: instead of the usual $scorechange = make - break$ a better score measure is $-(break + make/2)$.

The value of c_b determines the greediness of the algorithm. We concentrate on c_b in this discussion since it seems to be the more important parameter. The higher the value of c_b, the more greedy is the algorithm. A low value of c_b (in the extreme, $c_b = 1$ in the exp-algorithm) morphs the algorithm to a random walk algorithm with flip probabilities $(\frac{1}{k}, \dots \frac{1}{k})$ like the one considered in [12]. Examining Figure 2, almost a phase-transition can be observed. If c_b falls under some critical value, like 2.0, the expected run time increases tremendously. Turning towards the other side of the scale, increasing the value of c_b, i.e. making the algorithm more greedy, also degrades the performance but not with such an abrupt rise of the running time as in the other case.

3   Experimental Analysis of the Functions

To determine the performance of our probability distribution based solver we have designed a wide variety of experiments. In the first part of our experiments we try to determine good settings for the parameters c_b and c_m by means of automatic configuration procedures. In the second part we will compare our solver to other state-of-the-art solvers.

3.1   The Benchmark Formulae

All random instances used in our settings are uniform random k-SAT problems generated with different clause to variable ratios, which we denote with α. The class of random 3-SAT problems is the best studied class of random problems and because of this reason we have four different sets of 3-SAT instances.

1. 3sat1k[15]: 10^3 variables at $\alpha = 4.26$ (500 instances)
2. 3sat10k[15]: 10^4 variables at $\alpha = 4.2$ (500 instances)
3. 3satComp[16]: all large 3-SAT instances from the SAT Competition 2011 category random with variables range $2 \cdot 10^3 \dots 5 \cdot 10^4$ at $\alpha = 4.2$ (100 instances)
4. 3satExtreme: $10^5 \dots 5 \cdot 10^5$ variables at $\alpha = 4.2$ (180 instances)

The 5-SAT and 7-SAT problems used in our experiments come from [15]: 5sat500 (500 variables at $\alpha = 20$) and 7sat90 (90 variables at $\alpha = 85$). The 3sat1k, 3sat10k, 5sat500 and 7sat90 instance classes are divided into two equal sized classes called train and test. The train set is used to determine good parameters for c_b and c_m and the second class is used to report the performance. Further we also include the set of satisfiable random and crafted instances from the SAT Competition 2011.

Table 1. Parameter setting for c_b and c_m: Each cell represents a good setting for c_b and c_m dependent on the function used by the solver. Parameters values around these values have similar good performance.

	3sat1k	3sat10k	5sat500	7sat90
$exp(c_b, c_m)$	3.6 0.5	3.97 0.3	3.1 1.3	3.2 1.4
$poly(c_b, c_m)$	3.1 -0.8	2.86 -0.81	-	-
$exp(c_b)$	2.50	2.33	3.6	4.4
$poly(c_b)$	2.38	2.16	-	-

3.2   Good Parameter Setting for c_b and c_m

The problem that every solver designer is confronted with is the determination of good parameters for its solvers. We have avoided to accomplish this task by manual tuning but instead have used an automatic procedure.

As our search space is relatively small, we have opted to use a modified version of the iterated F-race [5] configurator, which we have implemented in Java. The idea of F-race is relatively simple: good configurations should be evaluated more often than poor ones which should be dropped as soon as possible. F-race uses a family Friedman test to check if there is a significant performance difference between solver configurations. The test is conducted every time the solvers have been run on an instance. If the test is positive poor configurations are dropped, and only the good ones are further evaluated. The configurator ends when the number of solvers left in the race is less than 2 times the number of parameters or if there are no more instances to evaluate on.

To determine good values for c_b and c_m we have run our modified version of F-race on the training sets 3sat1k, 3sat10k, 5sat500 and 7sat90. The cutoff time for the solvers were set to 10 seconds for 3sat1k and to 100 seconds for the rest. The best values returned by this procedure are listed in Table 1. Values for the class of 3sat1k problems were also included, because the preliminary analysis of the parameter search space was done on this class. The best parameter of the break-only-exp-algorithm for k-SAT can be roughly described by the formula $c_b = k^{0.8}$.

For the 3sat10k instance set the parameter space performance plots in Figure 2 looks similar to that of 3sat1k (Figure 1), though the area with good configurations is narrower, which can be explained by the short cutoff limit of 100 seconds used for this class (SLS solvers from the SAT Competition 2011 had an average runtime of 180 seconds on this type of instances).

In case of 5sat500 and 7sat90 we have opted to analyze only the exponential function because the polynomial function, other than in the 3SAT case, exhibited poor performance on these sets. Figure 3 shows the parameter space performance plot for the 5sat500 and 7sat90 sets. When comparing these plots with those for 3-SAT, the area with good configurations is much larger. For the 7-SAT instances the promising area seems to take almost half of the parameter space. The performance curve of the break-exp-only algorithm is also wider than that of 3-SAT and in the case of 7-SAT no clear curve is recognizable.

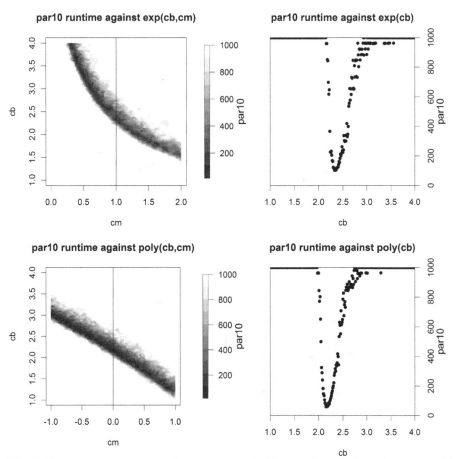

Fig. 2. Parameter space performance plot: The runtime of the solver using different function and for varying c_b and c_m on a the 3sat10k instances set

4   Evaluations

In the second part of our experiments we compare the performance of our solvers to that of the SAT Competition 2011 winners and also to WalkSAT SKC. An additional comparison to a survey propagation algorithm will show how far our probSAT local search solver can get.

4.1   Soft- and Hardware

The solvers were run on a part of the bwGrid clusters [4] (Intel Harpertown quad-core CPUs with 2.83 GHz and 8 GByte RAM). The operating system was Scientific Linux. All experiments were conducted with EDACC, a platform that distributes solver execution on clusters [1].

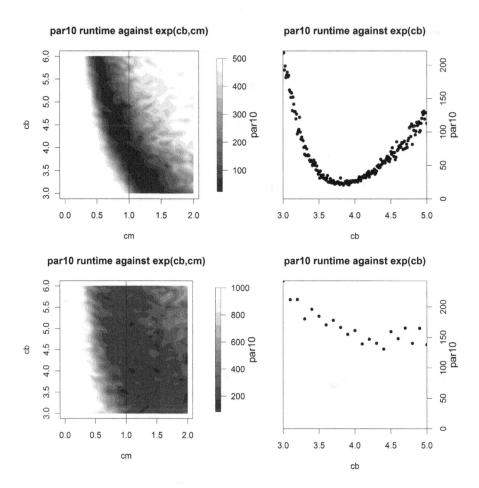

Fig. 3. Parameter space performance plot: The runtime of the exp-solvers with different functions and varying c_b and c_m on a the 5sat500 instances at the top and on the 7sat90 instances bottom

4.2 The Competitors

The WalkSAT SKC solver is implemented within our own code basis. We use our own implementation and not the original code provided by Henry Kautz, because our implementation is approximately 1.35 times faster. We have used version 1.4 of the survey propagation solver provided by Zecchina[1], which was changed to be DIMACS conform. For all other solvers we have used the binaries from the SAT Competition 2011[2].

---

[1] http://users.ictp.it/~zecchina/SP/

[2] http://www.cril.univ-artois.fr/SAT11/
solvers/SAT2011-static-binaries.tar.gz

Parameter Settings for Competitors. Sparrow is highly tuned on our target set of instances and incorporates optimal settings for each set within its code. WalkSAT has only one single parameter, the walk probability wp. In case of 3-SAT we took the optimal values for $wp = 0.567$ computed in [7]. Because we could not find any settings for 5-SAT and 7-SAT problems we have run our modified version of F-race to find good settings. For 5sat500 the configurator reported $wp = 0.25$ and for 7sat90 $wp = 0.1$. The survey propagation solver was evaluated with the default settings reported in [17].

4.3   Results

We have evaluated our solvers and the competitors on the test set of the instance sets 3sat1k, 3sat10k, 5sat500 and 7sat90 (note that the training set was used only for finding good parameters for the solvers). The parameter setting for c_b and c_m are those from Table 1 (in case of 3-SAT we have always used the parameters for 3sat10k). The results of the evaluations are listed in Table 2.

Table 2. Evaluation results: Each cell represents the par10 runtime and the number of successful runs for the solvers on the given instance set. Results are highlighted if the solver succeeded in solving all instances within the cutoff time, or if it has the best par10 runtime. Cutoff times are 600 seconds for 3sat1k, 5sat500 and 7sat90 and 5000 seconds for the rest.

	3sat10k		3satComp		3satExtreme		5sat500		7sat90	
$exp(c_b, c_m)$	46.6	(998)	93.84	**(500)**	-		12.49	(10^3)	201.68	(974)
$poly(c_b, c_m)$	46.65	(996)	76.81	**(500)**	-		-		-	
$exp(c_b)$	53.02	(997)	126.59	**(500)**	-		**7.84**	(10^3)	134.06	(984)
$poly(c_b)$	**22.80**	**(1000)**	**54.37**	**(500)**	1121.34	**(180)**	-		-	
Sparrow2011	199.78	(973)	498.05	(498)	47419	(10)	9.52	(10^3)	**14.94**	(10^3)
WalkSAT	61.74	(995)	172.21	(499)	1751.77	(178)	14.71	(10^3)	69.34	(994)
sp 1.4	3146.17	(116)	18515.79	(63)	**599.01**	**(180)**	5856	(6)	6000	(0)

On the 3-SAT insatances, the polynomial function yields the overall best performance. On the 3-SAT competition set all of our solver variants exhibited the most stable performance, being able to solve all problems within cutoff time. The survey propagation solver has problems with the 3sat10k and the 3satComp problems (probably because of the relatively small number of variables). The good performance of the break-only-poly-solver remains surprisingly good even on the 3satExtreme set where the number of variables reaches $5 \cdot 10^5$ (ten times larger than that from the SAT Competition 2011). From the class of SLS solvers it exhibits the best performance on this set and is only approx. 2 times slower than survey propagation. Note that a value of $c_b = 2.165$ for the break-only-poly solver further improved the runtime of the solver by approximately 30% on the 3satExtreme set.

On the 5-SAT instances the exponential break-only-exp solver yields the best performance being able to beat even Sparrow, which was the best solver for

5-SAT within the SAT Competition 2011. On the 7-SAT instances though the performance of our solvers is relatively poor. We observed a very strong variance of the run times on this set and it was relatively hard for the configurator to cope with such high variances.

Overall the performance of our simple probability based solvers reaches state-of-the-art performance and can even get into problem size regions where only survey propagation could catch ground.

Scaling Behavior with N. The survey propagation algorithm scales linearly with N on formulas generated near the threshold ratio. The same seems to hold for WalkSAT with optimal noise as the results in [7] shows. The 3satExtreme instance set contains very large instances with varying $N \in \{10^5 \ldots 5 \cdot 10^5\}$. To analyze the scaling behavior of our probSAT solver in the break-only-poly variant we have computed for each run the number of flips per variable performed by the solver until a solution was found. The number of flips per variable remains constant at about $2 \cdot 10^3$ independent of N. The same holds for WalkSAT, though WalkSAT seems to have a slight larger variance of the run times.

Results on the SAT Competition 2011 Random Set. We have compiled an adaptive version of our probSAT solver and of WalkSAT, that first checks the size of the clauses (i.e. k) and then sets the parameters accordingly (like Sparrow2011 does). We have ran this solvers on the complete satisfiable instances set from the SAT Competition 2011 random category along with all other competition winning solvers from this category: Sparrow2011, sattime2011 and EagleUp. Cutoff time was set to 5000 seconds. We report only the results on the large set, as the medium set was completely solved by all solvers and the solvers had a median runtime under one second. As can be seen from the results of the cactus plot in Figure 4, the adaptive version of probSAT would have been able to win the competition. Interestingly is to see that the adaptive version of WalkSAT would have ranked third.

Results on the SAT Competition 2011 Satisfiable Crafted Set. We have also ran the different solvers on the satisfiable instances from the crafted set of SAT Competition 2011 (with a cutoff time of 5000 seconds). The results are listed in Table 3. We have also inculded the results of the best three complete solvers from the crafted category. The probSAT solver and the WalkSAT solver performed best in their 7-SAT break-only configuration solving 81 respectively 101 instances. The performance of WalkSAT could not be improved by changing the walk probability. The probSAT solver though exhibited better performance with $cb = 7$ and a switch to the polynomial break-only scheme, being then able to solve 93 instances. With such a high cb value (very greedy) the probability of getting stuck in local minima is very high. By adding a static restart strategy after $2 \cdot 10^4$ flips per variable the probSAT solver was then able to solve 99 instances (as listed in the table).

The high greediness level needed for WalkSAT and probSAT to solve the crafted instances indicates that this instances might be more similar to the 7-SAT instances (generally to higher k-SAT). A confirmation of this conjecture is that

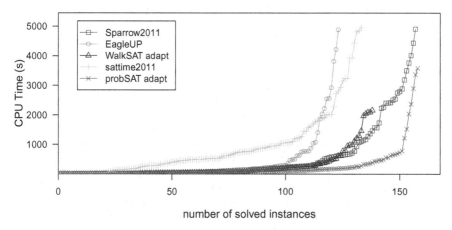

Fig. 4. Results on the "large" set of the SAT Competition 2011 random instances

Sparrow with fixed parameters for 7-SAT instances could solve 103 instances vs. 104 in the default setting. We suppose that improving SLS solvers for random instances with large clause length would also yield improvements for non random instances.

To check weather the performance of SLS solvers can be improved by preprocessing the instances first, we have run the preprocessor of lingeling [3], which incorporates all main preprocessing techniques, to simplify the instances. The results unluckily show the contrary of what would have been expected (see Table 3). None of the SLS solvers could benefit from the preprocessing step, solving equal or less instances.

Table 3. Results on the crafted satisfiable instances: Each cell reports the number of solved instances within the cutoff time (5000 seconds). The first line shows the results on the original instances and the second on the preprocessed instances.

	sattime	Sparrow	WalkSAT	probSAT	MPhaseSAT (complete)	clasp (complete)	SArTagnan (complete)
Crafted	**107**	104	101	99	93	81	46
Crafted pre.	86	97	**101**	95	98	80	48

5   Comparison with WalkSAT

In principle, WalkSAT [10] also uses a certain pattern of probabilities for flipping one of the variables within a non-satisfied clause. But the probability distribution does not depend on a single continuous function f as in our algorithms described above, but uses some non-continuous if-then-else decisions as described in [10].

In Table 3 we compare the flipping probabilities in WalkSAT (using the noise value 0.57) with the break-only-poly-algorithm (with $c_b = 2.3$) and the break-only-exp-algorithm (with $c_b = 2.5$) using a few examples of *break*-values that might occur within a 3-CNF clause. Even though the probabilities look very similar, we think that the small differences renders our approach to be more robust in case of 3-SAT and 5-SAT.

Table 4. Probability comparison of WalkSAT and probSAT: The first columns show some typical *break*-value combinations that occur within a clause in a 3-SAT formula during the search. For the different solvers considered here the probabilities for the each of the 3 variables to be flipped are listed.

breaks			WalkSAT			break-only-poly			break-only-exp		
0	0	0	0.33	0.33	0.33	0.33	0.33	0.33	0.33	0.33	0.33
0	0	1	0.5	0.5	0	0.46	0.46	0.08	0.42	0.42	0.16
0	1	1	1.0	0	0	0.72	0.14	0.14	0.56	0.22	0.22
0	1	2	1.0	0	0	0.79	0.15	0.06	0.64	0.26	0.1
0	2	2	1.0	0	0	0.88	0.06	0.06	0.76	0.12	0.12
1	1	1	0.33	0.33	0.33	0.33	0.33	0.33	0.33	0.33	0.33
1	1	2	0.4	0.4	0.19	0.42	0.42	0.16	0.42	0.42	0.16
1	2	2	0.62	0.19	0.19	0.56	0.22	0.22	0.56	0.22	0.22
1	2	3	0.62	0.19	0.19	0.63	0.24	0.13	0.64	0.26	0.1

6   Summary and Future Work

We introduced a simple algorithmic design principle for a SLS solver which does its job without heuristics and "tricks". It just relies on the concept of probability distribution and focused search. It is though flexible enough to allow plugging in various functions f which guide the search.

Using this concept we were able to discover a non-symmetry regarding the importance of the *break* and *make*-values: the *break*-value is the more important one; one can even do without the *make*-value completely.

We have systematically used an automatic configurator to find the best parameters and to visualize the mutual dependency and impact of the parameters.

Furthermore, we observe a large variation regarding the running times even on the same input formula. Therefore the issue of introducing an optimally chosen restart point arises. Some initial experiments show that performing restarts, even after a relatively short period of flips (e.g. $20N$) does give favorable results on hard instances. It seems that the probability distribution of the number of flips until a solution is found, shows some strong heavy tail behavior (cf. [9],[13]).

Plugging in the *age* property into the distribution function and analyze how strong its influence should be is also of interest.

Finally, a theoretical analysis of the Markov chain convergence and speed of convergence underlying this algorithm would be most desirable, extending the results in [12].

Acknowledgments. We would like to thank the BWGrid [4] project for providing the computational resources. This project was funded by the Deutsche Forschungsgemeinschaft (DFG) under the number SCHO 302/9-1. We thank

Daniel Diepold and Simon Gerber for implementing the F-race configurator and providing different analysis tools within the EDACC framework. We would also like to thank Andreas Fröhlich for fruitful discussions on this topic.

References

1. Balint, A., Diepold, D., Gall, D., Gerber, S., Kapler, G., Retz, R.: EDACC - An Advanced Platform for the Experiment Design, Administration and Analysis of Empirical Algorithms. In: Coello, C.A.C. (ed.) LION 2011. LNCS, vol. 6683, pp. 586–599. Springer, Heidelberg (2011)
2. Balint, A., Fröhlich, A.: Improving Stochastic Local Search for SAT with a New Probability Distribution. In: Strichman, O., Szeider, S. (eds.) SAT 2010. LNCS, vol. 6175, pp. 10–15. Springer, Heidelberg (2010)
3. Biere, A.: Lingeling and Friends at the SAT Competition 2011. Tehnical report 11/1, FMV Reports Series (2011),
   http://fmv.jku.at/papers/Biere-FMV-TR-11-1.pdf
4. bwGRiD, member of the German D-Grid initiative, funded by the Ministry for Education and Research (Bundesministerium für Bildung und Forschung) and the Ministry for Science, Research and Arts Baden-Württemberg (Ministerium für Wissenschaft, Forschung und Kunst Baden-Württemberg), http://www.bw-grid.de
5. Birattari, M., Yuan, Z., Balaprakash, P., Stützle, T.: F-Race and Iterated F-Race: An Overview. In: Bartz-Beielstein, T., et al. (eds.) Empirical Methods for the Analysis of Optimization Algorithms, pp. 311–336. Springer, Berlin (2010)
6. Hoos, H.H.: An adaptive noise mechanism for WalkSAT. In: Proceedings of AAAI 2002, pp. 635–660 (2002)
7. Kroc, L., Sabharwal, A., Selman, B.: An Empirical Study of Optimal Noise and Runtime Distributions in Local Search. In: Strichman, O., Szeider, S. (eds.) SAT 2010. LNCS, vol. 6175, pp. 346–351. Springer, Heidelberg (2010)
8. Li, C.-M., Huang, W.Q.: Diversification and Determinism in Local Search for Satisfiability. In: Bacchus, F., Walsh, T. (eds.) SAT 2005. LNCS, vol. 3569, pp. 158–172. Springer, Heidelberg (2005)
9. Luby, M., Sinclair, A., Zuckerman, D.: Optimal speedup of Las Vegas algorithms. Information Proc. Letters 47, 173–180 (1993)
10. McAllester, D., Selman, B., Kautz, H.: Evidence for invariant in local search. In: Proceedings of AAAI 1997, pp. 321–326 (1997)
11. Papadimitriou, C.H.: On selecting a satisfying truth assignment. In: Proceedings FOCS 1991, pp. 163–169. IEEE (1991)
12. Schöning, U.: A probabilistic algorithm for k-SAT and constraint satisfaction problems. In: Proceedings FOCS 1991, pp. 410–414. IEEE (1999)
13. Schöning, U.: Principles of Stochastic Local Search. In: Akl, S.G., Calude, C.S., Dinneen, M.J., Rozenberg, G., Wareham, H.T. (eds.) UC 2007. LNCS, vol. 4618, pp. 178–187. Springer, Heidelberg (2007)
14. Seitz, S., Alava, M., Orponen, P.: Focused local search for random 3-satisfiability. arXiv:cond-mat/051707v1 (2005)
15. Tompkins, D.A.D., Balint, A., Hoos, H.H.: Captain Jack: New Variable Selection Heuristics in Local Search for SAT. In: Sakallah, K.A., Simon, L. (eds.) SAT 2011. LNCS, vol. 6695, pp. 302–316. Springer, Heidelberg (2011)
16. The SAT Competition Homepage, http://www.satcompetition.org
17. Braunstein, A., Mezard, M., Zecchina, R.: Survey propagation: an algorithm for satisfiability. Random Structures and Algorithms 27, 201–226 (2005)

Off the Trail: Re-examining the CDCL Algorithm*

Alexandra Goultiaeva and Fahiem Bacchus

Department of Computer Science
University of Toronto
{alexia,fbacchus}@cs.toronto.edu

Abstract. Most state of the art SAT solvers for industrial problems are based on the Conflict Driven Clause Learning (CDCL) paradigm. Although this paradigm evolved from the systematic DPLL search algorithm, modern techniques of far backtracking and restarts make CDCL solvers non-systematic. CDCL solvers do not systematically examine all possible truth assignments as does DPLL.

Local search solvers are also non-systematic and in this paper we show that CDCL can be reformulated as a local search algorithm: a local search algorithm that through clause learning is able to prove UNSAT. We show that the standard formulation of CDCL as a backtracking search algorithm and our new formulation of CDCL as a local search algorithm are equivalent, up to tie breaking.

In the new formulation of CDCL as local search, the trail no longer plays a central role in the algorithm. Instead, the ordering of the literals on the trail is only a mechanism for efficiently controlling clause learning. This changes the paradigm and opens up avenues for further research and algorithm design. For example, in QBF the quantifier places restrictions on the ordering of variables on the trail. By making the trail less important, an extension of our local search algorithm to QBF may provide a way of reducing the impact of these variable ordering restrictions.

1 Introduction

The modern CDCL algorithm has evolved from DPLL, which is a systematic search through variable assignments [4]. CDCL algorithms have evolved through the years, various features and techniques have been added [10] that have demonstrated empirical success. These features have moved CDCL away from exhaustive search, and, for example, [9] has argued that modern CDCL algorithms are better thought of as guided resolution rather than as exhaustive backtracking search.

New features have been added as we have gained a better understanding of CDCL both through theoretical developments and via empirical testing. For example, the important technique of restarts was originally motivated by theoretical and empirical studies of the effect of heavy-tailed run-time distributions [7] on solver run-times.

* Supported by Natural Sciences and Engineering Research Council of Canada.

A. Cimatti and R. Sebastiani (Eds.): SAT 2012, LNCS 7317, pp. 30–43, 2012.
© Springer-Verlag Berlin Heidelberg 2012

Combinations of features, however, can sometimes interact in complex ways that can undermine the original motivation of individual features. For example, phase saving, also called light-weight component caching, was conceived as a progress saving technique, so that backtracking would not retract already discovered solutions of disjoint subproblems [12] and then have to spend time rediscovering these solutions. However, when we add phase saving to restarts, we reduce some of the randomization introduced by restarts, potentially limiting the ability of restarts to short-circuit heavy-tailed run-times. Nevertheless, even when combined, restarts and phase saving both continue to provide a useful performance boost in practice and are both commonly used in CDCL solvers.

When combined with a strong activity-based heuristic, phase saving further changes the behavior of restarts. In this context it is no longer obvious that restarts serve to move the solver to a different part of the search space. Instead, it can be shown empirically that after a restart a large percentage of the trail is re-created exactly as it was prior to the restart, indicating that the solver typically returns to the same part of the search space. In fact, there is evidence to support the conclusion that the main effect of restarts in current solvers is simply to update the trail with respect to the changed heuristic scores. For example, [14] show that often a large part of the trail can be reused after backtracking. With the appropriate implementation techniques reusing rather than reconstructing the trail can speed up the search by reducing the computational costs of restarts.

In this paper we examine another feature of modern SAT solvers that ties them with the historical paradigm of DPLL: the trail used to keep track of the current set of variable assignments. We show that modern SAT solvers, in which phase savings causes an extensive recreation of the trail after backtracking, can actually be reformulated as local search algorithms.

Local search solvers work with complete truth assignments [15], and a single step usually consists of picking a variable and flipping its value. Local search algorithms have borrowed techniques from CDCL. For example, unit propagation has been employed [6,8,2], and clause learning as also been used [1]. However, such solvers are usually limited to demonstrating satisfiability, and often cannot be used to reliably prove UNSAT. Our reformulation of the CDCL algorithm yields a local search algorithm that is able to derive UNSAT since it can perform exactly the same steps as CDCL would. It also gives a different perspective on the role of the trail in CDCL solvers. In particular, we show that the trail can be viewed as providing an ordering of the literals in the current truth assignment, an ordering that can be used to guide clause learning. This view allows more flexible clause learning techniques to be developed, and different types of heuristics to be supported. It also opens the door for potentially reformulating QBF algorithms, which suffer from strong restrictions on the ordering of the variables on the trail.

Section 2 examines the existing CDCL algorithm and describes our intuition in more detail. Section 3 presents a local search formulation of the modern CDCL algorithm and proves that the two formulations are equivalent. Section 4 presents some simple experiments which suggest further directions for research. Section 5 concludes the paper.

Algorithm 1: Modern CDCL algorithm

Data: ϕ—a formula in CNF
Result: TRUE if ϕ is SAT, FALSE if ϕ is UNSAT

```
1   π ← ∅ ; C ← ∅ while TRUE do
2   |   π ← unitPropagate(φ ∪ C, π)
3   |   if reduce(φ ∪ C, π) contains an empty clause then
4   |   |   c' ← clauseLearn(π, φ ∪ C)
5   |   |   if c' = ∅ then return FALSE
6   |   |   C = C ∪ {c'}
7   |   |   π = backtrack(c')
8   |   else if φ is TRUE under π then  return TRUE
9   |   else
10  |   |   v ← unassigned variable with largest heuristic value
11  |   |   v ← phase[v]
12  |   |   π.append(v)
13  |   end
14  |   if timeToRestart() then backtrack(0)
15  end
```

---

2   Examining the CDCL Algorithm

A modern CDCL algorithm is outlined in Algorithm 1. Each iteration starts by adding literals implied by unit propagation to the trail π. If a conflict is discovered clause learning is performed to obtain a new clause $c' = (\alpha \rightarrow y)$. The new clause is guaranteed to be **empowering**, which means that it is able to produce unit implications in situations when none of the old clauses can [13]. In this case, c' generates a new implication y earlier in the trail, and the solver backtracks to the point where the new implication would have been made if the clause had previously been known. Backtracking removes part of the trail in order to add the new implication in the right place. On the next iteration unit propagation will continue adding implications, starting with the newly implied literal y. If all variables are assigned without a conflict, the formula is satisfied. Otherwise, the algorithm picks a decision variable to add to the trail. It picks an unassigned variable with the largest heuristic value, and restores its value to the value it had when it was last assigned. The technique of restoring the variable's value is called **phase saving**. We will say that the **phase** of a variable v, $phase[v]$, is the most recent value it had; if v has never been assigned, $phase[v]$ will be an arbitrary value set at the beginning of the algorithm; if v is assigned, $phase[v]$ will be its current value.

Lastly, sometimes the solver restarts: it removes everything from the trail except for literals unit propagated at the top level. This might be done according to a set schedule, or some heuristic [3].

As already mentioned, after backtracking or restarting, the solver often recreates much of the trail. For example, we found that the overwhelming majority of assignments Minisat makes simply restore a variable's previous value. We

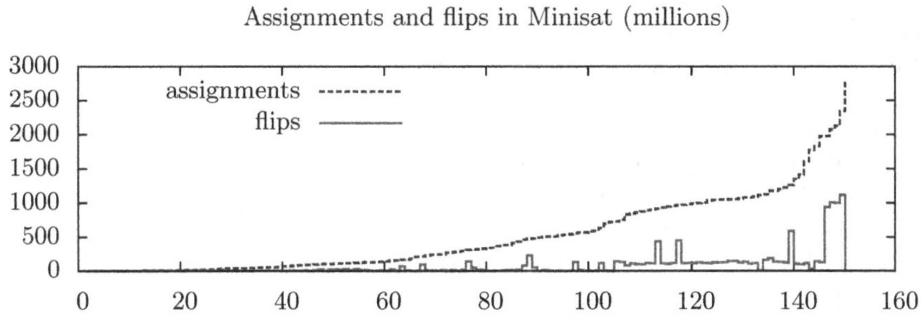

Assignments and flips in Minisat (millions)

Fig. 1. Assignments and flips on both solved and unsolved (after a 1000s timeout) instances of SAT11 dataset. Sorted by the number of assignments.

have ran Minisat on the 150 problems from the SAT11 dataset of the SAT competition, with a timeout of 1000 seconds. Figure 1 shows the distribution of assignments Minisat made, and the number of "flips" it made, where flips are when a variable is assigned a different value than it had before. On average, the solver performed 165.08 flips per conflict, and 3530.4 assignments per conflict. It has already been noted that flips can be correlated with the progress that the solver is making [3].

Whenever the solver with phase saving backtracks, it removes variable assignments, but unless something forces the variable to get a different value, it would restore the old value when it gets to it. So, we can imagine that the solver is working with a complete assignment, which is the phase settings for all the variables $phase[v]$, and performing a flip from $\neg l$ to l only in one of the following cases. (1) l is implied by a new conflict clause. (2) l is implied by a variable that was moved up in the trail because its heuristic value was upgraded. Or (3) l is implied by another "flipped" variable. Phase saving ensures that unforced literals, i.e., decisions, cannot be flipped.

In all of these cases l is part of some clause c that is falsified by the current "complete" assignment (consisting of the phase set variables); c would then become its reason clause; at the point when l is flipped, c is the earliest encountered false clause; and l is the single unassigned variable in c (i.e., without c, l would have been assigned later in the search). As we will see below, we can use these conditions to determine which variable to flip in a local search algorithm.

Note that we will not consider the randomization of decision variables in this paper, although this could be accommodated by making random flips in the local search algorithm. The benefits of randomizing the decision variables are still poorly understood. In our experiments we found that turning off randomization does not noticeably harm performance of Minisat. Among ten runs with different seeds, Minisat solved between 51 and 59 instances, on average 55. With randomization turned off, it solved 56.

Algorithm 2: Local Search

    Data: ϕ - a formula in CNF
    Result: TRUE if ϕ is SAT, FALSE if ϕ is UNSAT

1 **while** TRUE **do**
2 | $I \leftarrow$ initValues()
3 | **while** $\phi|_I$ *contains* FALSE *clauses* **do**
4 | | **if** *timeToRestart()* **then** break
5 | | $v \leftarrow$ pickVar(I)
6 | | flip(v)
7 | **end**
8 **end**
9 **return** TRUE

---

3 Local Search

Algorithm 2 presents a generic local search algorithm. A local search solver works with a complete assignment I. At each stage in the search, it picks a variable and flips its value. There are different techniques for choosing which variable to flip, from simple heuristics such as minimizing the number of falsified clauses [15], to complicated multi-stage filtering procedures [16].

Typically, the algorithm tries to flip a variable that will reduce the distance between the current complete assignment and a satisfying assignment. However, estimating the distance to a solution is difficult and unreliable, and local search solvers often get stuck in local minima. It was noted that it is possible to escape the local minimum by generating new clauses that would steer the search. Also, if new non-duplicated clauses are being generated at every local minimum, the resulting algorithm can be shown to be complete. An approach exploiting this fact was proposed, using a single resolution step to generate one new clause at each such point [5]. The approach was then extended to utilize an implication graph, and incorporate more powerful clause learning into a local search solver, resulting in the CDLS algorithm [1]. However, as we will see below, CDLS cannot ensure completeness because the clause learning scheme it employs can generate redundant clauses.

The main difficulty for such an approach is the generation of an implication graph from the complete assignment I. The first step consists of identifying **once-satisfied** clauses. A clause c is considered to be **once-satisfied** by a literal x and a complete assignment I if there is exactly one literal $x \in c$ that is true in I ($c \cap I = \{x\}$).

Theoretically, any clause c_f with $\neg x \in c_f$ that is false under I can be resolved with any clause c_o that is once-satisfied by literal x. This resolution would produce a non-tautological clause c_R which is false under I and which can potentially be further resolved with other once-satisfied clauses. However, in order to be useful, the algorithm performing such resolutions needs to ensure that it does not follow a cycle or produce a subsumed clause.

In order to avoid cycles, it is sufficient to define some ordering ψ on variables in I, and only allow the resolution of falsified clauses c_f and once-satisfied clauses c_o when all of the false literals in c_o precede the satisfying literal in the ordering ψ. However, a simple ordering does not ensure that the new clauses are useful.

Clause learning can be guided more effectively by considering the effects of unit propagation. We define an **ordering** ψ on the complete assignment I to be a sequence of literals $\psi = \{x_1, x_2, \ldots, x_k\}$ from I ($\forall i.x_i \in I$). A literal $x_i \in \psi$ is **implied** in ψ if there is some clause $(\neg x_{j_1}, \ldots, \neg x_{j_n}, x_i)$ with $j_1 < j_2 < \cdots < j_n < i$. In this case j_n is called an **implication point** for x_i (the implication point is 0 if the clause is unit). x_i is said to be **implied at** k if k is the smallest implication point for x_i.

Finally, an ordering ψ will be said to be **UP-compatible** if for any $x_i \in \psi$, if x_i is implied at j_n, then it must appear in the ordering ψ as soon after j_n as possible. In particular, UP-compatibility requires that any literals between x_i and its smallest implication point j_n, i.e., the literals $x_{j_n+1}, \ldots, x_{i-1}$, also be implied in ψ. For example, for a set of clauses (a), $(\neg a, \neg b, \neg c)$, $(\neg c, d)$, the orderings $\{a, c, d, b\}$ and $\{c, d, b, \neg a\}$ are UP-compatible, but $\{c, d, b, a\}$ or $\{c, b, d, \neg a\}$ are not. In the first case, a is implied by the clause (a), but follows non-implied c. In the second, d is implied by c with $(\neg c, d)$, but follows non-implied b.

A CDCL solver that ignores all conflicts would produce a UP-compatible ordering. However, not every UP-compatible ordering can be produced by a CDCL solver. This is because the definition considers only the given assignment, and does not take into account falsified clauses. So, it is possible that for some literal x_i, $\neg x_i$ is implied by a smaller prefix of the assignment, but this implication is ignored because it disagrees with the current assignment.

Given a complete assignment I and a UP-compatible ordering ψ we can define a **decision literal** to be any literal in ψ that is not implied. For each $x_i \in \psi$, we can define the decision level of x_i to be the number of decision literals in $\{x_1, x_2, \ldots, x_i\}$. For each implied literal, we can say that its reason is the clause that implied it. Note that the reason clauses are always once-satisfied by I. So, the ordering ψ gives us an implication graph over which clause learning can be performed as in a standard CDCL solver.

Consider a false clause $c = (\neg x_{c_1}, \neg x_{c_2}, \ldots, \neg x_{c_n})$ with $c_1 < c_2 < \ldots < c_n$. If $\neg x_{c_n}$ had been in I, it would have been implied at the same decision level as $x_{c_{n-1}}$. We will call such $\neg x_{c_n}$ a **failed implication**. We will say that x_{c_n} is **f-implied** at a decision level i if it is a failed implication at the decision level i but not earlier.

The scheme used by CDLS [1] is to construct a **derived partial interpretation** I'. Let i be the first decision at which a failed implication $\neg x_f$ occurs due to some clause $(\beta, \neg x_f)$. I' is then the prefix of ψ up to and including all variables with decision level i. If $x_f \in I'$, then x_f and $\neg x_f$ are implied at the same decision level, and clause learning can be performed as usual. We will call this kind of failed implication **conflicting**. In this case the execution is identical to a corresponding run of a CDCL solver, so the resulting clause is subject to all the guarantees a CDCL solver provides. In particular, the new clause is guaranteed to be empowering [13].

If $x_f \notin I'$ then $\neg x_f$ does not cause a conflict. It is a failed implication simply because it is incompatible with the current assignment. We shall call this kind of failed implication a **non-conflicting** implication. In CDLS the learning scheme is only applied when no variable flip is able to reduce the number of falsified clauses. So, there must be some clause (α, x_f) that is once-satisfied by x_i. CDLS then extends I' by including decisions $\{\alpha - I'\}$ as assumptions. In the new I', both α and β are falsified, so both x_f and $\neg x_f$ are implied, which causes a conflict that can be used as a starting point for clause learning. However, no guarantees apply to the new clause in this case. Because the added assumptions are not propagated, it is possible that the newly generated clause is not only not empowering, but is actually subsumed by existing clauses. For example, suppose the formula contains clauses (x_1, x_2, c), $(x_1, x_2, \neg x_3)$ and $(\neg x_3, \neg c)$. Suppose that the current assignment contains $(\neg x_1, \neg x_2, x_3, c)$. If x_3 is chosen as the first decision, the conflict immediately occurs because $\neg c$ is a failed implication at the first level. The implication graph, after adding the necessary assumptions, contains only two clauses, $(\neg x_3, \neg c)$ and (x_1, x_2, c). The resulting clause, $(x_1, x_2, \neg x_3)$, repeats a clause already in the database.

Instead of stopping at the first failed implication, we could use a larger prefix of ψ. Namely, we could apply learning to the first conflicting failed implication. However, this would not guarantee an unsubsumed new clause. It is possible that clause learning generates a clause (α, x) implying x that is the same as one of the previously ignored clauses causing a (non-conflicting) failed implication.

The problem arises because, from the point of view of CDCL, x_f is not a conflict. Instead of doing clause learning, a CDCL algorithm would have flipped x_f and continued with the search. Picking a correct ordering ψ would not help either. The problem here is with objective functions used to guide local search.

The following example is for the objective function that minimizes the number of satisfied clauses. Suppose we have the following clauses: (a, b), (c, d), $(\neg a, c)$, $(\neg b, d)$, $(\neg c, a)$, $(\neg d, b)$. An assignment $\pi = (\neg a, \neg b, \neg c, \neg d)$ is a local minimum: it falsifies two clauses. No literal flip falsifies less, and no ordering of π produces a conflicting implication. If we initially set $\neg a$, the two implications are b and $\neg c$. The first is a failed implication because it disagrees with π. The two possible implications from $\neg c$ are d and $\neg a$. The first is a failed implication, and the second is already set. All the other variables are completely symmetrical.

To avoid this problem, the flips need to be guided using some notion of unit propagation. Intuitively, a non-conflicting failed implication does not give enough information to clause learning, and thus would not produce a useful conflict. So, it should be resolved using a flip rather than clause learning, and should not constitute a local minimum.

Algorithm 3 demonstrates a strategy to guide the local search outlined in Algorithm 2. It selects a UP-compatible ordering ψ on I (of course, this could be updated incrementally and not generated from scratch every time). It then picks the first failed implication on ψ. If it is conflicting, clause learning is performed.

Algorithm 3: pickVar(I)

Data: I - a complete assignment
Result: y - next variable to flip
1 $\psi \leftarrow$ UP-compatible ordering on I
2 $y \leftarrow$ first failed implication in ψ
3 **if** $\neg y$ *is conflicting* **then**
4 $\quad$ $c \leftarrow firstUIP(\psi)$
5 $\quad$ **if** $c = \emptyset$ **then** EXIT(FALSE)
6 $\quad$ $attachClause(c)$; $y \leftarrow c.implicate$
7 **end**
8 **return** y

Note that the resultant clause c is guaranteed to be FALSE under I, and it would produce a failed implication at an earlier level. If an empty clause is derived, the formula is proven unsatisfiable. Otherwise, the new failed implication is now the earliest, and is non-conflicting, so the new implicate needs to be flipped.

One detail that is left out of the above algorithm is how to pick the ordering ψ. Note that if we are given some base ordering ψ_b, we can construct a UP-compatible ordering ψ in which decision literals respect ψ_b. In this case ψ_b plays the role of the variable selection heuristic. Of course, the heuristic must be chosen carefully so as not to lead the algorithm in cycles. An easy sufficient condition is when ψ_b is only updated after clause learning, as VSIDS is.

3.1   Connection to CDCL

In this section, we will focus on Algorithm 2 guided by the variable selection and clause learning technique presented in Algorithm 3 and with no restarts. We will refer to this as A2. We will refer to Algorithm 1 as A1.

Define a **trace** of an algorithm A to be a sequence of flips performed and clauses learned by A. Note that this definition applies to both A2 and A1: recall that for A1 a flip is an assignment where the variable's new value is different from its phase setting.

Theorem 1. *For any heuristic h there is a heuristic h' such that for any input formula ϕ, A1 with h would produce the same trace as A2 with h' (provided they make the same decisions in the presence of ties).*

Proof. We will say that a heuristic h is **stable** for (a version of) CDCL algorithm A if during any execution of A with h we have $h(v_1) \geq h(v_2)$ for some decision variable v_2 only if $h(v_1) \geq h(v_2)$ also held just before v_2 was last assigned.

Intuitively, a heuristic is stable if the ordering of decision variables is always correct with respect to the heuristic, and is not simply historical. One way to ensure that a heuristic is stable is to restart after every change to the heuristic. For example, the VSIDS heuristic is stable for a version of A1 which restarts after every conflict.

We will first prove the claim for a heuristic h that is stable for A1. Then, we will show that for any h, we can find an equivalent h' that is stable for A1 and such that A1 with h' produces the same trace as with h.

Let the initial assignment of A1 be the same as the initial phase setting of A2, and let both algorithms use the same heuristic h. It is easy to verify that if a partial execution of A1 has the same trace as a partial execution of A2, that means that the phase setting of any variable in A1 matches its value in A2.

To show that A1 and A2 would produce the same trace when run on the same formula ϕ, we will consider partial executions. By induction on n, we can show that if we stop each algorithm just after it has produced a trace of length n, the traces will be identical.

If $n = 0$, the claim trivially holds. Also note that if one algorithm halts after producing trace T, so will the other, and their returned values will match. Both algorithms will return FALSE iff T ends with an empty clause. If A2 has no failed implications, then A1 will restore all variables to their phase values and obtain a solution, and vice versa. Suppose the algorithms have produced a trace T.

Let S be the Next Flip or Learned Clause of A1. Let π be the trail of A1 just before it produced S.

Because h is stable for A1, then the heuristic values of the decision variables in π are non-increasing. That is because if $h(v_1) > h(v_2)$ for two decision variables, then the same must have held when v_2 was assigned. If v_1 had been unassigned at that point, it would have been chosen as the decision variable instead. So, v_1 must have been assigned before v_2.

So, π is a UP-compatible ordering respecting h over the partial assignment: any implication is placed as early as possible in π, and non-implied (decision) literals have non-increasing heuristic value. Because unit propagation was performed to completion (except for possibly the last decision level), and because the heuristic value of all unassigned literals is less than that of the last decision literal, π can always be extended to a UP-compatible ordering ψ on I.

Let $C = \{\alpha, v\}$ be the clause that caused v to be flipped to TRUE if S is a flip; otherwise, let it be the conflicting clause that started clause learning, with v being the trail-deepest of its literals. In both cases, C is FALSE at P_1, so v is a failed implication in ψ. This is the first conflict encountered by A1, so there are no false clauses that consist entirely of literals with earlier decision levels. So, v is the first failed implication in ψ.

If S is a flip, then v is non-conflicting, and A2 would match the flip. Otherwise, v is a conflicting failed implication, and will cause clause learning. For the ψ which matches π, clause learning would produce a clause identical to that produced by A1. So, the next entry in the trace of A2 will also be S.

Let S be the Next Flip or Learned Clause of A2. Let v be the first failed implication just before S was performed, and let ψ be the corresponding UP-compatible ordering. Let π be the trail of A1 just before it produces its next flip or a learned clause. We will show that whenever π differs from ψ, A1 could have broken ties differently to make them match. Let $v_1 \in \pi$ and $v_2 \in \psi$ be the first pair of literals that are different between π and ψ. Suppose v_1 is implied.

Then, because ψ is UP-compatible, v_2 must also be implied by preceding literals. So, A1 could have propagated v_2 before v_1. If v_1 is a decision literal, then so is v_2. Otherwise, v_2 should have been unit propagated before v_1 was assigned. If $h(v_2) < h(v_1)$, this would break the fact that ψ respects the heuristic. So, $h(v_2) \geq h(v_1)$. Then the same must have been true at the time v_1 was assigned. So, v_2 was at worst an equal candidate for the decision variable, and could have been picked instead.

So, provided A1 breaks ties accordingly, it would have the trail π that is a prefix of ψ. It can continue assigning variables until the trail includes all the variables at the decision level at which v is f-implied. Because v is the first failed implication in ψ, no conflicts or flip would be performed up to that point. At this point, there will be some clause $C = (\alpha, v)$. If S is a flip, then v is not conflicting, C will be unit, and a flip will be performed. If S is a learned clause, then v is conflicting, which means that v was among the implied literals at this level. So, clause learning will be performed. Either way, the next entry in the trace of A1 will also be S.

So, we have shown that A2 and A1 would produce the same trace given the same heuristic h' which is stable for A1. Now we will sketch a proof that given any variable heuristic h, we can construct a heuristic h' which is stable for A1 and such that A1 with h would produce the same trace as with h'.

We will define $h'(v) = h(v)$ whenever v is unassigned. Otherwise, we will set $h'(v) = M + V - D + 0.5d$, where M is some value greater than the maximum $h(v)$ of all non-frozen variables, V is the number of variables in the problem, and D is the decision level at which v was assigned when it became frozen, and d is 1 if v is decision and 0 otherwise.

Because a heuristic is only considered for unassigned variables, then the behavior of the algorithm is unaffected, and it will produce the same traces. Also, unassigned values always have a smaller heuristic value than those that are assigned; those assigned later always have a smaller heuristic value than earlier decision literals. So, the heuristic is stable for A1.

As a corollary: because Algorithm 1 is complete, so is Algorithm 2.

3.2   Other Failed Implications

In Algorithm 3 we always choose the first failed implication. However, it is not a necessary condition to generate empowering clauses.

Theorem 2. *Suppose that ψ is UP-compatible ordering on I. Let c be a clause generated by 1UIP on some failed implication x. Suppose $c = (\alpha, y)$ where y is the new implicant. If no failed implication that is earlier than x can be derived by unit propagation from α, then c is empowering.*

Proof. Suppose that c is not empowering. Then y can be derived by unit propagation from α. Because y was not implied by α at that level, then the unit propagation chain contains at least one literal that contradicts the current assignment. Let p be such a literal which occurs first during unit propagation. Then p is a failed implication that can be derived from α.

Note that this is sufficient, but not a necessary condition. It is possible that an earlier failed implication x can be derived from α, but $\alpha \cap x$ still do not allow the derivation of y.

3.3 Potential Extension to QBF Solving

The trail has always played a central role in the formalization of the SAT algorithm. It added semantic meaning to the the chronological sequence of assignments by linking it to the way clause learning is performed.

In SAT, this restriction has no major consequences, since the variables can be assigned in any order. However, in an extension of SAT, Quantified Boolean Formula (QBF) solving, this restriction becomes important.

In QBF variables are either existentially or universally quantified, and the inner variables can depend on the preceding ones. Clause learning utilizes a special **universal reduction** step, which allows a universal to be dropped from a clause if there are no existential variables that depend on it. In order to work, clause learning requires the implication graph to be of a particular form, with deeper variables having larger decision levels. Because of the tight link between the trail and clause learning, the same restriction is applied to the order in which the algorithm was permitted to consider variables. Only outermost variables were allowed to be picked as decision literals.

This restriction is a big impediment to performance in QBF. One illustration of this fact is that there is still a big discrepancy between search-based and expansion based solvers in QBF. The former are constrained to consider variables according to the quantifier prefix, while the latter are constrained to consider the variables in reverse of the quantifier prefix. The fact that the two approaches are incomparable, and that there are sets of benchmarks challenging for one but not the other, suggests that the ordering restriction plays a big role in QBF. Another indication of this is the success of dependency schemes, which are attempts to partially relax this restriction [11].

The reformulation presented here is a step towards relaxing this restriction. We show that the chronological sequence of assignments does not have any semantic meaning, and thus should not impose constraints on the solver. Extending the present approach to QBF should allow one to get an algorithm with the freedom to choose the order in which the search is performed.

To extend to QBF, the definition of UP-compatible ordering would need to be augmented to allow for universal reduction. One way to do this would be to constrain the ordering by quantifier level, to ensure that universal reduction is possible and any false clause would have an implicate. However, this ordering is no longer linked to the chronological sequence of variables considered by the solver, and will be well-defined after any variable flip. At each step, the solver will be able to choose which of the failed implications to consider, according to some heuristic not necessarily linked to its UP-compatible ordering.

So, decoupling the chronological variable assignments from clause learning would allow one to construct a solver that would be free to consider variables in any order, and would still have well-defined clause learning procedure when it encounters a conflict.

Table 1. Problems from SAT11 solved within a 1000s timeout by Minisat with phase saving, and by the modified versions C_n, where n is the number of full runs the solver performs at each restart. C_All is the version that performs exclusively full runs. The number of problems solved is shown for All, True and False instances.

Family	Minisat			C_1			C_5			C_10			C_100			C_1000			C_All		
	A	T	F	A	T	F	A	T	F	A	T	F	A	T	F	A	T	F	A	T	F
fuhs (34)	10	**9**	1	10	8	2	9	8	1	9	8	1	10	7	**3**	10	8	2	7	7	0
manthey (9)	3	**3**	0	2	2	0	2	2	0	3	**3**	0	3	**3**	0	1	1	0	0	0	0
jarvisalo (47)	24	8	**16**	24	8	**16**	24	9	15	24	**10**	14	25	9	**16**	17	6	11	15	6	9
leberre (17)	11	4	7	13	**6**	7	14	**6**	**8**	13	**6**	7	13	**6**	7	12	5	7	7	1	6
rintanen (30)	9	**7**	2	8	5	**3**	7	4	**3**	8	5	**3**	8	5	**3**	2	1	1	1	0	1
kullmann (13)	2	2	0	3	3	0	3	3	0	2	2	0	**4**	**4**	0	3	3	0	3	3	0
Total (150)	59	33	26	60	32	28	59	32	27	59	**34**	25	**63**	**34**	**29**	45	24	21	33	17	16

4   Experiments

We have investigated whether subsequent failed implications, mentioned in Section 3.2, can be useful in practice. To evaluate this, we have equipped Minisat with the ability to continue the search ignoring conflict clauses. Note: here we use a version of Minisat with phase saving turned on.

This is equivalent to building a UP-compatible assignment with no non-conflicting failed implications.

For each decision level, it would only store the first conflict clause encountered, because learning multiple clauses from the same decision level is likely to produce redundant clauses. After all the variables are assigned, it would backtrack, performing clause learning on each stored conflict, and adding the new clauses to the database. We will say that one iteration of this cycle is a **full run**.

Obviously, not yet having any method of guiding the selection, the algorithm could end up producing many unhelpful clauses. To offset this problem, and to evaluate whether the other clauses are *sometimes* helpful, we constructed an algorithm that performs a full run only some of the time.

We have added a parameter n so that at every restart, the next n runs of the solver would be full runs. We experimented with $n \in \{1, 5, 10, 100, 1000\}$ and with a version which only performs full runs.

We ran the modified version on the 150 benchmarks from SAT11 set of the Sat Competition, with timeout of 1000 seconds. The tests were run on a 2.8GHz machine with 12GB of RAM.

Table 1 summarizes the results. As expected with an untuned method, some families show improvement, while for others the performance is reduced. However, we see that the addition of the new clauses can improve the results for both satisfiable instances (as in benchmarks sets leberre and kullmann), and unsatisfiable ones (as in fuhs and rintanen).

Figure 2 compares the number of conflicts learned while solving the problems in Minisat and C_100. For instances which only one solver solved, the other solver's value is set to the number of conflicts it learned within the 1000s timeout.

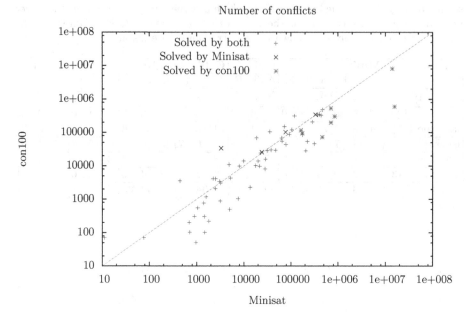

Fig. 2. The number of conflicts needed to solve the problem. Below the line are instances on which C_100 encountered fewer conflicts than Minisat.

We note that in the conflict count for C_100 we include all the conflicts it ever learned, so a single full run might add many conflicts at once. These are unfiltered, so we expect that good heuristics and pruning methods can greatly reduce this number. However, even with all the extra conflicts C_100 encounters, there is a fair number of cases where it needs fewer conflicts to solve the problem than Minisat.

5   Conclusion

We have presented a reformulation of the CDCL algorithm as local search. The trail is shown to be simply an efficient way to control clause learning. By decoupling clause learning from the chronological sequence in which variables are considered, we introduce new flexibility to be studied.

One potential application of this flexibility would be to produce QBF solvers whose search space is not so heavily constrained by the variable ordering. Another is to find good heuristics to choose which conflict clauses are considered during search.

Current CDCL solvers effectively maintain a UP-compatible ordering on the trail by removing the order up to the place affected by a flip, and recomputing it again. An interesting question worth investigating is whether it is possible to develop algorithms to update the order more efficiently.

References

1. Audemard, G., Lagniez, J.-M., Mazure, B., Sais, L.: Learning in local search. In: ICTAI, pp. 417–424 (2009)
2. Belov, A., Stachniak, Z.: Improved Local Search for Circuit Satisfiability. In: Strichman, O., Szeider, S. (eds.) SAT 2010. LNCS, vol. 6175, pp. 293–299. Springer, Heidelberg (2010)
3. Biere, A.: Adaptive Restart Strategies for Conflict Driven SAT Solvers. In: Kleine Büning, H., Zhao, X. (eds.) SAT 2008. LNCS, vol. 4996, pp. 28–33. Springer, Heidelberg (2008)
4. Davis, M., Logemann, G., Loveland, D.: A machine program for theorem-proving. Commun. ACM 5, 394–397 (1962)
5. Fang, H.: Complete local search for propositional satisfiability. In: Proceedings of AAAI, pp. 161–166 (2004)
6. Gableske, O., Heule, M.J.H.: EagleUP: Solving Random 3-SAT Using SLS with Unit Propagation. In: Sakallah, K.A., Simon, L. (eds.) SAT 2011. LNCS, vol. 6695, pp. 367–368. Springer, Heidelberg (2011)
7. Gomes, C.P., Selman, B., Crato, N.: Heavy-Tailed Distributions in Combinatorial Search. In: Smolka, G. (ed.) CP 1997. LNCS, vol. 1330, pp. 121–135. Springer, Heidelberg (1997)
8. Hirsch, E.A., Kojevnikov, A.: Unitwalk: A new sat solver that uses local search guided by unit clause elimination. Ann. Math. Artif. Intell. 43(1), 91–111 (2005)
9. Huang, J.: A Case for Simple SAT Solvers. In: Bessière, C. (ed.) CP 2007. LNCS, vol. 4741, pp. 839–846. Springer, Heidelberg (2007)
10. Katebi, H., Sakallah, K.A., Marques-Silva, J.P.: Empirical Study of the Anatomy of Modern Sat Solvers. In: Sakallah, K.A., Simon, L. (eds.) SAT 2011. LNCS, vol. 6695, pp. 343–356. Springer, Heidelberg (2011)
11. Lonsing, F., Biere, A.: Depqbf: A dependency-aware qbf solver. JSAT 7(2-3), 71–76 (2010)
12. Pipatsrisawat, K., Darwiche, A.: A lightweight component caching scheme for satisfiability solvers. In: 10th International Conference on Theory and Applications of Satisfiability Testing, pp. 294–299 (2007)
13. Pipatsrisawat, K., Darwiche, A.: On the Power of Clause-Learning SAT Solvers with Restarts. In: Gent, I.P. (ed.) CP 2009. LNCS, vol. 5732, pp. 654–668. Springer, Heidelberg (2009)
14. Ramos, A., van der Tak, P., Heule, M.J.H.: Between Restarts and Backjumps. In: Sakallah, K.A., Simon, L. (eds.) SAT 2011. LNCS, vol. 6695, pp. 216–229. Springer, Heidelberg (2011)
15. Selman, B., Kautz, H., Cohen, B.: Local Search Strategies for Satisfiability Testing. In: Tamassia, R., Tollis, I.G. (eds.) GD 1994. LNCS, vol. 894, pp. 521–532. Springer, Heidelberg (1995)
16. Tompkins, D.A.D., Balint, A., Hoos, H.H.: Captain Jack: New Variable Selection Heuristics in Local Search for SAT. In: Sakallah, K.A., Simon, L. (eds.) SAT 2011. LNCS, vol. 6695, pp. 302–316. Springer, Heidelberg (2011)

An Improved Separation of Regular Resolution from Pool Resolution and Clause Learning

Maria Luisa Bonet[1],[*] and Sam Buss[2],[**]

[1] Lenguajes y Sistemas Informáticos, Universidad Politécnica de Cataluña,
Barcelona, Spain
bonet@lsi.upc.edu
[2] Department of Mathematics, University of California, San Diego, La Jolla, CA
92093-0112, USA
sbuss@math.ucsd.edu

Abstract. We prove that the graph tautology principles of Alekhnovich, Johannsen, Pitassi and Urquhart have polynomial size pool resolution refutations that use only input lemmas as learned clauses and without degenerate resolution inferences. These graph tautology principles can be refuted by polynomial size DPLL proofs with clause learning, even when restricted to greedy, unit-propagating DPLL search.

1 Introduction

DPLL algorithms with clause learning have been highly successful at solving real-world instances of satisfiability (SAT), especially when extended with techniques such as clause learning, restarts, variable selection heuristics, etc. The basic DPLL procedure without clause learning or restarts is equivalent to tree-like resolution. The addition of clause learning makes DPLL considerably stronger. In fact, clause learning together with unlimited restarts is capable of simulating general resolution proofs [12]. However, the exact power of DPLL with clause learning but without restarts is unknown. This question is interesting both for theoretical reasons and for the potential for better understanding the practical performance of DPLL with clause learning.

Beame, Kautz, and Sabharwal [3] gave the first theoretical analysis of DPLL with clause learning. Among other things, they noted that clause learning with restarts simulates full resolution. Their construction required the DPLL algorithm to ignore some contradictions, but this was rectified by Pipatsrisawat and Darwiche [12] who showed that SAT solvers which do not ignore contradictions can also simulate resolution. (See [2] for the bounded width setting.)

[3] also studied DPLL clause learning without restarts. Using "proof trace extensions", they were able to show that DPLL with clause learning and no restarts is strictly stronger than any "natural" proof system strictly weaker than

---

[*] Supported in part by grant TIN2010-20967-C04-02.
[**] Supported in part by NSF grants DMS-0700533 and DMS-1101228, and by a grant from the Simons Foundation (#208717 to Sam Buss).

A. Cimatti and R. Sebastiani (Eds.): SAT 2012, LNCS 7317, pp. 44–57, 2012.
© Springer-Verlag Berlin Heidelberg 2012

resolution. Here, a *natural* proof system is one in which proofs do not increase in length when variables are restricted to constants. However, the proof trace method and the improved constructions of [9,7] have the drawback of introducing extraneous variables and clauses, and using contrived resolution refutations.

There have been two approaches to formalizing DPLL with clause learning as a static proof system rather than as a proof search algorithm. The first approach was pool resolution with a degenerate resolution inference [16,9]. Pool resolution requires proofs to have a depth-first regular traversal similarly to the search space of a DPLL algorithm. Degenerate resolution allows resolution inferences in which the hypotheses may lack occurrences of the resolution literal. Van Gelder [16] argued that pool resolution with degenerate resolution inferences simulates a wide range of DPLL algorithms with clause learning. He also gave a proof, based on [1], that pool resolution with degenerate inferences is stronger than regular resolution, using extraneous variables similar to proof trace extensions.

The second approach [7] is the proof system regWTRI that uses a "partially degenerate" resolution rule called w-resolution, and clause learning of input lemmas. [7] showed that regWRTI exactly captures non-greedy DPLL with clause learning. By "non-greedy" is meant that contradictions may need to be ignored.

It remains open whether any of DPLL with clause learning, pool resolution, or the regWRTI proof system can polynomially simulate general resolution. One approach to answering these questions is to try to separate pool resolution (say) from general resolution. So far, however, separation results are known only for the weaker system of regular resolution; namely, Alekhnovitch et al. [1], gave an exponential separation between regular resolution and general resolution based on two families of tautologies, variants of the graph tautologies GT′ and the "Stone" pebbling tautologies. Urquhart [15] subsequently gave a related separation.[1] In the present paper, we call the tautologies GT′ the *guarded* graph tautologies, and henceforth denote them GGT instead of GT′.

The obvious next question is whether pool resolution (say) has polynomial size proofs of the GGT tautologies or the Stone tautologies. The main result of the present paper resolves the first question by showing that pool resolution does indeed have polynomial size proofs of the graph tautologies GGT. Our proofs apply to the original GGT principles, without the use of extraneous variables in the style of proof trace extensions; and our refutations use only the traditional resolution rule, not degenerate resolution inferences or w-resolution inferences. In addition, we use only learning of input clauses; thus, our refutations are also regWRTI proofs (and in fact regRTI proofs) in the terminology of [7]. As a corollary of the characterization of regWRTI by [7], the GGT principles have polynomial size refutations that can found by a DPLL algorithm with clause learning and without restarts (under the appropriate variable selection order).

It is still open if there are polynomial size pool resolution refutations for the Stone principles. A much more ambitious project would be to show that pool

---

[1] Huang and Yu [10] also gave a separation of regular resolution and general resolution, but only for a single set of clauses. Goerdt [8] gave a quasipolynomial separation of regular resolution and general resolution.

resolution or regWRTI can simulate full resolution, or that DPLL with clause learning and without restarts can simulate full resolution. It is far from clear that this holds, but, if so, our methods may represent a step in that direction.

The first idea for constructing our pool resolution or regRTI proofs might be to try to follow the regular refutations of the graph tautology clauses GT_n as given by [14,5,17]: however, these refutations cannot be used directly since the transitivity clauses of GT_n are "guarded" in the GGT_n clauses and this yields refutations which violate the regularity/pool property. So, the second idea is that the proof search process branches as needed to learn transitivity clauses. This generates additional clauses that must be proved: to handle these, we develop a notion of "bipartite partial order" and show that the refutations of [14,5,17] can still be used in the presence of a bipartite partial order. The tricky part is to be sure that exactly the right set of clauses is derived by each subproof.

Our refutations of the GGT_n tautologies can be modified so that they are "greedy" and "unit-propagating". This means that, at any point in the proof search process, if it is possible to give an "input" refutation of the current clause, then that refutation is used immediately. The greedy and unit-propagating conditions correspond well to actual implemented DPLL proof search algorithms which backtrack whenever a contradiction can be found by unit propagation (c.f., [9]). The paper concludes with a short description of a greedy, unit-propagating DPLL clause learning algorithm for GGT_n.

For space reasons, only the main constructions for our proofs are included in this extended abstract. Complete proofs are in the full version of the paper available at the authors' web pages and at `http://arxiv.org/abs/1202.2296`.

2   Preliminaries and Main Results

Propositional variables range over the values *True* and *False*. The notation $\overline{x}$ expresses the negation of x. A *literal* is either a variable x or a negated variable $\overline{x}$. A *clause* C is a set of literals, interpreted as the disjunction ($\vee$) of its members.

Definition 1. *The various forms of resolution take two premise clauses A and B and a resolution literal x, and produce a new clause C called the* resolvent.

$$\frac{A \qquad B}{C}$$

It is required that $\overline{x} \notin A$ and $x \notin B$. The different forms of resolution are:

Resolution rule. *Here $A := A' \vee x$ and $B := B' \vee \overline{x}$, and C is $A' \vee B'$.*

Degenerate resolution rule. [9,16] *If $x \in A$ and $\overline{x} \in B$, we apply the resolution rule to obtain C. If A contains x, and B doesn't contain $\overline{x}$, then the resolvent C is B. If A doesn't contain x, and B contains $\overline{x}$, then the resolvent C is A. If neither A nor B contains the literal x or $\overline{x}$, then C is the lesser of A or B according to some tiebreaking ordering of clauses.*

w-resolution rule. [7] *Here $C := (A \setminus \{x\}) \vee (B \setminus \{\overline{x}\})$. If the literal $x \notin A$ (resp., $\overline{x} \notin B$), then it is called a* phantom literal *of A (resp., B).*

A *resolution derivation* of a clause C from a set F of clauses is a sequence of clauses that derives C from the clauses of F using resolution. Degenerate and w-resolution derivations are defined similarly. A *refutation* of F is a derivation of the empty clause. A refutation is *tree-like* if its underlying graph is a tree. A resolution derivation is *regular* provided that, along any path in the directed acyclic graph, each variable is resolved at most once and provided that no variable appearing in the final clause is used as a resolution variable.

Resolution is well-known to be sound and complete; in particular, C is a consequence of F iff there is a derivation of some $C' \subseteq C$ from F.

We define pool resolution using the conventions of [7], who called this concept "tree-like regular resolution with lemmas". The idea is that any clause appearing in the proof is a learned lemma and can be used freely from then on.

Definition 2. *The* postorder *ordering $<_T$ of the nodes in a tree T is defined so that if u is a node of T, v a node in the subtree rooted at the left child of u, and w a node in the subtree rooted at the right child of u, then $v <_T w <_T u$.*

Definition 3. *A pool resolution* proof from *a set of initial clauses F is a resolution proof tree T that fulfills the following conditions: (a) each leaf is labeled either with a clause of F or with a clause (called a "lemma") that appears earlier in the tree in the $<_T$ ordering; (b) each internal node is labeled with a clause and a literal, and the clause is obtained by resolution from the clauses labeling the node's children, by resolving on the given literal; (c) the proof tree is regular; (d) the root is labeled with the conclusion clause (the empty clause in the case of a pool refutation).*

The notions of *degenerate pool resolution* proof and *pool w-resolution* proof are defined similarly. Note that [16,9] defined pool resolution to be the degenerate pool resolution system, so our notion of pool resolution is more restrictive than theirs. (Our definition is equivalent to the one in [6], however.)

A "lemma" in part (a) of Definition 3 is called an *input lemma* if it is derived by *input* subderivation, namely by a subderivation in which each inference has at least one hypothesis which is a member of F or is a lemma.

The various graph tautologies, sometimes also called "ordering principles" use a size parameter $n > 1$, and variables $x_{i,j}$ with $i,j \in [n]$ and $i \neq j$, where $[n] = \{0,1,2,\ldots,n-1\}$. A variable $x_{i,j}$ will intuitively represent the condition that $i \prec j$ with $\prec$ intended to be a total, linear order. We thus adopt the convention that $x_{i,j}$ and $\overline{x}_{j,i}$ are the identical literal. This identification makes no essential difference to the complexity of proofs of the tautologies, but reduces the number of literals and clauses, and simplifies definitions.

The following tautologies are based on Krishnamurthy [11]. These tautologies, or similar ones, have also been studied by [14,5,1,4,13,17].

Definition 4. *Let $n > 1$. Then* GT_n *is the following set of clauses:*

$(\alpha_\emptyset)$ *The clauses* $\bigvee_{j \neq i} x_{j,i}$, *for each value $i < n$.*
$(\gamma_\emptyset)$ *The transitivity clauses* $T_{i,j,k} := \overline{x}_{i,j} \vee \overline{x}_{j,k} \vee \overline{x}_{k,i}$ *for all distinct i,j,k in $[n]$.*

Note that the clauses $T_{i,j,k}$, $T_{j,k,i}$ and $T_{k,i,j}$ are identical.

The next definition is from [1] who used the notation GT'_n. They used particular functions r and s for their lower bound proof, but since our upper bound proof does not depend on the details of r and s we leave them unspecified. We require that $r(i,j,k) \neq s(i,j,k)$ and that the set $\{r(i,j,k), s(i,j,k)\} \not\subseteq \{i,j,k\}$. W.l.o.g., $r(i,j,k) = r(j,k,i) = r(k,i,j)$, and similarly for s.

Definition 5. *Let $n \geq 1$, and let $r(i,j,k)$ and $s(i,j,k)$ be functions mapping $[n]^3 \to [n]$ as above. The* guarded graph tautology GGT_n *consists of:*

$(\alpha_\emptyset)$ *The clauses* $\bigvee_{j \neq i} x_{j,i}$, *for each value $i < n$.*
$(\gamma'_\emptyset)$ *The* guarded transitivity clauses $T_{i,j,k} \vee x_{r,s}$ and $T_{i,j,k} \vee \overline{x}_{r,s}$, *for all distinct i,j,k in $[n]$, where $r = r(i,j,k)$ and $s = s(i,j,k)$.*

Theorem 1. *The guarded graph tautology principles GGT_n have polynomial size pool resolution refutations.*

Theorem 2. *The guarded graph tautology principles GGT_n have polynomial size, tree-like regular resolution refutations with input lemmas.*

A consequence of Theorem 2 is that the GGT_n clauses can be shown unsatisfiable by non-greedy polynomial size DPLL searches using clause learning. This follows via Theorem 5.6 of [7]. Even better, we can improve the constructions of Theorems 1 and 2 to show that the GGT_n principles can be refuted also by greedy, unit-propagating polynomial size DPLL searches with clause learning.

Definition 6. *Let R be a tree-like regular resolution (or w-resolution) refutation with input lemmas from the initial clauses Γ. Let C be a clause in R. Define $\Gamma(C)$ to be Γ plus every clause $D <_R C$ in R that is derived by an input subproof. Define C^+ to be the set of literals that occur as a literal (or as a literal or phantom literal) in any clause on the path from C down to the root of R.*

The refutation R is greedy *and* unit-propagating *provided that, for each clause C of R, if there is an input derivation from $\Gamma(C)$ of some clause $C' \subseteq C^+$ which does not resolve on any literal in C^+, then C is derived in R by such a derivation.*

Note that, as proved in [3], the condition that there is a input derivation from $\Gamma(C)$ of some $C' \subseteq C^+$ which does not resolve on literals in C^+ is equivalent to the condition that if all literals of C^+ are set false then unit propagation yields a contradiction from $\Gamma(C)$. (In [3], these are called "trivial" proofs.) This justifies the terminology "unit-propagating".

Theorem 3. *The guarded graph tautology principles GGT_n have greedy, unit-propagating, polynomial size, tree-like, regular w-resolution refutations with input lemmas.*

A similar theorem holds for greedy, unit-propagating pool resolution refutations with degenerate resolution inferences.

Theorem 4. *There are DPLL search procedures with clause learning which are greedy, unit-propagating, but do not use restarts, that refute the GGT_n clauses in polynomial time.*

3   Proof of Main Theorems

The following theorem is an important ingredient of our upper bound proof.

Theorem 5. (Stålmarck [14], Bonet-Galesi [5], Van Gelder [17]) *The sets* GT_n *have regular resolution refutations* P_n *of polynomial size* $O(n^3)$.

The refutations P_n can be modified to give refutations of GGT_n by first deriving each transitive clause $T_{i,j,k}$ from the two guarded transitivity clauses of $(\gamma'_\emptyset)$. This however destroys the regularity property, and in fact no polynomial size regular refutations exist for GGT_n [1].

As usual, a *partial order* on $[n]$ is an antisymmetric, transitive relation binary relation on $[n]$. We shall be mostly interested in "partial specifications" of partial orders: partial specifications are not required to be transitive.

Definition 7. *A partial specification, τ, of a partial order is a set of ordered pairs $\tau \subseteq [n] \times [n]$ which are consistent with some (partial) order. The minimal partial order containing τ is the transitive closure of τ. We write $i \prec_\tau j$ to denote $\langle i, j \rangle \in \tau$, and write $i \prec_\tau^* j$ to denote that $\langle i, j \rangle$ is in the transitive closure of τ.*
The τ-minimal elements are the i's such that $j \prec_\tau i$ does not hold for any j.

We are primarily interested in particular kinds of partial orders, called "bipartite" partial orders, which do not have any chain of inequalities $x \prec y \prec z$.

Definition 8. *A bipartite partial order is a binary relation π on $[n]$ with disjoint domain and range. The set of π-minimal elements is denoted M_π.*

Figure 1 shows an example. The bipartiteness of π arises from the fact that M_π and $[n] \setminus M_\pi$ partition $[n]$ into two sets. Note that if $i \prec_\pi j$, then $i \in M_\pi$ and $j \notin M_\pi$. In addition, M_π contains the isolated points of π.

Definition 9. *Let τ be a specification of a partial order. The bipartite partial order π that is associated with τ is defined by letting $i \prec_\pi j$ hold for precisely those i and j such that i is τ-minimal and $i \prec_\tau^* j$.*

It is easy to check that π is a bipartite partial order. The intuition is that π retains only the information about whether $i \prec_\tau^* j$ for minimal elements i, and forgets the ordering that τ imposes on non-minimal elements. (See Fig. 1.)

Definition 10. *Let π be a bipartite partial order on $[n]$. Then $\mathrm{GT}_{\pi,n}$ is the set of clauses containing:*

(α) The clauses $\bigvee_{j \neq i} x_{j,i}$, for each value $i \in M_\pi$.
(β) The transitivity clauses $T_{i,j,k} := \overline{x}_{i,j} \vee \overline{x}_{j,k} \vee \overline{x}_{k,i}$ for all distinct i, j, k in M_π.
(Vertices i, j, k' in Fig. 2 show an example.)
(γ) The transitivity clauses $T_{i,j,k}$ for all distinct i, j, k such that $i, j \in M_\pi$ and $i \not\prec_\pi k$ and $j \prec_\pi k$. (As shown in Fig. 2.)

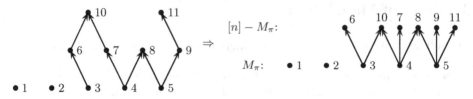

Fig. 1. Example of a partial specification of a partial order (left) and the associated bipartite partial order (right)

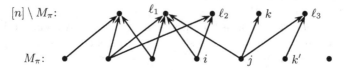

Fig. 2. A bipartite partial order π is pictured, with the ordered pairs of π shown as directed edges. (For instance, $j \prec_\pi k$ holds.) The nodes i, j, k shown are an example of nodes used for a transitivity axiom $\overline{x}_{i,j} \vee \overline{x}_{j,k} \vee \overline{x}_{k,i}$ of type (γ). The nodes i, j, k' are an example of the nodes for a transitivity axiom of type (β).

$\mathrm{GT}_{\pi,n}$ is satisfiable if π is nonempty, for example by the assignment that sets $x_{j,i}$ true for some fixed $j \notin M_\pi$ and every $i \in M_\pi$, and sets all other variables false. However, there is no assignment which satisfies $\mathrm{GT}_{\pi,n}$ and is consistent with π. This fact is proved by the regular derivation P_π of Lemma 1.

Definition 11. *For π a bipartite partial order, the clause $(\bigvee \overline{\pi})$ is defined by*

$$\left(\bigvee \overline{\pi} \right) := \{\overline{x}_{i,j} : i \prec_\pi j\},$$

Lemma 1. *Let π be a bipartite partial order on $[n]$. Then there is a regular derivation P_π of $(\bigvee \overline{\pi})$ from the set $\mathrm{GT}_{\pi,n}$.*
The only variables resolved on in P_π are the following: the variables $x_{i,j}$ such that $i, j \in M_\pi$, and the variables $x_{i,k}$ such that $k \notin M_\pi$, $i \in M_\pi$, and $i \not\prec_\pi k$.

Lemma 1 implies that if π is the bipartite partial order associated with a partial specification τ of a partial order, then the derivation P_π does not resolve on any literal whose value is set by τ. This is proved by noting that if $i \prec_\tau j$, then $j \notin M_\pi$.

If π is empty, $M_\pi = [n]$ and there are no clauses of type (γ). In this case, $\mathrm{GT}_{\pi,n}$ is identical to GT_n, and P_π is the refutation of GT_n of Theorem 5.

Lemma 1 is proved similarly to Theorem 5, taking care to resolve on variables in the correct order. The proof is left to the full version of the paper.

Proof (of Theorem 1). We will construct a series of "LR partial refutations", denoted $R_0, R_1, R_2, \ldots$; this process eventually terminates with a pool refutation of GGT_n. The terminology "LR partial" indicates that the refutation is being constructed in left-to-right order, with the left part of the refutation properly formed, but with many of the remaining leaves being labeled with bipartite partial orders instead of with valid learned clauses or initial clauses from GGT_n.

An LR partial refutation R is a tree with nodes labeled with clauses that form a correct pool resolution proof, except possibly at the leaves (the initial clauses). Furthermore, it must satisfy the following conditions.

a. R is a tree. The root is labeled with the empty clause. Each non-leaf node in R has a left child and right child; the clause labeling the node is derived by resolution from the clauses on its two children.
b. For C a clause occurring in R, define $\tau(C)$ to be the set of ordered pairs $\langle i, j \rangle$ such that $\overline{x}_{i,j} \in C^+$. Note that $C \subseteq C^+$ by definition. In many cases, $\tau(C)$ will be a partial specification of a partial order, but this is not always true. For instance, if C is a transitivity axiom, $\tau(C)$ has a 3-cycle and is not consistent as a specification of a partial order.
c. Leaves of R are flagged as "finished" or "unfinished".
d. Each finished leaf L is labeled with either a clause from GGT_n or a clause that occurs to the left of L in the postorder traversal of R.
e. For an unfinished leaf labeled with clause C, the set $\tau(C)$ is a partial specification of a partial order. Furthermore, letting π be the bipartite partial order associated with $\tau(C)$, the clause C is equal to $(\bigvee \overline{\pi})$.

Property e. is crucial for avoiding degenerate resolution inferences, and is a novel part of our construction. As shown below, each unfinished leaf, labeled with a clause $C = (\bigvee \overline{\pi})$, will be replaced by a derivation S. The derivation S often will be based on P_π, and thus might be expected to end with exactly the clause C; however, some of the resolution inferences needed for P_π might be disallowed by the pool property. So S will instead be a derivation of a clause C' such that $C \subseteq C' \subseteq C^+$. The condition $C' \subseteq C^+$ is required because any literal $x \in C' \setminus C$ will be handled by modifying the refutation R by propagating x downward in R until reaching a clause that already contains x. The condition $C' \subseteq C^+$ ensures that such a clause exists. The fact that $C' \supseteq C$ means that enough literals are present for the derivation to use only (non-degenerate) resolution inferences — indeed our constructions will pick C so that it contains the literals that must be present for use as resolution literals.

The construction begins by letting R_0 be the "empty" refutation, containing just the empty clause. Of course, this clause is an unfinished leaf, and $\tau(\emptyset) = \emptyset$.

Assume R_i has been already constructed, with C the leftmost unfinished clause. R_{i+1} will be formed by replacing C by a refutation S of some clause C' such that $C \subseteq C' \subseteq C^+$.

We need to describe the (LR partial) refutation S. By e., C is $(\bigvee \overline{\pi})$. The intuition is that we would like to let S be the derivation P_π of C from Lemma 1. The first difficulty with this is that P_π is dag-like, and the LR-refutation is intended to be tree-like. This difficulty, however, can be circumvented by just expanding P_π, which is regular, into a tree-like regular derivation with lemmas by the simple expedient of using a depth-first traversal of P_π. The second, and more serious, difficulty is that P_π is a derivation from GT_n, not GGT_n; namely, P_π uses the transitivity clauses of GT_n instead of the guarded transitivity clauses of GGT_n. These transitivity clauses $T_{i,j,k}$ are handled one at a time treating them, as needed, with four separate cases. Case (i) requires no change to P_π; cases (ii) and (iii) require a small change; and case (iv) abandons the subproof P_π and instead "learns" the transitivity clause.

By the remark made after Lemma 1, no literal in C^+ is used as a resolution literal in P_π.

(i) If an initial transitivity clause of P_π already appears earlier in R_i (that is, to the left of C), then it is already *learned*, and can be used freely in P_π.

In the remaining cases (ii)-(iv), the transitivity clause $T_{i,j,k}$ is not yet learned. Let the guard variable for $T_{i,j,k}$ be $x_{r,s}$, so $r = r(i, j, k)$ and $s = s(i, j, k)$.

(ii) Suppose case (i) does not apply and that the guard variable $x_{r,s}$ or its negation $\overline{x}_{r,s}$ is a member of C^+. The guard variable thus is used as a resolution variable somewhere along the branch from the root to clause C. Then, as just argued above, Lemma 1 implies that $x_{r,s}$ is not resolved on in P_π. Therefore, we can add the literal $x_{r,s}$ or $\overline{x}_{r,s}$ (respectively) to the clause $T_{i,j,k}$ and to every clause on any path below $T_{i,j,k}$ until reaching a clause that already contains that literal. This replaces $T_{i,j,k}$ with one of the initial clauses $T_{i,j,k} \vee x_{r,s}$ or $T_{i,j,k} \vee \overline{x}_{r,s}$ of GGT_n. By construction, it preserves the validity of the resolution inferences of R_i as well as the regularity property. Note this adds the literal $x_{r,s}$ or $\overline{x}_{r,s}$ to the final clause C' of the modified P_π. This maintains the property that $C \subseteq C' \subseteq C^+$.

(iii) Suppose case (i) does not apply and that $x_{r,s}$ is not used as a resolution variable anywhere below $T_{i,j,k}$ in P_π and is not a member of C^+. In this case, P_π is modified so as to derive the clause $T_{i,j,k}$ from the two GGT_n clauses $T_{i,j,k} \vee x_{r,s}$ and $T_{i,j,k} \vee \overline{x}_{r,s}$ by resolving on $x_{r,s}$. This maintains the regularity of the derivation. And, henceforth $T_{i,j,k}$ will be learned.

If all of the transitivity clauses in P_π can be handled by cases (i)-(iii), then we use P_π to define R_{i+1}. Namely, let P'_π be the derivation P_π as modified by the applications of cases (ii) and (iii). The derivation P'_π is regular and dag-like, so we can recast it as a tree-like derivation S with lemmas, by using a depth-first traversal of P'_π. The size of S is linear in the size of P'_π, since the only new clauses in S are clauses which are repeated as lemmas and, as an overestimate, there are at most two lemmas per clause in P'_π. The final line of S is the clause C', namely C plus the literals introduced by case (ii). The derivation R_{i+1} is formed from R_i by replacing the clause C with the derivation S of C', and then propagating each new literal $x \in C' \setminus C$ downward, adding x to clauses below S until reaching a clause that already contains x. Since S contains no unfinished leaf, R_{i+1} contains one fewer unfinished leaves than R_i.

On the other hand, if even one transitivity axiom $T_{i,j,k}$ in P_π is not covered by the above three cases, then case (iv) must be used instead. This introduces a completely different construction to form S:

(iv) Let $T_{i,j,k}$ be any transitivity axiom in P_π that is not covered by cases (i)-(iii). The guard variable $x_{r,s}$ is used as a resolution variable in P_π somewhere below $T_{i,j,k}$; in general, this means we cannot use resolution on $x_{r,s}$ to derive $T_{i,j,k}$ while maintaining the pool property. Hence, P_π is no longer used, and we instead form S with a short left-branching path that "learns" $T_{i,j,k}$. This will generate two or three new unfinished leaf nodes. Since unfinished leaf nodes in a LR partial derivation must be labeled with clauses from bipartite partial orders, it is also necessary to attach short derivations to these unfinished leaf nodes to make the unfinished leaf clauses of S correspond correctly to bipartite partial orders. These unfinished leaf nodes are then kept in R_{i+1} to be handled at later stages. There are separate constructions depending on whether $T_{i,j,k}$ is a clause of type (β) or (γ); some of the details are given below.

First suppose $T_{i,j,k}$ is of type (γ), and thus $\overline{x}_{j,k}$ appears in C. (Refer to Fig. 2.) Let $x_{r,s}$ be the guard variable for the transitivity axiom $T_{i,j,k}$. The derivation S will have the form

$$
\frac{
\dfrac{T_{i,j,k}, x_{r,s} \qquad T_{i,j,k}, \overline{x}_{r,s}}{T_{i,j,k}}
\qquad
\dfrac{S_1 \cdots \vdots \cdots \qquad}{\overline{x}_{i,j}, \overline{x}_{i,k}, \overline{\pi}_{-[jk;jR(i)]}}
}{
\dfrac{\overline{x}_{i,j}, \overline{x}_{j,k}, \overline{\pi}_{-[jk;jR(i)]} \qquad \dfrac{S_2 \cdots \vdots \cdots}{\overline{x}_{j,i}, \overline{x}_{j,k}, \overline{\pi}_{-[jk;iR(j)]}}}{\overline{x}_{j,k}, \overline{\pi}_{-[jk]}}
}
$$

The notation $\overline{\pi}_{-[jk]}$ denotes the disjunction of the negations of the literals in π omitting the literal $\overline{x}_{j,k}$. We write "$iR(j)$" to indicate literals $x_{i,\ell}$ such that $j \prec_\pi \ell$. (The "$R(j)$" means "range of j".) Thus $\overline{\pi}_{-[jk;iR(j)]}$ denotes the clause containing the negations of the literals in π, omitting $\overline{x}_{j,k}$ and any literals $\overline{x}_{i,\ell}$ such that $j \prec_\pi \ell$. The clause $\overline{\pi}_{-[jk;jR(i)]}$ is defined similarly, and the notation extends in the obvious way.

The upper leftmost inference of S is a resolution inference on the variable $x_{r,s}$. Since $T_{i,j,k}$ is not covered by either case (i) or (ii), the variable $x_{r,s}$ does not appear in or below clause C in R_i. Thus, this use of $x_{r,s}$ as a resolution variable does not violate regularity. Furthermore, since $T_{i,j,k}$ is of type (γ), we have $i \not\prec_{\tau(C)} j$, $j \not\prec_{\tau(C)} i$, $i \not\prec_{\tau(C)} k$, and $k \not\prec_{\tau(C)} i$. Thus the literals $x_{i,j}$ and $x_{i,k}$ do not appear in or below C, so they also can be resolved on without violating regularity.

Let C_1 and C_2 be the final clauses of S_1 and S_2, and let C_1^- be the clause below C_1 and above C. The set $\tau(C_2)$ is obtained by adding $\langle j, i \rangle$ to $\tau(C)$, and similarly $\tau(C_1^-)$ is $\tau(C)$ plus $\langle i, j \rangle$. Since $T_{i,j,k}$ is type (γ) we have $i, j \in M_\pi$. Therefore, since $\tau(C)$ is a partial specification of a partial order, $\tau(C_2)$ and $\tau(C_1^-)$ are also both partial specifications of partial orders. Let π_2 and π_1 be the bipartite orders associated with these two partial specifications (respectively). We will form the subproof S_1 so that it contains the clause $(\bigvee \overline{\pi}_1)$ as its only unfinished clause. This will require adding inferences in S_1 which add and remove the appropriate literals. The first step of this type already occurs in going up from C_1^- to C_1 since this has removed $\overline{x}_{j,k}$ and added $\overline{x}_{i,k}$, reflecting the fact that j is not π_1-minimal and thus $x_{i,k} \in \pi_1$ but $x_{j,k} \notin \pi_1$. Similarly, we will form S_2 so that its only unfinished clause is $(\bigvee \overline{\pi}_2)$.

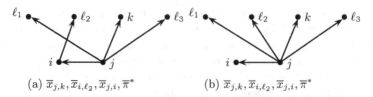

(a) $\overline{x}_{j,k}, \overline{x}_{i,\ell_2}, \overline{x}_{j,i}, \overline{\pi}^*$ (b) $\overline{x}_{j,k}, \overline{x}_{i,\ell_2}, \overline{x}_{j,i}, \overline{\pi}^*$

Fig. 3. The partial orders for the fragment of S_2 shown in (1)

The situation for the subproof S_2 is shown in Fig. 3, which shows an extract from Fig. 2: the edges shown in part (a) of the figure correspond to the literals in the final line C_2 of S_2. Recall that literals $\overline{x}_{i,\ell}$ such that $j \prec_\tau \ell$ are omitted from the last line of S_2. (Correspondingly, the edge from i to ℓ_1 is omitted from Fig. 3.) C_2 may not correspond to a bipartite partial order as it may not partition $[n]$ into minimal and non-minimal elements; thus, C_2 may not qualify to be an unfinished node of R_{i+1}. (An example of this in Fig. 3(a) is that $j \prec_{\tau(C_2)} i \prec_{\tau(C_2)} \ell_2$, corresponding to $\overline{x}_{j,i}$ and $\overline{x}_{i,\ell_2}$ being in C_2.) The bipartite partial order π_2 associated with $\tau(C_2)$ is equal to the bipartite partial order that agrees with π except that each $i \prec_\pi \ell$ condition is replaced with the condition $j \prec_{\pi_2} \ell$. (This is represented in Fig. 3(b) by the fact that the edge from i to ℓ_2 has been replaced by the edge from j to ℓ_2. Note that the vertex i is no longer a minimal element of π_2; that is, $i \notin M_{\pi_2}$.) We wish to form S_2 to be a regular derivation of the clause $\overline{x}_{j,i}, \overline{\pi}_{-[jk;iR(j)]}$ from the clause $(\bigvee \overline{\pi}_2)$.

The subproof of S_2 for replacing $\overline{x}_{i,\ell_2}$ in $\overline{\pi}$ with $\overline{x}_{j,\ell_2}$ in $\overline{\pi}_2$ is

$$\frac{S_2' \cdots \vdots \cdots \qquad\qquad \cdots \vdots \cdots \text{rest of } S_2}{\overline{x}_{j,i}, \overline{x}_{i,\ell_2}, \overline{x}_{\ell_2,j} \qquad \overline{x}_{j,k}, \overline{x}_{j,\ell_2}, \overline{x}_{j,i}, \overline{\pi}^*}{\overline{x}_{j,k}, \overline{x}_{i,\ell_2}, \overline{x}_{j,i}, \overline{\pi}^*} \tag{1}$$

where $\overline{\pi}^*$ is $\overline{\pi}_{-[jk;iR(j);i\ell_2]}$. The part labeled "rest of S_2" will handle similarly the other literals ℓ such that $i \prec_\pi \ell$ and $j \not\prec_\pi \ell$. The final line of S_2' is T_{j,i,ℓ_2}. This is a GT_n axiom, not a GGT_n axiom; however, it can be handled by the methods of cases (i)-(iii). Namely, if T_{j,i,ℓ_2} has already been learned by appearing somewhere to the left in R_i, then S_2' is just this single clause. Otherwise, let the guard variable for T_{j,i,ℓ_2} be $x_{r',s'}$. If $x_{r',s'}$ is used as a resolution variable below T_{j,i,ℓ_2}, then replace T_{j,i,ℓ_2} with $T_{j,i,\ell_2} \vee x_{r',s'}$ or $T_{j,i\ell_2} \vee \overline{x}_{r',s'}$, and propagate the $x_{r',s'}$ or $\overline{x}_{r',s'}$ to clauses down the branch leading to T_{j,i,ℓ_2} until reaching a clause that already contains that literal. Finally, if $x_{r',s'}$ has not been used as a resolution variable in R_i below C, then let S_2' consist of a resolution inference deriving (and learning) T_{j,i,ℓ_2} from the clauses $T_{j,i,\ell_2}, x_{r',s'}$ and $T_{j,i,\ell_2}, \overline{x}_{r',s'}$.

To complete the construction of S_2, the inference (1) is repeated for each value of ℓ such that $i \prec_\pi \ell$ and $j \not\prec_\pi \ell$. The result is that S_2 has one unfinished leaf clause, and it is labelled with the clause $(\bigvee \overline{\pi}_2)$.

We next describe the subproof S_1 of S. The situation is shown in Fig. 4. As in the formation of S_2, the final clause C_1 in S_1 may need to be modified in order to correspond to the bipartite partial order π_1 which is associated with $\tau(C_1)$.

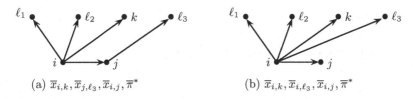

Fig. 4. The partial orders for the fragment of S_1 shown in (2)

First, note that the literal $\overline{x}_{j,k}$ is already replaced by $\overline{x}_{i,k}$ in the final clause of S_1. The other change that is needed is that, for every ℓ such that $j \prec_\pi \ell$ and $i \not\prec_\pi \ell$, we must replace $\overline{x}_{j,\ell}$ with $\overline{x}_{i,\ell}$ since we have $j \not\prec_{\pi_1} \ell$ and $i \prec_{\pi_1} \ell$. Vertex ℓ_3 in Fig. 4 is an example of a such a value ℓ. The ordering in the final clause of S_1 is shown in part (a), and the desired ordered pairs of π_1 are shown in part (b). Note that j is no longer a minimal element in π_1.

The replacement of $\overline{x}_{j,\ell_3}$ with $\overline{x}_{i,\ell_3}$ is effected by the following inference, letting $\overline{\pi}^*$ now be $\overline{\pi}_{-[jk;jR(i);j\ell_3]}$.

$$
\frac{S_1' \cdots \vdots \cdots \qquad\qquad \cdots \vdots \cdots \text{rest of } S_1}{\overline{x}_{i,j}, \overline{x}_{j,\ell_3}, \overline{x}_{\ell_3,i} \qquad \overline{x}_{i,k}, \overline{x}_{i,\ell_3}, \overline{x}_{i,j}, \overline{\pi}^*}
$$
$$
\overline{x}_{i,k}, \overline{x}_{j,\ell_3}, \overline{x}_{i,j}, \overline{\pi}^*
$$

$$(2)$$

The "rest of S_1" will handle similarly the other literals ℓ such that $j \prec_\pi \ell$ and $i \not\prec_\pi \ell$. Note that the final clause of S_1' is the transitivity axiom T_{i,j,ℓ_3}. The subproof S_1' is formed in the same way that S_2' was formed above. Namely, depending on the status of the guard variable $x_{r',s'}$ for T_{i,j,ℓ_3}, one of the following is done: (i) the clause T_{i,j,ℓ_3} is already learned and can be used as is, or (ii) one of $x_{r',s'}$ or $\overline{x}_{r',s'}$ is added to the clause and propagated down the proof, or (iii) the clause T_{i,j,ℓ_3} is inferred using resolution on $x_{r',s'}$ and becomes learned.

To complete the construction of S_1, the inference (2) is repeated for each value of ℓ such that $j \prec_\pi \ell$ and $i \not\prec_\pi \ell$. The result is that S_1 has one unfinished leaf clause, and it corresponds to the bipartite partial order π_1.

That completes the construction of the subproof S for the subcase of (iv) where $T_{i,j,k}$ is of type (γ). Now suppose $T_{i,j,k}$ is of type (β). (For instance, the values i, j, k' of Fig. 2.) In this case the derivation S will have the form

$$
\frac{\dfrac{T_{i,j,k}, x_{r,s} \quad T_{i,j,k}, \overline{x}_{r,s}}{T_{i,j,k}} \quad \dfrac{S_3 \cdots \vdots \cdots}{\overline{x}_{i,j}, \overline{x}_{i,k}, \overline{\pi}_{-[jR(i),kR(i\cup j)]}}}{\overline{x}_{i,j}, \overline{x}_{j,k}, \overline{\pi}_{-[jR(i),kR(i\cup j)]}} \quad \dfrac{S_4 \cdots \vdots \cdots}{\overline{x}_{i,j}, \overline{x}_{k,j}, \overline{\pi}_{-[jR(i\cap k)]}} \quad \dfrac{S_5 \cdots \vdots \cdots}{\overline{x}_{j,i}, \overline{\pi}_{-[iR(j)]}}}{\dfrac{\overline{x}_{i,j}, \overline{\pi}_{-[jR(i\cap k)]}}{\overline{\pi}}}
$$

where $x_{r,s}$ is the guard variable for $T_{i,j,k}$. We write $[\overline{\pi}_{-[jR(i\cap k)]}]$ to mean the negations of literals in π omitting any literal $\overline{x}_{j,\ell}$ such that $i \prec_\pi \ell$ and $k \prec_\pi \ell$. Similarly, $\overline{\pi}_{-[jR(i),kR(i\cup j)]}$ indicates the negations of literals in π, omitting the literals $\overline{x}_{j,\ell}$ such that $i \prec_\pi \ell$ and the literals $\overline{x}_{k,\ell}$ such that $i \prec_\pi \ell$ or $j \prec_\pi \ell$.

Note that the resolution on $x_{r,s}$ used to derive $T_{i,j,k}$ does not violate regularity, since otherwise $T_{i,j,k}$ would have been covered by case (ii). Likewise, the resolutions on $x_{i,j}$, $x_{i,k}$, and $x_{j,k}$ do not violate regularity since $T_{i,j,k}$ is of type (β).

The subproofs S_3, S_4, and S_5 are handled similarly to the way the subproofs S_1 and S_2 were handled above, albeit with some extra complications in the S_4 case. The detailed constructions are in the full version of the paper.

Once some R_i has no unfinished clauses, we have the desired pool refutation. We claim that the process stops after polynomially many stages.

To prove this, recall that R_{i+1} is formed by handling the leftmost unfinished clause using one of cases (i)-(iv). In the first three cases, the unfinished clause is replaced by a derivation based on P_π. Since P_π has size $O(n^3)$, this means that the number of clauses in R_{i+1} is at most the number of clauses in R_i plus $O(n^3)$. Also, by construction, R_{i+1} has one fewer unfinished clauses than R_i. In case (iv) however, R_{i+1} is formed by adding up to $O(n)$ many clauses to R_i plus adding either two or three new unfinished leaf clauses. However, case (iv) always causes at least one transitivity axiom $T_{i,j,k}$ to be learned. Therefore, case (iv) can occur at most $2\binom{n}{3} = O(n^3)$ times. Consequently at most $3 \cdot 2\binom{n}{3} = O(n^3)$ many unfinished clauses are added throughout the entire process. It follows that the process stops with R_i having no unfinished clauses for some $i \leq 6\binom{n}{3} = O(n^3)$. Therefore there is a pool refutation of GGT_n with $O(n^6)$ lines.

By inspection, each clause in the refutation contains $O(n^2)$ literals. This is because the largest clauses are those corresponding to (small modifications of) bipartite partial orders, and because bipartite partial orders can contain at most $O(n^2)$ many ordered pairs. Furthermore, the refutations P_n for the graph tautology GT_n contain only clauses of size $O(n^2)$.   Q.E.D. Theorem 1

The proofs of Theorems 2 and 3 are left to the full version of the paper, but use similar methods. Theorem 4 follows from the algorithm implicit in the proof of Theorem 3. The following gives a sketch of the algorithm for DPLL search with clause learning which always succeeds in finding a refutation of the GGT_n clauses. At each point in the DPLL search procedure, there is a partial assignment τ, and the search algorithm must do one of the following:

(1) If unit propagation yields a contradiction, then learn a clause $T_{i,j,k}$ if possible, and backtrack.
(2) Otherwise, if there are any literals in the bipartite partial order π associated with τ which are not assigned a value, branch on one of these literals to set its value.
(3) Otherwise, determine whether there is a clause $T_{i,j,k}$ which is used in the proof P_π whose guard literals are resolved on in P_π. (See Lemma 1.) If not, do a DPLL traversal of P_π, eventually backtracking from the assignment τ.
(4) Otherwise, let $T_{i,j,k}$ block P_π from being traversed, and branch on its variables in the order given in the above proof. From this, learn the clause $T_{i,j,k}$.

Acknowledgements. We thank J. Hoffmann and J. Johannsen for a correction to an earlier version of the proof of Theorem 2, and A. Van Gelder, A. Beckmann, and T. Pitassi for encouragement, suggestions, and comments.

References

1. Alekhnovich, M., Johannsen, J., Pitassi, T., Urquhart, A.: An exponential separation between regular and general resolution. Theory of Computation 3(4), 81–102 (2007)
2. Atserias, A., Fichte, J.K., Thurley, M.: Clause-learning algorithms with many restarts and and bounded-width resolution. Journal of Artificial Intelligence Research 40, 353–373 (2011)
3. Beame, P., Kautz, H.A., Sabharwal, A.: Towards understanding and harnessing the potential of clause learning. J. Artificial Intelligence Research 22, 319–351 (2004)
4. Beckmann, A., Buss, S.R.: Separation results for the size of constant-depth propositional proofs. Annals of Pure and Applied Logic 136, 30–55 (2005)
5. Bonet, M.L., Galesi, N.: A study of proof search algorithms for resolution and polynomial calculus. In: 40th Annual IEEE Symp. on Foundations of Computer Science, pp. 422–431. IEEE Computer Society (1999)
6. Buss, S.R.: Pool resolution is NP-hard to recognise. Archive for Mathematical Logic 48(8), 793–798 (2009)
7. Buss, S.R., Hoffmann, J., Johannsen, J.: Resolution trees with lemmas: Resolution refinements that characterize DLL-algorithms with clause learning. Logical Methods of Computer Science 4, 4:13(4:13), 1–18 (2008)
8. Goerdt, A.: Regular resolution versus unrestricted resolution. SIAM Journal on Computing 22(4), 661–683 (1993)
9. Hertel, P., Bacchus, F., Pitassi, T., Van Gelder, A.: Clause learning can effectively p-simulate general propositional resolution. In: Proc. 23rd AAAI Conf. on Artificial Intelligence (AAAI 2008), pp. 283–290. AAAI Press (2008)
10. Huang, W., Yu, X.: A DNF without regular shortest consensus path. SIAM Journal on Computing 16(5), 836–840 (1987)
11. Krishnamurthy, B.: Short proofs for tricky formulas. Acta Informatica 22(3), 253–275 (1985)
12. Pipatsrisawat, K., Darwiche, A.: On the power of clause-learning sat solvers as resolution engines. Artificial Intelligence 172(2), 512–525 (2011)
13. Segerlind, N., Buss, S.R., Impagliazzo, R.: A switching lemma for small restrictions and lower bounds for k-DNF resolution. SIAM Journal on Computing 33(5), 1171–1200 (2004)
14. Stålmarck, G.: Short resolution proofs for a sequence of tricky formulas. Acta Informatica 33(3), 277–280 (1996)
15. Urquhart, A.: A near-optimal separation of regular and general resolution. SIAM Journal on Computing 40(1), 107–121 (2011)
16. Van Gelder, A.: Pool Resolution and Its Relation to Regular Resolution and DPLL with Clause Learning. In: Sutcliffe, G., Voronkov, A. (eds.) LPAR 2005. LNCS (LNAI), vol. 3835, pp. 580–594. Springer, Heidelberg (2005)
17. Van Gelder, A.: Preliminary Report on Input Cover Number as a Metric for Propositional Resolution Proofs. In: Biere, A., Gomes, C.P. (eds.) SAT 2006. LNCS, vol. 4121, pp. 48–53. Springer, Heidelberg (2006)

Computing Resolution-Path Dependencies in Linear Time[*],[**]

Friedrich Slivovsky and Stefan Szeider

Institute of Information Systems, Vienna University of Technology, A-1040 Vienna, Austria
friedrich.slivovsky@tuwien.ac.at, stefan@szeider.net

Abstract. The alternation of existential and universal quantifiers in a quantified boolean formula (QBF) generates dependencies among variables that must be respected when evaluating the formula. Dependency schemes provide a general framework for representing such dependencies. Since it is generally intractable to determine dependencies exactly, a set of potential dependencies is computed instead, which may include false positives. Among the schemes proposed so far, resolution path dependencies introduce the fewest spurious dependencies. In this work, we describe an algorithm that detects resolution-path dependencies in linear time, resolving a problem posed by Van Gelder (CP 2011).

1 Introduction

Deciding the satisfiability of quantified boolean formulas (QBF) is a canonical PSPACE-complete problem [14]. Under standard complexity theoretic assumptions, that means it is much harder than testing satisfiability of propositional formulas. The source of this discrepancy can be found in variable dependencies introduced by the alternation of universal and existential quantifiers in a QBF. The kind of dependencies we consider can be illustrated with the following example:

$$\mathcal{F} = \forall x \exists y \, (x \vee \neg y) \wedge (\neg x \vee y)$$

While $\mathcal{F}$ is satisfiable, there is no single satisfying assignment to y. Instead, the value of y that satisfies $\mathcal{F}$ *depends* on the value of x.

For formulas in prenex normal form, it is safe to assume that a variable depends on all variables to its left in the quantifier prefix, but this assumption may result in a large number of spurious dependencies. More accurate representations of the dependency structure in a formula can be exploited for various purposes, and variable dependencies have been studied in a series of works, including [1–4, 9–12, 15].

Unfortunately, the problem of computing variable dependencies exactly is PSPACE-complete [12]. In practice one therefore computes an over-approximation of dependencies that may contain false positives. This leads to a trade-off between tractability and generality.

---

[*] Research supported by the European Research Council (ERC), project COMPLEX REASON 239962, and WWTF grant WWTF016.

[**] Dedicated to the memory of Marko Samer.

A. Cimatti and R. Sebastiani (Eds.): SAT 2012, LNCS 7317, pp. 58–71, 2012.

In a recent paper, Van Gelder [15] introduced *resolution-path dependencies* and argued that they generate fewer spurious dependencies than all previously considered notions of variable dependency (see Figure 1).

Van Gelder stated as an open problem whether resolution-path dependencies can be computed in polynomial time [15]. In this work, we solve this problem by describing a linear-time algorithm that identifies resolution-path dependencies. We obtain this result by a reduction to the problem of finding properly colored walks in edge-colored graphs, which is in turn solved using a variant of breadth-first search. We thus show that the most general dependency relation among those considered so far is tractable.

Dependency schemes are a generic framework for representing variable dependencies [12] that are useful in various settings. In particular, they have recently been built into state-of-the-art QBF solvers, with beneficial effects [9, 10]. We prove that resolution-path dependencies give rise to a dependency scheme, thereby providing a basis for their use across a variety of applications.

The proofs of statements marked with ($\star$) have been omitted due to space constraints. They can be found in the full version of this paper, which is available on arXiv:1202.3097.

2 Preliminaries

2.1 Quantified Boolean Formulas

In this section, we cover basic definitions and notation used throughout the paper. For an in-depth treatment of theoretical and practical aspects of QBFs, we refer the reader to [6] and [5], respectively.

We consider quantified boolean formulas in *quantified conjunctive normal form* (QCNF). A QCNF formula consists of a (quantifier) *prefix* and a CNF formula, called the *matrix*. A CNF formula is a finite conjunction of *clauses*, where each clause is a finite disjunction of *literals*. We identify a CNF formula with the set of its clauses, and a clause with the set of its literals. Literals are negated or unnegated propositional *variables*. If x is a variable, we put $\overline{x} = \neg x$ and $\overline{\neg x} = x$, and let $var(x) = var(\neg x) = x$.

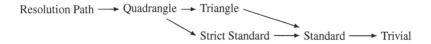

Fig. 1. Various notions of variable dependency ordered by generality [15]. An arrow from A to B should be read as "A is strictly more general than B." *Trivial dependencies* include all pairs of variables not contained in the same quantifier block as dependent and serve as a baseline. *Standard dependencies* [12] identify dependencies based on a notion of local connectivity of clauses, extending ideas introduced in work on universal expansion [2, 3]. *Triangle dependencies* generalize standard dependencies without increasing the worst-case asymptotic runtime [12]. *Quadrangle dependencies* in turn refine triangle dependencies, and *strict standard dependencies* refine standard dependencies [15]. *Resolution path dependencies* are based on a sophisticated notion of connectivity motivated by properties of Q-resolution [15].

If X is a set of literals, we write $\overline{X}$ for the set $\{\overline{x} : x \in X\}$. For a clause C, we let $var(C)$ be the set of variables occuring (negated or unnegated) in C. For a QCNF formula $\mathcal{F}$ with matrix F, we put $var(\mathcal{F}) = var(F) = \bigcup_{C \in F} var(C)$, and $lit(\mathcal{F}) = var(\mathcal{F}) \cup \overline{var(\mathcal{F})}$. We call a clause *tautological* if it contains the same variable negated as well as unnegated. Unless otherwise stated, we assume that the matrix of a formula does not contain tautological clauses (tautological clauses can be deleted without changing satisfiability of a formula). The prefix of a QCNF formula $\mathcal{F}$ is a sequence $Q_1 x_1 \ldots Q_n x_n$ of *quantifications* $Q_i x_i$, where $x_1, \ldots, x_n$ are pairwise distinct variables in $var(\mathcal{F})$ and $Q_i \in \{\forall, \exists\}$ for $1 \leq i \leq n$. We define the *depth* of variable x_p as $\delta_{\mathcal{F}}(x_p) = p$, and let $q_{\mathcal{F}}(x_p) = Q_p$. A QCNF formula $\mathcal{F}'$ is obtained from $\mathcal{F}$ by *quantifier reordering* if there is a permutation $i_1, \ldots, i_n$ of $1, \ldots, n$ such that $\mathcal{F}' = Q_{i_1} x_{i_1}, \ldots, Q_{i_n} x_{i_n} F$, where F denotes the matrix of $\mathcal{F}$. The sets of *existential* and *universal* variables occurring in $\mathcal{F}$ are given by $var_\exists(\mathcal{F}) = \{x \in var(\mathcal{F}) : q_{\mathcal{F}}(x) = \exists\}$ and $var_\forall(\mathcal{F}) = \{x \in var(\mathcal{F}) : q_{\mathcal{F}}(x) = \forall\}$, respectively. We call a literal ℓ existential (universal) if $var(\ell)$ is existential (universal). We assume that every variable in $var(\mathcal{F})$ appears in the prefix of $\mathcal{F}$, and – conversely – that every variable quantified in the prefix appears in F. The *size* of a QCNF formula $\mathcal{F}$ with matrix F is defined as $|\mathcal{F}| = \sum_{C \in F} |C|$.

For a set X of variables, a *truth assignment* is a mapping $\tau : X \to \{0, 1\}$. We extend τ to literals by setting $\tau(\neg x) = 1 - \tau(x)$, for $x \in X$. Let $\tau : X \to \{0, 1\}$ be a truth assignment and F a CNF formula. By $F[\tau]$ we denote the formula obtained from F by removing all clauses containing a literal ℓ such that $\tau(\ell) = 1$, and removing from every clause all literals ℓ for which $\tau(\ell) = 0$; moreover, if $\mathcal{F}$ is a QCNF formula, we write $\mathcal{F}[\tau]$ for the formula obtained from $\mathcal{F}$ by replacing its matrix F with $F[\tau]$ and deleting all superfluous quantifications in its prefix.

The evaluation function ν on QCNF formulas is recursively defined by $\nu(\exists x \mathcal{F}) = \max(\nu(\mathcal{F}[x \mapsto 0]), \nu(\mathcal{F}[x \mapsto 1]))$, $\nu(\forall x \mathcal{F}) = \min(\nu(\mathcal{F}[x \mapsto 0]), \nu(\mathcal{F}[x \mapsto 1]))$, $\nu(\emptyset) = 1$, and $\nu(\{\emptyset\}) = 0$, where $x \mapsto \varepsilon$ denotes the assignment $\tau : \{x\} \to \{0, 1\}$ such that $\tau(x) = \varepsilon$. A QCNF formula $\mathcal{F}$ is *satisfiable* if $\nu(\mathcal{F}) = 1$ and *unsatisfiable* if $\nu(\mathcal{F}) = 0$. Two formulas $\mathcal{F}$ and $\mathcal{F}'$ are *equivalent* if $\nu(\mathcal{F}) = \nu(\mathcal{F}')$. We call a clause *ternary* if it contains at most three literals. A QCNF formula is ternary if all of the clauses in its matrix are ternary. We denote the class of ternary QCNF formulas by Q3CNF.

2.2 Q-Resolution

Q-resolution [7] is an extension of propositional resolution. Let $\mathcal{F}$ be QCNF formula with matrix F. A *tree-like Q-resolution derivation* of clause D from $\mathcal{F}$ is a pair $\pi = (T, \lambda)$ of a rooted binary tree T and a labeling λ satisfying the following properties. The labeling λ assigns to each node a clause, and to each edge a variable. The leaves of T are labeled with clauses of F, and the root of T is labeled with D. Whenever a node t has two children t' and t'', then there is an existential literal ℓ such that $\ell \in \lambda(t')$, $\overline{\ell} \in \lambda(t'')$, and $\lambda(tt') = \lambda(tt'') = var(\ell)$. Moreover, $\lambda(t) = (\lambda(t') \setminus \{\ell\}) \cup (\lambda(t'') \setminus \{\overline{\ell}\})$ and $\lambda(t)$ is non-tautological. We call $\lambda(t)$ the *(Q-)resolvent* of $\lambda(t')$ and $\lambda(t'')$, and say that $\lambda(t)$ is obtained by *resolution* of $\lambda(t')$ and $\lambda(t'')$ on variable $var(\ell)$. If a node t has a single child t', then $\lambda(t) = \lambda(t') \setminus \{\ell\}$ and $\lambda(tt') = var(\ell)$ for some *tailing* universal literal ℓ in $\lambda(t')$. A universal literal ℓ is *tailing* in $\lambda(t')$ if for all existential variables $x \in var(\lambda(t'))$, we have $\delta_{\mathcal{F}}(x) < \delta_{\mathcal{F}}(var(\ell))$. The clause $\lambda(t)$ is the result of *universal reduction* of $\lambda(t')$ on variable $var(\ell)$. We call an instance of resolution or universal reduction in π a *derivation step* in π. We say π is *strict* if for every path $t_1, \ldots, t_n$ from the root of T to one of its leaves we have $\delta_{\mathcal{F}}(\lambda(t_i t_{i+1})) < \delta_{\mathcal{F}}(\lambda(t_{i+1} t_{i+2}))$, for all $i \in \{1, \ldots, n-2\}$. We call π *regular* if every existential variable appears at most once as an edge-label on a path from the root of T to one of its leaves. For a tree-like Q-resolution derivation $\pi = (T, \lambda)$, we define the set of *resolved variables of* π as $resvar(\pi) = \{ y \in var_{\exists}(\mathcal{F}) : \text{there is an edge } e \in T \text{ such that } \lambda(e) = y \}$. We define the *height* of a tree-like Q-resolution derivation $\pi = (T, \lambda)$ as the height of T. A tree-like Q-resolution derivation of the empty clause from $\mathcal{F}$ is called a *Q-resolution refutation* of $\mathcal{F}$.

Theorem 1. *A QCNF formula $\mathcal{F}$ is unsatisfiable if and only if it has a strict, tree-like Q-resolution refutation.*

Proof. Completeness of "ordinary" Q-resolution is proved in [7]. It is straightforward to turn the derivations used in this proof into strict, tree-like derivations. □

3 Dependency Schemes

For a binary relation $\mathcal{R}$ over some set V we write $\mathcal{R}^*$ to denote the reflexive and transitive *closure* of $\mathcal{R}$, i.e., the smallest set $\mathcal{R}^*$ such that $\mathcal{R}^* = \mathcal{R} \cup \{(x, x) : x \in V\} \cup \{(x, y) : \exists z \text{ such that } (x, z) \in \mathcal{R}^* \text{ and } (z, y) \in \mathcal{R}\}$. Moreover, we let $\mathcal{R}(x) = \{ y : (x, y) \in \mathcal{R} \}$ for $x \in V$, and $\mathcal{R}(X) = \bigcup_{x \in X} \mathcal{R}(x)$ for $X \subseteq V$. For a QCNF formula $\mathcal{F}$, we define the binary relation $R_{\mathcal{F}}$ over $var(\mathcal{F})$ as $R_{\mathcal{F}} = \{ (x, y) : x, y \in var(\mathcal{F}), \delta_{\mathcal{F}}(x) < \delta_{\mathcal{F}}(y) \}$. That is to say, $R_{\mathcal{F}}$ assigns to each variable x the variables on the right of x in the prefix.

Definition 1 (Shifting). *Let $\mathcal{F}$ be a QCNF formula and $X \subseteq var(\mathcal{F})$. We say the QCNF formula $\mathcal{F}'$ is obtained from $\mathcal{F}$ by* down-shifting X, *in symbols $\mathcal{F}' = S^{\downarrow}(\mathcal{F}, X)$, if $\mathcal{F}'$ is obtained from $\mathcal{F}$ by quantifier reordering such that the following conditions hold:*

1. $X = R_{\mathcal{F}'}(x)$ *for some* $x \in var(\mathcal{F}) = var(\mathcal{F}')$.
2. $\delta_{\mathcal{F}'}(x) < \delta_{\mathcal{F}'}(y)$ *if and only if* $\delta_{\mathcal{F}}(x) < \delta_{\mathcal{F}}(y)$ *for all* $x, y \in X$.
3. $\delta_{\mathcal{F}'}(x) < \delta_{\mathcal{F}'}(y)$ *if and only if* $\delta_{\mathcal{F}}(x) < \delta_{\mathcal{F}}(y)$ *for all* $x, y \in var(\mathcal{F}) \setminus X$.

For example, let $\mathcal{F} = \exists x \forall y \exists z \forall u \forall w\ F$, and $X = \{x, z, u\}$. Then $S^{\downarrow}(\mathcal{F}, X) = \forall y \forall w \exists x \exists z \forall u\ F$. Note that the result of shifting is unique. In general, shifting does not yield an equivalent formula.

Definition 2 (Dependency scheme). *A dependency scheme D assigns to each QCNF formula $\mathcal{F}$ a binary relation $D_{\mathcal{F}} \subseteq R_{\mathcal{F}}$ such that $\mathcal{F}$ and $S^{\downarrow}(\mathcal{F}, D_{\mathcal{F}}^*(x))$ are equivalent for all $x \in var(\mathcal{F})$. A dependency scheme D is* tractable *if $D_{\mathcal{F}}$ can be computed in time that is polynomial in $|\mathcal{F}|$.*

Intuitively, for a QCNF formula $\mathcal{F}$, variable $x \in var(\mathcal{F})$, and dependency scheme D, the set $D_{\mathcal{F}}(x)$ consists of variables that *may* depend on x. More specifically, if we want to simplify $\mathcal{F}$ by moving the variable x to the rightmost position in the prefix, we can use a dependency scheme to identify a set X so that down-shifting of $X \cup \{x\}$ preserves satisfiability. Typically, we are interested in dependency schemes that allow us to identify sound shifts for entire sets of variables.

Definition 3 (Cumulative). *A dependency scheme D is* cumulative *if for every QCNF formula $\mathcal{F}$ and set $X \subseteq var(\mathcal{F})$, $\mathcal{F}$ and $S^{\downarrow}(\mathcal{F}, D_{\mathcal{F}}^{*}(X))$ are equivalent.*

Cumulative dependency schemes play a crucial role in the context of backdoor sets [12], and have been integrated in search-based QBF solvers [10].

It is easy to verify that we can transpose adjacent quantifications $Q_x x Q_y y$ in the prefix of a QCNF $\mathcal{F}$ as long as $y \notin D_{\mathcal{F}}(x)$ for some dependency scheme D. In other words, every dependency scheme satisfies the property defined below.

Definition 4 (Sound for Transpositions). *Let D be a function that assigns to each QCNF formula $\mathcal{F}$ a binary relation $D_{\mathcal{F}} \subseteq R_{\mathcal{F}}$. We say D is* sound for transpositions *if any two QCNF formulas $\mathcal{F} = Q_1 x_1 \ldots Q_r x_r Q_{r+1} x_{r+1} \ldots Q_n x_n F$ and $Q_1 x_1 \ldots Q_{r+1} x_{r+1} Q_r x_r \ldots Q_n x_n F$ are equivalent given that $(x_r, x_{r+1}) \notin D_{\mathcal{F}}$.*

Further restrictions are required when going beyond individual transpositions: let $\mathcal{F} = \forall x \exists y \exists z\, F$, where F is the CNF encoding of $z \leftrightarrow (x \vee y)$, and let D be a mapping such that $D(\mathcal{F}) = D_{\mathcal{F}} = \emptyset$ and $D(\mathcal{F}') = R_{\mathcal{F}'}$ for $\mathcal{F}' \neq \mathcal{F}$. $\mathcal{F}$ is satisfiable and remains satisfiable after transposing y and x (or y and z) in the prefix. However, the formula $S^{\downarrow}(\mathcal{F}, D_{\mathcal{F}}^{*}(x)) = \exists y \exists z \forall x\, F$ is unsatisfiable. So D is sound for transpositions but not a dependency scheme.

Definition 5 (Continuous). *Let D be a function that maps each QCNF formula $\mathcal{F}$ to a binary relation $D_{\mathcal{F}} \subseteq R_{\mathcal{F}}$. We say D is* continuous *if the following holds for every pair $\mathcal{F} = Q_1 x_1 \ldots Q_r x_r Q_{r+1} x_{r+1} \ldots Q_n x_n F$ and $\mathcal{F}' = Q_1 x_1 \ldots Q_{r+1} x_{r+1} Q_r x_r \ldots Q_n x_n F$ of QCNF formulas: $D_{\mathcal{F}}(v) = D_{\mathcal{F}'}(v)$ for $v \in var(\mathcal{F}) \setminus \{x_r, x_{r+1}\}$, and $D_{\mathcal{F}'}(x_r) \subseteq D_{\mathcal{F}}(x_r)$ as well as $D_{\mathcal{F}'}(x_{r+1}) \supseteq D_{\mathcal{F}}(x_{r+1})$.*

Lemma 1. $(\star)$ *Let D be a function that maps each QCNF formula $\mathcal{F}$ to a binary relation $D_{\mathcal{F}} \subseteq R_{\mathcal{F}}$. If D is sound for transpositions and continuous, then D is a cumulative dependency scheme.*

Lemma 2. $(\star)$ *Let D' be a function that maps each QCNF formula $\mathcal{F}$ to a binary relation $D'_{\mathcal{F}} \subseteq R_{\mathcal{F}}$, and let D be a cumulative dependency scheme. If $D_{\mathcal{F}} \subseteq D'_{\mathcal{F}}$ for all formulas $\mathcal{F}$, then D' is a cumulative dependency scheme as well.*

4 Resolution-Path Dependencies

In this section, we will define the resolution path dependency *scheme*, which corresponds to the resolution-path dependency relation proposed by Van Gelder [15]. We justify this change of name by proving that the resolution path dependency scheme is indeed a cumulative dependency scheme.

Van Gelder [15] gives two definitions for resolution paths (Definitions 4.1 and 5.2), the former being more restrictive than the latter. The former definition is problematic as we will explain in Example 2 below. Hence we will base our considerations on the latter definition, which defines resolution paths as certain walks in a graph associated with a QBF formula. However, to avoid clashes with graph-theoretic terminology introduced below, we simply define resolution paths as particular sequences of clauses and literals.

Definition 6 (Resolution Path). *Let $\mathcal{F}$ be a QCNF formula with clause set F and $X \subseteq var_\exists(\mathcal{F})$. An X-resolution path in $\mathcal{F}$ is a sequence of clauses and literals $\ell_1, C_1, \ell_1', \ell_2, C_2, \ell_2', \ldots, \ell_n, C_n, \ell_n'$, satisfying the following properties:*

1. *$C_i \in F$ and $\ell_i, \ell_i' \in lit(\mathcal{F})$ for $i \in \{1, \ldots, n\}$.*
2. *$\ell_i, \ell_i' \in C_i$ for $i \in \{1, \ldots, n\}$.*
3. *$\ell_{i+1} = \overline{\ell_i'}$ and $\ell_i', \ell_{i+1} \in X \cup \overline{X}$, for $i \in \{1, \ldots, n-1\}$.*
4. *$var(\ell_i) \neq var(\ell_i')$ for $i \in \{1, \ldots, n\}$, and $\ell_1 \neq \ell_n'$.*

If $\ell_1, \ldots, \ell_n'$ is an X-resolution path in $\mathcal{F}$, we say that ℓ_1 and ℓ_n' are resolution connected in $\mathcal{F}$ with respect to X.

Example 1. Let $\mathcal{F} = \exists y_1 \exists y_2 \forall x_1 \exists y_3 \forall x_2\, C_1 \wedge C_2 \wedge C_3 \wedge C_4$, where $C_1 = (x_1 \vee x_2 \vee y_2 \vee y_1)$, $C_2 = (\neg x_1 \vee \neg y_2 \vee \neg y_1)$, $C_3 = (\neg y_1 \vee \neg y_3)$, and $C_4 = (\neg y_1 \vee y_3)$.

The sequence $x_1, C_1, y_1, \neg y_1, C_4, y_3$ is a $\{y_1\}$-resolution path in $\mathcal{F}$, and so the literals x_1 and $\neg y_3$ are resolution connected with respect to $\{y_1\}$. By contrast, the sequence $\neg x_1, C_2, \neg y_1, C_3, \neg y_3$ is not a resolution path in $\mathcal{F}$, because $\neg y_1$ is followed by a clause instead of the complementary literal y_1.   □

Resolution path dependencies are induced by a pair of resolution paths that connect the same two variables in reverse polarities:

Definition 7 (Dependency Pair). *Let $\mathcal{F}$ be a QCNF formula and $x, y \in var(\mathcal{F})$. We say (x, y) is a resolution-path dependency pair in $\mathcal{F}$ with respect to $X \subseteq var_\exists(\mathcal{F})$ if at least one of the following conditions holds:*

- *x and y, as well as $\neg x$ and $\neg y$, are resolution connected in $\mathcal{F}$ with respect to X.*
- *x and $\neg y$, as well as $\neg x$ and y, are resolution connected in $\mathcal{F}$ with respect to X.*

Definition 8 (Resolution-Path Dependency Scheme). *The resolution-path dependency scheme is a mapping D^{res} that assigns to each QCNF formula $\mathcal{F}$ the relation $D_{\mathcal{F}}^{res} = \{(x, y) \in R_{\mathcal{F}} : q_{\mathcal{F}}(x) \neq q_{\mathcal{F}}(y)$ and (x, y) is a resolution-path dependency pair in $\mathcal{F}$ with respect to $R_{\mathcal{F}}(x) \setminus (var_\forall(\mathcal{F}) \cup \{x, y\})\}$.*

In the formula $\mathcal{F}$ of Example 1 above, (y_1, x_1) is resolution-path dependency pair with respect to $\emptyset$, and (x_1, y_3) is a resolution-path dependency pair with respect to $\{y_1, y_2\}$. But while $(y_1, x_1) \in D_{\mathcal{F}}^{res}$, we have $(x_1, y_3) \notin D_{\mathcal{F}}^{res}$, because $\neg x_1$ is not resolution connected in $\mathcal{F}$ to either of y_3 or $\neg y_3$ with respect to $R_{\mathcal{F}}(x_1) \setminus \{y_3\} = \emptyset$.

The next lemma will be needed in the proof of Theorem 2 below.

Lemma 3 ([15]). *($\star$) Let $\mathcal{F}$ be QCNF formula, $\ell, \ell' \in lit(\mathcal{F})$ where $\ell \neq \ell'$, and $\pi = (T, \lambda)$ a regular, tree-like Q-resolution derivation of a clause D such that $\ell, \ell' \in D$. Then ℓ and ℓ' are resolution connected in $\mathcal{F}$ with respect to $resvar(\pi)$.*

The following result corresponds to Theorem 4.7 in [15].

Theorem 2 ([15]). (⋆) *Let $\mathcal{F}$ be a QCNF formula where $\forall u$ is followed by $\exists e$ in the quantifier prefix, so that $\delta_{\mathcal{F}}(e) = \delta_{\mathcal{F}}(u) + 1$. Suppose $(u, e) \notin D_{\mathcal{F}}^{\mathrm{res}}$. Let $\mathcal{F}'$ be the result of transposing $\exists e$ and $\forall u$ in the quantifier prefix. Then $\mathcal{F}'$ and $\mathcal{F}$ are equivalent.*

With the next example, we illustrate the importance of allowing consecutive clauses with a tautological Q-resolvent in the definition of resolution paths.

Example 2. Let $\mathcal{G} = \forall u \exists e \exists v \forall x \exists y \exists z \ C_1' \wedge C_2' \wedge C_3' \wedge C_4' \wedge C_5'$, where $C_1' = (u \vee y)$, $C_2' = (\neg y \vee \neg x \vee v)$, $C_3' = (\neg v \vee x \vee z)$, $C_4' = (\neg z \vee e)$, and $C_5' = (\neg u \vee \neg e)$. Figure 2 shows a Q-resolution derivation of the clause $(u \vee e)$ from $\mathcal{G}$. By Lemma 3, there must be a $\{v, y, z\}$-resolution path in $\mathcal{G}$ connecting u and e, and indeed it is straightforward to check that $u, C_1', y, \neg y, C_2', v, \neg v, C_3', z, \neg z, C_4', e$ is a resolution path. The literals $\neg u$ and $\neg e$ are trivially resolution connected, so (u, e) is a resolution path dependency pair with respect to $\{v, y, z\}$, and $(u, e) \in D_{\mathcal{F}}^{\mathrm{res}}$. This is a genuine dependency: it is easily verified that switching $\forall u$ and $\exists e$ in the prefix of $\mathcal{G}$ results in a formula that is unsatisfiable, while $\mathcal{G}$ itself is satisfiable.

Note that the clauses C_2' and C_3' do not have a non-tautological resolvent. All resolution paths in $\mathcal{G}$ between u and e lead through C_2' and C_3'. Consequently, if we would restrict Definition 6 so as to require consecutive clauses in a resolution path to have a non-tautological Q-resolvent (as in Definition 4.1 of [15]), u and e would no longer be resolution connected in $\mathcal{G}$, and e would not be identified as dependent on u.   ☐

Theorem 3. D^{res} *is a cumulative dependency scheme.*

Proof. We prove that D^{res} is (a) continuous and (b) sound for transpositions. The result then follows by Lemma 1. (a) Let $\mathcal{F}$ and $\mathcal{F}'$ be QCNF formulas such that $\mathcal{F}'$ is obtained from $\mathcal{F}$ by quantifier reordering. Let $x \in var(\mathcal{F}) = var(\mathcal{F}')$, and $P = R_{\mathcal{F}}(x) \setminus (var_{\forall}(\mathcal{F}) \cup \{x\})$, $P' = R_{\mathcal{F}'}(x) \setminus (var_{\forall}(\mathcal{F}') \cup \{x\})$. The set of P-resolution paths in $\mathcal{F}$ starting from x is identical to the set of P'-resolution paths in $\mathcal{F}'$ starting from x unless $R_{\mathcal{F}}(x) \neq R_{\mathcal{F}'}(x)$. If $R_{\mathcal{F}}(x) \subseteq R_{\mathcal{F}'}(x)$, every P-resolution path in $\mathcal{F}$ is a P'-resolution path in $\mathcal{F}'$. It is an easy consequence that D^{res} is continuous.

(b) Let $\mathcal{F}$ be a QCNF formula and $x, y \in var(\mathcal{F})$ so that $\delta_{\mathcal{F}}(y) = \delta_{\mathcal{F}}(x) + 1$ and $(x, y) \notin D_{\mathcal{F}}^{\mathrm{res}}$. If $x \in var_{\forall}(\mathcal{F})$ and $y \in var_{\exists}(\mathcal{F})$, the result follows from Theorem 2.

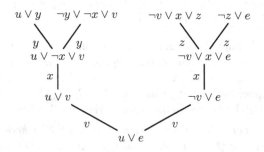

Fig. 2. Q-resolution derivation of $u \vee e$ from $\mathcal{G}$

Suppose $x \in var_\exists(\mathcal{F})$ and $y \in var_\forall(\mathcal{F})$. Let $\mathcal{F}'$ be the result of transposing $\exists x$ and $\forall y$ in the quantifier prefix of $\mathcal{F}$. Because of $(x, y) \notin D_\mathcal{F}^{res}$, we must have $(y, x) \notin D_{\mathcal{F}'}^{res}$, so we can again apply Theorem 2 and conclude that $\mathcal{F}$ and $\mathcal{F}'$ are equivalent. If $q_\mathcal{F}(x) = q_\mathcal{F}(y)$, equivalence is trivial. $\qquad\qquad\qquad\qquad\qquad\qquad\qquad\qquad\qquad\qquad\qquad\qquad\qquad$ □

Using Lemma 2, we can conclude that all dependency relations appearing in Figure 1 are cumulative dependency schemes.

5 Computing Resolution-Path Dependencies

This section will be devoted to proving that D^{res} is tractable. More specifically, we will show that the set of literals that are resolution connected to a given literal in a QCNF formula $\mathcal{F}$ with respect to a set $X \subseteq var_\exists(\mathcal{F})$ can be computed in linear time. This result in turn establishes linear time-tractability of deciding whether a pair of variables is contained in $D_\mathcal{F}^{res}$.

We will reduce the problem of finding resolution paths to the task of finding properly edge-colored walks in certain edge-colored graphs. A *graph* G consists of a finite set $V(G)$ of *vertices* and a set $E(G)$ of *edges*, where the edge between two vertices u and v is denoted by uv or equivalently vu. All graphs we consider are undirected and simple (i.e., without self-loops or multi-edges). If G is a graph and $v \in V(G)$, elements of the set $N_G(v) = \{w \in V(G) : vw \in E(G)\}$ are called *neighbors of v in G*. In a *c-edge-colored graph* G, every edge $e \in E(G)$ is assigned a *color* $\chi_G(e) \in \{1, \ldots, k\}$. Given a (not necessarily edge-colored) graph G, a *walk* from s to t in G is a sequence of vertices $\pi = v_1, v_2, \ldots, v_n$, where $v_1 = s$, $v_n = t$, and $v_i v_{i+1} \in E(G)$ for $i = 1, \ldots, n-1$. If further $v_i \neq v_{i+2}$ for all $i \in \{1, \ldots, n-2\}$, π is said to be *retracting-free*. A walk $\pi = v_1, \ldots, v_n$ in a *c-edge-colored* graph G is *properly edge-colored (PEC)* if $\chi_G(v_i v_{i+1}) \neq \chi_G(v_{i+1}v_{i+2})$ for all $i \in \{1, \ldots, n-2\}$. A walk $v_1, \ldots, v_n$ satisfying $v_i \neq v_j$ for distinct $i, j \in \{1, \ldots, n\}$ is a *path*. A PEC walk which is a path is called a *PEC path*. The *length* of a walk $v_1, \ldots, v_{n+1}$ is n. For 2-edge-colored graphs, we use the names *red* and *blue* to denote the colors 1 and 2, respectively.

Note that there can be a PEC walk from a vertex s to a vertex t without there being a PEC path from s to t. For instance, consider a 2-edge-colored graph with vertex set $\{s, u, v, w, t\}$ and edge set $\{su, ut, uv, uw, vw\}$, such that uv and uw are red and the remaining edges are blue. The sequence s, u, v, w, u, t is a PEC walk from s to t, but there is no PEC path from s to t.

Construction. Let $\mathcal{F}$ be a QCNF formula with matrix F, and let $X \subseteq var_\exists(\mathcal{F})$. We construct two graphs $G_{\mathcal{F},X}$ and $G'_{\mathcal{F},X}$:

- For the set of vertices of $G_{\mathcal{F},X}$, we choose $F \cup lit(\mathcal{F})$. Its edge set consists of all edges $\neg zz$ for $z \in X$, and all edges $C\ell$ where $\ell \in C$.
- We define $G'_{\mathcal{F},X}$ to be a 2-edge-colored graph with vertex set $lit(\mathcal{F})$ and edge set $E_r \cup E_b$, where the set E_r consists of all edges $\neg zz$ for $z \in X$, and E_b consists of all edges $\ell\ell'$ such that there is a clause $C \in F$ with $\ell, \ell' \in C$. The edges in E_r are red, while those in E_b are blue.

For general QCNF formulas $\mathcal{F}$, the size of $G'_{\mathcal{F},X}$ can be quadratic in the size of $\mathcal{F}$, since every clause of size n gives rise to a clique with n vertices. This can be avoided by using the following trick: we first convert $\mathcal{F}$ to a Q3CNF formula $\mathcal{F}'$ and then carry out the construction. For any set $X' \subseteq var(\mathcal{F}')$, we can clearly compute $G'_{\mathcal{F}',X'}$ in time $\mathcal{O}(|\mathcal{F}'|)$. Furthermore, it is well known that SAT can be reduced to 3SAT in linear time [8]. We show that this reduction preserves resolution connectedness.

Lemma 4. $(\star)$ *Let $\mathcal{F}$ be an arbitrary QCNF formula and $X \subseteq var_{\exists}(\mathcal{F})$. In time $\mathcal{O}(|\mathcal{F}|)$, one can construct a Q3CNF formula $\mathcal{F}'$ and a set $X' \subseteq var_{\exists}(\mathcal{F}')$ satisfying the following property: two literals $\ell, \ell' \in lit(\mathcal{F})$ are resolution connected in $\mathcal{F}$ with respect to X if and only if ℓ and ℓ' are r-connected in $\mathcal{F}'$ with respect to X'.*

Proposition 1. *Given a Q3CNF formula $\mathcal{F}$ and a set $X \subseteq var_{\exists}(\mathcal{F})$, the graph $G'_{\mathcal{F},X}$ can be constructed in time $\mathcal{O}(|\mathcal{F}|)$.*

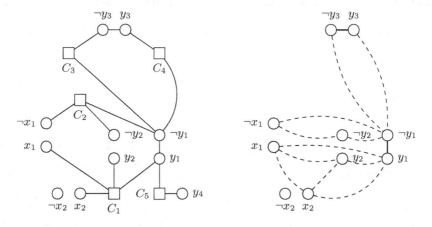

Fig. 3. The graphs $G_{\mathcal{F},X}$ (left) and $G'_{\mathcal{F},X}$ (right) for the formula $\mathcal{F}$ of Example 1 and $X = \{y_1, y_3\}$. Red edges of $G'_{\mathcal{F},X}$ are represented by solid lines, and blue edges by dashed lines.

Lemma 5. *Let $\mathcal{F}$ be a QCNF formula, $X \subseteq var_{\exists}(\mathcal{F})$, and $\ell, \ell' \in lit(\mathcal{F})$ such that $\ell \neq \ell'$. The following statements are equivalent:*

1. *ℓ and ℓ' are resolution connected in $\mathcal{F}$ with respect to X.*
2. *There is a retracting-free walk $\ell_1, C_1, \ell'_1, \ell_2, C_2, \ell'_2, \dots, \ell_n, C_n, \ell'_n$ in $G_{\mathcal{F},X}$ from ℓ to ℓ', where $C_i \in F$ and $\ell_i, \ell'_i \in lit(\mathcal{F})$ for $i \in \{1, \dots, n\}$.*
3. *There is a PEC walk in $G'_{\mathcal{F},X}$ from ℓ to ℓ' whose first and last edges are blue.*

Proof. $(1 \Rightarrow 2)$ Suppose ℓ and ℓ' are resolution connected in $\mathcal{F}$ with respect to X. Then there exists an X-resolution path $\pi = \ell_1, C_1, \ell'_1, \ell_2, C_2, \dots, \ell_n, C_n, \ell'_n$ in $\mathcal{F}$ from ℓ to ℓ'. We claim that π is already a retracting-free walk in $G_{\mathcal{F},X}$ of the desired form. Because π is a resolution path, we have $\ell_{i+1} = \overline{\ell'_i}$ and therefore $\ell'_i \ell_{i+1} \in E(G_{\mathcal{F},X})$ for all $i \in \{1, \dots, n-1\}$. Moreover, because $\ell_i, \ell'_i \in C_i$ for $i \in \{1, \dots, n\}$, we have $\ell_i C_i, \ell'_i C_i \in E(G_{\mathcal{F},X})$ as well. So π is indeed a walk in $G_{\mathcal{F},X}$. Since $var(\ell_i) \neq var(\ell'_i)$ for $i \in \{1, \dots, n\}$, π must be retracting-free.

$(2 \Rightarrow 3)$ Let $\pi = \ell_1, C_1, \ell_1', \ldots, \ell_n, C_n, \ell_n'$ be a retracting-free walk from ℓ to ℓ' in $G_{\mathcal{F},X}$ so that $C_i \in F$ and $\ell_i, \ell_i' \in lit(\mathcal{F})$ for $i \in \{1, \ldots, n\}$. We show that the sequence $\pi' = \ell_1, \ell_1', \ldots, \ell_n, \ell_n'$ is a PEC walk from ℓ to ℓ' in $G_{\mathcal{F},X}'$ whose first and last edges are blue. Let $\ell_i C_i$, $C_i \ell_i'$ be a pair of consecutive edges in π where $i \in \{1, \ldots, n\}$. By construction of $G_{\mathcal{F},X}$, we have $\ell_i, \ell_i' \in C_i$. Because π is retracting-free, $\ell_i \neq \ell_i'$, and thus there is a blue edge $\ell_i \ell_i'$ in $G_{\mathcal{F},X}'$. For all $i \in \{1, \ldots, n-1\}$, the edge $\ell_i' \ell_{i+1}$ of π is a red edge in $G_{\mathcal{F},X}'$. So π' is a walk in $G_{\mathcal{F},X}'$. Moreover, the first and last edges of π are blue, and it is easily to verified that π' is PEC.

$(3 \Rightarrow 1)$ Now let $\pi = \ell_1, \ell_1', \ell_2, \ell_2', \ldots, \ell_n, \ell_n'$ be a PEC walk from ℓ to ℓ' in $G_{\mathcal{F},X}'$ whose first and last edges are blue. By construction of $G_{\mathcal{F},X}'$, for every blue edge $\ell_i \ell_i'$ traversed by π, there is a clause C_i in F such that $\ell_i, \ell_i' \in C_i$, for $i \in \{1, \ldots, n\}$. For every red edge $\ell_i' \ell_{i+1}$, where $i \in \{1, \ldots, n-1\}$, we have $\ell_{i+1} = \overline{\ell_i'}$ and $\ell_i', \ell_{i+1} \in X \cup \overline{X}$. Let π' be the sequence $\ell_1, C_1, \ell_1', \ldots, \ell_n, C_n, \ell_n'$. π' is an X-resolution path in $\mathcal{F}$: we already know that π' satisfies conditions 1-3 of Definition 6. To verify condition 4, we must show that $var(\ell_i) \neq var(\ell_i')$ for all $i \in \{1, \ldots, n\}$. Suppose to the contrary that $var(\ell_i) = var(\ell_i')$ for some $i \in \{1, \ldots, n\}$. Because $G_{\mathcal{F},X}'$ does not contain self-loops, this implies $\ell_i' = \overline{\ell_i}$. But then $\ell_i, \overline{\ell_i} \in C_i$, contrary to the assumption that F does not contain tautological clauses. This concludes the proof that π' is an X-resolution path in $\mathcal{F}$. It follows that ℓ and ℓ' are resolution connected in $\mathcal{F}$ with respect to X. $\square$

Algorithm PEC-Walk. We now describe the algorithm *PEC-Walk* that takes as input a 2-edge-colored graph G and a vertex $s \in V(G)$, and computes the set of vertices t such that there is a PEC walk from s to t whose first and last edges are blue. We maintain a set Q containing (ordered) pairs of vertices (v, w) joined by edges that can be traversed by a PEC walk starting from s. Initially, Q is empty. For each vertex v, we store a set $\psi(v) \subseteq \{red, blue\}$, where $c \in \psi(v)$ indicates that there is a PEC walk from s to v ending in an edge with color c. In an initialization phase, we first set $\psi(u) = \emptyset$ for all vertices u. We then add all pairs (s, v) to Q such that v is a neighbor of s and sv is a blue edge, inserting *blue* into $\psi(v)$ at the same time. In the main procedure, we repeat the following steps until Q is empty: we remove a pair (v, w) from Q and add all pairs (w, u) to Q such that u is a neighbor of w, wu is an edge with color c different from the color of vw, and c is not already in $\psi(w)$. For every pair (v, w) we put into Q, we add its color to $\psi(w)$.

Lemma 6. *Let G be a 2-edge-colored graph and $s \in V(G)$. On input (G, s), PEC-Walk runs in time $\mathcal{O}(|E(G)| + |V(G)|)$.*

Proof. Every ordered pair of vertices joined by an edge is examined at most twice and added to Q at most once. The algorithm terminates when Q is empty, and an element is removed from Q in each iteration. Initialization can take at most $\mathcal{O}(|E(G)| + |V(G)|)$ steps. So the time required by the entire algorithm is $\mathcal{O}(|E(G)| + |V(G)|)$. $\square$

Lemma 7. *Let G be a 2-edge-colored graph, $s, t \in V(G)$, $s \neq t$, and let ψ be a vertex labeling generated by running PEC-Walk on input (G, s). There is a PEC walk from s to t whose first edge is blue and whose last edge has color $c \in \{red, blue\}$ if and only if $c \in \psi(t)$.*

Proof. By the preceding lemma, the algorithm always terminates and produces a labeling ψ.

($\Leftarrow$) Let t be a vertex of G different from s. We show that if $c \in \psi(t)$, there is a PEC walk from s to t whose first edge is blue and whose final edge has color c. We proceed by induction on the number n of times the algorithm enters the main loop with $c \notin \psi(t)$. If $n = 0$, color c is added to $\psi(t)$ during the initialization phase, so there must be a blue edge st. Assume the statement holds for all $0 \leq k \leq n$, and c is added to $\psi(t)$ in iteration $n+1$. Then there must be a pair (v,t) with $\chi_G(vt) = c$ which is added to Q in this iteration. That is the case only if a pair (u,v) is removed from Q during the same iteration with $\chi_G(uv) = c'$, where $c' \neq c$. The pair (u,v) must have been inserted into Q before iteration $n + 1$, at which point c' was added to $\psi(v)$. Applying the induction hypothesis, we can conclude there must be a PEC walk from s to v such that its first edge is blue and its last edge has color c'. By appending vt to this walk, we obtain a PEC walk from s to t with the desired properties.

($\Rightarrow$) Suppose there is a PEC walk from s to t whose first edge is blue and whose last edge has color c. Let n be the smallest integer that is the length of such a walk. We will show by induction on n that $c \in \psi(t)$. The case $n = 1$ is taken care of by the initialization phase of the algorithm. Suppose the statement holds for all $n \in \{1, \ldots, m\}$. Let $v_0, \ldots, v_{m+1}$ be a PEC walk from s to t with the property that its first edge is blue and its last edge has color c, and assume there is no shorter PEC walk with this property. Then $v_0, \ldots, v_m$ is a PEC walk from s to v_m so that v_0v_1 is blue, and $\chi_G(v_{m-1}v_m) = c'$ where $c \neq c'$. There can be no $k < m$ such that there is a PEC walk of length k from s to v_m whose first edge is blue and whose last edge has color c': otherwise, one could append v_mv_{m+1} to this path to obtain a PEC walk from s to v_{m+1} whose initial edge is blue and whose final edge has color c of length $k + 1 < m + 1$, a contradiction. We can therefore apply the induction hypothesis and conclude that $c' \in \psi(v_m)$. Let (w, v_m) be the pair that was removed from Q in the iteration of the main loop in which c' was added to $\psi(v_m)$. Because $c' \neq c$, in the same iteration the pair (v_m, v_{m+1}) must have been added to Q and c put into to $\psi(v_{m+1})$, unless already $c \in \psi(v_{m+1})$. $\square$

The next result is immediate from Lemmas 6 and 7.

Proposition 2. *Given a 2-edge-colored graph G, a vertex $s \in V(G)$, and some $c \in \{red, blue\}$, the set of vertices reachable from s along some PEC walk in G whose first edge is blue and whose last edge has color c can be computed in time $\mathcal{O}(|E(G)| + |V(G)|)$.*

With all the pieces in place, it is now straightforward to prove our main result.

Theorem 4. *Given a QCNF formula $\mathcal{F}$ and a pair of variables $x, y \in var(\mathcal{F})$, one can decide whether $(x, y) \in D_{\mathcal{F}}^{res}$ in time $\mathcal{O}(|\mathcal{F}|)$. Hence the resolution-path dependency scheme is tractable.*

Proof. We prove that there is a linear time decision algorithm. We first check whether $q_{\mathcal{F}}(x) \neq q_{\mathcal{F}}(y)$ and (x, y) is in $R_{\mathcal{F}}$. Using Lemma 4, we can then in linear time compute a QCNF formula $\mathcal{F}'$ and a set R' from $\mathcal{F}$ and $R_{\mathcal{F}}(x) \setminus (var_\forall(\mathcal{F}) \cup \{x, y\})$ so that two literals are resolution connected in $\mathcal{F}'$ with respect to R' if and only if they are resolution connected in $\mathcal{F}$ with respect to $R_{\mathcal{F}}(x) \setminus (var_\forall(\mathcal{F}) \cup \{x, y\})$. We can then construct the graph $G'_{\mathcal{F}',R'}$ and determine for all pairs ℓ_x, ℓ_y with $\ell_x \in \{x, \neg x\}$ and $\ell_y \in \{y, \neg y\}$ whether there is a properly edge-colored walk from ℓ_x to ℓ_y whose first and last edges are blue, which by Lemma 5 is equivalent to ℓ_x and ℓ_y being resolution connected in $\mathcal{F}'$ with respect to R' (according to Propositions 1 and 2, this can be done in linear time). Using this information, it is straightforward to decide whether (x, y) is a resolution-path dependency pair in $\mathcal{F}$ with respect to $R_{\mathcal{F}}(x) \setminus (var_\forall(\mathcal{F}) \cup \{x, y\})$. Each of these steps requires linear time, so we need $\mathcal{O}(|\mathcal{F}|)$ time in total.   □

Samer and Szeider [12] generalized the notion of a strong backdoor set from CNF formulas to QCNF formulas, by adding the requirement that the backdoor set is closed under a cumulative dependency scheme. They showed that evaluating QCNF formulas is fixed-parameter tractable (fpt) when parameterized by the size of a smallest strong backdoor set (with respect to the classes QHORN or Q2CNF) provided that the considered cumulative dependency scheme is tractable. By Theorems 3 and 4, one can use the resolution path dependency scheme here and thus get an fpt result that is stronger than the results achieved by using any of the other dependency schemes appearing in Fig. 1.

For an existentially quantified variable y in a QCNF $\mathcal{F}$, the entire set $D^{\text{res}}_{\mathcal{F}}(y)$ can be computed in linear time: we first determine the sets $D = \{ \ell \in lit(\mathcal{F}) : y$ is resolution connected to ℓ in $\mathcal{F}$ with respect to $R_{\mathcal{F}}(y) \setminus var_\forall(\mathcal{F}) \}$ and $D_\neg = \{ \ell \in lit(\mathcal{F}) : \neg y$ is resolution connected to ℓ in $\mathcal{F}$ with respect to $R_{\mathcal{F}}(y) \setminus var_\forall(\mathcal{F}) \}$ and store them in a data structure that allows us to decide membership of literals in constant time (say, an array). To determine $D^{\text{res}}_{\mathcal{F}}(y)$, we simply check for each element x of $R_{\mathcal{F}} \cap var_\forall(\mathcal{F})$ whether $x \in D$ and $\neg x \in D_\neg$, or $\neg x \in D$ and $x \in D_\neg$.

Unfortunately we cannot use the same approach to compute the set of dependent variables $D^{\text{res}}_{\mathcal{F}}(x)$ for a universal variable $x \in var_\forall(\mathcal{F})$. For every existential variable $y \in var_\exists(\mathcal{F})$, resolution paths that entail $(x, y) \in D^{\text{res}}_{\mathcal{F}}$ cannot contain y or $\neg y$. Hence the relevant resolution paths are subject to different constraints for each y, and it is not sufficient in general to construct $G'_{\mathcal{F},X}$ for a single set X.

6 Minimal Dependency Schemes

The fact that the resolution-path dependency scheme is the bottom element of the lattice represented in Figure 1 gives reason to wonder whether it is the most general dependency scheme. However, computing a minimal dependency scheme is complete for PSPACE [12]. Since the resolution path dependency scheme is tractable, it follows that it cannot be minimal. Can we instead prove that D^{res} is minimal relative to a class of "natural" dependency schemes? At the very least, such a class should include all the dependency schemes considered so far, which have the following feature in common: whether a pair of variables is considered dependent is determined almost entirely in terms of the matrix. We use this property to define a candidate class.

Definition 9. *A dependency scheme D is called a* matrix dependency scheme *if it satisfies the following property: Let $\mathcal{F}$ and $\mathcal{F}'$ be QCNF formulas such that $\mathcal{F}'$ is obtained from $\mathcal{F}$ by quantifier reordering. Moreover, let $x \in var(\mathcal{F})$ such that $R_{\mathcal{F}}(x) = R_{\mathcal{F}'}(x)$. Then for any $y \in var(\mathcal{F})$, we have $(x, y) \in D_{\mathcal{F}}$ if and only if $(x, y) \in D_{\mathcal{F}'}$.*

The next proposition can be easily verified by inspecting Definition 8.

Proposition 3. *The resolution-path dependency scheme D^{res} is a matrix dependency scheme.*

Unfortunately, D^{res} is not even the most general matrix dependency scheme. We now show that there is a cumulative matrix dependency scheme which is strictly more general than D^{res}. Let $\mathcal{F}$ be an arbitrary QCNF formula.

Definition 10. *We let $D^{\text{mat}} : \mathcal{F} \mapsto D_{\mathcal{F}}^{\text{mat}}$, where $D_{\mathcal{F}}^{\text{mat}} = \{ (x,y) \in R_{\mathcal{F}} : \text{there is a}$ formula $\mathcal{F}' = Q_1 x_1 \ldots Q_x x Q_y y \ldots Q_n x_n F$ obtained from $\mathcal{F}$ by quantifier reordering, such that $R_{\mathcal{F}}(x) \supseteq R_{\mathcal{F}'}(x)$ and $\nu(\mathcal{F}') \neq \nu(\mathcal{F}'')$, where $\mathcal{F}'' = Q_1 x_1 \ldots Q_y y Q_x x \ldots Q_n x_n F \}$.*

Proposition 4. $(\star)$ *D^{mat} is a cumulative matrix dependency scheme.*

Proposition 5. $(\star)$ *For every QCNF formula $\mathcal{F}$, the relation $D_{\mathcal{F}}^{\text{mat}}$ is contained in $D_{\mathcal{F}}^{\text{res}}$, and containment is strict in some cases.*

The reduction applied in the proof of the following result essentially corresponds to the one used by Samer and Szeider to establish PSPACE-hardness of computing minimal dependency schemes [12].

Proposition 6. $(\star)$ *Let $\mathcal{F}$ be a QCNF formula with matrix F and $x, y \in var(\mathcal{F})$. The problem of deciding whether there exists a matrix dependency scheme D such that $(x, y) \notin D_{\mathcal{F}}$ is Σ_2^P-hard.*

One may object that these considerations do not rule out the possibility that D^{res} is the most general *tractable* matrix dependency scheme. That this is not the case can be seen from the following simple argument. For any nonnegative integer k, we define a mapping D^k such that for any QCNF formula $\mathcal{F}$ we have $D_{\mathcal{F}}^k = D_{\mathcal{F}}^{\text{mat}}$ if $|\mathcal{F}| \leq k$, and $D_{\mathcal{F}}^k = D_{\mathcal{F}}^{\text{res}}$ otherwise. As both D^{mat} and D^{res} are cumulative matrix dependency schemes and the relevant properties are defined pointwise, any such function D^k must be a cumulative matrix dependency scheme as well. Moreover, each scheme D^k is clearly tractable and from the proof of Proposition 5 we know that D^k is strictly more general than D^{res} for $k \geq 5$.

7   Conclusion

We have shown that resolution path dependencies give rise to a cumulative dependency scheme that can be decided in linear time. While the latter result is optimal for the decision problem, we see at least two obstacles for an efficient implementation. First, computing the entire relation $D_{\mathcal{F}}^{\text{res}}$ using our current algorithm requires $\mathcal{O}(|\mathcal{F}|^3)$ time, which is prohibitive for practical purposes. Second, it is unclear whether one can find succinct representations of the relation $D_{\mathcal{F}}^{\text{res}}$ similar to those used for the standard dependency scheme [9]. We leave this issues for future work.

To capture the kind of variable dependencies relevant for expansion-based QBF solvers, Samer considered an alternative definition of dependency schemes based on variable *independence* [11]. It might be interesting to study resolution path dependencies in this context as well.

References

1. Ayari, A., Basin, D.: Qubos: Deciding Quantified Boolean Logic Using Propositional Satisfiability Solvers. In: Aagaard, M.D., O'Leary, J.W. (eds.) FMCAD 2002. LNCS, vol. 2517, pp. 187–201. Springer, Heidelberg (2002)
2. Biere, A.: Resolve and Expand. In: Hoos, H., Mitchell, D.G. (eds.) SAT 2004. LNCS, vol. 3542, pp. 59–70. Springer, Heidelberg (2005)
3. Bubeck, U., Kleine Büning, H.: Bounded Universal Expansion for Preprocessing QBF. In: Marques-Silva, J., Sakallah, K.A. (eds.) SAT 2007. LNCS, vol. 4501, pp. 244–257. Springer, Heidelberg (2007)
4. Egly, U., Tompits, H., Woltran, S.: On quantifier shifting for quantified Boolean formulas. In: Proc. SAT 2002 Workshop on Theory and Applications of Quantified Boolean Formulas. Informal Proceedings, pp. 48–61 (2002)
5. Giunchiglia, E., Marin, P., Narizzano, M.: Reasoning with quantified boolean formulas. In: Biere, A., Heule, M., van Maaren, H., Walsh, T. (eds.) Handbook of Satisfiability, vol. 185, pp. 761–780. IOS Press (2009)
6. Kleine Büning, H., Bubeck, U.: Theory of quantified boolean formulas. In: Biere, A., Heule, M., van Maaren, H., Walsh, T. (eds.) Handbook of Satisfiability, ch. 23, pp. 735–760. IOS Press (2009)
7. Kleine Büning, H., Karpinski, M., Flögel, A.: Resolution for quantified Boolean formulas. Information and Computation 117(1), 12–18 (1995)
8. Kleine Büning, H., Lettman, T.: Propositional logic: deduction and algorithms. Cambridge University Press, Cambridge (1999)
9. Lonsing, F., Biere, A.: A Compact Representation for Syntactic Dependencies in QBFs. In: Kullmann, O. (ed.) SAT 2009. LNCS, vol. 5584, pp. 398–411. Springer, Heidelberg (2009)
10. Lonsing, F., Biere, A.: Integrating Dependency Schemes in Search-Based QBF Solvers. In: Strichman, O., Szeider, S. (eds.) SAT 2010. LNCS, vol. 6175, pp. 158–171. Springer, Heidelberg (2010)
11. Samer, M.: Variable Dependencies of Quantified CSPs. In: Cervesato, I., Veith, H., Voronkov, A. (eds.) LPAR 2008. LNCS (LNAI), vol. 5330, pp. 512–527. Springer, Heidelberg (2008)
12. Samer, M., Szeider, S.: Backdoor sets of quantified Boolean formulas. Journal of Automated Reasoning 42(1), 77–97 (2009)
13. Stockmeyer, L.J.: The polynomial-time hierarchy. Theoretical Computer Science 3(1), 1–22 (1976)
14. Stockmeyer, L.J., Meyer, A.R.: Word problems requiring exponential time. In: Proc. Theory of Computing, pp. 1–9. ACM (1973)
15. Van Gelder, A.: Variable Independence and Resolution Paths for Quantified Boolean Formulas. In: Lee, J. (ed.) CP 2011. LNCS, vol. 6876, pp. 789–803. Springer, Heidelberg (2011)

Strong Backdoors to Nested Satisfiability[*]

Serge Gaspers[1,2] and Stefan Szeider[2]

[1] School of Computer Science and Engineering, The University of New South Wales,
Sydney, Australia
gaspers@kr.tuwien.ac.at
[2] Institute of Information Systems, Vienna University of Technology, Vienna, Austria
stefan@szeider.net

Abstract. Knuth (1990) introduced the class of nested formulas and showed that their satisfiability can be decided in polynomial time. We show that, parameterized by the size of a smallest strong backdoor set to the base class of nested formulas, computing the number of satisfying assignments of any CNF formula is fixed-parameter tractable. Thus, for any $k > 0$, the satisfiability problem can be solved in polynomial time for any formula F for which there exists a set B of at most k variables such that for every truth assignment τ to B, the reduced formula $F[\tau]$ is nested; moreover, the degree of the polynomial is independent of k.

Our algorithm uses the grid-minor theorem of Robertson and Seymour (1986) to either find that the incidence graph of the formula has bounded treewidth—a case that is solved by model checking for monadic second order logic—or to find many vertex-disjoint obstructions in the incidence graph. For the latter case, new combinatorial arguments are used to find a small backdoor set. Combining both cases leads to an approximation algorithm producing a strong backdoor set whose size is upper bounded by a function of the optimum. Going through all assignments to this set of variables and using Knuth's algorithm, the satisfiability of the input formula can be decided. With a similar approach, one can also count the number of satisfying assignments of the given formula.

1 Introduction

In a 1990 paper [20] Knuth introduced the class of nested CNF formulas and showed that their satisfiability can be decided in polynomial time. A CNF formula is *nested* if its variables can be linearly ordered such that there is no pair of clauses that *straddle* each other; a clause c straddles a clause c' if there are variables $x, y \in \mathsf{var}(c)$ and $z \in \mathsf{var}(c')$ such that $x < z < y$ in the linear ordering under consideration. NESTED denotes the class of nested CNF formulas. For an example see Figure 1. Since nested formulas have incidence graphs of bounded treewidth [2], one can use treewidth-based algorithms [10,34] to even compute the number of satisfying truth assignments of nested formulas in polynomial time (incidence graphs are defined in Section 2). Hence the problems SAT and #SAT are polynomial for nested formulas.

---

[*] The full version of the paper is available on arXiv [16].

A. Cimatti and R. Sebastiani (Eds.): SAT 2012, LNCS 7317, pp. 72–85, 2012.
© Springer-Verlag Berlin Heidelberg 2012

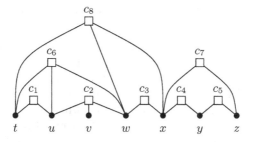

Fig. 1. Incidence graph of the nested formula $F = \bigwedge_{i=1}^{8} c_i$ with $c_1 = t \vee \neg u$, $c_2 = u \vee v \vee w$, $c_3 = w \vee x$, $c_4 = x \vee \neg y$, $c_5 = y \vee \neg z$, $c_6 = t \vee u \vee \neg w$, $c_7 = \neg x \vee z$, $c_8 = \neg t \vee w \vee x$

The aim of this paper is to extend the nice computational properties of nested formulas to formulas that are not nested but are of small distance from being nested. We measure the distance of a CNF formula F from being nested as the size of a smallest set B of variables, such that for all partial truth assignments τ to B, the reduced formula $F[\tau]$ is nested. Such a set B is called a *strong backdoor set* with respect to the class of nested formulas [37], or strong NESTED-backdoor set, for short. Once we have found such a backdoor set of size k, we can decide the satisfiability of F by checking the satisfiability of 2^k nested formulas, or for model counting, we can take the sum of the number of models of the 2^k nested formulas. Thus the problems SAT and #SAT can be solved in time $O(2^k |F|^c)$ where $|F|$ denotes the length of F and k denotes the size of the given strong NESTED-backdoor set; c is a small constant. In other words, the problems SAT and #SAT are *fixed-parameter tractable* for parameter k (for background on fixed-parameter tractability see Section 2). However, in order to use the backdoor set we must find it first. Is the detection of strong NESTED-backdoor sets fixed-parameter tractable as well?

Let $\mathbf{sb_N}(F)$ denote the size of a smallest strong NESTED-backdoor set of a CNF formula F. To find a strong backdoor set of size at most $k = \mathbf{sb_N}(F)$ one can try all possible sets of variables of size at most k, and check for each set whether it is a strong backdoor set. However, for a formula with n variables we have to check $\binom{n}{k} = \Omega(n^k)$ such sets. Thus, this brute-force approach scales poorly in k and does not provide fixed-parameter tractability, as the order of the polynomial increases with k.

In this paper we show that one can overcome this limitation with a more sophisticated algorithm. *We show that the problems SAT and #SAT are fixed-parameter tractable when parameterized by* $\mathbf{sb_N}$, *the size of a smallest strong* NESTED-*backdoor set, even when the backdoor set is not provided as an input.*

Our algorithm is constructive and uses the Grid Minor Theorem of Robertson and Seymour [32] to either find that the incidence graph of the formula has bounded treewidth—a case that is solved using model checking for monadic second order logic [1]—or to find many vertex-disjoint obstructions in the incidence graph. For the latter case, new combinatorial arguments are used to find a small strong backdoor set. Combining both cases leads to an algorithm producing a strong backdoor set of a given formula F of size at most 2^k for $k = \mathbf{sb_N}(F)$. Solving all the 2^{2^k} resulting nested formulas provides a solution to F.

Our result provides a new parameter $\mathbf{sb_N}$ that makes SAT and #SAT fixed-parameter tractable. The parameter $\mathbf{sb_N}$ is *incomparable* with other known parameters that make SAT and #SAT fixed-parameter tractable. Take for instance the treewidth of the incidence graph of a CNF formula F, denoted $\mathbf{tw}^*(F)$. As mentioned above, SAT and #SAT are fixed-parameter tractable for parameter $\mathbf{tw}^*$ [10,34], and $\mathbf{tw}^*(F) \leq 3$ holds if $\mathbf{sb_N}(F) = 0$ (i.e., if $F \in$ NESTED) [2]. However, by allowing only $\mathbf{sb_N}(F) = 1$ we already get formulas with arbitrarily large $\mathbf{tw}^*(F)$. This can be seen as follows. Consider an $n \times n$ grid whose vertices represent variables of a CNF formula F_n and subdivide each edge of the grid with a clause of F_n. It is well known that the $n \times n$ grid, $n \geq 2$, has treewidth n and that subdividing edges does not decrease the treewidth of a graph (folklore). Now take a new variable x and add it positively to all horizontal clauses and negatively to all vertical clauses, where a clause is *horizontal* (resp. *vertical*) if it subdivides a horizontal (resp. vertical) edge of the $n \times n$ grid in a natural layout. Let F_n^x denote the new formula. Since the incidence graph of F_n is a subgraph of the incidence graph of F_n^x, we have $\mathbf{tw}^*(F_n^x) \geq \mathbf{tw}^*(F_n) \geq n$. However, setting x to true removes all horizontal clauses and thus yields a formula whose incidence graph is a disjoint union of paths, which is easily seen to be nested. Similarly, setting x to false yields a nested formula as well. Hence $\{x\}$ forms a strong NESTED-backdoor set, and so $\mathbf{sb_N}(F) = 1$. One can also construct formulas where $\mathbf{sb_N}$ is large and $\mathbf{tw}^*$ is small, for example by taking the variable-disjoint union F of formulas $F_i = (x_i \vee y_i \vee z_i) \wedge (\neg x_i \vee y_i \vee z_i)$ with $\mathbf{sb_N}(F_i) = 1$ and $\mathbf{tw}^*(F_i) = 2$, $1 \leq i \leq n$. Then $\mathbf{tw}^*(F) = \mathbf{tw}^*(F_i) = 2$, but $\mathbf{sb_N}(F) = \sum_{i=1}^n \mathbf{sb_N}(F_i) = n$.

One can also define *deletion backdoor sets* of a CNF formula F with respect to a base class of formulas by requiring that deleting all literals $x, \neg x$ with $x \in B$ from F produces a formula that belongs to the base class [29]. For many base classes it holds that every deletion backdoor set is a strong backdoor set, but in most cases, including the base class NESTED, the reverse is not true. In fact, it is easy to see that if a CNF formula F has a NESTED-deletion backdoor set of size k, then $\mathbf{tw}^*(F) \leq k + 3$. In other words, the parameter "size of a smallest deletion NESTED-backdoor set" is dominated by the parameter incidence treewidth and therefore of limited interest. We note in passing, that one can use the algorithm from [24] to show that the detection of deletion NESTED-backdoor sets is fixed-parameter tractable.

Related Work. Williams *et al.* [37] introduced the notion of backdoor sets to explain favorable running times and the heavy-tailed behavior of SAT and CSP solvers on practical instances. The parameterized complexity of finding small backdoor sets was initiated by Nishimura *et al.* [28] who showed that with respect to the classes of Horn formulas and of 2CNF formulas, the detection of strong backdoor sets is fixed-parameter tractable. Their algorithms exploit the fact that for these two base classes strong and deletion backdoor sets coincide. For other base classes, deleting literals is a less powerful operation than applying partial truth assignments. This is the case for the class NESTED but also for the class RHORN of renamable Horn formulas [7]. In fact, finding a deletion RHORN-backdoor set is fixed-parameter tractable [30], but it is open whether this is the case for the detection of strong RHORN-backdoor sets. For clustering formulas, detection of deletion backdoor sets is fixed-parameter tractable, detection of strong backdoor sets is most probably not [29]. Very recently, the authors of the present paper showed that for the base class FOREST, of formulas whose incidence graph is acyclic, there is a fixed-parameter approximation algorithm for strong backdoor sets. That is, the following problem is fixed-parameter tractable: find a strong FOREST-backdoor set of size at most k or decide that there is no strong FOREST-backdoor set of size at most 2^k [13]. The present paper extends the ideas from [13] to the significantly more involved case with NESTED as the base class, which is a strict superclass of FOREST.

We conclude this section by referring to a recent survey on the parameterized complexity of backdoor sets [14].

2 Preliminaries

Parameterized Complexity. Parameterized Complexity is a two-dimensional framework to classify the complexity of problems based on their input size n and some additional parameter k [8,11,27]. It distinguishes between running times of the form $f(k)n^{g(k)}$ where the degree of the polynomial depends on k and running times of the form $f(k)n^{O(1)}$ where the exponential part of the running time is independent of n. A parameterized problem is *fixed-parameter tractable* (FPT) if there exists an algorithm that solves an input of size n and parameter k in time bounded by $f(k)n^{O(1)}$. In this case we say that the *parameter dependence* of the algorithm is f and we call it an *FPT algorithm*. Parameterized Complexity has a hardness theory, similar to the theory of NP-completeness to show that certain problems have no FPT algorithm under complexity-theoretic assumptions.

Graphs. Let $G = (V, E)$ be a simple, finite, undirected graph. Let $S \subseteq V$ and $v \in V$. We denote by $G - S$ the graph obtained from G by removing all vertices in S and all edges incident to vertices in S. We denote by $G[S]$ the graph $G - (V \setminus S)$. The *(open) neighborhood* of v is $N(v) = \{u \in V : uv \in E\}$, the *(open) neighborhood* of S is $N(S) = \bigcup_{u \in S} N(u) \setminus S$, and their *closed neighborhoods* are $N[v] = N(v) \cup \{v\}$ and $N[S] = N(S) \cup S$, respectively. A v_1-v_k *path* P of length k in G is a sequence of k pairwise distinct vertices $(v_1, v_2, \cdots, v_k)$ such that $v_i v_{i+1} \in E$ for each $i \in \{1, \ldots, k-1\}$. The vertices v_1 and v_k are the *endpoints* of P and all other vertices from P are *internal*. An edge is *internal* to P if it is incident to two internal vertices from P. Two or more paths are *independent* if none of them contains an inner vertex of another.

A *tree decomposition* of G is a pair $(\{X_i : i \in I\}, T)$ where $X_i \subseteq V$, $i \in I$, and T is a tree with elements of I as nodes such that (i) for each edge $uv \in E$, there is an $i \in I$ such that $\{u, v\} \subseteq X_i$, and (ii) for each $v \in V$, $T[\{i \in I : v \in X_i\}]$

is connected and has at least one node. The *width* of a tree decomposition is $\max_{i \in I} |X_i| - 1$. The *treewidth* [31] of G is the minimum width taken over all tree decompositions of G and it is denoted $\mathbf{tw}(G)$. A graph is *planar* if it can be drawn in the plane with no crossing edges. For other standard graph-theoretic notions not defined here, we refer to [6].

CNF Formulas and Satisfiability. We consider propositional formulas in conjunctive normal form (CNF) where no clause contains a complementary pair of literals. For a clause c, we write $\text{lit}(c)$ and $\text{var}(c)$ for the sets of literals and variables occurring in c, respectively. For a CNF formula F we write $\text{cla}(F)$ for its set of clauses, $\text{lit}(F) = \bigcup_{c \in \text{cla}(F)} \text{lit}(c)$ for its set of literals, and $\text{var}(F) = \bigcup_{c \in \text{cla}(F)} \text{var}(c)$ for its set of variables.

For a set $X \subseteq \text{var}(F)$ we denote by 2^X the set of all mappings $\tau : X \to \{0,1\}$, the *truth assignments* on X. A truth assignment on X can be extended to the literals over X by setting $\tau(\neg x) = 1 - \tau(x)$ for all $x \in X$. Given a CNF formula F and a truth assignment $\tau \in 2^X$ we define $F[\tau]$ to be the formula obtained from F by removing all clauses c such that τ sets a literal of c to 1, and removing the literals set to 0 from all remaining clauses.

A CNF formula F is *satisfiable* if there is some $\tau \in 2^{\text{var}(F)}$ with $F[\tau] = \emptyset$. SAT is the NP-complete problem of deciding whether a given CNF formula is satisfiable [4,22]. #SAT is the #P-complete problem of determining the number of distinct $\tau \in 2^{\text{var}(F)}$ with $F[\tau] = \emptyset$ [35].

Nested Formulas. Consider a linear order $<$ of the variables of a CNF formula F. A clause c *straddles* a clause c' if there are variables $x, y \in \text{var}(c)$ and $z \in \text{var}(c')$ such that $x < z < y$. Two clauses *overlap* if they straddle each other. A CNF formula F is *nested* if there exists a linear ordering $<$ of $\text{var}(F)$ in which no two clauses of F overlap [20]. The satisfiability of a nested CNF formula can be determined in polynomial time [20].

The *incidence graph* of F is the bipartite graph $\text{inc}(F) = (V, E)$ with $V = \text{var}(F) \cup \text{cla}(F)$ and for a variable $x \in \text{var}(F)$ and a clause $c \in \text{cla}(F)$ we have $xc \in E$ if $x \in \text{var}(c)$. The *sign* of the edge xc is *positive* if $x \in \text{lit}(c)$ and *negative* if $\neg x \in \text{lit}(c)$. Recall that $\mathbf{tw}^*(F)$ denotes the treewidth of $\text{inc}(F)$.

The graph $\text{inc+u}(F)$ is $\text{inc}(\text{univ}(F))$, where $\text{univ}(F)$ is obtained from F by adding a *universal* clause c^* containing all variables of F. By a result of Kratochvíl and Křivánek [21], F is nested if and only if $\text{inc+u}(F)$ is planar. Since $\mathbf{tw}^*(F) \leq 3$ if F is nested [2], the number of satisfying assignments of F can also be counted in polynomial time [10,34].

Backdoors. Backdoor sets are defined with respect to a fixed class $\mathcal{C}$ of CNF formulas, the *base class*. Let F be a CNF formula and $B \subseteq \text{var}(F)$. B is a *strong $\mathcal{C}$-backdoor set* of F if $F[\tau] \in \mathcal{C}$ for each $\tau \in 2^B$. B is a *deletion $\mathcal{C}$-backdoor set* of F if $F - B \in \mathcal{C}$, where $\text{cla}(F - B) = \{c \setminus \{x, \neg x : x \in B\} : c \in \text{cla}(F)\}$.

If we are given a strong $\mathcal{C}$-backdoor set of F of size k, we can reduce the satisfiability of F to the satisfiability of 2^k formulas in $\mathcal{C}$. Thus SAT becomes FPT in k if $\mathcal{C}$ is polynomial-time solvable. If $\mathcal{C}$ is clause-induced (i.e., $F \in \mathcal{C}$

implies $F' \in \mathcal{C}$ for every F' such that $\mathsf{cla}(F') \subseteq \mathsf{cla}(F)$), any deletion $\mathcal{C}$-backdoor set of F is a strong $\mathcal{C}$-backdoor set of F. The interest in deletion backdoor sets is motivated for base classes where they are easier to detect than strong backdoor sets. The challenging problem is to find a strong or deletion $\mathcal{C}$-backdoor set of size at most k if it exists. We denote by $\mathbf{sb_N}(F)$ the size of a smallest strong NESTED-backdoor set.

Minors and Grids. The *r-grid* is the graph $L_r = (V, E)$ with vertex set $V = \{(i,j) : 1 \leq i \leq r, 1 \leq j \leq r\}$ in which two vertices (i,j) and (i',j') are adjacent if and only if $|i - i'| + |j - j'| = 1$. We say that a vertex $(i,j) \in V$ has horizontal index i and vertical index j.

A graph H is a *minor* of a graph G if H can be obtained from a subgraph of G by contracting edges. The *contraction* of an edge uv makes u adjacent to all vertices in $N(v) \setminus \{u\}$ and removes v. If H is a minor of G, then one can find a model of H in G. A *model* of H in G is a set of vertex-disjoint connected subgraphs of G, one subgraph C_u for each vertex u of H, such that if uv is an edge of H, then there is an edge of G with one endpoint in C_u and the other in C_v. We will use Robertson and Seymour's grid-minor theorem.

Theorem 1 ([32]). *For every positive integer r, there exists a constant $f(r)$ such that if a graph G has treewidth at least $f(r)$, then G contains an r-grid as a minor.*

It is known that $f(r) \leq 20^{2r^5}$ [33]. A linear-time FPT algorithm (parameterized by k) by Bodlaender [3] finds a tree decomposition of width at most k of a graph G if $\mathbf{tw}(G) \leq k$. A quadratic FPT algorithm (parameterized by r) by Kawarabayashi *et al.* [17] finds an r-grid minor in a graph G if G contains an r-grid as a minor.

By Wagner's theorem [36], a graph is planar if and only if it has no $K_{3,3}$ and no K_5 as a minor. Here, K_5 denotes the complete graph on 5 vertices and $K_{3,3}$ the complete bipartite graph with 3 vertices in both independent sets of the bipartition.

3   Detection of Strong Nested-Backdoor Sets

Let F be a CNF formula and k be an integer. Our FPT algorithm will count the number of satisfying truth assignements of F if F has a strong NESTED-backdoor set of size at most k.

The first step of the algorithm is to find a good approximation for a smallest strong NESTED-backdoor set. Specifically, it will either determine that F has no strong NESTED-backdoor set of size at most k, or it will compute a strong NESTED-backdoor set of size at most 2^k. An algorithm of that kind is called an *FPT-approximation algorithm* [23], as it is an FPT algorithm that computes a solution that approximates the optimum with an error bounded by a function of the parameter. In case F has no strong NESTED-backdoor set of size at most k, the algorithm stops, and if it finds a strong NESTED-backdoor set B

of size at most 2^k, for every truth assignment τ to B, a tree decomposition of $\text{inc}(F[\tau])$ of width ast most 3 can be computed in linear time [2,3], and treewidth-based dynamic programming algorithms can be used to compute the number of satisfying assignments of $F[\tau]$ in polynomial time [10,34]. We will arrive at our main theorem.

Theorem 2. *The problems SAT and #SAT are fixed-parameter tractable parameterized by* $\mathbf{sb_N}(F)$.

It only remains to design the FPT-approximation algorithm for strong NESTED-backdoor set detection. Consider the incidence graph $G = (V, E) = \text{inc}(F)$ of F. By [33], it either has treewidth at most $\text{tw}(k)$, or it has a $\text{grid}(k)$-grid as a minor. We will treat both cases separately. Here,

$$\text{tw}(k) := 20^{2\text{grid}(k)^5},$$
$$\text{grid}(k) := 4 \cdot \sqrt{\text{obs}(k) + 1},$$
$$\text{obs}(k) := 2^k \cdot \text{same}(k) + k, \text{ and}$$
$$\text{same}(k) := 15 \cdot 2^{2k+2}.$$

The functions $\text{obs}(k)$ and $\text{same}(k)$ will be used in the next subsection.

Lemma 1. *There is an FPT algorithm that, given a CNF formula F, a positive integer parameter k, and a $\text{grid}(k)$-grid as a minor in $\text{inc}(F)$, computes a set $S^* \subseteq \text{var}(F)$ of size $2^{O(k^{10})}$ such that every strong NESTED-backdoor set of size at most k contains a variable from S^*.*

Lemma 2. *There is an FPT algorithm that takes as input a CNF formula F, a positive integer parameter k, and a tree decomposition of $G = \text{inc}(F)$ of width at most $\text{tw}(k)$, and finds a strong NESTED-backdoor set of F of size k if one exists.*

The proof of Lemma 2 [16] relies on Arnborg et al.'s extension [1] of Courcelle's theorem [5]. Lemma 1 is proven in Subsection 3.1 and contains the main combinatorial arguments of this paper. These two lemmas can now be used to compute a strong NESTED-backdoor set of F.

Theorem 3. *There is an FPT algorithm, which, for a CNF formula F and a positive integer parameter k, either concludes that F has no strong NESTED-backdoor set of size at most k or finds a strong NESTED-backdoor set of F of size at most 2^k.*

Proof. If $k \leq 1$, our algorithm solves the problem exactly in polynomial time. Otherwise, it runs Bodlaender's FPT algorithm [3] with input G and parameter $\text{tw}(k)$ to either find a tree decomposition of G of width at most $\text{tw}(k)$ or to determine that $\mathbf{tw}(G) > \text{tw}(k)$. In case a tree decomposition of width at most $\text{tw}(k)$ is found, the algorithm uses Lemma 2 to compute a strong NESTED-backdoor set of F of size k if one exists, and it returns the answer.

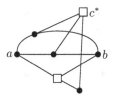

Fig. 2. A NESTED-obstruction leading to a $K_{3,3}$-minor with the universal clause c^*

In case Bodlaender's algorithm determines that $\mathbf{tw}(G) > \mathsf{tw}(k)$, by [33] we know that G has a $\mathsf{grid}(k)$-grid as a minor. Such a $\mathsf{grid}(k)$-grid is found by running the FPT algorithm of Kawarabayashi *et al.* [17] with input G and parameter $\mathsf{grid}(k)$. The algorithm now executes the procedure from Lemma 1 to find a set S^* of $2^{O(k^{10})}$ variables from $\mathsf{var}(F)$ such that every strong NESTED-backdoor set of size at most k contains a variable from S^*. The algorithm considers all possibilities that the backdoor set contains some $x \in S^*$; there are $2^{O(k^{10})}$ choices for x. For each such choice, recurse on $F[x = 1]$ and $F[x = 0]$ with parameter $k - 1$. If, for some $x \in S^*$, both recursive calls return backdoor sets B_x and $B_{\neg x}$, then return $B_x \cup B_{\neg x} \cup \{x\}$, otherwise, return NO. As $2^k - 1 = 2 \cdot (2^{k-1} - 1) + 1$, the solution size is upper bounded by $2^k - 1$. On the other hand, if at least one recursive call returns NO for every $x \in S^*$, then F has no strong NESTED-backdoor set of size at most k.   □

In particular, this proves Theorem 2. For showing Theorem 2 one could also avoid the use of Lemma 2 and directly apply the algorithms from [10,34] to count the number of satisfying assignments in case $\mathbf{tw}^*(F) \leq \mathsf{tw}(k)$.

3.1   Large Grid Minor

The goal of this subsection is to prove Lemma 1. Suppose G has a $\mathsf{grid}(k)$-grid as a minor.

Definition 1. *An a–b NESTED-obstruction is a subgraph of* $\mathsf{inc}(F)$ *consisting of*

- *five distinct vertices* a, b, p_1, p_2, p_3, *such that* p_1, p_2, p_3 *are variables,*
- *three independent a–b paths* P_1, P_2, P_3, *and*
- *an edge between* p_i *and an internal vertex from* P_i *for each* $i \in \{1, 2, 3\}$.

In particular, if a path P_i has a variable v as an interior vertex, we can take $p_i := v$. See Figure 2.

Lemma 3. *If F' is a CNF formula such that* $\mathsf{inc}(F')$ *contains a NESTED-obstruction, then* $F' \notin$ NESTED.

Lemma 3 can easily be proven by exhibiting a $K_{3,3}$-minor in $\mathsf{inc+u}(F')$. The lemma implies that for each assignment to the variables of a strong NESTED-backdoor set, at least one variable from each NESTED-obstruction vanishes in the

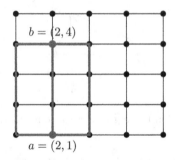

Fig. 3. The 5-grid and a highlighted NESTED-obstruction

reduced formula. Using the r-grid, we now find a set $\mathcal{O}$ of $\mathsf{obs}(k)$ vertex-disjoint NESTED-obstructions in G.

Lemma 4. *Given a $\mathsf{grid}(k)$-grid minor of $G = \mathsf{inc}(F)$, a set of $\mathsf{obs}(k)$ vertex-disjoint NESTED-obstructions can be found in polynomial time.*

The proof of Lemma 4 (see [16]) packs vertex-disjoint NESTED-obstructions, like the one highlighted in Figure 3, into the $\mathsf{grid}(k)$-grid.

Denote by $\mathcal{O}$ a set of $\mathsf{obs}(k)$ vertex-disjoint NESTED-obstructions obtained via Lemma 4. A backdoor variable can destroy a NESTED-obstruction either because it participates in the NESTED-obstruction, or because every setting of the variable satisfies a clause that participates in the NESTED-obstruction.

Definition 2. *Let x be a variable and O a NESTED-obstruction in G. We say that x kills O if neither $\mathsf{inc}(F[x = 1])$ nor $\mathsf{inc}(F[x = 0])$ contains O as a subgraph. We say that x kills O internally if $x \in \mathsf{var}(O)$, and that x kills O externally if x kills O but does not kill it internally. In the latter case, O contains a clause c containing x and a clause c' containing $\neg x$ and we say that x kills O (externally) in c and c'.*

By Lemma 3, for every strong NESTED-backdoor set B of F and every NESTED-obstruction O, there is at least one $x \in B$ that kills O.

We will describe an algorithm that first performs a constant number of non-deterministic steps (*guesses*) to determine some properties about the strong NESTED-backdoor set. Each such guess is made out of a number of choices that is upper bounded by a function of k. The algorithm will then only search for strong NESTED-backdoor sets that have the determined properties, and such backdoor sets are called *valid*. Finally, the algorithm will be made deterministic by executing each possible combination of nondeterministic steps.

For a fixed series of guesses, the algorithm will compute a set $S \subseteq \mathsf{var}(F)$ such that every valid strong NESTED-backdoor set of size at most k contains a variable from S. The union of all such S, taken over all possible series of guesses, forms a set S^* and each strong NESTED-backdoor set of size at most k contains a variable from S^*. Bounding the size of each S by a function of k enables us to bound $|S^*|$ by a function of k, and S^* can then be used in a bounded search tree algorithm (see Theorem 3).

For every strong NESTED-backdoor set of size at most k, at most k NESTED-obstructions from $\mathcal{O}$ are killed internally since they are vertex-disjoint. The algorithm guesses k NESTED-obstructions from $\mathcal{O}$ that may be killed internally. Let $\mathcal{O}'$ denote the set of the remaining NESTED-obstructions, which need to be killed externally.

Suppose F has a strong NESTED-backdoor set B of size k killing no NESTED-obstruction from $\mathcal{O}'$ internally. Then, B defines a partition of $\mathcal{O}'$ into 2^k parts where for each part, the NESTED-obstructions contained in this part are killed externally by the same set of variables from B. Since $|\mathcal{O}'| = \mathsf{obs}(k) - k = 2^k \cdot \mathsf{same}(k)$, at least one of these parts contains at least $\mathsf{same}(k)$ NESTED-obstructions from $\mathcal{O}'$. The algorithm guesses a subset $\mathcal{O}_s \subseteq \mathcal{O}'$ of $\mathsf{same}(k)$ NESTED-obstructions from this part and it guesses how many variables from the strong NESTED-backdoor set kill the obstructions in this part externally.

Suppose each NESTED-obstruction in $\mathcal{O}_s$ is killed externally by the same set of ℓ backdoor variables, and no other backdoor variable kills any NESTED-obstruction from $\mathcal{O}_s$. Clearly, $1 \leq \ell \leq k$. Compute the set of external killers for each NESTED-obstruction in $\mathcal{O}_s$. Denote by Z the common external killers of the NESTED-obstructions in $\mathcal{O}_s$. The presumed backdoor set contains exactly ℓ variables from Z and no other variable from the backdoor set kills any NESTED-obstruction from $\mathcal{O}_s$.

We will define three rules for the construction of S, and the algorithm will execute the first applicable rule.

Rule 1 (Few Common Killers). *If $|Z| < |\mathcal{O}_s|$, then set $S := Z$.*

The correctness of this rule follows since any valid strong NESTED-backdoor set contains ℓ variables from Z and $\ell \geq 1$.

For each $O \in \mathcal{O}_s$ we define an auxiliary graph $G_O = (Z, E_O)$ whose edge set is initially empty. As long as G_O has a vertex v with degree 0 such that v and some other vertex in Z have a common neighbor from O in G, select a vertex u of minimum degree in G_O such that u and v have a common neighbor from O in G and add the edge uv to E_O. As long as G_O has a vertex v with degree 0, select a vertex u of minimum degree in G_O such that v has a neighbor $v' \in V(O)$ in G and u has a neighbor $u' \in V(O)$ in G and there is a v'–u' path in O in which no internal vertex is adjacent to a vertex from $Z \setminus \{v\}$; add the edge uv to E_O.

Fact 1. *For each $O \in \mathcal{O}_s$, the graph G_O has minimum degree at least 1.*

Recall that no clause contains complimentary literals. Consider two variables $u, v \in Z$ that share an edge in G_O. By the construction of G_O, there is a u–v path P in G whose internal edges are in O, such that for each variable $z \in Z$, all edges incident to z and a clause from P have the same sign. Moreover, since no variable from a valid strong NESTED-backdoor set kills O externally, unless it is in Z, for each potential backdoor variable $x \in \mathsf{var}(F) \setminus Z$, all edges incident to x and a clause from P have the same sign. Thus, we have the following fact.

Fact 2. *If $u, v \in Z$ share an edge in G_O, then for every valid strong NESTED-backdoor set that does not contain u and v, there is a truth assignment τ to B such that $\mathrm{inc}(F[\tau])$ contains a u–v path whose internal edges are in O.*

Consider the multigraph $G_m(\mathcal{O}_s) = (Z, \biguplus_{O \in \mathcal{O}_s} E_O)$, i.e., the union of all G_O over all $O \in \mathcal{O}_s$, where the multiplicity of an edge is the number of distinct sets E_O where it appears, $O \in \mathcal{O}_s$.

Rule 2 (Multiple Edges). *If there are two vertices $u, v \in Z$ such that $G_m(\mathcal{O}_s)$ has a u–v edge with multiplicity at least $2 \cdot 2^k + 1$, then set $S := \{u, v\}$.*

Consider any valid strong NESTED-backdoor set B of size k. Then, by Fact 2, for each u–v edge there is some truth assignment τ to B such that $\mathrm{inc}(F[\tau])$ contains a u–v path in G. Moreover, since each u–v edge comes from a different $O \in \mathcal{O}_s$, all these u–v paths are independent. Since there are 2^k truth assignments to B but at least $2 \cdot 2^k + 1$ u–v edges, for at least one truth assignment τ to B, there are 3 independent u–v paths P_1, P_2, P_3 in $\mathrm{inc}(F[\tau])$. We obtain a u–v NESTED-obstruction choosing as $p_i, 1 \leq i \leq 3$, a variable from P_i or a variable neighboring a clause from P_i and belonging to the same NESTED-obstruction in $\mathcal{O}_s$. Thus, any valid strong NESTED-backdoor set contains u or v.

Now, consider the graph $G(\mathcal{O}_s)$ obtained from the multigraph $G_m(\mathcal{O}_s)$ by merging multiple edges, i.e., we retain each edge only once.

Rule 3 (No Multiple Edges). *Set S to be the $2k$ vertices of highest degree in $G(\mathcal{O}_s)$ (ties are broken arbitrarily).*

For the sake of contradiction, suppose F has a valid strong NESTED-backdoor set B of size k with $B \cap S = \emptyset$. First, we show a lower bound on the number of edges in $G(\mathcal{O}_s) - B$. Since $G_m(\mathcal{O}_s)$ has at least $\frac{|Z|}{2} \mathsf{same}(k)$ edges and each edge has multiplicity at most 2^{k+1}, the graph $G(\mathcal{O}_s)$ has at least $\frac{|Z|\mathsf{same}(k)}{2 \cdot 2^{k+1}} = 3 \cdot 5 \cdot 2^k \cdot |Z|$ edges. Let d be the sum of the degrees in $G(\mathcal{O}_s)$ of the vertices in $B \cap Z$. Now, the sum of degrees of vertices in S is at least $2d$ in $G(\mathcal{O}_s)$, and at least d in $G(\mathcal{O}_s) - B$. Therefore, $G(\mathcal{O}_s) - B$ has at least $d/2$ edges. On the other hand, the number of edges deleted to obtain $G(\mathcal{O}_s) - B$ from $G(\mathcal{O}_s)$ is at most d. It follows that the number of edges in $G(\mathcal{O}_s) - B$ is at least a third the number of edges in $G(\mathcal{O}_s)$, and thus at least $5 \cdot 2^k \cdot |Z|$.

Now, we iteratively build a truth assignment τ for B. Set $H := G(\mathcal{O}_s) - B$. Order the variables of B as $b_1, \ldots, b_k$. For increasing i, we set $\tau(b_i) = 0$ if in G, the vertex $v \in B$ is adjacent with a positive edge to more paths that correspond to an edge in H than with a negative edge and set $\tau(b_i) = 1$ otherwise; if $\tau(b_i) = 0$, then remove each edge from H that corresponds to a path in G that is adjacent with a negative edge to b_i, otherwise remove each edge from H that corresponds to a path in G that is adjacent with a positive edge to b_i.

Observe that for a variable $v \in B$ and a path P in G that corresponds to an edge in $G(\mathcal{O}_s) - B$, v is not adjacent with a positive and a negative edge to P. If $v \in Z$ this follows by the construction of G_O, and if $v \notin Z$, this follows since v does not kill any NESTED-obstruction from $\mathcal{O}_s$. Therefore, each of the k

iterations building the truth assignment τ has removed at most half the edges of H. In the end, H has at least $5|Z|$ edges.

Next, we use the following theorem of Kirousis et al. [19].

Theorem 4 ([19]). *If a graph has n vertices and $m > 0$ edges, then it has an induced subgraph that is $\lceil \frac{m+n}{2n} \rceil$-vertex-connected.*

We conclude that H has an induced subgraph H' that is 3-vertex-connected. Let $x, y \in V(H')$. We use Menger's theorem [25].

Theorem 5 ([25]). *Let $G = (V, E)$ be a graph and $x, y \in V$. Then the size of a minimum x, y-vertex-cut in G is equal to the maximum number of independent x–y paths in G.*

Since the minimum size of an x, y-vertex cut is at least 3 in H', there are 3 independent x–y paths in H'. Replacing each edge by its corresponding path in G, gives rise to 3 walks from x to y in G. Shortcutting cycles, we obtain three x–y paths P_1, P_2, P_3 in G. By construction, each edge of these paths is incident to a vertex from a NESTED-obstruction in $\mathcal{O}_s$. We assume that P_1, P_2, P_3 are edge-disjoint. Indeed, by the construction of the G_O, $O \in \mathcal{O}_s$, they can only share the first and last edges. In case P_1 shares the first edge with P_2, replace x by its neighbor on P_1, remove the first edge from P_1 and P_2, and replace P_3 by its symmetric difference with this edge. Act symmetrically for the other combinations of paths sharing the first or last edge.

Lemma 5 ([12]). *Let $G = (V, E)$ be a graph. If there are two vertices $x, y \in V$ with 3 edge-disjoint x–y paths in G, then there are two vertices $x', y' \in V$ with 3 independent x'–y' paths in G.*

By Lemma 5 we obtain two vertices x', y' in G with 3 independent x'–y' paths P_1', P_2', P_3' in G. Since the lemma does not presuppose any other edges in G besides those from the edge-disjoint x–y paths, P_1', P_2', P_3' use only edges from the paths P_1, P_2, P_3. Thus, each edge of P_1', P_2', P_3' is incident to a vertex from a NESTED-obstruction in $\mathcal{O}_s$. Thus, we obtain an x'–y' NESTED-obstruction with the paths P_1', P_2', P_3', and for each path P_i', we choose a variable from this path or a variable from $\mathcal{O}_s$ neighboring a clause from this path. We arrive at a contradiction for B being a valid strong NESTED-backdoor set. This proves the correctness of Rule 3.

The number of possible guesses the algorithm makes is upper bounded by $\binom{\mathsf{obs}(k)}{k} \cdot \binom{\mathsf{obs}(k)-k}{\mathsf{same}(k)} \cdot k = 2^{O(k^8)}$, and each series of guesses leads to a set S of at most $\mathsf{same}(k)$ variables. Thus, the set S^*, the union of all such S, contains at most $2^{O(k^8)} \cdot \mathsf{same}(k) = 2^{O(k^{10})}$ variables. This completes the proof of Lemma 1.

4   Conclusion

We have classified the problems SAT and #SAT as fixed-parameter tractable when parameterized by the size of a smallest strong backdoor set with respect to the base class of nested formulas. As argued in the introduction, this parameter is incomparable with incidence treewidth.

The parameter dependence makes our algorithm impractical. However, the class of fixed-parameter tractable problems has proven to be quite robust: Once a problem is shown to belong to this class, one can start to develop faster and more practical algorithms. For many cases in the past this has been successful. For instance, the problem of recognizing graphs of genus k was originally shown to be fixed-parameter tractable by means of non-constructive tools from graph minor theory [9]. Later a linear-time algorithm with doubly exponential parameter dependence was found [26], and more recently, an algorithm with a single exponential parameter dependence [18]. It would be interesting to see whether a similar improvement is possible for finding or FPT-approximating strong backdoor sets with respect to nested formulas.

We would like to point out that the results of this paper have been recently generalized to the base class of formulas with bounded incidence treewidth [15].

Acknowledgments. The authors acknowledge support from the European Research Council (COMPLEX REASON, 239962). Serge Gaspers acknowledges partial support from the Australian Research Council (DE120101761).

References

1. Arnborg, S., Lagergren, J., Seese, D.: Easy problems for tree-decomposable graphs. J. Algorithms 12(2), 308–340 (1991)
2. Biedl, T., Henderson, P.: Nested SAT graphs have treewidth three. Technical Report CS-2004-70. University of Waterloo (2004)
3. Bodlaender, H.L.: A linear-time algorithm for finding tree-decompositions of small treewidth. SIAM J. Comput. 25(6), 1305–1317 (1996)
4. Cook, S.A.: The complexity of theorem-proving procedures. In: Proc. of STOC 1971, pp. 151–158 (1971)
5. Courcelle, B.: Graph rewriting: an algebraic and logic approach. In: Handbook of Theoretical Computer Science, vol. B, pp. 193–242. Elsevier (1990)
6. Diestel, R.: Graph Theory, 4th edn. Graduate Texts in Mathematics, vol. 173. Springer (2010)
7. Dilkina, B.N., Gomes, C.P., Sabharwal, A.: Tradeoffs in the Complexity of Backdoor Detection. In: Bessière, C. (ed.) CP 2007. LNCS, vol. 4741, pp. 256–270. Springer, Heidelberg (2007)
8. Downey, R.G., Fellows, M.R.: Parameterized Complexity. Monographs in Computer Science. Springer (1999)
9. Fellows, M.R., Langston, M.A.: Nonconstructive tools for proving polynomial-time decidability. J. ACM 35(3), 727–739 (1988)
10. Fischer, E., Makowsky, J.A., Ravve, E.R.: Counting truth assignments of formulas of bounded tree-width or clique-width. Discr. Appl. Math. 156(4), 511–529 (2008)
11. Flum, J., Grohe, M.: Parameterized Complexity Theory. Texts in Theoretical Computer Science. An EATCS Series, vol. XIV. Springer (2006)
12. Gaspers, S.: From edge-disjoint paths to independent paths. Technical Report 1203.4483, arXiv (2012)

13. Gaspers, S., Szeider, S.: Backdoors to Acyclic SAT. In: Czumaj, A., et al. (eds.) ICALP 2012, Part I. LNCS, vol. 7391, pp. 363–374. Springer, Heidelberg (2012)
14. Gaspers, S., Szeider, S.: Backdoors to Satisfaction. In: Bodlaender, H.L., Downey, R.G., Fomin, F.V., Marx, D. (eds.) Fellows Festschrift. LNCS, vol. 7370, pp. 287–317. Springer, Heidelberg (2012)
15. Gaspers, S., Szeider, S.: Strong backdoors to bounded treewidth SAT. Technical Report 1204.6233, arXiv (2012)
16. Gaspers, S., Szeider, S.: Strong backdoors to nested satisfiability. Technical Report 1202.4331, arXiv (2012)
17. Kawarabayashi, K.-I., Kobayashi, Y., Reed, B.: The disjoint paths problem in quadratic time. J. Comb. Theory, Ser. B 102(2), 424–435 (2012)
18. Kawarabayashi, K.-I., Mohar, B., Reed, B.A.: A simpler linear time algorithm for embedding graphs into an arbitrary surface and the genus of graphs of bounded tree-width. In: Proc. of FOCS 2008, pp. 771–780 (2008)
19. Kirousis, L.M., Serna, M.J., Spirakis, P.G.: Parallel complexity of the connected subgraph problem. SIAM J. Comput. 22(3), 573–586 (1993)
20. Knuth, D.E.: Nested satisfiability. Acta Informatica 28(1), 1–6 (1990)
21. Kratochvíl, J., Křivánek, M.: Satisfiability and co-nested formulas. Acta Inf. 30, 397–403 (1993)
22. Levin, L.: Universal sequential search problems. Problems of Information Transmission 9(3), 265–266 (1973)
23. Marx, D.: Parameterized complexity and approximation algorithms. Comput. J. 51(1), 60–78 (2008)
24. Marx, D., Schlotter, I.: Obtaining a planar graph by vertex deletion. Algorithmica 62(3-4), 807–822 (2012)
25. Menger, K.: Zur allgemeinen Kurventheorie. Fundamenta Mathematicae 10, 96–115 (1927)
26. Mohar, B.: Embedding graphs in an arbitrary surface in linear time. In: Proc. of STOC 1996, pp. 392–397 (1996)
27. Niedermeier, R.: Invitation to Fixed-Parameter Algorithms. Oxford Lecture Series in Mathematics and its Applications. Oxford University Press (2006)
28. Nishimura, N., Ragde, P., Szeider, S.: Detecting backdoor sets with respect to Horn and binary clauses. In: Proc. of SAT 2004, pp. 96–103 (2004)
29. Nishimura, N., Ragde, P., Szeider, S.: Solving #SAT using vertex covers. Acta Inf. 44(7-8), 509–523 (2007)
30. Razgon, I., O'Sullivan, B.: Almost 2-SAT is fixed parameter tractable. J. Comput. Syst. Sci. 75(8), 435–450 (2009)
31. Robertson, N., Seymour, P.D.: Graph minors. II. Algorithmic aspects of tree-width. J. Algorithms 7(3), 309–322 (1986)
32. Robertson, N., Seymour, P.D.: Graph minors. V. Excluding a planar graph. J. Combin. Theory, Ser. B 41(1), 92–114 (1986)
33. Robertson, N., Seymour, P., Thomas, R.: Quickly excluding a planar graph. J. Combin. Theory, Ser. B 62(2), 323–348 (1994)
34. Samer, M., Szeider, S.: Algorithms for propositional model counting. J. Discrete Algorithms 8(1), 50–64 (2010)
35. Valiant, L.G.: The complexity of computing the permanent. Theor. Comput. Sci. 8(2), 189–201 (1979)
36. Wagner, K.: Über eine Eigenschaft der ebenen Komplexe. Mathematische Annalen 114(1), 570–590 (1937)
37. Williams, R., Gomes, C., Selman, B.: Backdoors to typical case complexity. In: Proc. of IJCAI 2003, pp. 1173–1178 (2003)

Extended Failed-Literal Preprocessing
for Quantified Boolean Formulas*

Allen Van Gelder[1], Samuel B. Wood[1], and Florian Lonsing[2]

[1] University of California, Santa Cruz
[2] Johannes Kepler University

Abstract. Building on recent work that adapts failed-literal analysis (FL) to Quantified Boolean Formulas (QBF), this paper introduces extended failed-literal analysis (EFL). FL and EFL are both preprocessing methods that apply a fast, but incomplete reasoning procedure to abstractions of the underlying QBF. EFL extends FL by remembering certain binary clauses that are implied by the same reasoning procedure as FL when it assumes one literal and that implies a second literal. This extension is almost free because the second literals are implied anyway during FL, but compared to analogous techniques for propositional satisfiability, its correctness involves some subtleties. For the first time, application of the universal pure literal rule is considered without also applying the existential pure literal rule. It is shown that using both pure literal rules in EFL is unsound. A modified reasoning procedure for QBF, called Unit-clause Propagation with Universal Pure literals (UPUP) is described and correctness is proved for EFL based on UPUP. Empirical results on the 568-benchmark suite of QBFEVAL-10 are presented.

Keywords: quantified boolean formulas, QBF, failed literals, extended failed literals, 1-saturation, look-ahead, preprocessing.

1   Introduction

With the advent of capable solvers for Quantified Boolean Formulas (QBFs), their use for encoding problems from industrial applications is increasing rapidly. As with propositional satisfiability, preprocessors have been found to be an important part of the QBF solving toolkit. Preprocessors typically do a predictable (polynomially bounded) amount of work to simplify the original formula, making it more amenable to the complete solver. The newer complete QBF solvers are typically based on search, a form of back-chaining, whereas preprocessors use forward reasoning. The two approaches often complement each other nicely.

The essence of failed-literal analysis is to add an assumption that some literal is true to a given formula and use incomplete (but usually fast) forward reasoning to see if the formula can now be proven false; if so, then the *negation* of the assumed literal can soundly be added to the formula. This idea was introduced for propositional SAT solving by Jon Freeman [Fre93].

---

* http://www.cse.ucsc.edu/{~avg,~sbwood}, http://fmv.jku.at/lonsing

A. Cimatti and R. Sebastiani (Eds.): SAT 2012, LNCS 7317, pp. 86–99, 2012.

This paper builds upon recent work of Lonsing and Biere [LB11] that adapts failed-literal analysis to QBF solving. That work uses the QBF pure-literal rule [GNT04] heavily in its incomplete forward reasoning, which they call *QBCP*. We re-examine the QBF pure-literal rule, which applies to both existential and universal literals, and consider those parts separately. We observe that the two parts operate quite differently and have different properties. We show that neither part is a *super-sound inference rule* (in the sense that tree models are preserved, as specified in Section 2); they are only *safe heuristics* (in the sense that the truth value of a closed QBF is not changed by their use). (This fact is well-known for propositional formulas and the *existential* pure-literal rule, but seems not to have been considered for the *universal* pure-literal rule in QBF.) We show that using them both as though they were super-sound logical inferences can lead to fallacious reasoning in QBF. We believe this is the first time that application of the universal pure literal rule has been considered without also applying the existential pure literal rule.

Then we develop an enhancement of failed literal analysis based on propositional techniques called *1-saturation*, described by Groote and Warners [GW00] and *double unit-propagation look-ahead* described by Le Berre [LB01]. We show that using the universal pure-literal rule *and not* the existential pure-literal rule is safe for 1-saturation in QBF.

The essence of *1-saturation* in propositional logic is the observation that, with a given formula, if assuming some literal q allows some other literal p to be soundly derived, and if assuming $\overline{q}$ also allows p to be soundly derived, then p can soundly be added to the formula as a unit clause. Unfortunately for QBF, literals derived with *QBCP* are not necessarily super-soundly derived.

We introduce an incomplete forward reasoning procedure that employs unit-clause propagation (including universal reduction) and the *universal pure-literal rule*. We show that this procedure super-soundly derives literals. We call it *UPUP* for Unit-clause Propagation with Universal Pure literals.

Although *UPUP* is weaker than *QBCP* in the sense that it assigns values to fewer literals, we found experimentally that it is considerably faster, for simple failed-literal analysis. In addition, it serves as the basis for adapting *1-saturation* to QBF, and it can log proof steps that are verifiable as *Q-resolution* steps [KBKF95]. We call our adaptation of 1-saturation to QBF **extended failed literal analysis** (**EFL**).

One reason for our interest in EFL is that it enables a significant fraction of the popular QBFLIB benchmark suite to be solved with preprocessing alone. The first QBF preprocessor to solve a significant number of benchmarks in this suite was sQueezeBF, described by Giunchiglia *et al.* [GMN10]. With their publicly available binary code, QuBE-7.2, we solved 40 of the 568 QBFLIB benchmarks. Lonsing and Biere reported [LB11] that their QBF failed-literal tool, publicly available as qxbf, processing the output of sQueezeBF 7.1 when it did not solve the instance, solved an additional 25 benchmarks. We confirmed the same result with sQueezeBF 7.2.

Another reason for our interest in EFL is that it can be used without any pure-literal rule to simplify QBFs without changing the set of tree models. This can be important in applications where the tree models themselves are important. Other popular preprocessors make changes that preserve the value (true or false) but add and delete tree models as they operate on a true QBF.

The paper is organized as follows.[1] Section 2 sets forth the notation and basic definitions. Section 3 reviews QBF forms of pure-literal rules. Section 4 reviews QBF abstraction and its combination with failed literal analysis. Extended failed literal analysis (EFL) is introduced. Existential and universal pure-literal rules are separated and soundness issues are examined. The main theoretical result is that EFL with abstraction and the universal pure-literal rule is safe. Experimental results are presented in Section 5. The paper concludes with Section 6.

2   Preliminaries

In general, *quantified boolean formulas* (QBFs) generalize propositional formulas by adding operations consisting of universal and existential quantification of boolean variables. See [KBL99] for a thorough introduction. This paper uses standard notation as much as possible. One minor variation is that we consider resolution and universal reduction separately, although some papers combine them. Also, we use tree models, which are not found in all QBF papers, to distinguish between *super-sound* and *safe* operations, and we define ordered assignments.

A *closed* QBF evaluates to either *invalid* (false) or *valid* (true), as defined by induction on its principal operator. We use 0 and 1 for truth values of literals and use true and false for semantic values of formulas.

1. $(\exists x \, \varPhi(x))$ is true if and only if ($\varPhi(0)$ is true or $\varPhi(1)$ is true).
2. $(\forall x \, \varPhi(x))$ is false if and only if ($\varPhi(0)$ is false or $\varPhi(1)$ is false).
3. Other operators have the same semantics as in propositional logic.

This definition emphasizes the connection of QBF to two-person games, in which player E (Existential) tries to set existential variables to make the QBF evaluate to true, and player A (Universal) tries to set universal variables to make the QBF evaluate to false. Players set their variable when it is outermost, or for non-prenex, when it is the root of a subformula (see [KSGC10] for more details). Only one player has a winning strategy.

We say that a QBF is in *prenex conjunction normal form* if all the quantifiers are outermost operators (the prenex, or quantifier prefix), and the quantifier-free portion (also called the matrix) is in CNF; i.e., $\varPsi = \vec{Q}.\mathcal{F}$ consists of prenex $\vec{Q}$ and matrix $\mathcal{F}$. Clauses in $\mathcal{F}$ are called *input clauses*. For this paper QBFs are in prenex conjunction normal form.

For this paper a *clause* is a *disjunctively* connected set of literals. Literals are variables or negated variables, with overbar denoting negation. Clauses may be written as literals enclosed in square brackets (e.g., $[p, q, \overline{r}]$), and $[]$ denotes the

---

[1] See http://www.cse.ucsc.edu/~avg/EFL/ for a longer version of this paper and other supplementary materials.

empty clause. Where the context permits, letters e and others near the beginning of the alphabet denote existential literals, while letters u and others near the end of the alphabet denote universal literals. Letters like p, q, r denote literals of unspecified quantifier type. The variable underlying a literal p is denoted by $|p|$ where necessary.

The quantifier prefix is partitioned into maximal contiguous subsequences of variables of the same quantifier type, called *quantifier blocks*. Each quantifier block has a unique *qdepth*, with the outermost block having qdepth $= 1$. The *scope* of a quantified variable is the qdepth of its quantifier block. We say scopes are *outer* or *inner* to another scope to avoid any confusion about the direction, since there are varying conventions in the literature for numbering scopes.

Definition 2.1 (Assignment). *An* assignment *is a partial function from variables to truth values, usually represented as the set of literals mapped to 1. A* total assignment *is an assignment to all variables. Assignments are denoted by ρ, σ, τ. Application of an assignment σ to a logical expression is called a* **restriction** *and is denoted by $q\lceil_\sigma$, $C\lceil_\sigma$, $\mathcal{F}\lceil_\sigma$, etc. Quantifiers for assigned variables are deleted in $\Psi\lceil_\sigma$.*

An **ordered assignment** *is a special term that denotes a total assignment that is represented by a sequence of literals that are assigned 1 and are in the same order as their variables appear in the quantifier prefix.* ☐

A *winning strategy* can be presented as an unordered directed tree. If it is a winning strategy for the E player, it is also called a **tree model**, which we now describe. We shorten *unordered directed tree* to *tree* throughout this paper. The qualifier "unordered" means that the children of a node do not have a specified order; they are a set. Recall that a **branch** in a tree is a path from the root node to some leaf node. A tree can be represented as the set of its branches. We also define a a **branch prefix** to be a path from the root node that might terminate before reaching a leaf.

Definition 2.2 (Tree Model). *Let a QBF $\Phi = \vec{Q} \cdot \mathcal{F}$ be given. In this definition, σ denotes a (possibly empty) branch prefix of some ordered assignment for Φ. A tree model M for Φ is a nonempty set of ordered assignments for Φ that defines a tree, such that*

1. *Each ordered assignment makes $\mathcal{F}$* **true***, i.e., satisfies $\mathcal{F}$ in the usual propositional sense.*
2. *If e is an existential literal in Φ and some branch of M has the prefix (σ, e), then no branch has the prefix $(\sigma, \overline{e})$; that is, treating σ as a tree node in M, it has only one child and the edge to that child is labeled e.*
3. *If u is an universal literal in Φ and some branch of M has the prefix (σ, u), then some branch of M has the prefix $(\sigma, \overline{u})$; that is, treating σ as a tree node in M, it has two children and the edges to those children are labeled u and $\overline{u}$.*

Although the wording is different, if M is a tree model by this definition, it is also a tree model by definitions found in other papers [SB07,LB11]. If τ is a *partial assignment to all variables outer to existential variable e, then require-ment 2 ensures that the "Skolem function" $e(\tau)$ is well defined as the unique literal following τ in a branch of M. If the formula evaluates to* false, *the set of tree models is empty.* □

Definition 2.3 (Safe, Super Sound). *For this paper purposes, an operation on a closed QBF is said to be* **safe** *if it does not change the truth value of the formula. An operation on a closed QBF is said to be* **super sound** *if it preserves (i.e., does not change) the set of tree models. Clearly, preserving the set of tree models is a sufficient condition for safety.* □

The proof system known as *Q-resolution* consists of two operations, *resolution* and *universal reduction*. Q-resolution is of central importance for QBFs because it is a *refutationally* complete proof system [KBKF95]. Unlike resolution for propositional logic, Q-resolution is not *inferentially* complete. That is, a (new) clause C might be a super-sound addition to a closed QBF Φ (i.e., C evaluates to true in every tree-model of Φ), yet no subset of C is derivable by Q-resolution (see [LB11], example 6).

Definition 2.4 (Resolution, Universal Reduction). *Resolution is defined as usual. Let clauses $C_1 = [q, \alpha]$ and $C_2 = [\overline{q}, \beta]$, where α and β are literal sequences without conflicting literals among them and without q and $\overline{q}$. (Either or both of α and β may be empty.) Also, let the clashing literal q be an existential literal. Then* $\mathsf{res}_q(C_1, C_2) = \alpha \cup \beta$ *is the resolvent, which cannot be tautologous in Q-resolution.*

Universal reduction is special to QBF. Let clause $C_1 = [q, \alpha]$, where α is a literal sequence without conflicting literals and without q and $\overline{q}$, the reduction literal q is a universal literal, and q is tailing for α. A universal literal q is said to be tailing for α if its quantifier depth is greater than that of any existential literal in α. Then $\mathsf{unrd}_q(C_1) = \alpha$. □

Lemma 2.5. *Resolution and universal reduction are super-sound operations. Proof: Straightforward application of the definitions.* ∎

3   Pure Literals in QBF

A literal is called *pure* or *monotone* if it appears in (the matrix of) the formula and its negation does not appear in the formula. In QBF the *pure-literal rule* consists of setting any existential pure literal to 1 and setting any universal pure literal to 0. As far as we know, the two parts of this rule have not been considered separately. It is well known that the *existential* pure-literal rule does not preserve tree models, since it does not preserve models in the propositional case. That is, if e is an existential pure literal, there may be tree models in which e is assigned 0 on some branches. In such a case, e is not a necessary assignment. Similarly, if u is a universal pure literal, deleting all occurrences of u in the matrix eliminates some tree models, in general. Combining abstraction, defined next, with the pure-literal rule can lead to fallacious conclusions if done carelessly.

4 QBF Abstraction

The idea of formula abstraction was introduced by Lonsing and Biere [LB11]. The idea is to create a formula that is easier to reason about for preprocessing purposes.

Definition 4.1. *Let Φ be a closed QBF.*

1. *Let e be an existential variable in Φ. Then the **abstraction of Φ with respect to** e, denoted $abst(\Phi, e)$, is the formula in which all universally quantified variables with scopes outer to e are changed to existential variables.*
2. *Let u be a universal variable in Φ. Then the **abstraction of Φ with respect to** u, denoted $abst(\Phi, u)$, is the formula obtained as follows. First, transpose u to be outermost within its own quantifier block and change it to an existential variable. Then, change all universally quantified variables with scopes outer to u to existential variables. Any other variables with the same original scope as u remain universal in $abst(\Phi, u)$.*

Thus the outermost scope of $abst(\Phi, p)$ is existential and contains p, whether p is existential or universal.[2] □

Theorem 4.2 [LB11] Let Φ be a closed QBF. Let p be a variable in Φ. Then the set of tree models of $abst(\Phi, p)$ is a superset of the set of tree models of Φ. □

We define a **necessary assignment** to be an assignment to a variable that occurs in every branch of every tree model. Adding a necessary assignment as a unit clause in the matrix is clearly super sound. The main idea is to find necessary assignments for existential variables in the outermost scope of abstractions of Φ. By Theorem 4.2, these are also necessary assignments for Φ when the variable is existential in Φ, as well. In the case that a necessary assignment is found for a universal variable of Φ that became existential due to abstraction, Φ must be false because every tree model has branches for both assignments to any universal variable.

Lonsing and Biere detect necessary assignments by using **failed literal** analysis: If the assumption that literal $p = 1$ in the outermost scope of $abst(\Phi, p)$ derives the empty clause using incomplete forward reasoning, then $\overline{p}$ is a necessary assignment for both $abst(\Phi, p)$ and Φ. For incomplete forward reasoning, they use the *QBCP* procedure, which consists of unit-clause propagation (including universal reduction), and pure literal propagation.

This paper shows how to extend this approach to include **1-saturation**: Separately, assume p and assume $\overline{p}$ in the outermost scope of $abst(\Phi, p)$. If neither assumption derives the empty clause by incomplete forward reasoning, intersect the sets of *variables* that were assigned during the two propagations

---

[2] Our definition is worded slightly differently from [LB11], but is the same in practice. If p is universal, by their definition p remains universal in $abst(\Phi, p)$, whereas in our definition p becomes outer to other variables in its scope and switches to being existential. But p is assigned a truth value immediately after forming $abst(\Phi, p)$, so it does not matter whether p is considered existential or universal.

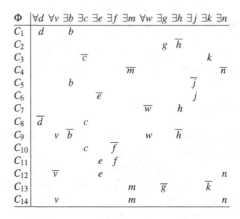

Φ	$\forall d$	$\forall v$	$\exists b$	$\exists c$	$\exists e$	$\exists f$	$\exists m$	$\forall w$	$\exists g$	$\exists h$	$\exists j$	$\exists k$	$\exists n$
C_1	d		b										
C_2									g	$\overline{h}$			
C_3				$\overline{c}$								k	
C_4							$\overline{m}$						$\overline{n}$
C_5			b								$\overline{j}$		
C_6					$\overline{e}$						j		
C_7								$\overline{w}$		h			
C_8	$\overline{d}$			c									
C_9		v	$\overline{b}$					w		$\overline{h}$			
C_{10}				c		$\overline{f}$							
C_{11}					e	f							
C_{12}		$\overline{v}$			e								n
C_{13}							m		$\overline{g}$			$\overline{k}$	
C_{14}		v					m						n

Fig. 1. QBF for Example 4.3

to find additional necessary assignments and equivalent literals. If a variable q was assigned the same value after both assumptions, we want to add q as an additional necessary assignment. If the assumption of p derives literal q and the assumption of $\overline{p}$ derives $\overline{q}$, we want to add the constraint $p = q$ (as two binary clauses, say). Our extension needs to be done in such a way that these additions to Φ are at least *safe* and preferably *super sound*.

The following example shows that the combination of *QBCP*, abstraction, and 1-saturation is *unsafe*.

Example 4.3 Consider $abst(\Phi, b)$, where Φ is shown in Figure 1. Variables d and v are temporarily existential. Only the crucial assignments are mentioned.

The assumption of $\overline{b}$ satisfies C_9. Now $\overline{w}$ is universal pure, which implies h.

The assumption of b satisfies C_5. Existential pure literal propagation assigns true to j, e, v in that order, satisfying C_9. Now $\overline{w}$ is again universal pure, which implies h.

Consequently, EFL using both pure-literal rules would derive $h = 1$ as a necessary assignment. However, adding $[h]$ to Φ changes it from true to false. In particular, if $h = 1$ the A player can choose $d = 0$, $v = 0$, and $w = 0$, after which C_1 and C_9 form a contradiction.

The original Φ is true, as shown by the E player's strategy: $b = 1$, $c = 1$, $e = 1$, $f = 0$, $m = 1$, $g = 1$, $j = 1$, $k = 1$, $n = 0$, $h = w$. (The A player must choose w before the E player chooses h.) $\qquad\square$

The main theoretical result of the paper is that the *universal* pure-literal rule *can* be combined safely with abstraction and 1-saturation. This is nontrivial because assigning a universal pure literal to 0 might *not* be a necessary assignment.

We now describe an incomplete forward reasoning procedure that employs unit-clause propagation (including universal reduction) and the *universal pure-literal rule*.

Definition 4.4. *The procedure **UPUP** (for Unit-clause Propagation with Universal Pure literals) consists of applying the following steps to a QBF until none are applicable or an empty clause is derived.*

1. *(**Unit Prop**) If a clause C has exactly one unassigned existential literal e and all other literals in C are 0 by previous assignment or universal reduction, then assign $e = 1$.*
2. *(**Empty Clause**) If all literals in clause C are 0 by previous assignment or universal reduction, derive the empty clause.*
3. *(**Univ Pure**) If u is an unassigned pure universal literal based on previous assignments, then u may be deleted from all clauses in which it occurs, as if by universal reduction. (For bookkeeping, $u = 0$ might be assigned on this computation path, but we avoid treating it as a derived literal).* ☐

Theorem 4.5 Let $\Phi = \overrightarrow{Q}.\mathcal{F}$ be a closed QBF, let e be an existential literal, and let p be a literal in Φ other than e or $\overline{e}$. Let $\Phi_1 = \overrightarrow{Q}.(\mathcal{F} \cup \{[p, \overline{e}]\})$. If assuming e in $abst(\Phi, |e|)$ derives p by UPUP, then Φ_1 has the same truth value as Φ. If the universal pure-literal rule is not used, then Φ_1 has the same set of tree models as Φ.

Proof: To show that Φ_1 has the same truth value as Φ, it suffices to show that if Φ_1 is false, then Φ is false, because Φ_1 has more constraints. To show that Φ_1 has the same set of tree models as Φ, it suffices to show that if M is a tree model of Φ, then M is a tree model of Φ_1. In this proof we identify the literal $\overline{p}$ with the assignment $p = 0$ and identify p with $p = 1$.

W.l.o.g., let e be innermost in its own quantifier block. If Φ_1 is false, then *every* tree model of Φ (if there are any) contains *some* branch on which $p = 0$ and $e = 1$. Let M be any tree model of Φ, represented as its set of ordered assignments, each ordered assignment being a branch in M (Definition 2.2). Let M_e be the subset of branches on which $e = 1$. By hypothesis, some branch of M_e has $p = 0$. Follow the steps by which p was derived in $abst(\Phi, |e|)$ after assuming e. Let the sequence of partial assignments σ_i, for $i \geq 1$, denote $e = 1$ followed by the derived assignments of UPUP, beginning with $\sigma_1 = \{e\}$. Each literal derived before any use of universal reduction or universal pure-literal rule is 1 on every branch of M_e. Universal reduction can only apply on literals with quantifier scope inner to e, and cannot falsify a clause or M would not be a tree model. If step i consists of universal reduction, $\sigma_i = \sigma_{i-1}$.

Consider the first application of the universal pure-literal rule, say at step $k + 1$. Suppose it deletes all occurrences of u in $\mathcal{F}$ because all occurrences of $\overline{u}$ are in satisfied clauses at this point. We treat this as an operation on $\mathcal{F}$, not an assignment, so $\sigma_{k+1} = \sigma_k$.

Group the branches of M_e by their literals outer to u. Let ρ be the (partial) assignment for any such group. Note that ρ specifies a branch prefix in M to a specific node whose children are u and $\overline{u}$; let us call this node N_ρ. Then $(\rho, u = 0)$ and $(\rho, u = 1)$ produce subtrees of N_ρ, which we denote as $T_{\rho,u=0}$ and $T_{\rho,u=1}$, respectively. That is, the set of branches in M_e consistent with ρ is equal to $((\rho, T_{\rho,u=0}) \cup (\rho, T_{\rho,u=1}))$, where ρ, T denotes a tree in which every branch has ρ as a prefix.

For each group replace $T_{\rho,u=1}$ by $T_{\rho,u=0}$, except that $u = 1$ replaces $u = 0$ in their ordered assignments. The assignment in the replacement branch still satisfies every clause of the matrix. Retain the branches in $M_{\overline{e}}$ unchanged. This gives another tree M'. This is a tree model of Φ because the variables that changed from universal to existential in $abst(\Phi, |e|)$ are all outer to e and u.

By the hypothesis that Φ_1 is false, some branch of M'_e has $p = 0$. Therefore, the $T_{\rho,u=0}$ corresponding to some ρ has $p = 0$ on some branch. Every branch of M' is consistent with σ_{k+1}. Use M' in the role of M for further tracking of the UPUP computation. If the universal pure-literal rule is not used, the original M is a tree model of Φ_1.

Continuing in this way, we see that after each stage i in the computation, there is some tree model of Φ that is consistent with σ_i on *every* branch that contains $e = 1$, and further, has $p = 0$ on *some* branch that has $e = 1$. Thus $p = 1$ cannot be derived by UPUP after assuming $e = 1$ if Φ is true and Φ_1 is false. (Recall that we do not treat p as "derived" if $\overline{p}$ is processed as a pure universal literal.) Similarly, if the universal pure-literal rule is not used, $p = 1$ cannot be derived by UPUP after assuming $e = 1$ if M is a tree model of Φ and not of Φ_1. This contradicts the hypothesis of the theorem, completing the proof. ∎

Several corollaries follow from Theorem 4.5 and Lemma 2.5.

Corollary 4.6. Let $\Phi = \overrightarrow{Q}.\mathcal{F}$ be a closed QBF, let v be a universal literal, and let p be a literal in Φ other than v or $\overline{v}$. Let $\Phi_1 = \overrightarrow{Q}.(\mathcal{F} \cup \{[p, \overline{v}]\})$. If assuming v in $abst(\Phi, |v|)$ derives p by UPUP, then Φ_1 has the same truth value as Φ. □

Corollary 4.7. Let $\Phi = \overrightarrow{Q}.\mathcal{F}$ be a closed QBF, let q be a literal, and let p be a literal in Φ other than q or $\overline{q}$. Let $\Phi_3 = \overrightarrow{Q}.(\mathcal{F} \cup \{[p]\})$. If the assumption that $q = 1$ in $abst(\Phi, |q|)$ derives p by UPUP and the assumption that $q = 0$ in $abst(\Phi, |q|)$ derives p by UPUP, then Φ_3 has the same truth value as Φ. □

Corollary 4.8. Let $\Phi = \overrightarrow{Q}.\mathcal{F}$ be a closed QBF, let q be a literal, and let p be a literal in Φ other than q or $\overline{q}$. Let $\Phi_4 = \overrightarrow{Q}.(\mathcal{F} \cup \{[p, \overline{q}], [\overline{p}, q]\})$. If the assumption that $q = 1$ in $abst(\Phi, |q|)$ derives p by UPUP and the assumption that $q = 0$ in $abst(\Phi, |q|)$ derives $\overline{p}$ by UPUP, then Φ_4 has the same truth value as Φ. □

5   Experimental Results

This section describes our experimental procedures and shows the results. Several procedural issues are discussed first.

As reported, qxbf monitors its own CPU time for certain operations and discontinues that kind of operation if its budget is used up. The budget can be supplied by the user on the command line; otherwise a default (e.g., 40 seconds) is used. This leads to unrepeatable behavior, even among runs on the same platform, and obviously gives different outcomes across platforms. To make runs repeatable and platform independent, we introduced functions to *estimate* CPU

Table 1. Replication of selected `qxbf` data in SAT 2011, Table 1, using `qxbfCntrs`

QBFEVAL-10: 568 formulas						
Preproc.: `qxbfCntrs`	*Solver*	*Solved*	*Time*	*(Preproc.)*	*SAT*	*UNSAT*
sQueezeBF+(ABST+SAT)	depqbf	434	238.24	(42.00)	201	233
SAT	depqbf	380	320.02	(8.34)	168	212
ABST+SAT	depqbf	377	321.84	(7.24)	167	210
ABST	depqbf	375	325.11	(2.84)	168	207
Preproc.: `qxbf`	*Solver*	*Solved*	*Time*	*(Preproc.)*	*SAT*	*UNSAT*
sQueezeBF+(ABST+SAT)	depqbf	434	239.84	(42.79)	201	233
SAT	depqbf	379	322.31	(7.17)	167	212
ABST+SAT	depqbf	378	323.19	(7.21)	167	211
ABST	depqbf	375	327.64	(3.33)	168	207

time based on various counters maintained in the program. The resulting program is called `qxbfCntrs`.

The user interface is unchanged; `qxbfCntrs` simply compares its (repeatable, platform independent) *estimated* CPU time to its budget. The estimation model was arrived at using linear regression, and was designed to give the same results as the published paper [LB11], or very close. The estimation model is not particularly accurate, but the budget amounts are only heuristics, so the overall performance might be about the same.

To validate this conjecture, we attempted to replicate parts of the published results for `qxbf`, using `qxbfCntrs` with all the same parameters. We chose those results that were most relevant for the topics of this paper. The replication is compared with the original in Table 1. For the replication to be meaningful, due to externally imposed time limits, it was carried out on the same platform, in the same environment, as the published table. Times shown in Table 1 are measured by the system; they are not estimates.

In addition, all counts in Tables 2 and 3 of the SAT 2011 paper were confirmed during replication. These tables analyze the preprocessor only, giving it 900 seconds, without internal budgets, so high correspondence is expected. Here, `qxbfCntrs` times were about 4% faster and one additional instance was completed in just under 900 seconds. This can be attributed to compiling with a newer compiler. One factor helping the close correspondence is that the limits imposed by the internal budget are often not reached.

In summary, we conclude that the functions for estimating time are adequate for producing useful repeatable experiments. For all remaining experiments we use `qxbfCntrs` for the baseline against which variations are compared, use the same estimation functions for monitoring against internal budgets, and drop the suffix `Cntrs` from here on.

The main goal of this research is to implement and evaluate 1-saturation for QBF. There are many preprocessing techniques known in the literature, so we are interested to know whether 1-saturation adds new capabilities, or just finds mostly the same inferences and simplifications as other techniques. Therefore, our general approach is to apply a strong existing preprocessor to

Table 2. Effect of pure-literal policies on `qxbf`, 420 preprocessed but unsolved instances

	qxbf			follow-on depqbf		
	N Solved		average	N Solved		average
Pure-lit policy	true	false	seconds	true	false	seconds
exist. and univ.	0	24	76.50	162	134	380
none	0	24	42.11	162	133	380
only univ.	0	24	53.53	162	134	380

the raw benchmarks, then apply failed literals or 1-saturation to the result. We chose the recently reported **bloqqer** since it is open source [BSL11].

There are (at least) two distinct ways to apply preprocessing. The usual way is to give limited budgets to the preprocessor(s) with the idea that a complete solver will do the bulk of the work. Another choice is to give the preprocessor(s) all the resources and see how many instances they can solve completely. Ideally, for the latter approach, the preprocessor(s) would run to completion, not be stopped by time limits. This ideal is not completely far-fetched because preprocessors are expected to stop in polynomial time. Unfortunately, for some larger benchmarks, the time to completion is impractically long.

Our computing resource for Tables 2, 3, and Figure 2 is a pair of 48-core AMD Opteron 6174 computers running Linux with 180 GB of memory each and a 2.2 GHz clock, managed by SunGrid queueing software. Our tests show that this platform is about two times slower than than the platform used in [LB11] and for Table 1, which is a dedicated cluster running one job per two-core processor.

The principal benchmark suite for QBF currently is the 568 instances used for QBFEVAL-10. We follow this tradition. The preprocessor **bloqqer** solved 148 instances (62 **true**, 86 **false**), leaving 420. This initial run averaged 8.89 seconds per instance; all runs completed normally.

The main purpose of this study was to evaluate extended failed-literal analysis (EFL). This necessitated changing how pure literals were handled. To be sure any experimental differences are attributable to EFL, we checked the influence of changes only in pure-literal processing.

The published program `qxbf` uses pure literals, although it has a command-line switch to disable all pure literal processing. To incorporate 1-saturation, it is only necessary to disable *existential* pure literal processing. To measure the effect of each change in the procedure, we created `qxbf_noepure`, which has the same logic as `qxbf` except that existential pure literal processing is disabled. Table 2 shows the effect of disabling all pure literal processing and the effect of disabling existential pure literal processing. Indeed, the "only univ." row in Table 2 demonstrates UPUP results without 1-saturation.

The data suggests that the existential pure-literal rule did not help `qxbf` and the universal pure-literal rule helped a little. These rules have not been examined separately before. Empirical data shows that pure-literal rules are important for QDPLL solvers, but possibly that most of the contribution comes from the universal pure-literal rule (also a source of substantial overhead).

Table 3. Additional completely solved instances among 420 instances that were pre-processed but unsolved by `bloqqer`. `qxbf` uses FL and both pure-literal rules. `eqxbf` uses EFL and the universal pure-literal rule.

| | bloqqer only | | | | qxbf, bloqqer | | | | eqxbf, bloqqer | | | |
| | N Solved | | | avg. | N Solved | | | avg. | N Solved | | | avg. |
Round	true	false	both	secs.	true	false	both	secs.	true	false	both	secs.
1	4	9	13	9	8	29	37	79	11	32	43	58
2	1	0	1	1	1	0	1	75	0	2	2	44
3	0	1	1	1	0	1	1	79	0	1	1	42
total	5	10	15	11	9	30	39	233	11	35	46	144

5.1   Complete Solutions with 1-Saturation

Table 3 compares the solving abilities of `bloqqer` alone, `qxbf` followed by `bloqqer`, and `eqxbf` followed by `bloqqer`. There are no time-outs in this table. The reason for following `qxbf` and `eqxbf` by `bloqqer` is that `qxbf` and `eqxbf` apply a single procedure and have no general simplification capabilities. In fact they can never detect that a formula is true. Our goal was to evaluate their solving capabilities in conjunction with other typical preprocessing techniques. They were run for several rounds in an attempt to reach a fixpoint. However, we discovered that `bloqqer` itself does not quickly reach a fixpoint, and stopped after three rounds.

By the criterion of complete solutions, we observe that extended failed literal analysis (EFL, `eqxbf`) produces moderate gains over failed literal analysis (FL, `qxbf`). We also confirm that FL produces about the same gains beyond the initial preprocessor as the 25 reported by Lonsing and Biere [LB11, Sec. 7]. We observed 24 additional solutions by `qxbf` alone. Overall, the data shows that 194 out of 568 QBFEVAL-10 benchmarks can be solved by preprocessing.

For completeness we ran `eqxbf` with all pure-literal processing disabled. The program slowed down slightly, but solved the same instances as `eqxbf` in round 1 of Table 3. This shows that super-sound preprocessing may be feasible, for applications in which preserving all tree models is important.

5.2   Timed Runs with 1-Saturation and a Complete Solver

This section evaluates whether extended failed literal analysis (EFL) makes the overall solving task faster. We established a total time limit of 1800 seconds, which is equivalent to about 900 seconds on the newest platforms. We ran `bloqqer` with adaptive command-line parameters so that it would not take too much of the 1800 seconds on very large benchmarks. The command-line parameters were set heuristically depending on the numbers of variables, clauses and literals in the instance. No `bloqqer` run on an original QBFEVAL-10 instance took more than 140 seconds for the experiments in this section. After this common start on the 568 QBFEVAL-10 benchmarks, the solving attempts used three strategies. All three strategies used `depqbf` as the complete solver.

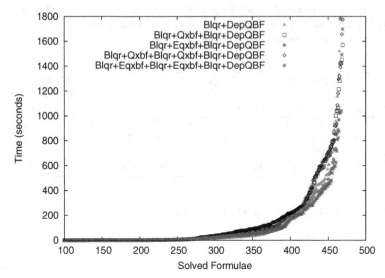

Fig. 2. Timed solving runs. (Larger figure at http://www.cse.ucsc.edu/~avg/EFL.)

To be considered successful, the sum of all preprocessing times and the solving time had to be under 1800 seconds. The outcomes are summarized in Figure 2.

The first strategy simply runs depqbf on the preprocessed instances with a time limit of 1800 seconds. Results should be comparable to runs reported in [BSL11] with 900 seconds time limits.

The second strategy uses one or two rounds of qxbf, bloqqer, as described in Section 5.1, then runs depqbf. Recall that qxbf uses FL and both pure-literal rules. For this table qxbf is given a budget of 80 "estimated seconds" and is run with an external time limit of 1800 seconds. Time-outs by qxbf occurred in three instances during round 2. The time limit for bloqqer was 300 seconds, but it never came close to timing out.

The third strategy uses one or two rounds of eqxbf, bloqqer, as described in Section 5.1, then runs depqbf. Recall that eqxbf uses EFL and the universal pure-literal rule. The details are the same as for the second strategy, except that eqxbf timed out on only one instance during round 2.

In Figure 2 we observe that eqxbf (EFL), combined with bloqqer for one or two rounds, helps depqbf slightly more than bloqqer alone. This observation is confirmed using the *Careful Ranking* procedures reported in SAT-11 and used unofficially in that competition [VG11].

6   Conclusion

This paper presents theoretical analysis of the universal pure-literal rule in combination with formula abstraction and extended failed literal analysis. It shows that binary clauses can be safely derived. It further shows that the use of the existential pure-literal rule in combination with formula abstraction is generally unsafe. An incomplete forward reasoning procedure called UPUP was implemented and tested. Another result is that binary clauses can be super soundly derived if no pure-literal rules are used.

Experimental results suggest that EFL provides some improvement in overall solving performance on the QBFEVAL-10 benchmarks. The previous mark of 148 solved by preprocessing alone now stands at 191, although EFL does not get all the credit. Preliminary data indicates that EFL does almost as well without any pure-literal rules, which is important if preserving model trees is important on true QBFs. Future work should integrate EFL with a general preprocessor, such as `bloqqer`.

Acknowledgment. We thank Craigslist Inc. for equipment donations that facilitated this research.

References

BSL11.   Biere, A., Lonsing, F., Seidl, M.: Blocked Clause Elimination for QBF. In: Bjørner, N., Sofronie-Stokkermans, V. (eds.) CADE 2011. LNCS, vol. 6803, pp. 101–115. Springer, Heidelberg (2011)

Fre93.   Freeman, J.W.: Failed literals in the Davis-Putnam procedure for SAT. In: DIMACS Challenge Workshop: Cliques, Coloring and Satisfiability (1993)

GMN10.   Giunchiglia, E., Marin, P., Narizzano, M.: sQueezeBF: An Effective Preprocessor for QBFs Based on Equivalence Reasoning. In: Strichman, O., Szeider, S. (eds.) SAT 2010. LNCS, vol. 6175, pp. 85–98. Springer, Heidelberg (2010)

GNT04.   Giunchiglia, E., Narizzano, M., Tacchella, A.: Monotone Literals and Learning in QBF Reasoning. In: Wallace, M. (ed.) CP 2004. LNCS, vol. 3258, pp. 260–273. Springer, Heidelberg (2004)

GW00.   Groote, J.F., Warners, J.P.: The propositional formula checker HeerHugo. J. Automated Reasoning 24(1), 101–125 (2000)

JBS+07.   Jussila, T., Biere, A., Sinz, C., Kroning, D., Wintersteiger, C.M.: A First Step Towards a Unified Proof Checker for QBF. In: Marques-Silva, J., Sakallah, K.A. (eds.) SAT 2007. LNCS, vol. 4501, pp. 201–214. Springer, Heidelberg (2007)

KBKF95.   Kleine Büning, H., Karpinski, M., Flögel, A.: Resolution for quantified boolean formulas. Information and Computation 117, 12–18 (1995)

KBL99.   Kleine Büning, H., Lettmann, T.: Propositional Logic: Deduction and Algorithms. Cambridge University Press (1999)

KSGC10.   Klieber, W., Sapra, S., Gao, S., Clarke, E.: A Non-prenex, Non-clausal QBF Solver with Game-State Learning. In: Strichman, O., Szeider, S. (eds.) SAT 2010. LNCS, vol. 6175, pp. 128–142. Springer, Heidelberg (2010)

LB01.   Le Berre, D.: Exploiting the real power of unit propagation lookahead. In: IEEE/ASL LICS Satisfiability Workshop, pp. 59–80. Elsevier (2001)

LB11.   Lonsing, F., Biere, A.: Failed Literal Detection for QBF. In: Sakallah, K.A., Simon, L. (eds.) SAT 2011. LNCS, vol. 6695, pp. 259–272. Springer, Heidelberg (2011)

SB07.   Samulowitz, H., Bacchus, F.: Dynamically Partitioning for Solving QBF. In: Marques-Silva, J., Sakallah, K.A. (eds.) SAT 2007. LNCS, vol. 4501, pp. 215–229. Springer, Heidelberg (2007)

VG11.   Van Gelder, A.: Careful Ranking of Multiple Solvers with Timeouts and Ties. In: Sakallah, K.A., Simon, L. (eds.) SAT 2011. LNCS, vol. 6695, pp. 317–328. Springer, Heidelberg (2011)

On Sequent Systems and Resolution for QBFs*

Uwe Egly

Institut für Informationssysteme 184/3, Technische Universität Wien,
Favoritenstrasse 9–11, A-1040 Vienna, Austria
uwe@kr.tuwien.ac.at

Abstract. Quantified Boolean formulas generalize propositional formulas by admitting quantifications over propositional variables. We compare proof systems with different quantifier handling paradigms for quantified Boolean formulas (QBFs) with respect to their ability to allow succinct proofs. We analyze cut-free sequent systems extended by different quantifier rules and show that some rules are better than some others.

Q-resolution is an elegant extension of propositional resolution to QBFs and is applicable to formulas in prenex conjunctive normal form. In Q-resolution, there is no explicit handling of quantifiers by specific rules. Instead the forall reduction rule which operates on single clauses inspects the global quantifier prefix. We show that there are classes of formulas for which there are short cut-free tree proofs in a sequent system, but any Q-resolution refutation of the negation of the formula is exponential.

1 Introduction

Quantified resolution (or Q-resolution) [10] is a relatively inconspicuous calculus. It was introduced as an elegant extension of resolution to process quantified Boolean formulas (QBFs) in prenex conjunctive normal form. Although there are only a few QBF solvers directly based on Q-resolution, it has gained an enormous practical importance as a subcalculus in modern DPLL solvers with clause learning. Moreover, an early proposal for a uniform proof format [9] is based on resolution. Nowadays many QBF solvers produce Q-resolution proofs and certificate generation [1] can be based on them.

Sequent calculi [7] are well explored proof systems, which are not restricted to specific normal forms. Variants of these calculi like tableau systems are widely used in (first-order) theorem proving for classical and non-classical logics, where often no clausal normal form is available. Variants of sequent calculi are available for QBFs and used for a variety of purposes [4,11]. Even some solvers not based on prenex conjunctive normal form like qpro [6] implement proof search in a restricted variant of a sequent calculus, and a look at a high-level description of its main procedure indicates that it is not too far away from DPLL.

* The work was partially supported by the Austrian Science Foundation (FWF) under grant S11409-N23. We thank the referees for valuable comments.

A. Cimatti and R. Sebastiani (Eds.): SAT 2012, LNCS 7317, pp. 100–113, 2012.

Initially driven by the $P =? NP$ question [5], propositional proof systems are well studied and compared with respect to their relative efficency, i.e., their ability to allow for succinct proofs. In this paper, we compare Q-resolution and sequent systems for QBFs. The crucial difference between systems for SAT and systems for quantified SAT (QSAT) is Boolean quantification in the latter, which allows for more succinct problem representations. As we will see later, there are different methods to handle quantifiers like the rules implementing semantics directly, rules inspired by first-order logic, or a completely different technique to handle quantifiers in Q-resolution. It turns out that the way how quantifiers are handled strongly influence proof complexity.

Contributions. First we consider *cut-free* propositional sequent systems extended by different quantifier rules. We show that these rules have increasing strength by providing formula classes which can be used for exponential separations. Second we partially solve the problem stated in [4] whether in sequent systems with restricted cuts, the quantifier rule introducing propositional formulas can be polynomially simulated by the one introducing variables. We show this for all tree-like systems except the one with only propositional cuts. Third we show an exponential separation between *cut-free tree-like* sequent systems and *arbitrary* Q-resolution. This result is surprising because the first system is extremely weak, whereas the second one does not have to obey the tree restriction and has an atomic cut (the resolution rule) in addition. It turns out that the relative strength comes from the more powerful quantifier rules of the sequent system.

Structure. In Sect. 2, we introduce necessary concepts. Section 3 presents sequent systems and Q-resolution. Different quantifier rules are compared in Sect. 4. In Sect. 5, we present an exponential separation between cut-free tree-like sequent systems and arbitrary Q-resolution. We show that the latter cannot polynomially simulate the former. Concluding remarks are presented in Sect. 6.

2   Preliminaries

We assume basic familiarity with the syntax and semantics of propositional logic. We consider a propositional language based on a set $\mathcal{PV}$ of Boolean variables and truth constants $\top$ (true) and $\bot$ (false), both of which are not in $\mathcal{PV}$. A variable or a truth constant is called *atomic*. We use connectives from $\{\neg, \wedge, \vee, \rightarrow\}$ and $A \leftrightarrow B$ is a shorthand for $(A \rightarrow B) \wedge (B \rightarrow A)$. A *clause* is a disjunction of literals. *Tautological clauses* contain a variable and its negation and the *empty clause* is denoted by $\square$. Propositional formulas are denoted by capital Latin letters like A, B, C possibly annotated with subscripts, superscripts or primes.

We extend the propositional language by Boolean quantifiers. Universal ($\forall$) and existential ($\exists$) quantification is allowed within a QBF. QBFs are denoted by Greek letters. Observe that we allow non-prenex formulas, i.e., quantifiers may occur deeply in a QBF and not only in an initial quantifier prefix. An example for a non-prenex formula is $\forall p\, (p \rightarrow \forall q \exists r\, (q \wedge r \wedge s))$, where p, q, r and s are variables. Moreover, free variables (like s) are allowed, i.e., there might be occurrences of variables in the formula for which we have no quantification. Formulas without free variables are called *closed*; otherwise they are called *open*. The *universal* (*existential*) closure of φ is $\forall x_1 \ldots \forall x_n \varphi$ ($\exists x_1 \ldots \exists x_n \varphi$), for which we often write $\forall \boldsymbol{X} \varphi$ ($\exists \boldsymbol{X} \varphi$) if $\boldsymbol{X} = \{x_1, \ldots, x_n\}$ is the set of all free variables in φ. A formula in *prenex conjunctive normal form* (PCNF) has the form $Q_1 p_1 \ldots Q_n p_n A$, where $Q_1 p_1 \ldots Q_n p_n$ is the *quantifier prefix*, $Q_i \in \{\forall, \exists\}$ and A is the (propositional) *matrix* which is in CNF. The *size* of a formula φ, $size(\varphi)$, is the number of occurrences of connectives or quantifiers.

Let Σ_0^q and Π_0^q both denote the set of propositional formulas. For $i > 0$, Σ_i^q is the set of all QBFs whose prenex forms starts with $\exists$ and which have $i - 1$ quantifier alternations. Π_i^q is the dual of Σ_i^q and $\Sigma_{i-1}^q \subseteq \Pi_i^q$ as well as $\Pi_{i-1}^q \subseteq \Sigma_i^q$ holds. We refer to [11] for more details.

The semantics of propositional logic is based on an *evaluation function* indexed by a *variable assignment* I for free variables. The semantics is extended to quantifiers by $\nu_I(Qp\,\varphi) = \nu_I(\varphi\{p/\top\} \circ \varphi\{p/\bot\})$, where $\circ = \vee$ if $Q = \exists$, and $\circ = \wedge$ if $Q = \forall$. We denote by $\varphi\{p/\psi\}$ the replacement of all (free) occurrences of p by ψ in φ.

A *quantified propositional proof system* is a surjective PTIME-computable function F from the set of strings over some finite alphabet to the set of valid QBFs. Every string α is then a proof of $F(\alpha)$. Let P_1 and P_2 be two proof systems. Then P_1 *polynomially simulates* (p-simulates) P_2 if there is a polynomial p such that for every natural number n and every formula φ, the following holds. If there is a proof of φ in P_2 of size n, then there is a proof of φ (or a suitable translation of it) in P_1 whose size is less than $p(n)$.

3   Calculi for Quantified Boolean Formulas

We first discuss sequent calculi [7] with different alternative quantifier rules. Later Q-resolution [10] is introduced which is applicable to QBFs in PCNF.

3.1   Sequent Calculi for Quantified Boolean Formulas

Sequent calculi do not work directly on formulas but on sequents. A *sequent* S is an ordered pair of the form $\Gamma \vdash \Delta$, where Γ and Δ are finite sequences of formulas. Γ is the *antecedent* of S, and Δ is the *succedent* of S. A formula occurring in one of Γ or Δ is called a *sequent formula* (of S). We write "$\vdash \Delta$" or "$\Gamma \vdash$" whenever Γ or Δ is empty, respectively. The meaning of a sequent $\Phi_1, \ldots, \Phi_n \vdash \Psi_1, \ldots, \Psi_m$ is the same as the meaning of $(\bigwedge_{i=1}^{n} \Phi_i) \rightarrow (\bigvee_{i=1}^{m} \Psi_i)$. The *size* of S, $size(S)$, is the sum over the size of all sequent formulas in S.

We introduce the axioms and the rules in Fig. 1. In the *strong quantifier rules* $\exists l_e$ and $\forall r_e$, q has to satisfy the *eigenvariable* (EV) *condition*, i.e., q does not occur as a free variable in the conclusion of these rules. In the *weak quantifier rules* $\forall l$ and $\exists r$, no free variable of Ψ is allowed to become bound in $\Phi\{p/\Psi\}$. For instance, this restriction forbids the introduction of x for y in the (false) QBF $\exists y \forall x\,(x \leftrightarrow y)$. Without this restriction, the true QBF $\forall x\,(x \leftrightarrow x)$ would result.

In the following, we instantiate the quantifier rules as follows. If the formula Ψ in $\forall l$ and $\exists r$ is restricted to a propositional formula, we call the quantifier

$$\Phi \vdash \Phi \ \ \text{Ax} \qquad\qquad \bot \vdash \ \bot l \qquad\qquad \vdash \top \ \text{Tr}$$

$$\frac{\Gamma \vdash \Delta}{\Phi^*, \Gamma \vdash \Delta} \ wl \qquad\qquad\qquad \frac{\Gamma \vdash \Delta}{\Gamma \vdash \Delta, \Phi^*} \ wr$$

$$\frac{\Gamma_1, \Phi^+, \Psi^+, \Gamma_2 \vdash \Delta}{\Gamma_1, \Psi^*, \Phi^*, \Gamma_2 \vdash \Delta} \ el \qquad\qquad \frac{\Gamma \vdash \Delta_1, \Phi^+, \Psi^+, \Delta_2}{\Gamma \vdash \Delta_1, \Psi^*, \Phi^*, \Delta_2} \ er$$

$$\frac{\Gamma_1, \Phi^+, \Phi^+, \Gamma_2 \vdash \Delta}{\Gamma_1, \Phi^*, \Gamma_2 \vdash \Delta} \ cl \qquad\qquad \frac{\Gamma \vdash \Delta_1, \Phi^+, \Phi^+, \Delta_2}{\Gamma \vdash \Delta_1, \Phi^*, \Delta_2} \ cr$$

$$\frac{\Gamma \vdash \Delta, \Phi^+}{(\neg\Phi)^*, \Gamma \vdash \Delta} \ \neg l \qquad\qquad \frac{\Phi^+, \Gamma \vdash \Delta}{\Gamma \vdash \Delta, (\neg\Phi)^*} \ \neg r$$

$$\frac{\Phi^+, \Psi^+, \Gamma \vdash \Delta}{(\Phi \wedge \Psi)^*, \Gamma \vdash \Delta} \ \wedge l \qquad\qquad \frac{\Gamma \vdash \Delta, \Phi^+ \quad \Gamma \vdash \Delta, \Psi^+}{\Gamma \vdash \Delta, (\Phi \wedge \Psi)^*} \ \wedge r$$

$$\frac{\Phi^+, \Gamma \vdash \Delta \quad \Psi^+, \Gamma \vdash \Delta}{(\Phi \vee \Psi)^*, \Gamma \vdash \Delta} \ \vee l \qquad\qquad \frac{\Gamma \vdash \Delta, \Phi^+, \Psi^+}{\Gamma \vdash \Delta, (\Phi \vee \Psi)^*} \ \vee r$$

$$\frac{\Gamma \vdash \Delta, \Phi^+ \quad \Psi^+, \Gamma \vdash \Delta}{(\Phi \to \Psi)^*, \Gamma \vdash \Delta} \ \to l \qquad\qquad \frac{\Phi^+, \Gamma \vdash \Delta, \Psi^+}{\Gamma \vdash \Delta, (\Phi \to \Psi)^*} \ \to r$$

$$\frac{\Gamma \vdash \Delta, \Phi\{p/q\}^+}{\Gamma \vdash \Delta, (\forall p\,\Phi)^*} \ \forall r_e \qquad\qquad \frac{\Phi\{p/q\}^+, \Gamma \vdash \Delta}{(\exists p\,\Phi)^*, \Gamma \vdash \Delta} \ \exists l_e$$

$$\frac{\Phi\{p/\Psi\}^+, \Gamma \vdash \Delta}{(\forall p\,\Phi)^*, \Gamma \vdash \Delta} \ \forall l \qquad\qquad \frac{\Gamma \vdash \Delta, \Phi\{p/\Psi\}^+}{\Gamma \vdash \Delta, (\exists p\,\Phi)^*} \ \exists r$$

Fig. 1. Axioms and inference rules for sequent calculi. *Principal formulas* are marked by *, *auxiliary formulas* by $^+$, the other (unmarked) formulas are *side formulas*.

rules $\forall l_f$ and $\exists r_f$. If only variables or truth constants are allowed, then the index f is replaced by v. Finally, if Ψ is further restricted to truth constants, then the index is s. We define three different sequent calculi Gqxe ($x \in \{s, v, f\}$) for QBFs possessing the quantifier rules with index x and $\forall r_e$ and $\exists l_e$. A fourth calculus, Gqss, is defined by adopting $\forall l_s$ and $\exists r_s$ together with the following two rules.

$$\frac{\Gamma \vdash \Delta, (\Phi\{p/\top\} \wedge \Phi\{p/\bot\})^+}{\Gamma \vdash \Delta, (\forall p\,\Phi)^*} \ \forall r_s \qquad\qquad \frac{(\Phi\{p/\top\} \vee \Phi\{p/\bot\})^+, \Gamma \vdash \Delta}{(\exists p\,\Phi)^*, \Gamma \vdash \Delta} \ \exists l_s$$

All the calculi introduced above are *cut-free*, i.e., the *cut rule*

$$\frac{\Gamma_1 \vdash \Delta_1, \Phi^+ \qquad \Phi^+, \Gamma_2 \vdash \Delta_2}{\Gamma_1, \Gamma_2 \vdash \Delta_1, \Delta_2} \ cut$$

is not part of the calculus. For $i \geq 0$ and $\mathsf{G} \in \{\mathsf{Gqss}, \mathsf{Gqse}, \mathsf{Gqve}, \mathsf{Gqfe}\}$, G_i is G extended by cut, where the cut formula Φ is restricted to be a $\Pi_i^q \cup \Sigma_i^q$ formula.

A *sequence* proof α of a sequent S (the *end sequent*) in G is a sequence $S_1, \ldots, S_m$ of sequents such that $S_m = S$ and, for every S_i ($1 \leq i \leq m$), S_i is either an axiom of G, the conclusion of an application of a unary inference from G with premise S_j, or the conclusion of an application of a binary inference from G with premises S_j, S_k $(j, k < i)$. Proofs in G are called G proofs. If α is a proof of $\vdash \Phi$, then α is a proof of the *formula* Φ. A proof α is called *tree-like* or a *tree proof*, if every sequent in α is used at most once as a premise. The *length*, $l(\alpha)$, of α is the number m of sequents occurring in α and its size is $\sum_{i=1}^{m} size(S_i)$.

We denote by G^* the version of G which permits only tree proofs. They are assumed to be in *free variable normal form* (FVNF) [2,4], to which they can be translated efficiently. A tree proof α is in FVNF, if (i) no free variable from the end sequent is used as an EV, and (ii) every other free variable z occurring in α is used exactly once as an EV and appears in α only in sequents above the application of $\exists l_e$ or $\forall r_e$ which introduced z.

Later, we have to trace formula occurrences through a tree proof. The means to do this is an *ancestor* relation between formula occurrences in a tree proof [2]. We first define *immediate descendants* (IDs). If Φ is an auxiliary formula of any rule R except exchange or cut, then Φ's ID is the principal formula of R. For the exchange rules el and er, the ID of Φ or Ψ in the premise is Φ or Ψ, respectively, in the conclusion. An occurrence of the cut formula does not have any ID. If Φ is a side formula at position i in $\Gamma, \Gamma_1, \Gamma_2, \Delta, \Delta_1, \Delta_2$ of the premise(s), then Φ's ID is the same formula at the same position of the same sequence in the conclusion. Now, Φ is an *immediate ancestor* of Ψ iff Ψ is an ID of Φ. The *ancestor relation* is the reflexive and transitive closure of the immediate ancestor relation.

G is *sound* and *complete*, i.e., a sequent S is valid iff it has a G proof. We will consider variants of our tree calculi without exchange rules and where sequents consists of multisets instead of sequences. Since the multiset and the sequence version are p-equivalent, it is sufficient to consider the multiset version.

The calculus in Fig. 1 is a cut-free variant of calculi proposed by Krajíček and Pudlák (KP) (cf, e.g., [11]). In the calculi KP_i, only $\Sigma_i^q \cup \Pi_i^q$ formulas can occur in a proof. Cook and Morioka [4] modified the KP calculi by allowing arbitrary QBFs as sequent formulas, but restricting cut formulas to $\Sigma_i^q \cup \Pi_i^q$ formulas. Moreover, $\forall l$ and $\exists r$ are replaced by $\forall l_f$ and $\exists r_f$.[1] They show in [4] that any of their system G_i $(i > 0)$ is p-equivalent to the corresponding system KP_i for proving formulas from $\Sigma_i^q \cup \Pi_i^q$. G_i is complete for QBFs (in contrast to KP_i).

3.2   The Q-resolution Calculus

The quantified resolution calculus, Q-res, is an extension of propositional resolution to QBFs [10]. There is no explicit handling of quantifiers by specific rules. Instead the $\forall$ reduction rule which operates on single clauses inspects the global quantifier prefix. As we will see, this processing of quantifiers results in a relatively weak calculus with respect to the ability to produce succinct refutations.

The input for Q-res is a (closed) QBF in PCNF. Quantifier blocks are numbered from left to right in increasing order and bound variables from quantifier

---

[1] The restriction to *propositional* formulas is necessary. For unrestricted QBFs, the hierarchy of calculi would "collapse" to G_1.

$$\frac{C_1 \vee x \vee C_2 \quad C_3 \vee \neg x \vee C_4}{C_1 \vee C_2 \vee C_3 \vee C_4} \ \exists\text{PR} \qquad \frac{C_1 \vee \ell \vee C_2 \vee \ell \vee C_3}{C_1 \vee \ell \vee C_2 \vee C_3} \ \text{PF} \qquad \frac{C_5 \vee k \vee C_6}{C_5 \vee C_6} \ \forall\text{R}$$

> C_1 to C_6 are clauses, x is an $\exists$ variable and ℓ a literal. $C_5 \vee k \vee C_6$ is non-tautological and k is a $\forall$ literal with level i. Any $\exists$ literal in $C_5 \vee C_6$ has a level smaller than i.

Fig. 2. The rules of the Q-resolution calculus

block i have level i. Literal occurrences in the CNF inherit the level from their variable in the quantifier prefix. Q-res consists of the *propositional resolution rule* $\exists$PR over existential literals, the *factoring rule* PF and the $\forall$ *reduction rule* $\forall$R, all of which are shown in Fig. 2. The following is Theorem 2.1 in [10].

Theorem 1. *A QBF φ in PCNF is false iff $\square$ can be derived from φ by Q-res.*

A Q-res refutation can be in tree form as well as in sequence form. The *length* of a Q-res refutation is the number of clauses in it. The *size* of a Q-res refutation is the sum of the sizes of its clauses.

4   Comparing Different Quantifier Rules

We compare Gqss, Gqse, Gqve and Gqfe with respect to p-simulation. Let $G \in \{\text{Gqse}, \text{Gqve}, \text{Gqfe}\}$. We reproduce Definition 6 and Lemma 3 from [4] below.

Definition 1. *Let φ be a quantified QBF in prenex form and let S be the sequent $\vdash \varphi$. Let $\alpha(S)$ be a G_0 proof of S. Then any quantifier-free formula A in $\alpha(S)$ that occurs as the auxiliary formula of a quantifier inference is called an α-prototype of φ. Define the Herbrand α-disjunction to be the sequent $\vdash A_1, \ldots, A_m$, where $A_1, \ldots, A_m$, are all the α-prototypes of φ.*

Lemma 1. *Let φ be a quantified QBF in prenex form and let S be the sequent $\vdash \varphi$. Let $\alpha(S)$ be a G_0 proof of S. Then the Herbrand α-disjunction is valid and it has a purely propositional sequent proof of size polynomial in the size of $\alpha(S)$.*

In the construction of the proof of the Herbrand α-disjunction in Lemma 1, no (new) cut is introduced and the form of the proof is retained. Consequently, if $\alpha(S)$ is cut-free and tree-like, then so is the resulting propositional proof.

Proposition 1. *(1) Gqss_0 cannot p-simulate Gqse^*, (2) Gqse_0 cannot p-simulate Gqve^* and (3) Gqve_0 cannot p-simulate Gqfe^*.*

We show (3) in detail. Let $(F_n)_{n>0}$ be a sequence of propositional formulas of the form $\bigwedge_{i=1}^{n}((\neg x_i) \leftrightarrow y_i)$ and let φ_n be $\forall \boldsymbol{X}_n \exists \boldsymbol{Y}_n F_n$ with $\boldsymbol{X}_n = \{x_1, \ldots, x_n\}$ and $\boldsymbol{Y}_n = \{y_1, \ldots, y_n\}$. The size of φ_n is linear in n and it has a short proof in Gqfe^* of length linear in n. It can be obtained by (i) introducing eigenvariable c_i for x_i for all i $(1 \leq i \leq n)$, (ii) introducing formula $\neg c_i$ for y_i for all i $(1 \leq i \leq n)$ and (iii) proving $\bigwedge_{i=1}^{n}((\neg c_i) \leftrightarrow (\neg c_i))$ with $O(n)$ sequents.

Next we show that any proof of φ_n in $\mathsf{Gqve_0}$ is exponential in n. The key observation is that only the introduction of truth constants for y_i makes sense. Otherwise we obtain conjunctive subformulas of the form $(\neg c_i) \leftrightarrow v_i$ which are unprovable. Consequently, all $\exists r$ inferences introduce truth constants.

Let α_n be an arbitrary $\mathsf{Gqve_0}$ proof of $\vdash \varphi_n$. By Lemma 1 we get a purely propositional $\mathsf{Gqve_0}$ proof β_n of the valid Herbrand α_n-disjunction

$$\vdash F_{n,1}, \ldots, F_{n,m} \ .$$

Moreover, the size of β_n is polynomially related to the size of α_n. We argue in the following that this disjunction consists of $m = 2^n$ formulas. Let S_n be the following set $\{F_n\{x_1/c_1, \ldots, x_n/c_n, y_1/t_1, \ldots, y_n/t_n\} \mid t_i \in \{\bot, \top\}\}$ of all possible substitution instances of F_n with 2^n elements. We show in the following that $\bigvee_{d \in D} d$ is not valid if $D \subset S_n$ holds. Then all elements of S_n have to occur in the Herbrand α_n-disjunction and the exponential lower bound follows.

Let C be an arbitrary instance $\bigwedge_{i=1}^n ((\neg c_i) \leftrightarrow t_i)$ of F_n which is in S_n but not in D. Let I be any assignment that makes C true, i.e., each c_i is assigned to the dual of t_i by I. Now take an arbitrary $d \in D$ of the form $\bigwedge_{i=1}^n ((\neg c_i) \leftrightarrow s_i)$. There must be an index k, $1 \leq k \leq n$, such that $s_k \neq t_k$. Then $(\neg c_k) \leftrightarrow s_k$ is false under I and so is d. Since d has been chosen arbitrarily, all elements of D are false under I and so is $\bigvee_{d \in D} d$. Consequently, all elements of S_n have to occur in the Herbrand α_n-disjunction and the exponential lower bound follows.

For (2), we can use a similar argumentation with $(G_n)_{n>0}$ instead of F_n, where G_n is of the form $\bigwedge_{i=1}^n (x_i \leftrightarrow y_i)$. For (1), the family of formula is $(\psi_n)_{n>1}$, where ψ_n is of the form $\exists x_n \forall y_n \ldots \exists x_1 \forall y_1 (x_n \lor y_n \lor \cdots \lor x_1 \lor y_1)$.

Looking at the structure of φ_n, one immediately realizes that the quantifiers can be pushed into the formula ("antiprenexed") in an equivalence-preserving way. This antiprenexed formula F_n': $\bigwedge_{i=1}^n (\forall x_i \exists y_i ((\neg x_i) \leftrightarrow y_i))$ has short proofs in $\mathsf{Gqve^*}$, $\mathsf{Gqse^*}$ and even in $\mathsf{Gqss^*}$, mainly because $\forall x_i \exists y_i ((\neg x_i) \leftrightarrow y_i)$ has a proof of constant length. A similar statement holds for the other two cases.

4.1   Using Eliminable Extensions to Simulate $\exists r_f / \forall l_f$ by $\exists r_v / \forall l_v$

We show in the following that the weak quantifier rules $\exists r_f$ and $\forall l_f$ in $\mathsf{Gqfe_i^*}$ can be simulated efficiently by $\exists r_v$ and $\forall l_v$ in $\mathsf{Gqve_i^*}$ for $i \geq 1$. The key idea is to use a quantified extension $\varepsilon(B)$ of the form $\exists x (x \leftrightarrow B)$ with B being a propositional formula. $\varepsilon(B)$ has a proof $\alpha(\varepsilon(B))$ in $\mathsf{Gqve^*}$ and $\mathsf{Gqse^*}$ of constant length.

Given a tree proof β_e of an end sequent S_e. For any occurrence of an inference $\forall l_f$ and $\exists r_f$ introducing non-atomic propositional formula B, we perform the following. Take an occurrence I of an inference $\exists r_f$ (the case of $\forall l_f$ is similar) and a globally new variable q, not occurring in β_e and not introduced as a new variable before. Employ the ancestor relation for I's auxiliary formula $\Phi\{p/B\}$ and get all highest sequents with occurrences of the sequent formula B originating from I. Start from the next lower sequent of these highest positions downwards until the conclusion of I and put $F(B) = q \leftrightarrow B$ into the antecedent of each sequent. If there is already a copy there, then do nothing. If there are strong quantifier rules, then there is *no* violation of the EV condition because we add only a globally new variable; all the variables from B have been already present in the sequent before.

Employing the ancestor relation again and starting from $\Phi\{p/B\}$, we replace any formula $\Psi\{p/B\}$ by $\Psi\{p/q\}$ in sequents containing $F(B)$. This includes a replacement of B by q. Perform the above procedure for each of the w occurrences of $\forall l_f$ and $\exists r_f$. We have not increased the number of sequents yet, but there are $O(w)$ additional formulas in any sequent.

We are going to correct the inference tree. We check all sequents with sequent formulas of the form $F(B)$ whether binary rules are violated, like, e.g., in the left inference figure below for the case of $\wedge r$. It is replaced by the correct right figure. ($F(B_1)$ and $F(B_2)$ are replaced by F_1, F_2 for space reasons).

$$\frac{F_1, F_2, \Gamma \vdash \Delta, \Phi_1\{p/q\} \quad \Gamma \vdash \Delta, \Phi_2}{F_1, F_2, \Gamma \vdash \Delta, \Phi_1\{p/q\} \wedge \Phi_2} \qquad \frac{F_1, F_2, \Gamma \vdash \Delta, \Phi_1\{p/q\} \quad \dfrac{\Gamma \vdash \Delta, \Phi_2}{F_1, F_2, \Gamma \vdash \Delta, \Phi_2} \; wl*}{F_1, F_2, \Gamma \vdash \Delta, \Phi_1\{p/q\} \wedge \Phi_2}$$

We have to perform two additional corrections, namely (i) to get rid of $F(B)$ immediately below the conclusion of I and (ii) to correct the situation when B originating from I occurs as a principal formula in a propositional inference or as a formula in an axiom of the original proof β_e. For the former, we use

$$\frac{\dfrac{\alpha(\varepsilon(B))}{\vdash \exists x\,(x \leftrightarrow B)} \qquad \dfrac{\dfrac{\dfrac{q \leftrightarrow B,\, \Gamma \vdash \Delta,\, \Phi\{p/q\}}{q \leftrightarrow B,\, \Gamma \vdash \Delta,\, \exists p\,\Phi} \; \exists r_v}{\exists x\,(x \leftrightarrow B),\, \Gamma \vdash \Delta,\, \exists p\,\Phi} \; \exists l_e}{} }{\Gamma \vdash \Delta, \exists p\,\Phi} \; cut$$

with a cut on a Σ_1^q-formula. Let us consider (ii) where B is the principal formula of a propositional inference. Below is one possible case for $B = B_1 \vee B_2$.

$$\frac{F(B), \Gamma \vdash \Delta, B_1, B_2}{F(B), \Gamma \vdash \Delta, q} \; \vee r \qquad\qquad \frac{\dfrac{\dfrac{F(B), \Gamma \vdash \Delta, B_1, B_2}{F(B), \Gamma \vdash \Delta, B} \; \vee r \qquad \dfrac{\alpha}{B, q \leftrightarrow B \vdash q}}{F(B), F(B), \Gamma \vdash \Delta, q} \; cut}{F(B), \Gamma \vdash \Delta, q} \; cl$$

cl is needed if $F(B)$ is required in the left branch. The case for the axiom is simpler. Finally, wl inferences are introduced to remove $q \leftrightarrow B$.

During the proof manipulations, we have added to each sequent $O(w)$ formulas. Moreover, by correcting the binary inferences, we added $O(w)$ sequents for any sequent in the original proof. For each occurrence of B and each of the w occurrences of the quantifier rules, we added a deduction of length $O(1)$ In total, we obtain a polynomial increase in length and size.

5   Exponential Separation of Q-res and Gqve*

We stepwisely construct a family $(\varphi_n)_{n>1}$ of closed QBFs φ_n for which (1) there exists short proofs in Gqve*, but (2) any Q-resolution refutation of $\neg\varphi_n$ has length exponential in n. We use the traditional approaches, namely a refutational approach with resolution and an affirmative approach with sequent systems.

5.1   The Construction of φ_n

We start with a version of the well-known pigeon hole formula in *disjunctive normal form*. The formula for n holes and $n + 1$ pigeons is given by

$$\left(\bigvee_{i=1}^{n+1} \bigwedge_{j=1}^{n} \neg x_{i,j} \right) \vee \left(\bigvee_{j=1}^{n} \bigvee_{1 \leq i_1 < i_2 \leq n+1} (x_{i_1,j} \wedge x_{i_2,j}) \right) .$$

Let $\mathsf{DPHP}_n^{X_n}$ denote this formula over the variables in $X_n = \{x_{1,1}, \ldots, x_{n+1,n}\}$. Variable $x_{i,j}$ is intended to denote that pigeon i is sitting in hole j. The usual (unsatisfiable) version of the pigeon hole formula in CNF, $\mathsf{CPHP}_n^{X_n}$, is given by

$$\left(\bigwedge_{i=1}^{n+1} \left(\bigvee_{j=1}^{n} x_{i,j} \right) \right) \wedge \left(\bigwedge_{j=1}^{n} \bigwedge_{1 \leq i_1 < i_2 \leq n+1} (\neg x_{i_1,j} \vee \neg x_{i_2,j}) \right) .$$

The number of clauses in $\mathsf{CPHP}_n^{X_n}$ is $l_n = (n+1) + n^2(n+1)/2$, $size(\mathsf{CPHP}_n^{X_n})$ is $O(n^3)$, and $\mathsf{CPHP}_n^{X_n}$ is obtained from $\neg\mathsf{DPHP}_n^{X_n}$ by shifting negations inwards using de Morgan's laws and eliminating double negations. Intuitively, we want to show that the refutation problem corresponding to the negation of the formula

$$\forall X_n \exists Y_n \left(\mathsf{DPHP}_n^{Y_n} \rightarrow \mathsf{DPHP}_n^{X_n} \right) \tag{1}$$

results only in Q-res refutations of length exponential in n. A short Gqve* proof of (1) exists which mainly relies on a unification property, namely that (i) $\forall r_e$ introduces eigenvariables C_n for X_n and (ii) $\exists r_v$ introduces exactly the same variables C_n for Y_n, therefore unifying the two versions of DPHP_n. As we will see later, this instantiation property of $\exists r_v$ is important to get a short proof.

A problem occurs if we want to translate the provability problem of (1) into a refutation problem of its negation. Clausifying the disjunctive normal form $\mathsf{DPHP}_n^{Y_n}$ using distributivity laws results in an exponential number of (tautological) clauses. We slightly modify the formula to be considered by introducing new variables of the form $z_{i_1,i_2,j}$ for disjuncts in $\mathsf{DPHP}_n^{Y_n}$. This procedure is in the spirit of the well-known Tseitin translation [13]. We use the "one polarity optimization" of [12]. For the first $n + 1$ disjuncts of the form $\bigwedge_{j=1}^{n} \neg y_{i,j}$ with $1 \leq i \leq n + 1$, we use variables $z_{1,0,0}, \ldots, z_{n+1,0,0}$. For the second part, for any $1 \leq j \leq n$ and the $n(n+1)/2$ disjuncts, we use

$$z_{1,2,j}, \ldots, z_{1,n+1,j}, z_{2,3,j}, \ldots, z_{2,n+1,j}, \ldots, z_{n,n+1,j} . \tag{2}$$

The set of these variables for DPHP_n is denoted by Z_n. Due to this construction, we can speak about the conjunction corresponding to the variable $z_{i_1,i_2,j}$.

We construct the conjunctive normal form $\mathsf{TPHP}_n^{Y_n,Z_n}$ of $\mathsf{DPHP}_n^{Y_n,Z_n}$ as follows. First, we take the clause $D_n^{Z_n} = \bigvee_{z \in Z_n} \neg z$ over all variables in Z_n. The formula $P_n^{Y_n,Z_n}$ for the first $(n+1)$ disjuncts of $\mathsf{DPHP}_n^{Y_n}$ is of the form

$$\bigwedge_{i=1}^{n+1} \bigwedge_{j=1}^{n} (z_{i,0,0} \vee \neg y_{i,j}) .$$

For the remaining $n^2(n+1)/2$ disjuncts of $\mathsf{DPHP}_n^{Y_n}$, we have the formula $Q_n^{Y_n,Z_n}$

$$\bigwedge_{j=1}^{n} \bigwedge_{1 \leq i_1 < i_2 \leq n+1} \left((z_{i_1,i_2,j} \vee y_{i_1,j}) \wedge (z_{i_1,i_2,j} \vee y_{i_2,j}) \right) \ .$$

Then $\mathsf{TPHP}_n^{Y_n,Z_n}$ is $D_n^{Z_n} \wedge P_n^{Y_n,Z_n} \wedge Q_n^{Y_n,Z_n}$ and $size(\mathsf{TPHP}_n^{Y_n,Z_n})$ is $O(n^3)$. The family of formulas we consider in the following is $(\varphi_n)_{n>1}$, where φ_n is

$$\forall X_n \exists Y_n \forall Z_n \left(\mathsf{TPHP}_n^{Y_n,Z_n} \to \mathsf{DPHP}_n^{X_n} \right). \tag{3}$$

Formula (1) is equivalent to formula (3) because $\mathsf{DPHP}_n^{X_n}$ is valid. We show that

$$\mathsf{DPHP}_n^{Y_n} \equiv \exists Z_n \, \mathsf{TPHP}_n^{Y_n,Z_n} \tag{4}$$

holds.

$\Longrightarrow$: Let I be a model of $\mathsf{DPHP}_n^{Y_n}$, i.e., $I \models \mathsf{DPHP}_n^{Y_n}$ holds.

Case 1: There exists an index i such that $I \models \bigwedge_{j=1}^{n} \neg y_{i,j}$ holds. Therefore, $I \models \neg y_{i,1}, \ldots, I \models \neg y_{i,n}$ as well as $I \models \bigwedge_{j=1}^{n} z_{i,0,0} \vee \neg y_{i,j}$ hold. Let us extend I to an interpretation J such that $\mathsf{TPHP}_n^{Y_n,Z_n}$ is true under J. We set all $z_{k,l,m}$ from Z_n to true under J except $z_{i,0,0}$ which is set to false. Then $J \models D_n^{Z_n}$, $J \models P_n^{Y_n,Z_n}$ and $J \models Q_n^{Y_n,Z_n}$ hold.

Case 2: There exist indices i_1, i_2 and j such that $I \models y_{i_1,j} \wedge y_{i_2,j}$ holds. Then $I \models (z_{i_1,i_2,j} \vee y_{i_1,j}) \wedge (z_{i_1,i_2,j} \vee y_{i_2,j})$ holds. Again, we extend I to J such that $J \models \mathsf{TPHP}_n^{Y_n,Z_n}$ holds. We set all $z_{k,l,m}$ from Z_n to true under J except $z_{i_1,i_2,j}$ which is set to false. Then $J \models D_n^{Z_n}$, $J \models P_n^{Y_n,Z_n}$ and $J \models Q_n^{Y_n,Z_n}$ hold.

In both cases, there exists an extension J of I (which interprets all variables in Z_n), such that $J \models \mathsf{TPHP}_n^{Y_n,Z_n}$. Hence, $\exists Z_n \mathsf{TPHP}_n^{Y_n,Z_n}$ is true under I.

$\Longleftarrow$: Let I be an interpretation such that $I \models \exists Z_n \mathsf{TPHP}_n^{Y_n,Z_n}$ holds. Then there exists an extension J of I (which interprets all variables in Z_n), such that $J \models \mathsf{TPHP}_n^{Y_n,Z_n}$. Consequently $J \models D_n^{Z_n}$ holds and at least one z variable has to be false under J.

Case 1: There exists an index i such that $J \models \neg z_{i,0,0}$ holds. Since J satisfies $\bigwedge_{j=1}^{n}(z_{i,0,0} \vee \neg y_{i,j})$, J and also I make $\bigwedge_{j=1}^{n} \neg y_{i,j}$ true. Then $I \models \mathsf{DPHP}_n^{Y_n}$ holds.

Case 2: There exist indices i_1, i_2 and j such that $J \models \neg z_{i_1,i_2,j}$ holds. Since $J \models (z_{i_1,i_2,j} \vee y_{i_1,j}) \wedge (z_{i_1,i_2,j} \vee y_{i_2,j})$ also holds, $y_{i_1,j} \wedge y_{i_2,j}$ has to be true under J and I. Then $I \models \mathsf{DPHP}_n^{Y_n}$ holds.

We continue in the next subsection with the construction of a short proof of φ_n in Gqve*. Afterwards, we show in Section 5.3 that any sequence Q-res refutation of $\neg\varphi_n$ possesses a number of clauses which is exponential in n.

5.2 Short Proofs of φ_n in Gqve*

We provide a short proof of φ_n in Gqve*. Observe that any proof of $\forall X_n \mathsf{DPHP}_n^{X_n}$ is exponential (see Theorem 5.3.5 in [3]).

Proposition 2. *Let* $(\varphi_n)_{n>1}$ *be a family of formulas where* φ_n *is given in (3).*
Then there exists a proof of $\vdash \varphi_n$ *in* Gqve* *of size polynomial in* n.

We first show that sequents $S_{i_1,i_2,j}$ of the form

$$\neg z_{i_1,i_2,j}, P_n^{C_n,Z_n}, Q_n^{C_n,Z_n} \vdash \bigvee_{i=1}^{n+1} \bigwedge_{j=1}^{n} \neg c_{i,j}, \bigvee_{j=1}^{n} \bigvee_{1 \leq i_1 < i_2 \leq n+1} (c_{i_1,j} \wedge c_{i_2,j})$$

are derivable using $O(n^3)$ sequents.

Case 1: $z_{i_1,i_2,j}$ is of the form $z_{i,0,0}$ for $1 \leq i \leq n+1$. Take axioms and derive

$$\neg c_{i,1}, \ldots, \neg c_{i,n} \vdash \bigwedge_{j=1}^{n} \neg c_{i,j}$$

by applications of $\wedge r$ and wl using $O(n^2)$ sequents. Continue with the derived
sequent by using axioms and applications of $\neg l$, weakening and $\vee l$ to generate

$$\neg z_{i,0,0}, z_{i,0,0} \vee \neg c_{i,1}, \ldots, z_{i,0,0} \vee \neg c_{i,n} \vdash \bigwedge_{j=1}^{n} \neg c_{i,j}$$

using $O(n^2)$ sequents. By applications of $\wedge l$ to the last sequent, we obtain

$$\neg z_{i,0,0}, \bigwedge_{j=1}^{n} (z_{i,0,0} \vee \neg c_{i,j}) \vdash \bigwedge_{j=1}^{n} \neg c_{i,j}$$

requiring further $O(n)$ sequents. Continue with weakening, $\wedge l$ and $\vee r$ to generate

$$\neg z_{i,0,0}, P_n^{C_n,Z_n}, Q_n^{C_n,Z_n} \vdash \bigvee_{i=1}^{n+1} \bigwedge_{j=1}^{n} \neg c_{i,j}, \bigvee_{j=1}^{n} \bigvee_{1 \leq i_1 < i_2 \leq n+1} (c_{i_1,j} \wedge c_{i_2,j})$$

from the sequent above using $O(n)$ sequents. In total, the derivation of each of
the $(n+1)$ sequents $S_{1,0,0}, \ldots, S_{n+1,0,0}$ requires $O(n^2)$ sequents, each of which
consists of $O(n)$ sequent formulas.

Case 2: $z_{i_1,i_2,j}$ occurs as an element in (2). Start from axioms and derive

$$c_{i_1,j}, c_{i_2,j} \vdash c_{i_1,j} \wedge c_{i_2,j}$$

by weakenings and $\wedge r$ using $O(1)$ sequents. Take axioms and apply $\neg l$, weakening, $\vee l$ and $\wedge l$ to get from the sequent above

$$\neg z_{i_1,i_2,j}, (z_{i_1,i_2,j} \vee c_{i_1,j}) \wedge (z_{i_1,i_2,j} \vee c_{i_2,j}) \vdash c_{i_1,j} \wedge c_{i_2,j}$$

with $O(1)$ further sequents. Using $O(n^3)$ weakenings, $\wedge l$ and $\vee r$, we obtain

$$\neg z_{i_1,i_2,j}, P_n^{C_n,Z_n}, Q_n^{C_n,Z_n} \vdash \bigvee_{i=1}^{n+1} \bigwedge_{j=1}^{n} \neg c_{i,j}, \bigvee_{j=1}^{n} \bigvee_{1 \leq i_1 < i_2 \leq n+1} (c_{i_1,j} \wedge c_{i_2,j}).$$

In total, we have to derive $n^2(n + 1)/2$ sequents using at most a cubic number of sequents in each derivation. Each sequent has $O(n^3)$ sequent formulas.

This completes the case analysis. The sequent

$$D_n^{Z_n}, P_n^{C_n,Z_n}, Q_n^{C_n,Z_n} \vdash \bigvee_{i=1}^{n+1} \bigwedge_{j=1}^{n} \neg c_{i,j}, \bigvee_{j=1}^{n} \bigvee_{1 \le i_1 < i_2 \le n+1} (c_{i_1,j} \wedge c_{i_2,j})$$

can be derived from the $O(n^3)$ different sequents $S_{i_1,i_2,j}$ by repeated applications of $\vee l$ using $O(n^3)$ sequents. Then we can continue as follows.

$$
\cfrac{
\cfrac{
\cfrac{
\cfrac{
D_n^{Z_n}, P_n^{C_n,Z_n}, Q_n^{C_n,Z_n} \vdash \bigvee_{i=1}^{n+1} \bigwedge_{j=1}^{n} \neg c_{i,j}, \bigvee_{j=1}^{n} \bigvee_{1 \le i_1 < i_2 \le n+1} (c_{i_1,j} \vee c_{i_2,j})
}{
D_n^{Z_n}, P_n^{C_n,Z_n}, Q_n^{C_n,Z_n} \vdash \mathsf{DPHP}_n^{C_n}
} \vee r
}{
\mathsf{TPHP}_n^{C_n,Z_n} \vdash \mathsf{DPHP}_n^{C_n}
} \wedge l, \wedge l
}{
\vdash \mathsf{TPHP}_n^{C_n,Z_n} \to \mathsf{DPHP}_n^{C_n}
} \to r
}{
\vdash \forall X_n \exists Y_n \forall Z_n (\mathsf{TPHP}_n^{Y_n,Z_n} \to \mathsf{DPHP}_n^{X_n})
} \forall r_e, \exists r_v, \forall r_e
$$

Hence the overall number of sequents used to derive the indicated end sequent is $O(n^6)$. There are $O(n^3)$ sequent formulas in each sequent and each such formula is a subformula of φ_n. Therefore, we have a polynomial size proof of φ_n in Gqve^*.

5.3 Q-resolution Refutations of $\neg \varphi_n$

We reconsider φ_n from above. Since φ_n is valid iff $\neg \varphi_n$ is unsatisfiable, we use the latter and show it by Q-resolution. As we will see, any Q-resolution refutation of $\neg \varphi_n$ is exponential in n. Take $\neg \varphi_n$ and push negation inwards. Then we get

$$\neg \varphi_n \text{ is unsat iff } \exists X_n \forall Y_n \exists Z_n (\mathsf{TPHP}_n^{Y_n,Z_n} \wedge \mathsf{CPHP}_n^{X_n}) \text{ is unsat.}$$

Proposition 3. *Any Q-res refutation of $\exists X_n \forall Y_n \exists Z_n (\mathsf{TPHP}_n^{Y_n,Z_n} \wedge \mathsf{CPHP}_n^{X_n})$ has exponential size.*

Since the two indicated CNFs $\mathsf{TPHP}_n^{Y_n,Z_n}$ and $\mathsf{CPHP}_n^{X_n}$ belong to completely different languages, no resolution is possible where one parent clause is from the one part and the other parent clause is from the other part. Therefore

$$\forall Y_n \exists Z_n (\mathsf{TPHP}_n^{Y_n,Z_n}) \text{ is unsat} \quad \text{or} \quad \exists X_n (\mathsf{CPHP}_n^{X_n}) \text{ is unsat.}$$

We first consider $\exists X_n (\mathsf{CPHP}_n^{X_n})$ which is the existential closure of the purely propositional pigeon hole formula $\mathsf{CPHP}_n^{X_n}$ in conjunctive normal form. Only the propositional resolution rule is applicable because no $\forall$ variable occurs. By Haken's famous result [8], any resolution refutation of $\mathsf{CPHP}_n^{X_n}$ is exponential in n. Consequently, the same holds for any Q-res refutation of the same formula. Hence, $\exists X_n (\mathsf{CPHP}_n^{X_n})$ is false and therefore unsatisfiable.

We next consider $\forall Y_n \exists Z_n \mathsf{TPHP}_n^{Y_n,Z_n}$. Above we proved the following equivalence $\mathsf{DPHP}_n^{Y_n} \equiv \exists Z_n \mathsf{TPHP}_n^{Y_n,Z_n}$. Since $\mathsf{DPHP}_n^{Y_n}$ is valid, so is $\exists Z_n \mathsf{TPHP}_n^{Y_n,Z_n}$

and therefore $\forall Y_n \exists Z_n (\mathsf{TPHP}_n^{Y_n, Z_n})$ is true. By the soundness and completeness of Q-resolution, no (non-tautological) clause with only universal literals can be derived. Hence, $\forall Y_n \exists Z_n \mathsf{TPHP}_n^{Y_n, Z_n}$ cannot provide any refutation.

In conclusion, any Q-res refutation of $\neg\varphi$ is exponential in n. Consider

$$\left(\mathsf{TPHP}_n^{X_n, Z_n} \wedge \mathsf{CPHP}_n^{X_n}\right) \tag{5}$$

which can be obtained by instantiating the quantifiers for Y_n properly. Interestingly, there exists a tree (Q-)resolution refutation of (the existential closure of) formula (5) of size polynomial in n, which identifies the simple way of handling quantifiers by $\forall$R to be the weak point in Q-res. Obviously, quantifier rules resulting an instantiation of the matrix formula can yield more succinct proofs.

From the above complexity analysis of Q-resolution refutations of $\neg\varphi$, a simple corollary can be drawn. Let us reconsider $\exists X_n \forall Y_n \exists Z_n (\mathsf{TPHP}_n^{Y_n, Z_n} \wedge \mathsf{CPHP}_n^{X_n})$ to which we apply the QDPLL algorithm with clause learning. The only variables which are processed are from X_n because $\mathsf{CPHP}_n^{X_n}$ is unsatisfiable. Finding the conflicts results in learned clauses, which can be used to construct a Q-res refutation of the input formula as a witness for unsatisfiability. Since any Q-resolution refutation is exponential in n, so is the QDPLL refutation.[2]

6   Conclusion

We studied different techniques to handle quantification in QBFs. Integrated into a sequent calculus for propositional logic, all discussed combinations of quantifier rules yield sound and complete calculi, differing in their non-deterministic strength, i.e., their ability to represent proofs succinctly. We have seen that Q-res is a weaker calculus than sequent systems with reasonable quantifier rules. Although this result seems to be of limited relevance for practical applications, one should keep in mind that certificates (or solutions) are extracted from Q-res refutations produced by QBF solvers [1]. Since the size of the certificate corresponds to the size of the Q-res refutation, a more succinct proof could be beneficial.

We have identified instantiation as *the* feature for obtaining short proofs for our formulas. Neither the quantifier handling in Q-res nor semantically motivated quantifier rules possess this feature. Strong quantifier rules based on semantics are essentially binary inferences and in general not powerful enough in a cut-free sequent system. These rules require additional techniques like propagation of values, formula simplification, dependency directed backtracking, etc. to compensate their weakness. Such techniques can be integrated in sequent systems via restricted versions of cut or as additional inferences, cf. [6] for examples.

Although $\forall l_f$ and $\exists r_f$ are the rules with most non-deterministic power, they are not necessarily required for our problem formulas. They were actually proved with weaker rules $\forall l_v$ and $\exists r_v$ allowing only the introduction of variables (and truth constants). We provided some indication that, at least in some variants of sequent calculi like Gqve_i^* ($i \geq 1$), the weaker rules are sufficient. But a closer

---

[2] We learned this argument from F. Lonsing (private communication).

look reveals the practical problem of the $\forall l_f$ and $\exists r_f$ inferences, the simulation by extension and the simulation by cut (not discussed here): How does a good formula for the quantifier, the extension step or the cut rule look like?

References

1. Balabanov, V., Jiang, J.-H.R.: Resolution Proofs and Skolem Functions in QBF Evaluation and Applications. In: Gopalakrishnan, G., Qadeer, S. (eds.) CAV 2011. LNCS, vol. 6806, pp. 149–164. Springer, Heidelberg (2011)
2. Buss, S.: An introduction to proof theory. In: Handbook of Proof Theory, pp. 1–78. Elsevier, Amsterdam (1998)
3. Clote, P., Kranakis, E.: Boolean Functions and Models of Computation. Springer, Heidelberg (2002)
4. Cook, S.A., Morioka, T.: Quantified propositional calculus and a second-order theory for NC^1. Arch. Math. Log. 44(6), 711–749 (2005)
5. Cook, S.A., Reckhow, R.A.: On the lengths of proofs in the propositional calculus (preliminary version). In: STOC, pp. 135–148 (1974)
6. Egly, U., Seidl, M., Woltran, S.: A solver for QBFs in negation normal form. Constraints 14(1), 38–79 (2009)
7. Gentzen, G.: Untersuchungen über das logische Schließen. Mathematische Zeitschrift 39, 176–210, 405–431 (1935)
8. Haken, A.: The intractability of resolution. Theor. Comput. Sci. 39, 297–308 (1985)
9. Jussila, T., Biere, A., Sinz, C., Kroning, D., Wintersteiger, C.M.: A First Step Towards a Unified Proof Checker for QBF. In: Marques-Silva, J., Sakallah, K.A. (eds.) SAT 2007. LNCS, vol. 4501, pp. 201–214. Springer, Heidelberg (2007)
10. Kleine Büning, H., Karpinski, M., Flögel, A.: Resolution for quantified Boolean formulas. Inf. Comput. 117(1), 12–18 (1995)
11. Krajíček, J.: Bounded Arithmetic, Propositional Logic, and Complexity Theory. Encyclopedia of Mathematics and its Application, vol. 60. Cambridge University Press (1995)
12. Plaisted, D.A., Greenbaum, S.: A structure-preserving clause form translation. J. Symb. Comput. 2(3), 293–304 (1986)
13. Tseitin, G.S.: On the Complexity of Derivation in Propositional Calculus. In: Slisenko, A.O. (ed.) Studies in Constructive Mathematics and Mathematical Logic, Part II. Seminars in Mathematics, vol. 8, pp. 234–259. Steklov Mathematical Institute, Leningrad (1968)

Solving QBF with Counterexample Guided Refinement[*]

Mikoláš Janota[1], William Klieber[3], Joao Marques-Silva[1,2],
and Edmund Clarke[3]

[1] IST/INESC-ID, Lisbon, Portugal
[2] University College Dublin, Ireland
[3] Carnegie Mellon University, Pittsburgh, PA, USA

Abstract. We propose two novel approaches for using Counterexample-Guided Abstraction Refinement (CEGAR) in Quantified Boolean Formula (QBF) solvers. The first approach develops a recursive algorithm whose search is driven by CEGAR (rather than by DPLL). The second approach employs CEGAR as an additional learning technique in an existing DPLL-based QBF solver. Experimental evaluation of the implemented prototypes shows that the CEGAR-driven solver outperforms existing solvers on a number of families in the QBF-LIB and that the DPLL solver benefits from the additional type of learning. Thus this article opens two promising avenues in QBF: CEGAR-driven solvers as an alternative to existing approaches and a novel type of learning in DPLL.

1 Introduction

Quantified Boolean formulas (QBFs) [8] naturally extend the SAT problem by enabling expressing PSPACE-complete problems, which can be found in a number of areas [13]. While nonrandom SAT solving has been dominated by the DPLL procedure, it has proven to be far from a silver bullet for QBF solving. Indeed, a number of solving techniques have been proposed for QBF [12,3,4,19,15], complemented by a variety of *preprocessing techniques* [7,14,21,5].

This paper extends the family of QBF solving techniques by employing the counterexample guided abstraction refinement (CEGAR) paradigm [10]. This is done in two different ways. The first approach develops a novel algorithm, named RAReQS, that gradually *expands* the given formula into a propositional one. In contrast to the existing expansion-based solvers [1,4,19], the use of CEGAR in RAReQS enables terminating before the formula is fully expanded and thus substantially mitigates the problems with memory blowup inherent to expansion-based solvers. The second approach employs CEGAR as an *additional learning technique* in an existing DPLL-based QBF solver. At the price of higher memory consumption, this learning technique enables more aggressive pruning of

---

[*] This work is partially supported by FCT grants ATTEST (CMU-PT/ELE/0009/-2009) and POLARIS (PTDC/EIA-CCO/123051/2010), by SFI grant BEACON (09/IN.1/I2618), and by Semiconductor Research Corporation contract 2005TJ1366.

A. Cimatti and R. Sebastiani (Eds.): SAT 2012, LNCS 7317, pp. 114–128, 2012.

the search space than the existing techniques [28]. The experimental evaluation carried out demonstrates that CEGAR-based techniques are useful for a large number of families in the QBF-LIB [25].

2 Preliminaries

Quantified Boolean formulas (QBF) are assumed, unless noted otherwise, to be in *prenex* form $Q_1 z_1 \ldots Q_n z_n.\phi$ where $Q_i \in \{\forall, \exists\}$, z_i are distinct variables, and ϕ is a propositional formula using only the variables z_i and the constants 0 (false), 1 (true). The sequence of quantifiers in a QBF is called the *prefix* and the propositional formula the *matrix*. The prefix is divided into *quantifier blocks*, each of which is a subsequence $\forall x_1 \ldots \forall x_n$ or resp. $\exists x_1 \ldots \exists x_n$, which we denote by $\forall X$ or resp. $\exists X$, where $X = \{x_1, ..., x_n\}$.

Notation. We write $\bar{Q}$ for "$\forall$" (if Q is "$\exists$") or "$\exists$" (if Q is "$\forall$").

Whenever convenient, parts of a prefix are denoted as P with possible subscripts, e.g. $P_1 \forall X P_2.\phi$ denotes a QBF with the matrix ϕ and a prefix that contains $\forall X$. If the quantifier of a block Y occurs within the scope of the quantifier of another block X, we say that variables in X are *upstream* of variables in Y and that variables in Y are *downstream* of variables in X.

Variable assignments are represented as sets of literals. In particular, an assignment τ to the set of variables X contains exactly one of x, $\neg x$ for each $x \in X$, with the meaning that if $x \in \tau$, the variable x has the value 1 in τ and if $\neg x \in \tau$, it has the value 0.

Notation. We write $\mathcal{B}^Y$ for the set of assignments to the variables Y.

For a Boolean formula ϕ and an assignment τ we write $\phi[\tau]$ for the substitution of τ in ϕ. In practice a substitution also performs basic simplifications, e.g. $(\neg x \vee y)[\{\neg x\}] = (\neg 0 \vee y) = 1$. We extend the notion of substitution to QBF so that it first removes the quantifiers of substituted variables and then substitutes all occurrences with their assigned values. E.g., if τ is an assignment to a block X, then $(P_1 Q X P_2.\ \phi)[\tau]$ results in $P_1 P_2.\ \phi[\tau]$.

A Boolean formula in *conjunctive normal form (CNF)* is a conjunction of *clauses*, where a clause is a disjunction of *literals*, and a literal is either a variable or its complement. Whenever convenient, a CNF formula is treated as a set of clauses. For a literal l, $\mathsf{var}(l)$ denotes the variable in l, i.e. $\mathsf{var}(\neg x) = \mathsf{var}(x) = x$.

The pseudocode throughout the paper uses the function $\mathsf{SAT}(\phi)$ to represent a call to a SAT solver on a propositional formula ϕ. The function returns a satisfying assignment for ϕ, if such exists, and returns NULL otherwise.

2.1 Game-Centric View

A QBF can be seen as a a *game* between the *universal player* and the *existential player*. During the game, the existential player assigns values to the existentially quantified variables and the universal player assigns values to the universally quantified ones. A player can assign a value to a variable only if all variables upstream of it already have a value. The existential player wins if the formula evaluates to 1 and the universal player wins if it evaluates to 0.

We note that the order in which values are given to variables in the same block is unimportant. Hence, by a *move* we mean an assignment to variables in a certain block. A concept useful throughout the paper are the *winning moves*.

Definition 1 (winning move). *Consider a (nonprenex) closed QBF $QX.\Phi$ and an assignment τ to X. Then τ is called a* winning move *for $QX.\Phi$ if $Q=\exists$ and $\Phi[\tau]$ is true or $Q=\forall$ and $\Phi[\tau]$ is false.*
Notation. *We write $\mathcal{M}(QX.\Phi)$ to denote the set of winning moves for $QX.\Phi$.*

Observation 1. *Let Φ be a QBF.*
  A closed QBF $\exists X.\Phi$ is true iff there exists a winning move for $\exists X.\Phi$.
  A closed QBF $\forall Y.\Phi$ is true iff there does not *exist a winning move for $\forall Y.\Phi$.*

3   Recursive CEGAR-Based Algorithm

Previous work on QBF shows how CEGAR can be used to solve formulas with 2 levels of quantifiers [17]. Here we generalize this approach to an arbitrary number of quantifiers by recursion. The recursion follows the prefix of the given formula starting with the most upstream variables progressing towards more downstream variables. It tries to find a winning move (Definition 1) for variables in a certain block by making recursive calls to obtain winning moves for the downstream variables. The base case of the recursion, i.e., a QBF with one quantifier, is handled by a SAT solver.

The algorithm is presented as a recursive function returning a winning move for the given formula, if such move exists. Following the CEGAR paradigm, the function builds an abstraction which provides *candidates* for the winning move. This abstraction is gradually refined as the algorithm progresses. Refinement is realized by *strengthening* the abstraction, which means reducing the set of winning moves; strengthening is achieved by applying conjunction and disjunction.

Observation 2. *Let $\Phi_1, \ldots, \Phi_n$ be QBFs with free variables in X.*
$$\mathcal{M}\left(\forall X.\ (\Phi_1 \vee \cdots \vee \Phi_n)\right) \subseteq \mathcal{M}\left(\forall X.\ \Phi_i\right),\ i \in 1..n.$$
$$\mathcal{M}\left(\exists X.\ (\Phi_1 \wedge \cdots \wedge \Phi_n)\right) \subseteq \mathcal{M}\left(\exists X.\ \Phi_i\right),\ i \in 1..n.$$
$$\mathcal{M}(\forall X \exists Y.\ \Phi) = \mathcal{M}(\forall X.\ \bigvee_{\mu \in \mathcal{B}^Y} \Phi[\mu])$$
$$\mathcal{M}(\exists X \forall Q Y.\ \Phi) = \mathcal{M}(\exists X.\ \bigwedge_{\mu \in \mathcal{B}^Y} \Phi[\mu])$$

The second half of the above observation gives us a recipe how to eliminate quantifiers by *expanding* them into the corresponding propositional operator. One could thus eliminate quantifiers one by one and eventually call a SAT solver if only one quantifier is left. The clear disadvantage of this approach is that the formula grows rapidly and therefore performing the expansion is often unfeasible. This is where CEGAR comes in; the algorithm expands quantifiers *carefully*, based on counterexamples that show that the current expansion is too weak. In this spirit, we define abstraction as a partial expansion of the given formula.

Algorithm 1. Basic recursive CEGAR algorithm for QBF

---

1 **Function** Solve $(QX.\Phi)$

    input   : $QX.\Phi$ is a closed QBF in prenex form with no adjacent blocks with the same quantifier

    output   : a winning move for $QX.\Phi$ if there is one, NULL otherwise

2 **begin**

3 | **if** Φ *has no quantifiers* **then**

4 | | **return** $(Q = \exists)$? SAT(ϕ) : SAT$(\neg\phi)$

5 | **end**

6 | $\omega \leftarrow \emptyset$

7 | **while** true **do**

8 | | $\alpha \leftarrow (Q = \exists)$? $\bigwedge_{\mu\in\omega} \Phi[\mu]$: $\bigvee_{\mu\in\omega} \Phi[\mu]$ `// build abstraction`

9 | | $\tau' \leftarrow$ Solve(Prenex($QX.\,\alpha$))   `// find a candidate solution`

10 | | **if** $\tau' = $ NULL **then return** NULL   `// no winning move`

11 | | $\tau \leftarrow \{l \mid l \in \tau' \wedge \mathrm{var}(l) \in X\}$ `// filter a move for X`

12 | | $\mu \leftarrow$ Solve($\Phi[\tau]$)   `// find a counterexample`

13 | | **if** $\mu = $ NULL **then return** τ

14 | | $\omega \leftarrow \omega \cup \{\mu\}$ `// refine`

15 | **end**

16 **end**

Definition 2 (ω-abstraction). *Let ω be a subset of $\mathcal{B}^Y$.*
The ω-abstraction of a closed QBF $\forall X \exists Y.\ \Phi$ is the formula $\forall X.\ \bigvee_{\mu\in\omega} \Phi[\mu]$.
The ω-abstraction of a closed QBF $\exists X \forall Y.\ \Phi$ is the formula $\exists X.\ \bigwedge_{\mu\in\omega} \Phi[\mu]$.

Observe that any winning move for $QX\bar{Q}Y.\ \Phi$ is also a winning move for its ω-abstraction (for arbitrary ω). The reverse, however, does not hold. Hence, following the CEGAR paradigm, we first find a winning move for the abstraction and then *verify* that it is also a winning move for the given formula. Verifying that a given assignment is a winning move entails solving another QBF.

Observation 3. *An assignment τ is a winning move for a closed $QX\bar{Q}Y.\ \Phi$ iff $\bar{Q}Y.\ \Phi[\tau]$ has no winning move.*

If a winning move for the abstraction is verified to be a winning move for the given formula, the move is returned. However, if this is not the case, the abstraction is strengthened. Observation 3 tells us that if an assignment τ is *not* a winning move for $QX\bar{Q}Y.\ \Phi$, then there *is* a winning move μ for the opposing quantifier $\bar{Q}$ for the QBF $\bar{Q}Y.\ \Phi[\tau]$. We say that this move μ is a *counterexample* to τ because it serves as a witness demonstrating that τ is not a winning move for $QX\bar{Q}Y.\ \Phi$. In accordance with the concept of counterexample *guided* abstraction refinement, if a counterexample μ is found, the current ω-abstraction is strengthened by adding μ to ω.

When we put these things together, we obtain Algorithm 1. The algorithm is given a closed QBF $QX.\Phi$. and returns a winning move for $QX.\Phi$, if such exists, and returns NULL otherwise. It is required that $QX.\Phi$ is in prenex form where no two adjacent blocks have the same quantifiers (the blocks are maximal). The algorithm starts with $\omega = \emptyset$; this represents an abstraction that can be won by any candidate. In each iteration of the CEGAR loop it first solves the abstraction (line 9) and then verifies whether the move winning the abstraction is also a winning move for the given problem (line 12). These operations are realized as recursive calls. If there is no winning move for the abstraction, then there is no winning move for the given problem and the function terminates. If there is no counterexample to the move winning the abstraction, then this move is also a winning move for the given problem and the function terminates. If there is a counterexample to the move winning the abstraction, the abstraction must be refined (line 14).

The precondition of the function that the input formula must be in prenex form with no adjacent blocks with the same quantifier poses some technical difficulty. When constructed directly according to its definition (Definition 2), the abstraction does not necessarily satisfy this condition.

Consider the case for $Q = \exists$ ($Q = \forall$ is analogous). The abstraction is of the form $\exists X. \bigwedge_{\mu \in \omega} \Phi[\mu]$. Prenexing the abstraction generates fresh variables for each of the conjuncts $\Phi[\mu]$, interleaves them into a single prefix, and merges adjacent blocks that start with the same quantifier. Since each $\Phi[\mu]$ starts with the existential quantifier (the substitution of μ eliminated the universal variables at the top), after prenexing, the abstraction's prefix starts with $\exists X X_1 \ldots X_k$ where X_i are the fresh variables for the conjuncts $\Phi[\mu]$. For this reason if a winning move for the abstraction is computed, only the assignments to the variables X are considered (line 11).

Example 1. Consider the QBF $\exists vw.\Phi$, where $\Phi = \forall u \exists xy. (v \lor w \lor x) \land (\bar{v} \lor y) \land (\bar{w} \lor y) \land (u \lor \bar{x}) \land (\bar{u} \lor \bar{y})$, and the candidates $\{v, w\}$ and $\{\bar{v}, \bar{w}\}$, and corresponding counterexamples $\{u\}$ and $\{\bar{u}\}$. Refinement yields the abstraction $\exists vw. \Phi[\{u\}] \land \Phi[\{\bar{u}\}]$, with the prenex form $\exists vwxyx'y'. (v \lor w \lor x) \land (\bar{v} \lor y) \land (\bar{w} \lor y) \land (\bar{y}) \land (v \lor w \lor x') \land (\bar{v} \lor y') \land (\bar{w} \lor y') \land (\bar{x}')$ with no winning move and the algorithm terminates with the return value NULL.

3.1 Improving Recursive CEGAR-Based Algorithm

Algorithm 1 clearly suffers from high memory consumption since in each iteration of the loop the abstraction is increased by the size of the input formula and the number of its variables is doubled (in the worst case). Recursive calls further amplify this unfavorable behavior. For the input formula $\exists X. \Phi$, performing n_1 iterations with the counterexamples $\mu_1^1, \ldots, \mu_{n_1}^1$ yields the abstraction $\Omega = \exists X. \phi[\mu_1^1] \land \cdots \land \phi[\mu_{n_1}^1]$. The algorithm subsequently invokes the recursive call Solve(Ω) on line 9. If within this recursive call the loop iterates n_2 times, its abstraction is of the form $\exists X. \Omega[\mu_1^2] \lor \cdots \lor \Omega[\mu_{n_2}^2]$ with the size $O(n_1 \times n_2 \times |\phi|)$. In general, if the algorithm iterates n_i times at a recursion level i, the abstraction at level k is of the size $O(n_1 \times \ldots \times n_k \times |\phi|)$.

To cope with this inefficiency, we exploit the form of the formulas that the algorithm handles. In the case of the existential quantifier, the abstraction is a conjunct, and it is a disjunct in the case of the universal quantifier. For the sake of uniformity, we bridge these two forms by introducing the notion of a *multi-game* where a player tries to find a move that wins multiple formulas simultaneously.

Definition 3 (multi-game). *A multi-game is denoted by $QX.\{\Phi_1, \ldots, \Phi_n\}$ where each Φ_i is a prenex QBF starting with $\bar{Q}$ or has no quantifiers. The free variables of each Φ_i must be in X and all Φ_i have the same number of quantifier blocks. We refer to the formulas Φ_i as subgames and QX as the top-level prefix.*

A winning move for a multi-game is an assignment to the variables X such that it is a winning move for each of the formulas $QX.$ Φ_i.

Observe that the set of winning moves of a multi-game $QX.\{\Phi_1, \ldots, \Phi_n\}$ is the same as the set of winning moves of the QBF $\forall X.(\Phi_1 \vee \cdots \vee \Phi_n)$ for $Q = \forall$ and it is the same as $\exists X.(\Phi_1 \wedge \cdots \wedge \Phi_n)$ for $Q = \exists$. And, any QBF $QX.$ Φ corresponds to a multi-game with a single subgame $QX.\{\Phi\}$

To solve multi-games we use Algorithm 2. The algorithm is given a multi-game to solve and the abstraction is again a multi-game. To determine whether the candidate τ is a winning move, it tests whether it is a winning move for the subgames in turn. If it finds a subgame Φ_i s.t. $\Phi_i[\tau]$ is won by the opponent $\bar{Q}$ by a move μ, then $\Phi_i[\mu]$ is used to strengthen the abstraction.

Since an abstraction is a multi-game, it seems natural to add $\Phi_i[\mu]$ to the set of its subgames. This, however, cannot be done right away because the formula is not in the right form. In particular, all the subgames must start with the opposite quantifier as the top-level prefix. Hence, if Φ_i is of the form $\bar{Q}YQX_1.\Psi_i$ and $\mu \in \mathcal{B}^Y$, then $\Phi_i[\mu] = QX_1.\Psi_i[\mu]$. To bring the formula into the right form, we introduce fresh variables for the variables X_1 and move them into the top-level prefix. More precisely, the function $\texttt{Refine}(\alpha, \Phi_l, \mu_l)$ is defined as follows (observe that the subgames remain in prenex form).

$\texttt{Refine}\big(QX.\{\Psi_1, \ldots, \Psi_n\}, \bar{Q}YQX_1.\Psi, \mu\big) := QXX_1'.\{\Psi_1, \ldots, \Psi_n, \Psi'[\mu]\}$
where X_1' are fresh duplicates of the variables X_1 and Ψ' is Ψ with X_1 replaced by X_1'

$\texttt{Refine}\big(QX.\{\Psi_1, \ldots, \Psi_n\}, \bar{Q}Y.\psi, \mu\big) := QX.\{\Psi_1, \ldots, \Psi_n, \psi[\mu]\}$
where ψ is a propositional formula (where no duplicates are needed)

Similarly to Algorithm 1, after the refinement, the abstraction's top-level prefix contains additional variables besides the variables X. Hence, values for these variables are filtered out if a winning move for the abstraction is found.

3.2   Properties of the Algorithms

In CEGAR loop of Algorithm 1 no candidate or counterexample repeats. Intuitively, this is because once a counterexample μ is found, the abstraction is strengthened so that in the future winning moves for the abstraction cannot be beaten by the move μ. Consequently, the loop is terminating and for a formula $QX\bar{Q}Y.\Phi$ the number of its iterations is bounded by the number of possible assignments to the variables X and Y, i.e. $\min(2^{|X|}, 2^{|Y|})$. In the worst case, in each iteration the abstraction grows by the size of Φ. For a multi-game $QX.\{\Phi_1, \ldots, \Phi_n\}$ in the CEGAR loop of Algorithm 2 no candidates repeat

---

Algorithm 2. Recursive CEGAR algorithm for multi-games

---

1 **Function** RAReQS $(QX. \{\Phi_1, \ldots, \Phi_n\})$
2 **output:** a winning move for $QX. \{\Phi_1, \ldots, \Phi_n\}$ if there is one; NULL
   otherwise

3 **begin**
4   **if** Φ_i *have no quantifiers* **then**
5 | **return** $Q = \exists$? SAT$(\bigwedge_i \Phi_i)$: SAT$(\neg(\bigvee_i \Phi))$
6     $\alpha \leftarrow QX. \{\}$
7   **while** true **do**
8        $\tau' \leftarrow$ RAReQS(α) // find a candidate solution
9   **if** $\tau' =$ NULL **then return** NULL
10      $\tau \leftarrow \{l \mid l \in \tau' \wedge \mathsf{var}(l) \in X\}$ // filter a move for X
11   **for** $i \leftarrow 1$ **to** n **do** $\mu_i \leftarrow$ RAReQS$(\Phi_i[\tau])$ // find a
       counterexample
12   **if** $\mu_i =$ NULL *for all* $i \in \{1..n\}$ **then return** τ
13   let $l \in \{1..n\}$ be s.t. $\mu_l \neq$ NULL
14      $\alpha \leftarrow$ Refine(α, Φ_l, μ_l) // refine
15   **end**
16 **end**

---

but counterexamples may. However, for a given $i \in 1..n$, a counterexample μ_i does not repeat. More precisely there are no two distinct iterations of the loop with the corresponding candidates and counterexamples $\tau_1, \mu_1, \tau_2, \mu_2$, such that $\mu_1 = \mu_2$ and μ_1 is a winning move for both $\Phi_i[\tau_1]$ and $\Phi_i[\tau_2]$ for some i. This demonstrates termination with the upper bound for the number of iterations as $\min(2^{|X|}, n \times 2^{|Y|})$. In the worst case, in each iteration the abstraction grows by the maximum of the sizes of the subgames $\Phi_1, \ldots, \Phi_n$. Soundness and completeness of the algorithms 1 and 2 are direct consequences of Observation 2.

3.3   Implementation Details

We have implemented a prototype of RAReQS in C++, supporting the QDIMACS format, with the underlying SAT solver minisat 2.2 [11].

The implementation has several distinctive features. In Algorithm 2, an abstraction computed within a sub-call is forgotten once the call returns. This may lead to repetition of work and hence the solver supports maintaining these abstractions and strengthening them gradually, similarly to the way SAT solvers provide *incremental* interface. This incremental approach, however, tends to lead to unwieldy memory consumption and therefore, it is used only when the given multigame's subgames have 2 or fewer quantification blocks.

If an assignment τ is a candidate for a winning move that turns out *not* to be a winning move, the refinement guarantees that τ is not a solution to the abstraction in the future iterations of the CEGAR loop. This knowledge enables

Algorithm 3. DPLL Algorithm with CEGAR Learning

```
1.    global π_cur = ∅;
2.    function dpll_solve(Φ_in) {
3.        while (true) {
4.            while (we don't know who has a winning strategy under π_cur) {
5.                decide_lit(); propagate();
7.            }
8.            Φ_in := dpll_learn(Φ_in);
9.            if (we learned who has a winning strategy under ∅) return;
10.           if (last decision literal is owned by winner) {
11.               Φ_in := cegar_learn(Φ_in);
12.           }
13.           backtrack();
14.           propagate(); // Learned information will force a literal.
15.       }
16.   }
```

us to make the subcall for solving the abstraction more efficient by explicitly disabling τ as a winning move for the abstraction. We refer to this technique as *blocking* and it is similar to the refinement used in certain SMT solvers [24,2].

Throughout its course, the algorithm may produce a large number of new formulas, either by substitution or refinement. Since these formulas tend to be simpler than the given one, they can be further simplified by standard QBF *preprocessing* techniques. The implementation uses unit propagation and *monotone* (pure) literal rule [9]. These simplifications introduce the complication that in a multi-game $QX.\{\Phi_1, \ldots, \Phi_n\}$ the individual subgames might not necessarily have the same number of quantifier levels. In such case, all games with no quantifiers are immediately put into the abstraction before the loop starts.

4   CEGAR as a Learning Technique in DPLL

The previous section shows that CEGAR can give rise to a complete and sound algorithm for QBF. In this section we show that CEGAR enables us to extend existing DPLL solvers with an additional learning technique. To illustrate the basic idea consider the QBF $\forall X. (\exists Y. \phi)$ and a situation when the solver assigned values to variables in X and Y such that ϕ is satisfied, i.e., the existential player won. This assignment has two disjoint parts, π_{cand} and π_{cex}, which are assignments to X and Y, respectively. Conceptually, π_{cand} corresponds the candidate assignment in RAReQS and π_{cex} to its counterexample. In this case, the CEGAR-based learning will correspond to disjoining the formula $\phi[\pi_{cex}]$ onto ϕ, resulting in $\forall X. (\exists Y. \phi) \vee \phi[\pi_{cex}]$, so that π_{cand} is avoided in the future.

The CEGAR learning in DPLL is most naturally described in the context of a non-prenex, non-clausal solver such as GhostQ [18]. Given an assignment

---

1. Let X_c be the quantifier block of the last decision literal.
   Let Q_c and Φ_c be such that $(Q_c X_c . \Phi_c)$ is a subformula of Φ_{in}.
2. Let π_c be a complete assignment for X_c created by extending the solver's current assignment with arbitrary values for the unassigned variables in X_c and removing variables in blocks other than X_c. This assignment π_c corresponds to the *counterexample* in the recursive CEGAR approach.
3. We modify Φ_{in} by:
   - substituting $(\exists X_c . \Phi_c)$ with $(\exists X_c . \Phi_c) \vee \Phi_c[\pi_c]$, if $Q_c = $ "$\exists$", or
   - substituting $(\forall X_c . \Phi_c)$ with $(\forall X_c . \Phi_c) \wedge \Phi_c[\pi_c]$, if $Q_c = $ "$\forall$".
4. All variables that are bound by a quantifier inside $\Phi_c[\pi_c]$ are renamed to preserve uniqueness of variable names.

Fig. 1. CEGAR Learning in DPLL

π, such a solver will tell us that either (1) the existential player has a winning stategy under π (i.e., $\Phi_{\text{in}}[\pi]$ is true), (2) the universal player has a winning stategy under π (i.e., $\Phi_{\text{in}}[\pi]$ is false), or (3) it is not yet known which player has a winning strategy under π.

We modify such a solver by inserting a call to a new CEGAR-learning procedure after performing standard DPLL learning, as shown in Algorithm 3. We write "Φ_{in}" to denote the current input formula, i.e., the input formula enhanced with what the solver has learned up to now. Both standard DPLL learning and CEGAR learning are performed by modifying Φ_{in}. As shown in Algorithm 3, CEGAR learning is performed only if the last decision literal is owned by the winner. (The case where the last decision literal is owned by the losing player corresponds to the conflicts that take place *within* the underlying SAT solver in RAReQS.) The CEGAR-learning procedure is shown in Figure 1. Step 3 is justified by Observation 5 below, which in turn is justified by Observation 4.

Observation 4. Consider an arbitrary QBF $(Q_c X_c . \Phi_c)$, possibly containing free variables, but where each bound variable is bound by at most one quantifier. Then it follows immediately from definition of quantification that:

$$\exists X_c . \Phi_c = \bigvee_{\pi \in \mathcal{B}^{X_c}} \Phi_c[\pi] \quad \text{and} \quad \forall X_c . \Phi_c = \bigwedge_{\pi \in \mathcal{B}^{X_c}} \Phi_c[\pi]$$

(Recall that "$\mathcal{B}^{X_c}$" denotes the set of all assignments to X_c.)

Observation 5. Since conjunction and disjunction are idempotent,

$$\exists X_c . \Phi_c = (\exists X_c . \Phi_c) \vee \Phi_c[\pi_c], \text{ where } \pi_c \in \mathcal{B}^{X_c}$$
$$\forall X_c . \Phi_c = (\forall X_c . \Phi_c) \wedge \Phi_c[\pi_c], \text{ where } \pi_c \in \mathcal{B}^{X_c}$$

4.1   Implementation Details

We have implemented a limited version of CEGAR learning in the solver GhostQ [18]. Our implementation uses a modified version of step 3 of Figure 1. We substitute π_c into the original version of the input formula Φ_{in}, not the current version of Φ_{in}. Although substituting into the original formula instead of the current formula potentially reduces the effectiveness of CEGAR learning (since we can't learn a refinement of a refinement), it reduces the memory consumed per refinement. Unit propagation and the Pure Literal Rule are applied to simplify the result of the substitution, among other optimizations.

Step 2 of Figure 1 extends the counterexample π_c to a complete assignment to the quantifier block X_c. This allows completely eliminating a quantifier block, which may cause two quantifier blocks of the same quantification type to become adjacent to each other. If so, the two adjacent blocks are merged together, providing greater freedom in selecting variable order.

5   Experimental Results

Our objective was to analyze the effect of CEGAR on the different families of available benchmarks. Due to do the large number of families in QBF-LIB [25], we have targeted families from *formal verification* and *planning* as two prominent applications of QBF. Several large and hard families were sampled with 150 files (terminator, tipfixpoint, Strategic Companies); the area of planning contains four classes for robot planning, each counting 1000 instances with similar characteristics and thus only one of these classes was selected (Robots2D). The solvers QuBE7.2, Quantor, and Nenofex were chosen for comparison. QuBE7.2 is a state-of-the-art DPLL-based solver; Quantor and Nenofex are expansion-based solvers (c.f. Section 6). The experimental results were obtained on an Intel Xeon 5160 3GHz, with 4GB of memory. The time limit was set to 800 seconds and the memory limit to 2GB.

All the instances were preprocessed by the preprocessor bloqqer [5] and instances solved by the preprocessor alone were excluded from further analysis. An exception was made for the family Debug where preprocessing turned out to be infeasible and the family was considered in its unpreprocessed form.

Unlike the other solvers, GhostQ's input format is not clause-based (QDI-MACS) but it is circuit-based. To enable running GhostQ on the targeted instances, the solver was prepended with a reverse-engineering front-end. Since this front-end cannot handle bloqqer's output, GhostQ was run directly on the instances without preprocessing. The other solvers were run on the preprocessed instances (further preprocessing was disabled for QuBE7.2).

The relation between solving times and instances is presented by a cactus plot in Figure 2; number of solved instances per family are shown in Table 2; a comparison of RAReQS with other solvers is presented in Table 1. More detailed information can be found at http://sat.inesc-id.pt/~mikolas/sat12.

On the considered benchmarks, RAReQS solved the most instances, approximately 33% more than the second solver QuBE7.2. RAReQS also turned out

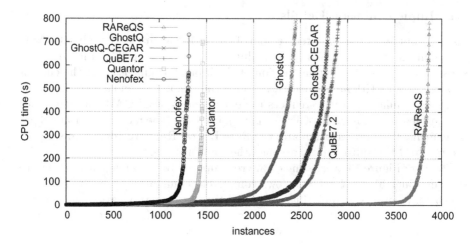

Fig. 2. Cactus plot of the overall results

Table 1. Number of instances solved by RAReQS but not by a competing solver, and *vice versa*

	GhostQ	GhostQ-CEGAR	QuBE7.2	Quantor	Nenofex
Only RAReQS	1661	1336	998	2436	2564
Only competitor	242	269	46	30	13

to be the best solver for most of the types of the considered instances. Table 1 further shows that for each of the other solvers, there is only a small portion of instances that the other solver can solve and RAReQS cannot. Out of the 801 instances when the solver was aborted, only 50 ran out of of memory.

In several families the addition of CEGAR learning to GhostQ worsened its performance. With the exception of Robots2D, however, the performance was worse only slightly. Overall, GhostQ benefited from the additional CEGAR learning and in particular for certain families. A family worth noting is irqlkeapclte, where no instances were solved by any of the solvers except for GhostQ-CEGAR.

The usefulness of CEGAR was in particular demonstrated by the families incrementer-encoder, conformant-planning, trafficlight-controller, Sorting-networks, and BMC where RAReQS solved significantly more instances than the existing solvers, and GhostQ-CEGAR improved significantly over GhostQ. Most notably, for incrementer-encoder (484) and RobotsD2 (700) only one instance was not solved by RAReQS, and for blackbox-01X-QBF (320) and trafficlight-controller (1459) RAReQS solved *all* instances.

6   Related Work

CEGAR has proven useful in number of areas, most notably in model checking [10] and SMT solving [24,2]; more recently it has been applied to handle

Table 2. Number of instances solved within 800 seconds by each solver. "Lev" indicates the number of quantifier blocks (min–max) in the family of instances, post-bloqqer.

Family	Lev.	RAReQS	GhostQ	GhostQ-Cegar	QuBE7.2	Quantor	Nenofex
trafficlight-ctlr (1459)	1–287	**1459**	806	1001	1092	955	863
RobotsD2 (700)	2–2	**699**	350	271	630	0	30
incrementer-encoder (484)	3–119	**483**	285	477	284	51	27
blackbox-01X-QBF (320)	2–21	**320**	138	126	224	3	4
Strat. Comp. (samp.) (150)	1–2	**107**	12	12	**107**	18	12
BMC (85)	1–3	**73**	26	48	37	65	64
Sorting-networks (84)	1–3	**72**	24	32	45	38	38
blackbox-design (27)	5–9	**27**	**27**	**27**	18	0	0
conformant-planning (23)	1–3	**17**	7	16	5	13	12
Adder (28)	3–7	**11**	2	2	4	5	9
Lin. Bitvec. Rank. Fun. (60)	3–3	**9**	0	0	0	0	0
Ling (8)	1–3	**8**	6	**8**	**8**	**8**	**8**
Blocks (7)	3–3	**7**	6	**7**	5	**7**	**7**
fpu (6)	1–3	**6**	0	0	**6**	**6**	**6**
RankingFunctions (4)	2–2	**3**	0	0	**3**	0	0
Logn (2)	3–3	**2**	**2**	**2**	**2**	**2**	**2**
Mneimneh-Sakallah (163)	1–3	110	**148**	141	89	3	22
tipfixpoint-sample (150)	1–3	26	**128**	127	22	5	6
terminator-sample (150)	2–2	98	**109**	103	9	25	0
tipdiam (121)	1–3	55	**99**	93	54	21	14
Scholl-Becker (55)	1–29	37	**43**	40	29	32	27
evader-pursuer (15)	5–19	10	**11**	8	**11**	2	2
uclid (3)	4–6	0	**2**	**2**	0	0	0
toilet-all (136)	1–1	134	133	131	131	**135**	133
Counter (58)	1–125	30	14	11	20	**33**	15
Debug (38)	3–5	3	0	0	0	**24**	6
circuits (63)	1–3	8	4	5	5	**9**	8
Gent-Rowley (205)	7–81	52	67	67	**70**	2	0
jmc-quant (+squaring) (20)	3–9	2	0	0	**6**	0	2
irqlkeapclte (45)	2–2	0	0	**44**	0	0	0
total (4669)		**3868**	2449	2801	2916	1462	1317

quantification in SMT [27,23]. Special cases of QBF, with limited number of quantifiers, have been targeted by CEGAR: computing vertex eccentricity [22], nonmonotonic reasoning [6,16], two-level quantification [17].

A SAT solver was used in [26] to guide DPLL search of a QBF solver and to cut out unsatisfiable branches. A notion of abstraction was also used in QBF preprocessing [21]. This notion, however, differs from the one used in RAReQS as it means treating universally quantified variables as existentially quantified.

An important feature of RAReQS is the expansion of the given QBF into a propositional formula, which is then solved by a SAT solver. This technique is used for preprocessing [7] but also several existing solvers tackle QBF solving in this way, most notably QUBOS [1], Quantor [4], and Nenofex [19]. Just as RAReQS uses multi-games, these solvers employ some various techniques to mitigate the blowup of the expansion (besides preprocessing). QUBOS uses *miniscoping*, Quantor *tree-like prefixes*, and Nenofex uses *negation normal form*. In these aspects, the solvers share similarities with RAReQS.

The way the expansion is carried out is significantly different. While the other solvers start the expansion from the innermost variables, RAReQS starts from the outermost variables. The main difference, however, lies in the *careful expansion* in RAReQS. In the aforementioned solvers, once a variable is scheduled to be expanded, both of its values are considered in the expansion. In contrast, in RAReQS only a particular assignment to a block of variables chosen in the expansion and the expansion is checked whether it is sufficient or not. This is an important factor for both time and space complexity. For large formulas, the traditional expansion-based solvers are bound to generate unwieldy formulas but the use of abstraction in RAReQS enables the solver to stop before this expansion is reached. This leads to generating easier formulas for the underlying SAT solver and dramatically mitigates the problems with memory blowup.

7   Conclusions and Future Work

Applying the CEGAR paradigm, this paper develops two novel techniques for QBF solving. The first technique is a CEGAR-driven solver RAReQS and the second an additional learning technique for DPLL solvers.

In its workings, RAReQS is close to expansion-based solvers (e.g. Quantor, Nenofex) but with the important difference that the expansion is done step-by-step, driven by counterexamples. Thus, the solver builds an abstraction of the given formula by constructing a partial expansion. The downside of this approach may be that if in the end a *full* expansion is needed, then RAReQS performs the same expansion as a traditional expansion-based solver but with the overhead of intermediate tests for whether or not the expansion is already sufficient.

However, the approach has important advantages. Whenever there is no winning move for the partial expansion, then there is no winning move for the given formula. This enables RAReQS to quickly stop for formulas with no winning moves. For formulas for which there *is* a winning move, RAReQS only needs to build a strong-enough partial expansion whose winning moves are also likely to be winning moves for the given formula. The experimental results demonstrate the ability of RAReQS to avoid the inherent memory blowup of expansion solvers, and, that careful expansion outperforms a traditional DPLL-based approach on a large number of practical instances.

We have shown that abstraction-refinement as used in RAReQS is also applicable within DPLL solvers as an additional learning mechanism. This provides a more powerful learning technique than standard clause/cube learning, although it requires more memory. Experimental evaluation indicates that this type of learning is indeed useful for DPLL-based solvers.

In the future we plan to further develop our DPLL solver so that it supports the full range of CEGAR learning exploited by RAReQS and to investigate how to fine-tune this learning in order to mitigate the speed penalty for the cases where the learning provides little information over the traditional learning. This can not only be done by better engineering of the solver but also devising schemata that disable the learning once deemed too costly. In RAReQS we plan to investigate how to integrate techniques used in other solvers. In particular, more aggressive preprocessing as used in Quantor and techniques for finding commonalities in formulas used in Nenofex and dependency detection [20].

References

1. Ayari, A., Basin, D.A.: QUBOS: Deciding Quantified Boolean Logic Using Propositional Satisfiability Solvers. In: Aagaard, M.D., O'Leary, J.W. (eds.) FMCAD 2002. LNCS, vol. 2517, pp. 187–201. Springer, Heidelberg (2002)
2. Barrett, C.W., Dill, D.L., Stump, A.: Checking Satisfiability of First-Order Formulas by Incremental Translation to SAT. In: Brinksma, E., Larsen, K.G. (eds.) CAV 2002. LNCS, vol. 2404, pp. 236–249. Springer, Heidelberg (2002)
3. Benedetti, M.: Evaluating QBFs via Symbolic Skolemization. In: Baader, F., Voronkov, A. (eds.) LPAR 2004. LNCS (LNAI), vol. 3452, pp. 285–300. Springer, Heidelberg (2005)
4. Biere, A.: Resolve and Expand. In: Hoos, H.H., Mitchell, D.G. (eds.) SAT 2004. LNCS, vol. 3542, pp. 59–70. Springer, Heidelberg (2005)
5. Biere, A., Lonsing, F., Seidl, M.: Blocked Clause Elimination for QBF. In: Bjørner, N., Sofronie-Stokkermans, V. (eds.) CADE 2011. LNCS, vol. 6803, pp. 101–115. Springer, Heidelberg (2011)
6. Browning, B., Remshagen, A.: A SAT-based solver for Q-ALL SAT. In: ACM Southeast Regional Conference (2006)
7. Bubeck, U., Kleine Büning, H.: Bounded Universal Expansion for Preprocessing QBF. In: Marques-Silva, J., Sakallah, K.A. (eds.) SAT 2007. LNCS, vol. 4501, pp. 244–257. Springer, Heidelberg (2007)
8. Büning, H.K., Bubeck, U.: Theory of quantified boolean formulas. In: Handbook of Satisfiability. IOS Press (2009)
9. Cadoli, M., Giovanardi, A., Schaerf, M.: An algorithm to evaluate Quantified Boolean Formulae. In: National Conference on Artificial Intelligence (1998)
10. Clarke, E.M., Grumberg, O., Jha, S., Lu, Y., Veith, H.: Counterexample-guided abstraction refinement for symbolic model checking. J. ACM 50(5) (2003)
11. Eén, N., Sörensson, N.: An Extensible SAT-solver. In: Giunchiglia, E., Tacchella, A. (eds.) SAT 2003. LNCS, vol. 2919, pp. 502–518. Springer, Heidelberg (2004)
12. Giunchiglia, E., Marin, P., Narizzano, M.: QuBE 7.0 system description. Journal on Satisfiability, Boolean Modeling and Computation 7 (2010)
13. Giunchiglia, E., Marin, P., Narizzano, M.: Reasoning with quantified boolean formulas. In: Handbook of Satisfiability. IOS Press (2009)
14. Giunchiglia, E., Marin, P., Narizzano, M.: sQueezeBF: An Effective Preprocessor for QBFs Based on Equivalence Reasoning. In: Strichman, O., Szeider, S. (eds.) SAT 2010. LNCS, vol. 6175, pp. 85–98. Springer, Heidelberg (2010)
15. Goultiaeva, A., Bacchus, F.: Exploiting QBF duality on a circuit representation. In: AAAI (2010)
16. Janota, M., Grigore, R., Marques-Silva, J.: Counterexample Guided Abstraction Refinement Algorithm for Propositional Circumscription. In: Janhunen, T., Niemelä, I. (eds.) JELIA 2010. LNCS, vol. 6341, pp. 195–207. Springer, Heidelberg (2010)
17. Janota, M., Marques-Silva, J.: Abstraction-Based Algorithm for 2QBF. In: Sakallah, K.A., Simon, L. (eds.) SAT 2011. LNCS, vol. 6695, pp. 230–244. Springer, Heidelberg (2011)
18. Klieber, W., Sapra, S., Gao, S., Clarke, E.M.: A Non-prenex, Non-clausal QBF Solver with Game-State Learning. In: Strichman, O., Szeider, S. (eds.) SAT 2010. LNCS, vol. 6175, pp. 128–142. Springer, Heidelberg (2010)
19. Lonsing, F., Biere, A.: Nenofex: Expanding NNF for QBF Solving. In: Kleine Büning, H., Zhao, X. (eds.) SAT 2008. LNCS, vol. 4996, pp. 196–210. Springer, Heidelberg (2008)

20. Lonsing, F., Biere, A.: DepQBF: A dependency-aware QBF solver. JSAT (2010)
21. Lonsing, F., Biere, A.: Failed Literal Detection for QBF. In: Sakallah, K.A., Simon, L. (eds.) SAT 2011. LNCS, vol. 6695, pp. 259–272. Springer, Heidelberg (2011)
22. Mneimneh, M., Sakallah, K.: Computing Vertex Eccentricity in Exponentially Large Graphs: QBF Formulation and Solution. In: Giunchiglia, E., Tacchella, A. (eds.) SAT 2003. LNCS, vol. 2919, pp. 411–425. Springer, Heidelberg (2004)
23. Monniaux, D.: Quantifier Elimination by Lazy Model Enumeration. In: Touili, T., Cook, B., Jackson, P. (eds.) CAV 2010. LNCS, vol. 6174, pp. 585–599. Springer, Heidelberg (2010)
24. de Moura, L., Rue, H., Sorea, M.: Lazy Theorem Proving for Bounded Model Checking over Infinite Domains. In: Voronkov, A. (ed.) CADE 2002. LNCS (LNAI), vol. 2392, pp. 438–455. Springer, Heidelberg (2002)
25. The Quantified Boolean Formulas satisfiability library, http://www.qbflib.org/
26. Samulowitz, H., Bacchus, F.: Using SAT in QBF. In: van Beek, P. (ed.) CP 2005. LNCS, vol. 3709, pp. 578–592. Springer, Heidelberg (2005)
27. Wintersteiger, C.M., Hamadi, Y., de Moura, L.: Efficiently Solving Quantified Bit-Vector Formulas. In: FMCAD (2010)
28. Zhang, L., Malik, S.: Conflict driven learning in a quantified Boolean satisfiability solver. In: ICCAD (2002)

Henkin Quantifiers and Boolean Formulae

Valeriy Balabanov, Hui-Ju Katherine Chiang,
and Jie-Hong Roland Jiang

Department of Electrical Engineering / Graduate Institute of Electronics Engineering
National Taiwan University, Taipei 10617, Taiwan
balabasik@gmail.com, b97901184@ntu.edu.tw, jhjiang@cc.ee.ntu.edu.tw

Abstract. Henkin quantifiers, when applied on Boolean formulae, yielding the so-called dependency quantified Boolean formulae (DQBF), offer succinct descriptive power specifying variable dependencies. Despite their natural applications to games with incomplete information, logic synthesis with constrained input dependencies, etc., DQBF remain a relatively unexplored subject however. This paper investigates their basic properties, including formula negation and complement, formula expansion, and prenex and non-prenex form conversions. In particular, the proposed DQBF formulation is established from a synthesis perspective concerned with Skolem-function models and Herbrand-function countermodels.

1 Introduction

Henkin quantifiers [9], also known as branching quantifiers among other names, generalize the standard quantification by admitting explicit specification, for an existentially quantified variable, about its dependence on universally quantified variables. In addition to mathematical logic, Henkin quantifiers appear not uncommonly in various contexts, such as natural languages [12], computation [2], game theory [11], and even system design. They permit the expression of (in)dependence in language, logic and computation, the modelling of incomplete information in noncooperative games, and the specification of partial dependencies among components in system design, which is the main motivation of this work.

When Henkin quantifiers are imposed on first-order logic (FOL) formulae, it results in the formulation of independence-friendly (IF) logic [10], which was shown to be more expressive than first-order logic and exhibit expressive power same as existential second-order logic. However one notable limitation among others of IF logic under the game-theoretical semantics is the violation of the law of the excluded middle, which states either a proposition or its negation is true. Therefore negating a formula can be problematic in terms of truth and falsity. From a game-theoretical viewpoint, it corresponds to undetermined games, where there are cases under which no player has a winning strategy. Moreover, the winning strategies of the semantic games do not exactly correspond to Skolem and Herbrand functions in synthesis applications although syntactic rules for negating IF logic formulae were suggested in [7,6].

A. Cimatti and R. Sebastiani (Eds.): SAT 2012, LNCS 7317, pp. 129–142, 2012.
© Springer-Verlag Berlin Heidelberg 2012

When Henkin quantifiers are imposed on Boolean formulae, it results in the so-called dependency quantified Boolean formulae (DQBF), whose satisfiability lies in the complexity class of NEXPTIME-complete [11]. In contrast to QBF, which is PSPACE-complete, DQBF offers more succinct descriptive power than QBF provided that NEXPTIME is not in PSPACE. By expansion on universally quantified variables, a DQBF can be converted to a QBF with the cost of exponential blow up in formula size [4,5].

This paper studies DQBF from a synthesis perspective. By distinguishing formula negation and complement, the connections between Skolem and Herbrand functions are established. While the law of the excluded middle holds for negation, it does not hold for complement. The special subset of the DQBF whose truth and falsity coincide with the existence of Skolem and Herbrand functions, respectively, is characterized. Our formulation provides a unified view on DQBF models and countermodels, which encompasses QBF as a special case. Some fundamental properties of DQBF are studied in Section 3, and the potential application of DQBF on Boolean relation determinization for input constrained function extraction is presented in Section 4. Discussions and conclusions are then given in Section 5 and Section 6, respectively.

2   Preliminaries

As conventional notation, a set is denoted with an upper-case letter, e.g., V; its elements are in lower-case letters, e.g., $v_i \in V$. The ordered version (i.e., vector) of $V = \{v_1, \ldots, v_n\}$ is denoted as $\boldsymbol{v} = (v_1, \ldots, v_n)$. Substituting a term t (respectively a vector of terms $\boldsymbol{t} = (t_1, \ldots, t_n)$) for some variable v (respectively a vector of variables $\boldsymbol{v} = (v_1, \ldots, v_n)$) in a formula ϕ is denoted as $\phi[v/t]$ (respectively $\phi[\boldsymbol{v}/\boldsymbol{t}]$ or $\phi[v_1/t_1, \ldots, v_n/t_n]$). A formula ϕ under some truth assignment α to its variables is denoted as $\phi|_\alpha$.

2.1   Quantified Boolean Formulae

A *quantified Boolean formula* (QBF) Φ over variables $V = \{v_1, \ldots, v_k\}$ in the *prenex form* is expressed as

$$Q_1 v_1 \cdots Q_k v_k . \phi,$$

where $Q_1 v_1 \cdots Q_k v_k$, with $Q_i \in \{\exists, \forall\}$, is called the *prefix*, denoted Φ_{pfx}, and ϕ, a quantifier-free formula over variables V, is called the *matrix*, denoted Φ_{mtx}. We call variable v_i in a QBF an *existential variable* if $Q_i = \exists$, or a *universal variable* if $Q_i = \forall$. A QBF is of *non-prenex form* if its quantifiers are scattered around the formula without a clean separation between the prefix and the matrix. Unless otherwise said, we shall assume that a QBF is in the prenex form and is totally quantified, i.e., with no free variables. As a notational convention, unless otherwise specified we shall let $X = \{x_1, \ldots, x_n\}$ be the set of universal variables and $Y = \{y_1, \ldots, y_m\}$ existential variables.

Given a QBF Φ over variables V, the *quantification level* $\ell : V \to \mathbb{N}$ of variable $v_i \in V$ is defined to be the number of quantifier alternations between $\exists$ and $\forall$ from the outermost variable to variable v_i in Φ_{pfx}, e.g., $\ell(v_1) = \ell(v_2) = 0$, $\ell(v_3) = 1$, and $\ell(v_4) = 2$ for QBF $\exists v_1 \exists v_2 \forall v_3 \exists v_4.\phi$.

Any QBF Φ over variables $X \cup Y$ can be converted into the well-known *Skolem normal form* [13]. In the conversion, every appearance of $y_i \in Y$ in Φ_{mtx} is replaced by its respective newly introduced function symbol F_{y_i} corresponding to the *Skolem function* of y_i, which refers only to the universal variables $x_j \in X$ with $\ell(x_j) < \ell(y_i)$. These function symbols are then existentially quantified before (on the left of) other universal quantifiers in Φ_{pfx}. This conversion, called *Skolemization*, is satisfiability preserving. Essentially a QBF Φ is true if and only if its Skolem functions exist such that substituting F_{y_i} for every appearance of y_i in Φ_{mtx} makes the new formula true (i.e., a tautology).

Example 1. Skolemizing the QBF

$$\forall x_1 \exists y_1 \forall x_2 \exists y_2.(x_1 \vee y_1 \vee \neg y_2)(\neg x_1 \vee \neg x_2 \vee y_2)$$

yields

$$\exists F_{y_1} \exists F_{y_2} \forall x_1 \forall x_2.(x_1 \vee F_{y_1} \vee \neg F_{y_2})(\neg x_1 \vee \neg x_2 \vee F_{y_2})$$

where F_{y_1} is a 1-ary function symbol referring to x_1, and F_{y_2} is a 2-ary function symbol referring to x_1 and x_2. Since the QBF is true, Skolem functions exist, for instance, $F_{y_1} = \neg x_1$ and $F_{y_2} = x_1 \wedge x_2$.

The notion of Skolem function has its dual form, known as the *Herbrand function*. For a QBF Φ, the Herbrand function F_{x_i} of variable $x_i \in X$ refers only to the existential variables $y_j \in Y$ with $\ell(y_j) < \ell(x_i)$. Essentially a QBF Φ is false if and only if Herbrand functions exist such that substituting F_{x_i} for every appearance of x_i in Φ_{mtx} makes the new formula false (i.e., unsatisfiable) [3].

2.2   Dependency Quantified Boolean Formulae

A *dependency quantified Boolean formula* (DQBF) generalizes a QBF in its allowance for explicit specification of variable dependencies. Syntactically, a DQBF Φ is the same as a QBF except that in Φ_{pfx} an existential variable y_i is annotated with the set $S_i \subseteq X$ of universal variables referred to by its Skolem function, denoted as $\exists y_{i(S_i)}$, or a universal variable x_j is annotated with the set $H_j \subseteq Y$ of existential variables referred to by its Herbrand function, denoted as $\forall x_{j(H_j)}$, where S_i and H_j are called the *support sets* of y_i and x_j, respectively. However, either the dependencies for the existential variables or the dependencies for the universal variables (but not both) shall be specified. That is, a prenex DQBF is in either of the two forms:

$$\text{S-form: } \forall x_1 \cdots \forall x_n \exists y_{1(S_1)} \cdots \exists y_{m(S_m)}.\phi, \text{ and} \tag{1}$$

$$\text{H-form: } \forall x_{1(H_1)} \cdots \forall x_{n(H_n)} \exists y_1 \cdots \exists y_m.\phi, \tag{2}$$

where ϕ is some quantifier-free formula. Note that *the syntactic quantification order in the prefix of a DQBF is immaterial and can be arbitrary* because the variable dependencies are explicitly specified by the support sets. Such quantification with dependency specification corresponds to the Henkin quantifier [9].[1]

By the above syntactic extension of DQBF, the inputs of the Skolem (respectively Herbrand) function of an existential (respectively universal) variable can be explicitly specified, rather than inferred from the syntactic quantification order. That is, an existential variable y_i (respectively universal variable x_j) can be specified to be semantically independent of a universal variable (respectively an existential variable) whose syntactic scope covers y_i (respectively x_j). Unlike the totally ordered set formed by those of a QBF, the support sets of the existential or universal variables of a DQBF form a partially ordered set in general. This extension makes DQBF more succinct in expressive power than QBF [11].

For the semantics, the truth and falsity of a DQBF can be interpreted by the existence of Skolem and Herbrand functions. Precisely an S-form (respectively H-form) DQBF is true (respectively false) if and only if its Skolem (respectively Herbrand) functions exist for the existential (respectively universal) variables while the specified variable dependencies are satisfied. Consequently, Skolem functions serve as the model to a true S-form DQBF whereas Herbrand functions serve as the countermodel to a false H-form DQBF.

Alternatively, the truth and falsity of a DQBF can be understood from a game-theoretic viewpoint. Essentially an S-form DQBF can be interpreted as a game played by one $\forall$-player and m noncooperative $\exists$-players [11]. An S-form DQBF is true if and only if the $\exists$-players have winning strategies, which correspond to the Skolem functions. Similarly an H-form DQBF can be interpreted as a game played by one $\exists$-player and n noncooperative $\forall$-players. An H-form DQBF is false if and only if the $\forall$-players have winning strategies, which correspond to the Herbrand functions.

As was shown in [4,5], an S-form DQBF Φ can be converted to a *logically equivalent*[2] QBF Φ' by formula expansion on the universal variables. Assume that universal variable x_1 is to be expanded in Formula (1) and $x_1 \notin S_1 \cup \cdots \cup S_{k-1}$ and $x_1 \in S_k \cap \cdots \cap S_m$. Then Formula (1) can be expanded to

$$\forall x_2 \cdots \forall x_n \exists y_{1(S_1)} \cdots \exists y_{k-1(S_{k-1})}$$

$$\exists y_{k(S_k[x_1/0])} \exists y_{k(S_k[x_1/1])} \cdots \exists y_{m(S_m[x_1/0])} \exists y_{m(S_m[x_1/1])} \cdot \phi|_{x_1=0} \wedge \phi|_{x_1=1},$$

where $S_i[x_1/v]$ denotes x_1 in S_i is substituted with logic value $v \in \{0,1\}$, and $\phi|_{x_1=v}$ denotes all appearances of x_1 in ϕ are substituted with v including those in the support sets of variables $y_{i(S_i)}$ for $i = k, \ldots, m$. (The subscript of the support set of an existential variable are helpful for tracing expansion paths. Different expansion paths of an existential variable result in distinct existential variables.) Such expansion can be repeatedly applied for every universal variables. The resultant formula after expanding all universal variables is a QBF,

---

[1] Henkin quantifiers in their original proposal [9] specify dependencies for existential variables only. The dependencies are extended in this paper to universal variables.

[2] That is, Φ and Φ' characterize the same set of Skolem-function models (by properly relating the existential variables of Φ' to those of Φ).

whose variables are all existentially quantified. As to be shown in Section 3.2, expansion can be applied also to H-form DQBF.

3   Properties of DQBF

3.1   Negation vs. Complement

In the light of QBF certification, where there always exists either a Skolem-function model or a Herbrand-function countermodel to a QBF, one intriguing question is whether or not the same property carries to DQBF as well. To answer this question, we distinguish two operators, *negation* (symbolized by "$\neg$") and *complement* (by "$\sim$"), for DQBF. Let Φ_S and Φ_H be Formulae (1) and (2), respectively. By negation, we define

$$\neg\Phi_S = \exists x_1 \cdots \exists x_n \forall y_{1(S_1)} \cdots \forall y_{m(S_m)}.\neg\phi \text{ and} \tag{3}$$

$$\neg\Phi_H = \exists x_{1(H_1)} \cdots \exists x_{n(H_n)} \forall y_1 \cdots \forall y_m.\neg\phi. \tag{4}$$

By complement, we define

$$\sim\Phi_S = \exists x_{1(H_1')} \cdots \exists x_{n(H_n')} \forall y_1 \cdots \forall y_m.\neg\phi \text{ and} \tag{5}$$

$$\sim\Phi_H = \exists x_1 \cdots \exists x_n \forall y_{1(S_1')} \cdots \forall y_{m(S_m')}.\neg\phi, \tag{6}$$

where $H_i' = \{y_j \in Y \mid x_i \notin S_j\}$ and $S_k' = \{x_l \in X \mid y_k \notin H_l\}$, which follow what we call the *complementary principle* of the Skolem and Herbrand support sets.

By the above definitions, one verifies that $\neg\neg\Phi = \Phi$, $\sim\sim\Phi = \Phi$, and $\neg\sim\Phi = \sim\neg\Phi$. Moreover, because the Skolem functions of Φ_S, if they exist, are exactly the Herbrand functions of $\neg\Phi_S$, and the Herbrand functions of Φ_H, if they exist, are exactly the Skolem functions of $\neg\Phi_H$, the following proposition holds.

Proposition 1. *DQBF under the negation operation obey the* law of the excluded middle. *That is, a DQBF is true if and only if its negation is false.*

Since any DQBF can be converted to a logically equivalent QBF by formula expansion, it also explains that the law of the excluded middle should hold under negation for DQBF as it holds for QBF.

A remaining question is whether or not the complement of DQBF obeys the law of the excluded middle. The answer to this question is in general negative as we show below. Based on the existence of Skolem and Herbrand functions, we classify DQBF into four categories:

$\mathcal{C}_S = \{\Phi \mid \Phi \text{ is true and } \sim\Phi \text{ is false}\}$,

$\mathcal{C}_H = \{\Phi \mid \Phi \text{ is false and } \sim\Phi \text{ is true}\}$,

$\mathcal{C}_{SH} = \{\Phi \mid \Phi \text{ and } \sim\Phi \text{ are true for S-form } \Phi, \text{ or false for H-form } \Phi\}$, and

$\mathcal{C}_\emptyset = \{\Phi \mid \Phi \text{ and } \sim\Phi \text{ are false for S-form } \Phi, \text{ or true for H-form } \Phi\}$.

Note that if $\Phi \in \mathcal{C}_S$, then $\sim\Phi \in \mathcal{C}_H$; if $\Phi \in \mathcal{C}_H$, then $\sim\Phi \in \mathcal{C}_S$; if $\Phi \in \mathcal{C}_{SH}$, then $\sim\Phi \in \mathcal{C}_{SH}$; if $\Phi \in \mathcal{C}_\emptyset$, then $\sim\Phi \in \mathcal{C}_\emptyset$.

Under the above DQBF partition, observe that the complement of DQBF obeys the law of the excluded middle if and only if $\mathcal{C}_{SH}$ and $\mathcal{C}_\emptyset$ are empty. In fact, as to be shown, for any QBF Φ, $\Phi \notin \mathcal{C}_{SH} \cup \mathcal{C}_\emptyset$. As a consequence, the complement and negation operations for any QBF Φ coincide, and thus $\neg{\sim}\Phi = \Phi$. However, for general DQBF, $\mathcal{C}_{SH}$ and $\mathcal{C}_\emptyset$ are not empty as the following two examples show.

Example 2. Consider the DQBF

$$\Phi = \forall x_1 \forall x_2 \exists y_{1(x_1)} \exists y_{2(x_2)}.((y_1 \oplus x_1) \wedge (y_2 \overline{\oplus} x_2)) \vee ((y_2 \oplus x_2) \wedge (y_1 \overline{\oplus} x_1)),$$

where symbols "$\oplus$" and "$\overline{\oplus}$" stand for Boolean XOR and XNOR operators, respectively. Φ has Skolem functions, e.g., x_1 and $\neg x_2$ for existential variables y_1 and y_2, respectively, and $\neg{\sim}\Phi$ has Herbrand functions, e.g., y_2 and y_1 for universal variables x_1 for x_2, respectively. That is, $\Phi \in \mathcal{C}_{SH}$.

Example 3. Consider the DQBF

$$\Phi = \forall x_1 \forall x_2 \exists y_{1(x_1)} \exists y_{2(x_2)}.(y_1 \vee \neg x_1 \vee x_2) \wedge (y_2 \vee x_1 \vee \neg x_2) \wedge (\neg y_1 \vee \neg y_2 \vee \neg x_1 \vee \neg x_2).$$

It can be verified that Φ has no Skolem functions, and $\neg{\sim}\Phi$ has no Herbrand functions. That is, $\Phi \in \mathcal{C}_\emptyset$.

By these two examples, the following proposition can be concluded.

Proposition 2. *DQBF under the complement operation do not obey the law of the excluded middle. That is, the truth (falsity) of a DQBF cannot be decided from the falsity (truth) of its complement.*

Nevertheless, if a DQBF $\Phi \notin \mathcal{C}_{SH} \cup \mathcal{C}_\emptyset$, then its truth and falsity can surely be certified by a Skolem-function model and a Herbrand-function countermodel, respectively.[3] That is, excluding $\Phi \in \mathcal{C}_{SH} \cup \mathcal{C}_\emptyset$, DQBF under the complement operation obeys the law of the excluded middle.

A sufficient condition for a DQBF not in $\mathcal{C}_{SH}$ (equivalently, a necessary condition for a DQBF in $\mathcal{C}_{SH}$) is presented in Theorem 1.

Theorem 1. *Let ϕ be a quantifier-free formula over variables $X \cup Y$, let $\Phi_1 = \forall x_1 \cdots \forall x_n \exists y_{1(S_1)} \cdots \exists y_{m(S_m)}.\phi$ and $\Phi_2 = \forall x_{1(H_1)} \cdots \forall x_{n(H_n)} \exists y_1 \cdots \exists y_m.\phi$ with $H_i = \{y_j \in Y \mid x_i \notin S_{y_j}\}$. Then there exist Skolem functions $\boldsymbol{f} = (f_1, \ldots, f_m)$ for Φ_1 and Herbrand functions $\boldsymbol{g} = (g_1, \ldots, g_n)$ for Φ_2 only if the composite function vector $\boldsymbol{g} \circ \boldsymbol{f}$ admits no fixed-point, that is, there exists no truth assignment α to variables $\boldsymbol{x} = (x_1, \ldots, x_n)$ such that $\alpha = \boldsymbol{g}(\boldsymbol{f}(\alpha))$.*

Proof. Since Φ_1 is true and has Skolem functions $\boldsymbol{f}$, formula $\phi[\boldsymbol{y}/\boldsymbol{f}]$ must be a tautology. On the other hand, since Φ_2 is false and has Herbrand functions $\boldsymbol{g}$, formula $\phi[\boldsymbol{x}/\boldsymbol{g}]$ must be unsatisfiable. Suppose that the fixed-point condition $\alpha = \boldsymbol{g}(\boldsymbol{f}(\alpha))$ holds under some truth assignment α to $\boldsymbol{x}$. Then $\phi[\boldsymbol{y}/\boldsymbol{f}]|_\alpha = \phi[\boldsymbol{x}/\boldsymbol{g}]|_\beta$ for $\beta = \boldsymbol{f}(\alpha)$ being the truth assignment to $\boldsymbol{y}$. It contradicts with the fact that $\phi[\boldsymbol{y}/\boldsymbol{f}]$ must be a tautology and $\phi[\boldsymbol{x}/\boldsymbol{g}]$ must be unsatisfiable. ∎

---

[3] In general a false S-form DQBF has no Herbrand-function countermodel, and a true H-form DQBF has no Skolem-function model.

The following corollary shows that $\Phi \notin \mathcal{C}_{SH}$ for any QBF Φ.

Corollary 1. *For any QBF Φ, the Skolem-function model and Herbrand-function countermodel cannot co-exist.*

Proof. If a QBF is false, its Skolem-function model does not exist and the corollary trivially holds. Without loss of generality, assume a true QBF is of the form $\Phi = \exists y_1 \forall x_1 \cdots \exists y_n \forall x_n.\phi$. Let $\{y_1 = f_1(), \ldots, y_n = f_n(x_1, \ldots, x_{n-1})\}$ be a model for Φ. Further by contradiction assume there exist a countermodel $\{x_1 = g_1(y_1), \ldots, x_n = g_n(y_1, \ldots, y_n)\}$. So the fixed-point condition is $\{x_1 = g_1(f_1()), \ldots, x_n = g_n(f_1(), \ldots, f_n(x_1, \ldots, x_{n-1}))\}$. Since no cyclic dependency presents in the fixed-point equations, the set of equations always has a solution. In other words, due to the complete ordering of the prefix of a QBF, a fixed-point exists. By Theorem 1, the Skolem-function model and Herbrand-function countermodel cannot co-exist. ■

A sufficient condition for a DQBF not in $\mathcal{C}_\emptyset$ can be characterized by procedure *HerbrandConstruct* as shown in Figure 1. Note that although the algorithm computes Herbrand functions of $\neg\sim\Phi_S$ for a false S-form DQBF Φ_S, it can be used to compute Skolem functions of $\neg\sim\Phi_H$ for a true H-form DQBF Φ_H by taking as input the negation of the formula.

Given a false S-form DQBF Φ with $n \geq 1$ universal variables, procedure *HerbrandConstruct* in line 1 collects the support set H_n for universal variable x_n. Let $H_n = \{y_{a_1}, \ldots, y_{a_k}\}$ and the rest be $\{y_{a_{k+1}}, \ldots, y_{a_m}\}$. It then recursively constructs the Herbrand functions of the formula expanded on x_n until $n = 1$. By formula expansion on x_n in line 3, variables $\{y_{a_{k+1}}, \ldots, y_{a_m}\}$, which depend on x_n, are instantiated in $\Phi_{\exp}$ into two copies, say, $\{y'_{a_{k+1}}, y''_{a_{k+1}}, \ldots, y'_{a_m}, y''_{a_m}\}$. Then the *VariableMerge* step in line 6 lets $g_i = g_i^\dagger[y'_{a_{k+1}}/y_{a_{k+1}}, y''_{a_{k+1}}/y_{a_{k+1}}, \ldots, y'_{a_m}/y_{a_m}, y''_{a_m}/y_{a_m}]$.[4] In constructing the Herbrand function g_n of x_n, each assignment α to H_n is examined. Since Herbrand function aims to falsity ϕ, the value of $g_n(\alpha)$ is set to the x_n value that makes $\phi[x_1/g_1, \ldots, x_{n-1}/g_{n-1}]|_\alpha$ unsatisfiable.

Theorem 2. *Given a false S-form DQBF Φ, algorithm HerbrandConstruct returns either nothing or correct Herbrand functions, which falsify $\neg\sim\Phi$.*

Proof. Observe first that the functions returned by the algorithm satisfy the support-set dependencies for the universal variables. It remains to show that $\phi[x_1/g_1, \ldots, x_n/g_n]$ is unsatisfiable. By contradiction, suppose there exists an assignment β to the existential variables Y such that $\phi[x_1/g_1, \ldots, x_n/g_n]|_\beta = 1$. Let $v \in \{0, 1\}$ be the value of $g_n|_\alpha$ for α being the projection of β on $H_n \subseteq Y$. Then $\phi[x_1/g_1, \ldots, x_{n-1}/g_{n-1}, x_n/v]|_\beta = 1$. However it contradicts with the way

---

[4] The method to perform *VariableMerge* in line 6 is not unique. In theory, as long as no violation of variable dependencies is incurred, any substitution can be applied. In practice, however the choice of substitution may affect the strength of the algorithm *HerbrandConstruct* in terms of the likelihood of returning (non-empty) Herbrand functions.

HerbrandConstruct

    input: a false S-form DQBF $\Phi = \forall x_1 \cdots \forall x_n \exists y_1{}_{(S_1)} \cdots \exists y_m{}_{(S_m)}.\phi$, and
          the number n of universal variables
    output: Herbrand-functions $(g_1, \cdots, g_n)$ of $\neg \sim \Phi$
01   $H_n := \{y_i \in Y \mid x_n \notin S_i\}$
02   **if** $(n > 1)$
03      $\Phi_{\exp} := FormulaExpand(\Phi, x_n)$;
04      $g^\dagger := HerbrandConstruct(\Phi_{\exp}, n-1)$;
05   **if** $(g^\dagger = \emptyset)$ **return** $\emptyset$;
06      $g := VariableMerge(g^\dagger)$;
07   **for each** assignment α to H_n
08   **if** $(\phi[x_1/g_1, \ldots, x_{n-1}/g_{n-1}]|_{\alpha,x_n=0}$ is unsatisfiable$)$
09            $g_n(\alpha) = 0$;
10   **if** $(\phi[x_1/g_1, \ldots, x_{n-1}/g_{n-1}]|_{\alpha,x_n=1}$ is unsatisfiable$)$
11            $g_n(\alpha) = 1$;
12   **else return** $\emptyset$;
13   **else**
14   **for each** assignment α to H_n
15   **if** $(\phi|_{\alpha,x_n=0}$ is unsatisfiable$)$
16            $g_n(\alpha) = 0$;
17   **if** $(\phi|_{\alpha,x_n=1}$ is unsatisfiable$)$
18            $g_n(\alpha) = 1$;
19   **else return** $\emptyset$;
20   **return** $(g_1, \ldots, g_n)$;
   end

Fig. 1. Algorithm: Herbrand-function Construction

how $g_n|_\alpha$ is constructed. Hence the returned Herbrand functions $(g_1, \ldots, g_n)$, if they are not empty, are indeed correct Herbrand functions. ∎

The following corollary shows that $\Phi \notin \mathcal{C}_\emptyset$ for any QBF Φ.

Corollary 2. *If Φ is a false QBF and its universal variables $x_1, \ldots, x_n$ follow the QBF's prefix order, algorithm HerbrandConstruct always returns non-empty Herbrand functions.*

Proof. We prove the statement by induction on the number of universal variables. For the base case, without loss of generality consider QBF $\Phi = \exists y_1 \cdots \exists y_k \forall x \exists y_{k+1} \cdots \exists y_m.\phi$. After line 1, *HerbrandConstruct* enters line 14. Since the QBF is false and has only one universal variable x, expanding on x yields a purely existentially quantified unsatisfiable formula: $\exists y_1 \cdots \exists y_k (\exists y'_{k+1} \cdots \exists y'_m.\phi|_{x=0} \land \exists y''_{k+1} \cdots \exists y''_m.\phi|_{x=1})$. By its unsatisfiability, for every assignment α to $y_1, \cdots, y_k$, formula $\exists y'_{k+1} \cdots \exists y'_m.\phi|_{\alpha,x=0} \land \exists y''_{k+1} \cdots \exists y''_m.\phi|_{\alpha,x=1}$ must be unsatisfiable. Since $\exists y'_{k+1} \cdots \exists y'_m.\phi|_{\alpha,x=0}$ and $\exists y''_{k+1} \cdots \exists y''_m.\phi|_{\alpha,x=1}$ share no common variables, at least one of them must be unsatisfiable. Hence the procedure returns a non-empty Herbrand function.

For the inductive step, assume the previous recursive calls for $k = 1, \ldots, n-1$ of *HerbrandConstruct* do not return $\emptyset$. We show that the current call for $k = n$ cannot return $\emptyset$. Expanding Φ on x_n yields $\Phi_{\exp} = \forall x_1 \cdots \forall x_{n-1} \exists y_{1(S_1)} \cdots \exists y_{k(S_k)}$ $(\exists y'_{k+1(S_{k+1})} \cdots \exists y'_{m(S_m)}.\phi|_{x_n=0} \wedge \exists y''_{k+1(S_{k+1})} \cdots \exists y''_{m(S_m)}.\phi|_{x_n=1})$. By the inductive hypothesis, functions $g_1^{\dagger}, \cdots, g_{n-1}^{\dagger}$ are returned. Moreover, $g_i^{\dagger}$ for any $i = 1, \ldots, n-1$ is independent of y'_j and y''_j for $j = k+1, \ldots, m$. So we construct $g_i = g_i^{\dagger}$. Since $g_1, \ldots, g_{n-1}$ have been constructed in a way such that $\exists y_1 \cdots \exists y_k (\exists y'_{k+1} \cdots \exists y'_m.\phi[x_1/g_1, \cdots, x_{n-1}/g_{n-1}]|_{x_n=0} \wedge \exists y''_{k+1} \cdots \exists y''_m.\phi[x_1/g_1, \cdots, x_{n-1}/g_{n-1}]|_{x_n=1})$ is unsatisfiable, under every assignment α to $y_1, \cdots, y_k$ formula $(\exists y'_{k+1} \cdots \exists y'_m.\phi[x_1/g_1, \cdots, x_{n-1}/g_{n-1}]|_{x_n=0} \wedge \exists y''_{k+1} \cdots \exists y''_m.\phi[x_1/g_1, \cdots, x_{n-1}/g_{n-1}]|_{x_n=1})$ is unsatisfiable. Moreover, since $\exists y'_{k+1} \cdots \exists y'_m. \phi[x_1/g_1, \cdots, x_{n-1}/g_{n-1}]|_{x_n=0}$ and $\exists y''_{k+1} \cdots \exists y''_m. \phi[x_1/g_1, \cdots, x_{n-1}/g_{n-1}]|_{x_n=1}$ do not share any variables, at least one of them must be unsatisfiable. So g_n is returned. ■

Note that the above proof does not explicitly perform the substitution $g_i = g_i^{\dagger}[y'_{a_{k+1}}/y_{a_{k+1}}, y''_{a_{k+1}}/y_{a_{k+1}}, \ldots, y'_{a_m}/y_{a_m}, y''_{a_m}/y_{a_m}]$ in *VariableMerge* because all g_i in fact do not depend on primed or double-primed variables in the QBF case.

Procedure *HerbrandConstruct* is useful in deriving Herbrand functions not only for QBF but also for general DQBF as the following example suggests.

Example 4. Consider the DQBF $\Phi = \forall x_1 \forall x_2 \exists y_{1(x_1)} \exists y_{2(x_2)}.\phi$ with $\phi = (y_1 \vee x_2) \wedge (y_2 \vee x_1) \wedge (\neg y_1 \vee \neg y_2 \vee \neg x_1 \vee \neg x_2)$. *HerbrandConstruct*$(\Phi, 2)$ computes Herbrand functions for $\neg \sim \Phi$ with the following steps. Expanding Φ on x_2 yields $\Phi_{\exp} = \forall x_1 \exists y_{1(x_1)} \exists y'_2 \exists y''_2.\phi|_{x_2=0} \wedge \phi|_{x_2=1}$ with $\phi|_{x_2=0} = (y_1) \wedge (y'_2 \vee x_1)$ and $\phi|_{x_2=1} = (y''_2 \vee x_1) \wedge (\neg y_1 \vee \neg y''_2 \vee \neg x_1)$. The recursive call to *HerbrandConstruct*$(\Phi_{\exp}, 1)$ determines the value of function $g_1^{\dagger}(y'_2, y''_2)$ under every assignment α to (y'_2, y''_2). In particular, $g_1^{\dagger}(0,0) = 0$ due to $\phi_{\exp} = (y_1) \wedge (x_1) \wedge (x_1)$; $g_1^{\dagger}(0,1) = 0$ (or 1) due to $\phi_{\exp} = (y_1) \wedge (x_1) \wedge (\neg y_1 \vee \neg x_1)$; $g_1^{\dagger}(1,0) = 0$ due to $\phi_{\exp} = (y_1) \wedge (x_1)$; $g_1^{\dagger}(1,1) = 1$ due to $\phi_{\exp} = (y_1) \wedge (\neg y_1 \vee \neg x_1)$. So $g_1^{\dagger}(y'_2, y''_2) = y'_2 y''_2$ (or y''_2), and $g_1(y_2) = g_1^{\dagger}[y'_2/y_2, y''_2/y_2] = y_2$.

Returning to *HerbrandConstruct*$(\Phi, 2)$, we have $\phi[x_1/g_1] = (y_1 \vee x_2) \wedge (y_2) \wedge (\neg y_1 \vee \neg y_2 \vee \neg x_2)$. The value of function g_2 for each assignment α to y_1 can be determined with $g_2(0) = 0$ due to $\phi[x_1/g_1]|_{y_1=0} = (x_2) \wedge (y_2)$ and $g_2(1) = 1$ due to $\phi[x_1/g_1]|_{y_1=1} = (y_2) \wedge (\neg y_2 \vee \neg x_2)$. That is, $g_2(y_1) = y_1$. The computed g_1 and g_2 indeed make $\phi[x_1/g_1, x_2/g_2] = (y_1) \wedge (y_2) \wedge (\neg y_1 \vee \neg y_2)$ unsatisfiable.

Since the DQBF subset $\mathcal{C}_S \cup \mathcal{C}_H$ obeys the law of the excluded middle under the complement operation, Theorems 1 and 2 provide a tool to test whether a DQBF Φ can be equivalently expressed as $\neg \sim \Phi$, that is, whether a DQBF has either a Skolem-function model or a Herbrand-function countermodel. Figure 2 shows the four DQBF categories and the regions characterized by Theorems 1 and 2.

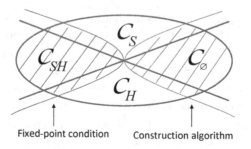

↑   ↑

Fixed-point condition   Construction algorithm

Fig. 2. Four DQBF categories and regions characterized by Theorems 1 and 2

3.2   Formula Expansion on Existential Variables

Formula expansion on existential variables for DQBF can be achieved by negation using De Morgan's law and expansion on universal variables. It leads to the following expansion rule, which is dual to expanding universal variables.

Proposition 3. *Given a DQBF* $\forall x_{1(H_1)} \cdots \forall x_{n(H_n)} \exists y_1 \cdots \exists y_m.\phi$, *assume without loss of generality that* y_1 *is to be expanded with* $y_1 \notin H_1 \cup \cdots \cup H_{k-1}$ *and* $y_1 \in H_k \cap \cdots \cap H_n$. *The formula can be expanded to*

$$\forall x_{1(H_1)} \cdots \forall x_{k-1(H_{k-1})} \forall x_{k(H_k[y_1/0])} \forall x_{k(H_k[y_1/1])} \cdots \forall x_{n(H_n[y_1/0])} \forall x_{n(H_n[y_1/1])}$$
$$\exists y_2 \cdots \exists y_m.\phi|_{y_1=0} \vee \phi|_{y_1=1},$$

where $H_i[y_1/v]$ *denotes* y_1 *in* H_i *is substituted with logic value* $v \in \{0, 1\}$, *and* $\phi|_{y_1=v}$ *denotes all appearances of* y_1 *in* ϕ *are substituted with* v *including those in the support sets of variables* $x_{i(H_i)}$ *for* $i = k, \ldots, n$.

Such expansion can be repeatedly applied for every existential variables. The resultant formula after expanding all existential variables is a QBF. Note that, when Skolem functions are concerned rather than Herbrand functions, the support sets of the existential variables should be listed and can be obtained from H_i by the aforementioned complementary principle.

Example 5. Consider expanding variable y_1 of DQBF

$$\Phi = \forall x_{1(y_1)} \forall x_{2(y_2)} \forall x_{3(y_3)} \exists y_1 \exists y_2 \exists y_3.\phi.$$

By De Morgan's law and expansion on a universal variable, we obtain

$$\neg\neg\Phi = \neg \exists x_{1(y_1)} \exists x_{2(y_2)} \exists x_{3(y_3)} \forall y_1 \forall y_2 \forall y_3.\neg\phi$$
$$= \neg \exists x_{1(0)} \exists x_{1(1)} \exists x_{2(y_2)} \exists x_{3(y_3)} \forall y_2 \forall y_3.\neg\phi|_{y_1=0} \wedge \neg\phi|_{y_1=1}$$
$$= \forall x_{1(0)} \forall x_{1(1)} \forall x_{2(y_2)} \forall x_{3(y_3)} \exists y_2 \exists y_3.\phi|_{y_1=0} \vee \phi|_{y_1=1}.$$

3.3   Prenex and Non-prenex Conversion

This section studies some syntactic rules that allow localization of quantifiers to sub-formulae. We focus on the truth (namely the Skolem-function model), while similar results can be concluded by duality for the falsity (namely the Herbrand-function countermodel), of a formula.

The following proposition shows the localization of existential quantifiers to the sub-formulas of a disjunction.

Proposition 4. *The DQBF*

$$\forall \boldsymbol{x} \exists y_{1(S_1)} \cdots \exists y_{m(S_m)} . \phi_A \vee \phi_B,$$

where $\forall \boldsymbol{x}$ denotes $\forall x_1 \cdots \forall x_n$, sub-formula ϕ_A (respectively ϕ_B) refers to variables $X_A \subseteq X$ and $Y_A \subseteq Y$ (respectively $X_B \subseteq X$ and $Y_B \subseteq Y$), is logically equivalent to

$$\forall \boldsymbol{x_c} \left(\forall \boldsymbol{x_a} \exists y_{a_1(S_{a_1} \cap X_A)} \cdots \exists y_{a_p(S_{a_p} \cap X_A)} \phi_A \vee \forall \boldsymbol{x_b} \exists y_{b_1(S_{b_1} \cap X_B)} \cdots \exists y_{b_q(S_{b_q} \cap X_B)} \phi_B \right),$$

where variables $\boldsymbol{x_c}$ are in $X_A \cap X_B$, variables $\boldsymbol{x_a}$ are in $X_A \backslash X_B$, variables $\boldsymbol{x_b}$ are in $X_B \backslash X_A$, $y_{a_i} \in Y_A$, and $y_{b_j} \in Y_B$.

Proof. A model to the former expression consists of every truth assignment to X and the induced Skolem function valuation to Y. Since every such combined assignment to $X \cup Y$ either satisfies ϕ_A or ϕ_B, by collecting those satisfying ϕ_A (respectively ϕ_B) and projecting to variables $X_A \cup Y_A$ (respectively $X_B \cup Y_B$) the model (i.e., the Skolem functions for $\boldsymbol{y_a}$ and $\boldsymbol{y_b}$) to the latter expression can be constructed. (Note that, for a quantifier $\exists y_i$ splitting into two, one for ϕ_A and the other for ϕ_B, in the latter expression, they are considered distinct and have their own Skolem functions.)

In addition, the Skolem functions for $\forall \boldsymbol{x_a} \exists y_{a_1(S_{a_1} \cap X_A)} \cdots \exists y_{a_p(S_{a_p} \cap X_A)} \phi_A|\alpha$ and those for $\forall \boldsymbol{x_b} \exists y_{b_1(S_{b_1} \cap X_B)} \cdots \exists y_{b_q(S_{b_q} \cap X_B)} \phi_B|\alpha$ under every assignment α to $\boldsymbol{x_c}$ can be collected and combined to form a model for the former expression. In particular the respective Skolem functions $f_{a_j}|\alpha$ and $f_{b_k}|\alpha$ under α for y_{a_j} and y_{b_k} originating from the same quantifier y_i in the former expression are merged into one Skolem function $f_i = \bigvee_\alpha \left(\chi_\alpha (f_{a_j}|\alpha \vee f_{b_k}|\alpha) \right)$, where χ_α denotes the characteristic function of α, e.g., $\chi_\alpha = x_1 x_2 \neg x_3$ for $\alpha = (x_1 = 1, x_2 = 1, x_3 = 0)$. ∎

Example 6. Consider the QBF

$$\Phi = \forall x_1 \exists y_1 \forall x_2 \exists y_2 \forall x_3 \exists y_3 . \phi_A \vee \phi_B$$

with ϕ_A refers to variables x_1, x_2, y_1, y_2 and ϕ_B refers to x_2, x_3, y_2, y_3. It has the following equivalent DQBF expressions.

$$\Phi = \forall x_1 \forall x_2 \forall x_3 \exists y_{1(x_1)} \exists y_{2(x_1, x_2)} \exists y_{3(x_1, x_2, x_3)} . \phi_A \vee \phi_B$$

$$= \forall x_1 \forall x_2 \forall x_3 \left(\exists y_{1(x_1)} \exists y_{2(x_1, x_2)} \phi_A \vee \exists y_{2(x_2)} \exists y_{3(x_2, x_3)} \phi_B \right)$$

$$= \forall x_2 \left(\forall x_1 \exists y_{1(x_1)} \exists y_{2(x_1, x_2)} \phi_A \vee \forall x_3 \exists y_{2(x_2)} \exists y_{3(x_2, x_3)} \phi_B \right)$$

In contrast, conventionally the quantifiers of the QBF can only be localized to

$$\forall x_1 \exists y_1 \forall x_2 \exists y_2 \left(\phi_A \vee \forall x_3 \exists y_3 \phi_B \right).$$

The following proposition shows the localization of universal quantifiers to a sub-formula of a conjunction.

Proposition 5. *The DQBF*

$$\forall \boldsymbol{x} \exists y_{1(S_1)} \cdots \exists y_{k(S_k)} . \phi_A \wedge \phi_B,$$

where $\forall \boldsymbol{x}$ denotes $\forall x_1 \cdots \forall x_n$, sub-formula ϕ_A (respectively ϕ_B) refers to variables $X_A \subseteq X$ and $Y_A \subseteq Y$ (respectively $X_B \subseteq X$ and $Y_B \subseteq Y$), is logically equivalent to

$$\forall \boldsymbol{x} \exists y_{2(S_2)} \cdots \exists y_{k(S_k)} . \left(\exists y_{1(S_1 \cap X_A)} \phi_A \right) \wedge \phi_B,$$

for $y_1 \notin Y_B$.

Proof. The proposition follows from the fact that the Skolem function of y_1 is purely constrained by ϕ_A only, and is the same for both expressions. Note that the former formula is equivalent to $\forall \boldsymbol{x} \exists y_{1(S_1 \cap X_A)} \cdots \exists y_{k(S_k)} . \phi_A \wedge \phi_B$. ∎

Essentially DQBF allow tighter localization of quantifier scopes than QBF. On the other hand, converting a non-prenex QBF to the prenex form may incur the size increase of support sets of existential variables due to the linear (or complete order) structure of the prefix. With DQBF, such spurious increase can be eliminated.

4   Applications

Although to date there is no DQBF solver, we note that the framework provided by QBF solver sKizzo [1], which is based on Skolemization, can be easily extended to DQBF solving. A natural application of DQBF is Boolean relation determinization [8,3] in logic circuit synthesis. Consider a Boolean relation $R(\boldsymbol{x}, \boldsymbol{y})$ as a characteristic function (quantifier-free Boolean formula) specifying the input and output behavior of some (possibly non-deterministic) combinational system with inputs X and outputs Y. To realize the outputs of the system, the Skolem functions of the QBF

$$\forall \boldsymbol{x} \exists \boldsymbol{y} . R(\boldsymbol{x}, \boldsymbol{y})$$

is to be solved. Often the inputs of some output y_i need to be restricted to depend only on a subset of X. This restriction can be naturally described by DQBF. Therefore DQBF can be exploited for topologically constrained logic synthesis [14].

5 Discussions

IF logic [10] with the game-theoretical semantics is known to violate the law of the excluded middle. A simple example is the IF logic formula $\forall x \exists y_{/x}.(x = y)$ for $x, y \in \{0, 1\}$, where $y_{/x}$ indicates the independence of y on x [7]. It assumes that not only y is independent of x, but also is x independent of y. That is, it is equivalent to $\forall x_{()} \exists y_{()}.(x = y)$ in our dependency notation. In a game-theoretic viewpoint, neither the $\exists$-player nor the $\forall$-player has a winning strategy. Therefore this formula is neither true nor false, and has no equivalent DQBF since any DQBF can always be expanded into a QBF, whose truth and falsity can be fully determined.

On the other hand, the game-theoretical semantics of IF logic, when extended to DQBF, does not provide a fully meaningful approach to synthesizing Skolem and Herbrand functions. Unlike the unimportance of the syntactic quantification order in our formulation, the semantic game of IF logic should be played with respect to the prefix order. Since different orders correspond to different games, the semantics is not directly useful in our considered synthesis application.

Henkin quantifiers in their original form [9] specified only the dependencies of existential variables on universal variables. Such restricted dependencies were assumed in early IF logic [10] research. As was argued in [7], the dependency of universal variables on existential variables are necessary to accomplish a symmetric treatment on the falsity, in addition to truth, of an IF logic formula. With such extension, IF logic formulae can be closed under negation. However, how the dependencies of existential variables and universal variables relate to each other was not studied. The essential notion of Herbrand functions was missing. In contrast, our formulation on DQBF treats Skolem and Herbrand functions on an equal footing. Unlike [7], we restrict a formula to be of either S-form or H-form, rather than simultaneous specification of dependencies for existential and universal variables. This restriction makes the synthesis of Skolem and Herbrand functions for DQBF more natural.

Prior work [11,5] assumed DQBF are of S-form only. In [11], a DQBF was formulated as a game played by a $\forall$-player and multiple noncooperative $\exists$-players. This game formulation is fundamentally different from that of IF-logic. The winning strategies, if they exist, of the $\exists$-players correspond to the Skolem functions of the DQBF. This game interpretation can be naturally extended to H-form DQBF.

6 Conclusions

The syntax and semantics of DQBF presented in this paper made DQBF a natural extension of QBF from a certification viewpoint. Basic DQBF properties, including formula negation, complement, expansion, and prenex and non-prenex form conversion, were shown. Our formulation is adequate for applications where Skolem/Herbrand functions are of concern.

Acknowledgments. This work was supported in part by the National Science Council under grants NSC 99-2221-E-002-214-MY3, 99-2923-E-002-005-MY3, and 100-2923-E-002-008.

References

1. Benedetti, M.: Evaluating QBFs via Symbolic Skolemization. In: Baader, F., Voronkov, A. (eds.) LPAR 2004. LNCS (LNAI), vol. 3452, pp. 285–300. Springer, Heidelberg (2005)
2. Blass, A., Gurevich, Y.: Henkin quantifiers and complete problems. Annals of Pure and Applied Logic 32, 1–16 (1986)
3. Balabanov, V., Jiang, J.-H.R.: Unified QBF certification and its applications. In: Formal Methods in System Design (2012)
4. Bubeck, U., Kleine Büning, H.: Dependency Quantified Horn Formulas: Models and Complexity. In: Biere, A., Gomes, C.P. (eds.) SAT 2006. LNCS, vol. 4121, pp. 198–211. Springer, Heidelberg (2006)
5. Bubeck, U.: Model-based Transformations for Quantified Boolean Formulas. IOS Press (2010)
6. Caicedo, X., Dechesne, F., Janssen, T.: Equivalence and quanftifier rules for logic with imperfect information. Logic Journal of the IGPL 17(1), 91–129 (2009)
7. Dechesne, F.: Game, Set, Maths: Formal Investigations into Logic with Imperfect Information. PhD thesis, Tilburg University (2005)
8. Jiang, J.-H.R., Lin, H.-P., Hung, W.-L.: Interpolating functions from large Boolean relations. In: Proc. Int'l Conf. on Computer-Aided Design (ICCAD), pp. 770–784 (2009)
9. Henkin, L.: Some remarks on infinitely long formulas. Infinitistic Methods, 167–183 (1961)
10. Hintikka, J., Sandu, G.: Informational independence as a semantical phenomenon. In: Logic, Methodology and Philosophy of Science, pp. 571–589 (1989)
11. Peterson, G., Reif, J., Azhar, S.: Lower bounds for multiplayer non-cooperative games of imcomplete information. Computers and Mathematics with Applications 41(7-8), 957–992 (2001)
12. Peters, S., Westerstahl, D.: Quantifiers in Language and Logic. Oxford University Press (2006)
13. Skolem, T.: Uber die mathematische Logik. Norsk. Mat. Tidsk. 10, 125–142 (1928); Translation in From Frege to Gödel, A Source Book in Mathematical Logic, J. van Heijenoort. Harvard Univ. Press (1967)
14. Sinha, S., Mishchenko, A., Brayton, R.K.: Topologically constrained logic synthesis. In: Proc. Int'l Conf. on Computer-Aided Design (ICCAD), pp. 679–686 (2002)

Lynx: A Programmatic SAT Solver
for the RNA-Folding Problem

Vijay Ganesh[1], Charles W. O'Donnell[1], Mate Soos[2], Srinivas Devadas[1],
Martin C. Rinard[1], and Armando Solar-Lezama[1]

[1] Massachusetts Institute of Technology
{vganesh,cwo,devadas,rinard,asolar}@csail.mit.edu
[2] Security Research Labs
mate@srlabs.de

Abstract. This paper introduces Lynx, an incremental programmatic SAT solver
that allows non-expert users to introduce domain-specific code into modern
conflict-driven clause-learning (CDCL) SAT solvers, thus enabling users to guide
the behavior of the solver.

The key idea of Lynx is a *callback interface* that enables non-expert users to
specialize the SAT solver to a class of Boolean instances. The user writes specialized code for a class of Boolean formulas, which is periodically called by Lynx's
search routine in its inner loop through the callback interface. The user-provided
code is allowed to examine partial solutions generated by the solver during its
search, and to respond by adding CNF clauses back to the solver dynamically
and incrementally. Thus, the user-provided code can specialize and influence the
solver's search in a highly targeted fashion. While the power of incremental SAT
solvers has been amply demonstrated in the SAT literature and in the context of
DPLL(T), it has not been previously made available as a programmatic API that
is easy to use for non-expert users. Lynx's callback interface is a simple yet very
effective strategy that addresses this need.

We demonstrate the benefits of Lynx through a case-study from computational biology, namely, the RNA secondary structure prediction problem. The
constraints that make up this problem fall into two categories: structural constraints, which describe properties of the biological structure of the solution,
and energetic constraints, which encode quantitative requirements that the solution must satisfy. We show that by introducing structural constraints on-demand
through user provided code we can achieve, in comparison with standard SAT approaches, upto 30x reduction in memory usage and upto 100x reduction in time.

1   Introduction

Conflict-driven clause-learning (CDCL) Boolean SAT solvers have had a huge impact
on a variety of domains ranging from program analysis to AI [3]. This success can
partly be attributed to their simple interface and powerful heuristics. In many cases, a
straightforward translation from a program analysis or AI problem into Boolean formulas in CNF (conjunctive normal form) format is sufficient to leverage the power of
the solver. Unfortunately, there are many other important domains (e.g., biology) where
straightforward translation of problems to CNF clauses leads to formulas that are too

A. Cimatti and R. Sebastiani (Eds.): SAT 2012, LNCS 7317, pp. 143–156, 2012.

large or complex for solvers to handle. For many of these domains, however, small domain-specific modifications to the solver can make SAT-based solution feasible. The challenge addressed by this paper is to enable users to make these small adaptations with minimal effort and without breaking subtle invariants in the solver implementation. The solution we provide allows for the resultant specialized solver to be adaptive, efficient for the problem-at-hand, and easy to build and maintain. Equally important, users are not burdened with knowing too much about the internals of SAT solvers and related technologies.

1.1 Our Contributions

– To address the problem described above, we created the solver Lynx that extends CryptoMiniSat [23] with an API allowing user-provided code to examine partial solutions generated by the SAT solver and add CNF clauses back to the solver in response. The added code is called inside the inner loop of the SAT solver, allowing the user to tightly integrate problem-specific clause-generation heuristics into the solver.

    We call solvers extended in this way *programmatic*, i.e., the user can programmatically influence solver behavior and adapt it to their specific problem domain in ways that are difficult to achieve otherwise. Programmatic solvers address the "solvers are unpredictable black boxes" problem by giving users more control over their search heuristics.
– Using Lynx we developed the first SAT based tool for solving the RNA-folding prediction problem. We present a detailed experimental evaluation of our technique in comparison with standard approaches. We use the above-mentioned callback interface in efficiently translating the RNA prediction problem into Boolean formulas. The interface allows Lynx to incrementally translate the RNA-folding structure inside the inner loop of the SAT solver, allowing a tighter, highly targeted and more efficient integration of the SAT solver and the translator.

1.2 Existing Approaches to Incremental and Adaptive Solving

Incremental solvers, that use some form of *abstraction-refinement* [3], have been proposed as a solution to the above-mentioned issue of *simple but inefficient translations* from problems to Boolean formulas. Instead of translating the entire input problem-instance into a potentially very large Boolean formula in one step, *abstraction-refinement approaches* translate the input instance into Boolean formulas incrementally and call the solver on these incrementally generated formulas. Such formulas are abstractions of the input instance and are often easier to solve than the entire input instance. The solver terminates if it gets the correct result to the input instance by solving an abstraction. Otherwise the solver iteratively refines the abstractions as necessary until it gets the correct result. Typically these abstractions and their refinements are performed by a layer outside the inner loop of the SAT solver. For an excellent reference on abstraction-refinement strategies refer to the Handbook of Satisfiability [3].

    Such incremental SAT solvers with an outside abstraction-refinement loop are relatively easy to build. However, the problem with such an approach is that it may not be the most efficient for the problem-at-hand. Indeed, Ohrimenko et al. [18] have proposed incremental translation of problems to SAT where the integration of the solver and the incremental translation is much tighter and more efficient than an outer layer translator. However, their implementation is non-adaptive, and is specific to a class of difference logic formulas — they do not provide an API for users to easily adapt or extend the solver for a previously unknown class of Boolean formulas.

An example of an API that allows users to adapt or extend solvers is the powerful idea of DPLL(T) [11] aimed at solving Boolean combination of formulas in rich theories such as integer linear arithmetic, uninterpreted functions and datatypes (aka SMT solvers [3]). In this approach, there is a tight integration of a CDCL SAT solver with a *theory solver* (aka a T-solver) that can handle conjunction of constraints represented in a rich logic. The CDCL SAT solver does the search on the Boolean structure of the formula without knowing the semantics of the literals, while the T-solver reasons about the literals themselves adding any new derived literals back to the Boolean CDCL solver appropriately. The tight integration enables the T-solver to influence the CDCL solver's behavior in ways not possible otherwise, and the resultant combination is typically a solver than can handle arbitrary Boolean combination of theory formulas efficiently.

A lay non-expert user could implement a "T-solver" using the DPLL(T) framework that reasons about a specific domain (say, theory of RNA folding) and adds constraints incrementally to the SAT solver. The resultant combination can be a powerful incremental domain-specific solver. However, the DPLL(T) API imposes strict requirements on the user-specified code (T-solver) to ensure that the resultant combination is sound and complete. Such requirements make perfect sense for constructing powerful SMT solvers with complex T-solvers, the problem for which the DPLL(T) approach was originally proposed. However, for the lay non-expert users such requirements may be onerous, and may not be essential. Lynx, by contrast provides a simple interface which is relatively easy to prove correct and is tailored for problem-specific extensions.

1.3   RNA-Folding with Lynx

To explore the benefits of using the Lynx's callback interface, we applied the technique to the problem of RNA folding. This is an application of significant practical relevance: understanding RNA folding is crucial to understanding a number of biological processes, including the replication of single-strand RNA viruses such as the poliovirus which causes polio in humans. Moreover, RNA prediction actually shares important similarities with other structure prediction problems of biological interest. This problem is particularly suitable to benchmark our approach. First, a SAT based solution to this problem is desirable because it gives researchers the ability to easily experiment with different formulations for the basic problem. Moreover, previous work in the literature has succeeded in formalizing the problem in a form that lends itself very naturally to solution with a Boolean SAT solver. SAT based solutions, however, have been elusive because the standard encoding leads to Boolean SAT instances that are too big for solvers to handle. Using Lynx's callback interface allowed us to encode instances of the RNA folding problem in a memory efficient manner, producing the first successful SAT based solution to this problem. The resultant incremental (or online abstraction-refinement) solver led to a 30-fold reduction in the amount of memory required to solve some of these problems compared to standard SAT approach, and demonstrated dramatic time improvements over standard abstraction refinement techniques.

Paper Layout. In Section 2 we provide a detailed overview of our incremental approach. In Section 3 we provide a self-contained description of the RNA-folding structure prediction problem. In Section 4 we provide detailed description of our experimental setup and results. We review the related work in Section 5, and conclude in Section 6.

2   Incrementality in Lynx

This section details how the callback interface in Lynx makes the solver incremental, what we sometimes also refer to as online abstraction-refinement or OAR. In order to facilitate the description, let us introduce a simple running example which shares some features with the more complex biology application.

The running example is a formula of the form $P(x) \land C(x)$ over a vector $x = \langle x_0, x_1, \ldots, x_N \rangle$ of Boolean variables, where $P(x)$ consists of some arbitrary set of constraints and $C(x)$ is a cardinality constraint that says that no more than 2 bits in x can be set to 1.

$$C(x) \equiv \forall_{i \neq j \neq k} (\neg x_i \lor \neg x_j \lor \neg x_k)$$

The above definition of $C(x)$ can be trivially encoded as a set of N^3 CNF clauses — too many for large values of N. For this specific case, more efficient encodings exist using only $O(N)$ clauses, but they are more complicated and require the introduction of additional SAT variables. By contrast, online abstraction refinement allows us to use the simple encoding without having to pay the price of introducing N^3 clauses.

The first step in using OAR is to divide the problem into a core set of clauses added to the solver from the very beginning, and a different set of *dynamic* clauses added to the solver incrementally by a *callback function*. The callback function is a user-provided function $\mathcal{M}$ producing a set of clauses given a partial assignment to the variables of the solver's input instance. A partial assignment sets each variable in the problem to either 1, 0, or $\perp$ (undefined), and is represented as a vector $t \in \{0, 1, \perp\}^N$.

In the case of the example, we define $P(x)$ to be the core clauses, and $C(x)$ to be the clauses added dynamically by a callback function defined as:

$$\mathcal{M}(t) \equiv \{(\neg x_i \lor \neg x_j \lor \neg x_k) \mid i \neq j \neq k \land t_i = t_j = t_k = 1\}$$

This callback function receives a partial assignment t, and returns a set of clauses of the form $(\neg x_i \lor \neg x_j \lor \neg x_k)$ where x_i, x_j and x_k are variables set to 1 in the partial assignment (i.e., $t_i = t_j = t_k = 1$). The clauses produced by the callback function eliminate those incorrect solutions that would have been eliminated by $C(x)$, so running the solver with constraints $P(x)$ and callback function $\mathcal{M}$ is the same as solving $P(x) \land C(x)$.

Lynx incorporates the callback function into the solution process by invoking it periodically with the current partial assignment. If the callback function returns any clause, these are incorporated into the problem. This process continues until an assignment q

is found such that: a) q satisfies all the core constraints, b) q satisfies all the constraints ever produced by the callback function, and c) the callback function produces an empty set of clauses when applied to q indicating that the process can be terminated. If the input problem is unsatisfiable, the solver with the callback function is guaranteed to report unsatisfiable and terminate. It is possible for the user-code, without any restrictions, to render the combination of base solver plus user-code incomplete. However, we can impose some minimal conditions on the user-code such that the combination is guarateed to be a complete decision procedure. In particular, one such condition is as follows: assume the desired input instance to be solved is $P(x) \wedge C(x)$, and $P(x)$ is input to the base solver. Then, the user-code must "encode" $C(x)$ exactly. Imposing this particular condition on the user-code is guaranteed to render the combination complete.

3   Biological Problem Overview

RNA is a versatile polymer essential to all of life. A chain of covalently bound nucleotides, RNA classically acts as a cellular messenger which duplicates DNA sequence information in the nucleus/nucleoid and transports that code to ribosomes for the construction of proteins. However, this chain can also fold in on itself into a 3-dimensional globular molecule which catalyzes biological reactions by itself. In fact, modern studies have suggested that such non-coding RNA (ncRNA) may play even a bigger cellular role than messenger RNA, with significant effects on metabolism, signal transduction, gene regulation, and chromosome inactivation. Such RNA function is determined by its nucleotide composition and 3-dimensional structure, however, relatively little ncRNA structural data is known [25], severely limiting our understanding of these mechanisms. Therefore, algorithmic prediction of RNA structure from its nucleotide sequence has been a longstanding computational goal.

3.1   Structure Prediction via SAT

The computational problem we address is "how to correctly attribute a unique structural state to each nucleic acid (or groups of nucleic acids) within an RNA polymer sequence". This problem has a long history of solutions based on many different algorithmic models — the most successful of which using a recursive, grammatical approach introduced by Zuker [26]. In this biophysical model, each nucleotide is allowed to form a pairwise bond with another, and each pair is assigned an energetic cost based on spatially adjacent nucleotide types [16]. The most likely structure is predicted by optimizing pairing configuration according to a fixed thermodynamic scoring system (energy minimization). Efficient computation is made possible through the imposition of specific, often biologically-inspired model restrictions — for example, limiting base-pairs to be sequentially nested (i.e. no "pseudoknots") and scoring only a subset of all potential energetic interactions (i.e. only Watson-Crick or wobble base-pairs). Unfortunately, this entangles the optimization techniques used with a particular set of biological assumptions. While these methods have shown good predictive accuracy, changes to the algorithm can be difficult to implement as new scientific data comes to light. For example, it has been shown that a more complex description of the RNA interaction energetics can lead to greatly improved results [19].

We propose a declarative approach for the structure prediction problem, providing a decoupled platform for reasoning about biological concepts in clear, succinct rules, backed by the powerful generic optimization of CDCL SAT solvers. This allows biological models to be tested and flexibly refined using a constraint-based philosophy, independent of performance improvements to the underlying solver.

To study this approach, we have implemented an RNA structure prediction algorithm using Lynx. Rather than comparing the benefits and disadvantages of different biological models, we base our implementation on an RNA scoring model recently proposed by Kato, et al. for integer programming optimization [20]. Although other models outperform this scoring system's accuracy, we believe our results are easily generalizable to greater classes of RNA structures [4] and more complex (non-RNA) structure prediction problems in general.

To implement energy minimization as a SAT-based decision procedure, we pose the question of whether an assignment exists that is lower than a certain energy threshold and perform iterative binary search. Despite this search routine, this approach can often be more efficient than the dynamic programming methods used by grammatical models as the problem can be finely partitioned into smaller jobs that are run in parallel. Further, when a sub-optimal solution is sufficient, this method quickly short-circuits, along with a guarantee of how near the solution is to optimality.

3.2 RNA Secondary Structure Prediction with Pseudoknots

The RNA prediction algorithm described here differentiates itself from classical prediction methods in its goal of predicting pseudoknots. Earlier grammar-based predictors allowed only base-pairs to occur in a recursively nested fashion (i.e. for every base-pair i-j there exists no base-pair k-l such that $i < k < j < l$) to enable highly efficient energy minimization via dynamic programming. However, pseudo-knotted structures which break this restriction are known to be essential to a number of functions, such as the Diels-Alder ribozyme and mouse mammary tumor virus [24]. However, predicting pseudoknotted structures is computationally much harder with fewer solutions [17,20,21]. In fact, the prediction of truly arbitrary pseudoknots has been shown NP-complete [14], and classes of pseudoknotted structures are often more easily defined by the algorithms which recognize them rather than their biological significance [7]. This motivates the use of a declarative approach, allowing easy exploration of different trade-offs between representation and optimization, especially if the underlying scoring system is changed from the standard Watson-Crick/wobble base-pair models to more complex interactions [19]. However, in the remainder of this work we restrict ourselves to the model proposed by Kato, et al. [20].

3.3 Encoding RNA Structure Prediction in SAT

Our SAT encoding is formulated by two sets of constraints, structural and energetic, that control the assignment of a vector of free variables which represent the final structural solution. The assignment of each free variable indicates whether two nucleotides are base-paired in the final RNA structure, fixed by structural constraints and an associated energetic score. Figure 1 depicts this formulation.

Solution Variables. The set of all properly-nested base-pairs within the final output RNA structure is represented by the variables $X_{i,j}$: where i and j indicate the sequence position of two nucleotides, a value $X_{i,j} = \mathbf{T}$ indicates a hydrogen bond base-pair exists between nucleotides at i and j, and $X_{i,j} = \mathbf{F}$ indicates that no base-pairing occurs between positions i and j. The set of pseudoknotted base pairs that cannot be properly nested are similarly represented by the independent variables $Y_{i,j}$. In this way pseudoknots are represented solely by the alignment of properly-nested $X_{i,j}$ pairs and properly-nested $Y_{i,j}$ pairs. Since RNA structure permits any nucleotide position i to pair with any other position j, a valid biological structure requires a complete assignment of all $X_{i,j}$s and $Y_{i,j}$s for every i, j ($0 \leq i, j < length(sequence)$). Therefore, the number of solution variables, the number of resultant constraints, and thus the difficulty of the SAT problem depends directly on the sequence length of the input RNA.

Structural Constraints. The structural representation places requirements on the assignment of the solution bits $X_{i,j}$ and $Y_{i,j}$ to ensure a biologically consistent structure. Therefore, we declare the following constraints, which must be satisfied in any valid solution:

- Every position i can at most pair with one other position j, independent of whether that pairing is properly-nested or a pseudoknot (Figure 1(a-d)). Four straightforward constraints ensure this:

$$\forall i, j, k, \quad i < j < k$$
$$(X_{i,j} \wedge X_{j,k}) = \mathbf{F} \wedge (Y_{i,j} \wedge Y_{j,k}) = \mathbf{F} \wedge$$
$$(X_{i,j} \wedge Y_{j,k}) = \mathbf{F} \wedge (Y_{i,j} \wedge X_{j,k}) = \mathbf{F}$$

- All base-pairs i, j are properly nested or a pseudoknot, but not both (Figure 1(e)):

$$\forall i, j \ (X_{i,j} \wedge Y_{i,j}) = \mathbf{F}$$

- We define all $X_{i,j}$ and $Y_{i,j}$ base-pairs to be independently knot-free (Figure 1(f-g)):

$$\forall i, j, k, l, \quad i < k < j < l$$
$$(X_{i,j} \wedge X_{k,l}) = \mathbf{F} \wedge (Y_{i,j} \wedge Y_{k,l}) = \mathbf{F}$$

- We only permit bifurcations within the "normal" base-pairs in $X_{i,j}$ since pseudoknots are rare and deserve distinct energetic treatment. Therefore (Figure 1(h)):

$$\forall i, j, k, l, i < k < j < l \ (Y_{i,j} \wedge Y_{k,l}) = \mathbf{F}$$

- Finally, the class of structures with "double-crossing" pseudoknots are rare and present unusual energetics which are not handled by the energy model we use, thus we constrain pseudoknots to only cross at most once (Figure 1(i-j)):

$$\forall i, j, k, l, m, n, \quad i < m < j < k < n < l$$
$$(X_{i,j} \wedge Y_{m,n}) \implies (X_{k,l} = \mathbf{F}) \wedge$$
$$(X_{k,l} \wedge Y_{m,n}) \implies (X_{i,j} = \mathbf{F})$$

Energetic Constraints. The total energy of an RNA structure is defined as the sum of experimentally-derived energy parameters [26,20] for every constituent base-pair *stack*, where a stack indicates two adjacent base pairs, e.g. $X_{i,j}$ and $X_{i+1,j-1}$. Energy parameters are given in terms of base-pair stacks because nucleotide π-orbital overlap serves as a dominant stabilizing factor in RNA structure. Thus, an energy value is assigned to every base-pair stack $X_{i,j}X_{i+1,j-1}$ according to the four nucleotide types at sequence positions i, j, $i+1$, and $j-1$ (Parameters found in [20]). By including a logical adder of all possible energetic assignments, we can then define a valid solution as an assignment of $X_{i,j}$ and $Y_{i,j}$ (subject to structural constraints), where the output of the adder overcomes some minimum threshold energy $E_{threshold}$ (the energy bound). As a logical declaration, we write:

$$\forall i, j, \; i < j \; (X_{i,j} \wedge X_{i+1,j-1}) = \mathbf{T} \Rightarrow (E_{X_{i,j}} = EnergyConstant_{(i,j,i+1,j-1)}) \quad \wedge$$
$$(Y_{i,j} \wedge Y_{i+1,j-1}) = \mathbf{T} \Rightarrow (E_{Y_{i,j}} = EnergyConstant_{(i,j,i+1,j-1)}) \quad \wedge$$
$$(X_{i,j} \wedge X_{i+1,j-1}) = \mathbf{F} \Rightarrow (E_{X_{i,j}} = 0) \quad \wedge$$
$$(Y_{i,j} \wedge Y_{i+1,j-1}) = \mathbf{F} \Rightarrow (E_{Y_{i,j}} = 0),$$

where $EnergyConstant(i, j, i+1, j-1)$ indicates the energy score of the four nucleotides found at positions i, j, $i+1$, and $j+1$ base-pairing and stacking, and

$$\sum_{\forall i,j} (E_{X_{i,j}} + E_{Y_{i,j}}) \geq E_{threshold}.$$

Finally, to enforce that all assigned base-pairs are accounted within the adder by stacking energy parameters, we require:

$$\forall i, j \quad s.t. \; i < j$$
$$(\overline{X_{i-1,j+1}} \wedge X_{i,j} \wedge \overline{X_{i+1,j-1}}) = \mathbf{F} \quad \wedge$$
$$(\overline{Y_{i-1,j+1}} \wedge Y_{i,j} \wedge \overline{Y_{i+1,j-1}}) = \mathbf{F}$$

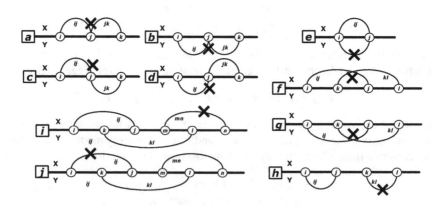

Fig. 1. RNA Constraints

4   Experimental Results

In this section we describe the results of our experimental evaluation of Lynx and competing approaches over input tests obtained from a set of RNA sequences. As described in detail in 3, we solve the two dimensional RNA optimum structure prediction problem (where the structures may have pseudoknots). We ran all experiments on a 3GHz Intel Xeon X5460 with 64GB of RAM and a 6MB L2 cache with 1 hour timeout per SAT instance.

4.1   Description of Input Tests

We acquired a set of benchmark RNA sequences and structures from the PseudoBase website [1]. These RNA sequences are widely used by computational biologists for a variety of structure prediction tasks. The biological accuracy of our lowest-energy structure predictions were verified to agree with Kato, et al. [20], whose scoring model we duplicate. Recall that the optimization problem is treated as a series of decision problems performing a binary search of the energy space. For each RNA sequence, a corresponding SAT instance is therefore constructed containing the energy and structural constraints along with an energetic bound that captures the minimum and maximum allowed energy for that step in the binary search. Given the precision of our energy model a search depth of 10 sufficed to identify the minimum energy structure of any structure tested.

4.2   Experimental Methodology

We solve the structure prediction problem using the following three methods:

- **Baseline Approach Using CryptoMiniSat (BA):** A standard encoding of our problem in SAT. We generate the complete SAT encoding (with XOR clauses as appropriate) of the RNA secondary structure prediction problem, then use CryptoMiniSat to solve this problem. We also used MiniSat2 [9], and found that for this problem its performance is similar to CryptoMiniSat [23].
- **Offline Abstraction Refinement (OFFA):** An encoding of our problem using established refinement techniques. Starting with only the energy constraints from the SAT encoding of the RNA structure prediction problem to form the abstracted constraint, we use offline abstraction refinement to obtain a solution to the complete structure prediction problem. Each refinement step uses CryptoMiniSat to solve the current SAT problem, computes the set of constraints from the complete structure prediction problem that are inconsistent with this solution, and generates a new problem by incrementally adding these constraints to the current problem in SAT. The refinement process continues until it produces a solution to the complete input problem.
- **Online Abstraction Refinement (ONA):** The methodology enabled by our tool Lynx. Starting with only the energy constraints from the SAT encoding of the RNA structure prediction problem to form the abstracted constraint, we use online abstraction refinement to obtain a solution to the complete structure prediction problem. After each CryptoMiniSat propagation step, the constraint manager examines

the current partial solution to find the set of constraints from the full structure prediction problem that conflict with the current solution. It then incrementally adds these constraints to the current problem before CryptoMiniSat takes the next partial solution step. The difference between the Offline (OFFA) and Online (ONA) approaches is the granularity of the refinement steps. Each refinement step in the OFFA version takes place only after CryptoMiniSat produces a complete solution to the current problem. Each refinement step in the ONA version, in contrast, takes place at the much finer granularity, every time CryptoMiniSat extends the current partial solution.

Table 1. Comparison of running times between Baseline (BA), Offline (OFFA), and Online (ONA) methods. Total cumulative time (across all solver instances during search) is reported, broken down by the amount of time spent in the SAT solver versus the amount of time spent in refinement. The number of refinement steps involved is also given. T.O. indicates that a timeout occured after 1hr of an individual SAT solver instance.

RNA	sequence length	Baseline (sec)	Offline Tot(sec)=SAT+Ref (# steps)	Online Tot(sec)=SAT+Ref (# steps)
PKB115	24	1.4	**1.7** = 1.3+0.4 (205)	**0.8** = 0.6+0.2 (2,538)
PKB102	24	1.3	**1.0** = 0.7+0.3 (129)	**0.6** = 0.5+0.1 (1,766)
PKB119	24	2.1	**3.6** = 3.0+0.6 (266)	**1.6** = 1.3+0.3 (4,108)
PKB103	25	3.1	**6.6** = 5.4+1.2 (417)	**3.5** = 3.1+0.4 (6,191)
PKB123	26	5.6	**24.7** = 22.7+2.0 (597)	**7.4** = 6.8+0.6 (8,980)
PKB154	26	2.5	**3.8** = 3.2+0.6 (236)	**1.9** = 1.7+0.2 (4,070)
PKB152	26	3.2	**6.2** = 5.2+1.0 (255)	**2.3** = 2.0+0.3 (5,528)
PKB126	27	4.0	**6.6** = 5.5+1.1 (384)	**2.8** = 2.5+0.3 (5,874)
PKB124	29	4.7	**5.1** = 4.4+0.7 (262)	**2.3** = 2.1+0.2 (4,635)
PKB100	31	11.0	**52.3** = 49.4+2.9 (315)	**6.8** = 6.0+0.8 (11,890)
PKB105	32	17.0	**58.3** = 54.0+4.3 (1004)	**18.1** = 17.0+1.1 (16,817)
PKB118	33	13.7	**32.8** = 29.6+3.2 (591)	**8.2** = 7.4+0.8 (12,878)
PKB120	36	36.1	**571.1** = 560.6+10.5 (652)	**24.1** = 21.9+2.2 (26,370)
PKB065	46	185.1	**11,341.9** = 11,298.7+43.2 (1,344)	**112.7** = 108.1+4.6 (50,508)
PKB205	48	388.6	T.O.	**391.6** = 381.9+9.7 (72,922)
PKB147	51	1,917.3	T.O.	**1,087.9** = 1,067.2+20.7 (131,321)
PKB248	66	T.O.	T.O.	T.O.
PKB072	67	5,352.6	T.O.	**2,414.1** = 2,367.6+46.5 (286,881)

4.3 Results

Table 1 presents the total execution times required for the different methods to solve the RNA structure prediction problems. We ran each method with a timeout of 3600 seconds for each SAT solution problem (i.e., each binary search step). Each row in the table corresponds to a single RNA. The first column is the number of base pairs in the RNA sequence. The next column presents the time (in seconds) required for the BA method to solve the problem. Recall that each problem requires the solution of 10 SAT instances; the reported total time is the sum of the 10 individual SAT solution times. The next column presents data from the OFFA method and is of the form $t = s + c(r)$.

Table 2. Comparison of memory usage between Baseline (BA), Offline (OFFA), and Online (ONA) methods. Given is the maximum memory (in MB) required throughout all SAT solver instances, along with the sum of the total number of clauses (in thousands) both input and generated during refinement. T.O. indicates that a timeout occured after 1hr of an individual SAT solver instance.

RNA	sequence length	Baseline Mem(MB) / Clauses	Offline Mem(MB) / Clauses	Online Mem(MB) / Clauses
PKB115	24	**5.0** / 3,223k	**5.0** / 94k	**72.1** / 82k
PKB102	24	**5.0** / 3,219k	**5.0** / 86k	**5.0** / 75k
PKB119	24	**5.0** / 3,240k	**5.0** / 130k	**5.0** / 104k
PKB103	25	**5.0** / 4,142k	**16.5** /174k	**5.0** / 136k
PKB123	26	**43.4** / 5,244k	**19.7** / 226k	**74.7** / 168k
PKB154	26	**5.0** / 5,204k	**5.0** / 128k	**5.0** / 106k
PKB152	26	**5.0** / 5,220k	**16.6** / 174k	**5.0** / 128k
PKB126	27	**72.1** / 6,544k	**74.5** / 171k	**5.0** / 129k
PKB124	29	**5.0** / 10,076k	**5.0** / 142k	**5.0** / 108k
PKB100	31	**90.5** / 16,937k	**23.9** / 376k	**90.0** / 231k
PKB105	32	**157.4** / 20,584k	**75.9** / 448k	**95.7** / 260k
PKB118	33	**131.9** / 24,870k	**23.2** / 355k	**22.8** / 227k
PKB120	36	**276.0** / 42,698k	**76.7** / 729k	**75.3** / 369k
PKB065	46	**1,011.8** / 196,236k	**150.6** / 341k	**122.9** / 595k
PKB205	48	**1,221.3** / 255,861k	**T.O.**	**145.0** / 808k
PKB147	51	**1,988.9** / 373,294k	**T.O.**	**188.7** / 1,322k
PKB248	66	**T.O.**	**T.O.**	**T.O.**
PKB072	67	**9,046.5** / 2,031,362k	**T.O.**	**313.1** / 2,652k

Here t is the total time required to solve the structure prediction problem (the sum of the solution times for the 10 SAT problems), s is the amount of time spent in the SAT solver, c is the amount of time spent in the constraint manager, and r is the total number of refinement steps (summed over all 10 SAT problems). The last column presents data from the ONA method and is also of the form $t = s + c(r)$.

Up to problem PKB124, the solution times for all of the methods are roughly comparable: each is less than ten seconds and within a factor of two for the same RNA sequence. For larger problems the OFFA approach starts to exhibit substantially larger solution times than either BA or ONA approaches; for the largest problems in our benchmark set OFFA times out. For two of the largest three problem sizes BA is roughly a factor of two slower than ONA; BA times out for PKB248.

We note that there is a substantial difference between the number of refinement steps that the ONA and OFFA methods perform — OFFA typically performs hundreds of (relatively coarse grain) refinement steps, while ONA performs thousands of (fine grain) refinement steps. These data indicate that, as expected, the SAT solver can respond much more quickly to fine grain than to coarse grain refinement steps, but that the ONA method requires more fine grain steps to reach a solution.

Table 2 presents the maximum amount of memory required to solve the structure prediction problem (this is the maximum over all runs of the SAT solver of the amount of memory that the SAT solver consumes) and the total number of clauses for each

RNA. For the OFFA and ONA methods, the total number of clauses is the sum over all binary search steps of the number of clauses in the problem at the final refinement step. Each entry of the table is in the form m/c, where m is the maximum memory and c is the number of clauses. Both the OFFA and ONA methods generate problems with substantially smaller numbers of clauses than the BA method (BA typically generates hundred to thousand times as many clauses OFFA and ONA typically generate). For larger RNA sequences, these larger clause sizes translate into substantially larger memory requirements for the BA method — OFFA and ONA never go above several hundred Mbytes, while BA starts requiring more than 1Gbyte of memory for the larger sequences.

4.4 Discussion

These data highlight how the ONA method is able to combine the benefit of small memory requirements, which it shares with OFFA, and feasible execution times, which it shares with BA (further note that ONA often exhibits roughly a factor of two performance advantage over BA). We attribute these characteristics to, first, the ability of the ONA method to effectively find relatively small problems whose solution also happens to be a solution of the complete structure prediction problem, and second, the ability of the ONA method to efficiently guide the SAT solver to the solution through fine-grain corrections to partial solution missteps. A comparison with the OFFA method illustrates how quickly correcting any missteps on the part of the SAT solver (by operating the refinement steps after every intermediate SAT solver decision rather than after every complete solution) can deliver very efficient solution times even in situations where the more coarse OFFA approach fails to solve the problem in an acceptable amount of time.

5   Related Work

There has been a lot of recent work on incremental SAT solvers [18], DPLL(T) [11], abstraction-refinement based techniques in the context of model-checking and decision procedures for SMT theories [2]. We summarize the related work, and contrast Lynx with other tools.

Incrementality, Extensibility and SAT Solvers. The work that is closest to ours is by Stuckey et al. [18] and the related idea of DPLL(T) [11]. Our work is different from Stuckey et al. in the mechanism employed to implement incrementality, namely, a callback interface. Our approach is more flexible in the sense that it can be used to expose other internals of SAT solvers (e.g., branching heuristics or restart triggers) to lay nonexpert users. While DPLL(T) is a very powerful idea, it places more requirements on user-code (to ensure completeness and soundness) and is probably best used by experts.

Abstraction-Refinement in Decision Procedures. The idea of counter-example guided abstraction refinement was originally developed in the context of model-checking [6]. Since then the basic idea has been adapted in different ways to solve the satisfiability problems of SMT theories [2]. Kroening, Ouaknine, Seshia, and Strichman [13] were the first to adapt CEGAR to deciding quantifier-free Presburger arithmetic. More recently, Brummayer and Biere give a new technique that allows early termination of an under-approximation refinement loop even when the original formula is unsatisfiable [5]. Ganesh and Dill proposed the use of abstraction-refinement for deciding the theory of arrays [10].

RNA Secondary Structure Prediction. Zuker introduced the first optimal algorithms for RNA secondary structure prediction based on a dynamic programming solution to energy minimization [26], although many improved predictors have followed [15]. Non-thermodynamic approaches have also met success through the use of phylogenetic relationships [12], or via machine learning [8]. The first efficient thermodynamic-based algorithm for predicting RNA pseudoknotted secondary structure was introduced by Rivas and Eddy (PKNOTS [22]). Subsequent algorithms have recognized alternate classes of pseudknots or improved upon the efficiency of solutions [17,4], including the IP formulation focused on in this paper [20], and heuristics such as HotKnots [21].

6   Conclusions

We present Lynx, a programmatic incremental SAT solver that allows non-expert users to easily introduce domain-specific or instance-specific code into modern CDCL SAT solvers, thus enabling users to control the behavior of the solver in ways not possible otherwise. While there has been work on incremental SAT [18] before and related ideas such as DPLL(T), Lynx's interface is simple to use and the requirements placed on user code are minimal. The approach is a template on how to expose other internals of the SAT solver to non-expert users in a easy-to-use and intuitive way. We demonstrate the benefits of Lynx through a first-of-its-kind solver case-study from computational biology, namely, RNA secondary structure prediction.

References

1. Pseudobase RNA sequence, Most widely used database for research on RNA sequences with Psuedoknots. website, http://pseudobaseplusplus.utep.edu/
2. SMTLIB website, http://combination.cs.uiowa.edu/smtlib/
3. Biere, A., Heule, M.J.H., van Maaren, H., Walsh, T. (eds.): *Handbook of Satisfiability.* Frontiers in Artificial Intelligence and Applications, vol. 185. IOS Press (February 2009)
4. Bon, M., Vernizzi, G., Orland, H., Zee, A.: Topological classification of RNA structures. J. Mol. Biol. 379(4), 900–911 (2008)
5. Brummayer, R., Biere, A.: Effective Bit-Width and Under-Approximation. In: Moreno-Díaz, R., Pichler, F., Quesada-Arencibia, A. (eds.) EUROCAST 2009. LNCS, vol. 5717, pp. 304–311. Springer, Heidelberg (2009)
6. Clarke, E., Grumberg, O., Jha, S., Lu, Y., Veith, H.: Counterexample-guided abstraction refinement for symbolic model checking. J. ACM 50(5), 752–794 (2003)
7. Condon, A., Davy, B., Rastegari, B., Chao, S., Tarrant, F.: Classifying RNA pseudoknotted structures. Theoretical Computer Science 320, 35–50 (2004)
8. Do, C., Woods, D., Batzoglou, S.: CONTRAfold: RNA secondary structure prediction without energy-based models. Bioinformatics 22(14), e90–e98 (2006)

9. Een, N., Sorensson, N.: An extensible SAT-solver. In: Proc. Sixth International Conference on Theory and Applications of Satisfiability Testing, pp. 78–92 (May 2003)

10. Ganesh, V., Dill, D.L.: A Decision Procedure for Bit-Vectors and Arrays. In: Damm, W., Hermanns, H. (eds.) CAV 2007. LNCS, vol. 4590, pp. 519–531. Springer, Heidelberg (2007)

11. Ganzinger, H., Hagen, G., Nieuwenhuis, R., Oliveras, A., Tinelli, C.: DPLL(*T*): Fast Decision Procedures. In: Alur, R., Peled, D.A. (eds.) CAV 2004. LNCS, vol. 3114, pp. 175–188. Springer, Heidelberg (2004)

12. Knudsen, B., Hein, J.: RNA secondary structure prediction using stochasatic context-free grammars and evolutionary history. Bioinformatics 15, 446–454 (1999)

13. Kroning, D., Ouaknine, J., Seshia, S.A., Strichman, O.: Abstraction-Based Satisfiability Solving of Presburger Arithmetic. In: Alur, R., Peled, D.A. (eds.) CAV 2004. LNCS, vol. 3114, pp. 308–320. Springer, Heidelberg (2004)

14. Lyngsø, R.B., Pedersen, C.N.S.: Pseudoknots in RNA secondary structures. In: Proc. Computational Molecular Biology, RECOMB 2000, pp. 201–209. ACM (2000)

15. Mathews, D., Disney, M., Childs, J., Schroeder, S., Zuker, M., Turner, D.: Incorporating chemical modification constraints into a dynamic programming algorithm for prediction of RNA secondary structure. Proc. Natl. Acad. Sci. 101, 7287–7292 (2004)

16. Mathews, D.H., Sabina, J., Zuker, M., Turner, D.H.: Expanded sequence dependence of thermodynamic parameters improves prediction of RNA secondary structure. Journal of Molecular Biology 288(5), 911–940 (1999)

17. Mathews, D.H., Turner, D.H.: Prediction of RNA secondary structure by free energy. Curr. Opin. Struct. Biol. 16, 270–278 (2006)

18. Ohrimenko, O., Stuckey, P.J., Codish, M.: Propagation = Lazy Clause Generation. In: Bessière, C. (ed.) CP 2007. LNCS, vol. 4741, pp. 544–558. Springer, Heidelberg (2007)

19. Parisien, M., Major, F.: The MC-Fold and MC-Sym pipeline infers RNA structure from sequence data. Nature 452, 51–55 (2008)

20. Poolsap, U., Kato, Y., Akutsu, T.: Prediction of RNA secondary structure with pseudoknots using integer programming. BMC Bioinformatics 10, S38 (2009)

21. Ren, J., Rastegari, B., Condon, A., Hoos, H.H.: HotKnots: Heuristic prediction of RNA secondary structures including pseudoknots. RNA 11, 1494–1504 (2005)

22. Rivas, E., Eddy, S.: A dynamic programming algorithm for RNA structure prediction including pseudoknots. J. Mol. Biol. 285, 2053–2068 (1999)

23. Soos, M., Nohl, K., Castelluccia, C.: Extending SAT Solvers to Cryptographic Problems. In: Kullmann, O. (ed.) SAT 2009. LNCS, vol. 5584, pp. 244–257. Springer, Heidelberg (2009)

24. Staple, D.W., Butcher, S.E.: Pseudoknots: RNA structures with diverse functions. PLoS Biol. 3(6), e213 (2005)

25. Washietl, S., Hofacker, I., Lukasser, M., Hüttenhofer, A., Stadler, P.: Mapping of conserved RNA secondary structures predicts thousands of functional noncoding RNAs in the human genome. Nat. Biotechnol. 23(11), 1383–1390 (2005)

26. Zuker, M., Stiegler, P.: Optimal computer folding of large RNA sequences using thermodynamics and auxiliary information. Nucleic Acids Research 9(1), 133–148 (1981)

Generalized Property Directed Reachability

Kryštof Hoder[1] and Nikolaj Bjørner[2]

[1] The University of Manchester
[2] Microsoft Research, Redmond

Abstract. The IC3 algorithm was recently introduced for proving properties of finite state reactive systems. It has been applied very successfully to hardware model checking. We provide a specification of the algorithm using an abstract transition system and highlight its dual operation: model search and conflict resolution. We then generalize it along two dimensions. Along one dimension we address nonlinear fixed-point operators (push-down systems) and evaluate the algorithm on Boolean programs. In the second dimension we leverage proofs and models and generalize the method to Boolean constraints involving theories.

1 Introduction

Efficient SAT and SMT solvers are at the heart of many program analysis, verification and test tools. Such tools reduce program representations and logics to first-order or propositional queries. An ongoing quest is how one can raise the level of abstraction and power of the logic engines. We pursue satisfiability modulo least fixed-points. The propositional fragment corresponds to Boolean Programs with procedure calls, equivalently monotone Datalog, and first-order existential fixed-points provide a precise match for Hoare Logic [3].

The IC3 algorithm [4] was recently used successfully for hardware model checking [4,6]. We use the current popular, and descriptive, terminology *Property Directed Reachability* (PDR) to refer to IC3 and its derivatives. PDR has several intriguing characteristics. It simultaneously strengthens an abstraction of reachable states and prunes a search for counter examples, but in contrast to predicate abstraction methods [2] and methods based on interpolants [12,8,1], it maintains precise transition relations and only refines state abstractions. Importantly, it leverages induction proofs to strengthen invariant candidates.

We are motivated by software analysis, where handling procedure calls and theories is relevant. In the pursuit of this goal, we contribute the following:

- provide an abstract account of the PDR algorithm;
- generalize PDR to *nonlinear* fixed-point operators;
- further generalize PDR to theories, specifically Linear Real Arithmetic.

The paper is organized as follows: Section 2 motivates satisfiability modulo least fixed-points. Sections 3 (4) present the abstract account of (nonlinear) PDR. Section 5 generalizes PDR to theories and Section 6 describes our implementation and experiments.

A. Cimatti and R. Sebastiani (Eds.): SAT 2012, LNCS 7317, pp. 157–171, 2012.

2 From Safety Verification to Least Fixed-points

To motivate the use of fixed-point operators and solving satisfiability modulo least fixed-points consider Lamport's two process Bakery algorithm. It ensures mutual exclusion between processes P_1 and P_2. They cannot simultaneously execute **critical**.

initially $y_1 := y_2 := 0$;

$$P_1 :: \begin{bmatrix} \textbf{loop forever do} \\ \begin{bmatrix} \ell_0 : y_1 := y_2 + 1; \\ \ell_1 : \textbf{await } y_2 = 0 \vee y_1 \leq y_2; \\ \ell_2 : \textbf{critical}; \\ \ell_3 : y_1 := 0; \end{bmatrix} \end{bmatrix} \quad || \quad P_2 :: \begin{bmatrix} \textbf{loop forever do} \\ \begin{bmatrix} \ell_0 : y_2 := y_1 + 1; \\ \ell_1 : \textbf{await } y_1 = 0 \vee y_2 \leq y_1; \\ \ell_2 : \textbf{critical}; \\ \ell_3 : y_2 := 0; \end{bmatrix} \end{bmatrix}$$

Mutual exclusion and other safety properties are proved by *induction* over the set of reachable states. The induction proof requires finding an *inductive invariant* that is true in the initial state, is maintained

$$\frac{\begin{array}{c} \Theta(\boldsymbol{x}) \rightarrow R(\boldsymbol{x}) \\ R(\boldsymbol{x}) \wedge \rho(\boldsymbol{x}, \boldsymbol{x}') \rightarrow R(\boldsymbol{x}') \\ R(\boldsymbol{x}) \rightarrow S(\boldsymbol{x}) \end{array}}{\mathcal{S} \models \Box S(\boldsymbol{x})} \text{ G-INV}$$

by each step of the system, and implies the safety property. The induction G-INV rule can be formalized following [11]. Programs denote *transition systems* $\mathcal{S} = \langle \boldsymbol{x}, \Theta, \rho(\boldsymbol{x}, \boldsymbol{x}') \rangle$, where $\boldsymbol{x}$ is a set of state variables, Θ a formula describing a set of initial configurations over $\boldsymbol{x}$ and $\rho(\boldsymbol{x}, \boldsymbol{x}')$ is a transition relation. For the Bakery algorithm, $\Theta := y_1 = y_2 = L = M = 0$ where L is program counter for P_1 and M is a program counter for P_2. The safety property for Bakery is $S := \neg(L = 2 \wedge M = 2)$. The predicate R serves as the inductive invariant in G-INV. Notice that the premises of G-INV are Horn clauses. We reformulate the Bakery verification problem on the left using Prolog convention of capitalization for universally quantified variables. The non-recursive predicate T is a shorthand for ρ. It

$$R(0,0,0,0)$$
$$R(L, M, Y_1, Y_2) \wedge T(L, L', Y_1, Y_2, Y_1') \rightarrow R(L', M, Y_1', Y_2)$$
$$R(L, M, Y_1, Y_2) \wedge T(M, M', Y_2, Y_1, Y_2') \rightarrow R(L, M', Y_1, Y_2')$$
$$T(0, 1, Y_1, Y_2, Y_2 + 1)$$
$$Y_1 \leq Y_2 \vee Y_2 = 0 \rightarrow T(1, 2, Y_1, Y_2, Y_1)$$
$$T(2, 3, Y_1, Y_2, Y_1)$$
$$T(3, 0, Y_1, Y_2, 0)$$
$$R(2, 2, Y_1, Y_2) \rightarrow \textit{false}$$

exploits the symmetry of P_1 and P_2. The real challenge is to find a solution to the recursive predicate R. A solution exists iff there is a solution to the least fixed-point of the *strongest post-condition* predicate transformer that defines R. For transition systems, the predicate transformer follows the template:

$$\mathcal{F}(R)(\boldsymbol{x}) := \exists \boldsymbol{x}_0 \; . \; \underbrace{\Theta(\boldsymbol{x}) \; \vee \; R(\boldsymbol{x}_0) \wedge \rho(\boldsymbol{x}_0, \boldsymbol{x})}_{\mathcal{T}[R(\boldsymbol{x}_0)]}, \tag{1}$$

where the quantifier-free body $\mathcal{T}$ of $\mathcal{F}$ is the *transition relation*, which now includes also the initial condition Θ. The predicate transformer (1) is *linear*; the argument R occurs in at most one (positive) position. Using the terminology of predicate transformers, finding an inductive invariant amounts to finding a *post fixed-point* R, such that $\mathcal{F}(R) \to R$ and $R \to S$. Recall that the least fixed-point $\mu R.\mathcal{F}(R)$ (the infinite disjunction $\bigvee_{i \geq 0} \mathcal{F}^i(\textit{false})$) is contained in any R that satisfies $\mathcal{F}(R) \to R$.

3   Abstract Property Directed Reachability

This section recalls the Property Directed Reachability algorithm using terminology of predicate transformers and we specify the algorithm as an abstract transition system. We build upon [6] as we arrive to a specification.

$$S \qquad R_{i+1} \atop \nwarrow \quad \nearrow \quad \nwarrow \qquad (2) \atop R_i \qquad\qquad \mathcal{F}(R_i)$$

Fig. 1. Invariant (2). Each arrow is an implication: $R_i \to S$, $R_i \to R_{i+1}$, $\mathcal{F}(R_i) \to R_{i+1}$.

The original IC3 algorithm verifies invariants of *linear* fixed-point operators. PDR maintains formulas $\Theta = R_0, \ldots, R_N$, such that for $0 \leq i < N$ invariant (2) holds. Initially we set $R_0 = \mathcal{F}(\textit{false})$ and $N = 0$, so the invariant holds trivially. N is incremented if $R_N \to S$ is established.

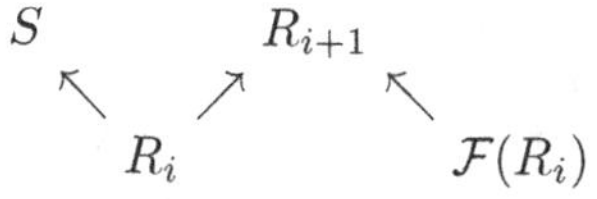

For Bakery, we have $R_0 := \Theta := (L = M = Y_1 = Y_2 = 0)$ and $S := \neg(L = M = 2)$ so $R_0 \to S$, and we can increment $N := 1$, $R_1 := \textit{true}$. The next action is to check if $R_1 \wedge \neg S$ is satisfiable. It is, and one *partial* model is denoted by $\mathcal{M} := L = 2 \wedge M = 2$ (it does not include assignments to Y_1, Y_2). The model $\mathcal{M}$ violates the safety property, but is $\mathcal{M}$ reachable from the current unfolding? They would be if $\mathcal{F}(R_0) \wedge \mathcal{M}$ was satisfiable. It is not satisfiable. One of several possible unsatisfiable cores is $L \neq 0 \wedge M \neq 0$. We then update R_1 with the negation: $R_1 := R_1 \wedge (L = 0 \vee M = 0)$ and now we can unfold again $N := 2$.

These steps give a taste of how PDR uses spurious counter examples, as partial models, to build up R_i. PDR also contains a clever mechanism for strengthening clauses in R_i by using *induction*, presented in the following. The R_i are sets of clauses. We shall however often (ab)use notation and use conjunction instead of set union and freely switch between viewing a set of formulas as a conjunction. The R_i denote sets of states (the set of models that satisfy R_i) and over-approximate the states reachable by unfolding the transition relation i times. All stages except the last imply the safety property. We can visualize the implications between the approximations using the picture below (for $N = 3$).

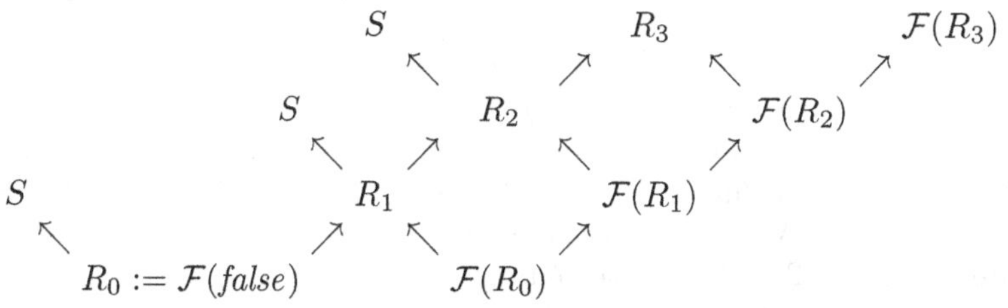

$$S \qquad\qquad R_3 \qquad\qquad \mathcal{F}(R_3)$$
$$\nwarrow \quad \nearrow \qquad \nwarrow \quad \nearrow$$
$$S \qquad\qquad R_2 \qquad\qquad \mathcal{F}(R_2)$$
$$\nwarrow \quad \nearrow \qquad \nwarrow \quad \nearrow$$
$$S \qquad\qquad R_1 \qquad\qquad \mathcal{F}(R_1)$$
$$\nwarrow \quad \nearrow \qquad \nwarrow \quad \nearrow$$
$$S \atop \nwarrow \qquad R_0 := \mathcal{F}(\textit{false}) \qquad\qquad \mathcal{F}(R_0)$$

PDR relies on refining counter examples that are models. A model is a conjunction of equalities between variables and values. For example the model

$a = true \wedge b = false \wedge x = 3$ (more compactly written $a \wedge \neg b \wedge x = 3$), assigns a to $true$, b to $false$ and x to 3. We use $\mathcal{M}$ as shorthand for models of the form $\boldsymbol{x} = \boldsymbol{c}$. A model $\mathcal{M}$ of a formula φ is allowed to be partial (omit assigning values to some variables) as long as φ is true under $\mathcal{M}$. When φ is a clause, then $\neg\varphi$ is treated as a conjunction of literals. So $\neg\varphi \subseteq \mathcal{M}$ means that all literals in φ are false in $\mathcal{M}$.

The following updates to R_i are made by the algorithm:

Valid. For $i < N$, if $R_i \subseteq R_{i+1}$, then R_i is an inductive invariant. Return *Valid*.

Unfold. If $R_N \to S$, then set $N \leftarrow N+1$, $R_N \leftarrow true$.

Induction. For $0 \leq i < N$, a clause $(\varphi \vee \psi) \in R_i$, $\varphi \notin R_{i+1}$, if $\mathcal{F}(R_i \wedge \varphi) \to \varphi$, then conjoin φ to R_j, for each $j \leq i+1$. While induction is a separate rule, it is useful to apply it immediately following Unfold and Conflict. The rule is sound because $\mathcal{F}$ is monotone and Invariant (2) ensures $R_j \to R_i$ for $j < i$. Therefore $\mathcal{F}(false) \to \varphi$ and $\mathcal{F}(R_j \wedge \varphi) \to \varphi$ for each $j < i$.

PDR includes a *dual* mode where it searches for a candidate counter-model to S. The candidate model is used to guide the strengthening of R_i.

Candidate. If $\mathcal{M} \models R_N(\boldsymbol{x}) \wedge \neg S(\boldsymbol{x})$, then produce candidate $\langle \mathcal{M}, N \rangle$.

Decide. If $\langle \boldsymbol{x} = \boldsymbol{c}, i+1 \rangle$ for $0 \leq i < N$ is a candidate model and there is a subset $\tilde{\boldsymbol{x}}_0$ of $\boldsymbol{x}_0$ and constants $\boldsymbol{c}_0$, such that $\boldsymbol{x} = \boldsymbol{c}, \tilde{\boldsymbol{x}}_0 = \boldsymbol{c}_0 \models \mathcal{T}[R_i(\boldsymbol{x}_0)]$, then add the candidate model $\langle \tilde{\boldsymbol{x}} = \boldsymbol{c}_0, i \rangle$ (renaming $\tilde{\boldsymbol{x}}_0$ to $\tilde{\boldsymbol{x}}$).

Model. If $\langle \mathcal{M}, 0 \rangle$ is a candidate model, then report that S is violated.

Conflict. For $0 \leq i < N$: given a candidate model $\langle \mathcal{M}, i+1 \rangle$ and clause φ, such that $\neg\varphi \subseteq \mathcal{M}$, if $\mathcal{F}(R_i) \to \varphi$, then conjoin φ to R_j, for $j \leq i+1$.

3.1 PDR as an Abstract Transition System

Figure 2 summarizes the PDR algorithm as an abstract transition system. It maintains states of the form $M \parallel A$, where M is a candidate counter example trace that is a stack of models labeled by a level i, and A is the current abstraction comprising of the maximal level N and sets of clauses $R_0, \ldots, R_N$.

Example 1. Suppose we are given the safety property $S(x, y, z) \equiv \neg y$ and the predicate transformer $\mathcal{F} = \lambda R \lambda xyz . \exists x_0 y_0 z_0 . (x, y, z) = (1, 0, 0) \vee ((x, y, z) = (y_0, z_0, x_0) \wedge R(x_0, y_0, z_0)))$ that corresponds to the rules: $R(1, 0, 0)$. $R(x, y, z) \to R(y, z, x)$. We use 0 for *false* and 1 for *true*. We can check that $\neg S$ is reachable.

Initialize	$\Longrightarrow$	$\epsilon \parallel [N \leftarrow 0, R_0 \leftarrow x \wedge \neg y \wedge \neg z]$
Unfold	$\Longrightarrow$	$\epsilon \parallel [N \leftarrow 1, R_0, R_1 \leftarrow true]$
Candidate	$\Longrightarrow$	$\langle y \wedge \neg z, 1 \rangle \parallel [N, R_0, R_1]$
Conflict	$\Longrightarrow$	$\epsilon \parallel [N, R_0, R_1 \leftarrow \neg y]$ since $y \wedge \neg z \models y$, $\mathcal{F}(R_0)(x, y, z) \to \neg y$
Unfold	$\Longrightarrow$	$\epsilon \parallel [N \leftarrow 2, R_0, R_1, R_2 \leftarrow true]$
Candidate	$\Longrightarrow$	$\langle y, 2 \rangle \parallel [N, R_0, R_1, R_2]$
Decide	$\Longrightarrow$	$\langle z, 1 \rangle \langle y, 2 \rangle \parallel [N, R_0, R_1, R_2]$
Decide	$\Longrightarrow$	$\langle x, 0 \rangle \langle z, 1 \rangle \langle y, 2 \rangle \parallel [N, R_0, R_1, R_2]$
Model	$\Longrightarrow$	$\langle x, 0 \rangle \langle z, 1 \rangle \langle y, 2 \rangle$ $\boxtimes$

Initialize	$\implies \epsilon \parallel [N \leftarrow 0, R_0 \leftarrow \mathcal{F}(false)]$	
Valid	$M \parallel A \implies$ Valid	$\textbf{if} \models R_{i-1} \subseteq R_i, i < N.$
Unfold	$M \parallel A \implies \epsilon \parallel A[R_{N+1} \leftarrow true, N \leftarrow N+1]$	$\textbf{if} \models R_N \to S,$
Induction	$M \parallel A \implies M \parallel A[R_j \leftarrow R_j \wedge \varphi]_{j=1}^{i+1}$	$\textbf{if } (\varphi \vee \psi) \in R_i, \varphi \notin R_{i+1},$ $\models \mathcal{F}(R_i \wedge \varphi) \to \varphi$
Candidate	$\epsilon \parallel A \implies \langle \mathcal{M}, N \rangle \parallel A$	$\textbf{if } \mathcal{M} \models R_N \wedge \neg S$
Decide	$\langle \mathcal{M}, i+1 \rangle M \parallel A \implies \langle \tilde{\boldsymbol{x}} = \boldsymbol{c_0}, i \rangle \langle \mathcal{M}, i+1 \rangle M \parallel A$	$\textbf{if } \mathcal{M}, \tilde{\boldsymbol{x}}_0 = \boldsymbol{c_0} \models \mathcal{T}[R_i(\boldsymbol{x_0})]$
Model	$\langle \mathcal{M}, 0 \rangle M \parallel A \implies$ Model $\langle \mathcal{M}, 0 \rangle M$	
Conflict	$\langle \mathcal{M}, i+1 \rangle M \parallel A \implies M \parallel A[R_j \leftarrow R_j \wedge \varphi]_{j=1}^{i+1}$	$\textbf{if } \neg\varphi \subseteq \mathcal{M}, \models \mathcal{F}(R_i) \to \varphi.$

Fig. 2. Abstract transition system specification of PDR

The example exercises several features of PDR. It starts with a Candidate counter example model to the last state and Decide pushes models down to the initial state. There is some freedom in choosing models to push down. Such models can be partial; they just need to force the transition relations. If models can be pushed all the way down, there is a counter example trace, otherwise a Conflict gets detected along the way. The induction rule is specified as a separate rule, but it can be applied immediately after Conflict to minimize the new clause. A good analogy is how subsumption is used when processing conflict clauses in modern SAT solvers. Induction also serves the purpose of pushing up clauses from $(\varphi \vee \psi) \in R_i$ to R_{i+1} by taking $\psi = false$. Such propagation can be applied immediately after Conflict and before Unfold.

Correctness of the algorithm follows from four observations:

Lemma 1 (Invariant (2)). *The rules from Figure 2 maintain Invariant (2).*

Lemma 2 (Validity). *When* $\models R_i \subseteq R_{i+1}$, *then* S *is invariant.*

Proof. Let us add this condition to the implications from invariant (2) and we get that R_i is a post-fixed point that is contained in S: $\mathcal{F}(R_i) \to R_{i+1} \to R_i \to S$. Thus, R_i satisfies the premises of G-INV and therefore S is invariant.

Lemma 3 (Satisfiability). *When* $\langle \mathcal{M}, 0 \rangle$ *is reached, then* S *is violated with a path of length* N.

Corollary 1 (Correctness of PDR). *If PDR terminates with* Valid, *then* S *is invariant. If PDR terminates with* Model M, *then* M *is a trace leading to a violation of* S.

It is also the case that each step makes progress by either extending models or strengthening states. The set of possible different states R_i is bounded by the set of possible models (assuming that clauses are normalized) so the algorithm terminates for finite domains. Therefore,

Theorem 1 (Termination on Finite Domains). *Any derivation sequence terminates with a verdict* Valid *or* Model *when* $\mathcal{F}$ *is finite domain.*

Note that PDR represents traces explicitly, so while reachability of Boolean systems is PSPACE, PDR may nevertheless consume exponential space.

4 Nonlinear PDR

Nonlinear transformers are important in the context of checking software with procedures. The Static Driver Verifier [2] implements a model checker for programs with procedure calls.

$$\texttt{mc(x)} = \texttt{if x} > \texttt{100 then x} - \texttt{10 else mc(mc(x} + \texttt{11))}$$
$$\texttt{assert } \forall \texttt{x.mc(x)} \geq \texttt{91}$$

$$X > 100 \;\to\; mc(X, X - 10)$$
$$X \leq 100 \land mc(X + 11, Y) \land mc(Y, R) \;\to\; mc(X, R)$$
$$mc(X, R) \;\to\; R \geq 91$$

Nonlinear predicate transformers correspond to general Horn clauses. An example program with procedure calls and resulting non-linear Horn clauses comes from McCarthy's 91 function and the accompanying assertion.

We therefore consider nonlinear predicate transformers of the form

$$\mathcal{F}(R)(\boldsymbol{x}) = \exists \boldsymbol{x_0}, \boldsymbol{x_1} \; . \; \underbrace{\Theta(\boldsymbol{x}) \; \lor \; [R(\boldsymbol{x_0}) \; \land \; R(\boldsymbol{x_1}) \; \land \; \rho(\boldsymbol{x_0}, \boldsymbol{x_1}, \boldsymbol{x})]}_{\mathcal{T}[R(\boldsymbol{x_0}), R(\boldsymbol{x_1})]} \qquad (3)$$

We use the template (3) when presenting algorithms for nonlinear PDR. The terminology of predicate transformers was useful for formulating the main invariant (2), and we find it particularly instrumental for generalizing PDR to general Horn clauses. The extension to nonlinear predicate transformers with more than two occurrences of R, and systems of nonlinear predicate transformers is relatively straight-forward.

4.1 State

In contrast to linear predicate transformers, counter examples for nonlinear transformers unfold into trees. A compressed view of counter examples is as DAGs, and the potential savings of using DAGs can be exponential. A challenge is to create and maintain such counter examples. We propose an approach where states that are known to be reachable are put in a cache, and PDR inserts nodes into a DAG. So it inspects the current DAG to see if a new (potentially) reachable state is already being expanded before creating a new node. States are compared syntactically. A more powerful alternative is to represent the cache as a formula and check cache containment semantically, but we found no practical use for such added power: counter examples for recursive predicates from programs can be expected to have a small tree unfolding. We present this approach in the following.

The state of the algorithm is maintained as a triple $\mathcal{D} \parallel A \parallel C$, where:

$\mathcal{D}$, *the model search DAG* represents a partial unfolding of a counter example. It is initially the empty DAG ϵ. Nodes are labeled with queries $\langle \mathcal{M}, i \rangle$, where i is a level and $\mathcal{M}$ is a partial model. We use L as a shorthand for $\langle \mathcal{M}, i \rangle$; use $\mathcal{D}[L \bullet \{\mathcal{D}' \, \mathcal{D}''\}]$ to refer to an internal node L with two children; and $\mathsf{model}(\mathcal{D})$ to access the model at the root of a DAG.

A, *the property state* of the form $[N, R_0, \ldots, R_N]$.

C, *the cache* of reachable states. It contains a set of partial interpretations $\mathcal{M}$ that imply $\mathcal{F}^n(\mathit{false})$ for some $n \geq 0$. Consequently, every completion of $\mathcal{M}$ is contained in the least fixed-point and is therefore reachable.

4.2   Algorithm Specification

Figure 3 contains the new rules we need for the nonlinear variant of PDR. Rules Initialize, Valid, Induction, Unfold, Candidate are unchanged from Figure 2, with the exception that we add a column for the cache C and we replace the stack of models M by a DAG $\mathcal{D}$.

Decide $\mathcal{D}[\langle \mathcal{M}, i+1 \rangle] \parallel A \parallel C \implies \mathcal{D}[\langle \mathcal{M}, i+1 \rangle] \bullet \{ \langle \tilde{\boldsymbol{x}} = \boldsymbol{c_0}, i \rangle \, \langle \tilde{\tilde{\boldsymbol{x}}} = \boldsymbol{c_1}, i \rangle \}] \parallel A \parallel C$
$\qquad\qquad$ if $\mathcal{M}, \tilde{\boldsymbol{x}}_0 = \boldsymbol{c_0}, \tilde{\tilde{\boldsymbol{x}}}_1 = \boldsymbol{c_1} \models \mathcal{T}[R_i(\boldsymbol{x_0}), R_i(\boldsymbol{x_1})]$

Model $\mathcal{D} \parallel A \parallel C \implies \mathsf{Model} \; \mathcal{D}$ if $\langle \mathcal{M}, N \rangle \in \mathcal{D}, \mathcal{M} \in C$

Conflict $\mathcal{D}[L \bullet \{\mathcal{D}' \, \mathcal{D}''\}] \parallel A \parallel C \implies \mathcal{D}[L] \parallel A[R_j \leftarrow R_j \wedge \varphi]_{j=1}^{i+1} \parallel C$
$\qquad\qquad$ if $\neg \varphi \subseteq \mathsf{model}(\mathcal{D}'), \models \mathcal{F}(R_i) \rightarrow \varphi$.

Base $\mathcal{D}[\langle \mathcal{M}, i \rangle] \parallel A \parallel C \implies \mathcal{D} \parallel A \parallel C \cup \{\mathcal{M}\}$ if $\mathcal{M} \models R_0$.

Close $\mathcal{D}[\langle \mathcal{M}, i+1 \rangle \bullet \{\mathcal{D}' \, \mathcal{D}''\}] \parallel A \parallel C \implies \mathcal{D} \parallel A \parallel C \cup \{\mathcal{M}\}$
$\qquad\qquad$ if $\mathsf{model}(\mathcal{D}'), \mathsf{model}(\mathcal{D}'') \in C$.

Fig. 3. Abstract nonlinear transitions

Decide extends a leaf L in $\mathcal{D}$ with two children. The nodes correspond to partial models for the variables that are arguments to the recursive predicates in $\mathcal{F}$. To differentiate two possibly different subsets of $\boldsymbol{x}$ we use $\tilde{\boldsymbol{x}}$ and $\tilde{\tilde{\boldsymbol{x}}}$. The children are possibly pointers to nodes that already exist in $\mathcal{D}$ (so that we don't expand the same model twice). Model declares a counter example when all the leaves and internal nodes have been validated. This amounts to that the root of $\mathcal{D}$ is in the cache C. Conflicts are similar, Conflict backtracks from a leaf when the (partial) model annotating the leaf contradicts the constraints at its level.

There are two new rules. The rules are Base and Close. Their role is to propagate cache hits upwards in the model DAG. At the base level, a model $\mathcal{M}$ is added to the cache C if it implies R_0. The Close rule removes children from an internal node if each child is reachable. The model annotating the internal node is then also reachable, so added to C.

Correctness follows analogously to the basic PDR algorithm, as we maintain the following properties for a state $\mathcal{D} \parallel A \parallel C$:

1. $R_0 \equiv \mathcal{F}(false)$.
2. Invariant (2) holds.
3. Every member $\mathcal{M} \in C$ is contained in $\mathcal{F}^N(false)$.
4. Every internal node $\langle \mathcal{M}, i+1 \rangle$ with children $\langle \tilde{x} = c_0, i \rangle, \langle \tilde{\tilde{x}} = c_1, i \rangle$, it is the case that $\mathcal{M}, \tilde{x}_0 = c_0, \tilde{\tilde{x}}_1 = c_1 \models \mathcal{T}[R_i(x_0), R_i(x_1)]$.

Example 2. Consider a nonlinear system $R(true, true)$. $R(x_0, y_0) \wedge R(x_1, y_1) \to R(x_0 \oplus x_1, y_0 \oplus y_1)$. $R(true, false) \to false$. A sample run of the algorithm proceeds as follows:

Initialize	$\implies \epsilon \parallel A_0 \parallel \{\}$	for $A_0 = [N \leftarrow 0, R_0 \leftarrow x \wedge y]$
Unfold	$\implies \epsilon \parallel A_1 \parallel \{\}$	for $A_1 = A_0[N \leftarrow 1, R_1 \leftarrow true]$
Candidate	$\implies \langle x \wedge \neg y, 1 \rangle \parallel A_1 \parallel \{\}$	
Conflict	$\implies \epsilon \parallel A_2 \parallel \{\}$	for $A_2 = A_1[R_1 \leftarrow R_1 \wedge (\neg x \vee y)]$
Unfold	$\implies \epsilon \parallel A_3 \parallel \{\}$	for $A_3 = A_2[N \leftarrow 2, R_2 \leftarrow true]$
Candidate	$\implies \langle x \wedge \neg y, 2 \rangle \parallel A_3 \parallel \{\}$	
Decide	$\implies \langle x \wedge \neg y, 2 \rangle \bullet \{\langle x \wedge y, 1 \rangle \langle \neg x \wedge y, 1 \rangle\} \parallel A_3 \parallel \{\}$	
Base	$\implies \langle x \wedge \neg y, 2 \rangle \bullet \{\langle x \wedge y, 1 \rangle \langle \neg x \wedge y, 1 \rangle\} \parallel A_3 \parallel \{x \wedge y\}$	
Conflict	$\implies \langle x \wedge \neg y, 2 \rangle \parallel A_4 \parallel \{x \wedge y\}$	for $A_4 = A_3[R_j \leftarrow R_j \wedge (x \vee \neg y)]_{j=1}^2$
Induction	$\implies \ldots \parallel A_4[R_j \leftarrow R_j \wedge (\neg x \vee y)]_{j=1}^2 \parallel \{x \wedge y\}$	
Valid	$\implies$ Valid	

Note how Decide develops two branches. When one child is in conflict then both children are collapsed. Note also how Induction is used to push $(\neg x \vee y)$ up to level 2. The property is inductive when combined with the property $(x \vee \neg y)$. At this point $R_2 \to R_1$ (e.g., $R_1 \subseteq R_2$) so the procedure terminates with Valid. $\boxtimes$

5   Theories - The Case of Linear Real Arithmetic

We generalize PDR to handle non-Boolean constraints. The problem goes from PSPACE to highly intractable. Nevertheless, we identify a subclass, timed pushdown systems, that are handled by our generalization. Our approach is to lift the Conflict and Decide rules and instantiate the generalization to the theory of linear real arithmetic. Central to our approach is the use of models for guiding the creation of conflict clauses as interpolants. The interpolants are a minimal set of constraints implied by the existing abstraction that suffice to exclude a spurious counter example. When iterated over all spurious counter examples, our procedure does in fact produce interpolants for systems of non-recursive Horn clauses [8]. Our incremental approach is appealing compared to an approach that computes interpolants in from an eager unfolding: intermediary results from spurious counter examples act as conflict clauses for future traversals. We use the calculus from Section 3 to keep definitions simpler.

5.1   Conflicts

Recall Conflict applies when there is a $\varphi \subseteq \neg \mathcal{M}$ such that $\mathcal{F}(R_i) \to \varphi$. The Conflict rule therefore applies when $\mathcal{F}(R_i) \to \neg \mathcal{M}$. The propositional version lets us add any subset of $\neg \mathcal{M}$ that is implied by $\mathcal{F}(R_i)$. The clause φ is also an interpolant by construction. A problem with using a subset of $\neg \mathcal{M}$ for infinite domains is that the number of potential counter-models is unbounded, so blocking one of an unbounded set of models does not help to ensure convergence. In principle one can take any clause $Post$ such that

$$\mathcal{F}(R_i) \ \to \ Post, \qquad Post \ \to \ \neg \mathcal{M} \ . \tag{4}$$

This suggests a G-Conflict rule (formulated for linear fixed-points) as:

G-Conflict $\langle \mathcal{M}, i+1 \rangle M \parallel A \ \Longrightarrow \ M \parallel A[R_j \leftarrow R_j \wedge Post]_{j=1}^{i+1}$
$\qquad\qquad\qquad\qquad$ **if** $\models \mathcal{F}(R_i) \to Post, \quad Post \to \neg \mathcal{M}.$

Where $Post$ is any clause that uses the variables x and implies $\neg \mathcal{M}$. Notice that we require $Post$ to be a single clause. At the other extreme, one could think of taking $Post := \mathcal{F}(R_i)$, the strongest post-condition that is independent of $\mathcal{M}$. The resulting algorithm would have to rely on quantifier elimination to convert the result into a set of clauses and for making effective use of Induction. The rule G-Conflict without further conditions is not informative.

Arithmetical Conflicts. We instantiate G-Conflict for the theory of Linear Real Arithmetic (LRA) and show that we obtain a decision procedure for safety properties of timed push-down systems. The main idea is to compute the *strongest* conflict clause modulo linear real arithmetic from unsatisfiability of $\mathcal{M} \wedge \mathcal{F}(R_i)$.

The conflict clause is by construction an interpolant and the way it is extracted can be described as a specialized interpolation procedure. On the right is the stage $N = 4$ where PDR pushes a counter example down for Bakery. It reaches a conflict because

$$L = 2 \wedge M = 2 \models \mathcal{F}(R_3) \wedge \neg S$$
$$\uparrow$$
$$L = 1 \wedge M = 2 \wedge Y_2 = 0 \models \mathcal{F}(R_2)$$
$$\uparrow$$
$$L = 1 \wedge M = 1 \wedge Y_1 = 1 \wedge Y_2 = 0 \models \mathcal{F}(R_1)$$
$$\vdots$$
$$L = 0 \wedge M = 1 \wedge Y_2 = 0 \models \neg \mathcal{F}(R_0)$$

$L = 0 \wedge M = 1 \wedge Y_2 = 0 \wedge \mathcal{T}[R_0(x_0)]$ is unsatisfiable. The justification includes the clause $\neg(Y_2 \leq 0 \wedge Y_1 \geq 0 \wedge Y_2 \geq Y_1 + 1)$. The last two literals are from $\mathcal{T}[R_0(\boldsymbol{x_0})]$. They resolve to $Y_2 > 0$, justifying the stronger conflict clause $\neg(L = 0 \land M = 1 \land Y_2 \leq 0)$.

In general, assume $\mathcal{M}$ is of the form $\bigwedge_i k_i \leq x_i \leq k_i$ where x_i are variables and k_i are numerals of type Real. The G-Conflict rule applies when $\mathcal{M} \land \mathcal{T}[R_i(\boldsymbol{x_0})]$ is unsatisfiable and there is a resolution proof Π that derives the empty clause. In the following we make two important assumptions for our construction, first we assume that all literals in Π are already in $\mathcal{M} \land \mathcal{T}[R_i(\boldsymbol{x_0})]$. This is the case for proofs produced by the DPLL(T) framework [13]. Second, we assume that all literals in $\mathcal{M}$ are used in unit-resolution with input clauses. This can be enforced by permuting $\mathcal{M}$ up in proofs. The leaves of Π comprise of the inequalities (unit-literals) from $\mathcal{M}$, clauses from $\mathcal{T}[R_i(\boldsymbol{x_0})]$ and T-axioms. In the theory of LRA, the T-axioms are of the form $\bigvee_i \neg(A_i\boldsymbol{x} - \boldsymbol{b_i} \leq 0)$, where A_i are row vectors and $\boldsymbol{b_i}$ are constants. Recall that we can represent strict inequalities $t > s$ using non-strict inequalities by using an infinitesimal ϵ constant for $t \geq s + \epsilon$. Let us write $A\boldsymbol{x} \leq \boldsymbol{b}$ for the conjunction $\bigwedge_i A_i\boldsymbol{x} \leq \boldsymbol{b_i}$. Farkas' lemma implies that there is a corresponding set of non-negative coefficients $\boldsymbol{\lambda}$, such that $\boldsymbol{\lambda} \cdot A \cdot \boldsymbol{x}$ is a numeric constant and $\boldsymbol{\lambda} \cdot A \cdot \boldsymbol{x} > \boldsymbol{\lambda b}$. These coefficients are produced as a side-effect of the Simplex procedure. Proof-objects exposed by Z3 [9] include the coefficients.

The method for creating *Post* is now as follows: conjoin every literal from $\mathcal{M}$ that resolves against a clause from $\mathcal{T}[R_i(\boldsymbol{x_0})]$ in Π. Furthermore, for every T-axiom we partition the literals into two groups, the first group contains the literals that resolve against a literal from $\mathcal{M}$, the second comprises of literals that resolve against clauses from $\mathcal{T}[R_i(\boldsymbol{x_0})]$. Rewrite the inequality as $\begin{bmatrix} C \\ D \end{bmatrix} x \leq \begin{bmatrix} \boldsymbol{c} \\ \boldsymbol{d} \end{bmatrix}$, where the inequalities with coefficients $C, \boldsymbol{c}$ resolve against $\mathcal{M}$ and the remaining inequalities resolve against $\mathcal{T}[R_i(\boldsymbol{x_0})]$. The coefficients from Farkas' lemma are $\boldsymbol{\lambda}_C$ and $\boldsymbol{\lambda}_D$ respectively, such that:

$$\boldsymbol{\lambda}_C C\boldsymbol{x} + \boldsymbol{\lambda}_D D\boldsymbol{x} > \boldsymbol{\lambda}_C \boldsymbol{c} + \boldsymbol{\lambda}_D \boldsymbol{d}, \tag{5}$$

and therefore:

$$D\boldsymbol{x} \leq \boldsymbol{d} \;\rightarrow\; \boldsymbol{\lambda}_D D\boldsymbol{x} \leq \boldsymbol{\lambda}_D \boldsymbol{d}, \quad \boldsymbol{\lambda}_D D\boldsymbol{x} \leq \boldsymbol{\lambda}_D \boldsymbol{d} \;\rightarrow\; \boldsymbol{\lambda}_C C\boldsymbol{x} > \boldsymbol{\lambda}_C \boldsymbol{c} \;. \tag{6}$$

Then replace the theory axiom in Π by

$$\neg(D\boldsymbol{x} \leq \boldsymbol{d}) \;\lor\; \boldsymbol{\lambda}_D D\boldsymbol{x} \leq \boldsymbol{\lambda}_D \boldsymbol{d} \tag{7}$$

and conjoin $\boldsymbol{\lambda}_D D\boldsymbol{x} > \boldsymbol{\lambda}_D \boldsymbol{d}$ to *Post*. This literal is implied by the original literals $C\boldsymbol{x} > \boldsymbol{c}$ from $\mathcal{M}$. Denote by Farkas-Conflict the rule that extracts formula *Post* corresponding to a weakening of $\mathcal{M}$ determined by Π'.

5.2   Timed Push-Down Systems

Basic timed transition systems are of the form $\mathcal{S} = \langle \boldsymbol{x}, \mathcal{C}, \Theta, \rho \lor \rho_{tick} \rangle$, where $\boldsymbol{c} \subseteq \boldsymbol{x}$ is a designated set of clock variables, and $\boldsymbol{d} := \boldsymbol{x} \setminus \boldsymbol{c}$ are finite domain data-variables. There is a transition $\rho_{tick} : \exists \delta.\boldsymbol{c}' = \boldsymbol{c} + \delta \land \boldsymbol{d}' = \boldsymbol{d}$ that advances time on the clock variables. Other transitions are allowed to reset clocks to 0 and modify the data-variables. We consider a slight extension of timed transition systems with push-down capabilities. Reachable states can be described as:

$$R(\boldsymbol{c}, \boldsymbol{d}) \wedge \boldsymbol{c}' = \boldsymbol{c} + \delta \wedge \varphi(\boldsymbol{c}', \boldsymbol{d}) \rightarrow R(\boldsymbol{c}', \boldsymbol{d}) \tag{8}$$

$$R(\boldsymbol{c}, \boldsymbol{d}) \wedge \wedge_i \boldsymbol{c}'_i = reset?(\boldsymbol{c}_i) \rightarrow R(\boldsymbol{c}', \boldsymbol{d}) \tag{9}$$

$$R(\boldsymbol{c}, \boldsymbol{y}) \wedge R(\boldsymbol{c}, \boldsymbol{z}) \wedge \varphi(\boldsymbol{c}, \boldsymbol{d}, \boldsymbol{y}, \boldsymbol{z}) \rightarrow R(\boldsymbol{c}, \boldsymbol{d}) \tag{10}$$

where $reset?(c)$ is either c or 0 and the occurrences of clocks in φ is restricted to difference arithmetic formulas of the form $c_i - c_j \leq k$ for k a constant.

Theorem 2 (Timed Push-down System Reachability). *Generalized PDR with* Farkas-Conflict *decides timed push-down system reachability.*

Proof (Idea). Use the observation that Farkas-Conflict produces only literals in the transitive closure of the difference constraints from the timed push-down system. Assume $\mathcal{F}$ is a description of a timed push-down system that uses the difference constraints $\Delta = \{y_{i_1} - y_{j_1} \leq k_1, y_{i_2} \leq k_2, \ldots\}$ where each y_i is from $\boldsymbol{x_0}$. Add to Δ also the inequalities $y_i \geq 0, y_i \leq 0$ for each y_i from $\boldsymbol{x_0}$. As usual in difference arithmetic we can treat Δ as a directed graph whose edges are weighted by the difference constraints. Let Δ^* be the transitive closure that contains inequalities for every loop-free path in Δ. Suppose that $\boldsymbol{x} = \boldsymbol{c} \wedge \mathcal{T}[R_i(\boldsymbol{x_0})]$ is unsatisfiable (the premise of G-Conflict) with proof Π and let $C : \bigvee_i \neg(A_i\boldsymbol{x} - b_i \leq 0)$ be a clause in Π that is justified by Farkas lemma. Consider the most interesting case where C contains at most two literals $x_i \geq k_i$, $x_j \leq k_j$ from $\mathcal{M}[\boldsymbol{x}]$ and C contains the atoms $x_i = y_i + \delta, x_j = y_j + \delta$ (or $x_i = y_i, x_j = y_j$) together with literals from Δ^*. Since difference logic tautologies correspond to paths in a weighted graph, the literal $\lambda_D D\boldsymbol{x} > \lambda_D\boldsymbol{d}$ obtained from Farkas' lemma cancel out the coefficient δ and use a weight that corresponds to a directed path in Δ^*. In each case, every spurious counter example was blocked by a combination of literals in Δ^*. Since Δ^* is finite this process terminates.

The Farkas-Conflict rule also suffices for some non-timed transition systems. It can prove the mutual exclusion property of our initial Bakery algorithm example.

5.3   Decisions

Farkas-Conflict is useful for many scenarios, but it is easy to come up with Horn clauses where it is insufficient. For example, the inductive invariant $2x = y$ that is required to establish the satisfiability of the Horn clauses cannot be found using Farkas-Conflict.

$$R(x, y) \rightarrow R(x + 1, y + 2). \quad R(0, 0). \quad R(x, y) \wedge 2x \neq y \rightarrow false. \tag{11}$$

A remedy to this limitation is to generalize the Decide rule. The approach is motivated as a way of producing relevant predicates, similar to what predicate abstraction achieves. We cannot help to note some dualities between the Decide and the Conflict rules: Conflict strengthens invariants and uses over-approximations of strongest post-conditions; Decide weakens counter examples and uses under-approximations of pre-conditions. Recall the basic Decide rule:

Decide $\langle \mathcal{M}, i+1 \rangle M \parallel A \implies \langle \tilde{x} = c_0, i \rangle \langle \mathcal{M}, i+1 \rangle M \parallel A$ if $\mathcal{M}, \tilde{x}_0 = c_0 \models \mathcal{T}[R_i(x_0)]$

In order to retain predicates that are relevant to the counter example trace, we can use any pre-condition Pre such that

$$\tilde{x}_0 = c_0 \;\rightarrow\; Pre[\tilde{x}_0], \quad Pre[\tilde{x}_0] \;\rightarrow\; \exists x \;.\; \mathcal{M}[x] \wedge \mathcal{T}[R_i(x_0)] \;. \qquad (12)$$

Thus, the generalized Decide rule is:

G-Decide $\langle \mathcal{M}, i+1 \rangle M \parallel A \implies \langle \tilde{x} = c_0 \wedge Pre[\tilde{x}], i \rangle \langle \mathcal{M}, i+1 \rangle M \parallel A$
        if $\tilde{x}_0 = c_0 \;\rightarrow\; Pre[\tilde{x}_0], \quad Pre[\tilde{x}_0] \;\rightarrow\; \exists x \;.\; \mathcal{M}[x] \wedge \mathcal{T}[R_i(x_0)]$

A crucial insight in [6] is to use ternary simulation for computing the relevant subset $\tilde{x}_0$ of x. This reduces the set of literals in $\tilde{x}_0 = c_0$. We are not aware of a canonical approach to lifting model generalization to the first-order case. The following is a heuristic. For the first-order case we also leverage ternary simulation to minimize $\tilde{x}_0 = c_0$, and select the literals in $\mathcal{T}[R_i(x_0)]$ that contribute to making the formula true under $\tilde{x}_0 = c_0, \mathcal{M}[x]$. The goal is to produce a conjunction $\tilde{x}_0 = c_0 \wedge Pre[\tilde{x}_0]$ comprising of an assignment to $\tilde{x}_0$ and auxiliary literals over $\tilde{x}_0$ such that $\tilde{x}_0 = c_0 \models Pre[\tilde{x}_0]$. So by induction assume $\mathcal{M}[x]$ is also of this form: $\mathcal{M}^1 \wedge Pre^1$. When $\mathcal{F}$ is derived from guarded assignments, the variables x are typically given as a function of previous state variables, and the selected literals from $\mathcal{T}[R_i(x_0)]$ contains equalities of the form $x = t[x_0]$. We collect these equalities as a substitution θ. The condition for $Pre[\tilde{x}_0]$ is then reduced to:

$$\tilde{x}_0 = c_0 \;\rightarrow\; Pre[\tilde{x}_0], \quad Pre[\tilde{x}_0] \;\rightarrow\; \exists x \;.\; \mathcal{M}^1 \wedge (Pre^1 \wedge \mathcal{T}[R_i(x_0)])\theta \qquad (13)$$

Our current approach creates $Pre[\tilde{x}_0]$ as the conjunction of $\tilde{x}_0 = c_0$ and the selected literals from $(Pre^1 \wedge \mathcal{T}[R_i(x_0)])\theta$ that do not contain variables from x and that do not mix variables from different predecessor states.

Example 3. Assume a candidate counter example to (11) sets $x' = 3, y' = 1$. Then, $\exists x', y' . x' = 3 \wedge y' = 4 \wedge [(y' = y+2 \wedge x' = x+1 \wedge 2x' \neq y') \vee y' = x' = 0]$ yields the pre-condition $x = 2 \wedge y = 2 \wedge 2x \neq y$.   ⊠

It is now also necessary to generalize Conflict so that it can produce the necessary conflict clauses from either $\mathcal{M}$ or the predicates from the weakest pre-condition.

Multi-core Conflicts. Each unsatisfiable core for $\mathcal{F}(R_i) \wedge \mathcal{M}$ gives rise to a different conflict that can enable a different proof. A proper generalization of G-Conflict is therefore to allow multiple conflicts

MC-Conflict $\langle \mathcal{M}, i+1 \rangle M \parallel A \implies M \parallel A[R_{i+1} \leftarrow R_{i+1} \wedge Post_1 \wedge \ldots \wedge Post_k]$
        if $\models \mathcal{F}(R_i) \rightarrow Post_j, \quad Post_j \rightarrow \neg \mathcal{M}$, for $j = 1..k$.

Algorithms for unsatisfiable cores [10] and efficient integration of cores in PDR is beyond the scope of this paper.

6   Implementation and Experiments

We have implemented Generalized PDR in μZ [9] and we have performed a number of experiments to validate the generalizations to nonlinear PDR and linear real arithmetic. Additional material is online http://rise4fun.com/muZ.

We tested our implementation on a set of 2906 Boolean programs that come with the Windows Driver Research Platform.[1] Most programs are checked for safety violations within a second by both the Bebop tool and μZ. We wrote a basic converter from Boolean programs into Horn clauses. It associates a recursive predicate with each program statement and therefore sometimes requires much more space than the Boolean program. We are therefore not surprised that our prototype is generally 3 times slower than Bebop.

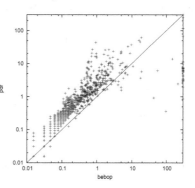

Nevertheless it prevails where it matters: *it was able to solve 32 programs that Bebop could not solve within a 5 minute timeout.* μZ times out on just one program where Bebop also times out.

In other experiments we use μZ successfully on instantiations of timed transition systems, the examples in this paper, and a set of device drivers provided by Ken McMillan. They use arithmetic for reasoning about pointer offsets so Farkas-Conflict also suffices for verifying safety properties of these programs.

7   Conclusions, Related and Future Work

We generalized PDR in two directions. To solve general Horn clauses we first developed an abstract account of PDR and leveraged it for nonlinear predicate transformers. We also provided a solution to lifting PDR to linear real arithmetic. The solution uses a generalization of unsatisfiable cores for theories. The idea is to compute an interpolant based on the unit literals from a spurious counter example. We applied it to timed automata (with push-down capabilities). This is a new algorithm for timed automata, but not a new decidability result. Other extensions such as vector addition systems can be formulated in Datalog [14]. These extensions are not addressed here.

PDR can be seen as an instance of a Counter Example Guided Abstraction Refinement [5]. It refines state abstractions while avoiding approximating or unfolding the transition relation. Related approaches [2,12,8,7,1] also refine transition relations. In several cases (and in contrast to PDR), the abstraction refinement loop relies on unfolding the transition relation up to a certain depth. Of particular interest is [7], which explicates the connection between proof rules and solving Horn clauses.

---

[1] http://research.microsoft.com/slam

Generalizing PDR to theories is an open-ended enterprise. The experiments so far indicate that Generalized PDR is attractive as a tool for satisfiability modulo fixed-points. Nevertheless, several extensions and optimizations should be pursued and there are several avenues for future work. A study of *weakest* T-*unsat cores* deserves attention from both an algorithmic point of view and from a point of view of commonly used theories. Our implementation in μZ works with algebraic data-types, but not yet with general uninterpreted functions. We believe uninterpreted functions can be handled by extending models to carry also a congruence class of terms. The corresponding version of Farkas-Conflict is then super-position on T-conflicts from congruence closure. We would also like to generalize other parts of PDR, in particular the crucial Induction rule. The implementations of PDR we are aware of use cheap strategies, they pick random literals in clauses and try to drop them one-by-one until a limit (of 4) failed strengthening attempts is reached. It is tempting to speculate of other generalizations for strengthening clauses. For example, $(\varphi \lor \neg(x \le y + 1) \lor \neg(z + 2 \le x)) \in R_i$ could be strengthened to $(\varphi \lor \neg(x + 1 \le y))$, and $(x \not\simeq y \lor \varphi[x]) \in R_i$ could be strengthened to $\varphi[y]$.

Acknowledgments. Thanks to Natarajan Shankar, Josh Berdine, Bruno Dutertre, Sam Owre and the reviewers for significant constructive feedback. Also thanks to Andrey Rybalchenko and Ken McMillan for numerous discussions.

References

1. Albarghouthi, A., Gurfinkel, A., Chechik, M.: WHALE: An Interpolation-Based Algorithm for Inter-procedural Verification. In: Kuncak, V., Rybalchenko, A. (eds.) VMCAI 2012. LNCS, vol. 7148, pp. 39–55. Springer, Heidelberg (2012)
2. Ball, T., Rajamani, S.K.: The SLAM project: debugging system software via static analysis. SIGPLAN Not. 37(1), 1–3 (2002)
3. Blass, A., Gurevich, Y.: Existential Fixed-Point Logic. In: Börger, E. (ed.) Computation Theory and Logic. LNCS, vol. 270, pp. 20–36. Springer, Heidelberg (1987)
4. Bradley, A.R.: SAT-Based Model Checking without Unrolling. In: Jhala, R., Schmidt, D. (eds.) VMCAI 2011. LNCS, vol. 6538, pp. 70–87. Springer, Heidelberg (2011)
5. Clarke, E.M., Grumberg, O., Jha, S., Lu, Y., Veith, H.: Counterexample-guided abstraction refinement. Journal of the ACM 50(5), 752–794 (2003)
6. Een, N., Mishchenko, A., Brayton, R.: Efficient implementation of property-directed reachability. In: FMCAD (2011)
7. Grebenshchikov, S., Lopes, N.P., Popeea, C., Rybalchenko, A.: Synthesizing software verifiers from proof rules. In: PLDI (2012)
8. Gupta, A., Popeea, C., Rybalchenko, A.: Solving Recursion-Free Horn Clauses over LI+UIF. In: Yang, H. (ed.) APLAS 2011. LNCS, vol. 7078, pp. 188–203. Springer, Heidelberg (2011)
9. Hoder, K., Bjørner, N., de Moura, L.: μZ– An Efficient Engine for Fixed Points with Constraints. In: Gopalakrishnan, G., Qadeer, S. (eds.) CAV 2011. LNCS, vol. 6806, pp. 457–462. Springer, Heidelberg (2011)

10. Liffiton, M.H., Sakallah, K.A.: Algorithms for computing minimal unsatisfiable subsets of constraints. J. Autom. Reasoning 40(1), 1–33 (2008)
11. Manna, Z., Pnueli, A.: Temporal verification of reactive systems - safety (1995)
12. McMillan, K.L.: Interpolants from Z3 proofs. In: FMCAD (2011)
13. Nieuwenhuis, R., Oliveras, A., Tinelli, C.: Solving SAT and SAT Modulo Theories: From an abstract DPLL procedure to DPLL(T). J. ACM 53(6) (2006)
14. Revesz, P.Z.: Safe Datalog Queries with Linear Constraints. In: Maher, M.J., Puget, J.-F. (eds.) CP 1998. LNCS, vol. 1520, pp. 355–369. Springer, Heidelberg (1998)

SMT-Aided Combinatorial Materials Discovery[*]

Stefano Ermon[1], Ronan Le Bras[1], Carla P. Gomes[1],
Bart Selman[1], and R. Bruce van Dover[2]

[1] Dept. of Computer Science
[2] Dept. of Materials Science and Engr.,
Cornell University, Ithaca, NY 14853

Abstract. In combinatorial materials discovery, one searches for new materials with desirable properties by obtaining measurements on hundreds of samples in a single high-throughput batch experiment. As manual data analysis is becoming more and more impractical, there is a growing need to develop new techniques to automatically analyze and interpret such data. We describe a novel approach to the phase map identification problem where we integrate domain-specific scientific background knowledge about the physical and chemical properties of the materials into an SMT reasoning framework. We evaluate the performance of our method on realistic synthetic measurements, and we show that it provides accurate and physically meaningful interpretations of the data, even in the presence of artificially added noise.

Keywords: SMT, Combinatorial Materials Discovery, Automated Reasoning.

1   Introduction

In recent years, we have witnessed an unprecedented growth in data generation rates in many fields of science [10]. For instance, in combinatorial materials discovery, one searches for materials with new desirable properties by obtaining measurements on hundreds of samples in a single batch experiment [7,14]. These are referred to as 'high-throughput' experiments, and are common to many other fields such as molecular biology or astronomy, where there is a need to optimize the data throughput of high-cost equipment [2]. As manual data analysis is becoming more and more impractical, there is a growing need to develop new techniques to automatically analyze and interpret such vast amount of data for important trends and results. Modern statistical machine learning and data-mining approaches have been quite effective in extracting relevant information from the ever increasing streams of raw digital data. However, in scientific data analysis, there is a large amount of rather complex domain-specific background knowledge that needs to be taken into account, such as the physical and chemical properties of the materials in the combinatorial materials discovery domain.

---

[*] This work was supported by NSF Grant 0832782.

A. Cimatti and R. Sebastiani (Eds.): SAT 2012, LNCS 7317, pp. 172–185, 2012.

In this paper, we describe a novel approach to the phase map identification problem, a key step towards understanding the properties of new materials created and examined using the combinatorial materials discovery method. The process of identifying a phase map has been traditionally carried out manually by domain-experts, but a completely automatic solution for the phase map identification problem would open the way for even more automation in the combinatorial approach pipeline. Further, a scalable and reliable automatic data interpretation procedure would allow us to analyze larger datasets that go beyond the capabilities of human experts.

In our approach, we integrate domain-specific scientific background knowledge about the physical and chemical properties of the materials into an SMT reasoning framework based on linear arithmetic. The problem has a hybrid nature, with continuous measurement data, discrete decision variables and combinatorial constraints at the same time. We show that using our novel encoding, state-of-the-art SMT solvers can automatically analyze large synthetic datasets, and generate interpretations that are physically meaningful and very accurate, even in the presence of artificially added noise. Moreover, our approach scales to realistic-sized problem instances, outperforming a previous approach based on Constraint Programming and a set-variables encoding [11]. Further, we show that SMT solving outperforms both Constraint Programming and Mixed Integer Programming translations of our SMT formulation. This suggests that the improvements come from the SMT solving procedure rather than from the new arithmetic-based encoding, opening a novel application area for SMT solving technology beyond the traditional verification domains [4,5].

We see this work as a first step towards using automated reasoning technology to aid the scientific discovery process. While several aspects of our method are specific to the materials discovery application, the approach we take to scientific data analysis is general. Given the flexibility and reasoning power of modern day SMT solvers, we expect to see more applications of this technology to other fields of science.

2   Combinatorial Materials Discovery

The combinatorial method is a general experimentation setting where many simultaneous experiments are performed and analyzed in batch at each step. This experimental methodology is intended to speed up the scientific discovery process, and is becoming common in a number of areas, including catalyst discovery, drug discovery, polymer optimization, and chemical synthesis. For example, new catalysts have been discovered 10 to 30 times faster using the combinatorial approach rather than conventional methodology [7,14]. This is an important research direction in the field of Computational Sustainability, for instance because new materials with improved catalytic activity can be used for fuel cell applications [8].

In this paper, we consider a combinatorial materials discovery approach called *composition-spread*, that has been recently applied with success to speed up the discovery of new catalysts [15]. In the composition spread approach, three metals (or oxides) are sputtered onto a silicon wafer using guns pointed at three distinct locations, resulting in a so-called *thin film*. Different locations on the silicon wafer correspond to different concentrations of the sputtered materials, depending on their distance from the gunpoints. During experimentation, a number of locations (samples) on the thin film are examined using an x-ray diffraction technique, obtaining a diffraction pattern for each sampled point that gives the intensity of the electromagnetic waves as a function of the scattering angle of radiation. The observed diffraction pattern is closely related to the underlying crystal structure, which provides important insights into chemical and physical properties of the corresponding composite material.

A key step towards understanding the chemical and physical properties of the composite materials on a *thin film* is to obtain a so-called *phase map*, that is used to identify regions of the silicon wafer that share the same underlying crystal structure (see Figure 2 for an example). Intuitively, the idea is that the different diffraction patterns observed across the *thin film* can all be explained as combinations of a small number (typically, less than 6) of diffraction patterns called *basis patterns* or *phases*. Finding the phase map corresponds to identifying these *basis patterns* and their location on the silicon wafer. This is a challenging task because we only observe combinations of the *basis patterns*, and the measurements are affected by noise. Furthermore, due to a fairly complicated physical process dealing with the expansion of crystals on the lattice, *basis patterns* can appear scaled (contracted to a smaller or larger frequency range), and they must satisfy a number of physical constraints (for instance, basis patterns must appear in contiguous locations on the *thin film* and there is a maximum number of *basis patterns* that can appear in each sample diffraction pattern).

2.1   Phase Map Identification

Formally, we are given P diffraction patterns $\mathbf{D}_0, \cdots, \mathbf{D}_{P-1}$, one for each of the P points sampled on the *thin film*, where each vector $\mathbf{D}_i = (d_{0,i}, \cdots, d_{B-1,i}) \in (\mathbb{R}_{\geq 0})^B$ represents the intensity of the electromagnetic waves for a fixed set of B scattering angles of radiation. The sample points are embedded into a graph $\mathcal{G}$, such that there is a vertex for every point and edges connect points that are close on the *thin film* (for instance, based on Delaunay triangulation). Given a norm $||\cdot||$ (for instance, an L_∞ norm), we want to find K basis patterns $\mathbf{B}_0, \cdots, \mathbf{B}_{K-1}$ where $\mathbf{B}_i \in (\mathbb{R}_{\geq 0})^B$, coefficients $a_{i,j} \in \mathbb{R}$ and scaling factors $s_{i,j} \in \mathbb{R}$ for $i = 0, \cdots, P-1, j = 0, \cdots, K-1$ that minimize

$$\sum_{i=0}^{P-1} ||\mathbf{D}_i - \sum a_{i,j} S(\mathbf{B}_j, s_{i,j})|| \tag{1}$$

where $S(\cdot)$ is an operator modeling the scaling phenomena (see below), and the coefficients $a_{i,j}$ must satisfy

$$a_{i,j} \geq 0 \quad i = 0, \cdots, P-1, j = 0, \cdots, K-1$$
$$|\{j | a_{i,j} > 0\}| \leq M \quad i = 0, \cdots, P-1$$

that is, they are non-negative and no more than M basis patterns can be used to explain a point i. Furthermore, the subgraph induced by $\{i | a_{i,j} > 0\}$ must be connected for $j = 0, \cdots, K - 1$ (so that the basis patterns appear in contiguous locations on the *thin film*). The scaling operator $S(\cdot)$ models the potential expansion of the crystals on the lattice. Specifically, a peak appearing at scattering angle a in the k-th basis pattern might appear respectively at scattering angles $s_{p,k} \cdot a$ and $s_{p',k} \cdot a$ at points p, p' of the silicon wafer because of the scaling effect. For each basis pattern k, the corresponding scaling coefficients $s_{i,k}$ must be continuous and monotonic as a function of the corresponding location i on the *thin film*. Further, the presence of 3 or more basis patterns in the same point prevents any significant expansion of the crystals, and therefore scalings do not occur.

Notice that this formulation is closely related to a principal component analysis (PCA) of the data, but includes additional constraints needed to ensure that the solution is physically meaningful, such as the non-negativity of eigenvectors, connectivity, and phase usage limitations.

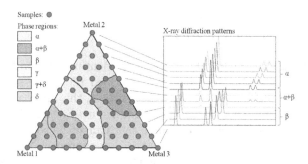

Fig. 1. Left: Pictorial depiction of the problem, showing a set of sampled points on a *thin film*. Each sample corresponds to a different composition, and has an associated measured x-ray diffraction pattern. Colors correspond to different combinations of the basis patterns $\alpha, \beta, \gamma, \delta$. On the right: Scaling (shifting) of the diffraction patterns as one moves from one point to a neighboring one.

3   Prior Work

There have been several attempts to automate the *phase map* identification process. Most of the solutions in the literature are based on unsupervised machine learning techniques, such as clustering and non-negative matrix factorization [13,12]. While these approaches are quite effective at extracting information from large amounts of noisy data, their major limitation is that it is hard to enforce the physical constraints of the problem at the same time. As a result, the interpretations obtained with these techniques are often not physically meaningful, for instance because regions corresponding to some basis patterns are not connected [11].

To address these limitations, in [11] they used a Constraint Programming approach to enforce the constraints on the phase maps, defining a new problem called *Pattern Decomposition with Scaling*. They propose an encoding based on set variables, but the main limitation of their work is that current state-of-the-art CP solvers cannot scale to realistic size instances (e.g., with at least 40 sample points). To overcome this limitation, the authors used a heuristic preprocessing step based on clustering to fix the value of certain variables before attempting to solve the problem. While the solutions they found are empirically shown to be accurate, their strategy cannot provide any guarantee because it only explores part of the search space.

Our approach is similar to the one proposed in [11], but in this work we introduce a novel SMT encoding based on arithmetic to formulate the phase map identification problem. The SMT formalism nicely captures the hybrid nature of the problem, which involves discrete decision variables and continuous measurement data at the same time. Furthermore, we show that the ability to reason at the level of arithmetic operations of SMT solvers allows our approach to scale to instances of realistic size without need for Machine Learning-based heuristics.

4   SMT-Aided Phase Map Identification

In our first attempt to model the phase map identification problem, we constructed an SMT-based model where we described the entire spectrum of all the unknown basis patterns $\mathbf{B}_0, \cdots, \mathbf{B}_{K-1}$. However, this approach requires too many variables to obtain a sufficiently fine-grained description of the diffraction patterns, and ultimately leads to instances that cannot be solved using current state-of-the art solvers. We therefore use the same approach presented in [11], and we preprocess the diffraction patterns $\mathbf{D}_0, \cdots, \mathbf{D}_{P-1}$ using a peak detection algorithm, extracting the locations of the *peaks* $\mathcal{Q}(p)$ in the x-ray diffraction pattern of each point p (see Figure 1). This is justified by the nature of the diffraction patterns, as constructive interference of the scattered x-rays occurs at specific angles (thus creating peaks of intensities) that characterize the underlying crystal. Furthermore, matching the locations of the peaks is what human experts do when they try to manually solve these problems.

Given the sets of observed peaks $\{\mathcal{Q}(p)\}_{p=0}^{P-1}$ extracted from the measured diffraction patterns $\mathbf{D}_0, \cdots, \mathbf{D}_{P-1}$, our goal is to find a set of peaks $\{\mathcal{E}_k\}_{k=0}^{K-1}$ for the K basis patterns that can explain the observed sets of peaks $\{\mathcal{Q}(p)\}_{p=0}^{P-1}$. The new variables $\{\mathcal{E}_k\}_{k=0}^{K-1}$ therefore replace the original variables $\mathbf{B}_0, \cdots, \mathbf{B}_{K-1}$ in the problem described earlier in Section 2. For each peak $c \in \mathcal{Q}(p)$ we want to have at least one peak $e \in \mathcal{E}_k$ that can explain it, i.e.

$$\forall c \in \mathcal{Q}(p) \exists e \in \mathcal{E}_k \ s.t. \ (a_{p,k} > 0 \land |c - s_{p,k} \cdot e| \le \epsilon)$$

where ϵ is a parameter that depends on how accurate the peak-detection algorithm is. Notice that we match the location of the peak, which can be measured accurately, but not its intensity, which can be very noisy. At the same time, we want to limit the number of missing peaks, i.e. peaks that should appear because they belong to some basis pattern but have not actually been measured. Therefore, instead of optimizing the objective in equation (1), we consider an approximation given by

$$\sum_{p=0}^{P-1} \sum_{k=0}^{K-1} \mathbb{1}_{[a_{p,k}>0]} \sum_{e \in \mathcal{E}_k} \mathbb{1}_{[\forall c \in \mathcal{Q}(p), |c - s_{p,k} \cdot e| > \epsilon]}$$

that gives the total number of missing peaks. All the other constraints of the problem previously introduced are not affected and still need to be satisfied. Note that we can avoid the use of expensive non-linear arithmetic by using a logarithmic scale for the x-ray data, so that multiplicative scalings become linear operations. We refer to these effects (corresponding to the scalings in the original problem formulation) as *shifts*. For each point, we therefore define a set $\mathcal{A}(p) = \{\log q, q \in \mathcal{Q}(p)\}$ of peak positions in log-scale and similarly we represent the positions of the peaks of the basis patterns using the same logarithmic scale.

After a preliminary investigation where we evaluated the performance of real-valued arithmetic, we decided to discretize the problem and use Integer variables to represent peak locations (with a user-defined discretization step). Since the diffraction data is measured using digital sensors, there is no actual loss of information if we use a small enough discretization step, and it significantly improves the efficiency of the solvers. In the resulting SMT model we therefore use a *quantifier-free linear integer arithmetic theory*.

4.1   Model Parameters

Let P be the number of sampled points on the *thin film*. We define L as the maximum number of peaks per point, i.e. $L = \max_p |\mathcal{A}_p|$. Based on the observed patterns, we precompute an upper and lower bound e_{max} and e_{min} for the positions of the peaks: $e_{max} = \max_p \max_{a \in \mathcal{A}(p)} a$, $e_{min} = \min_p \min_{a \in \mathcal{A}(p)} a$. There are also a number of user-defined parameters. K is the total maximum number of basis patterns used to explain the observed diffraction patterns, while M is the maximum number of basis patterns that can appear in any point p. ϵ is a tolerance level such that two peaks within an interval of size 2ϵ are considered to be overlapping. ϵ_S is a bound on the maximum allowed difference in the shifts of neighboring locations on the thin film, while S_{max} is a bound on the maximum possible shift. Furthermore, the user specifies a parameter T which gives a bound on the total number of peaks that should appear because they belong to some basis pattern but have not actually been measured (we will refer to them as *missing peaks*).

4.2   Variables

We use a set of Boolean variables

$$r_{p,k}, \quad p = 0, \cdots, P-1, k = 0, \cdots, K-1$$

where $r_{p,k} = TRUE$ means that phase (basis pattern) k appears in point p (i.e., $a_{p,k} > 0$). We also have the following *Integer* variables:

$$e_{k,\ell}, \quad k = 0, \cdots, K-1, \ell = 0, \cdots, L-1$$
$$S_{p,k}, \quad p = 0, \cdots, P-1, k = 0, \cdots, K-1$$
$$I_{p,k}, \quad p = 0, \cdots, P-1, k = 0, \cdots, K-1$$
$$t_p, \quad p = 0, \cdots, P-1$$

where $e_{k,\ell}$ represents the position of the ℓ-th peak of the k-th basis pattern. $S_{p,k}$ represents the shift of the k-th basis pattern at point p. The variables $I_{p,k}$ are redundant and used to count the number of phases used at point p. The variables t_p represent the number of unexplained peaks at point p, i.e. the number of missing peaks at point p. These are peaks that should appear according to the values of $\{r_{p,k}\}_{k=0}^{K-1}, \{e_{k,\ell}\}_{\ell=0}^{L-1}$, and $\{S_{p,k}\}_{k=0}^{K-1}$, but are not present, i.e. they do not belong to $\mathcal{Q}(p)$.

4.3   Constraints

The variables $I_{p,k}$ are Integer indicators for the Boolean variables $r_{p,k}$ that must satisfy

$$0 \leq I_{p,k} \leq 1 \; k = 0, \cdots, K-1, p = 0, \cdots, P-1$$
$$r_{p,k} \Leftrightarrow (I_{p,k} = 1) \; k = 0, \cdots, K-1, p = 0, \cdots, P-1$$

Peak locations $e_{k,\ell}$ in the basis patterns are bounded by what we observe in the x-ray diffraction pattern:

$$e_{min} \leq e_{k,\ell} \leq e_{max}, \quad k = 0, \cdots, K-1, \ell = 0, \cdots, L-1$$

Shifts are bounded by the maximum allowed shift, and can be assumed to be non-negative without loss of generality:

$$0 \leq S_{p,k} \leq S_{max}, \quad k = 0, \cdots, K-1, p = 0, \cdots, P-1$$

Every peak $a \in \mathcal{A}(p)$ appearing at point p must be explained by at least one peak belonging to one phase k, which can appear shifted by $S_{p,k}$:

$$\bigvee_{k=0}^{K-1} \bigvee_{\ell=0}^{L-1} (r_{p,k} \wedge (|e_{k,\ell} + S_{p,k} - a| \leq \epsilon)) \forall p, \forall a \in \mathcal{A}(p)$$

Inequalities involving the absolute value of an expression of the form $|e| < c$ where c is a positive constant are encoded as $(e < c) \wedge (e > -c)$.

If a phase k is chosen for point p (i.e., $r_{p,k} = TRUE$), then most of the peaks $e_{k,0}, \cdots, e_{k,L-1}$ should belong to $\mathcal{Q}(p)$. We count the number of missing peaks as follows:

$$t_p = \sum_{k=0}^{K-1} \sum_{\ell=0}^{L-1} ITE\left(r_{p,k} \wedge \neg \left(\bigvee_{a \in \mathcal{A}(p)} (|e_{k,\ell} + S_{p,k} - a| \leq \epsilon) \right), 1, 0\right), \forall p$$

where ITE is an if-then-else expression. Here we assume that each phase contains at least one peak, but since peaks can be overlapping (e.g., $e_{k,\ell} = e_{k,\ell+1}$) a basis pattern is allowed to contain less than L distinct peaks.

Missing Peaks Bound. We limit the number of total missing peaks (across all points p) with the user-defined parameter T

$$\sum_{p=0}^{P-1} t_p \leq T$$

Intuitively, the smaller T is, the better an interpretation of the data.

Phase Usage. There is a bound M on the total number of phases that can be used to explain the peaks observed at any location p:

$$\sum_{k=0}^{K-1} I_{p,k} \leq M, p = 0, \cdots, P-1$$

For instance, when three metals or oxides are used to obtain the thin film, we have a *ternary system*, where no more than three phases can appear in each point p, that is $M = 3$.

Shift Continuity. Phase shifting is a continuous process over the *thin film*. We therefore have the following constraint:

$$|S_{p,k} - S_{p',k}| < \epsilon_S, \forall p, \forall p' \in \mathcal{N}(p)$$

where $\mathcal{N}(p)$ is the set of neighbors of p according to the connectivity graph $\mathcal{G}$ (i.e., points that lie close to p on the *thin film*).

Shift Monotonicity. Let $\mathcal{D} = (d_0, \cdots, d_t)$ where $d_i \in \{0, \cdots, P-1\}$ be a sequence of points that lie in a straight line on the thin film. Shifting is a monotonic process, i.e. it must satisfy the following constraint

$$\left(\bigwedge_{i=0}^{t-1} (S_{d_i,k} \geq S_{d_{i+1},k}) \right) \vee \left(\bigwedge_{i=0}^{t-1} (S_{d_i,k} \leq S_{d_{i+1},k}) \right), k = 0, \cdots, K-1$$

Since points are usually collected on a grid lattice on the silicon wafer, we enforce shift monotonicity on the lines forming the grid.

Ternary Phases Shift. Ternary phases (where 3 basis patterns are used) are not affected by shifting:

$$\left(\left(\sum_{k=0}^{K-1} I_{p,k} = 3 \right) \wedge \bigwedge_{k=0}^{K-1} (r_{p,k} \Leftrightarrow r_{p',k}) \right) \Rightarrow (S_{p,k} = S_{p',k}), \forall p, \forall p' \in \mathcal{N}(p)$$

where $\mathcal{N}(p)$ is the set of neighbors of p.

Connectivity Constraint. Each of the basis patterns must be connected. Formally, for every pair of points p, p' such that $r_{p,k} \wedge r_{p',k}$, there must exist a path $\mathbb{P}$ from p to p' such that $r_{j,k} = TRUE$ for all $j \in \mathbb{P}$. Since it would require too many constraints, we use a lazy approach to enforce connectivity. If we find a solution where a basis pattern k is not connected, i.e. there exists p, p' such that $r_{p,k} \wedge r_{p',k}$ but there is no path $\mathbb{P}$ with p, p' as endpoints such that $r_{j,k} = TRUE$ for all $j \in \mathbb{P}$, then we consider the smallest cut C between p and p' such that $r_{j,k} = FALSE$ for all $j \in C$ and we add a new constraint

$$(r_{p,k} \wedge r_{p',k}) \Rightarrow \bigvee_{c \in C} r_{c,k}$$

Symmetry Breaking. Without loss of generality, we can impose an ordering on the peak locations within every phase k:

$$e_{k,\ell} \leq e_{k,\ell+1}, \ell = 0, \cdots, L - 2, k = 0, \cdots, K - 1$$

Furthermore, notice that the problem is symmetric with respect to permutations of the phase indexes $k = 0, \cdots, K - 1$. We therefore enforce an ordering on the way phases are assigned to points

$$\bigwedge_{k=1}^{K-1} \left(r_{0,k} \Rightarrow r_{0,k-1} \right)$$

$$\cdots$$

$$\bigwedge_{j=1}^{K-1} \left(\left(\bigwedge_{i=0}^{Y} \neg r_{i,j} \right) \Rightarrow \bigwedge_{k=j}^{K-1} \left(r_{Y+1,k} \Rightarrow r_{Y+1,k-1} \right) \right)$$

where we set $Y = 4$.

5   Experimental Results

We evaluate the performance of our approach on a benchmark set of synthetic instances for which the ground truth is known (namely, what the true basis patterns are and how they are combined to form the observed diffraction patterns). All the systems we consider are ternary, where three metals are combined, so that M is set to 3 in the entire experimental section. For all experiments, two peaks are considered to be overlapping if they are within 1% of each other, and the maximum allowed shift is 15%.

We compare our SMT-based approach with the Constraint Programming based solution presented in [11]. Since their CP-based formulation does not scale to realistic-sized instances, they integrate a Machine Learning based component to simplify the problem that the CP solver needs to solve to improve scalability. Note that by doing this they lose the completeness of the search, because they only explore a subtree (suggested by the ML part) of the original search space. In contrast, our approach scales to instances of realistic size (with over 40 points) without need for the ML component. Note however that if desired, the ML heuristic component could be easily integrated with our method.

Synthetic Data. We consider the known Al-Li-Fe system [1] previously used in [11], represented with a ternary diagram in Figure 2. A ternary diagram is a simplex where each point corresponds to a different concentration of the three constituent elements, in this case Al, Li, and Fe. The composition of a point depends on its distance from the corners. For a fixed value of the parameter P, synthetic instances are generated by sampling P points in the ternary diagram, each corresponding to different concentrations of the three constituent elements. For each point, synthetic x-ray diffraction patters are generated starting from known diffraction patterns of the constituent phases (taken from the JCPDS database [1]), that are combined according to the concentrations of the elements in that point. A peak detection algorithm is then used to generate a discrete set of peaks.

We first consider a set of instances without any noise, for which we have the exact location of all the peaks for every sample (the maximum number of peaks per sample is $L = 12$), without any outlier or missing peak. Starting from the diffraction patterns and the corresponding peaks, we generate the corresponding instance using the formulation described in the previous section, encoded in the SMTLibV2 language [3]. In this case, we set $K = 6$, the true number of underlying unknown basis patterns, and we try to recover a solution with $T = 0$ missing peaks. We also consider a set of simplified instances, where we fix some of the six unknown basis patterns to their true values. We solved these instances on a 3 Ghz Intel Core2Duo machine running Windows, using the SMT solvers Z3 [6] and MathSAT5 [9]. However, MathSAT is significantly slower (for instance, it takes over 50 minutes to solve a small instance with $P = 10$ points that Z3 solves in about 15 seconds) and it does not scale to larger problems. We therefore report only times obtained with Z3.

Running Time. We compare our method with previous CP-based approach presented in [11] on the same set of benchmark instances. The runtime for the CP solver are taken from [11], and were obtained on a comparable 3.8 GHz Intel Xeon machine. In Table 1a we show runtime as a function of the instance size P and the number of basis patterns left unknown K' (e.g., $K' = 3$ when the instance has been simplified by fixing three out of the six unknown basis patterns).

As we can see from the runtimes reported in Table 1a, our approach based on SMT and Z3 is always considerably faster, except for the smallest simplified problems where the difference is in the order of a few seconds. More importantly, our SMT-based approach shows a significantly improved scaling behavior, and can solve problems of realistic size with 6 unknown phases and over 40 points within an hour. In contrast, the previous CP-based approach can only solve simplified problems and cannot solve any problem with 6 unknown basis patterns [11].

Solving Strategy. In order to understand whether the improvement comes from the new problem encoding (based on integer arithmetic and not on set variables as the one in [11]) or from the SMT solving strategy, we translated

Table 1. P is the number of sampled points. K' is the number of basis patterns left unknown. e is the number of peaks removed (simulating measurement errors).

(a) Running time.			(b) Accuracy.		
Dataset	Z3 (s)	ILOG Solver (s)	Dataset	Precision (%)	Recall (%)
P=10 K'=3	8	0.5	P=10, e=0	95.8	100
K'=6	12	timeout at 1200	P=15, e=0	96.6	100
P=15 K'=3	13	0.5	P=18, e=0	97.2	96.6
K'=6	20	timeout at 1200	P=28, e=0	96.1	92.8
P=18 K'=3	29	384.8	P=45, e=0	95.8	91.6
K'=6	125	timeout at 1200	P=15, e=1	96.1	99.6
P=29 K'=3	78	276	P=15, e=2	96.3	99.3
K'=6	186	timeout at 1200	P=15, e=3	96.7	99.5
P=45 K'=6	518	timeout at 1200	P=15, e=4	95.3	98.9
			P=15, e=4	94.8	99.7

our arithmetic-based encoding as a Constraint Satisfaction Problem and as a Mixed Integer Program. As our SMT model combines logical constraints and linear inequalities exclusively, a Mixed Integer Programming (MIP) approach is particularly appealing. Indeed, one can fairly naturally translate the logical constraints of our model, namely 'Or', 'And', 'Not', 'IfThenElse', into a system of linear inequalities by using additional binary variables, and be left with a MIP formulation. The ability of the MIP to handle continuous variables for both the peak locations and the shifts, as well as to reason in terms of an objective function (e.g., the total number of missing peaks) makes it an attractive option. Nevertheless, the translation of the logical constraints yields a high number of binary variables (e.g., over 23K binary variables for a synthetic instance with $P = 10$), which contrasts with a low total number of continuous variables (about 120 for the same instance) and thus, weakens the potential of the MIP formulation. Empirically, none of the instances could be solved by the MIP formulation within the time limit of one hour. Similarly, we were not able to solve any of the instances (not even when simplified) obtained from translating our SMT formulation (symmetry breaking constraints included) to a CSP using the state-of-the-art IBM ILOG Cplex Solver within one hour. This suggests that the improvement over CP based solutions is not achieved thanks to the different problem encoding, but is due to the SMT solving procedure itself, which is stronger in the reasoning part and can handle well the intricate combinatorial constraints of the problem.

Accuracy. We evaluate the accuracy of our method by comparing the solutions we find (i.e., the phase map given by the values of $r_{p,k}$ for $p = 0, \cdots, P-1, k = 0, \cdots, K-1$) with the ground truth in terms of precision/recall scores, reported in Table 1b. Precision is defined as the fraction of the number of points correctly identified as belonging to phase k (true positives), over the total number of points identified as belonging to phase k (true positives + false positives). Recall is defined as the fraction of points correctly identified as belonging to phase k (true positives) over the true number of points belonging to phase k (true positives + false negatives). These values are obtained by comparing with ground truth all $K!$ permutations of the phases we obtain, and taking the one with the smallest number of errors (recall that the problem is symmetric with respect to permutations of the phase indexes k). Further, the values in Table 1b are the precision/recall scores obtained for each single phase k averaged over the $K = 6$ phases. The results show that the phase maps we identify are always very accurate, with precision and recall values always larger than 90%.

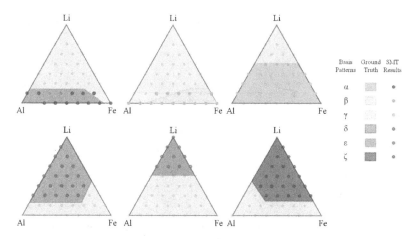

Fig. 2. Phase map for the synthetic Al-Li-Fe system with 45 sampled points, no errors. Each of the six colored areas represents one of the basis patterns $(\alpha, \beta, ..., \zeta)$ of the ground truth, while the colored dots correspond to the solution of our SMT model. The SMT results closely delimit each phase of the ground truth, which is quantitatively validated by the high precision/recall score of our approach.

Robustness. To evaluate the robustness of our method to experimental noise, we also consider another dataset from [11] where peaks are removed from the observed diffraction patterns with probability proportional to the square of the inverse peak height, in order to simulate the fact that low-intensity peaks might not be detected or they can be discarded by the peak detection algorithm. This situation is common for real-world instances, where measurements are affected by noise. We consider instances generated by removing exactly e peaks from the observed diffraction patterns, and we solve them by setting the upper bound T on the number of missing peaks equal to e. In figure 3 we see the median running time as a function of the number of missing peaks T. This is averaged over 10 instances with $P = 15$ points, and 20 runs per instance, with a timeout set at 1 hour. As shown in figure 3, the problem becomes significantly harder as we introduce missing peaks, because the constraint on the total number of missing peaks allowed T becomes less and less effective at pruning the search space as T grows. However, the median running time appears to increase linearly, and we are still able to recover a phase map efficiently even for instances affected by noise.

In table 1b we show precision recall values for these instances affected by noise. We see that the phase maps we identify are still very accurate even in presence of noise, with precision/recall scores over 95%.

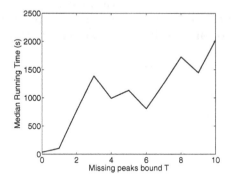

Fig. 3. Median running time as a function of the bound on the total number of missing peaks allowed T

6   Conclusions

We described a novel approach to the phase map identification problem, a key step towards automatically understanding the properties of new materials created and examined using the *composition spread* method. In our approach, we integrate domain-specific scientific background knowledge about the physical and chemical properties of the materials into an SMT reasoning framework based on linear arithmetic. Using state-of-the-art SMT solvers, we are able to automatically analyze large synthetic datasets, generating interpretations that are physically meaningful and very accurate, even in the presence of artificially added noise. Moreover, we showed that our solution outperforms in terms of scalability both Constraint Programming and Mixed Integer Programming approaches, allowing us to solve instances of realistic size. Our experiments show a novel application area for SMT technology, where we can exploit its reasoning power in a hybrid setting with continuous measurement data and rather intricate combinatorial constraints.

As there is an ever-growing amount of data in many fields of science, the grand challenge for computing and information science is how to provide efficient methods for interpreting such data, a process that generally requires integration with domain-specific scientific background knowledge. As a first step towards this goal, in this work we demonstrated the use of automated reasoning technology to support the scientific data analysis process in materials discovery. While several aspects of our method are specific to the phase map identification problem, the approach we take for the data analysis problem is quite general. Given the flexibility and ever-growing reasoning power of modern day SMT solvers, we expect to see more applications of this technology to other areas of scientific exploration that require sophisticated reasoning to interpret experimental data.

References

1. Powder Diffract. File, JCPDS Internat. Centre Diffract. Data, PA (2004)
2. Altschul, S., Gish, W., Miller, W., Myers, E., Lipman, D.: Basic local alignment search tool. Journal of Molecular Biology 215(3), 403–410 (1990)
3. Barrett, C., Stump, A., Tinelli, C.: The SMT-LIB Standard: Version 2.0. In: Gupta, A., Kroening, D. (eds.) Proceedings of the 8th International Workshop on Satisfiability Modulo Theories (Edinburgh, Edinburgh, England (2010)
4. Biere, A.: Sat, smt and applications. In: Logic Programming and Nonmonotonic Reasoning, pp. 1–1 (2009)
5. Brummayer, R., Biere, A.: Boolector: An Efficient SMT Solver for Bit-Vectors and Arrays. In: Kowalewski, S., Philippou, A. (eds.) TACAS 2009. LNCS, vol. 5505, pp. 174–177. Springer, Heidelberg (2009)
6. de Moura, L., Bjørner, N.S.: Z3: An Efficient SMT Solver. In: Ramakrishnan, C.R., Rehof, J. (eds.) TACAS 2008. LNCS, vol. 4963, pp. 337–340. Springer, Heidelberg (2008)
7. Ginley, D., Teplin, C., Taylor, M., van Hest, M., Perkins, J.: Combinatorial materials science. In: AccessScience. McGraw-Hill Companies (2005)
8. Gregoire, J.M., Tague, M.E., Cahen, S., Khan, S., Abruna, H.D., DiSalvo, F.J., van Dover, R.B.: Improved fuel cell oxidation catalysis in pt1-xtax. Chem. Mater. 22(3), 1080 (2010)
9. Griggio, A.: A Practical Approach to Satisfiability Modulo Linear Integer Arithmetic. JSAT 8, 1–27 (2012)
10. Halevy, A., Norvig, P., Pereira, F.: The unreasonable effectiveness of data. IEEE Intelligent Systems 24(2), 8–12 (2009)
11. Le Bras, R., Damoulas, T., Gregoire, J.M., Sabharwal, A., Gomes, C.P., van Dover, R.B.: Constraint Reasoning and Kernel Clustering for Pattern Decomposition with Scaling. In: Lee, J. (ed.) CP 2011. LNCS, vol. 6876, pp. 508–522. Springer, Heidelberg (2011)
12. Long, C.J., Bunker, D., Karen, V.L., Li, X., Takeuchi, I.: Rapid identification of structural phases in combinatorial thin-film libraries using x-ray diffraction and non-negative matrix factorization. Rev. Sci. Instruments 80(103902) (2009)
13. Long, C.J., Hattrick-Simpers, J., Murakami, M., Srivastava, R.C., Takeuchi, I., Karen, V.L., Li, X.: Rapid structural mapping of ternary metallic alloy systems using the combinatorial approach and cluster analysis. Rev. Sci. Inst. 78 (2007)
14. Narasimhan, B., Mallapragada, S., Porter, M.: Combinatorial materials science. John Wiley and Sons (2007)
15. Van Dover, R.B., Schneemeyer, L., Fleming, R.: Discovery of a useful thin-film dielectric using a composition-spread approach. Nature 392(6672), 162–164 (1998)

Faulty Interaction Identification
via Constraint Solving and Optimization*

Jian Zhang, Feifei Ma, and Zhiqiang Zhang

State Key Laboratory of Computer Science
Institute of Software, Chinese Academy of Sciences

Abstract. Combinatorial testing (CT) is an important black-box testing method. In CT, the behavior of the system under test (SUT) is affected by several parameters/components. Then CT generates a combinatorial test suite. After the user executes a test suite and starts debugging, some test cases fail and some pass. From the perspective of a black box, the failures are caused by interaction of several parameters. It will be helpful if we can identify a small set of interacting parameters that caused the failures. This paper proposes a new automatic approach to identifying faulty interactions. It uses (pseudo-Boolean) constraint solving and optimization techniques to analyze the execution results of the combinatorial test suite. Experimental results show that the method is quite efficient and it can find faulty combinatorial interactions quickly. They also shed some light on the relation between the size of test suite and the ability of fault localization.

1 Introduction

For many complex software systems, there are usually different components (or options or parameters) which interact with each other. Combinatorial testing (CT) [1,7,9] is an important black-box testing technique for such systems. It is a generic technique and can be applied to different testing levels such as unit testing, integration testing and system testing.

Using some automatic tools, we can generate a reasonable test suite (e.g., a covering array) which achieves a certain kind of coverage. For instance, pair-wise testing covers all different pairs of values for any two parameters.

After the system is tested, we get the execution results. Typically some test cases pass, while a few others fail. We would like to identify the cause of the failure. In CT, we adopt a parametric model of the system under test (SUT). We assume that the system fails due to the interaction of some parameters/components. Such failure-causing parameter interactions are called *faulty combinatorial interactions* (FCIs).

There are already some works on identifying FCIs, e.g., [10,12,2,8,11,14,5]. This paper proposes a new approach which is based on constraint solving and

---

* Supported in part by the National Natural Science Foundation of China under grants No. 61070039 and No. 61100064.

A. Cimatti and R. Sebastiani (Eds.): SAT 2012, LNCS 7317, pp. 186–199, 2012.

optimization techniques. Our approach is a test result analysis technique. The test results of the combinatorial test suite are used as input, and no additional test cases are generated. One benefit of our approach is that it can identify all possible solutions for the combinatorial test suite. Here, a solution is a set of suspicious FCIs, such that once the SUT has exactly these FCIs, the execution of the combinatorial test suite will be the same as the real execution result. Also, as we can find all possible solutions, the number of possible solutions can be used as a criterion to measure the precision of the solutions. Specifically, if the number of possible solutions is large, then each possible solution has a low precision. Our approach also provides evidence of the tradeoff between reducing the size of the test suite and enhancing its fault localization ability. (Most traditional CT techniques aim at generating small test suites to reduce the cost of testing. They provide good fault detection, but do not support fault localization very well.)

The remainder of this paper is organized as follows: Section 2 gives some definitions. Section 3 introduces our approach, in which we formulate the problem as a constraint satisfaction problem (CSP). Section 4 discusses translating the problem into a pseudo-Boolean optimization problem. Section 5 presents some experimental results. Section 6 compares our approach with some related work. Section 7 gives the conclusions and some implication of our results for CT.

2   Definitions

Now we give some formal definitions related to CT.

Definition 1. *(SUT model) A model $SUT(k, s)$ has k parameters $p_1, p_2, \ldots, p_k$. The vector s is of length k, i.e. $\langle s_1, s_2, \ldots, s_k \rangle$, where s_i indicates the number of possible values of the parameter p_i. The domain of p_i is $D_i = \{1, 2, \ldots, s_i\}$. And s_i is called the* level *of p_i.*

Definition 2. *(Test case) A test case t is a vector of length k, representing an assignment to each parameter with a specific value within its domain. The execution result of a test case is either* pass *or* fail.

Definition 3. *(Test suite) A test suite T is a set of test cases $\{t_1, t_2, \ldots, t_m\}$.*

Definition 4. *(CI) A combinatorial interaction (CI) is a vector of length k, which assigns t parameters to specific values, leaving the rest $k - t$ parameter values undetermined (undetermined values are denoted by '-'). Here t is the size of the CI.*

A CI represents the interaction of the t parameters with assigned values. The undetermined parameters ('-') are just placeholders, and do not participate in the interaction.

Definition 5. *(CI containment) A CI P_1 is contained by another CI P_2 if and only if all determined parameters of P_1 are determined parameters of P_2, and these parameters are assigned the same values in P_1 and P_2. A CI P is contained by a test case t if and only if all determined parameters in P have the same values as those in t.*

Definition 6. *(FCI) A faulty combinatorial interaction (FCI) is a CI such that all possible test cases containing it will fail.*

Note that if a CI is an FCI, then all CIs containing it are FCIs. Suppose we have an FCI P_1, and P_2 contains P_1. Then all test cases containing P_2 will contain P_1, and all of these test cases will fail. Thus P_2 is an FCI. For example, suppose $(1,-,2,-)$ is an FCI, then $(1,1,2,-)$ and $(1,-,2,3)$ are also FCIs. In this paper, we only identify the minimal FCIs, i.e. FCIs containing no other FCIs.

Example 1. Table 1 shows a covering array for pair-wise testing of an online payment system. There are 9 test cases, and for two of them the SUT fails. The braces ("{}") indicate the value combinations causing the failure of the SUT, which we do not know in advance. The test results are shown in the last column. We can see that this covering array can detect all FCIs of size 2, but we cannot tell what they are just from the test results. (e.g. the failure of the 3rd test case may also be caused by the interaction of Firefox and Apache.)

Table 1. A Sample Covering Array

Client	Web Server	Payment	Database	Exec
Firefox	WebSphere	MasterCard	DB/2	pass
Firefox	.NET	UnionPay	Oracle	pass
Firefox	{Apache}	{Visa}	Access	fail
IE	WebSphere	UnionPay	Access	pass
{IE}	Apache	MasterCard	{Oracle}	fail
IE	.NET	Visa	DB/2	pass
Opera	WebSphere	Visa	Oracle	pass
Opera	.NET	MasterCard	Access	pass
Opera	Apache	UnionPay	DB/2	pass

3   Formulation as a Constraint Satisfaction Problem

Suppose the system under test (SUT) has k attributes or parameters or components. The ith attribute may take one value from the following set: $D_i = \{1, 2, \ldots\}$

Assume that there are already m test cases. Some of them failed, while the others passed. The CT fault model assumes that failures are caused by value combinations (i.e. several parameters are assigned to specific values). We would like to know which combinations in the failing test cases caused the failure (e.g., made the SUT crash).

For simplicity, we first assume that there is only one failing test case in the test suite. We would like to identify the value combination (i.e., FCI) in this test case. It can be represented by a vector like this: $\langle x_1, x_2, \ldots, x_k \rangle$, where each x_i can be 0 or a value in the set D_i. When $x_j = 0$, it means that the j'th attribute does not appear in the FCI.

Our way to solve the FCI identification problem is to formulate it as a constraint satisfaction problem (CSP) or satisfiability (SAT) problem. In a CSP, there are some variables, each of which can take values from a certain domain; and there are also some constraints. Solving a CSP means finding a suitable value (in the domain) for each variable, such that all the constraints hold.

We have already given the "variables" in the CSP. Now we describe the "constraints". Roughly, a failing test case should match the FCI vector; and a passing test case should not match it. More formally, for a passing test case $(v_1, v_2, \ldots, v_k)$, we need a constraint like this:

```
   ( NOT(x1=0) AND NOT(x1=v1) )
OR ( NOT(x2=0) AND NOT(x2=v2) )
OR ......
OR ( NOT(xk=0) AND NOT(xk=vk) )
```

For a failing test case $(w_1, w_2, \ldots, w_k)$, we need the following constraint:

```
    ( (x1=0) OR (x1=w1) )
AND ( (x2=0) OR (x2=w2) )
AND ......
AND ( (xk=0) OR (xk=wk) )
```

In this way, we obtain the "constraints" in the CSP. After solving the CSP, we get the values of the variables x_i $(1 \leq i \leq k)$. Then, deleting the 0's, we get the FCI.

Example 1. (Continued)   Assume that the values in the table can be represented by integers as follows:

- Client can be: 1. Firefox; 2. IE; 3. Opera.
- WebServer can be: 1. WebSphere; 2. .NET; 3. Apache.
- Payment can be: 1. MasterCard; 2. UnionPay; 3. VISA.
- Database can be: 1. DB/2; 2. Oracle; 3. Access.

Suppose that the FCI is denoted by the vector $\langle x_1, x_2, x_3, x_4 \rangle$. The domain of each variable x_i is $[0..3]$. We can derive the following constraints from the first 3 test cases:

```
   ( NOT(x1=0) AND NOT(x1=1) )
OR ( NOT(x2=0) AND NOT(x2=1) )
OR ( NOT(x3=0) AND NOT(x3=1) )
OR ( NOT(x4=0) AND NOT(x4=1) ) ;

   ( NOT(x1=0) AND NOT(x1=1) )
OR ( NOT(x2=0) AND NOT(x2=2) )
OR ( NOT(x3=0) AND NOT(x3=2) )
OR ( NOT(x4=0) AND NOT(x4=2) ) ;
```

```
      ( (x1=0) OR (x1=1) )
AND ( (x2=0) OR (x2=3) )
AND ( (x3=0) OR (x3=3) )
AND ( (x4=0) OR (x4=3) ) .
```

More constraints can be derived from the other test cases, which are omitted here.

More than One FCIs

Example 1. (Continued) Now suppose that there are two different FCIs, denoted by $\langle x_1, x_2, x_3, x_4 \rangle$ and $\langle y_1, y_2, y_3, y_4 \rangle$, respectively. From the first test case (which passed), we obtain the following constraint:

```
(     ( NOT(x1=0) AND NOT(x1=1) )
  OR ( NOT(x2=0) AND NOT(x2=1) )
  OR ( NOT(x3=0) AND NOT(x3=1) )
  OR ( NOT(x4=0) AND NOT(x4=1) ))
AND
(     ( NOT(y1=0) AND NOT(y1=1) )
  OR ( NOT(y2=0) AND NOT(y2=1) )
  OR ( NOT(y3=0) AND NOT(y3=1) )
  OR ( NOT(y4=0) AND NOT(y4=1) )).
```

From the third test case (which failed), we obtain the following constraint:

```
(     ( (x1=0) OR (x1=1) )
AND ( (x2=0) OR (x2=3) )
AND ( (x3=0) OR (x3=3) )
AND ( (x4=0) OR (x4=3) ))
OR
(     ( (y1=0) OR (y1=1) )
AND ( (y2=0) OR (y2=3) )
AND ( (y3=0) OR (y3=3) )
AND ( (y4=0) OR (y4=3) )).
```

Symmetry Breaking

For the above example with two FCIs, suppose by solving the constraints we get $\langle x_1 = v_1, x_2 = v_2, x_3 = v_3, x_4 = v_4 \rangle$ and $\langle y_1 = v_1', y_2 = v_2', y_3 = v_3', y_4 = v_4' \rangle$. If we exchange the respective values of the two vectors, we get $\langle x_1 = v_1', x_2 = v_2', x_3 = v_3', x_4 = v_4' \rangle$ and $\langle y_1 = v_1, y_2 = v_2, y_3 = v_3, y_4 = v_4 \rangle$. Obviously it is another solution to the constraints. However, the two solutions represent the same set of FCIs. They are symmetric solutions.

A symmetry is a one to one mapping (bijection) on decision variables that preserves solutions and non-solutions. Symmetries can generate redundant search space, so it is very important to eliminate symmetries while solving the problem.

For a system with n FCIs, there are $n!$ permutations of the variable vectors of FCIs in the aforementioned constraints, hence results in as many symmetries. To break such symmetries, we introduce lexicographic order on the variable vectors. In the above example, we add the following constraints which ensure that one vector is lexicographically smaller than the other.

```
    ( (x1<=y1) )
AND ( (x1<y1) OR (x2<=y2) )
AND ( (x1<y1) OR (x2<y2) OR (x3<=y3) )
AND ( (x1<y1) OR (x2<y2) OR (x3<y3) OR (x4<y4) )
```

Apparently all symmetries caused by permutations of FCIs can be eliminated with the above constraints.

Determining the Number of FCIs

To use our method, one should specify the number of FCIs in advance. But we do not know the actual number. So we gradually increase the number. We start from $n = 1$; if the problem is unsatisfiable, then n is increased by 1; ... We repeat this procedure until the problem becomes satisfiable.

4   Translation to Pseudo-Boolean Optimization Problem

By checking the satisfiability of the aforementioned constraints, one or a set of FCIs can be discovered. However, there might be many solutions to the constraint satisfaction problem, and it is desirable to find the optimal one according to some criterion. A reasonable objective is to minimize the size of FCI, i.e., to maximize the number of '0's, so that the FCI is the most general one. As a result, in this paper we are investigating FCI identification as an optimization problem instead of a decision problem.

This problem can be naturally formulated as a Pseudo-Boolean Optimization (PBO) problem. In its broadest sense, a pseudo-Boolean function is a function that maps Boolean values to a real number.

A linear pseudo-Boolean constraint has the following form

$$\Sigma_i a_i x_i \geq b$$

where $x_i \in \{0, 1\}$ is a Boolean variable and a_i, b are integers.

A pseudo-Boolean constraint is nonlinear if it contains the product of Boolean variables. Such a constraint has the following form

$$\Sigma_i a_i (\Pi_k x_{i,k}) \geq b$$

A pseudo-Boolean optimization problem is to maximize (minimize) a pseudo-Boolean expression subject to a set of pseudo-Boolean constraints.

4.1   Encoding

Suppose there are w failing cases for the SUT, and we are looking for n ($n \leq w$) FCIs. We use Boolean variable $P_{i,j,v}$ to indicate that the jth ($1 \leq j \leq k$) parameter of the ith ($1 \leq i \leq n$) FCI $x_{i,j}$ takes value v, or formally, $P_{i,j,v} \equiv (x_{i,j} = v)$. When there are more than one FCIs, we also introduce auxiliary Boolean variable $E_{t,i}$ to indicate that the tth ($1 \leq t \leq w$) failing case is caused by the ith FCI. The P-variables are called *primary variables*, and the E-variables are called *auxiliary variables*.

Assume that our goal is to maximize the total number of zero values of all FCIs, we have the following objective function:

$Minimize \quad -\Sigma_i \Sigma_j P_{i,j,0}$

There are four types of pseudo-Boolean constraints:

1. Basic constraints: Constraints that guarantee the validity of the encoding. We have to make sure that each parameter of each FCI can take only one value. So we add constraints $\Sigma_v P_{i,j,v} = 1$ for all $1 \leq i \leq n, 1 \leq j \leq k$. Here the variable v ranges over $D_i \cup \{0\}$.
2. Constraints for passing cases: To facilitate encoding, we make slight changes to the original constraints by removing the innermost AND operator. For instance, in *Example 1* we replace (NOT(x1=0) AND NOT(x1=1)) with (x1=2 OR x1=3), and (NOT(x2=0) AND NOT(x2=1)) with (x2=2 OR x2=3), and so on. Generally, for each passing case $V = \{v_1, v_2, \ldots, v_k\}$, we add:

$$\bigwedge_{1 \leq i \leq n} \Sigma_{j=1}^{k} \Sigma_{v, v \neq 0, v \neq v_j} P_{i,j,v} \geq 1$$

3. Constraints for failing cases: If there is only one FCI, we simply translate the original constraints into pseudo-Boolean constraints. For each failing case $V = \{v_1, v_2, \ldots, v_k\}$, we add:

$$\bigwedge_{1 \leq j \leq k} P_{1,j,0} + P_{1,j,v_j} = 1$$

If the number of target FCIs n is more than 1, each failing case must match at least one FCI, which results in the constraints

$$\Sigma_{i=1}^{n} E_{t,i} \geq 1$$

for all $1 \leq t \leq w$.

In addition, for the tth failing case $V = \{v_1, v_2, \ldots, v_k\}$, we have

$$\bigwedge_{1 \leq i \leq n} \bigwedge_{1 \leq j \leq k} -E_{t,i} + P_{i,j,0} + P_{i,j,v_j} \geq 0$$

And

$$\bigwedge_{1 \leq i \leq n} E_{t,i} + \Sigma_{1 \leq j \leq k} \Sigma_{v, v \neq 0, v \neq v_j} P_{i,j,v} \geq 1$$

which means if the tth failing case matches the ith FCI, then the jth parameter of the ith FCI either takes value 0 or takes value v_j, and vice versa.

Besides, we must make sure that there is no useless FCI. Each FCI must match at least one failing case. So for all $1 \leq i \leq n$, we add:

$$\Sigma_{t=1}^{w} E_{t,i} \geq 1$$

4. Symmetry breaking constraints: A direct way to encode the inequality (e.g. x1<y1) is to enumerate all assignments allowed by the inequality and assert that at least one assignment is true. Therefore, the constraints that state the ith FCI is lexicographically smaller than the $(i+1)^{\text{th}}$ FCI are encoded as follows:

$$(\bigwedge_{1 \leq j \leq k-1} \Sigma_{l=1}^{j} \Sigma_{v_1 < v_2} P_{i,l,v1} P_{i+1,l,v2} + \Sigma_v P_{i,j,v} P_{i+1,j,v} \geq 1)$$
$$\bigwedge \quad \Sigma_{l=1}^{k} \Sigma_{v_1 < v_2} P_{i,l,v1} P_{i+1,l,v2} \geq 1$$

Unlike the other three classes of PB constraints, these PB constraints are nonlinear.

4.2 Tool Support

There are quite some tools for solving pseudo-Boolean constraints. In our work, we use the tool clasp [4], a conflict-driven nogood learning answer set solver, to solve the optimization problem. clasp can be applied as an ASP solver, as a SAT solver, or as a PB solver. We have developed a prototype tool which translates the original problem into the input of clasp.

5 Experiments

All the experiments in this paper are performed on a laptop with Intel CPU: Core i5 M540, 2.53GHz, running Ubuntu 11.10.

5.1 Simulation Results

In this part, we present some experimental simulation results. The results are shown in Table 2. For each row, an experiment is performed as follows:

1. We give an SUT model, as well as a set of FCIs in the SUT. Suppose the set of FCIs is unknown before our tool is used.
2. A combinatorial test suite is generated according to the SUT model and a given coverage strength t.
3. Label the test suite results with the set of FCIs. Any test case containing one or more FCIs will be labeled as fail; while other test cases are labeled as pass. This step simulates the testing procedure for real systems.
4. We apply our tool to the test results to generate possible solutions. Then we analyze the results.

Table 2. Results for Locating FCIs on Test Suites

Exp #	Test Suite	nTC	nFCI	sFCI	nFT	nSol/nSol_NSB	tSB (s)	tNSB (s)
1	$CAM(3^4,2)$	9	1	1(1/1)	3	1/1	0.001	=
2			1	2(1/2)	1	6/6	0.001	=
3			2	1(2/1)	5	1/6	0.001	0.001
4			2	2(2/2)	2	36/72	0.002	0.003
5	$CA(3^4,2)$	13	1	1(1/1)	5	1/1	0.001	=
6			1	2(1/2)	2	1/1	0.001	=
7			2	1(2/1)	7	2/36	0.001	0.002
8			2	2(1/2)	2	1/1	0.001	=
9	$CA(2^8,2)$	7	1	1(1/1)	4	1/1	0.001	=
10			1	2(1/2)	1	13/13	0.001	=
11			1	3(1/2)	1	6/6	0.001	=
12			2	1(2/1)	6	4/24	0.002	0.002
13			2	2(2/1.5)	3	12/48	0.003	0.004
14				1&2(2/1)	5	3/90	0.002	0.005
15	$CA(2^{20},2)$	10	1	1(1/1)	6	1/1	0.001	=
16			1	2(1/2)	3	4/4	0.001	=
17			1	3(1/2)	2	9/9	0.001	=
18			2	1(2/1)	7	4/216	0.006	0.022
19			2	2(1/2)	4	1/1	0.001	=
20	$CA(3^{20},2)$	24	1	1(1/1)	9	1/1	0.001	=
21			1	2(1/2)	3	1/1	0.001	=
22			1	3(1/2)	1	3/3	0.003	=
23			2	1(2/1)	16	1/18	0.017	0.007
24			2	2(2/2)	5	5/10	0.015	0.008
25			2	3(1/2)	2	3/3	0.001	=
26	$TS1(2^8)$	10	1	3(1/3)	6	1/1	0.001	=
27	$TS2(2^8)$	18	2	1&2(2/1.5)	14	1/54	0.002	0.004
28	$CA(6^{10},3)$	526	4	1-4(3/2)	97	1/162	0.177	0.157
29			8	1-4(5/1.8)	180	6/-	7.177	NA
30	$CA(3^{50},3)$	133	2	2&3(2/2.5)	20	1/6	0.137	0.052
31			4	1-4(3/2)	56	30/-	1.854	NA
32	$CA(3^{50},4)$	579	2	2&3(2/2.5)	83	1/192	0.302	0.279
33			4	1-4(4/2.5)	256	1/-	53.185	NA
34	$CA(3^{100},4)$	169	2	2&3(2/2.5)	22	1/6	0.920	0.201
35			4	1-4(3/2)	74	1/39366	11.975	51.781

In Table 2, column "nTC" shows the number of test cases, "nFCI" shows the number of FCIs we give in advance, column "sFCI" shows the size of FCIs we give in advance (the "a/b" in the parenthesis shows the number and average size of identified FCIs), column "nFT" shows the number of failing test cases, column "nSol / nSol_NSB" shows the number of solutions found when symmetry breaking is used or not used, columns "tSB" and "tNSB" show the running times when symmetry breaking is used and not used, respectively. An "=" in the tNSB column means that when the number of FCIs is 1, there is actually no symmetry breaking constraint, so there is no significant difference between tSB and tNSB.

"NA" stands for "not available", which means that the solver did not find an optimal solution within 300 seconds.

$CA(s^k, t)$ represents a covering array of strength t, having k parameters of level s (i.e. each parameter has s possible values). A covering array of strength t covers all possible value combinations among t parameters. A $CAM(s^k, t)$ is a covering array having the minimum number of test cases among all $CA(s^k, t)$'s. In Table 2, the CAs are generated by PICT of Microsoft [3]; and $CAM(3^4, 2)$ is the covering array in Example 1. $TS1(2^8)$ and $TS2(2^8)$ are two test suites generated using the algorithms in [14]. Each of the two test suites tries to localize FCIs in one failing test case.

Our technique only analyzes the execution results of the test suite. Details about running the SUT are not considered in our evaluation. From Table 2, we can see the average time for finding each possible solution is less than 0.010s for most of the small cases (exp # 1-27). And our method scales up to some large cases (exp # 28-35).

Applying symmetry breaking sometimes works worse than not adopting symmetry breaking, since the additional nonlinear constraints slows the constraint solving. But for some large cases, symmetry breaking can greatly reduce the solving time.

We also have the following observations:

- The number of solutions may be greater than 1. The reason is that the test suite may not be sufficient to localize the faults. The number of possible solutions can be used to measure the fault localization ability of the test suite. The more solutions we get, the less accurate the FCI localization is.
- From the results of $CAM(3^4, 2)$ and $CA(3^4, 2)$, we can see the number of solutions decreases with increase of the number of test cases. This is because a larger number of test cases will provide more information about the faults.
- From the results of $TS1$ and $TS2$, we can see that the test cases used by the methods in [14] are sufficient to localize the fault, and the number of solutions is 1 for each experiment.
- The results show that when the number and size of the FCIs of the SUT (which we declared in advance) grows, the number of solutions grows, or the number and size of identified FCIs are more likely to be inconsistent with that of the FCIs of the SUT. Both provide evidence that the increasing of the number and size of FCIs of the SUT will make fault localization more difficult. Also, when the number and size of FCIs is 1, the number of solution for $CA(s^k, 2)$ is always 1. These conclusions conform to that of [8].

5.2 Experiment on a Real System

We also applied our technique to a real system. The experiment subject is a module of the Traffic Collision Avoidance System (TCAS) benchmark.

First, we build a parameterized model for TCAS. Here we used the same SUT model as used in [6]. The model has some input parameters: 6 of the parameters are of level 2, 3 parameters are of level 3, 1 parameter is of level 4, and 2 parameters are of level 10. So there are $3 \times 2^5 \times 4 \times 10^2 \times 3 \times 2 \times 3 = 691200$ possible test cases in total.

The TCAS benchmark has 41 faulty versions and 1 correct version. We perform our experiment in the following steps:

1. Select a faulty version among the 41 versions.
2. Set a coverage strength t, and generate a covering array of strength t.
3. Execute the covering array on the selected faulty version. The execution result of a test case is determined by comparing the output of the faulty version with that of the correct version.
4. Apply our technique to the execution results of the covering array, in order to identify the FCIs.

In our evaluation, we conducted experiments on the first 15 versions of TCAS. The results are shown in Table 3. In the table, row "nTC" shows the number of test cases in the covering array. Row "nFTC" shows the number of failing test cases. Rows "nSol" and "nSol_NSB" show the number of solutions when symmetry breaking is used and not used, respectively. In these rows, "AP" means that all test cases passed and our technique is not applied. "NA" stands for "not available", i.e., the solver cannot reach an optimal solution within 300 seconds. Row "nFCI" shows the number of FCIs. Row "asFCI" shows the average size of identified FCIs.

In our experiments, the covering arrays of strength 5 and 6 are very large (4294 and 11333 test cases respectively). These large test suites are likely to trigger more FCIs, which makes it less possible to find a solution if we assume a small number of FCIs in advance. So we assume that the number of FCIs is large too. This will result in a large number of (pseudo-Boolean) variables and constraints, and make it difficult for the solver to find a solution. Indeed, most experiments for strength 5 and 6 got an "NA", so we do not show them in Table 3.

From the results, we can see that the performance of our technique is not so good for the TCAS example. The reason is that the size of FCIs is relatively large. The consequence is two-fold. On one hand, covering arrays of lower strengths are not likely to trigger SUT failures, and most experiments with lower strengths got an "AP". On the other hand, covering arrays of higher strengths will need more test cases, and more likely there are quite some FCIs. Both will make the number of constraints large, resulting an "NA".

6   Related Works

There are different kinds of FCI identification methods. Some adaptive approaches like [13,10,14] take one failing test case as input, then generate and execute additional test cases to identify the FCIs in the failing test case. Other adaptive approaches like [11,5] process the whole test suite. These methods proceed in an iterative way. In each iteration, they analyze the test suite to identify suspicious FCIs. Then they generate and execute some additional test cases to refine the suspicious FCIs.

Table 3. Experimental Results for TCAS

Version	v1			v2			v3		
t	2	3	4	2	3	4	2	3	4
nTC	100	402	1410	100	402	1410	100	402	1410
nFTC	0	0	0	0	0	0	0	3	1
nSol	AP	AP	AP	AP	AP	AP	AP	2	7
nSol_NSB	AP	AP	AP	AP	AP	AP	AP	4	7
nFCI	-	-	-	-	-	-	-	2	1
asFCI	-	-	-	-	-	-	-	4	4
tSB (s)	-	-	-	-	-	-	-	0.024	0.038
tNSB (s)	-	-	-	-	-	-	-	0.022	=

Version	v4			v5			v6		
t	2	3	4	2	3	4	2	3	4
nTC	100	402	1410	100	402	1410	100	402	1410
nFTC	0	0	2	0	6	21	0	6	9
nSol	AP	AP	1	AP	18	$\geq$10	AP	15	182
nSol_NSB	AP	AP	1	AP	432	$\geq$10	AP	90	7056
nFCI	-	-	1	-	4	-	-	3	4
asFCI	-	-	5	-	4	-	-	4	5
tSB (s)	-	-	0.030	-	1.014	-	-	0.072	1.256
tNSB (s)	-	-	=	-	5.213	-	-	0.113	6.836

Version	v7			v8			v9		
t	2	3	4	2	3	4	2	3	4
nTC	100	402	1410	100	402	1410	100	402	1410
nFTC	0	0	1	0	0	1	0	0	1
nSol	AP	AP	4	AP	AP	3	AP	AP	4
nSol_NSB	AP	AP	4	AP	AP	3	AP	AP	4
nFCI	-	-	1	-	-	1	-	-	1
asFCI	-	-	4	-	-	4	-	-	4
tSB (s)	-	-	0.032	-	-	0.032	-	-	0.032
tNSB (s)	-	-	=	-	-	=	-	-	=

Version	v10			v11			v12		
t	2	3	4	2	3	4	2	3	4
nTC	100	402	1410	100	402	1410	100	402	1410
nFTC	1	7	16	1	9	23	3	14	57
nSol	1	12	700	1	60	NA	1	NA	NA
nSol_NSB	1	72	NA	1	1440	NA	6	NA	NA
nFCI	1	3	7	1	4	$\geq$ 10	3	$\geq$ 8	$\geq$ 10
asFCI	2	5	5.57	2	4.75	-	2	-	-
tSB (s)	0.002	0.045	14.035	0.002	0.227	-	0.244	-	-
tNSB (s)	=	0.049	NA	=	0.865	-	0.257	-	-

Version	v13			v14			v15		
t	2	3	4	2	3	4	2	3	4
nTC	100	402	1410	100	402	1410	100	402	1410
nFTC	0	2	11	1	3	4	0	6	21
nSol	AP	5	34	1	12	30	AP	18	NA
nSol_NSB	AP	5	360	1	24	180	AP	432	NA
nFCI	-	1	3	1	2	3	-	4	$\geq$ 10
asFCI	-	5	5.67	2	3.5	4.67	-	4	-
tSB (s)	-	0.006	0.168	0.002	0.025	0.375	-	0.996	-
tNSB (s)	-	=	0.259	=	0.021	0.823	-	5.313	-

There are also some works aiming at generating covering arrays having fault localization abilities, such as LDA [2] and ELA [8]. One problem with these approaches is that the test suite is large and will cost a lot of resources during testing period. Besides, the generation of these covering arrays is still a problem.

Another kind of approaches aim at analyzing all the test execution results to identify the FCIs. Our technique belongs to this kind. The benefits of these methods are that only the test results are used as input, and no additional test cases are generated. However, these methods suffer from insufficient test results, which provide insufficient information about FCIs and will lower the precision of FCI identification. Previous works of this kind include [12]. It uses classification tree analysis (CTA) on the test results. The FCIs are represented as a tree structure, and a path from the root to each failing leaf node corresponds to an FCI. (An example of classification trees is shown in Fig. 1. The p_i's represent the parameters, and the numbers on the branches are their values.) This approach is fast and insensitive to occasional failures. However, in some situations, only a very small set of test cases fail, which means the input data for CTA is highly unbalanced. Then CTA will have very bad performance when processing this kind of data. Another point is that all the FCIs should contain the same parameter on the root (it can be observed from Fig. 1), but this is not always the case. Compared with the CTA approach, our technique will not face the difficulties of CTA, and we can find all possible solutions of FCIs. Each of these solutions will lead to exactly the same test result as the input test results.

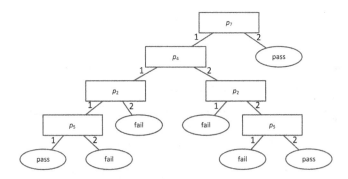

Fig. 1. An Example Classification Tree

7 Concluding Remarks

This paper proposes a new automated approach for identifying FCIs. It is based on pseudo-Boolean constraint solving and optimization techniques, and it is quite effective. The method has been implemented as a prototype tool. Preliminary results are encouraging. In most cases, it can provide a few FCIs which are helpful to the user when debugging. Of course, there are still some works to be done. For example, there can be different encodings of the problem, which are worth investigating.

Sometimes there may be a number of possible solutions for the generated constraints, which means that the localized FCIs may not be unique. This is because the test suite is insufficient to provide enough information about the failure. The more solutions we get, the less accurate the FCI localization is. One direction of our future works is to make a balance between reducing the size of the test suite and increasing the ability of fault localization.

Our work in this paper is another application of SAT/constraint solving (and optimization) techniques in software engineering. On the other hand, some benchmarks can be generated from such applications. They might be interesting and challenging to the SAT community.

References

1. Cohen, D.M., Dalal, S.R., Fredman, M.L., Patton, G.C.: The AETG system: An approach to testing based on combinatorial design. IEEE Trans. on Softw. Eng. 23(7), 437–444 (1997)
2. Colbourn, C., McClary, D.: Locating and detecting arrays for interaction faults. J. of Combinatorial Optimization 15(1), 17–48 (2011)
3. Czerwonka, J.: Pairwise testing in realworld: Practical extensions to test case generator. In: Proc. PNSQC 2006, pp. 419–430 (2006)
4. Gebser, M., Kaufmann, B., Neumann, A., Schaub, T.: Conflict-driven answer set solving. In: Proc. IJCAI 2007, pp. 386–392 (2007), http://www.cs.uni-potsdam.de/clasp/
5. Ghandehari, L.S.G., Lei, Y., Xie, T., Kuhn, D.R., Kacker, R.: Identifying failure-inducing combinations in a combinatorial test set. In: Proc. ICST 2012, pp. 370–379 (2012)
6. Kuhn, D.R., Okun, V.: Pseudo-exhaustive testing for software. In: Proc. SEW 2006, pp. 153–158 (2006)
7. Kuhn, D.R., Wallace, D.R., Gallo, A.M.: Software fault interactions and implications for software testing. IEEE Trans. on Softw. Eng. 30(6), 418–421 (2004)
8. Martínez, C., Moura, L., Panario, D., Stevens, B.: Algorithms to Locate Errors Using Covering Arrays. In: Laber, E.S., Bornstein, C., Nogueira, L.T., Faria, L. (eds.) LATIN 2008. LNCS, vol. 4957, pp. 504–519. Springer, Heidelberg (2008)
9. Nie, C., Leung, H.: A survey of combinatorial testing. ACM Computing Surveys 43(2) (2011)
10. Shi, L., Nie, C., Xu, B.: A Software Debugging Method Based on Pairwise Testing. In: Sunderam, V.S., van Albada, G.D., Sloot, P.M.A., Dongarra, J. (eds.) ICCS 2005, Part III. LNCS, vol. 3516, pp. 1088–1091. Springer, Heidelberg (2005)
11. Wang, Z., Xu, B., Chen, L., Xu, L.: Adaptive interaction fault location based on combinatorial testing. In: Proc. QSIC 2010, pp. 495–502 (2010)
12. Yilmaz, C., Cohen, M.B., Porter, A.A.: Covering arrays for efficient fault characterization in complex configuration spaces. IEEE Trans. on Softw. Eng. 32(1), 20–34 (2006)
13. Zeller, A., Hildebrandt, R.: Simplifying and isolating failure-inducing input. IEEE Trans. on Soft. Eng., 183–200 (2002)
14. Zhang, Z., Zhang, J.: Characterizing failure-causing parameter interactions by adaptive testing. In: Proc. ISSTA 2011, pp. 331–341 (2011)

Revisiting Clause Exchange in Parallel SAT Solving

Gilles Audemard[1], Benoît Hoessen[1], Saïd Jabbour[1],
Jean-Marie Lagniez[2], and Cédric Piette[1,*]

[1] Université Lille-Nord de France
CRIL - CNRS UMR 8188
Artois, F-62307 Lens
{audemard,hoessen,jabbour,piette}@cril.fr
[2] Institute for Formal Models and Verification Johannes Kepler
University, AT-4040 Linz, Austria
jean-marie.lagniez@jku.at

Abstract. Managing learnt clause database is known to be a tricky task in SAT solvers. In the portfolio framework, the collaboration between threads through learnt clause exchange makes this problem even more difficult to tackle. Several techniques have been proposed in the last few years, but practical results are still in favor of very limited collaboration, or even no collaboration at all. This is mainly due to the difficulty that each thread has to manage a large amount of learnt clauses generated by the other workers. In this paper, we propose new efficient techniques for clause exchanges within a parallel SAT solver. In contrast to most of the current clause exchange methods, our approach relies on both export and import policies, and makes use of recent techniques that proves very effective in the sequential case. Extensive experimentations show the practical interest of the proposed ideas.

1 Introduction

The practical resolution of the SAT problem has received major attention these two last decades. Particularly, due to the wide availability of cheap multicore architectures, the focus is now on the development of efficient parallel engines, able to solve large real-world problems. Several of them have been recently proposed, e.g. ManySat [10], SArTagnan [12], Plingeling [4], ppfolio [16], part-tree-learn [11].

Portfolio schema is a possible approach to tackle parallelism. One idea of portfolio algorithms is the collaboration between the different workers. In the SAT case, their joint effort is mainly achieved through the exchange of learnt clauses. Unfortunately, it is very hard to predict whether a clause generated by a working thread will be useful for the others, or not. To deal with this problem, ManySAT proposes a dynamic clause sharing policy which uses pairwise size limits to control the exchange between threads [9]. However, most of implementations (Plingeling [4], SArTagnan [12], etc.) only share unit clauses with the other threads. Moreover, in the last SAT competition, a new portfolio called ppfolio obtained very good results; ppfolio is actually a simple script that runs different state-of-art sequential solvers in an independent way.

---

* This work has been supported by CNRS and OSEO, under the ISI project "Pajero".

A. Cimatti and R. Sebastiani (Eds.): SAT 2012, LNCS 7317, pp. 200–213, 2012.
© Springer-Verlag Berlin Heidelberg 2012

Accordingly, no collaboration is achieved within this solver, and yet it proves very efficient during this competition. This shows that the current clause exchange techniques are not mature and may be improved.

The problem of predicting the usefulness of a given learnt clause is also known in the sequential case; recently, a new measure called *psm* [1] has been proposed to dynamically manage learnt clauses. Roughly, it consists in comparing the current (partial) interpretation to the set of literals of each learnt clause. The main idea is the following: if the set-theoretical intersection of the current interpretation and the clause is large, then the clause is unlikely useful in the current part of the search space. On the contrary, if this intersection is small, then the clause has a lot of chance to be useful for unit-propagation, reducing the search space. This measure has been used in a new strategy to manage the learnt clauses database which enables to *freeze* a clause, namely to remove it from the set of learnt clauses on a temporary basis, when it is considered "useless". Periodically all clauses are reevaluated in other to be frozen or activated. This technique proves very efficient in an empirical point of view, and succeeds to select relevant learnt clauses.

In this paper, we extend the results of [1] to the portfolio framework, proposing different efficient heuristical policies for exporting, importing and selecting relevant clauses for the different threads of a portfolio. This paper is structured as follows: in the next Section, we present the background knowledge about parallel SAT solving and learnt clause management. In the Section 3, we present some preliminaries about the behavior of a portfolio solver. Next, in Section 4, our case study solver called PeneLoPe is presented and it is compared to the best parallel SAT solvers in Section 5. Finally, we conclude with some perspectives.

2   Technical Background

First of all, we assume that the reader is familiar with Satisfiability notions (variables, literal, clause, unit clause, interpretations, CNF formula). Note that clauses and interpretations will be equally interpreted as set of literals. We just want to recall the global schema of CDCL (Conflict-Driven, Clause Learning) solvers: a typical branch of a CDCL solver can be seen as a sequence of decisions followed by propagations, repeated until a conflict is reached. Each decision literal, chosen by some heuristic, usually activity-based ones, is assigned at its own level, shared with all propagated literals assigned at the same level. Each time a conflict is reached, a *nogood* is extracted using a particular method, usually the First UIP (Unique Implication Point) one [14,19]. The learnt clause is then added to the learnt clause database and a *backjumping* level is computed from it. The interested reader can refer to [7] for more details. In the rest of the section, we give background about important notions for the paper understanding.

Control of the Learnt Clauses Database

The size of the learnt clauses database is clearly crucial to the solver performance. Indeed, keeping too many learnt clauses slows down the unit propagation process, while deleting too many of them breaks the overall learning benefit. To avoid such drawbacks, solvers periodically remove some clauses considered to be useless. Consequently, identifying good learnt clauses - relevant to the proof derivation - is clearly an important challenge. The first proposed quality measure follows the success of the activity-based VSIDS heuristic. More precisely, a learnt clause is considered relevant to the proof, if it is often involved in recent conflicts *i.e.* frequently used to derive asserting clauses. Clearly, this deletion strategy supposes that a useful clause in the past could be useful in the future. In [2], another measure called *lbd* is used to estimate the quality of a learnt clause (we denote $lbd(c)$ the *lbd* value of a clause c). This new measure is based on the number of different decision levels appearing in a learnt clause and is computed when the clause is generated. Extensive experiments demonstrates that clauses with small *lbd* values are used more often than those with higher *lbd* ones. Note also that *lbd* of clauses can be recomputed when they are used for unit propagations, and updated if the it becomes smaller. This update process is important to get many good clauses. However, these both measures are obviously heuristical ones and solvers are not safe from regularly eliminating relevant learnt clauses.

Parallel SAT Solving

Two approaches are commonly explored to parallelize SAT solvers. The first one is mainly a *divide-and-conquer* idea, which divides the search space into subspaces, successively allocated to SAT workers. Each time a worker finish its job (whereas the other ones are still doing their task), a load balancing strategy is invoked, and dynamically transfers subspaces to this idle worker [5,6]. A closely related approach is the iterative partitioning one [11]. Note that some of these approaches are able to share clauses [17,11] between workers. On the other hand, the parallel portfolio strategy exploits the complementarity between different sequential CDCL strategies to let them compete and cooperate on the same formula [10,4,12]. Since each worker deals with the whole formula, there is no need to introduce load balancing overheads, and cooperation is only achieved through the exchange of learnt clauses. With this approach, the crafting of the strategies is important, especially with a small number of workers. In general, the objective is to cover the space of good search strategies in the best possible way. Such strategies are efficient on multicore architectures.

As we said above, the size of the learnt clause database is crucial for sequential solvers. Then, for a parallel portfolio SAT solver, it is not desirable to share all learnt clauses between all threads. To deal with this problem, one has to select carefully the clauses that a thread wants to share with the others. A natural solution is to take into account the size of learnt clauses and share only the smallest ones (size less than 8 for example [10]). Based on the observation that small clauses appears less and less during the search, authors of ManySAT propose a dynamic clause sharing policy which uses pairwise size limits to control the exchange between threads [9].

However, it is surprising that Plingeling, one of the winner of the SAT'11 competition shares between threads only unit clauses. Furthermore, ppfolio which have obtained good results in that competition runs different state-of-art sequential solvers in an independent way without any sharing. This last observations show that current clause exchange techniques are not mature and may be improved. This is one of the goal of this paper.

3   Parallelism, Collaboration and Clause Exchange: A Premilinary Experimentation

To illustrate the current behavior of portfolio solvers with respect to clause exchanges, we first have conducted preliminaries experiments on a state-of-the-art solver. For a sequential solver, a "good" learnt clause is a clause that is used during the unit propagation process and the conflict analysis. For portfolio solvers, one can quite safely state the same idea: a "good" shared clause is a clause that helps at least one other thread reducing its search space, namely *propagating*.

Accordingly, we wanted to know how useful are the clauses shared in a portfolio-based solver. To this end, we ran some experiments[1] using a state-of-the-art portfolio-based SAT solver. We choose the solver ManySAT 2.0 (based on Minisat 2.2), because in this solver, the only difference between the working threads are caused by the first decision variables which are selected randomly. Except this initial interpretation, each CDCL worker exhibits the exact same behavior (in terms of restart strategy, branching variable heuristics, etc.), which allows us to make a fair comparison about clause exchange without any side effect. Hence, it represents a good framework to deal with parallel SAT solvers and clauses sharing. By default all clauses of size less than 8 are shared. Moreover, ManySAT provides a deterministic mode [8]. Let us emphasize that we have activated this option to make the obtained results fully reproducible and we report the detailled results online[2].

Let SC be the set of shared clauses, namely the set-theoretical union of each clause exported by a given thread to all the other ones. In this experiment, for each thread, we have considered two particular kinds of shared clauses. First, the shared clauses that are actually used (at least once) by a working thread to propagate. We denote this set $used(SC)$. Second, we have also focused on the set of shared clauses that are deleted without having been from any help during the search. This set is denoted $unused(SC)$. Clearly, $SC \setminus (used(SC) \cup unused(SC))$ represents the set of clauses that have neither been used nor been deleted, yet.

Figure 1 synthetizes the results obtained during this first experiment. The results are reported in the following way: each point of Figure 1 is associated with an instance, and the x-axys corresponds to the rate $\#used(SC)/\#SC$, whereas the y-axys corresponds to the rate $\#unused(SC)/\#SC$, and we report the average rate over the different threads. Figure 1(a) gives the results for ManySAT. First of all, we can remark that the rate of useful shared clauses differs greatly over different instances. We can also note, that in a lot of cases, ManySAT keeps shared clauses during the entire search (dots near the x-axys). This is due to the non-aggressive cleaning strategy of Minisat where in many instances no cleaning are performed. Threads can keep useless clauses a long time and have to support an over cost without any benefit.

---

[1] All experimentations of this paper have been conducted on a dual socket Intel XEON X5550 quad-core 2.66 GHz with 8 MB of cache and a RAM limit of 32GB, under Linux CentOS 6 (kernel 2.6.32). All solvers use 8 threads. The timeout was set to 1200 seconds WC for each instance. If no answer was delivered within this amount of time, the instance was considered unsolved. We used the application instances (300) of the SAT competition 2011.

[2] http://www.cril.fr/~hoessen/penelope.html

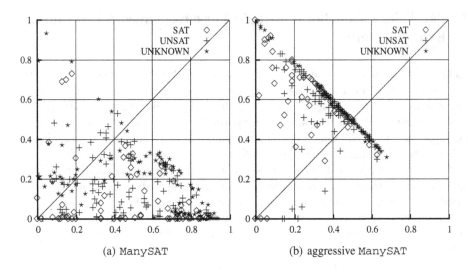

Fig. 1. Comparison between useful share clauses and unused deleted clauses. Each dot corresponds to an instances. x-axis gives the rate of useful shared clauses $\#used(\mathcal{SC})/\#\mathcal{SC}$, whereas the y-axis gives the rate of unused deleted shared clauses $\#unused(\mathcal{SC})/\#\mathcal{SC}$.

We have conducted the same experiment with a much more agressive cleaning strategy. We have choosed the one presented in [1] (see Section 4) and we report the result in Figure 1(b). Here, in many cases, shared clauses are deleted without any usage and the percentage of shared clauses that are used at least one time decreases with respect to the basic version of ManySAT. These results can be explained quite easily. If only few cleanings are done, the threads have to manage a lot of useful *and* useless shared clauses. In one hand, it owns a lot of information about the problem to solve, and propagates many units clauses. On the other hand, such a solver has to maintain a large number of clauses uselessly, which greatly slows down its exploration.

Conversely, when many cleanings are achieved, another problem occurs. Indeed, if a given clause is not used in conflict analysis and/or unit propagation very often, it has then a lot of chances to be quickly removed. Therefore, threads using an agressive strategy spend a lot of time importing clauses that are never used. We can also notice that only using the *lbd* measure for clause usefulness seems not efficient. Indeed, shared clauses are here small clauses, so they have small *lbd* values. Even if we can try to tune the cleaning strategy to obtain a stronger solver, we think that the classical strategy used to manage learnt clauses is not appropriate in the case of clauses sharing and multicore architectures. We propose a new scheme in the next section.

4   Selecting, Sharing and Activating Good Clauses

Managing learnt clauses is known to be difficult in the sequential case. Furthermore, dealing with imported clauses from other threads leads to additional problems:

- Imported clauses can be subsumed by clauses already present in the database. Since subsumption computation is time consuming, it is necessary to give the possibility to remove periodically learnt clauses.
- Imported clauses may be useless during a long time, and suddenly become useful.
- Each thread has to manage many more clauses.
- Characterizing good imported clauses is a real challenge.

For all of these reasons, we propose to use the dynamic management policy of learnt clauses proposed in [1] inside each thread. This recent technique enables to activate or freeze some learnt clauses, imported or locally generated. The advantage is twofold. The overhead caused by imported clauses is greatly reduced since clauses can be frozen. Nevertheless, clauses estimated useful in the next future of the search are activated. Let us present more precisely this method in the next Section.

Freezing Clauses

The strategy proposed in [1] differs from the other ones proposed in the past (see Section 2). Indeed, it is based on a dynamic freezing and activation principle of learnt clauses. At a given search state, it activates the most promising learnt clauses while freezing irrelevant ones. In this way, learned clauses can be discarded from the current step, but may be activated again in future steps of the search process. This strategy cannot be used with the other known measures such as activity or *lbd*-based ones. Indeed, the activity (VSIDS-based) measure is dynamic but can only be used to update the activity of learnt clauses currently in the database, while the *lbd* value of a given learnt clause is either static (and does not change during search) or dynamic but, in this case, the same problem as VSIDS-based one occurs. Then, this strategy is associated with another measure, defined in the following, for identifying good learnt clauses [1].

Let Σ be a CNF formula, c be a clause learnt by the solver, and ω the current interpretation saved from the polarity choice of decision variables [15]. The *psm* value of the clause c w.r.t. ω, denoted $psm_\omega(c)$, is equal to:

$$psm_\omega(c) = |\omega \cap c|$$

psm is a highly dynamic measure, since it is mainly based on the current interpretation. It aims at selecting relevant context (i.e. learnt clauses) with respect to the search in progress. To this end, the clauses that exhibit a low *psm* are considered relevant. Indeed, the lower is a *psm* value, the more likely the related clause is about to unit-propagate some literal, or to be falsified. On the opposite, a clause with a large *psm* value has a lot of chance to be satisfied by many literals, making it irrelevant for the search in progress.

Thus, only clauses that exhibit a low *psm* are selected and currently used by the solver, the other clauses being *frozen*. When a clause is frozen, it is removed from the list of the watched literals of the solver, in order to avoid the computational overhead of maintaining the data structure of the solver for this useless clause. Nevertheless, a frozen clause is not erased but it is kept in memory, since this clause may be useful in the next future of the search. As the current interpretation evolves, the set of learnt clauses actually used by the solver evolves, too. In this respect, the *psm* value is computed periodically, and sets of clauses are frozen or unfrozen with respect to their freshly computed new value.

Let P_k be a sequence where $P_0 = 500$ and $P_{i+1} = P_i + 500 + 100 \times i$. A function "*updateDB* " is called each time the number of conflict reaches P_i conflicts (where $i \in [0..\infty]$). This function computes new *psm* values for every learnt clauses (frozen or activated). A clause that has a *psm* value less than a given limit l is activated in the next part of the search. If its *psm* does not hold this condition, then it is frozen. Moreover, a clause that is not activated after k (equal to 7 by default) time steps is deleted. Similarly, a clause remaining active more than k steps without participating to the search is also permanently deleted.

Given the *psm* and *lbd* measures, we now define different policies for clause exchange. In a typical CDCL procedure, a nogood clause is learnt after each conflict. It appears that all clauses cannot be shared, especially because some of them are not useful in a long term. So, when collaboration is achieved, this is limited through some criterion. To the best of our knowledge, in all current portfolio solvers, this criterion is only based on the information from the sender of the clause, the receiver having to accept any clause judged locally relevant by another worker.

We present in the next Section a technique where both the sender and the receiver of a clause havea strategy. Obviously, any sender (export strategy) tries to find in its own learnt clause database the most relevant information to help the other workers. However, the receiver (import strategy) here does not accept the shared clauses in a blind way. We have called our case study solver PeneLoPe[3] (**Parallel Lbd Psm** solver.).

Importing Clause Policy. When a clause is imported, we can consider different cases, depending on the moment the clause is attached for participating to the search.

- *no-freeze*: each imported clause is actually stored with the current learnt database of the thread, and will be evaluated (and possibly frozen) during the next call to *updateDB* .
- *freeze-all*: each imported clause is *frozen* by default, and is only used later by the solver if it is evaluated relevant w.r.t. unfreezing conditions.
- *freeze*: each imported clause is evaluated as it would have been if locally generated. If the clause is considered relevant, it is added to the learnt clauses, otherwise it is frozen.

Exporting Clause Policy. Since PeneLoPe can freeze clauses, each thread can import more clauses than it would with a classical management of clauses, where all of them are attached. Then, we propose different strategies, more or less restrictive, to select which clauses have to be shared:

- *unlimited*: any generated clause is exported towards the different threads.
- *size limit*: only clauses whose size is less than a given value (8 in our experiments) are exported [9].

---

[3] In reference to Odysseus's faithful wife who wove a burial shroud, linking many *threads* together.

Table 1. Comparison between import, export & restart strategies using deterministic mode

psm used	export strategy	restart strategy	import strategy	#SAT	#UNSAT	#SAT + #UNSAT
✓	lbd limit	lbd	no freeze	94	111	**205**
✓	lbd limit	lbd	freeze	89	**113**	202
✓	size limit	lbd	freeze	93	107	200
✓	size limit	lbd	no freeze	89	107	196
✓	size limit	luby	no freeze	**97**	98	195
✓	lbd limit	lbd	freeze all	89	102	191
✓	size limit	luby	freeze all	96	92	188
✓	unlimited	lbd	freeze	86	102	188
✓	size limit	luby	freeze	92	96	188
✓	lbd limit	luby	freeze	91	97	188
ManySAT	-	-	-	95	93	188
✓	lbd limit	luby	no freeze	90	94	184
✓	unlimited	luby	freeze	91	92	183
	size limit	luby	no freeze	92	90	182
✓	unlimited	luby	no freeze	89	88	177
✓	size = 1	lbd	freeze	89	88	177

– *lbd limit*: a given clause c is exported to other threads if its *lbd* value $lbd(c)$ is less than a given limit value d (8 by default). Let us also note that the *lbd* value can vary over time, since it is computed with respect to the current interpretation. Therefore, as soon as $lbd(c)$ is less than d, the clause is exported.

Restarts Policy. Beside exchange policies, we define two restart strategies.

– *Luby*: Let l_i be the ith term of the Luby serie[13]. The ith restart is achieved after $l_i \times \alpha$ conflicts (α is set to 100 by default).
– *LBD* [2]: Let LBD_g be the average value of the LBD of each learnt clause since the beginning. Let LBD_{100} be the same value computed only for the last 100 generated learnt clause. With this policy, a restart is achieved as soon as $LBD_{100} \times \alpha > LBD_g$ (α is set to 0.7 by default). In addition, the VSIDS score of variables that are unit-propagated thank for a learnt clause whose *lbd* is equal to 2 are increased, as detailed in [2].

We have conducted experiments to compare these different import/export/restart strategies. We ran these different versions and Table 1 presents a sample of the obtained results This table report for each strategy the number of SAT instances solved (#SAT), together with the number of UNSAT instances solved (#UNSAT) and total (#SAT + #UNSAT).

Let us take a first look at the export strategy. Unsurprisingly, the "unlimited" policy obtained the worst results. Indeed, none of these versions have been able to solve more than 190 instances, regardless all other policies (export, restart). Here, every generated clause is exported, and we reach the maximum level of communication. As expected, with the multiplicity of the workers, the solvers are soon overwhelmed by clauses and their performances drop.

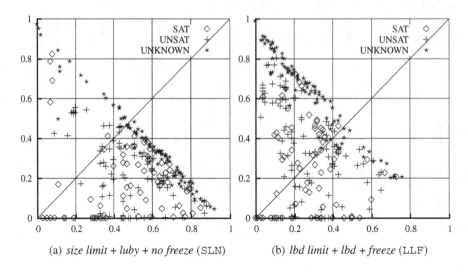

(a) *size limit + luby + no freeze* (SLN)   (b) *lbd limit + lbd + freeze* (LLF)

Fig. 2. Comparison between usefull share clauses and useless deleted clauses. Each dot corresponds to an instances. x-axis gives the rate of useful shared clauses $\frac{\#used(SC)}{\#SC}$, whereas the y-axis gives the rate of unused deleted shared clauses $\frac{\#unused(SC)}{\#SC}$.

This was the reason why a size-based limit was introduced with the idea that the smallest clauses produce the best syntactic filtering, and therefore are preferable. Indeed, in Table 1, it appears clearly that "size limit" (clauses containing less than 8 literals) policy outperforms the "unlimited" one. This simple limit shows its usefulness, but a main drawback is that it has been shown [3] that longer clauses may greatly reduce the size of the proof.

Using the *lbd* value $lbd(c)$ of a clause c can improve the situation as $lbd(c) \leq size(c)$. Hence, if the same value v is used for both the size and the *lbd* limits, more clauses are exported with the *lbd* policy. So, specifying a limit on the *lbd* allows us to import larger clauses if those ones are heuristically considered as promising. This could represent a problem for a parallel solver without the ability to freeze some clauses. Nevertheless, as PeneLoPe contains such mechanism, the impact is greatly reduced. From an empirical point of view, Table 1 shows that the "lbd limit" obtains the best results among all exporting strategies. We have also tried to limit the export to unary clauses (line *size=1*) like most current portfolio solvers do, but this does not lead to good performance, since only 177 instances are solved.

Let us now focus on the restart strategy. Quite obviously, the "luby" technique obtains overall worst results than the "*lbd* " one. This clearly shows the particular interest of this *lbd* concept introduced in [2]. About import strategy, no clear winner appears when looking at the results in Table 1. Indeed, the best results in term of number of solved SAT instances is obtained with *no freeze* (97) when associated with the "luby" restart and the "size limit" export strategy, whereas the best number of solved UNSAT instances is obtained with the *"freeze"* strategy (113). Furthermore, *no freeze* enables to obtain the best overall result solving 205 instances out of the 300 used ones. Hence, it would be audacious to plead for one of the 3 proposed techniques. However, a large number of our proposed policies performs in practice better than "classical" clause exchange techniques, represented in Table 1 by ManySat.

In a second experiment, we wanted to assess the behavior of the solver when using some of our proposed policies. To this end, we have conducted the exact same experiments than the one presented in Section 3; the obtained results are reported in the Figure 2. First, we have tried with *size limit*, *luby*, and *no freeze* policies (denoted SLN, see Figure 2(a)). Clearly, this version behaves very well, since most of the dots are located under the diagonal. Moreover, for most instances, $\frac{\#usedSC + \#unused(SC)}{\#SC}$ is close to 1 (dots near the second diagonal), which indicates that the solver does not carry useless clauses without deleting them. Most of them proves useful, and the other ones are deleted.

Then, the experimentation was conducted with the *lbd* limit, *lbd* and *freeze* combo (denoted LLF, see Figure 2(b)). At first sight, the behavior is here less satisfying than the SLN version, since for most instances at least half of imported clauses are deleted without being from any help. Actually, in this version, a much larger number of clauses are exported due to the "*lbd* limit" export policy, which leads to a lower rate of useful clauses. Fine-tuning parameters (*lbd* limit values, number of time a clause has to be frozen before being permanently deleted, etc.) might improve this behavior.

Looking at some detailed statistics provided in Table 2, it indeed appears that the LLF version shares a lot more clauses than the SLN one (column nb_u). Note that this Table contains some other very interesting information. For instance, it enables to see that for some benchmarks (e.g. AProVE07-21), about 90% of imported clauses are actually frozen and do not immediately participate to the search, whereas for other instances, we face the opposite situation (hwmcc10...) with only 10% of clauses that are frozen when imported. This reveals the high adaptability of the *psm* measure. Let us focus now on the number of imported clauses, compared to the number of conflicts needed to solve the instance. The SLN version produces very often more conflict clauses than it imports from other working threads ($nb_c/nb_i < 1$), even though this is not true with some benchmarks (e.g. AProVE07-21, hwmcc10...). Note that the nb_c/nb_i rate of the LLF version exhibits a very high variability, from 0.58 for the smallest value in Table 2 (velev-pipe-o-uns...) to more than 4, meaning that in such cases, each time the solver produces a conflict (and consequently a clause), it imports more than 4 clauses on average. Let us also emphasize that the computationnal cost of the *psm* measure is not major (see "*psm* time" column). During all our experiments, PeneLoPe have spent at most 5% of the solving time to compute *psm*.

On a more general view, even if the *no-freeze* policy seems to be the best in terms of efficiency in communication between threads of the solver, it has the disadvantage of adding every imported clause in the set of active clauses. This leads to a lower number of propagation per second until the next re-examination of the whole clause database. This might be a problem if we want to increase the number of threads of the solver. On the other hand, the *freeze-all* policy does not slow down the solver. Yet, such solver is not able to use the imported clauses as soon as they are available, and therefore explores subspaces that would have been pruned with the *no-freeze* policy.

Table 2. Statistics about some unsatisfiable instance solving. For each instance and each version, we report the WC time needed to solve it, the number of conficts (nb_c, in thousands), the number of imported clauses (nb_i, in thousands) with between brackets the rate between nb_i and nb_c, the percentage of clauses frozen at the import (nb_f), the percentage of useful imported clauses (nb_u) and the percentage of unused deleted clauses (nb_d). Finally, we provide the rate of time (w.r.t. the overall solving time) spent on computing the *psm* value. Except for time, we compute the average between the 8 threads for these statistics.

instance	version	time	nb_c	$nb_i(nb_c/nb_i)$	nb_f	nb_u	nb_d	*psm* time
dated-10-17-u	SLN	TO	1771	278 (0.15)	0%	45%	49%	2%
	LLF	949	1047	1251 (0.83)	64%	20%	60%	4%
hwmcc10-...	SLN	TO	5955	7989 (1.34)	0%	35%	60%	3%
k50-eijkbs6669-tseitin	LLF	766	3360	15299 (4.55)	10%	11%	80%	5%
velev-pipe-o-uns-1.1-6	SLN	150	981	69 (0.07)	0%	60%	24%	2%
	LLF	48	296	173 (0.58)	41%	31%	33%	3%
sokoban-sequential-p145-	SLN	TO	182	86 (0.47)	0%	92%	4%	0.1%
microban-sequential.040	LLF	530	74	155 (2.09)	5%	58%	17%	0.4%
AProVE07-21	SLN	10	78	83 (1.06)	0%	35%	16%	3%
	LLF	31	143	506 (3.53)	89%	9%	57%	5%
slp-synthesis-aes-bottom13	SLN	445	1628	194 (0.11)	0%	58%	30%	3%
	LLF	91	309	298 (0.96)	71%	24%	49%	4%
velev-vliw-uns-4.0-9-i1	SLN	TO	1664	262 (0.15)	0%	55%	40%	2%
	LLF	906	1165	824 (0.70)	35%	37%	48%	5%
x1mul.miter...-359	SLN	819	2073	421 (0.20)	0%	51%	37%	5%
	LLF	280	680	1134 (1.66)	76%	16%	59%	5%

5   Comparison with State of the Art Solvers

In this Section, we propose a comparison of two of our proposed prototypes against state-of-the-art parallel SAT solvers. We have selected solvers that prove the most effective during the last competitive events: ppfolio [16], cryptominisat [18], plingeling [4] and ManySat [10].

For PeneLoPe, we choose for both versions the *lbd* restart strategy and the *lbd limit* for the export policy. These two versions only differ from their import policies: *freeze* and *no freeze*. Let us precise that contrary to previous experiments, we do not use the deterministic mode in these experiments, in order to obtain the best possible performance.

Figure 3 shows the obtained results through different representations; Table 3(a) provides the number of solved instances for the different solvers, Figure 3(b) details the comparison of PeneLoPe and Plingeling through a scatter plot, and a cactus plot in Figure 3(c) gives the number of solved instances w.r.t. the time (in seconds) needed to solve them. PeneLoPe outperforms all other parallel solvers; indeed, it succeeds to solve 216 instances while no other solver is able to exceed 200 (Table 3(a)). Note that only considering SAT instances, the best results come from plingeling which solves 99 instances. This is particularly noticeable in Figure 3(b) where PeneLoPe and plingeling are more precisely compared; indeed, most of "SAT dots" are located

Solver	#SAT	#UNSAT	#SAT+#UNSAT
PeneLoPe *freeze*	97	**119**	**216**
PeneLoPe *no freeze*	96	**119**	215
plingeling [4]	**99**	97	196
ppfolio [16]	91	103	194
cryptominisat [18]	89	104	193
ManySat [10]	95	92	187

(a) PeneLoPe VS state-of-the-art parallel solvers

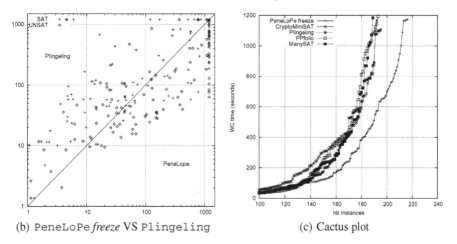

(b) PeneLoPe *freeze* VS Plingeling   (c) Cactus plot

Fig. 3. Comparison on 8 cores

above the diagonal, illustrating the strength of plingeling on these instances. However, results for SAT instances are closer from each other (97 for PeneLoPe *freeze*, 95 for ManySat, etc.), the gap being more important for UNSAT problems.

In addition, we have compared the same solvers on a 32 cores architecture. More precisely, the considered hardware configuration is now Intel Xeon CPU X7550 (4 processors, 32 cores) 2.00GHz with 18 MB of cache and a RAM limit of 256GB. The software framework is the same as with previous experiments. Each solver is run using 32 threads, and the obtained results are displayed in Figure 4 in a similar way than previously. First, let us remark that except for plingeling, all solvers improve their results when they are run with a larger number of threads. The benefit is limited for certain solvers, however. For example, cryptominisat solves 193 instances with 8 threads, and 201 instances with 32 threads. The improvement is stronger with PeneLoPe whose both versions solve 15 extra instances when 32 threads are used, and especially for ManySAT with a gain of 29 instances. The gap can be more remarkable looking at the cactus plot in Figure 4(c), since our 3 competitors solve about the same number of instances within the same time (curves very close to each other), whereas the curves of PeneLoPe and ManySAT clearly show their ability to solve a larger number of instances within a more restricted time. Besides, it is worth noting that PeneLoPe solves the same number of instances as Plingeling, ppfolio and cryptominisat with a (virtual) time limit of only 400 seconds. Finally, we can also notice than PeneLoPe can be improved on SAT instances. Indeed, it appears that

Solver	#SAT	#UNSAT	#SAT+#UNSAT
PeneLoPe *freeze*	104	127	**231**
PeneLoPe *no freeze*	99	**131**	230
ManySAT [10]	105	111	216
ppfolio [16]	**107**	97	204
cryptominisat [18]	96	105	201
Plingeling [4]	100	95	195

(a) PeneLoPe VS state-of-the-art parallel solvers

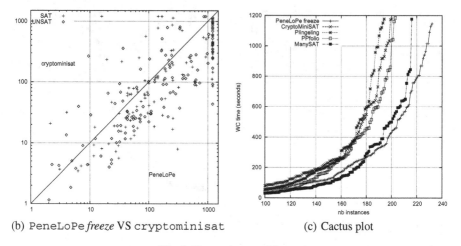

(b) PeneLoPe *freeze* VS cryptominisat   (c) Cactus plot

Fig. 4. Comparison on 32 cores

Luby restarts are more efficient for SAT than for UNSAT, whereas the exact opposite phenomenon happens for UNSAT instances with the *lbd* restart strategy.

Adding computing units has different impacts. For instance, for ppfolio and plingeling the gain is not major, since augmenting the number of working threads "just" improves the number of CDCL sequential solvers that explore the search space; each worker does not benefit from the exploration of the other ones, since with these solvers, little (if any) collaboration is done. PeneLoPe benefits more from more computing units because the number of exchanged clauses coming from different search subspaces is greater. This leads to a wider knowledge for each thread without being slowed down too much, thanks to the freezing mechanism.

Finally, let us emphasize that during all our experiments with PeneLoPe, all working threads share the exact same parameters and strategies, just like in our preliminary experimentation in Section 3. Improving diversification in the different sequential CDCL searches should probably boost even more our case study solver.

6 Conclusion

In this paper, we have proposed new strategies to manage clause exchange within parallel SAT solvers. Based on the recent *psm* and *lbd* concepts, the idea is to adopt different strategies for import and export of clauses. We have carefully studied different empirical aspects of our proposed ideas and compared our solver to best known parallel SAT engines, showing that it appears to be a highly competitive prototype.

Clearly, diversifying the different working threads should improve the performance of our case study solver `PeneLoPe`, since this technique is known to be the cornerstone of the efficiency of some portfolio solvers, like `ppfolio`. We plan to study this point in the next future.

References

1. Audemard, G., Lagniez, J.-M., Mazure, B., Saïs, L.: On Freezing and Reactivating Learnt Clauses. In: Sakallah, K.A., Simon, L. (eds.) SAT 2011. LNCS, vol. 6695, pp. 188–200. Springer, Heidelberg (2011)
2. Audemard, G., Simon, L.: Predicting learnt clauses quality in modern SAT solvers. In: Proceedings of IJCAI, pp. 399–404 (2009)
3. Beame, P., Kautz, H., Sabharwal, A.: Towards understanding and harnessing the potential of clause learning. Journal of Artificial Intelligence Research 22, 319–351 (2004)
4. Biere, A.: (p)lingeling, http://fmv.jku.at/lingeling
5. Chrabakh, W., Wolski, R.: GrADSAT: A parallel SAT solver for the grid. Technical report, UCSB (2003)
6. Chu, G., Stuckey, P.J., Harwood, A.: Pminisat: a parallelization of minisat 2.0. Technical report, SAT Race (2008)
7. Darwiche, A., Pipatsrisawat, K.: Complete Algorithms, ch. 3, pp. 99–130. IOS Press (2009)
8. Hamadi, Y., Jabbour, S., Piette, C., Saïs, L.: Deterministic parallel DPLL. Journal on Satisfiability, Boolean Modeling and Computation 7(4), 127–132 (2011)
9. Hamadi, Y., Jabbour, S., Saïs, L.: Control-based clause sharing in parallel SAT solving. In: Proceedings of IJCAI, pp. 499–504 (2009)
10. Hamadi, Y., Jabbour, S., Saïs, L.: Manysat: a parallel SAT solver. Journal on Satisfiability, Boolean Modeling and Computation 6, 245–262 (2009)
11. Hyvärinen, A.E.J., Junttila, T., Niemelä, I.: Grid-Based SAT Solving with Iterative Partitioning and Clause Learning. In: Lee, J. (ed.) CP 2011. LNCS, vol. 6876, pp. 385–399. Springer, Heidelberg (2011)
12. Kottler, S., Kaufmann, M.: SArTagnan - a parallel portfolio SAT solver with lockless physical clause sharing. In: Pragmatics of SAT (2011)
13. Luby, M., Sinclair, A., Zuckerman, D.: Optimal speedup of Las Vegas algorithms. In: Proceedings of ISTCS, pp. 128–133 (1993)
14. Marques-Silva, J., Sakallah, K.: GRASP - A New Search Algorithm for Satisfiability. In: Proceedings of ICCAD, pp. 220–227 (1996)
15. Pipatsrisawat, K., Darwiche, A.: A Lightweight Component Caching Scheme for Satisfiability Solvers. In: Marques-Silva, J., Sakallah, K.A. (eds.) SAT 2007. LNCS, vol. 4501, pp. 294–299. Springer, Heidelberg (2007)
16. Roussel, O.: ppfolio, http://www.cril.univ-artois.fr/~roussel/ppfolio
17. Schubert, T., Lewis, M., Becker, B.: Pamiraxt: Parallel SAT solving with threads and message passing. Journal on Satisfiability, Boolean Modeling and Computation 6(4), 203–222 (2009)
18. Soos, M.: Cryptominisat, http://www.msoos.org/cryptominisat2/
19. Zhang, L., Madigan, C., Moskewicz, M., Malik, S.: Efficient conflict driven learning in boolean satisfiability solver. In: Proceedings of ICCAD, pp. 279–285 (2001)

Designing Scalable Parallel SAT Solvers

Antti E.J. Hyvärinen[1,*] and Norbert Manthey[2]

[1] Aalto University
Department of Information and Computer Science
P.O. Box 15400, FI-00076 AALTO, Finland
antti.hyvarinen@aalto.fi
[2] Knowledge Representation and Reasoning Group
Technische Universität Dresden, 01062 Dresden, Germany
norbert@janeway.inf.tu-dresden.de

Abstract. Solving instances of the propositional satisfiability problem (SAT) in parallel has received a significant amount of attention as the number of cores in a typical workstation is steadily increasing. With the increase of the number of cores, in particular the scalability of such approaches becomes essential for fully harnessing the potential of modern architectures. The best parallel SAT solvers have, until recently, been based on algorithm portfolios, while search-space partitioning approaches have been less successful. We prove, under certain natural assumptions on the partitioning function, that search-space partitioning can always result in an increased expected run time, justifying the success of the portfolio approaches. Furthermore, we give first controlled experiments showing that an approach combining elements from partitioning and portfolios scales better than either of the two approaches and succeeds in solving instances not solved in a recent solver competition.

1   Introduction

The satisfiability problem (SAT) of determining whether a given propositional formula has a satisfying truth assignment has been a target of intense research efforts due to its theoretical significance [1] and the numerous applications, such as scheduling [2], termination analysis [3], configuration [4], and bioinformatics [5], where SAT solvers have proven successful. Parallelism seems now to dominate the performance of future computer systems, as already current computers provide more than ten CPU cores. As a result, the research on how to parallelize SAT solvers for an increasing number of cores is of high practical relevance.

This paper uses rigorous analysis and experiments to find a novel explanation to the effects certain well-known parallelization techniques have on the expected run time of solving SAT instances. The time a SAT solver S requires to solve a given formula ϕ is known to be highly erratic and might vary significantly as the formula or the solver is modified even slightly. Hence, given a solver S and

* Financially supported by the Academy of Finland under the Finnish Centre of Excellence in Computational Inference (COIN) and the project number 122399.

A. Cimatti and R. Sebastiani (Eds.): SAT 2012, LNCS 7317, pp. 214–227, 2012.

a formula ϕ, the run time is a random variable. The variance in run times has two important implications in parallel solving of formulas. Firstly, assume that the run time of S on ϕ is one second with probability 0.8 and ten seconds with probability 0.2. The expected run time of the solver S is then $1 \cdot 0.8 + 10 \cdot 0.2 = 2.8$ seconds. Running ten (randomized) solvers in parallel gives the solution in the expected time $(1 - 0.2^{10}) \cdot 1 + 0.2^{10} \cdot 10 \approx 1.000001$ seconds. Hence the use of this approach results in speed-up of 2.8. Secondly, assume there is a way of partitioning ϕ's search space into ten separately in parallel solvable, equally difficult parts. It is reasonable to assume that such a partitioning is not perfect so that the run time of each partition is, say, half of the original formula instead of tenth. Proving unsatisfiability of the formula would in this case require ensuring that there are no solutions in any of the partitions, and the expected run time with the partitioning approach is thus $0.8^{10} \cdot 1 \cdot 1/2 + (1 - 0.8^{10}) \cdot 10 \cdot 1/2 \approx$ 4.5 seconds, resulting in speedup of 0.6. This artificial example provides some insight to why the *portfolio solvers*, corresponding to the first case, perform often better than the *search space partitioning solvers* which correspond to the second case. The portfolio approach provides a substantially better speed-up, while the partitioning approach results in fact in a higher expected run time than solving with the underlying solver S.

In this paper we prove, under reasonable assumptions, that there is always a distribution which results in a similar increased expected run time as in the example above, unless the process of constructing partitions is *ideal* in the sense described later. Earlier it has been shown that by organizing the partitioning as a *partition tree* it can be guaranteed that not only the expected run time does not increase above that of S, but that increasing the number of parallel resources never increases the expected run time [6]. We experimentally confirm this using realistic and comparable implementations by showing that the partition tree based *iterative partitioning approach* scales better than either the portfolio approach or the partitioning approach. The implementation is able to solve four instances that were not solved in the SAT Competition 2011.

The run times of randomized tree-based searches have been studied analytically both for sequential solving [7] and in parallel cases [8,9,10,11]. Our analytic discussion differs from these by studying unsatisfiability proofs with a model of partitioning function that is an extension of [11]. Much work has been invested in studying search space partitioning solvers [12,13,14,15] and algorithm portfolios [16,17]. In this work our aim is to build understanding between the two approaches by implementing similar systems in as comparable manner as possible, omitting for instance the most sophisticated clause sharing mechanisms [18,19,20]. The iterative partitioning approach discussed in this work is introduced in [21] and further developed in [6] and [22]. We extend these studies by implementing the approach for multi-core architectures instead of computational grids, which enables us to provide a much more reliable comparison of the iterative partitioning approach to the portfolio and partitioning approaches.

The work is organized as follows: Section 2 defines our notation. Section 3 presents the proof for increasing run times, and formalizes the three parallelization approaches. Section 4 provides the multi-core implementations of the approaches and discusses how clause sharing is implemented in them. The implementations are experimentally evaluated in Sect. 5, and conclusions are given in Sect. 6.

2   Preliminaries

Let V be a finite set of Boolean variables. The set of *literals* $\{x, \neg x \mid x \in V\}$ consists of negative and positive Boolean variables, a *clause* is a disjunction of literals and a *formula* (in conjunctive normal form) is a conjunction of clauses. Whenever convenient, we denote clauses by sets of literals and formulas by sets of clauses. Clauses of size one are called *unit clauses*. An *assignment* σ is a set of literals, is consistent if for no variable x both $x \in \sigma$ and $\neg x \in \sigma$, and is inconsistent otherwise. If an assignment σ does not contain a literal l for each variable $v \in V$, it is called partial. A consistent assignment σ satisfies a clause C if C contains a literal in σ, and satisfies a formula if it satisfies all its clauses. A formula is *satisfiable* if there is a consistent truth assignment satisfying it, and *unsatisfiable* otherwise. A formula ψ is a logical consequence of ϕ, denoted $\phi \models \psi$, if ψ is satisfied by all satisfying assignments of ϕ. The formulas are *logically equivalent*, denoted $\phi \equiv \psi$, if they are logical consequences of each other.

3   Parallel Solving Approaches

This work studies the parallel SAT solver designs that have recently proved successful. In particular, we will discuss

- the *Simple Parallel SAT Solving* (SPSAT) approach, which is a simplified variant of the portfolio approach;
- the *plain partitioning* approach, which again is a simplified version of the search space partitioning approaches such as those based on *guiding paths* [23]; and
- the *iterative partitioning* approach, again a simple approach where partitioning is recursive and the solving is attempted on the search spaces related to all recursive levels until satisfiability is proved.

In the following, we describe the approaches and the concepts related to them in more detail.

The SPSAT Approach is based on solving a given formula ϕ with several SAT solvers in parallel. As all solvers are working on the same instance, the solution is obtained from the solver finishing first. The underlying solvers are slightly randomized so that they correspond to a straightforward portfolio of seemingly infinite different algorithms. The approach is known to be efficient for a wide range of application originated formulas, in particular if they are satisfiable and have several solutions [11].

The Plain Partitioning Approach first divides the search space of the formula, and then solves the resulting partitions separately in parallel. The search space is divided using a *partitioning function* $P(\phi, n)$, which maps a formula ϕ to n partitioning constraints $\kappa_1 \ldots, \kappa_n$ such that (i) $\phi \equiv (\phi \wedge \kappa_1) \vee \ldots \vee (\phi \wedge \kappa_n)$, and (ii) $\kappa_i \wedge \kappa_j \wedge \phi$ is unsatisfiable when $i \neq j$. The formulas $\phi_i = \phi \wedge \kappa_i$, $1 \leq i \leq n$, are called the *derived formulas* of ϕ. The satisfiability of ϕ can be determined by either showing all derived formulas unsatisfiable or showing ϕ_i satisfiable for some $1 \leq i \leq n$. By (i), in the former case also ϕ is unsatisfiable, and in the latter case the assignment satisfying ϕ_i satisfies also ϕ.

The Iterative Partitioning Approach is based on solving a hierarchical *partition tree* in a breadth-first order. Given a formula ϕ, iteratively constructed derived formulas can be presented by a *partition tree* T_ϕ. Each node ν_i is labeled with a set of clauses $Co(\nu_i)$ so that the root ν_0 is labeled with $Co(\nu_0) = \phi$, and given a node ν_k and a rooted path $\nu_0, \ldots, \nu_{k-1}$ to its parent, the label of ν_k is $Co(\nu_k) = \kappa_i$, where κ_i is one of the constraints given by $P(\bigwedge_{j=0}^{k-1} Co(\nu_j), n)$. Each node ν_k with a rooted path $\nu_0, \ldots, \nu_k$ represents the formula $\phi_{\nu_k} = \bigwedge_{i=0}^{k} Co(\nu_i)$. Solving is attempted for each ϕ_{ν_k} in the tree in a breadth-first order. The approach terminates if a satisfying assignment is found, or all rooted paths to the leaves contain a node ν_j such that ϕ_{ν_j} is shown unsatisfiable. In practice, we also limit the run time of each solving attempt to ensure that a reasonably large portion of the search tree will be covered.

The Partitioning Function Model.   A partitioning function $P(\phi, n)$ should produce derived formulas ϕ_i which are increasingly faster to solve as the number of derived formulas n increases. We will use an *efficiency function* to formalize how well P accomplishes this. Assume that the solver S performs with the same probability a given search that takes time t_ϕ in the formula ϕ but, due to the partitioning constraints, a shorter time t_{ϕ_i} in the derived formulas ϕ_i. The efficiency function $\epsilon(n)$ depends on the number n of derived formulas and gives the ratio of the two times, that is, $\epsilon(n) = t_\phi / t_{\phi_i}$.

We use a cumulative run time distribution $q_{S,\phi}(t)$ to describe the probability that a solver S determines the satisfiability of a formula ϕ in time t. This reasoning results in a model where, given a formula ϕ with the run time distribution $q_{S,\phi}(t)$ on a solver S, the n derived formulas ϕ_i all have the distributions $q_{S,\phi_i}(t) = q_{S,\phi}(\epsilon(n)t)$.

We will only consider efficiency functions of the form $\epsilon(n) = n^\alpha$ where $0 \leq \alpha \leq 1$ is a constant depending on the partitioning function. The function satisfies the following natural properties:

(1) $1 \leq \epsilon(n) \leq n$,
(2) $\epsilon(n) \leq \epsilon(n+1)$, and
(3) $\epsilon(n)^p = \epsilon(n^p)$ for all $p \in \mathbb{N}$

The first condition states that the partitioning function should not make a particular search of S super-linearly faster or slow the search down. The second condition requires that the efficiency does not decrease as more derived formulas are created. The last condition states that if a partitioning function $P(\phi, n)$ is used to produce n^p derived formulas recursively, the resulting efficiency must equal the efficiency of $P(\phi, n^p)$ where the derived formulas are all generated at once. Hence, given a partitioning function P with efficiency $\epsilon(n) = n^\alpha$, the cumulative run time distributions for the derived formulas ϕ_i of ϕ are

$$q_{\mathcal{S},\phi_i}(t) = q_{\mathcal{S},\phi}(n^\alpha t) \text{ for some } \alpha \text{ in the range } 0 \leq \alpha \leq 1, \tag{1}$$

where the partitioning function is called *ideal* if $\alpha = 1$, that is, $\epsilon(n) = n$.

3.1   Plain Partitioning Can Increase Expected Run Time

A run time distribution $q_{\mathcal{S},\phi}(t)$ for an unsatisfiable ϕ completely determines the run time distribution $q^n_{\text{Plain-Part}(\alpha),\phi}(t)$ for the plain partitioning approach with a partitioning function P with efficiency $\epsilon(n) = n^\alpha$. In particular, since the formula ϕ is shown unsatisfiable once all derived formulas have been shown unsatisfiable, by (1) we have

$$q^n_{\text{Plain-Part}(\alpha),\phi}(t) = q^n_{\mathcal{S},\phi_i}(t) = q^n_{\mathcal{S},\phi}(n^\alpha t). \tag{2}$$

In this section we are interested in studying the expected value of the random variables $T_{\mathcal{S},\phi}$ and $T^n_{\text{Plain-Part}(\alpha),\phi}$ describing the times required to solve ϕ with the solver $\mathcal{S}$ and the plain partitioning approach using n derived formulas, respectively. In particular, we wish to prove the somewhat surprising claim that for non-ideal partitioning functions there are distributions for unsatisfiable formulas such that the expected run time of the solver $\mathcal{S}$ is less than the expected run time of the plain partitioning approach, stated more formally as follows:

Proposition 1. *Let ϕ be unsatisfiable, $P(\phi, n)$ a partitioning function with efficiency $\epsilon(n) = n^\alpha$, and $\mathcal{S}$ a SAT solver. Then for sufficiently large n and every $0 \leq \alpha < 1$ there exists a distribution $q^n_{\mathcal{S},\phi}(t)$ such that the expected run time $\mathbb{E}T_{\mathcal{S},\phi}$ of $\mathcal{S}$ is lower than the expected run time $\mathbb{E}T^n_{Plain-Part(\alpha),\phi}$ of the plain partitioning approach.*

Proof. The family of distributions $q^n_{\mathcal{S},\phi}(t)$ we will use in the proof is

$$q^n_{\mathcal{S},\phi}(t) = \begin{cases} 0 & \text{if } t < t_1, \\ 1 - \frac{1}{n} & \text{if } t_1 \leq t < t_2, and \\ 1 & \text{if } t \geq t_2, \end{cases} \tag{3}$$

where $t_1 < t_2$. Thus the probabilities that the formula is solved by $\mathcal{S}$ exactly in time t_1 is $1 - 1/n$ and exactly in time t_2 is $1/n$. The expected run time for a formula following the distribution is

$$\mathbb{E}T_{\mathcal{S},\phi} = (1 - \frac{1}{n})t_1 + \frac{1}{n}t_2. \tag{4}$$

The expected run time of the plain partitioning approach using the partition function $\epsilon(n) = n^\alpha$ can be derived by noting that all derived formulas need to be solved before the result can be determined. This means that either all solvers are "lucky", and determine the unsatisfiability in time t_1/n^α, or at least one of the solvers runs for time t_2/n^α, which will then become the run time of the approach. This results in

$$\mathbb{E}T^n_{\text{Plain-Part}(\alpha),\phi} = \left(1 - \frac{1}{n}\right)^n \frac{t_1}{n^\alpha} + \left(1 - (1 - \frac{1}{n})^n\right) \frac{t_2}{n^\alpha}. \tag{5}$$

We claim that for every α, there are values for n, t_1 and t_2 such that $\mathbb{E}T_{\mathcal{S},\phi} < \mathbb{E}T^n_{\text{Plain-Part}(\alpha),\phi}$. Dividing both sides of the resulting inequality by t_2 and setting $k = t_1/t_2$ results in

$$(1 - \frac{1}{n})k + \frac{1}{n} < \frac{(1 - \frac{1}{n})^n}{n^\alpha}k + \frac{1 - (1 - \frac{1}{n})^n}{n^\alpha},$$

which can be reordered to

$$k\left((1 - \frac{1}{n}) - \frac{(1 - \frac{1}{n})^n}{n^\alpha}\right) < \frac{1 - (1 - \frac{1}{n})^n}{n^\alpha} - \frac{1}{n}.$$

We note that $(1 - \frac{1}{n}) > (1 - \frac{1}{n})^n/n^\alpha$ when $n \geq 2$, and therefore the left side of the inequality is positive and can be made arbitrarily small by setting k small. It remains to show that the right side of the inequality is positive for sufficiently large n, i.e.,

$$\frac{n - (1 - \frac{1}{n})^n n - n^\alpha}{n^{\alpha+1}} > 0.$$

Since $n^{\alpha+1}$ is always positive, we may simplify this and factor n from the nominator, resulting in

$$1 - (1 - \frac{1}{n})^n - n^{\alpha-1} > 0. \tag{6}$$

Noting that $\lim_{n\to\infty}(1 - \frac{1}{n})^n = \frac{1}{e} \approx 0.3$, and that $\lim_{n\to\infty} 1 - n^{\alpha-1} = 1$ if $\alpha < 1$, we get the desired result, that is, for sufficiently large n, there are values t_1 and t_2 such that $t_1 < t_2$ and $\mathbb{E}T_{\mathcal{S},\phi} < \mathbb{E}T^n_{\text{Plain-Part}(\alpha),\phi}$.

Note that the proof does not hold if the partitioning function is ideal, since the left hand side of the inequality (6) is negative if $\alpha = 1$. In fact, we have the following proposition proved in [11]:

Proposition 2. *Let $n \geq 1$, $\epsilon(n) = n^1 = n$ be the efficiency of an ideal partitioning function, and $q_T(t)$ be the run time distribution of an unsatisfiable formula ϕ with a randomized solver. Then $\mathbb{E}T^n_{\text{Plain-Part}(1),\phi} \geq \mathbb{E}T^{n+1}_{\text{Plain-Part}(1),\phi}$.*

The distribution $q^n_{\mathcal{S},\phi}(t)$ used in proof of Prop. 1 is clearly not a common distribution for any solver and unsatisfiable formula. Furthermore, many search space partitioning solvers are based on *guiding paths*, an approach designed to increase dynamically the number of derived formulas as the instance is being solved. Nevertheless we believe that the observation helps to understand the performance of parallel SAT solvers and thereby gives guidelines how to design better parallel solvers. To further evaluate the effect in practice, we compare the plain partitioning approach against the SPSAT approach and the iterative partitioning approach, both of which provably do not suffer from the increasing expected run times (see [11] and [24], respectively, for proofs).

4   Multi-core Implementations

The partitioning approaches discussed in Sect. 3 can be implemented in a relatively straightforward manner using the efficient off-the-shelf SAT solvers that are readily available. Our implementations use the POSIX threads library to enable multi-threaded computing. The SPSAT implementation is straightforward, and the plain partitioning approach can be seen as a special case of the iterative partitioning, where only the original formula is partitioned, and only the derived formulas are solved. Hence this section concentrates on describing the iterative partitioning approach. In interpreting the experimental results it is useful to keep in mind that the low-level SAT solving, corresponding to the underlying solver S in the analytical model, is performed by the same code in all three approach. This allows us to compare the results more reliably.

4.1   The Iterative Partitioning Approach

The iterative partitioning approach is implemented as a master-worker architecture, where the master maintains a tree of derived formulas and the workers both compute the partitioning function and run the underlying solvers. Communication is handled via shared memory and the locking primitives available from the library. The master thread takes care of the following tasks:

1. Maintaining the partition tree
2. Maintaining the queue of nodes to partition
3. Submitting partitioning tasks
4. Submitting solving tasks
5. Determining whether the search can be terminated

There are two kinds of workers: the *partitioner* and the *solver*. The maximum run times of both workers are limited. The partitioner takes as input a node ν_i in the partition tree and a number n, and produces the derived formulas computed by the partitioning function $P(\phi_{\nu_i}, n)$ upon reaching the time limit. The solver receives a node ν_i from the partition tree and tries to solve the corresponding formula ϕ_{ν_i}. At success, the solver returns either a satisfying truth assignment or concludes that ϕ_{ν_i} is unsatisfiable. Otherwise, if the run time limit is reached, no solution is returned and the corresponding node is marked *unknown*. Such nodes are subject to at most one partitioning and their satisfiability will be determined by attempting to solve recursively the formulas corresponding to child nodes.

A node ν_i is solved by first constructing the corresponding formula ϕ_{ν_i}. After successful solving of ϕ_{ν_i}, the master either updates the state of ν_i to *unsatisfiable* or receives the satisfying truth assignment depending on the outcome of the solver. Otherwise, if the solving of ϕ_{ν_i} failed due to a timeout, the state of ν_i remains *unknown*.

In case of receiving an unsatisfiable result on a node ν_i, the master checks the states of the sibling nodes. In case they all are already in the state *unsatisfiable*, also the parent of ν_i, if one exists, is marked *unsatisfiable*. This process is repeated recursively upwards. This way a node in the tree is marked *unsatisfiable* if and only if all paths from the node to the leaves pass through a node corresponding to a formula that is shown unsatisfiable with the solver.

The partitioner implements the *vsids* scattering function [21], where the partitioning constraints are in general clauses consisting of literals with a high *vsids* [25] score.

4.2   Clause Learning

To keep the discussion and the results generalizable, the underlying solvers of the approaches are only allowed a limited form of learned clause sharing. In particular, the sharing of only unit clauses is allowed, since sharing longer clauses might have negative impact on the overall performance [20,22].

The SPSAT implementation synchronizes its units with a centralized database at every restart and when learning a new unit clause. This operation can be performed with no locks with a Compare-and-Swap instruction, and has no noticeable negative performance effect. Clause sharing is less straightforward in the partitioning approaches. A clause learned in one derived formula is not, in general, a logical consequence of another derived formula, and hence the learned clauses are not transferred between derived formulas. The iterative partitioning approach shares the unit clauses only "downwards", that is, clauses learned in a node are shared with the formulas in the subtree rooted at that node, by storing the units learned while solving a formula ϕ_{ν_i} to the constraints $Co(\nu_i)$. In plain partitioning, the unit clauses are similarly saved to the constraints of the derived formula from which they are learned, and are hence shared between two consecutive solvers if the first solver fails.

It is possible to maintain more complicated data structures which allow tracking to some extent from which constraints a given clause depends. This usually involves an overhead some times high enough to completely ruin the speed-up obtained from the parallelization [22]. For simplicity and to help in interpreting the comparison, such data structures are not implemented in our experiments.

5   Results

This section analyzes the performance of the SPSAT, the plain partitioning and the iterative partitioning approaches using the application category instances from the 2009 and 2011 SAT competitions[1]. We first compare the wall clock run time of each solving approach to that of the underlying solver, and then study the scalability of the approaches with four and 12 cores. We continue by comparing the plain and iterative partitioning approaches, by showing how the iterative partitioning approach scales when moving to a grid-based system, and finally report on solving the instances that were not solved in SAT competition 2011. The reliability of the results is addressed shortly by solving repeatedly certain randomly chosen instances.

---

[1] See http://www.satcompetition.org/

5.1   Experimental Setup

All three approaches use MiniSAT 2.2.0 [26] as the underlying solver[2]. We use preprocessing only in the last experiment. The experiments are run on a cluster consisting of nodes with two six-core AMD Opteron 2435 processors. Each instance was solved on an exclusively reserved computing node. The memory usage for each instance was limited to 30 GB and the duration to four hours of wall clock time. Each thread was allocated an equal amount of memory, that depended on the number of threads used. For instance, when running 12 threads, each thread had approximately 2.5 GB of memory. If the thread ran out of memory, the unit clauses learned by the thread were collected and, in case of the SPSAT and plain partitioning approaches, the thread was restarted with the same formula.

The measurement of memory usage is always an estimate, and therefore the system may nevertheless run out of memory resulting in an early termination of the search. The run time of each solver thread in the SPSAT and the plain partitioning approaches was limited to four hours, while run times of the solver threads in the iterative partitioning approach was limited to 2400 seconds wall clock time (however, the master thread still had the time limit of four hours).

The partitioning function used in the iterative partitioning approach constructed eight derived instances, that is, it was the function $P(\cdot, 8)$. The plain partitioning used the function $P(\cdot, 1000)$ to obtain roughly the same amount of formulas in total for both approaches. Increasing the number of derived formulas increases the probability that some of them are trivially unsatisfiable. In 50% of the 2009 benchmark formulas the number of non-trivial derived formulas was over 200, and in 25% of the formulas the number was over 600. In total 70 seconds were allocated for computing the partitioning function in both cases.

In most of the experiments, we illustrate the results with scatter plots with two solving approaches on the axis. Satisfiable instances are denoted by × and unsatisfiable instances by □. The instances that timed out are plotted on the lines on the top and the right of the graphs, whereas the instances that ran out of memory despite the restart-forcing limitations are drawn at the edges of the graph. The dashed line in the figures correspond to the linear speedup.

5.2   Scalability of the Multi-core Implementation

Figure 1 shows scatter plots of the SPSAT approach, the iterative partitioning (*Iter-Part*) approach and the plain partitioning (*Plain-Part*) approach against the underlying solver. All three approaches are able to solve more formulas and are usually faster than MiniSAT 2.2.0. The SPSAT approach does not reach a linear speedup for unsatisfiable formulas, but works well for many of the satisfiable instances. The plain partitioning approach shows a noticeable slowdown for many of the instances where the run time is between hundred and thousand seconds. This could result from two factors; firstly, in multi-core computing the threads interfere between each other causing a slowdown. Secondly, as shown in Prop. 1, it is possible that the slowdown results from the shape of the distributions of the

---

[2] Solvers and data are available at http://tools.computational-logic.org/

original and the derived instances. Interestingly, the wall clock run time of plain partitioning for unsatisfiable instances is in 41 cases higher than that of the underlying solver, and lower in 69 cases (excluding instances not solved by both plain partitioning and the underlying solver). The corresponding numbers for iterative partitioning are substantially more convincing, 18 and 94.

As discussed in Sect. 3, the run time of a solver given a formula is essentially a random variable, and therefore a single run time pair on a single instance is inherently unreliable in comparing the performance of two algorithms. To estimate the quality of the results, we randomly selected ten unsatisfiable and ten satisfiable instances and repeated their solving with the iterative partitioning approach ten times. The average variation coefficient c_v, that is, the ratio of the standard deviation to the mean, itself averaged over the ten instances, is $c_v = 0.10$ for the unsatisfiable instances and $c_v = 0.31$ for the satisfiable instances.

5.3   Selecting a Scalable Algorithm

The use of more cores increases the memory access times and causes memory outs in the solvers as the data structures are replicated for each thread. A parallel solving approach should provide speed-up despite these adverse effects. Table 1 summarizes scalability using the instances that the approach solved both with four and 12 cores. In Fig. 2 we concentrate more on studying the run times of the unsatisfiable instances. The table distinguishes the results for satisfiable and unsatisfiable instances; the columns *slower* and *faster* denote the number of instances solved slower and faster, respectively, with 12 cores than with four cores. Hence if the number under *slower* is lower than the number under *faster*, this indicator shows that the approach scales. We also report the sum of the wall-clock run times for the approaches on the last four columns.

Based on the results we can make several interesting observations. Firstly, the wall-clock solving time for most instances in nearly all cases increases when the number of cores increases. The only exception is the iterative partitioning when solving satisfiable instances. Secondly, the total wall clock run time required to solve the instances decreases for almost all the approaches, here the exception being the SPSAT approach in unsatisfiable instances. The SPSAT approach scales

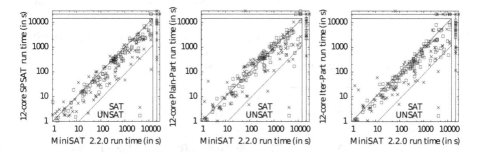

Fig. 1. The SPSAT, iterative and plain partitioning approaches with unit sharing

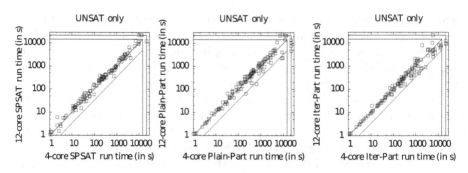

Fig. 2. The scalability of the parallel approaches

badly in unsatisfiable instances, while the plain and iterative partitioning approaches show better scalability, the iterative partitioning being clearly the best as it reduces the run time by 15% and almost never slows down the solving of an instance. As shown in Fig. 2, increasing the number of cores results in solving more instances in iterative partitioning, whereas in the other two approaches the number of solved instances either decreases or stays the same.

The scalability above suggest that the partitioning approaches scale better than SPSAT. Our three approaches are deliberately as simple as possible while still being interesting from the practical point of view, and hence none of the parallel solvers competing in the recent SAT competitions correspond exactly to any of the approaches. Nevertheless it is interesting to try to relate the observations here to the recent competitions. Of the four solvers competing in the 32 core track in 2011, three were variants of the SPSAT approach, one being implemented with the guiding path approach [23] related to the plain partitioning.

As can be seen in the comparisons in Figs. 1 and 2, the partitioning approaches are especially advantageous for the harder instances. We will next give some more insight into this. First, we show in left of Fig. 3 that the iterative partitioning approach compares favorably to the plain partitioning on unsatisfiable instances except for a handful of instances. The iterative partitioning approach also solves a significantly larger number of formulas than the plain partitioning approach. Again, there are two reasons for this. Firstly, in the light of the propositions 1 and 2 and results in [11], it is unlikely that the plain partitioning approach would obtain even close to linear speed-up. The iterative partitioning behaves analytically much

Table 1. Comparison on instances that the respective approaches could solve both with four and 12 cores. Column *slower* (resp. *faster*) denotes the number of instances solved slower (resp. faster) with 12 cores than with four cores.

| | SAT | | UNSAT | | SAT runtime | | UNSAT runtime | |
Approach	slower	faster	slower	faster	4-core	12-core	4-core	12-core
SPSAT	**47**	36	**93**	23	61784	**57380**	**111152**	127462
Plain-Part	**45**	34	**77**	39	61681	**60934**	121432	**119925**
Iter-Part	33	**45**	61	58	53918	**50642**	153726	**131521**

Table 2. Instances not solved in SAT 2011 competition

Name	Solution	w/ preproc	w/o preproc
aes_32_4_keyfind_1	SAT	—	6299
gus-md5-12	UNSAT	4367	6022
rbcl_xits_09_UNKNOWN	UNSAT	—	9635
smtlib-qfbv-aigs-VS3-benchmark-S2-tseitin	UNSAT	4732	8163

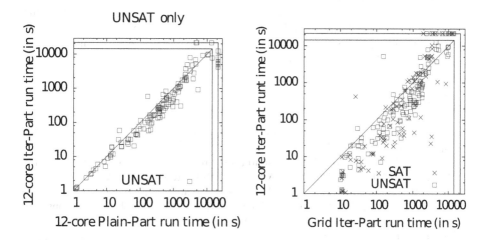

Fig. 3. Iterative partitioning in grid using at most 64 cores [22], and in multi-core approach using 12 cores

nicer [24]. Secondly, the iterative partitioning approach is able to adjust to the problem difficulty due to the dynamic construction of the partition tree.

We also give some insight into how the iterative partitioning approach scales beyond 12 cores in right of Fig. 3 by using a computing grid based implementation running on at most 64 cores. In this system the communication latencies are several orders of magnitude higher and the hardware is older than in the multi-core environment, rendering the results not directly comparable. We still note that increasing the number of cores helps in many hard unsatisfiable instances and results in solving roughly ten more instances.

Finally, we ran the iterative partitioning approach on the 2011 competition instances on 12 cores with and without the SatElite preprocessing techniques. The wall-clock run times are reported in Table 2. We only report the four instances that we could solve but were not solved by any solver in the competition.

6   Conclusions and Future Work

This work addresses some of the central questions in designing scalable parallel SAT solvers using a novel analysis based on a realistic model of search-space partitioning and an efficient uniform implementation based on widely used techniques. The analysis shows that partitioning inherently involves a risk that the expected run time increases compared to sequential solving. An earlier result [24], showing that organizing the search spaces as a tree instead of a set avoids this problem, motivates the experimental comparison of these approaches as well as the widely used *portfolio* approach. Our results confirm that the partition tree based *iterative partitioning approach* performs well compared to the set-based *plain partitioning*, both of which perform better in the unsatisfiable formulas than the portfolio approach. Surprisingly, the iterative partitioning approach over-performs portfolio also in satisfiable formulas. Finally we demonstrate the performance of the iterative partitioning approach by solving four instances that could not be solved in the SAT competition 2011.

References

1. Cook, S.A.: The complexity of theorem-proving procedures. In: Proc. STOC 1971, pp. 151–158. ACM (1971)
2. Béjar, R., Manyà, F.: Solving the round robin problem using propositional logic. In: Proc. AAAI 2000, pp. 262–266. AAAI Press/MIT Press (2000)
3. Fuhs, C., Giesl, J., Middeldorp, A., Schneider-Kamp, P., Thiemann, R., Zankl, H.: SAT Solving for Termination Analysis with Polynomial Interpretations. In: Marques-Silva, J., Sakallah, K.A. (eds.) SAT 2007. LNCS, vol. 4501, pp. 340–354. Springer, Heidelberg (2007)
4. Kautz, H.A., Selman, B.: Planning as satisfiability. In: Proc. ECAI 1992, pp. 359–363. John Wiley and Sons (1992)
5. Lynce, I., Marques-Silva, J.: SAT in Bioinformatics: Making the Case with Haplotype Inference. In: Biere, A., Gomes, C.P. (eds.) SAT 2006. LNCS, vol. 4121, pp. 136–141. Springer, Heidelberg (2006)
6. Hyvärinen, A.E.J., Junttila, T., Niemelä, I.: Partitioning SAT Instances for Distributed Solving. In: Fermüller, C.G., Voronkov, A. (eds.) LPAR-17. LNCS, vol. 6397, pp. 372–386. Springer, Heidelberg (2010)
7. Luby, M., Sinclair, A., Zuckerman, D.: Optimal speedup of Las Vegas algorithms. Information Processing Letters 47(4), 173–180 (1993)
8. Rao, V.N., Kumar, V.: On the efficiency of parallel backtracking. IEEE Transactions on Parallel and Distributed Systems 4(4), 427–437 (1993)
9. Luby, M., Ertel, W.: Optimal Parallelization of Las Vegas Algorithms. In: Enjalbert, P., Mayr, E.W., Wagner, K.W. (eds.) STACS 1994. LNCS, vol. 775, pp. 463–474. Springer, Heidelberg (1994)
10. Segre, A.M., Forman, S.L., Resta, G., Wildenberg, A.: Nagging: A scalable fault-tolerant paradigm for distributed search. Artificial Intelligence 140(1/2), 71–106 (2002)
11. Hyvärinen, A.E.J., Junttila, T., Niemelä, I.: Partitioning search spaces of a randomized search. Fundamenta Informaticae 107(2-3), 289–311 (2011)
12. Böhm, M., Speckenmeyer, E.: A fast parallel SAT-solver — efficient workload balancing. Annals of Mathematics and Artificial Intelligence 17(3-4), 381–400 (1996)
13. Sinz, C., Blochinger, W., Küchlin, W.: PaSAT — parallel SAT-checking with lemma exchange: Implementation and applications. In: Proc. SAT 2001. Electronic Notes in Discrete Mathematics, vol. 9. Elsevier (2001)

14. Schubert, T., Lewis, M., Becker, B.: PaMiraXT: Parallel SAT solving with threads and message passing. Journal on Satisfiability, Boolean Modeling and Computation 6(4), 203–222 (2009)
15. Heule, M.J., Kullmann, O., Wieringa, S., Biere, A.: Cube and conquer: Guiding CDCL SAT solvers by lookaheads. In: Proc. HVC 2001. Springer (2011) (to appear)
16. Huberman, B.A., Lukose, R.M., Hogg, T.: An economics approach to hard computational problems. Science 275(5296), 51–54 (1997)
17. Hamadi, Y., Jabbour, S., Saïs, L.: ManySAT: a parallel SAT solver. Journal on Satisfiability, Boolean Modeling and Computation 6(4), 245–262 (2009)
18. Hamadi, Y., Jabbour, S., Saïs, L.: Control-based clause sharing in parallel SAT solving. In: Proc. IJCAI 2009, pp. 499–504. IJCAI/AAAI (2009)
19. Guo, L., Hamadi, Y., Jabbour, S., Sais, L.: Diversification and Intensification in Parallel SAT Solving. In: Cohen, D. (ed.) CP 2010. LNCS, vol. 6308, pp. 252–265. Springer, Heidelberg (2010)
20. Hyvärinen, A.E.J., Junttila, T., Niemelä, I.: Incorporating clause learning in grid-based randomized SAT solving. Journal on Satisfiability, Boolean Modeling and Computation 6, 223–244 (2009)
21. Hyvärinen, A.E.J., Junttila, T.A., Niemelä, I.: A Distribution Method for Solving SAT in Grids. In: Biere, A., Gomes, C.P. (eds.) SAT 2006. LNCS, vol. 4121, pp. 430–435. Springer, Heidelberg (2006)
22. Hyvärinen, A.E.J., Junttila, T., Niemelä, I.: Grid-Based SAT Solving with Iterative Partitioning and Clause Learning. In: Lee, J. (ed.) CP 2011. LNCS, vol. 6876, pp. 385–399. Springer, Heidelberg (2011)
23. Zhang, H., Bonacina, M.P., Hsiang, J.: PSATO: a distributed propositional prover and its application to quasigroup problems. Journal of Symbolic Computation 21(4-6), 543–560 (1996)
24. Hyvärinen, A.E.J.: Grid Based Propositional Satisfiability Solving. PhD thesis, Aalto University (2011)
25. Moskewicz, M.W., Madigan, C.F., Zhao, Y., Zhang, L., Malik, S.: Chaff: Engineering an efficient SAT solver. In: Proc. DAC 2001, pp. 530–535. ACM (2001)
26. Eén, N., Sörensson, N.: An Extensible SAT-solver. In: Giunchiglia, E., Tacchella, A. (eds.) SAT 2003. LNCS, vol. 2919, pp. 502–518. Springer, Heidelberg (2004)

Evaluating Component Solver Contributions to Portfolio-Based Algorithm Selectors

Lin Xu, Frank Hutter, Holger Hoos, and Kevin Leyton-Brown

Department of Computer Science, University of British Columbia
{xulin730,hutter,hoos,kevinlb}@cs.ubc.ca

Abstract. Portfolio-based methods exploit the complementary strengths of a set of algorithms and—as evidenced in recent competitions—represent the state of the art for solving many NP-hard problems, including SAT. In this work, we argue that a state-of-the-art method for constructing portfolio-based algorithm selectors, SATzilla, also gives rise to an automated method for quantifying the importance of each of a set of available solvers. We entered a substantially improved version of SATzilla to the inaugural "analysis track" of the 2011 SAT competition, and draw two main conclusions from the results that we obtained. First, automatically-constructed portfolios of sequential, non-portfolio competition entries perform substantially better than the winners of all three sequential categories. Second, and more importantly, a detailed analysis of these portfolios yields valuable insights into the nature of successful solver designs in the different categories. For example, we show that the solvers contributing most to SATzilla were often *not* the overall best-performing solvers, but instead solvers that exploit novel solution strategies to solve instances that would remain unsolved without them.

1 Introduction

The propositional satisfiability problem (SAT) is among the most widely studied NP-complete problems, with applications to problems as diverse as scheduling [4], planning [15], graph colouring [30], bounded model checking [1], and formal verification [25]. The resulting diversity of SAT instances fueled the development of a multitude of solvers with complementary strengths.

Recent results, reported in the literature as well as in solver competitions, have demonstrated that this complementarity can be exploited by so-called "algorithm portfolios" that combine different SAT algorithms (or, indeed, algorithms for solving other hard combinatorial problems). Such algorithm portfolios include methods that select a single algorithm on a per-instance basis [21,7,10,31,24,26], methods that make online decisions between algorithms [16,3,23], and methods that run multiple algorithms independently on one instance, either in parallel or sequentially [13,9,20,6,27,14,22,11].

To show that such methods work in practice as well as in theory—that despite their overhead and potential to make mistakes, portfolios can outperform

A. Cimatti and R. Sebastiani (Eds.): SAT 2012, LNCS 7317, pp. 228–241, 2012.

their constituent solvers—we submitted our `SATzilla` portfolio-based algorithm selectors to the international SAT competition (`www.satcompetition.org`), starting in 2003 and 2004 [18,19]. After substantial improvements [31], `SATzilla` won three gold and two other medals (out of nine categories overall) in each of the 2007 and 2009 SAT competitions. Having established `SATzilla`'s effectiveness, we decided not to compete in the main track of the 2011 competition, to avoid discouraging new work on (non-portfolio) solvers. Instead, we entered `SATzilla` in a new "analysis track". However, other portfolio-based methods did feature prominently among the winners of the main track: the algorithm selection and scheduling system 3S [14] and the simple, yet efficient parallel portfolio `ppfolio` [22] won a total of seven gold and 16 other medals (out of 18 categories overall).

One of the main reasons for holding a SAT competition is to answer the question, *What is the current state of the art (SOTA) in SAT solving?*. The traditional answer to this question has been *the winner of the respective category of the SAT competition*; we call such a winner a *single best solver (SBS)*. However, as clearly demonstrated by the efficacy of algorithm portfolios, different solver strategies are (at least sometimes) complementary. This fact suggests a second answer to the SOTA question: *the virtual best solver (VBS)*, defined as the best SAT competition entry on a per-instance basis. The VBS typically achieves much better performance than the SBS, and does provide a useful theoretical upper bound on the performance currently achievable. However, this bound is typically not tight: the VBS is not an actual solver, because it only tells us which underlying solver to run after the performance of each solver on a given instance has been measured, and thus the VBS cannot be run on new instances. Here, we propose a third answer to the SOTA question: *a state-of-the-art portfolio that can be constructed in a fully automatically fashion from available solvers* (using an existing piece of code); we call such a portfolio a *SOTA portfolio*. Unlike the VBS, a SOTA portfolio is an actual solver that can be run on novel instances. We show how to build a SOTA portfolio based on an improved version of our `SATzilla` procedure, and then demonstrate techniques for analyzing the extent to which its performance depends on each of its component solvers. While (to the best of our knowledge) `SATzilla` is the only fully automated and publicly available portfolio construction method, our measures of solver contributions may also be applied to other portfolio approaches, including parallel portfolios.

In this paper, we first verify that our automatically constructed `SATzilla 2011` portfolios yielded cutting-edge performance (they closed 27.7% to 90.1% of the gap between the winners of each sequential category and the VBS). Next, we perform a detailed, quantitative analysis of the performance of `SATzilla 2011`'s component solvers and their pairwise correlations, their frequency of selection and success within `SATzilla 2011`, and their overall benefit to the portfolio. Overall, our analysis reveals that the solvers contributing most to `SATzilla 2011` were often *not* the overall best-performing solvers (SBSs), but instead solvers that exploit novel solution strategies to solve instances that would remain unsolved without them. We also provide a quantitative basis for the folk knowledge that—due to the dominance of MiniSAT-like architectures—the performance of most solvers for `Application` instances is tightly correlated. Our results suggest a shift away from rewarding solvers that perform well on average and towards rewarding creative approaches that can solve instance types not solved well by other solvers (even if they perform poorly on many other types of instances).

2 Improved Automated Construction of Portfolio-Based Algorithm Selectors: SATzilla 2011

In this work, we use an improved version of the construction method underlying the portfolio-based algorithm selection system SATzilla. We first describe SATzilla's automated construction procedure (which, at a high level, is unchanged since 2008; see [31] for details) and then discuss the improvements to its algorithm selection core we made in 2011.

How SATzilla Works. We first describe the automated SATzilla construction procedure and then describe the pipeline that the resulting portfolio executes on a new instance. First, SATzilla's construction procedure uses a set of very cheap features (*e.g.*, number of variables and clauses) to learn a classifier $\mathcal{M}$ that predicts whether the computation of a more comprehensive set of features will succeed within a time threshold t_f; SATzilla also selects a backup solver B that has the best performance on instances with large feature cost ($> t_f$ or failed). Second, it chooses a set of candidate presolvers and candidate presolving time cutoffs based on solver performance on the training set. Then, for all possible combinations of up to two presolvers and given presolving time cutoffs, it learns an algorithm selector based on all training instances with feature cost $\leq t_f$ that cannot be solved by presolvers. The final portfolio-based selector is chosen as the combination of presolvers, presolving cutoffs, and the corresponding algorithm selector with the best training performance.

When asked to solve a given instance π, SATzilla first inspects π to gather very cheap features and uses them with $\mathcal{M}$ to predict feature computation time. If this prediction exceeds t_f, it runs solver B for the remaining time. Otherwise, it sequentially runs its presolvers up to their presolving cutoffs. If a presolver succeeds, SATzilla terminates. Otherwise, it computes features (falling back on B if feature computation fails or exceeds t_f), uses them in its algorithm selector to pick the most promising algorithm and runs that for the remaining time.

Improvements to SATzilla's Algorithm Selection Core. For our entry to the 2011 SAT competition analysis track, we improved SATzilla's models by basing them on cost-sensitive decision forests, rather than linear or quadratic ridge regression. (This improvement is described in [32].) Our new selection procedure uses an explicit cost-sensitive loss function—punishing misclassifications in direct proportion to their impact on portfolio performance—without predicting runtime. Such an approach has never before been applied to algorithm selection: all existing classification approaches use a simple 0–1 loss function that penalizes all misclassifications equally (e.g., [23,10,12]), whereas previous versions of SATzilla used regression-based runtime predictions. Our cost-sensitive classification approach based on decision forests (DFs) has the advantage that it effectively partitions the feature space into qualitatively different parts; furthermore, in contrast to clustering methods, DFs take the response variable (here "algorithm A performs better/worse than algorithm B") into account when determining that partitioning.

We construct cost-sensitive DFs as collections of 99 cost-sensitive decision trees [29], following standard random forest methodology [2]. Given n training data points with k features each, for each tree we construct a bootstrap sample of n training data points sampled uniformly at random with replacement. During tree construction, we sample a random subset of $\lceil \log_2(k) + 1 \rceil$ features at each internal node to be considered for splitting the data at that node. Predictions are based on majority votes across all trees. Given a set of m algorithms $\{A_1, \ldots, A_m\}$, an $n \times k$ matrix holding the values of k features for each of n training instances, and an $n \times m$ matrix P holding the performance of the m algorithms on the n instances, we construct our selector based on $m(m-1)/2$ pairwise cost-sensitive decision forests, determining the labels and costs as follows. For any pair of algorithms (A_i, A_j), we train a cost-sensitive decision forest $DF(i,j)$ on the following weighted training data: we label an instance q as i if $P(q,i)$ is better than $P(q,j)$, and as j otherwise; the weight for that instance is $|P(q,i) - P(q,j)|$. For each test instance, we apply each $DF(i,j)$ to vote for either A_i or A_j and select the algorithm with the most votes as the best algorithm for that instance. Ties are broken by only counting the votes from those decision forests that involve algorithms which received equal votes; further ties are broken randomly. Our implementation of SATzilla 2011, integrating cost-sensitive decision forests based on Matlab R2010a's implementation of cost-sensitive decision trees, is available online at http://www.cs.ubc.ca/labs/beta/Projects/SATzilla. Our experimental results show that SATzilla 2011 always outperformed SATzilla 2009; this gap was particularly substantial in the Application category (see Table 1).

3   Measuring the Value of a Solver

We believe that the SAT community could benefit from rethinking how the value of individual solvers is measured. The most natural way of assessing the performance of a solver is by means of some statistic of its performance over a set (or distribution) of instances, such as the number of instances solved within a given time budget, or its average runtime on an instance set. While we see value in these performance measures, we believe that they are not sufficient for capturing the value a solver brings to the community. Take, for example, two solvers MiniSAT'++ and NewSAT, where MiniSAT'++ is based on MiniSAT [5] and improves some of its components, while NewSAT is a (hypothetical) radically different solver that performs extremely well on a limited class of instances and poorly elsewhere.[1] While MiniSAT'++ has a good chance of winning medals

---

[1] In the 2007 SAT competition, the solver closest to NewSAT was TTS: it solved 12 instances unsolved by all others and thus received a silver medal under the purse scoring scheme [8] (discussed below), even though it solved many fewer instances than the bronze medalist Minisat (39 instances vs 72). Purse scoring was abandoned for the 2009 competition.

in the SAT competition's `Application` track, `NewSAT` might not even be submitted, since (due to its poor average performance) it would be unlikely even to qualify for the final phase of the competition. However, `MiniSAT'++` might only very slightly improve on the previous (MiniSAT-based) incumbent's performance, while `NewSAT` might represent deep new insights into the solution of instances that are intractable for all other known techniques.

One way of evaluating the value a solver brings to the community is through the notion of *state-of-the-art (SOTA) contributors* [28]. It ranks the contribution of a constituent solver A by the performance decrease of VBS when omitting A and reflects the added value due to A much more effectively than A's average performance. A related method for scoring algorithms is the "purse score" [8], which rewards solving instances that cannot be solved by other solvers; it was used in the 2005 and 2007 SAT competitions. However, both of these methods describe idealized solver contributions rather than contributions to an actual executable method. Instead, we propose to measure the SOTA contribution of a solver as its contribution to a SOTA portfolio that can be automatically constructed from available solvers. This notion resembles the prior notion of SOTA contributors, but directly quantifies their contributions to an *executable* portfolio solver, rather than to an abstract virtual best solver (VBS). Thus, we argue that an additional scoring rule should be employed in SAT competitions to recognize solvers (potentially with weak overall performance) for the contributions they make to a *SOTA portfolio*.

We must still describe exactly how we should assess a solver A's contribution to a portfolio. We might measure the frequency with which the portfolio selects A, or the number of instances the portfolio solves using A. However, neither of these measures accounts for the fact that if A were not available other solvers would be chosen instead, and might perform nearly as well. (Consider again `MiniSAT'++`, and assume that it is chosen frequently by a portfolio-based selector. However, if it had not been created, the set of instances solved might be the same, and the portfolio's performance might be only slightly less.) We argue that a solver A should be judged by its *marginal contribution* to the SOTA: the difference between the SOTA portfolio's performance including A and the portfolio's performance excluding A. (Here, we measure portfolio performance as the percentage of instances solved, since this is the main performance metric in the SAT competition.)

4   Experimental Setup

Solvers. In order to evaluate the SOTA portfolio contributions of the SAT competition solvers, we constructed `SATzilla` portfolios using all sequential, non-portfolio solvers from Phase 2 of the 2011 SAT Competition as component solvers: 9, 15, and 18 candidate solvers for the `Random`, `Crafted`, and `Application` categories, respectively. (These solvers and their individual performance are shown in Figures 1(a), 2(a), and 3(a); see [17] for detailed information.) We hope that in the future, construction procedures will also be made publicly available for other portfolio builders, such as 3S [14]; if so, our analysis could be easily and automatically repeated for them. For each category, we also computed the performance of an *oracle* over sequential non-portfolio solvers (an idealized algorithm selector that picks the best solver for each instance) and the *virtual best solver* (*VBS*, an oracle over all 17, 25 and 31 entrants for the Random, Crafted and Application categories, respectively). These oracles do not represent the current state of the art in SAT solving, since they cannot be run on new instances; however, they serve as upper bounds on the performance that any portfolio-based selector over these solvers could achieve. We also compared to the performance of the winners of all three categories (including other portfolio-based solvers).

Features. We used 115 features similar to those used by SATzilla in the 2009 SAT Competition. They fall into 9 categories: problem size, variable graph, clause graph, variable-clause graph, balance, proximity to Horn formula, local search probing, clause learning, and survey propagation. Feature computation averaged 31.4, 51.8 and 158.5 CPU seconds on Random, Crafted, and Application instances, respectively; this time counted as part of SATzilla's runtime budget.

Methods. We constructed SATzilla 2011 portfolios using the improved procedure described in Section 2. We set the feature computation cutoff t_f to 500 CPU seconds (a tenth of the time allocated for solving an instance). To demonstrate the effectiveness of our improvement, we also constructed a version of SATzilla 2009 (which uses ridge regression models), using the same training data.

We used 10-fold cross-validation to obtain an unbiased estimate of SATzilla's performance. First, we eliminated all instances that could not be solved by any candidate solver (we denote this instance set as U). Then, we randomly partitioned the remaining instances (denoted S) into 10 disjoint sets. Treating each of these sets in turn as the test set, we constructed SATzilla using the union of the other 9 sets as training data and measured SATzilla's runtime on the test set. Finally, we computed SATzilla's average performance across the 10 test sets.

To evaluate how important each solver was for SATzilla, for each category we quantified the marginal contribution of each candidate solver, as well as the percentage of instances solved by each solver during SATzilla's presolving (Pre1 or Pre2), backup, and main stages. Note that our use of cross-validation means that we constructed 10 different SATzilla portfolios using 10 different subsets ("folds") of instances. These 10 portfolios can be qualitatively different (e.g., selecting different presolvers); we report aggregates over the 10 folds.

Data. Runtime data was provided by the organizers of the 2011 SAT competition. All feature computations were performed by Daniel Le Berre on a quad-core computer with 4GB of RAM and running Linux, using our code. Four out of 1200 instances (from the Crafted category) had no feature values, due to a database problem caused by duplicated file name. We treated these instances as timeouts for SATzilla, thus obtaining a lower bound on SATzilla's true performance.

Table 1. *Comparison of SATzilla 2011 to the VBS, an Oracle over its component solvers, SATzilla 2009, the 2011 SAT competition winners, and the best single SATzilla 2011 component solver for each category. We counted timed-out runs as 5000 CPU seconds (the cutoff). Because the construction procedure for 3S portfolios is not publicly available, we used the original 2011 SAT-competition-winning version of 3S, which was trained on pre-competition data; it is likely that 3S would have achieved better performance if we had been able to retrain it on the same data we used to train SATzilla.*

Solver	Application Runtime (Solved)	Crafted Runtime (Solved)	Random Runtime (Solved)
VBS	1104 (84.7%)	1542 (76.3%)	1074 (82.2%)
Oracle	1138 (84.3%)	1667 (73.7%)	1087 (82.0%)
SATzilla 2011	1685 (75.3%)	2096 (66.0%)	1172 (80.8%)
SATzilla 2009	1905 (70.3%)	2219(63.0%)	1205 (80.3%)
Gold medalist	Glucose2: 1856 (71.7%)	3S: 2602 (54.3%)	3S: 1836 (68.0%)
Best comp.	Glucose2: 1856 (71.7%)	Clasp2: 2996 (49.7%)	Sparrow: 2066 (60.3%)

5   Experimental Results

We begin by assessing the performance of our SATzilla portfolios, to confirm that they did indeed yield SOTA performance. Table 1 compares SATzilla 2011 to the other solvers mentioned above. SATzilla 2011 outperformed all of its component solvers in all categories . It also always outperformed SATzilla 2009, which in turn was slightly worse than the best component solver on Application.

SATzilla 2011 also outperformed each category's gold medalist (including portfolio solvers, such as 3S and ppfolio). Note that this does not constitute a fair comparison of the underlying portfolio construction procedures, as SATzilla had access to data and solvers unavailable to portfolios that competed in the competition. This finding does, however, give us reason to believe that SATzilla portfolios either represent or at least closely approximate the best performance reachable by current methods. Indeed, in terms of instances solved, SATzilla 2011 reduced the gap between the gold medalists and the (upper performance bound defined by the) VBS by 27.7% on Application, by 53.2% on Crafted, and by 90.1% on Random. The remainder of our analysis concerns the contributions of each component solver to these portfolios. To substantiate our previous claim that marginal contribution is the most informative measure, here we contrast it with various other measures.

Random. Figure 1 presents a comprehensive visualization of our findings for the Random category. (Table 2 in supplemental material at http://www.cs.ubc.ca/labs/beta/Projects/SATzilla/SAT12-EvaluatingSolverContributions.pdf shows the data underlying all figures.) First, Figure 1(a) considers the set of instances that could be solved by at least one solver, and shows the percentage that each component solver is able to solve. By this measure, the two best solvers were Sparrow and MPhaseSAT_M. The former is a local search algorithm; it solved 362 + 0 satisfiable and unsatisfiable instances, respectively. The latter is a complete search algorithm; it solved 255 + 104 = 359 instances. Neither of these solvers

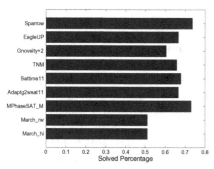

(a) Percentage of solvable instances solved by each component solver alone

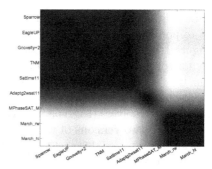

(b) Correlation matrix of solver runtimes (darker = more correlated)

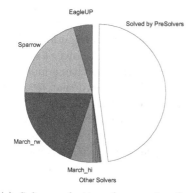

(c) Solver selection frequencies for SATzilla 2011; names only shown for solvers picked for > 5% of the instances

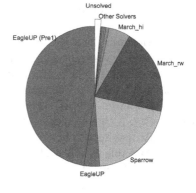

(d) Fraction of instances solved by SATzilla 2011 pre-solver(s), backup solver, and main stage solvers

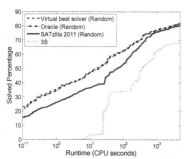

(e) Runtime CDF for SATzilla 2011, VBS, Oracle, and the gold medalist 3S

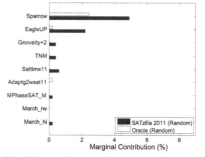

(f) Marginal contribution to Oracle and SATzilla 2011

Fig. 1. Visualization of results for category Random. In Figure 1(e), we used the original 2011 SAT-competition-winning version of 3S, which was trained on pre-competition data; thus, this figure is not a fair comparison of the SATzilla and 3S portfolio-building strategies. The data underlying this figure can be found in Table 2 of the supplemental material at http://www.cs.ubc.ca/labs/beta/Projects/SATzilla/SAT12-EvaluatingSolverContributions.pdf.

won medals in the combined SAT + UNSAT Random category, since all medals went to portfolio solvers that combined local search and complete solvers.

Figure 1(b) shows a correlation matrix of component solver performance: the entry for solver pair (A, B) is computed as the Spearman rank correlation coefficient between A's and B's runtime, with black and white representing perfect correlation and perfect independence respectively. Two clusters are apparent: six local search solvers (EagleUP, Sparrow, Gnovelty+2, Sattime11, Adaptg2wsat11, and TNM), and two versions of the complete solver March, which achieved almost identical performance. MPhaseSAT_M performed well on both satisfiable and unsatisfiable solvers; it was strongly correlated with local search solvers on the satisfiable instance subset and very strongly correlated with the March variants on the unsatisfiable subset.

Figure 1(c) shows the frequency with which different solvers were selected in SATzilla 2011. The main solvers selected in SATzilla 2011's main phase were the best-performing local search solver Sparrow and the best-performing complete solver March. As shown in Figure 1(d), the local search solver EagleUP was consistently chosen as a presolver and was responsible for more than half (51.3%) of the instances solved by SATzilla 2011 overall. We observe that MPhaseSAT_M did not play a large role in SATzilla 2011: it was only run for 2 out of 492 instances (0.4%). Although MPhaseSAT_M achieved very strong overall performance, its versatility appears to have come at the price of not excelling on either satisfiable or unsatisfiable instances, being largely dominated by local search solvers on the former and by March variants on the latter. Figure 1(e) shows that SATzilla 2011 closely approximated both the Oracle over its component solvers and the VBS, and stochastically dominated the category's gold medalist 3S.

Finally, Figure 1(f) shows the metric that we previously argued is the most important: each solver's marginal contribution to SATzilla 2011's performance. The most important portfolio contributor was Sparrow, with a marginal contribution of 4.9%, followed by EagleUP with a marginal contribution of 2.2%. EagleUP's low marginal contribution may be surprising at first glance (recall that it solved 51.3% of the instances SATzilla 2011 solved overall); however, most of these instances (49.1% out of 51.3%) were also solvable by other local search solvers. Similarly, both March variants had very low marginal contribution (0% and 0.2%, respectively) since they were essentially interchangeable (correlation coefficient 0.9974). Further insight can be gained by examining the marginal contribution of *sets* of highly correlated solvers. The marginal contribution of the set of both March variants was 4.0% (MPhaseSAT_M could still solve the most instances), while the marginal contribution of the set of six local search solvers was 22.5% (nearly one-third of the satisfiable instances were not solvable by any complete solver).

Crafted. Overall, sufficiently many solvers were relatively uncorrelated in the Crafted category (Figure 2) to yield a portfolio with many important contributors. The most important of these was Sol, which solved all of the 13.7% of the instances for which SATzilla 2011 selected it; without it, SATzilla 2011 would have solved 8.1% fewer instances! We observe that Sol was not identified as an important solver in the SAT competition results, ranking 11th of 24 solvers in

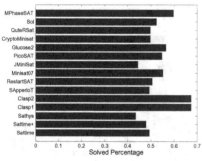
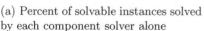

(a) Percent of solvable instances solved by each component solver alone

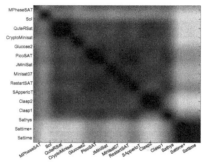

(b) Correlation matrix of solver runtimes (darker = more correlated)

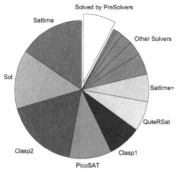

(c) Solver selection frequencies for SATzilla 2011; names only shown for solvers picked for > 5% of the instances

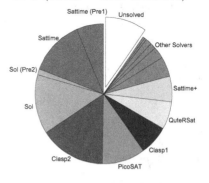

(d) Fraction of instances solved by SATzilla 2011 pre-solver(s), backup solver, and main stage solvers

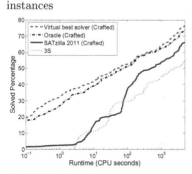

(e) Runtime CDF for SATzilla 2011, VBS, Oracle, and the gold medalist 3S

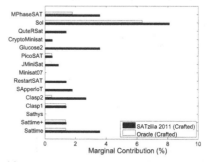

(f) Marginal contribution for Oracle and SATzilla 2011

Fig. 2. Visualization of results for category Crafted. In Figure 2(e), we used the original 2011 SAT-competition-winning version of 3S, which was trained on pre-competition data; thus, this figure is not a fair comparison of the SATzilla and 3S portfolio-building strategies. The data underlying this figure can be found in Table 2 of the supplemental material at http://www.cs.ubc.ca/labs/beta/Projects/SATzilla/SAT12-EvaluatingSolverContributions.pdf.

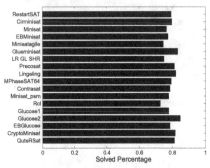

(a) Percent of solvable instances solved by each component solver alone

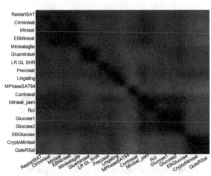

(b) Correlation matrix of solver runtimes (darker = more correlated)

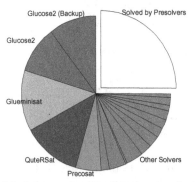

(c) Solver selection frequencies for SATzilla 2011; names only shown for solvers picked for > 5% of the instances

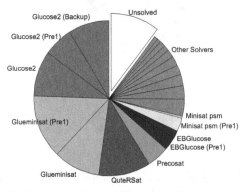

(d) Fraction of instances solved by SATzilla 2011 pre-solver(s), backup solver, and main stage solvers

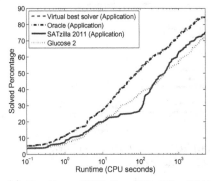

(e) Runtime CDF for SATzilla 2011, VBS, Oracle, and the gold medalist Glucose2

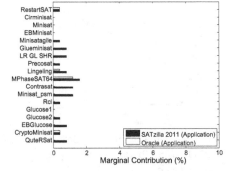

(f) Marginal contribution for Oracle and SATzilla 2011

Fig. 3. Visualization of results for category Application. The data underlying this figure can be found in Table 2 of the supplemental material, available at http://www.cs.ubc.ca/labs/beta/Projects/SATzilla/SAT12-EvaluatingSolverContributions.pdf.

the SAT+UNSAT category. Similarly, MPhaseSAT_M, Glucose2, and Sattime each solved a 3.6% fraction of instances that would have gone unsolved without them. (This is particularly noteworthy for MPhaseSAT_M, which was only selected for 5% of the instances in the first place.) Considering the marginal contributions of sets of highly correlated solvers, we observed that {Clasp1, Clasp2} was the most important at 6.3%, followed by {Sattime, Sattime11} at 5.4%. {QuteRSat, CryptoMiniSat} and {PicoSAT, JMiniSat, Minisat07, RestartSAT, SApperloT} were relatively unimportant even as sets, with marginal contributions of 0.5% and 1.8% respectively.

Application. All solvers in the Application category (Figure 3) exhibited rather highly correlated performance. It is thus not surprising that in 2011, no medals went to portfolio solvers in the sequential Application track, and that in 2007 and 2009, SATzilla versions performed worst in this track, only winning a single gold medal in the 2009 satisfiable category. As mentioned earlier, SATzilla 2011 did outperform all competition solvers, but here the margin was only 3.6% (as compared to 12.8% and 11.7% for Random and Crafted, respectively). All solvers were rather strongly correlated, and each solver could be replaced in SATzilla 2011 without a large decrease in performance; for example, dropping the competition winner only decreased SATzilla 2011's percentage of solved instances by 0.4%. The highest marginal contribution across all 18 solvers was four times larger: 1.6% for MPhaseSAT64. Like MPhaseSAT in the Crafted category, it was selected infrequently (only for 3.6% of the instances) but was the only solver able to solve about half of these instances. We conjecture that this was due to its unique phase selection mechanism. Both MPhaseSAT64 and Sol (in the Crafted category) thus come close to the hypothetical solver NewSAT mentioned earlier: they showed outstanding performance on certain instances and thus contributed substantially to a portfolio, but achieved unremarkable rankings in the competition (9th of 26 for MPhaseSAT64, 11th of 24 for Sol). We did observe one set of solvers that achieved a larger marginal contribution than that of MPhaseSAT64: 2.3% for {Glueminisat, LR GL SHR}. The other three highly correlated clusters also gave rise to relatively high marginal contributions: 1.5% for {CryptoMiniSat, QuteRSat}, 1.5% for {Glucose1,Glucose2,EBGlucose}, and 1.2% for {Minisat,EBMiniSAT, MiniSATagile}.

6   Conclusions

In this work, we investigated the question of assessing the contributions of individual SAT solvers by examining their value to SATzilla, a portfolio-based algorithm selector. SATzilla 2011 is an improved version of this procedure based on cost-based decision forests, which we entered into the new analysis track of the 2011 SAT competition. Its automatically generated portfolios achieved state of the art performance across all competition categories, and consistently outperformed its constituent solvers, other competition entrants, and our previous version of SATzilla. We observed that the frequency with which a component solver was selected is a poor measure of that solver's contribution to SATzilla 2011's performance. Instead, we advocate assessing solvers in terms of their marginal contributions to the state of the art in SAT solving.

We found that the solvers with the largest marginal contributions to SATzilla were often not competition winners (*e.g.*, MPhaseSAT64 in Application SAT+ UNSAT; Sol in Crafted SAT+UNSAT). To encourage improvements to the state of the art in SAT solving and taking into account the practical effectiveness of portfolio-based approaches, we suggest rethinking the way future SAT competitions are conducted. In particular, we suggest that all solvers that are able to solve instances not solved by any other entrant pass Phase 1 of the competition, and that solvers contributing most to the best-performing portfolio-based approaches be given formal recognition, for example by means of "portfolio contributor" or "uniqueness" medals. We also recommend that portfolio-based solvers be evaluated separately—and with access to all submitted solvers as components—rather than competing with traditional solvers. We hope that our analysis serves as an encouragement to the community to focus on creative approaches to SAT solving that complement the strengths of existing solvers, even though they may (at least initially) be effective only on certain classes of instances.

Acknowledgments. This research was supported by NSERC and Compute Canada / Westgrid. We thank Daniel Le Berre and Olivier Roussel for organizing the new SAT competition analysis track, for providing us with competition data and for running our feature computation code.

References

1. Biere, A., Cimatti, A., Clarke, E.M., Fujita, M., Zhu, Y.: Symbolic model checking using SAT procedures instead of BDDs. In: Proc. of DAC 1999, pp. 317–320 (1999)
2. Breiman, L.: Random forests. Machine Learning 45(1), 5–32 (2001)
3. Carchrae, T., Beck, J.C.: Applying machine learning to low-knowledge control of optimization algorithms. Computational Intelligence 21(4), 372–387 (2005)
4. Crawford, J.M., Baker, A.B.: Experimental results on the application of satisfiability algorithms to scheduling problems. In: Proc. of AAAI 1994, pp. 1092–1097 (1994)
5. Eén, N., Sörensson, N.: An Extensible SAT-solver. In: Giunchiglia, E., Tacchella, A. (eds.) SAT 2003. LNCS, vol. 2919, pp. 502–518. Springer, Heidelberg (2004)
6. Gagliolo, M., Schmidhuber, J.: Learning dynamic algorithm portfolios. Annals of Mathematics and Artificial Intelligence 47(3-4), 295–328 (2007)
7. Gebruers, C., Hnich, B., Bridge, D.G., Freuder, E.C.: Using CBR to Select Solution Strategies in Constraint Programming. In: Muñoz-Ávila, H., Ricci, F. (eds.) ICCBR 2005. LNCS (LNAI), vol. 3620, pp. 222–236. Springer, Heidelberg (2005)
8. Van Gelder, A., Le Berre, D., Biere, A., Kullmann, O., Simon, L.: Purse-based scoring for comparison of exponential-time programs. In: Proc. of SAT 2005 (2005)
9. Gomes, C.P., Selman, B.: Algorithm portfolios. Artificial Intelligence 126(1-2), 43–62 (2001)
10. Guerri, A., Milano, M.: Learning techniques for automatic algorithm portfolio selection. In: Proc. of ECAI 2004, pp. 475–479 (2004)
11. Helmert, M., Róger, G., Karpas, E.: Fast downward stone soup: A baseline for building planner portfolios. In: Proc. of ICAPS-PAL 2011, pp. 28–35 (2011)

12. Horvitz, E., Ruan, Y., Gomes, C.P., Kautz, H., Selman, B., Chickering, D.M.: A Bayesian approach to tackling hard computational problems. In: Proc. of UAI 2001, pp. 235–244 (2001)
13. Huberman, B.A., Lukose, R.M., Hogg, T.: An economics approach to hard computational problems. Science 265, 51–54 (1997)
14. Kadioglu, S., Malitsky, Y., Sabharwal, A., Samulowitz, H., Sellmann, M.: Algorithm Selection and Scheduling. In: Lee, J. (ed.) CP 2011. LNCS, vol. 6876, pp. 454–469. Springer, Heidelberg (2011)
15. Kautz, H.A., Selman, B.: Unifying SAT-based and graph-based planning. In: Proc. of IJCAI 1999, pp. 318–325 (1999)
16. Lagoudakis, M.G., Littman, M.L.: Learning to select branching rules in the DPLL procedure for satisfiability. Electronic Notes in Discrete Mathematics, pp. 344–359 (2001)
17. Le Berre, D., Roussel, O., Simon, L.: The international SAT Competitions web page (2012), http://www.satcompetition.org (last visited on January 29, 2012)
18. Leyton-Brown, K., Nudelman, E., Andrew, G., McFadden, J., Shoham, Y.: A portfolio approach to algorithm selection. In: Proc. of IJCAI 2003, pp. 1542–1543 (2003)
19. Nudelman, E., Leyton-Brown, K., Devkar, A., Shoham, Y., Hoos, H.: Satzilla: An algorithm portfolio for SAT. In: Solver Description, SAT Competition 2004 (2004)
20. Petri, M., Zilberstein, S.: Learning parallel portfolios of algorithms. Annals of Mathematics and Artificial Intelligence 48(1-2), 85–106 (2006)
21. Rice, J.R.: The algorithm selection problem. Advances in Computers 15, 65–118 (1976)
22. Roussel, O.: Description of ppfolio (2011), http://www.cril.univ-artois.fr/~roussel/ppfolio/solver1.pdf, Solver description (last visited on May 1, 2012)
23. Samulowitz, H., Memisevic, R.: Learning to solve QBF. In: Proc. of AAAI 2007, pp. 255–260 (2007)
24. Smith-Miles, K.: Cross-disciplinary perspectives on meta-learning for algorithm selection. ACM Computing Surveys 41(1) (2008)
25. Stephan, P., Brayton, R., Sangiovanni-Vencentelli, A.: Combinational test generation using satisfiability. IEEE Transactions on Computer-Aided Design of Integrated Circuits and Systems 15, 1167–1176 (1996)
26. Stern, D., Herbrich, R., Graepel, T., Samulowitz, H., Pulina, L., Tacchella, A.: Collaborative expert portfolio management. In: Proc. of AAAI 2010, pp. 210–216 (2010)
27. Streeter, M.J., Smith, S.F.: New techniques for algorithm portfolio design. In: Proc. of UAI 2008, pp. 519–527 (2008)
28. Sutcliffe, G., Suttner, C.B.: Evaluating general purpose automated theorem proving systems. Artificial Intelligence Journal 131(1-2), 39–54 (2001)
29. Ting, K.M.: An instance-weighting method to induce cost-sensitive trees. IEEE Transactions on Knowledge and Data Engineering 14(3), 659–665 (2002)
30. van Gelder, A.: Another look at graph coloring via propositional satisfiability. In: Proc. of COLOR 2002, pp. 48–54 (2002)
31. Xu, L., Hutter, F., Hoos, H.H., Leyton-Brown, K.: SATzilla: portfolio-based algorithm selection for SAT. Journal of Artificial Intelligence Research 32, 565–606 (2008)
32. Xu, L., Hutter, F., Hoos, H.H., Leyton-Brown, K.: Hydra-MIP: Automated algorithm configuration and selection for mixed integer programming. In: Proc. of IJCAI-RCRA 2011 (2011)

Efficient SAT Solving under Assumptions

Alexander Nadel[1] and Vadim Ryvchin[1,2]

[1] Intel Corporation, P.O. Box 1659, Haifa 31015 Israel
{alexander.nadel,vadim.ryvchin}@intel.com
[2] Information Systems Engineering, IE, Technion, Haifa, Israel

Abstract. In incremental SAT solving, assumptions are propositions that hold solely for one specific invocation of the solver. Effective propagation of assumptions is vital for ensuring SAT solving efficiency in a variety of applications. We propose algorithms to handle assumptions. In our approach, assumptions are modeled as unit clauses, in contrast to the current state-of-the-art approach that models assumptions as first decision variables. We show that a notable advantage of our approach is that it can make preprocessing algorithms much more effective. However, our initial scheme renders assumption-dependent (or temporary) conflict clauses unusable in subsequent invocations. To resolve the resulting problem of reduced learning power, we introduce an algorithm that transforms such temporary clauses into assumption-independent pervasive clauses. In addition, we show that our approach can be enhanced further when a limited form of look-ahead information is available. We demonstrate that our approach results in a considerable performance boost of the SAT solver on instances generated by a prominent industrial application in hardware validation.

1 Introduction

A variety of SAT applications require the ability to solve incrementally generated SAT instances online [1–7]. In such settings the solver is expected to be invoked multiple times. Each time it is asked to check the satisfiability status of all the available clauses under *assumptions* that hold solely for one specific invocation. The naïve algorithm which solves the instances independently is inefficient, since all learning is lost [1–4].

The current state-of-the-art approach to this problem was proposed in [4] and implemented in the MiniSat SAT solver [8]. MiniSat reuses a single SAT solver instance for all the invocations. Each time after solving is completed, the user can add new clauses to the solver and reinvoke it. The user is also allowed to provide the solver a set of *assumption literals*, that is, literals that are always picked as the first decision literals by the solver. In this scheme, all the conflict clauses generated are *pervasive*, that is, assumption-independent. We call this approach to the problem of incremental SAT solving under assumptions the *Literal-based Single instance (LS)* approach, since it reuses a single SAT solver instance and models assumptions as decision literals. The approach of [1] to our problem would use a separate SAT solver instance for each invocation, where

A. Cimatti and R. Sebastiani (Eds.): SAT 2012, LNCS 7317, pp. 242–255, 2012.

each assumption would be encoded as a unit clause. To increase the efficiency of learning, it would store and reuse the set of assumption-independent *pervasive* conflict clauses throughout all the SAT invocations. We call this approach the *Clause-based Multiple instances (CM)* approach, since it uses multiple SAT solver instances and models assumptions as unit clauses.

It was shown in [4] that LS outperforms CM in the context of model checking. As a result, LS is currently widely applied in practice (e.g. [5–7]). The goal of this paper is to demonstrate its limitations and to propose an efficient alternative.

This study springs from the authors' experiences, described herein, in tuning Intel's formal verification flow. Verification engineers reported to us that a critical property could not be solved by the SAT solver within two days. Our default flow used the LS approach, where to check a property the property's negation is provided as an assumption. The property holds iff the instance is unsatisfiable. Surprisingly, we discovered that providing the negation of the property as a unit clause, rather than as an assumption, rendered the property solvable within 30 minutes. The reason for this was that the unit clause triggered a huge simplification chain for our SatELite [9]-like preprocessor that drastically reduced the number of clauses in the formula.

Our experience highlights a drawback of LS: preprocessing techniques cannot propagate assumptions in LS, since they are modeled as decision variables, while assumptions can be propagated in CM, where they are modeled as unit clauses. Section 3 of this work demonstrates how to incorporate the SatELite algorithm within CM and shows why the applicability of SatELite for LS is an open problem.

LS has important advantages over CM related to the efficiency of learning. First, in LS all the conflict clauses are pervasive and can be reused, while CM cannot reuse *temporary* conflict clauses, that is, clauses that depend on assumptions. Second, LS reuses all the information relevant to guiding the SAT solver's heuristics, while CM has to gather relevant information from scratch for each new incremental invocation of the solver. Section 4 of this paper proposes an algorithm that overcomes the first of the above-mentioned drawbacks of CM: our algorithm transforms temporary clauses into pervasive clauses as a postprocessing step. Section 5 introduces an algorithm that mitigates the second of the above-mentioned advantages of plain LS over CM, given that limited lookahead information is available to the solver. In fact, we propose an algorithm that combines LS and CM to achieve the most efficient results.

We study the performance of algorithms for incremental SAT solving under assumptions on instances generated by a prominent industrial application in hardware validation, detailed in Section 2. Section 2 also provides some defintions and background information. Experimental results demonstrating the efficiency of our algorithms are provided in Section 6. We would like to emphasize that all the SAT instances used in this paper are publicly available from the authors. Section 7 concludes our work.

2 Background

An incremental SAT solver is provided with the input $\{F_i, A_i\}$ at each invocation i, where for each i, F_i is a formula in Conjunctive Normal Form (CNF) and $A_i = \{l_1, l_2, \ldots, l_n\}$ is a set (conjunction) of *assumptions*, where each assumption l_j is a unit clause (it is also a literal). Invocation i of the solver decides the satisfiability of $(\bigwedge_{j=1}^{i} F_j) \wedge A_i$. Intuitively, before each invocation the solver is provided with a new block of clauses and a set of assumptions. It is asked to solve a problem comprising all the clauses it has been provided with up to that moment under the set of assumptions relevant only to a single invocation of the solver. Modern SAT solvers generate *conflict clauses* by resolution over input clauses and previously generated conflict clauses. A clause α is *pervasive* if $(\bigwedge_{j=1}^{i} F_j) \rightarrow \alpha$, otherwise it is *temporary*.

The Clause-based Multiple instances (CM) approach [1] to incremental SAT solving under assumptions operates as follows. CM creates a new instance of a SAT solver for each invocation. Each invocation decides the satisfiability of $(\bigwedge_{j=1}^{i} F_j) \wedge (\bigwedge_{l=1}^{i-1} P_l) \wedge A_i$, where P_l is the set of pervasive conflict clauses generated at invocation l of the solver. To keep track of temporary and pervasive conflict clauses, the algorithm marks all the assumptions as temporary clauses and marks a newly generated conflict clause as temporary iff one or more temporary clauses participated in its resolution derivation.

The Literal-based Single instance (LS) approach [4] to incremental SAT solving under assumptions reuses the same SAT instance for all the invocations. The instance is always updated with a new block of clauses. The key idea is in providing the assumptions as *assumption literals*, that is, literals that are always picked as the first decision literals by the solver. Conflict-clause learning algorithms ensure that any conflict clause that depends on a set of assumptions will contain the negation of these assumptions. Hence, in LS all the conflict clauses are pervasive.

While all the algorithms for incremental SAT solving under assumptions discussed in this paper are application-independent, the experimental results section studies the performance of various algorithms on instances generated by the following prominent industrial application in hardware validation.

Assume that a verification engineer needs to formally verify a set of properties in some circuit up to a certain bound. Formal verification techniques cannot scale to large modern circuits, hence the engineer needs to select a sub-circuit and mimic the environment of the larger circuit by imposing assumptions (also called constraints) [10]. The engineer then invokes SAT-based Bounded Model Checking (BMC) [11] to verify a property under the assumptions. If the result is satisfiable, then either the environment is not set correctly, that is, assumptions are incorrect or missing, or there is a real bug. In practice the first reason is much more common than the second. To discover which of the possibilities is the correct one, the engineer needs to analyze the counter-example. If the reason for satisfiability lies in incorrect modeling of the environment, the assumptions must be modified and BMC invoked again. When one property has been verified, the engineer can move on to another. Practice shows that most of the validation time is spent in this process, which is known as the *debug loop*.

In the standard industrial BMC-based formal validation flow the model checker instance is built from scratch for each iteration of the debug loop. The key idea behind our solution is to take advantage of incremental SAT solving under assumptions *across multiple invocations of the model checker*. We keep only one instance of the model checker. For each invocation of BMC, given a transition system Ψ, a safety property Δ, and a set of assumptions Λ, we check whether Ψ satisfies Δ given Λ at each bound up to a given bound k using incremental SAT solving under assumptions, as follows. At each bound i, the transition system Ψ unrolled to i is translated to CNF and comprises the formula, while the set comprising the negation of Δ unrolled to i and the assumptions Λ unrolled to i is the set of assumptions provided to the SAT solver. We call our model checking algorithm *incremental Bounded Model Checking (BMC) under assumptions*.

Some recent works dedicated to BMC propose taking advantage of look-ahead information that is available, since the instance can be unrolled beyond the current bound [12,13]. In particular, it is proposed in [13] to apply preprocessing, including SatELite [9], for LS-based BMC, where complete look-ahead information is required to ensure soundness, as variables that are expected to appear in the future must not be eliminated. The technique of [13] cannot be applied in our application, since it is unknown a priori how the user would update the formula before subsequent invocations of the incremental model checker. The in-depth BMC algorithm of [12], which uses a limited form of look-ahead to boost BMC, served as an inspiration for our algorithm for incremental SAT solving under assumptions with step look-ahead, presented in Section 5.

3 Preprocessing under Assumptions

Preprocessing refers to a family of algorithms whose goal is to simplify the input CNF formula prior to the CDCL-based search in SAT. Preprocessing has commonly been applied in modern SAT solvers since the introduction of the SatELite preprocessor [9]. This section first explains why even a rather straightforward form of preprocessing, known as database simplification, is expected to be much more effective when used with CM as compared to LS. We then show that, unmodified, SatELite cannot be used with either CM or LS, and demonstrate how it can be modified so as to be safely used with CM.

Consider the following algorithm, which we call *database simplification* following MiniSat [8] notation: First, propagate unit clauses with Boolean Constraint Propagation (BCP). Second, eliminate satisfied clauses and falsified literals.

Database simplification is applied as an inprocessing step (that is, as an on-the-fly simplification procedure, applied at the global decision level) in modern SAT solvers [8,14,15]. It can be applied during preprocessing and inprocessing with either LS or CM without further modification. A key observation is that the efficiency of the first application of database simplification after a new portion of the incremental problem becomes available can be dramatically higher when assumptions are modeled as unit clauses (as in CM) rather than as assumption literals (as in LS). Indeed, database simplification takes full advantage of unit clauses by propagating them and eliminating resulting redundancies, while it does not take any advantage of assumption literals. In addition, variables representing assumptions are eliminated by database simplification with CM, but not with LS. Our experimental data, presented in Section 6, demonstrates that database simplification eliminates significantly more clauses for CM than for LS, and that the average conflict clause length for LS is much greater than it is for CM. These two factors favor CM as compared to LS as they have a significant impact on the efficiency of BCP and the overall efficiency of SAT solving.

Consider now the preprocessing algorithm of SatELite [9]. SatELite is a highly efficient algorithm used in leading SAT solvers [8,14,15]. SatELite is composed of the following three techniques:

1. *Variable elimination*: for each variable v, the algorithm performs resolution between clauses containing v (denoted by V^+) and $\neg v$ (denoted by V^-). Let U be the set of resulting clauses. If the number of clauses in U is less than or equal to the number of clauses in $V^+ \cup V^-$, then the algorithm eliminates v by replacing $V^+ \cup V^-$ by U.
2. *Subsumption*: if a clause α is subsumed by the clause β, that is, $\beta \subseteq \alpha$, α is removed.
3. *Self-subsuming resolution*: if $\alpha = \alpha_1 \vee l$ and $\beta = \beta_1 \vee \neg l$, where α_1 is subsumed by β_1, then α is replaced by α_1.

It is unclear how to apply SatELite with LS, let alone make its performance efficient. It is currently unknown how to apply SatELite for incremental SAT solving, since eliminated variables may be reintroduced (unless full look-ahead information is available [13], which is not always the case). However, even if the problem of incremental SatELite is solved, it is still unclear how to efficiently propagate assumptions when SatELite is applied with LS. One cannot apply SatELite as is, since eliminating assumption literals would render the algorithm unsound. A simple solution for ensuring soundness would be *freezing* the assumption literals [4,13], that is, not carrying out variable elimination for them. However, this solution has the same potential severe performance drawback as database simplification applied with LS as compared to CM: freezing assumptions is expected to have a significant negative impact on the preprocessor's ability to simplify the instance.

It is also unknown how SatELite can be applied with CM. The problem is that one has to keep track of pervasive and temporary clauses. Fortunately, we can propose a simple solution for this problem, based on the observation that SatELite uses nothing but resolution. SatELite can be updated as follows to keep track of pervasive and temporary clauses. If a variable is eliminated, each new clause $\alpha = \beta_1 \otimes \beta_2$ is marked as temporary iff one of the clauses β_1 or β_2 is temporary (where $\otimes$ corresponds to an application of the resolution rule). Whenever self-subsuming resolution is applied, the new clause α_1 is temporary iff either α or β is temporary (this operation is sound since α_1 is a resolvent of α and β). No changes are required for subsumption.

4 Transforming Temporary Clauses to Pervasive Clauses

We saw in Section 3 that CM has an important advantage over LS: preprocessing is expected to be much more efficient for CM. However, LS has its own
advantages. An important advantage is efficiency of learning: all the conflict
clauses learned by LS are pervasive, hence they can always be reused. In CM, all
the temporary conflict clauses are lost. In this section we propose an algorithm
that converts temporary clauses to pervasive clauses as a post-processing step
after the SAT solver is invoked. Our algorithm overcomes the above-mentioned
disadvantage of CM as compared to LS.

We start by providing some resolution-related definitions. The *resolution rule*
states that given clauses $\alpha_1 = \beta_1 \vee v$ and $\alpha_2 = \beta_2 \vee \neg v$, where β_1 and β_2
are also clauses, one can derive the clause $\alpha_3 = \beta_1 \vee \beta_2$. The resolution rule
application is denoted by $\alpha_3 = \alpha_1 \otimes^v \alpha_2$. A *resolution derivation* of a target clause α from a CNF formula $G = \{\alpha_1, \alpha_2, \ldots, \alpha_q\}$ is a sequence $\pi =
(\alpha_1, \alpha_2, \ldots, \alpha_q, \alpha_{q+1}, \alpha_{q+2}, \ldots, \alpha_p \equiv \alpha)$, where each clause α_i for $i \leq q$ is *ini-
tial* and α_i for $i > q$ is *derived* by applying the resolution rule to α_j and α_k,
where $j, k < i$.[1] A *resolution refutation* is a resolution derivation of the empty
clause $\square$. Modern SAT solvers are able to generate resolution refutations given
an unsatisfiable formula.

A resolution derivation π can naturally be considered as a directed acyclic
graph (dag) whose vertices correspond to all the clauses of π and in which there
is an edge from a clause α_j to a clause α_i iff $\alpha_i = \alpha_j \otimes \alpha_k$ (an example of such a
dag appears in Fig. 1). A clause $\beta \in \pi$ is *backward reachable* from $\gamma \in \pi$ if there
is a path (of 0 or more edges) from β to γ.

Assume now that the SAT solver is invoked over the CNF formula $A =
\{\alpha_1 = l_1, \ldots, \alpha_n = l_n\} \wedge F = \{\alpha_{n+1}, \ldots, \alpha_r\}$ (where the first n clauses are tem-
porary unit clauses corresponding to assumptions and the rest of the clauses are
pervasive). Assume that the solver generated a resolution refutation π of $A \wedge F$.
Let $\beta \in \pi$ be a clause. We denote by $\Gamma(\pi, \beta)$ the conjunction (set) of all the
backward reachable assumptions from β, that is, the conjunction (set) of all the
assumptions whose associated unit clauses are backward reachable from $\beta \in \pi$.
Let $\Gamma(\beta)$ be short for $\Gamma(\pi, \beta)$. To transform any clause $\beta \in \pi \setminus A$ to a pervasive
clause we propose applying the following operation:

$$\boxed{T2P(\beta) = \beta \vee \neg\Gamma(\beta)}$$

That is to say, we propose to update each temporary derived clause with the
negations of the assumptions that were required for its derivation, while pervasive clauses are left untouched. Consider the example in Fig. 1. The proposed
operation would transform α_7 to $c \vee d \vee \neg a$; α_8 to $\neg d \vee \neg b$; α_{10} to $c \vee \neg a \vee \neg b$; and
α_{11} to $\neg a \vee \neg b$. The pervasive clauses $\alpha_3, \alpha_4, \alpha_5, \alpha_6,$ and α_9 are left untouched.

Alg. 1 shows how to transform a resolution refutation π of $A \wedge F$ to a resolution
derivation $T2P(\pi)$ from F, such that every clause $\beta \in \pi \setminus A$ is mapped to a clause
$T2P(\beta) = \beta \vee \neg\Gamma(\beta) \in T2P(\pi)$. The pre- and post-conditions that must hold

---

[1] We force the resolution derivation to start with all the initial clauses, since such a
convention is more convenient for the subsequent discussion.

for Alg. 1 appear at the beginning of its text. The second pre-condition is not necessary, but it makes the algorithm's formulation and correctness proof easier. The algorithm's correctness is proved below.

Algorithm 1. Transform π to $T2P(\pi)$

Require: $\pi = (A = \{\alpha_1 = l_1, \ldots, \alpha_n = l_n\}, F = \{\alpha_{n+1}, \ldots, \alpha_r\}, \alpha_{r+1}, \ldots, \alpha_p)$ is a resolution refutation of $A \wedge F$
Require: All the assumptions in A are distinct and non-contradictory
Ensure: $T2P(\pi) = (T2P(\alpha_{n+1}), T2P(\alpha_{n+2}), \ldots, T2P(\alpha_r), T2P(\alpha_{r+1}), \ldots, T2P(\alpha_p))$ is a resolution derivation from F
Ensure: For each $i \in \{n+1, n+2, \ldots, r, \ldots, p\}$: $T2P(\alpha_i) = \alpha_i \vee \neg\Gamma(\alpha_i)$
 1: **for** $i \in \{n+1, n+2, \ldots, p\}$ **do**
 2:    **if** $\alpha_i \in F$ **then**
 3:       $T2P(\alpha_i) := \alpha_i$
 4:    **else**
 5:       Assume $\alpha_i = \alpha_j \otimes^v \alpha_k$
 6:       **if** α_j or α_k is an assumption **then**
 7:          Assume without limiting the generality that α_j is the assumption
 8:          $T2P(\alpha_i) := T2P(\alpha_k)$
 9:       **else**
10:          $T2P(\alpha_i) := T2P(\alpha_j) \otimes^v T2P(\alpha_k)$
11:    Append $T2P(\alpha_i)$ to $T2P(\pi)$

Proposition 1. *Algorithm 1 is sound, that is, its pre-conditions imply its post-conditions.*

Proof. The proof is by induction on i, starting with $i = r + 1$. Both post-conditions hold when the "for" loop condition is reached when $i = r + 1$, since $T2P(\pi)$ comprises precisely the clauses of F at that stage. Indeed, every clause α_i visited until that point is initial and is mapped to $T2P(\alpha_i) = \alpha_i$ by construction. It is left to prove that both post-conditions hold each time after a derived clause $\alpha_i \in \pi$ is translated to $T2P(\alpha_i)$ and $T2P(\alpha_i)$ is appended to $T2P(\pi)$. We divide the proof into three cases depending on the status of α_i.

When α_i is a pervasive derived clause, its sources α_j and α_k must also be pervasive by definition. By induction, we have $T2P(\alpha_j) = \alpha_j$ and $T2P(\alpha_k) = \alpha_k$, since $\Gamma(\alpha_j)$ and $\Gamma(\alpha_k)$ are empty. Hence, $T2P(\alpha_i) = T2P(\alpha_j) \otimes^v T2P(\alpha_k) = \alpha_j \otimes^v \alpha_k$. Thus, it holds that $T2P(\alpha_i)$ is derived from F by resolution, so the first post-condition holds. We also have the second post-condition, since we have seen that $T2P(\alpha_i) = \alpha_j \otimes^v \alpha_k = \alpha_i$, while $\Gamma(\alpha_i)$ is empty.

Consider the case where α_i is temporary and α_j is an assumption. The second pre-condition of the algorithm ensures that α_k will not be an assumption. The algorithm's flow ensures that $T2P(\alpha_i) = T2P(\alpha_k)$. By induction, $T2P(\alpha_k)$ is derived from F by resolution, hence $T2P(\alpha_i)$ is also derived from F by resolution and the first post-condition holds. The induction hypothesis yields that $T2P(\alpha_i) = T2P(\alpha_k) = \alpha_k \vee \neg\Gamma(\alpha_k)$. It must hold that $\alpha_k = \alpha_i \vee \neg l_j$, otherwise the resolution rule application $\alpha_i = (\alpha_j = l_j) \otimes^v \alpha_k$ would not be correct. Substituting the equation $\alpha_k = \alpha_i \vee \neg l_j$ into $T2P(\alpha_i) = \alpha_k \vee \neg\Gamma(\alpha_k)$

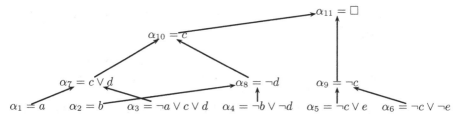

Fig. 1. An example of a resolution refutation for illustrating the *T2P* transformation. The pervasive input clauses are $F = \alpha_3 \wedge \alpha_4 \wedge \alpha_5 \wedge \alpha_6$; the assumptions are $\alpha_1 = a$ and $\alpha_2 = b$. The only pervasive derived clause is α_9; the rest of the derived clauses are temporary.

gives us $T2P(\alpha_i) = \alpha_i \vee \neg l_j \vee \neg\Gamma(\alpha_k) = \alpha_i \vee \neg(l_j \wedge \Gamma(\alpha_k))$. It must hold that $\Gamma(\alpha_i) = l_j \wedge \Gamma(\alpha_k)$ by resolution derivation construction. Substituting the latter equation into $T2P(\alpha_i) = \alpha_i \vee \neg(l_j \wedge \Gamma(\alpha_k))$ gives us precisely the second post-condition.

Finally consider the case where α_i is temporary and neither α_j nor α_k is an assumption. The first post-condition still holds after $T2P(\pi)$ is updated with $T2P(\alpha_i)$, since $T2P(\alpha_i) = T2P(\alpha_j) \otimes^v T2P(\alpha_k)$ by construction and both $T2P(\alpha_j)$ and $T2P(\alpha_k)$ are derived from F by resolution by the induction hypothesis. The induction hypothesis yields that $T2P(\alpha_i) = T2P(\alpha_j) \otimes^v T2P(\alpha_k) = (\alpha_j \vee \neg\Gamma(\alpha_j)) \otimes^v (\alpha_k \vee \neg\Gamma(\alpha_k))$. We have $\alpha_i = \alpha_j \otimes^v \alpha_k$. Hence, it holds that $T2P(\alpha_i) = (\alpha_j \otimes^v \alpha_k) \vee \neg\Gamma(\alpha_j) \vee \neg\Gamma(\alpha_k) = \alpha_i \vee \neg\Gamma(\alpha_j) \vee \neg\Gamma(\alpha_k)$. By resolution derivation construction, it holds that $\Gamma(\alpha_i) = \Gamma(\alpha_j) \wedge \Gamma(\alpha_k)$. Hence, $T2P(\alpha_i) = \alpha_i \vee \neg\Gamma(\alpha_i)$ and we have proved the second post-condition.   □

We implemented our method as follows. After SAT solving is completed, we go over the derived clauses in the generated resolution refutation π and associate each derived clause α with the set $\Gamma(\alpha)$. This operation can be applied independently of the SAT solving result, even if the problem is satisfiable. After that, we update each remaining temporary conflict clause α with $\neg\Gamma(\alpha)$ and mark the resulting clause as pervasive. In practice, there is no need to create a new resolution derivation $T2P(\pi)$.

Note that one needs to store and maintain the resolution derivation in order to apply our transformation. This may have a negative impact on performance. To mitigate this problem, we store only a subset of the resolution derivation, where each clause's associated set of backward reachable assumptions is non-empty. The idea of holding and maintaining only the relevant parts of the resolution derivation was proposed and proved useful in [16].

Finally, when the number of assumptions is large, our transformation might create pervasive clauses which are too large. To cope with this problem we use a user-given threshold n. Whenever the number of backward reachable assumptions for a clause is higher than n, that clause is not transformed into a pervasive clause, and thus is not reused in subsequent SAT invocations.

5   Incremental SAT Solving under Assumptions with Step Look-Ahead

In some applications of incremental SAT solving under assumptions, look-ahead information is available. Specifically, before invocation number i, the solver may already know the clauses F_j and assumptions A_j for some or all future invocations $j > i$. In this section, we propose an algorithm for incremental SAT solving under assumptions given a limited form of look-ahead, which we call step look-ahead. The reason for choosing this form of look-ahead is inspired by step-based approaches to BMC [12].

Given an integer step $s > 1$, an invocation i is *step-relevant* iff i modulo $s = 0$ (invocations are numbered starting with 0). Given an invocation q, its *step interval* is a set of successive invocations $SI(q) = [n * s, \ldots, q, \ldots, ((n+1) * s) - 1]$, where $n * s$ is the largest step-relevant invocation smaller than or equal to q. For example, for $s = 3$, invocations $0, 3, 6, 9, 12, \ldots$ are step-relevant; and $SI(3) = SI(4) = SI(5) = [3, 4, 5]$. In *step look-ahead*, at each step-relevant invocation i, the solver can access all the clauses and assumptions associated with invocations within $SI(i)$. In addition, in step look-ahead, given a step-relevant invocation i, it holds that $F_j \wedge A_j$ is satisfiable iff $F_j \wedge A_j \wedge F_k$ is satisfiable for every $j, k \in SI(i)$. That is to say, we assume that all the clauses available within the step interval hold for every invocation within that step interval.

One can adjust LS to take advantage of the fact that the solver has a wider view of the problem as follows. At a step-relevant invocation i, LS can be provided the problem $\bigwedge_{j=i}^{i+s-1} F_j$ and solve it s times, each time under a new set of assumptions A_j for each $j \in SI(i)$ (in this scheme non-step-relevant invocations are ignored). We call this approach the *Single instance Literal-based with Step look-ahead (LSS)* approach. LSS was proved to have advantages over the plain LS algorithm (which has a narrower view of the problem) in the context of standard BMC [12]. However, it suffers from the same major drawback as plain LS: preprocessing does not take advantage of assumptions.

We need to refine the semantics of the problem before proposing our solution. Given a step-relevant invocation i, an assumption $l \in A_i$ is *invocation-generic* iff $l \in A_j$ for every $j \in SI(i)$. Any assumption that is not invocation-generic is *invocation-specific*. That is, an assumption is invocation-generic iff it can be assumed for every invocation within the given step interval. In our application of incremental BMC under assumptions, described in Section 2, the negation of the property for each bound is invocation-specific, while the unrolled temporal assumptions are invocation-generic.

We propose an algorithm, called *Multiple instances Clause/Literal-based with Step look-ahead (CLMS)* (shown in Alg. 2), that combines LS and CM. The algorithm is applied at each step-relevant invocation. It creates the instance $\bigwedge_{j=i}^{i+s-1} F_j$ once as in LS. The key idea is that invocation-generic assumptions can be provided as unit clauses, since assuming them does not change the satisfiability status of the problem for any invocation within the current step interval. To ensure the soundness of solving subsequent step intervals, the unit clauses corresponding to invocation-generic assumptions must be temporary as in CM.

After creating the instance the solver is invoked s times for each invocation in the step interval, each time under the corresponding invocation-specific assumptions. To combine SatELite with Alg. 2 in a sound manner, all the invocation-specific assumptions must be frozen. Finally, note that our *T2P* transformation is applicable for CLMS.

Algorithm 2. CLMS Algorithm

1: **if** i is step-relevant **then**
2:    Let $G = \bigcap_{j=i}^{i+s-1} A_j$ be the set of all invocation-generic assumptions
3:    Create a SAT solver instance with pervasive clauses $\bigwedge_{j=i}^{i+s-1} F_j$ and temporary clauses G
4:    Optionally, apply SatELite, where all the invocation-specific assumptions in $\bigcup_{j=i}^{i+s-1} A_j$ must be frozen
5:    **for** $j \in \{i, i+1, \ldots, i+s-1\}$ **do**
6:       Invoke the SAT solver under the assumptions $A_j \setminus G$
7:       Optionally, transform temporary clauses to pervasive clauses using *T2P*
8:    Store the pervasive clauses and delete the SAT instance

---

6   Experimental Results

This section analyzes the performance of various algorithms for incremental SAT solving under assumptions on instances generated by incremental BMC under assumptions. In our analysis, we consider an instance satisfiable iff a certain invocation over that instance by one of the algorithms under consideration was satisfiable within a time-out of one hour. We picked instances from three satisfiable families comprising satisfiable instances only (128 instances) and four unsatisfiable families comprising unsatisfiable instances only (81 instances). We measured the number of completed incremental invocations for unsatisfiable families and the solving time until the first time an invocation was proved to be satisfiable for satisfiable families (the time-out was used as the solving time when an algorithm could not prove the satisfiability of a satisfiable instance). Each pair of invocations corresponds to a BMC bound (a clock transition and a real bound), where the complexity of SAT invocations in BMC grows exponentially with the bound. We implemented the algorithms in Intel's internal state-of-the-art Eureka SAT solver and used machines with Intel® Xeon® processors with 3Ghz CPU frequency and 32Gb of memory for the experiments.

We checked the performance of LS and CM as well as of LSS and CLMS with steps 10 and 50. We tested CM and CLMS with and without SatELite and with different thresholds for applying *T2P* transformation (0, 100, 100000). Our solver uses database minimization during inprocessing by default.

The graph on the left-hand side of Fig. 2 provides information about the number of variables and assumptions (satisfiable and unsatisfiable instances appear separately). For each instance we measured these numbers at the last invocation completed by both CM and LS (the basic algorithms). Note that the distribution of variables and assumptions for the satisfiable instances is more diverse.

This is explained by the fact that for satisfiable instances, the last invocation is sometimes very low or very high, while for unsatisfiable instances it is moderate. Overall, our satisfiable instances are easier to solve.

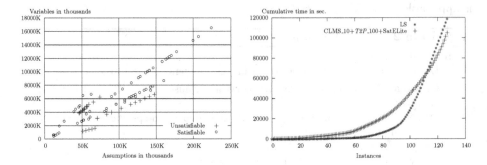

Fig. 2. Left-hand side: variables to assumptions ratio; Right-hand side: a comparison between plain LS and CLMS_10+*T2P*_100+SatELite with respect to the number of satisfiable instances solved within a given time.

Table 1. The number of invocations completed within an hour for the unsatisfiable instances from four families. The algorithms are sorted by the sum of completed invocations in decreasing order.

| \multicolumn{4}{c}{Algorithms} | | | | \multicolumn{5}{c}{Completed Invocations} | | | | |
LS?	SatELite?	Step	T2P Thr.	Overall	Fam. 1	Fam. 2	Fam. 3	Fam. 4
-	+	50	0	2967	1443	470	562	492
-	+	10	100	2934	1413	472	563	486
-	+	10	0	2932	1408	474	568	482
-	+	50	100	2927	1427	462	552	486
-	+	50	100000	2927	1427	462	552	486
-	+	1	0	2828	1365	468	539	456
-	+	1	100	2813	1363	462	535	453
-	-	10	100000	2806	1378	442	528	458
-	-	50	0	2801	1375	444	526	456
-	-	50	100	2795	1373	442	522	458
-	-	50	100000	2795	1373	442	522	458
-	-	10	100	2779	1357	440	528	454
-	-	10	0	2775	1353	438	530	454
-	-	1	100000	2736	1335	432	537	432
-	-	1	100	2734	1339	436	526	433
-	-	1	0	2732	1339	436	524	433
+	-	10	N/A	2579	1295	380	494	410
+	-	1	N/A	2575	1295	378	494	408
+	-	50	N/A	2563	1291	376	488	408
-	+	10	100000	2525	1245	390	507	383
-	+	1	100000	2250	1133	296	493	328

Consider Table 1, which compares the number of completed invocations for unsatisfiable instances. Compare basic CM and LS (configurations [-,-,1,0] and [+,-,1], respectively). CM significantly outperforms LS. As we discussed in Section 3, the reasons for this are related to the relative efficiency of database simplification and the average clause length for both algorithms. Fig. 3 demonstrates the huge difference between the two algorithms in these parameters in favor of CM. Note that when SatELite is not applied, the best performance is achieved by CLMS_10 (CLMS with step 10) with *T2P*_100000 (*T2P* with threshold 100000). Hence, without SatELite, both CLMS and *T2P* are helpful. SatELite increases the number of completed invocations considerably, while the absolutely best result is achieved by combining SatELite with CLMS_50 when *T2P* is turned off. Fig. 4 demonstrates that the reason for the inefficiency of the combination of *T2P* and SatELite is related to the fact that the time spent in preprocessing increases drastically when *T2P* is applied with threshold 100000. The degradation still exists, but is not that critical when the threshold is 100.

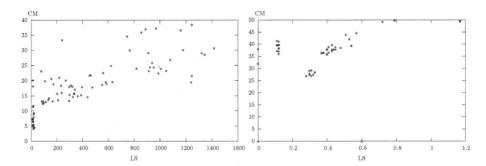

Fig. 3. Comparison of CM and LS with respect to average conflict cause length (left-hand side) and the percent of clauses removed by database simplification (right-hand side). Note the difference in the scales of the axes.

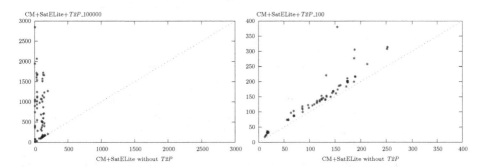

Fig. 4. Comparison between CM and CM+*T2P*_100000 (left-hand side) and between CM and CM+*T2P*_100 (right-hand side) in terms of time in seconds spent in SatELite

Consider now Table 2, which compares the run-time for satisfiable instances. Note that, unlike in the case of unsatisfiable instances, the default LS is one of the best algorithms. The advantage of LS over CM-based algorithms is that it maintains all the information relevant to the decision heuristic. This advantage proves to be very important in the context of relatively easy falsifiable instances. Still, the absolutely best configuration is the combination of CLMS_10 with SatELite and *T2P*_100, which uses all the algorithms proposed in this paper. The graph on the right-hand side of Fig. 2 shows that the advantage of our approach over LS becomes apparent as the run-time increases, while LS is still preferable for easier instances.

One can also see that the combination of CLMS_10 with SatELite and *T2P*_100 ([-,+,10,100]) is the most robust approach overall: it is the second best for unsatisfiable instances and the absolute best for satisfiable instances.

Table 2. Solving time in seconds for instances from three falsifiable families. The algorithms are sorted by overall solving time in increasing order.

Algorithms				Time			
LS?	SatELite?	Step	T2P Thr.	Overall	Fam. 1	Fam. 2	Fam. 3
-	+	10	100	104845	10843	35083	58919
+	-	1	N/A	118954	18005	41624	59325
-	+	10	0	134917	16886	40965	77067
+	-	10	N/A	139787	21726	53304	64757
-	+	10	100000	154437	22280	53436	78721
-	+	50	0	172104	10496	56087	105521
-	+	50	100	189965	11649	69373	108943
-	+	50	100000	192790	15220	68475	109096
-	-	10	100000	196784	12521	126153	58110
+	-	50	N/A	200261	22832	93635	83794
-	-	10	100	205124	16133	125529	63462
-	-	10	0	206390	14991	125400	65999
-	+	1	100	213278	31628	83009	98641
-	-	1	100	216714	20889	118703	77122
-	-	1	100000	220054	20639	128871	70545
-	+	1	0	219346	34447	89040	95859
-	-	1	0	228404	23642	121608	83154
-	-	50	0	244202	18996	138971	86235
-	+	1	100000	244826	34735	111862	98229
-	-	50	100000	247347	18514	138552	90281
-	-	50	100	250937	18897	141524	90516

7   Conclusion

This paper introduced efficient algorithms for incremental SAT solving under assumptions. While the currently widely-used approach (which we called LS) models assumptions as first decision variables, we proposed modeling assumptions as unit clauses. The advantage of our approach is that we allow the preprocessor to use assumptions while simplifying the formula. In particular, we demonstrated that the efficient SatELite preprocessor can easily be modified for use in our scheme, while it cannot be used with LS. Furthermore, we proposed an enhancement to our algorithm that transforms temporary clauses into pervasive clauses as a post-processing step, thus improving learning efficiency. In addition, we developed an algorithm which improves the performance further by taking advantage of a limited form of look-ahead information, which we called step look-ahead, when available. We showed that the combination of our algorithms outperforms LS on instances generated by a prominent industrial application. The empirical gap is especially significant for difficult unsatisfiable instances.

Acknowledgments. The authors would like to thank Amit Palti for supporting this work, Paul Inbar for editing the paper, and Niklas Eén, Armin Biere, and Mate Soos for their clarifications and suggestions that helped us to improve it.

References

1. Silva, J.P.M., Sakallah, K.A.: Robust search algorithms for test pattern generation. In: FTCS, pp. 152–161 (1997)
2. Shtrichman, O.: Pruning Techniques for the SAT-Based Bounded Model Checking Problem. In: Margaria, T., Melham, T.F. (eds.) CHARME 2001. LNCS, vol. 2144, pp. 58–70. Springer, Heidelberg (2001)
3. Whittemore, J., Kim, J., Sakallah, K.A.: SATIRE: A new incremental satisfiability engine. In: DAC, pp. 542–545. ACM (2001)
4. Eén, N., Sörensson, N.: Temporal induction by incremental SAT solving. Electr. Notes Theor. Comput. Sci. 89(4) (2003)
5. Cabodi, G., Lavagno, L., Murciano, M., Kondratyev, A., Watanabe, Y.: Speeding-up heuristic allocation, scheduling and binding with SAT-based abstraction/refinement techniques. ACM Trans. Design Autom. Electr. Syst. 15(2) (2010)
6. Franzén, A., Cimatti, A., Nadel, A., Sebastiani, R., Shalev, J.: Applying SMT in symbolic execution of microcode. [17], 121–128
7. Eén, N., Mishchenko, A., Amla, N.: A single-instance incremental SAT formulation of proof- and counterexample-based abstraction. [17], 181–188
8. Eén, N., Sörensson, N.: An Extensible SAT-solver. In: Giunchiglia, E., Tacchella, A. (eds.) SAT 2003. LNCS, vol. 2919, pp. 502–518. Springer, Heidelberg (2004)
9. Eén, N., Biere, A.: Effective Preprocessing in SAT Through Variable and Clause Elimination. In: Bacchus, F., Walsh, T. (eds.) SAT 2005. LNCS, vol. 3569, pp. 61–75. Springer, Heidelberg (2005)
10. Khasidashvili, Z., Kaiss, D., Bustan, D.: A compositional theory for post-reboot observational equivalence checking of hardware. In: FMCAD, pp. 136–143. IEEE (2009)
11. Biere, A., Cimatti, A., Clarke, E.M., Fujita, M., Zhu, Y.: Symbolic model checking using SAT procedures instead of BDDs. In: DAC, pp. 317–320 (1999)
12. Khasidashvili, Z., Nadel, A.: Implicative simultaneous satisfiability and applications. In: HVC 2011 (2011) (to appear)
13. Kupferschmid, S., Lewis, M.D.T., Schubert, T., Becker, B.: Incremental preprocessing methods for use in BMC. Formal Methods in System Design 39(2), 185–204 (2011)
14. Biere, A.: Lingeling and Plingeling, http://fmv.jku.at/lingeling/
15. Soos, M.: Cryptominisat2, http://www.msoos.org/cryptominisat2
16. Ryvchin, V., Strichman, O.: Faster Extraction of High-Level Minimal Unsatisfiable Cores. In: Sakallah, K.A., Simon, L. (eds.) SAT 2011. LNCS, vol. 6695, pp. 174–187. Springer, Heidelberg (2011)
17. Bloem, R., Sharygina, N. (eds.): Proceedings of 10th International Conference on Formal Methods in Computer-Aided Design, FMCAD 2010, Lugano, Switzerland, October 20-23. IEEE (2010)

Preprocessing in Incremental SAT

Alexander Nadel[1], Vadim Ryvchin[1,2], and Ofer Strichman[2]

[1] Design Technology Solutions Group, Intel Corporation, Haifa, Israel
`alexander.nadel@intel.com`
[2] Information Systems Engineering, IE, Technion, Haifa, Israel
`rvadim@tx.technion.ac.il, ofers@ie.technion.ac.il`

Abstract. Preprocessing of CNF formulas is an invaluable technique when attempting to solve large formulas, such as those that model industrial verification problems. Unfortunately, the best combination of preprocessing techniques, which involve variable elimination combined with subsumption, is incompatible with incremental satisfiability. The reason is that soundness is lost if a variable is eliminated and later reintroduced. *Look-ahead* is a known technique to solve this problem, which simply blocks elimination of variables that are expected to be part of future instances. The problem with this technique is that it relies on knowing the future instances, which is impossible in several prominent domains. We show a technique for this realm, which is empirically far better than the known alternatives: running without preprocessing at all or applying preprocessing separately at each step.

1 Introduction

Whereas CNF preprocessing techniques have been known for a long time (e.g., [1,2]), most are not cost-effective when it comes to formulas with millions of clauses – a typical size for industrial verification problems that are being routinely solved these days in the EDA industry. In that respect one of the major breakthroughs in practical SAT solving in the last few years has been the combined preprocessing techniques that were suggested by Een and Biere [3]: non-increasing variable elimination through resolution, coupled with subsumption and self-subsumption. These three techniques remove variables, clauses and literals, respectively. They are implemented in MiniSat [4] and the stand-alone preprocessor SatELite, and are in common use by many SAT solvers. Our experience with industrial verification instances shows that these techniques frequently remove more than half of the formula, and enable the solving of large instances that otherwise cannot be solved within a reasonable time limit. We will describe these techniques in more detail in Sect. 2.

A known problem with variable elimination is the fact that it is incompatible, at least in its basic form as published, with incremental SAT solving [4,9,10]. The reason, as was pointed out already in [3], is that variables that are eliminated may reappear in future instances. Soundness is not maintained in this scenario. For example, suppose that a formula contains the two clauses $(a \lor v), (b \lor \bar{v})$.

A. Cimatti and R. Sebastiani (Eds.): SAT 2012, LNCS 7317, pp. 256–269, 2012.

Eliminating v results in removing these two clauses and adding the resolvent $(a \lor b)$. Suppose, now, that in the next instance the clauses $(\bar{a}), (\bar{v})$ are added, which clearly contradict $(a \lor v)$. Yet since we erased that clause and since there is no contradiction between the resolvent and the new clauses, the new formula is possibly satisfiable — soundness is lost.

A possible remedy to this problem which was already suggested in [3] and experimented with in [7], is *look-ahead*. This means that variables that are known to be added in future instances are not eliminated. The problem with look-ahead is that it is not always possible, because information about future instances is not always available. Examples of such problem domains are:

- Some applications require interactive communication with the user for determining the next portion of the problem. For example, a recent article from IBM [3] describes a process in which the verification engineer may re-invoke the same instance of the SAT-based model checker for verifying a new property, which is not known a-priory (it depends on the result of the previous property). In such a case only a small part of the formula is changed, and hence incremental satisfiability may be crucial for performance.
- In some applications the calculation of the next portion of the problem depends on the results of the previous invocation of the SAT solver. For example, various tasks in MicroCode validation [6] are solved by using a symbolic execution engine to explore the paths of the program. The generated proof obligations are solved by an incremental SAT-based SMT solver. In this application, the next explored path of the program is determined based on the result of the previous computation.
- In Intel, the conversion of BMC problems to CNF is done after applying a 'saturation' optimization at the circuit level. Saturation divides all the variables into equivalence classes and tries to unite them by propagating short clauses that were learned in a previous instance — hence the dependency that prevents precalculating the instances. The SAT solver is provided only with the representatives of the equivalence classes. As a result, simple unrolling cannot predict those variables that will be present or absent in future instances.

Another possible remedy is called *full preprocessing*. It was briefly mentioned in [7] as an option that is expected not to scale, although in our experiments it is occasionally competitive. The idea is to perform full preprocessing before each instance. This means that all variables that were previously eliminated are returned to the formula and resolvents are removed, other than those that subsumed other clauses and hence cannot be removed. Therefor preprocessing is performed independently of past or future instances, other than the fact that it marks subsuming resolvents. The disadvantage of this approach comparing to *incremental preprocessing* — the main contribution of this article — is that it repeats a lot of work that has already been done in previous instances. Our experiments with large instances show that this extra overhead can add more than an hour to the preprocessing time.

In this article we suggest a method for combining the method of [3] with assumptions-based incremental SAT [4]. Our experiments show that it is much better than either running without preprocessing at all or full preprocessing. Look-ahead is still better overall, however, when possible. The solution we suggest is simple and rather easy to implement. Basically we eliminate variables regardless of future instances, and every time a variable is reintroduced into the formula we choose whether to *reeliminate*, or *reintroduce* it. An exception is made for the assumptions variables, which must be reintroduced. For both routes we need to save the clauses that were erased in the process of elimination: these need to be resolved with the new clauses for the former, and returned to the formula for the latter. As we show, the *order* in which variables are reeliminated or reintroduced matters for correctness. Specifically, the order must be *consistent* between instances. The order also changes the resulting reduced formula and hence the solving time. Our experiments show that in most cases the consistent order reduces the solving time.

We continue in the next section by describing the technical details of variable elimination, subsumption and self-subsumption. In Sect. 3 we present incremental preprocessing, which is an adaptation of these algorithms to the setting of incremental SAT. In Sect. 4 we summarize the results of our extensive experiments with industrial verification benchmarks from Intel.

2   Preliminaries

Let φ be a CNF formula. We denote by $vars(\varphi)$ the variables used in φ. For a clause c we write $c \in \varphi$ to denote that c is a clause in φ. For $v \in vars(\varphi)$ we define $\varphi_v = \{c \mid c \in \varphi \land v \in c\}$ and $\varphi_{\bar{v}} = \{c \mid c \in \varphi \land \bar{v} \in c\}$ (somewhat abusing notation, as we refer here to v as both a variable and a literal). Our setting includes the use of *assumptions* [5].

Variable Elimination
Input: formula φ and a variable $v \in vars(\varphi)$.
Output: formula φ' such that $v \notin vars(\varphi')$ and φ' and φ are equisatisfiable.

Typically this preprocessing is applied only if the number of clauses in φ' is not larger than in φ. More generally one may define a positive limit on the growth in the number of clauses, but for simplicity we will assume here that this limit is 0. Alg. 1 presents a variable elimination algorithm, where the eliminated variable v is the parameter. The variable v must be unassigned.

The function RESOLVE computes the set of non-tautological resolvents of two sets of clauses given to it as input (the check in line 5 excludes tautological resolvents). Function ELIMINATEVAR uses RESOLVE to compute the set Res of such resolvents of φ_v and $\varphi_{\bar{v}}$. If this set is larger than $|\varphi_v| + |\varphi_{\bar{v}}|$ it simply returns, and hence v is not eliminated. Otherwise in line 4 it adds the resolvents Res and discards the resolved clauses. All the variables in the resolvents are added to a list $TouchedVars$ in line 6. This list will be used later, in Alg. 2, for driving further subsumption and self-subsumption.

Algorithm 1. A variable elimination algorithm similar to the one implemented in MiniSat and in [3]

```
1: function RESOLVE(clauseset pos, clauseset neg)
2:     clauseset res = ∅;
3:     for each clause p ∈ pos do
4:         for each clause n ∈ neg do
5:             if p and n have a single possible pivot then
6:                 res = res ∪ resolution(p, n);
7:     return res;
```

```
1: function ELIMINATEVAR(var v)
2:     clauseset Res = RESOLVE (φᵥ, φᵥ̄);
3:     if |Res| > |φᵥ| + |φᵥ̄| then return ∅;          ▷ no variable elimination
4:     φ = (φ ∪ Res) \ (φᵥ ∪ φᵥ̄);
5:     ClearDataStructures(v);          ▷ clearing occurrence list, watch-list, scores-list
6:     TouchedVars = TouchedVars ∪ vars(Res);          ▷ used in Alg. 2
7:     return Res;
```

Subsumption

Input: $\varphi \wedge (l_1 \vee \cdots \vee l_i) \wedge (l_1 \vee \cdots \vee l_i \vee l_{i+1} \vee \cdots \vee l_j)$.
Output: $\varphi \wedge (l_1 \vee \cdots \vee l_i)$.

Self-subsumption

Input: $\varphi \wedge (l_1 \vee \cdots \vee l_i \vee l) \wedge (l_1 \vee \cdots \vee l_i \vee l_{i+1} \vee \cdots \vee l_j \vee \bar{l})$.
Output: $\varphi \wedge (l_1 \vee \cdots \vee l_i \vee l) \wedge (l_1 \vee \cdots \vee l_i \vee l_{i+1} \vee \cdots \vee l_j)$.

Preprocessing. The preprocessing algorithm described in Alg. 2 is similar to that implemented in MiniSat 2.2 [4] (based on the stand-alone preprocessor SatELite [3]). *SubsumptionQ* is a global queue of clauses. For each $c \in$ *SubsumptionQ*, and each $c' \in \varphi$, REMOVESUBSUMPTIONS (1) checks if $c \subset c'$ and if yes performs subsumption, and otherwise (2) if c self-subsumes c' then it performs self-subsumption. Essentially it is similar to the implementation suggested in [3]. Self-subsumption is followed by adding the reduced clause back to the queue. The function runs until the queue is empty. Note that assumptions are not eliminated. Eliminating assumptions would render the algorithm unsound.

In line 5 the variables are scanned in an increasing order of occurrences count. Note that in line 7 REMOVESUBSUMPTIONS is applied only to the set of newly generated resolvents.

3   Incremental Preprocessing

We now describe an incremental version of the preprocessing algorithm. In contrast to the full-preprocessing algorithm that was briefly described in the introduction (performing preprocessing of the new formula, together with learned clauses from previous instances), our suggested algorithm does not repeat preprocessing work that was done in previous instances.

Algorithm 2. Preprocessing, similar to the algorithm implemented in MiniSat 2.2

1: **function** PREPROCESS
2:     $SubsumptionQ = \varphi$;
3:     **while** $SubsumptionQ \neq \emptyset$ **do**
4:         REMOVESUBSUMPTIONS ();
5:         **for** each unassigned non-assumption variable v **do**   ▷ order heuristically
6:             $SubsumptionQ =$ ELIMINATEVAR (v);
7:             **if** $SubsumptionQ \neq \emptyset$ **then** REMOVESUBSUMPTIONS ();
8:         $SubsumptionQ = \{c \mid vars(c) \cap TouchedVars \neq \emptyset\}$;
9:         $TouchedVars = \emptyset$;

In our setting of incremental SAT, each instance is given as a set of clauses that should be added to the formula accumulated thus far. Removal of clauses is done indirectly, by using assumptions that are clause selectors. For example, if v is an assumption variable, then we can add $\bar{v}$ to a set of clauses. Assigning this variable FALSE is equivalent to removing this set.

Let φ^0 denote the initial formula, and Δ^i denote the set of clauses added at step i. Step i for $i > 0$ begins with a formula denoted φ^i, initially assigned the conjunction of φ^{i-1} at the *end* of the solving process (i.e., after being preprocessed and with additional learned clauses), and Δ^i. This formula changes during the solving process.

Preprocessing in an incremental SAT setting requires various changes. In step i, the easy case is when we wish to eliminate a variable v that is *not* eliminated in step $i - 1$. ELIMINATEVAR-INC, shown in Alg. 3 is a slight variation of ELIMINATEVAR that we saw in Alg. 1. The only difference is that if v is eliminated, then it saves additional data that will be used later on, as we will soon see. Specifically, it saves φ^i_v and $\varphi^i_{\bar{v}}$ in clause-sets denoted respectively by S_v and $S_{\bar{v}}$, and in the next line also the number of resolvents in a queue called $ElimVarQ$. This queue holds tuples of the form $\langle$variable v, int $resolvents\rangle$.

Algorithm 3. Variable elimination for φ^i, where the eliminated variable v was *not* eliminated in φ^{i-1}

1: **function** ELIMINATEVAR-INC(var v, int i)
2:     clauseset $Res =$ RESOLVE $(\varphi^i_v, \varphi^i_{\bar{v}})$;
3:     **if** $|Res| > |\varphi^i_v| + |\varphi^i_{\bar{v}}|$ **then return** $\emptyset$;   ▷ no variable elimination
4:     $S_v = \varphi^i_v$; $S_{\bar{v}} = \varphi^i_{\bar{v}}$;   ▷ Save for possible reintroduction
5:     $ElimVarQ$.push($\langle v, |Res|\rangle$);   ▷ Save #resolvents in queue
6:     $\varphi^i = (\varphi^i \cup Res) \setminus (\varphi^i_v \cup \varphi^i_{\bar{v}})$;
7:     CLEARDATASTRUCTURES (v);
8:     $TouchedVars = TouchedVars \cup vars(Res)$;   ▷ used in Alg. 5
9:     **return** Res;

The more difficult case is when v is already eliminated at step $i-1$. In that case we invoke REELIMINATE-OR-REINTRODUCE, as shown in Alg. 4. This function decides between reintroduction and reelimination.

- *Reelimination.* In Line 6 the algorithm computes the set of resolvents Res that need to be added in case v is reeliminated. Note that φ^i may contain v because of two separate reasons. First, $vars(\Delta^i)$ may contain v; Second, variables that were reintroduced in step i *prior* to v may have led to reintroduction of clauses that contain v. The total number of resolvents resulting from eliminating v is $|Res|$ + the number of resolvents incurred by eliminating v up to step i, which, recall, is saved in $ElimVarQ$.
- *Reintroduction.* In case we decide to cancel elimination, the previously removed clauses S_v and $S_{\bar{v}}$ have to be reintroduced. The total number of clauses resulting from reintroducing v is thus $|S_v \cup S_{\bar{v}} \cup \varphi_v^i \cup \varphi_{\bar{v}}^i|$. Note that the algorithm reintroduces variables that appear in the assumption list.

The decision between the two options is made in line 7. If reintroduction results in a smaller number of clauses, we simply return the saved clauses S_v and $S_{\bar{v}}$ by calling REINTRODUCEVAR, which also removes its entry from $ElimVarQ$ because v is no longer eliminated. The rest of the code is self-explanatory.

Algorithm 4. Variable elimination for φ^i, where the eliminated variable (located in $ElimVarQ[loc].v$) was already eliminated in φ^{i-1}

```
1: function REINTRODUCEVAR(var v, int loc, int i)
2:     φⁱ += S_v ∪ S_v̄;
3:     erase ElimVarQ[loc];                    ▷ v is not eliminated, hence 0 resolvents
```

```
1: function REELIMINATEVAR(clauseset Res, var v, int loc, int i)
2:     S_v = S_v ∪ φ_v^i; S_v̄ = S_v̄ ∪ φ_v̄^i;
3:     ElimVarQ[loc].resolvents += |Res|;
4:     φⁱ = (φⁱ ∪ Res) \ (φ_v^i ∪ φ_v̄^i);
5:     CLEARDATASTRUCTURES (v);
6:     TouchedVars = TouchedVars ∪ vars(Res);
```

```
1: function REELIMINATE-OR-REINTRODUCE(int loc, int i)
2:     var v = ElimVarQ[loc].v;                     ▷ The variable to eliminate
3:     if v is an assumption then
4:         REINTRODUCEVAR(v, loc, i);
5:         return ∅;
6:     clauseset Res =  RESOLVE(φ_v^i, φ_v̄^i) ∪
                        RESOLVE(φ_v^i, S_v̄) ∪ RESOLVE(S_v, φ_v̄^i);
7:     if (|Res| + ElimVarQ[loc].resolvents) > |S_v ∪ S_v̄ ∪ φ_v^i ∪ φ_v̄^i| then
8:         REINTRODUCEVAR(v, loc, i);
9:         return ∅;
10:    REELIMINATEVAR (Res, v, loc, i);
11:    return Res
```

Given ELIMINATEVAR-INC and REELIMINATE-OR-REINTRODUCE we can now focus on PREPROCESS-INC in Alg. 5, which is parameterized by the instance number i. The difference from Alg. 2 is twofold: First, variables that are already eliminated in the end of step $i - 1$ are processed by REELIMINATE-OR-REINTRODUCE; Second, other variables are processed in ELIMINATEVAR-INC. The crucial point here is the *order* in which variables are eliminated. Note that 1) elimination is consistent between instances, and 2) variables that are not currently eliminated are checked for elimination only at the end. These two conditions are necessary for correctness, because, recall, REINTRODUCEVAR may return clauses that were previously erased. These clauses may contain any variable that was not eliminated at the time they were erased.

Example 1. Suppose that in step $i - 1$, v_1 was eliminated, and as a result a clause $c = (v_1 \vee v_2)$ was removed. Then v_2 was eliminated as well. Suppose now that in step i we first reeliminate v_2, and then decide to reintroduce v_1. The clause c above is added back to the formula. But c contains v_2 which was already eliminated.   □

Algorithm 5. Preprocessing in an incremental SAT setting

1: **function** PREPROCESS-INC(int i)   ▷ preprocessing of φ^i
2:     $SubsumptionQ = \{c \mid \exists v. \ v \in c \wedge v \in vars(\Delta^i)\}$;
3:     REMOVESUBSUMPTIONS ();
4:     **for** $(j = 0 \ldots |ElimVarQ| - 1)$ **do**   ▷ scanning eliminated vars *in order*
5:         $v = ElimVarQ[j].v$;
6:         **if** $|\varphi_v^i| = |\varphi_{\bar{v}}^i| = 0$ **then continue**;
7:         REELIMINATE-OR-REINTRODUCE (j, i);
8:     **while** $SubsumptionQ \neq \emptyset$ **do**
9:         **for** each non-assumption variable $v \notin ElimVarQ$ **do**   ▷ scanning the rest
10:            $SubsumptionQ = $ ELIMINATEVAR-INC (v, i);
11:            REMOVESUBSUMPTIONS ();
12:        $SubsumptionQ = \{c \mid vars(c) \cap TouchedVars \neq \emptyset\}$;
13:        $TouchedVars = \emptyset$;

Let $\psi^n = \varphi^0 \wedge \bigwedge_{i=1}^{n} \Delta^i$, i.e., ψ_n is the n-th formula without preprocessing at all. We claim that:

Proposition 1. *Algorithm* PREPROCESS-INC *is correct, i.e., for all n*

$$\psi^n \text{ is equisatisfiable with } \varphi^n .$$

Proof. The full proof is given in a technical report [8]. Here we only sketch its main steps. The proof is by induction on n. The base case corresponds to standard (i.e., non-incremental) preprocessing. Proving the step of the induction relies on another induction, which proves that the following two implications hold right after line 7 at the j-th iteration of the first loop in PREPROCESS-INC, for $j \in [0 \ldots |ElimVarQ| - 1]$:

$$\psi^n \implies \left(\varphi^n \wedge \bigwedge_{k=j+1}^{|ElimVarQ|-1} \bigwedge_{c \in S_{v_k} \cup S_{\bar{v}_k}} c\right) \implies \exists v_1 \ldots v_j. \ \psi^n ,$$

The implication on the right requires some attention: existential quantification is necessary because of variable elimination via resolution (in the same way that $Res(x \vee A)(\bar{x} \vee B) = (A \vee B)$ and $(A \vee B) \implies \exists x. \ (x \vee A)(\bar{x} \vee B)$). The crucial point in the proof of this implication is to show that if a variable is eliminated at step j, it cannot reappear in the formula in later iterations. This is indeed guaranteed by the order in which the first loop processes the variables.

Note that at the last iteration $j = |ElimVarQ| - 1$ and the big conjunctions disappear. This leaves us with

$$\psi^n \implies \varphi^n \implies \exists v_1 \ldots v_j. \ \psi^n \ ,$$

which implies that ψ^n is equisatisfiable with the formula after the last iteration. The second loop of PREPROCESS-INC is non-incremental preprocessing, and hence clearly maintains satisfiability. $\qquad\square$

Removal of Resolvents. Recall that REINTRODUCEVAR returns the clause sets S_v and $S_{\bar{v}}$ to the formula. So far we ignored the question of what to do with the resolvents: should we remove them given that we canceled the elimination of v? These clauses are implied by the original formula, so keeping them does not hinder correctness. *Removing* them, however, is not so simple, because they may have participated in subsumption / self-subsumption of other clauses. Removing them hinders soundness, as demonstrated by the following example.

Example 2. Consider the following four clauses:

$$c_1 = (l_1 \vee l_2 \vee l_3) \quad c_2 = (l_4 \vee l_5 \vee \bar{l}_3)$$
$$c_3 = (l_1 \vee l_2 \vee \bar{l}_4) \quad c_4 = (l_1 \vee l_2 \vee \bar{l}_5) \ ,$$

and the following sequence:

- elimination of $var(l_3)$:
    - $c_5 = res(c_1, c_2) = (l_1 \vee l_2 \vee l_4 \vee l_5)$ is added;
    - c_1 and c_2 are removed and saved.
- self-subsumption between c_3 and c_5: $c_5 = (l_1 \vee l_2 \vee l_5)$.
- self-subsumption between c_4 and c_5: $c_5 = (l_1 \vee l_2)$.
- subsumption of c_3 and c_4 by c_5.
- removal of the resolvent c_5 and returning of c_1 and c_2.

We are left with only a subset of the original clauses (c_1 and c_2), which do not imply the rest. In this case the original formula is satisfiable, but it is not hard to see that the subsumed clauses (c_3, c_4) could have been part of an unsatisfiable set of clauses, and hence that their removal could have changed the result from unsat to sat. Soundness is therefore not secured if resolvents that participated in subsumption are removed. $\qquad\square$

In our implementation we solve this problem as follows. When eliminating v, we associate all the resolvent clauses with v. In addition, we mark all clauses that subsumed other clauses. We then change REINTRODUCEVAR as can be seen in Alg. 6. Note that in line 3 we guarantee that unit resolvents remain: it does not affect correctness and is likely to improve performance.

Algorithm 6. REINTRODUCEVAR with removal of resolvents that did not participate in subsumption

1: **function** REINTRODUCEVAR(var v, int loc, int i)
2:     φ^i += $S_v \cup S_{\bar{v}}$;
3:     **for** each non-unit clause c associated with v **do**
4:         **if** c is not marked **then** Remove c from φ^i;
5:     erase $ElimVarQ[loc]$;

4   Experimental Results

We implemented incremental preprocessing on top of FIVER[1], and experimented with hundreds of large processor Bounded Model-checking instances from Intel, categorized to four different families. In each case the problem is defined as performing BMC up to a given bound[2] in increments of size 1, or finding a

Table 1. The number of time-outs and the average total run time (incl. preprocessing) achieved by the four compared methods

Method	Time-outs	Avg. total run-time
full-preprocessing	68	2465.5
no-preprocessing	42	1784.7
incremental-preprocessing	2	1221.3
look-ahead	0	1064.9

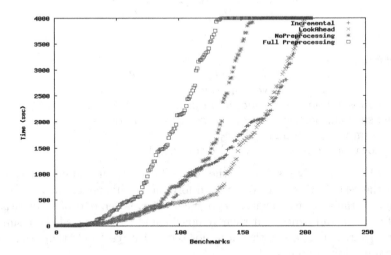

Fig. 1. Overall run-time of the four compared methods

---

[1] FIVER is a new SAT solver that was developed in Intel. It is a CDCL solver, combining techniques from EUREKA, MINISAT, and other modern solvers.
[2] Internal customers in Intel are typically interested in checking properties up to a given bound.

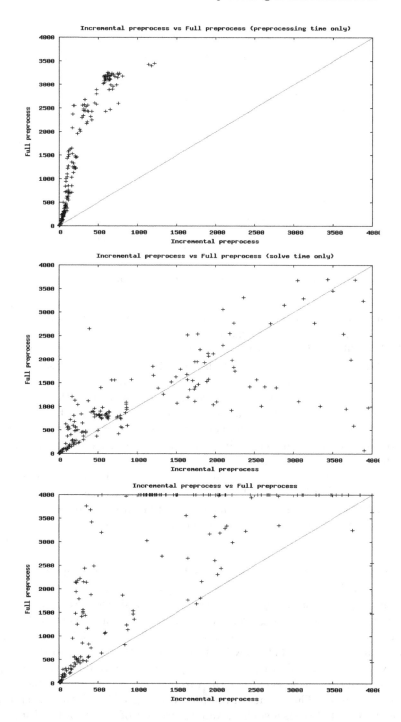

Fig. 2. Incremental preprocessing vs. full preprocessing: (top) preprocessing time, (middle) SAT time, and (bottom) total time

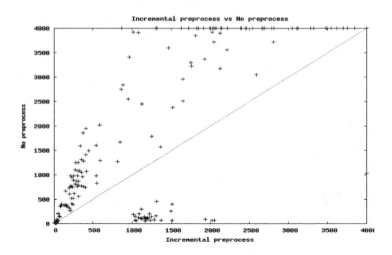

Fig. 3. Incremental preprocessing vs. no-preprocessing

satisfying assignment on the way to that bound. The time out was set to 4000 sec. After removing those benchmarks that cannot be solved by any of the tested methods within the time limit we were left with 206 BMC problems.[3] We turned off the 'saturation' optimization at the circuit level that was described in the introduction, in order to be able to compare our results to look-ahead. Overall in about half of the cases there is no satisfying assignment up to the given bound.

The first graph, in Fig. 1, summarizes the overall results of the four compared methods: full-preprocessing, no-preprocessing, incremental-preprocessing, and look-ahead. The number of time-outs and the average total run-time with these four methods is summarized in Table 1.

Look-ahead wins overall, but recall that in this article we focus on scenarios in which lookahead is impossible. Also note that it only has an advantage in a setting in which there is a short time-out. Incremental-preprocessing is able to close the gap and become almost equivalent once the time-out is set to a high value. It seems that the reason for the advantage of incremental preprocessing over look-ahead in hard instances is that unlike the latter, it does not force each variable to stay in the formula until it is known that it will not be added from thereon.

We now examine the results in more detail. Fig. 2 shows the consistent benefit of incremental preprocessing over full preprocessing. The generated formula is not necessarily the same because of the order in which the variables are examined. Recall that it is consistent between instances in PREPROCESS-INC and gives priority to those variables that are currently eliminated. In full preprocessing, on the other hand, it checks each time the variable that is contained in the minimal number of clauses. The impact of the preprocessing order on the search time is inconsistent, but there is a slight advantage to that of PREPROCESS-INC, as can

---

[3] The benchmarks are available upon request from the authors.

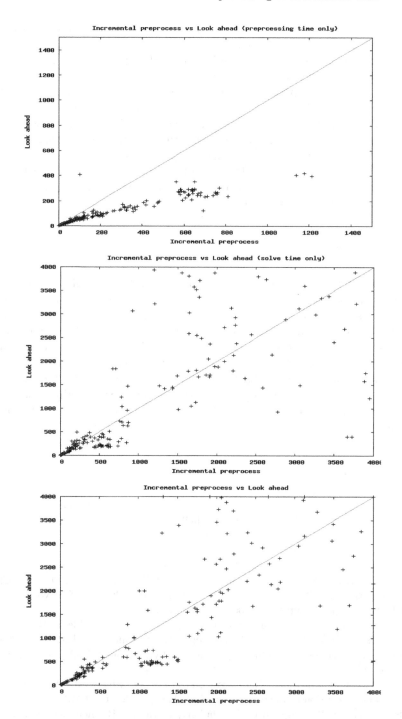

Fig. 4. Incremental preprocessing vs. look-ahead: (top) preprocessing time, (middle) SAT time, and (bottom) total time

be seen in the middle figure. The overall run time favors PREPROCESS-INC, as can be seen at the bottom figure.

Fig. 3 compares incremental preprocessing and no preprocessing at all. Again, the advantage of the former is very clear.

Finally, Fig. 4 compares incremental preprocessing and look-ahead, which shows the benefit of knowing the future. The fact that the preprocessing time of the latter is smaller is very much expected, because it does not have the overhead incurred by the checks in Alg. 3 and the multiple times that each variable can be reeliminated and reintroduced. The last graph shows that a few more instances were solved overall faster with look-ahead, but recall that according to Fig. 1 with a long-enough timeout the two methods have very similar results in terms of the number of solved instances.

5   Conclusion

In various domains there is a need for incremental SAT, but the sequence of instances cannot be computed apriori, because of dependance on the result of previous instances. In such scenarios applying preprocessing with look-ahead, namely preventing elimination of variables that are expected to be reintroduced, is impossible. Incremental preprocessing, the method we suggest here, is an effective algorithm for solving this problem. Our experiments with hundreds of industrial benchmarks show that it is much faster than the two known alternatives, namely full-preprocessing and no-preprocessing. Specifically, with a timeout of 4000 sec. it is able to reduce the number of time-outs by a factor of four and three, respectively.

References

1. Bacchus, F., Winter, J.: Effective Preprocessing with Hyper-Resolution and Equality Reduction. In: Giunchiglia, E., Tacchella, A. (eds.) SAT 2003. LNCS, vol. 2919, pp. 341–355. Springer, Heidelberg (2004)
2. Le Berre, D.: Exploiting the real power of unit propagation lookahead. Electronic Notes in Discrete Mathematics 9, 59–80 (2001)
3. Eén, N., Biere, A.: Effective Preprocessing in SAT Through Variable and Clause Elimination. In: Bacchus, F., Walsh, T. (eds.) SAT 2005. LNCS, vol. 3569, pp. 61–75. Springer, Heidelberg (2005)
4. Eén, N., Sörensson, N.: An Extensible SAT-solver. In: Giunchiglia, E., Tacchella, A. (eds.) SAT 2003. LNCS, vol. 2919, pp. 502–518. Springer, Heidelberg (2004)
5. Eén, N., Sörensson, N.: Temporal induction by incremental SAT solving. Electr. Notes Theor. Comput. Sci. 89(4), 543–560 (2003)
6. Franzén, A., Cimatti, A., Nadel, A., Sebastiani, R., Shalev, J.: Applying smt in symbolic execution of microcode. In: FMCAD, pp. 121–128 (2010)
7. Kupferschmid, S., Lewis, M.D.T., Schubert, T., Becker, B.: Incremental preprocessing methods for use in BMC. Formal Methods in System Design 39(2), 185–204 (2011)

8. Nadel, A., Ryvchin, V., Strichman, O.: Preprocessing in incremental SAT. Technical Report IE/IS-2012-02, Industrial Engineering, Technion (2012), http://ie.technion.ac.il/~ofers/publications/sat12t.pdf
9. Shtrichman, O.: Pruning Techniques for the SAT-Based Bounded Model Checking Problem. In: Margaria, T., Melham, T.F. (eds.) CHARME 2001. LNCS, vol. 2144, pp. 58–70. Springer, Heidelberg (2001)
10. Whittemore, J., Kim, J., Sakallah, K.: SATIRE: a new incremental satisfiability engine. In: IEEE/ACM Design Automation Conference, DAC (2001)

On Davis-Putnam Reductions
for Minimally Unsatisfiable Clause-Sets

Oliver Kullmann[1] and Xishun Zhao[2,*]

[1] Computer Science Department, Swansea University
http://cs.swan.ac.uk/~csoliver
[2] Institute of Logic and Cognition
Sun Yat-sen University, Guangzhou, 510275, P.R.C.

Abstract. DP-reduction $F \rightsquigarrow \mathrm{DP}_v(F)$, applied to a clause-set F and a variable v, replaces all clauses containing v by their resolvents (on v). A basic case, where the number of clauses is decreased (i.e., $c(\mathrm{DP}_v(F)) < c(F)$), is *singular DP-reduction* (sDP-reduction), where v must occur in one polarity only once. For minimally unsatisfiable $F \in \mathcal{MU}$, sDP-reduction produces another $F' := \mathrm{DP}_v(F) \in \mathcal{MU}$ with the same deficiency, that is, $\delta(F') = \delta(F)$; recall $\delta(F) = c(F) - n(F)$, using $n(F)$ for the number of variables. Let $\mathrm{sDP}(F)$ for $F \in \mathcal{MU}$ be the set of results of complete sDP-reduction for F; so $F' \in \mathrm{sDP}(F)$ fulfil $F' \in \mathcal{MU}$, are *non-singular* (every literal occurs at least twice), and we have $\delta(F') = \delta(F)$. We show that for $F \in \mathcal{MU}$ all complete reductions by sDP must have the same length, establishing the *singularity index* of F. In other words, for $F', F'' \in \mathrm{sDP}(F)$ we have $n(F') = n(F'')$. In general the elements of $\mathrm{sDP}(F)$ are not even (pairwise) isomorphic. Using the fundamental characterisation by Kleine Büning, we obtain as application of the singularity index, that we have *confluence modulo isomorphism* (all elements of $\mathrm{sDP}(F)$ are pairwise isomorphic) in case $\delta(F) = 2$. In general we prove that we have confluence (i.e., $|\mathrm{sDP}(F)| = 1$) for saturated F (i.e., $F \in \mathcal{SMU}$). More generally, we show confluence modulo isomorphism for *eventually saturated* F, that is, where we have $\mathrm{sDP}(F) \subseteq \mathcal{SMU}$, yielding another proof for confluence modulo isomorphism in case of $\delta(F) = 2$.

1 Introduction

Minimally unsatisfiable clause-sets ("MU's") are a fundamental form of irredundant unsatisfiable clause-sets. Regarding the subset relation, they are the hardest examples for proof systems. A substantial amount of insight has been gained into their structure, as witnessed by the handbook article [6]. A related area of MU, which gained importance in recent industrial applications, is the study of "MUS's", that is minimally unsatisfiable *sub*-clause-sets $F' \in \mathcal{MU}$ with $F' \subseteq F$ as the "cores" of unsatisfiable clause-sets F; see [16] for a recent overview. Now the investigations of this paper relate two areas: The structure of MU (see below), and the study of DP-reduction as started with [7,12,13]:

- A fundamental result shown there is that DP-reduction is commutative modulo subsumption (see Subsection 5.1 for the precise formulation).

* Partially supported by NSFC Grant 60970040 and MOE grant 11JD720020.

A. Cimatti and R. Sebastiani (Eds.): SAT 2012, LNCS 7317, pp. 270–283, 2012.

- Singular DP-reduction is a special case of length-reducing DP-reduction (while in general one step of DP-reduction can yield a quadratic blow-up).
- Confluence *modulo isomorphism* was shown in [7] (Theorem 13, Page 52) for a combination of subsumption elimination with special cases of length-reducing DP-reductions, namely DP-reduction in case no (non-tautological) resolvent is possible, and singular DP-reduction in case there is only one side clause, or the main clause is of length at most 2 (see Definition 2).

The basic questions for this paper are:

- When does singular DP-reduction, applied to MU, yield unique (non-singular) results (i.e., we have confluence)?
- And when are the results at least determined up to isomorphism (i.e., we have confluence modulo isomorphism)?

Investigating the structure of $\mathcal{MU}(k)$ We give now an overview on characterising the levels $\mathcal{MU}_{\delta=k} := \{F \in \mathcal{MU} : \delta(F) = k\}$; see [6] for further information.

The field of the combinatorial study of minimally unsatisfiable clause-sets was opened by [1], showing the fundamental insight $\delta(F) \geq 1$ for $F \in \mathcal{MU}$ (see [9,6] for generalisations of the underlying method, based on autarky theory). Also $\mathcal{SMU}_{\delta=1}$ was characterised there, where $\mathcal{SMU} \subset \mathcal{MU}$ is the set of "saturated" minimally unsatisfiable clause-sets, which are minimal not only w.r.t. having no superfluous clauses, but also w.r.t. that no clause can be further weakened. The fundamental "saturation method" $F \in \mathcal{MU} \rightsquigarrow F' \in \mathcal{SMU}$ was introduced in [4] (see Definition 1). Basic for all studies of MU is detailed knowledge on minimal number of occurrences of a (suitable) variable (yielding a suitable splitting variable): see [14] for the current state-of-art. The levels $\mathcal{MU}_{\delta=k}$ are decidable in polynomial time by [3,8]; see [17,10] for further extensions.

"Singular" variables v in $F \in \mathcal{MU}$, that is, variables occurring in at least one polarity only once, play a fundamental role — they are degenerations which (usually) need to be eliminated by *singular DP-reduction*. Let $\mathcal{MU}' \subset \mathcal{MU}$ be the set of non-singular minimally unsatisfiable clause-sets (not having singular variables), that is, the results of applying singular DP-reduction to the elements of $\mathcal{MU}$ as long as possible. The fundamental problem is the characterisation of $\mathcal{MU}'_{\delta=k}$ for arbitrary $k \in \mathbb{N}$. Up to now only $k \leq 2$ has been solved: $\mathcal{MU}'_{\delta=1}$ has been determined in [2], while $\mathcal{MU}'_{\delta=2} = \mathcal{SMU}'_{\delta=2}$ has been determined in [5]. Regarding higher deficiencies, until now only (very) partial results in [18] exist. Regarding singular minimally unsatisfiable clause-sets, also $\mathcal{MU}_{\delta=1}$ is very well known (with further extensions and generalisations in [8], and generalised to non-boolean clause-sets in [11]), while for $\mathcal{MU}_{\delta=2}$ not much is known.

For characterising $\mathcal{MU}'_{\delta=k}$, we need (very) detailed insights into (arbitrary) $\mathcal{MU}_{\delta<k}$, since the basic method to investigate $F \in \mathcal{MU}'_{\delta=k}$ is to split F into smaller parts from $\mathcal{MU}_{\delta<k}$ (usually containing singular variables). Assuming that we know $\mathcal{MU}'_{\delta<k}$, such insights can be based on some classification of $F \in \mathcal{MU}_{\delta<k}$ obtained from the set $\text{sDP}(F) \subseteq \mathcal{MU}'_{\delta<k}$ of singular-DP-reduction results. The easiest case is when $|\text{sDP}(F)| = 1$ holds (confluence), the second-easiest case is where all elements of $\text{sDP}(F)$ are pairwise isomorphic. This is the basic motivation for the questions raised and partially solved in this article. For

general k we have no conjecture yet how the classification of $\mathcal{MU}'_{\delta=k}$ could look like (besides the basic conjecture that enumeration of the isomorphism types can be done efficiently). However for unsatisfiable hitting clause-sets (two different clauses clash in at least one variable) we have the conjecture stated in [14], that for every fixed deficiency $k \in \mathbb{N}$ there are only finitely many isomorphism types.

Overview on results Section 3 introduces the basic notions regarding singularity, and the basic characterisations of singular DP-reduction on minimally unsatisfiable clause-sets are given. In Section 4 we consider the question of confluence of singular DP-reduction, with the first main result Theorem 12, showing confluence for saturated clause-sets. Section 5 mainly considers the question of changing the order of DP-reductions without changing the result. The second main result of this article is Theorem 35, establishing the singularity index. Section 6 is devoted to show confluence modulo isomorphism on eventually saturated clause-sets (Theorem 40), the third main result. As an application we determine the "types" of (possibly singular) minimally unsatisfiable clause-sets of deficiency 2 via Theorem 44. We conclude with a collection of open problems in Section 7. The underlying report [15] contains some additional material.

2   Preliminaries

We follow the general notations and definitions as outlined in [6]. Complementation of literals x is denoted by $\overline{x}$, while for a set L of literals we define $\overline{L} := \{\overline{x} : x \in L\}$. A **clause** C is a finite and clash-free set of literals (i.e., $C \cap \overline{C} = \emptyset$), while a **clause-set** $F \in \mathcal{CLS}$ is a finite set of clauses. We denote by $\mathrm{var}(F)$ the set of (occurring) variables, by $n(F) := |\mathrm{var}(F)|$ the number of variables, by $c(F) := |F|$ the number of clauses, and finally by $\delta(F) := c(F) - n(F)$ the deficiency. For clause-sets F, G we denote by $\boldsymbol{F \cong G}$ that both clause-sets are **isomorphic**, that is, the variables of F can be renamed and potentially flipped so that F is turned into G; more precisely, an isomorphism α from F to G is a bijection α on literal-sets which preserves complementation and which maps the clauses of F precisely to the clauses of G. The *literal-degree* $\mathrm{ld}_F(x) \in \mathbb{N}_0$ of a literal x for a clause-set F is the number of clauses the literal appears in, i.e., $\mathrm{ld}_F(x) := |\{C \in F : x \in C\}|$. The *variable-degree* $\mathrm{vd}_F(v) \in \mathbb{N}_0$ for a variable v is the number of clauses the variable appears in, i.e., $\mathrm{vd}_F(v) := \mathrm{ld}_F(v) + \mathrm{ld}_F(\overline{v})$.

For a clause-set F and a variable v by $\mathbf{DP}_v(\boldsymbol{F})$ we denote the result of applying DP-reduction on v, that is, removing all clauses containing v and adding all resolvents on v. More formally $\mathrm{DP}_v(F) := \{C \in F : v \notin \mathrm{var}(C)\} \cup \{C \diamond D : C, D \in F, C \cap \overline{D} = \{v\}\}$, where clauses C, D are resolvable iff they clash in exactly one literal, i.e., iff $|C \cap \overline{D}| = 1$, while for resolvable clauses C, D the resolvent $\boldsymbol{C \diamond D} := (C \cup D) \setminus \{x, \overline{x}\}$ for $C \cap \overline{D} = \{x\}$ is defined as the union minus the resolution literals. $\mathrm{DP}_v(F)$ is logically equivalent to the existential quantification of F by v, and thus F and $\mathrm{DP}_v(F)$ are satisfiability-equivalent.

The set of minimally unsatisfiable clause-sets (which are unsatisfiable, while removal of any clause makes them satisfiable) is $\mathcal{MU} \subset \mathcal{CLS}$. Note that for $v \in \mathrm{var}(F)$ with $F \in \mathcal{MU}$ we have $\mathrm{vd}_F(v) \geq 2$. Furthermore the set of saturated minimally unsatisfiable clause-sets is $\mathcal{SMU} \subset \mathcal{MU}$, which is the set of minimally unsatisfiable clause-sets such that addition of any literal to any clause renders them satisfiable. We recall the fact ([4], Lemma 5.1 in [11], and [15]) that every minimally unsatisfiable clause-set $F \in \mathcal{MU}$ can be **saturated**, i.e., by adding literal occurrences to F we obtain $F' \in \mathcal{SMU}$ with $\text{var}(F') = \text{var}(F)$ such that there is a bijection $\alpha : F \to F'$ with $C \subseteq \alpha(C)$ for all $C \in F$.

Definition 1. *The operation* $\mathbf{S}(F, C, x) := (F \setminus \{C\}) \cup (C \cup \{x\}) \in \mathcal{CLS}$ *(adding literal x to clause C in F) is defined if $F \in \mathcal{CLS}$, $C \in F$, and x is a literal with $\text{var}(x) \in \text{var}(F) \setminus \text{var}(C)$. A **saturation** $F' \in \mathcal{SMU}$ of $F \in \mathcal{MU}$ is obtained by a sequence $F = F_0, \ldots, F_m = F'$, $m \in \mathbb{N}_0$, such that for $0 \le i < m$ there are C_i, x_i with $F_{i+1} = \mathbf{S}(F_i, C_i, x_i)$, such that for all $1 \le i \le m$ we have $F_i \notin \mathcal{SAT}$, and such that the sequence cannot be extended. Note that $n(F') = n(F)$ and $c(F') = c(F)$ holds (and thus $\delta(F') = \delta(F)$). More generally, a **partial saturation** of a clause-set $F \in \mathcal{MU}$ is a clause-set $F' \in \mathcal{MU}$ such that $\text{var}(F') = \text{var}(F)$ and there is a bijection $\alpha : F \to F'$ such that for all $C \in F$ we have $C \subseteq \alpha(C)$.*

3 Singularity

In this section we present basic results on singular variables in minimally unsatisfiable clause-sets. Lemmas 5, 7 yield basic characterisations of singular DP-reduction for minimally resp. saturated minimally unsatisfiable clause-sets.

Definition 2. *We call a variable v **singular** for a clause-set $F \in \mathcal{CLS}$ if we have $\min(\text{ld}_F(v), \text{ld}_F(\overline{v})) = 1$; the set of singular variables of F is denoted by $\mathbf{var_s}(F) \subseteq \text{var}(F)$. F is called **nonsingular** if F does not contain singular variables. We use $\mathcal{MU}' := \{F \in \mathcal{MU} : \text{var}_s(F) = \emptyset\}$ for the set of nonsingular MU's, and $\mathcal{SMU}' := \mathcal{SMU} \cap \mathcal{MU}'$ for the set of nonsingular saturated MU's. More precisely, we call v **m-singular** for F for some $m \in \mathbb{N}$, if v is singular for F with $m = \text{vd}_F(v) - 1$. The set of 1-singular variables of F is denoted by $\mathbf{var_{1s}}(F) \subseteq \text{var}_s(F)$. That a variable is m-singular for some $m \ge 2$ is simply called **non-1-singular** (so "non-1-singular" variables are singular); the set of non-1-singular variables of F is denoted by $\mathbf{var_{\neg 1s}}(F) := \text{var}_s(F) \setminus \text{var}_{1s}(F)$. A **singular literal** for a singular variable v is a literal x with $\text{var}(x) = v$ and $\text{ld}_F(x) = 1$; if the underlying variable is 1-singular, then some choice is applied, so that we can speak of "the" singular literal of a singular variable. For a singular literal x we call the clause $C \in F$ with $x \in C$ the **main clause**, while the **side clauses** are the clauses $D_1, \ldots, D_m \in F$ with $\overline{x} \in D_i$ (here v is m-singular).*

Example 3. In clause-set $\{\{a\}, \{\overline{a}, b\}, \{\overline{a}, \overline{b}\}\}$, variable a is 2-singular, while variable b is 1-singular, and thus $\text{var}_s(F) = \{a, b\}$, $\text{var}_{1s}(F) = \{b\}$ and $\text{var}_{\neg 1s}(F) = \{a\}$. The main clause of a is $\{a\}$, its side clauses are $\{\overline{a}, b\}, \{\overline{a}, \overline{b}\}$, while for the main clause of b there is the choice between $\{\overline{a}, b\}$ and $\{\overline{a}, \overline{b}\}$. In general, if $F \in \mathcal{MU}$ contains a unit-clause $\{x\} \in F$, then $\text{var}(x)$ is singular for F. Thus the clause-sets $\{\bot\}$ and $\{\{a, b\}, \{a, \overline{b}\}, \{\overline{a}, b\}, \{\overline{a}, \overline{b}\}\}$ are the two smallest elements of $\mathcal{MU}'$ and $\mathcal{SMU}'$ regarding the number of clauses.

Singular DP-reduction The following special application of DP-reduction is fundamental for investigations of minimally unsatisfiable clause-sets (see [5], or Appendix B in [8] and subsequent [17,10]):

Definition 4. A **singular DP-reduction** is a reduction $F \rightsquigarrow \mathrm{DP}_v(F)$, where v is singular for $F \in \mathcal{MU}$. For $F, F' \in \mathcal{MU}$ by $F \xrightarrow{sDP} F'$ we denote that F' is obtained from F by one step of singular DP-reduction; i.e., there is a singular variable v for F with $F' = \mathrm{DP}_v(F)$, where v is called the **reduction variable**. We write $F \xrightarrow{sDP}_* F'$ if F' is obtained from F by an arbitrary number of steps of singular DP-reductions. The set of all nonsingular clause-sets obtainable from F by singular DP-reduction is denoted by $\mathbf{sDP}(F) := \{F' \in \mathcal{MU}' : F \xrightarrow{sDP}_* F'\}$.

The following lemma is kind of "folklore", but apparently the only place where its assertions are (partially) stated in the literature (in a more general form) is [10], Lemma 6.1. We add here various details; for the proof see [15]:

Lemma 5. Consider a clause-set F and a singular variable v for F. Then the following assertions are equivalent:

1. F is minimally unsatisfiable.
2. $\delta(\mathrm{DP}_v(F)) = \delta(F)$ and $\mathrm{DP}_v(F)$ is minimally unsatisfiable.
3. $\mathrm{DP}_v(F)$ is minimally unsatisfiable, and for the main clause C and the side clauses $D_1, \ldots, D_m$ for v (in F) we have:
   (a) Every D_i clashes with C in exactly one variable (namely in v).
   (b) For $1 \leq i \neq j \leq m$ we have $C \diamond D_i \neq C \diamond D_j$.
   (c) For $E \in F$ with $v \notin \mathrm{var}(E)$ and for all $1 \leq i \leq m$ we have $C \diamond D_i \neq E$.

Corollary 6. Consider $F \in \mathcal{MU}$ and a singular variable v with singular literal x, with main clause C and side clauses $D_1, \ldots, D_m$. Then adding $C \setminus \{x\}$ to D_i for all $i \in \{1, \ldots, m\}$ is a partial saturation of F (recall Definition 1).

Lemma 5 can be strengthened for saturated F by requiring special conditions for the occurrences of the singular variable; for the proof see [15]:

Lemma 7. Consider a clause-set F and a singular variable v for F. For a singular literal x for v consider the main clause C and the side clauses $D_1, \ldots D_m \in F$. Let $C' := C \setminus \{x\}$ and $D'_i := D_i \setminus \{\overline{x}\}$. The following assertions are equivalent:

1. F is saturated minimally unsatisfiable.
2. $\mathrm{DP}_v(F)$ is saturated minimally unsatisfiable, $C' = \bigcap_{i=1}^m D'_i$, and for every $E \in F$ with $v \notin \mathrm{var}(E)$ we have $C' \not\subseteq E$.

Corollary 8. The class $\mathcal{SMU}$ is stable under singular DP-reduction.

4   Confluence of Singular DP-Reduction

In this section we introduce the question of confluence of singular DP-reduction, and we obtain our first major result, namely confluence for $\mathcal{SMU}$ (Theorem 12).

Definition 9. Let $\mathbf{CFMU} := \{F \in \mathcal{MU} \,|\, |\mathrm{sDP}(F)| = 1\}$ be the set of $F \in \mathcal{MU}$ where singular DP-reduction is confluent, and let $\mathbf{CFIMU} := \{F \in \mathcal{MU} \,|\, \forall\, F', F'' \in \mathrm{sDP}(F) : F' \cong F''\}$ be the set of $F \in \mathcal{MU}$ where singular DP-reduction is confluent modulo isomorphism.

In [2] it is shown that every $F \in \mathcal{MU}_{\delta=1}$ contains a 1-singular variable (see [8,14] for further generalisations). So singular DP-reduction on $\mathcal{MU}_{\delta=1}$ must end in $\{\{\bot\}\}$, and we have $\mathcal{MU}'_{\delta=1} = \{\{\bot\}\}$. It follows $\mathcal{MU}_{\delta=1} \subseteq \mathcal{CFMU}$. For examples showing $\mathcal{MU}_{\delta=2} \not\subseteq \mathcal{CFMU}$ and $\mathcal{MU}_{\delta=3} \not\subseteq \mathcal{CFIMU}$ see [15].

Definition 10. *For clause-sets F, G we write $F \subseteq^{\mapsto} G$ if for all $C \in F$ there is $D \in G$ with $C \subseteq D$.*

"Nonsingular saturated patterns" are not destroyed by singular DP-reduction:

Lemma 11. *Consider $F_0, F, F' \in \mathcal{MU}$ with $F \xrightarrow{sDP}_* F'$.*

1. *If F_0 is nonsingular, then $F_0 \subseteq^{\mapsto} F \Rightarrow F_0 \subseteq^{\mapsto} F'$.*
2. *If $F_0, F, F' \in \mathcal{SMU}$, then $F_0 \subseteq^{\mapsto} F' \Rightarrow F_0 \subseteq^{\mapsto} F$.*

Proof. W.l.o.g. we can assume for both parts that $F' = \mathrm{DP}_v(F)$ for a singular variable v of F. Part 1 follows from the facts that $v \notin \mathrm{var}(F_0)$ due to the nonsingularity of F_0, and that due to the minimal unsatisfiability of F no clause gets lost by an application of singular DP-reduction. For Part 2 assume $\mathrm{ld}_F(v) = 1$. Due to the saturatedness of F we have for the clause $C \in F$ with $v \in C$ and for every clause $D \in F$ with $\overline{v} \in D$ that $C \setminus \{v\} \subseteq D \setminus \{\overline{v}\}$; the assertion follows.   □

Theorem 12. $\mathcal{SMU} \subset \mathcal{CFMU}$.

Proof. Consider $F \in \mathcal{SMU}$ and two nonsingular $F', F'' \in \mathcal{SMU}$ with $F \xrightarrow{sDP}_*$ F' and $F \xrightarrow{sDP}_* F''$. From $F' \subseteq^{\mapsto} F'$ and $F \xrightarrow{sDP}_* F'$ by Lemma 11, Part 2 we get $F' \subseteq^{\mapsto} F$, and then by Part 1 we get $F' \subseteq^{\mapsto} F''$; in the same way we obtain $F'' \subseteq^{\mapsto} F'$ and thus $F' = F''$.   □

5   Permutations of Sequences of DP-Reductions

This section contains central technical results on (iterated) singular DP-reduction. After some preliminaries follows an interlude on iterated general DP-reduction in Subsection 5.1, stating "commutativity modulo subsumption" and deriving the basic fact in Corollary 18, that in case a sequence of DP-reductions as well as some permutation both yield minimally unsatisfiable clause-sets, then actually these MU's are the same. In Subsection 5.2 then conclusions for singular DP-reductions are drawn, obtaining various conditions under which sDP-reductions can be permuted without changing the final result. A good overview on all possible sDP-reductions is obtained in Subsection 5.3 in case no 1-singular variables are present. In Subsection 5.4 we introduce the "singularity index", the minimal length of a maximal sDP-reduction sequence. Our second major result is Theorem 35, showing that in fact all maximal sDP-reduction-sequences have the same length. First we analyse the changes for literal-degrees after one step of sDP-reduction (for the simple proof and further details see [15]):

Lemma 13. *Consider $F \in \mathcal{MU}$ and an m-singular variable v ($m \in \mathbb{N}$). Let C be the main clause, let $D_1, \ldots, D_m$ be the side clauses, and let $F' := \mathrm{DP}_v(F)$. Consider a literal x with $\mathrm{var}(x) \neq v$. The task is to compare $\mathrm{ld}_F(x)$ and $\mathrm{ld}_{F'}(x)$.*

1. If $\operatorname{ld}_{F'}(x) < \operatorname{ld}_F(x)$, then $\operatorname{ld}_{F'}(x) \geq m$.
2. If $m = 1$, then $\operatorname{ld}_{F'}(x) \leq \operatorname{ld}_F(x)$.
3. For $x \in C$ let $p := |\{i \in \{1, \ldots, m\} : x \notin D_i\}|$; then $\operatorname{ld}_{F'}(x) = \operatorname{ld}_F(x) - 1 + p$.
4. We have $\operatorname{ld}_{F'}(x) > \operatorname{ld}_F(x)$ iff $x \in C$ and $p \geq 2$.

By Lemma 13, Parts 1 and 2 we get that singular variables can only be created for 1-singular DP-reduction, while singular variables can only be destroyed for non-1-singular DP-reductions; the details are as follows:

Corollary 14. *Consider $F \in \mathcal{MU}$ and an m-singular variable v for F ($m \in \mathbb{N}$), and let $F' := \operatorname{DP}_v(F)$.*

1. *(a) If $m \geq 2$, then $\operatorname{vars}(F') \subseteq \operatorname{vars}(F)$ with $\operatorname{var}_{1s}(F') \subseteq \operatorname{var}_{1s}(F)$.*
   (b) If $m = 1$ then $\operatorname{vars}(F') \setminus \operatorname{vars}(F) \subseteq \operatorname{var}_{\neg 1s}(F')$.
2. *(a) If $m = 1$, then $\operatorname{vars}(F) \setminus \{v\} \subseteq \operatorname{vars}(F')$ with $\operatorname{var}_{1s}(F) \setminus \{v\} \subseteq \operatorname{var}_{1s}(F')$.*
   (b) If $m \geq 2$ then $\operatorname{vars}(F) \setminus \operatorname{vars}(F') \subseteq \operatorname{var}_{\neg 1s}(F)$.

5.1   Iterated DP-Reduction

Definition 15. *Consider $F \in \mathcal{CLS}$ and a sequence $v_1, \ldots, v_n$ of variables for $n \in \mathbb{N}_0$. Then $\mathbf{DP}_{v_1,\ldots,v_n}(F) := \operatorname{DP}_{v_n}(\operatorname{DP}_{v_1,\ldots,v_{n-1}}(F))$.*

Thus in "$\operatorname{DP}_{v_1,\ldots,v_n}$" DP-reduction is performed in order $v_1, \ldots, v_n$. We have $\operatorname{var}(\operatorname{DP}_{v_1,\ldots,v_n}(F)) \subseteq \operatorname{var}(F) \setminus \{v_1, \ldots, v_n\}$. In [12] (Lemma 7.4, page 33) as well as in [13] (Lemma 7.6, page 27) the following fundamental result on iterated DP-reduction is shown (for more details see [15]):

Lemma 16. *If performing subsumption-elimination at the end, then iterated DP-reduction does not depend on the order of the variables; and performing subsumption-elimination inbetween does not influence then the result.*

Definition 17. *Consider $F \in \mathcal{CLS}$ and variables $v_1, \ldots, v_n$ ($n \in \mathbb{N}_0$). Then a permutation $\pi \in S_n$ is called **equality-preserving** for F and $v_1, \ldots, v_n$ (for short: "eq-preserving"), if we have $\operatorname{DP}_{v_1,\ldots,v_n}(F) = \operatorname{DP}_{\pi(v_1),\ldots,\pi(v_n)}(F)$. The set of all eq-preserving $\pi \in S_n$ is denoted by $\mathbf{eqp}(F, (v_1, \ldots, v_n)) \subseteq S_n$.*

Note that if $\operatorname{var}(F) \subseteq \{v_1, \ldots, v_n\}$, then $\operatorname{eqp}(F, (v_1, \ldots, v_n)) = S_n$. Since minimally unsatisfiable clause-sets do not contain subsumptions, we obtain:

Corollary 18. *Consider $F \in \mathcal{CLS}$ and variables $v_1, \ldots, v_n$ ($n \in \mathbb{N}_0$) such that $\operatorname{DP}_{v_1,\ldots,v_n}(F) \in \mathcal{MU}$. Then we have for $\pi \in S_n$ that $\pi \in \operatorname{eqp}(F, (v_1, \ldots, v_n))$ holds iff $\operatorname{DP}_{v_{\pi(1)},\ldots,v_{\pi(n)}}(F) \in \mathcal{MU}$.*

5.2   Iterated sDP-Reduction via Singular Tuples

Generalising Definition 2 we consider "singular tuples":

Definition 19. *Consider $F \in \mathcal{MU}$. A tuple $(v_1, \ldots, v_n)$ of variables ($n \in \mathbb{N}_0$) is called **singular** for F if for all $i \in \{1, \ldots, n\}$ we have that v_i is singular for $\operatorname{DP}_{v_1,\ldots,v_{i-1}}(F)$. Note that for a singular $(v_1, \ldots, v_n)$ all variables must be different. We call variable v_i ($i \in \{1, \ldots, n\}$) m-**singular** ($m \in \mathbb{N}$) for $(v_1, \ldots, v_n)$ and F, if v_i is m-singular for $\operatorname{DP}_{v_1,\ldots,v_{i-1}}(F)$.*

Example 20. Consider $F := \{\{a\}, \{\bar{a}, b\}, \{\bar{a}, \bar{b}\}\}$ (recall Example 3). There are 5 singular tuples for F, namely $(), (a), (b), (a, b), (b, a)$. Considering $\boldsymbol{v} := (a, b)$, variable a is 2-singular for $\boldsymbol{v}$ and F, and b is 1-singular for $\boldsymbol{v}$ and F, while considering $\boldsymbol{v}' := (b, a)$, both a and b are 1-singular for $\boldsymbol{v}'$ and F.

For the understanding of sDP-reduction of $F \in \mathcal{MU}$, understanding the set of singular tuples for F is an important task.

Definition 21. *Consider $F \in \mathcal{MU}$ and a singular tuple $(v_1, \ldots, v_n)$ for F. A permutation $\pi \in S_n$ is called **singularity-preserving** for F and $(v_1, \ldots, v_n)$ (for short: "s-preserving"), if also $(v_{\pi(1)}, \ldots, v_{\pi(n)})$ is singular for F. The set of all s-preserving $\pi \in S_n$ is denoted by $\mathbf{sp}(\boldsymbol{F}, (\boldsymbol{v_1}, \ldots, \boldsymbol{v_n})) \subseteq S_n$.*

By Corollary 18 we obtain the fundamental lemma, showing that singularity-preservation implies equality-preservation:

Lemma 22. *For $F \in \mathcal{MU}$ and a singular tuple $\boldsymbol{v}$ we have $\mathrm{sp}(F, \boldsymbol{v}) \subseteq \mathrm{eqp}(F, \boldsymbol{v})$.*

The corollary, that singular tuples with the same variables yield the same final reduction-result, is used throughout in the following:

Corollary 23. *Consider two singular tuples $(v_1, \ldots, v_n), (v'_1, \ldots, v'_n)$ for $F \in \mathcal{MU}$. If $\{v_1, \ldots, v_n\} = \{v'_1, \ldots, v'_n\}$, then $\mathrm{DP}_{v_1, \ldots, v_n}(F) = \mathrm{DP}_{v'_1, \ldots, v'_n}(F)$.*

In preparation for our results on singularity-preserving permutation, we consider first "homogeneous" singular pairs in the following two lemmas. By Lemma 13, Parts 3 and 1 we get:

Lemma 24. *Consider $F \in \mathcal{MU}$ and two different non-1-singular variables v, w for F. Let C be the main clause for v, and let D be the main clause for w. There are precisely two cases now:*

1. *If $C = D$, then $w \notin \mathrm{var}_s(\mathrm{DP}_v(F))$ and $v \notin \mathrm{var}_s(\mathrm{DP}_w(F))$.*
2. *If $C \neq D$, then $w \in \mathrm{var}_{\neg 1s}(\mathrm{DP}_v(F))$ and $v \in \mathrm{var}_{\neg 1s}(\mathrm{DP}_w(F))$.*

Lemma 25. *Consider $F \in \mathcal{MU}$ and a singular sequence (v, w) for F such that v and w are 1-singular for it. Let $C, D \in F$ be the two occurrences of v.*

1. *Assume w is not 1-singular in F. Then w is 2-singular in F. Let $E_0 \in F$ be the main-clause of w, and let $E_1, E_2 \in F$ be the two side-clauses. We have $\{E_1, E_2\} = \{C, D\}$. So v is 1-singular in $\mathrm{DP}_w(F)$.*
2. *Otherwise w is 1-singular in F. Then v is 1-singular in $\mathrm{DP}_w(F)$. Let E_1, E_2 be the two occurrences of w in F. We have $|\{C, D\} \cap \{E_1, E_2\}| \leq 1$.*

We are now able to characterise singularity-preserving neighbour-exchanges as follows (note that every permutation is the composition of neighbour-exchanges; for the simple proof see [15]):

Lemma 26. *Consider $F \in \mathcal{MU}$ and a singular tuple $\boldsymbol{v} = (v_1, \ldots, v_n)$ with $n \geq 2$. Consider $i \in \{1, \ldots, n-1\}$, and let $\pi \in S_n$ be the neighbour-exchange $i \leftrightarrow i+1$ (i.e., $\pi(j) = j$ for $j \in \{1, \ldots, n\} \setminus \{i, i+1\}$, while $\pi(i) = i+1$ and $\pi(i+1) = i$). The task is to characterise when $\pi \in \mathrm{sp}(F, \boldsymbol{v})$ holds; we need also to be able to apply such s-preserving neighbour-exchanges consecutively. For this let*

v_i be m_i-singular w.r.t. $F, \boldsymbol{v}$, and in case of $\pi \in \mathrm{sp}(F, \boldsymbol{v})$ let $v_{\pi(i)}$ be m'_i-singular w.r.t. $F, \boldsymbol{v}'$, where $\boldsymbol{v}' := (v_{\pi(1)}, \ldots, v_{\pi(n)})$.

1. If $\pi \in \mathrm{sp}(F, \boldsymbol{v})$, then for $j \in \{1, \ldots, n\} \setminus \{i, i+1\}$ we have $m'_j = m_j$.
2. Assume $m_i \geq 2$. Then $\pi \in \mathrm{sp}(F, \boldsymbol{v})$. And if $m_{i+1} \geq 2$, then $m'_i, m'_{i+1} \geq 2$. While if $m_{i+1} = 1$, then $m'_i = 1$ and $m'_{i+1} \geq m_i - 1$.
3. Assume $m_i = 1$.
   (a) Assume $m_{i+1} = 1$. Then $\pi \in \mathrm{sp}(F, \boldsymbol{v})$ and $m'_{i+1} = 1$ and $m'_i \in \{1, 2\}$.
   (b) Assume $m_{i+1} \geq 2$. Then $\pi \in \mathrm{sp}(F, \boldsymbol{v})$ if and only if v_{i+1} is singular in $\mathrm{DP}_{v_1, \ldots, v_{i-1}}(F)$. And if $\pi \in \mathrm{sp}(F, \boldsymbol{v})$, then $m'_i \geq 2$.

The gist of Lemma 26 is that in most cases neighbours in a singular tuple can be exchanged safely (s-preserving), except when a 1-singular DP-reduction is followed by a non-1-singular DP-reduction (Case 3b). By Case 2:

Corollary 27. *Consider $F \in \mathcal{MU}$ and a singular tuple $\boldsymbol{v}$ such that each v_i is non-1-singular for $\boldsymbol{v}$. Then $\mathrm{sp}(F, \boldsymbol{v}) = S_n$ (all permutations are singular), and in each permutation of $\boldsymbol{v}$ each v_i is non-1-singular.*

We get some normal form of a singular tuple $\boldsymbol{v}$ for $F \in \mathcal{MU}$ by moving the singular variables from F to the front, followed by 1-singular DP-reductions:

Corollary 28. *Consider $F \in \mathcal{MU}$ and a singular tuple $\boldsymbol{v} = (v_1, \ldots, v_n)$. Let $V := \{v_1, \ldots, v_n\} \cap \mathrm{var}_{1s}(F)$ and $p := |V|$. Consider any $\pi_0 : \{1, \ldots, p\} \to \{1, \ldots, n\}$ such that $\{v_{\pi_0(i)} : i \in \{1, \ldots, p\}\} = V$. Then there exists $q \in \{p, \ldots, n\}$ and an s-preserving permutation π for $\boldsymbol{v}$ such that π extends π_0 and $v_{\pi(i)}$ is 1-singular for $(v_{\pi(1)}, \ldots, v_{\pi(n)})$ and $i \in \{1, \ldots, n\}$ if and only if $i \leq q$.*

Comparing two different singular tuples, they don't need to overlap, however they need to have a "commutable beginning" via appropriate permutations, given they contain at least two variables:

Lemma 29. *Consider $F \in \mathcal{MU}$ and singular tuples $(v_1, \ldots, v_p), (w_1, \ldots, w_q)$ for F with $p, q \geq 2$. Then there is an s-preserving permutation π for $(v_1, \ldots, v_p)$ and an s-preserving permutation π' for $(w_1, \ldots, w_q)$ such that both $(v_{\pi(1)}, w_{\pi'(1)})$ and $(w_{\pi'(1)}, v_{\pi(1)})$ are singular for F.*

Proof. If one of the two tuples contains a 1-singular variable v_i resp. w_i, then the assertion follows by Corollary 28 and Part 2 of Corollary 14. And otherwise the assertion follows by Corollary 27 and Lemma 24. □

5.3 Without 1-Singular Variables

If $F \in \mathcal{MU}$ has no 1-singular variables, then we know its maximal singular tuples, namely they are given by choosing exactly one singular literal from each clause which contains singular literals. A general concept is helpful here:

Definition 30. *For $F \in \mathcal{MU}$ we define the **singularity hypergraph** $\mathbf{S}(F)$ as follows (recall that a hypergraph G has vertex-set $V(G)$ and hyperedge-set $E(G)$):*

- *The vertex set is $\mathrm{var}(F)$ (the variables of F).*

- For every $v \in \mathrm{vars}(F)$ let x_v be the singular literal (which depends on the given choice in case v is 1-singular), and let $L := \{x_v : v \in \mathrm{vars}(F)\}$.
- Now the hyperedges are given by $\mathrm{var}(C \cap L)$ for $C \in F$ with $C \cap L \neq \emptyset$.

Note that the hyperedges of $\mathrm{S}(F)$ are non-empty and pairwise disjoint.

Lemma 31. Consider $F \in \mathcal{MU}$ with $\mathrm{var}_{1s}(F) = \emptyset$. The variable-sets of maximal singular tuples for F are precisely the minimal transversals of $\mathrm{S}(F)$ (minimal sets of vertices intersecting every hyperedge). And the maximal singular tuples of F are precisely obtained as (arbitrary) linear orderings of these variable-sets.

Proof. By Corollary 14, Part 1a, for each singular tuple $(v_1, \ldots, v_n)$ of F we have $\{v_1, \ldots, v_n\} \subseteq \mathrm{var}_{\neg 1s}(F) = \mathrm{vars}(F)$, and every v_i is non-1-singular for $(v_1, \ldots, v_n)$. So by Corollary 27 all permutations are singular. Finally, for $v \in \mathrm{vars}(F)$ let $F_v := \mathrm{DP}_v(F)$, let $C_v \in F$ be the main clause of v, and let $H_v := \mathrm{var}(C_v) \cap \mathrm{vars}(F)$. Then we have $\mathrm{S}(F_v) = (V(\mathrm{S}(F)) \setminus \{v\}, E(\mathrm{S}(F)) \setminus \{H_v\})$. The assertion of the lemma follows now easily by induction. $\qquad\square$

Example 32. Consider

$$F := \big\{\, \{a,b\}, \{\overline{a},x,v\}, \{\overline{a},y,v'\}, \{\overline{b},x,v\}, \{\overline{b},y,v'\}, \{\overline{x},v\}, \{\overline{y},v'\}, \{\overline{v},\overline{v'}\} \,\big\}.$$

By definition we have $\mathrm{var}(F) = \mathrm{var}_{\neg 1s}(F)$, and by sDP-reduction we see that $F \in \mathcal{MU}_{\delta=2} \setminus \mathcal{SMU}_{\delta=2}$. We have $\mathrm{S}(F) = (\{a,b,x,y,v,v'\}, \{\{a,b\}, \{x\}, \{y\}, \{v,v'\}\})$. There are $2 \cdot 1 \cdot 1 \cdot 2 = 4$ minimal transversals $\{a,x,y,v\}, \{b,x,y,v\}, \{a,x,y,v'\}$, $\{b,x,y,v'\}$. There are thus 4 elements in $\mathrm{sDP}(F)$; Theorem 44 will show that they are necessarily all isomorphic.

5.4   The Singularity Index

Definition 33. Consider $F \in \mathcal{MU}$. A singular tuple $(v_1, \ldots, v_n)$ for F is called **maximal**, if there is no singular tuple extending it (i.e., $\mathrm{DP}_{v_1,\ldots,v_n}(F)$ is non-singular). The **singularity index** of F, denoted by $\mathbf{si}(F) \in \mathbb{N}_0$, is the minimal $n \in \mathbb{N}_0$ such that a maximal singular sequence of length n exists for F.

So $\mathrm{si}(F) = 0 \Leftrightarrow F \in \mathcal{MU}'$. See Corollary 41, Part 1, for a characterisation of $F \in \mathcal{MU}$ with $\mathrm{si}(F) = 1$. The meaning of $\mathrm{si}(F)$ in general is that of the minimum number of sDP-reductions needed to reach non-singularity. The aim is to show that $\mathrm{si}(F)$ actually does not depend on the choice of a maximal singular sequence. First, by Lemma 31 we can solve a special case:

Lemma 34. Consider $F \in \mathcal{MU}$ not having 1-singular variables (i.e., $\mathrm{var}_{1s}(F) = \emptyset$). Then every maximal singular tuple has length $\mathrm{si}(F)$, which is the number of different clauses of F containing at least one singular literal.

We are now ready to show that, more general than Lemma 34 but with less details, for *all* minimally unsatisfiable clause-sets all maximal singular tuples (i.e., maximal sDP-reduction sequences) have the same length. The basic idea is to utilise the good commutativity properties of 1-singular variables, so that induction on the singularity index can be used.

Theorem 35. *For $F \in \mathcal{MU}$ and every maximal singular tuple $(v_1, \ldots, v_m)$ for F we have $m = \mathrm{si}(F)$.*

Proof. We prove the assertion by induction on $\mathrm{si}(F)$. For $\mathrm{si}(F) = 0$ the assertion is trivial, so assume $\mathrm{si}(F) > 0$. If F has no 1-singular variables, then the assertion follows by Lemma 34, and so we assume that F has a 1-singular variable v. First we show that we can choose v such that $\mathrm{si}(\mathrm{DP}_v(F)) = n - 1$.

Consider a maximal singular tuple $(v_1, \ldots, v_n)$ of length $n = \mathrm{si}(F)$. Note that $\mathrm{si}(\mathrm{DP}_{v_1}(F)) = n - 1$. If v_1 is 1-singular, then we can use $v := v_1$ and we are done, and so assume v_1 is not 1-singular. The induction hypothesis, applied to $\mathrm{DP}_{v_1}(F)$, yields $\mathrm{si}(\mathrm{DP}_{v_1,v}(F)) = n - 2$. Now by Corollary 14, Part 2, both tuples (v_1, v) and (v, v_1) are singular for F, whence $\mathrm{DP}_{v_1,v}(F) = \mathrm{DP}_{v,v_1}(F)$ holds (Corollary 23), and so $\mathrm{si}(\mathrm{DP}_{v,v_1}(F)) = n - 2$. We obtain $\mathrm{si}(\mathrm{DP}_v(F)) \leq n - 1$, and thus $\mathrm{si}(\mathrm{DP}_v(F)) = n - 1$ as claimed.

Now consider an arbitrary maximal singular tuple $(w_1, \ldots, w_m)$. It suffices to show that $\mathrm{si}(\mathrm{DP}_{w_1}(F)) \leq n - 1$, from which by induction hypothesis the assertion follows. The argument is now similar to above. The claim holds for $w_1 = v$, and so assume $w_1 \neq v$. By induction hypothesis we have $\mathrm{si}(\mathrm{DP}_{v,w_1}(F)) = n - 2$. By Corollary 14, Part 2, both tuples (v, w_1) and (w_1, v) are singular for F. Thus $\mathrm{si}(\mathrm{DP}_{w_1,v}(F)) = n - 2$. We obtain $\mathrm{si}(\mathrm{DP}_{w_1}(F)) \leq n - 1$ as claimed.   □

Corollary 36. *For $F \in \mathcal{MU}$ and $F', F'' \in \mathrm{sDP}(F)$ we have $n(F') = n(F'')$.*

6   Confluence Mod Isomorphism on Eventually $\mathcal{SMU}$

Finally we are able to show our third major result, confluence modulo isomorphism of singular DP-reduction in case all maximal sDP-reductions yield saturated clause-sets.

Definition 37. *A minimally unsatisfiable clause-set F is called **eventually saturated**, if all nonsingular F' with $F \xrightarrow{\mathrm{sDP}}_* F'$ are saturated; the set of all eventually saturated clause-sets is $\boldsymbol{\mathcal{ESMU}} := \{F \in \mathcal{MU} : \mathrm{sDP}(F) \subseteq \mathcal{SMU}\}$.*

By Corollary 8 we have $\mathcal{SMU} \subseteq \mathcal{ESMU}$. If $\mathcal{C} \subseteq \mathcal{MU}$ is stable under sDP-reduction, then we have $\mathcal{C} \subseteq \mathcal{ESMU}$ iff $\mathcal{C} \cap \mathcal{MU}' \subseteq \mathcal{SMU}$. In order to show $\mathcal{ESMU} \subseteq \mathcal{CFIMU}$ we show first that "divergence in one step" is enough:

Lemma 38. *Consider $F \in \mathcal{MU} \setminus \mathcal{CFIMU}$ (recall Definition 9). So $\mathrm{si}(F) \geq 1$. Then there is a singular tuple $(v_1, \ldots, v_{\mathrm{si}(F)-1})$ for F, such that for $F' := \mathrm{DP}_{v_1, \ldots, v_{\mathrm{si}(F)-1}}(F)$ we still have $\mathrm{sDP}(F') \in \mathcal{MU} \setminus \mathcal{CFIMU}$ (note $\mathrm{si}(F') = 1$).*

Proof. We prove the assertion by induction on $\mathrm{si}(F) \geq 1$. The assertion is trivial for $\mathrm{si}(F) = 1$, and so consider $n := \mathrm{si}(F) \geq 2$. If there is a singular variable $v \in \mathrm{vars}_\mathrm{s}(F)$ with $\mathrm{DP}_v(F) \in \mathcal{MU} \setminus \mathcal{CFIMU}$, then the assertion follows by induction hypothesis. So assume for the sake of contradiction, that for all singular variables v we have $\mathrm{DP}_v(F) \in \mathcal{CFIMU}$. Consider (maximal) singular tuples $(v_1, \ldots, v_n), (w_1, \ldots, w_n)$ for F such that $\mathrm{DP}_v(F)$ and $\mathrm{DP}_w(F)$ are not isomorphic. By Lemma 29 w.l.o.g. we can assume that (v_1, w_1) and (w_1, v_1) are both singular for F, whence $\mathrm{DP}_{v_1,w_1}(F) = \mathrm{DP}_{w_1,v_1}(F)$ by Corollary 23. We have $\mathrm{DP}_{v_1}(F), \mathrm{DP}_{w_1}(F) \in \mathcal{CFIMU}$ by assumption, and we obtain the contradiction that $\mathrm{DP}_v(F)$ and $\mathrm{DP}_w(F)$ are isomorphic, since $\mathrm{DP}_v(F)$ is isomorphic to the result obtained by reducing F via a (maximal) singular tuple $\boldsymbol{v'} = (v_1, w_1, \dots)$ of length n, where permuting the first two elements in $\boldsymbol{v'}$ yields the singular tuple $\boldsymbol{w'} = (w_1, v_1, \dots)$ with the same result, which is isomorphic to $\mathrm{DP}_w(F)$.   □

Corollary 39. *Consider a class $\mathcal{C} \subseteq \mathcal{MU}$ which is stable under application of singular DP-reduction. Then we have $\mathcal{C} \subseteq \mathcal{CFIMU}$ if and only if $\{F \in \mathcal{C} : \mathrm{si}(F) = 1\} \subseteq \mathcal{CFIMU}$.*

Now we analyse the main case where all sDP-reductions give saturated results:

Lemma 40. *Consider $F \in \mathcal{MU}$ and a clause $C \in F$. Let $C' := \{x \in C : \mathrm{ld}_F(x) = 1\}$ be the set of singular literals in C, establishing C as the main clause for the underlying singular variables $\mathrm{var}(x)$ (for $x \in C'$), and let $F_x := \{D \in F : \overline{x} \in D\}$ be the set of side clauses of $\mathrm{var}(x)$ for $x \in C'$. Due to $F \in \mathcal{MU}$ the sets F_x are non-empty and pairwise disjoint (note that $\mathrm{var}(x)$ is $|F_x|$-singular in F for $x \in C'$). Now assume $|C'| \geq 2$, and that for all $x \in C'$ we have $\mathrm{DP}_{\mathrm{var}(x)}(F) \in \mathcal{SMU}$. Then:*

1. $|C'| = 2$.
2. $\forall\, x \in C'\, \forall\, D \in F_x : (C \setminus C') \subseteq D$.
3. *For $x, y \in C'$ we have that $\mathrm{DP}_{\mathrm{var}(x)}(F)$ and $\mathrm{DP}_{\mathrm{var}(y)}(F)$ are isomorphic.*

Proof. Consider (any) literals $x, y \in C'$ with $x \neq y$. Then for $D \in F_x$ we have $(C \setminus \{x, y\}) \subseteq D$ by Corollary 6, since otherwise the corollary can be applied to $\mathrm{var}(x)$, replacing D by $D \cup (C \setminus \{x, y\})$, which yields the partial saturation $F' \in \mathcal{MU}$ of F with singular variable $\mathrm{var}(y)$, and where then $\mathrm{DP}_{\mathrm{var}(y)}(F')$ would yield a proper partial saturation G of $\mathrm{DP}_{\mathrm{var}(y)}(F)$, contradicting that the latter is saturated. It follows that actually $C' = \{x, y\}$ must be the case, since if there would be $z \in C' \setminus \{x, y\}$, then $\mathrm{ld}_F(z) \geq 2$ contradicting the definition of C'. It follows Part 2. Finally for Part 3 we note that now $F \rightsquigarrow \mathrm{DP}_x(F)$ just replaces $\overline{x}$ in the clauses of F_x by y, while $F \rightsquigarrow \mathrm{DP}_y(F)$ just replaces $\overline{y}$ in the clauses of F_y by x, and thus renaming y in $\mathrm{DP}_x(F)$ to $\overline{x}$ yields $\mathrm{DP}_y(F)$.   □

Corollary 41. *For $F \in \mathcal{MU}$ with $\mathrm{si}(F) = 1$ we have:*

1. *If $|\mathrm{var}_s(F)| \geq 2$:*
   (a) *$\mathrm{var}_s(F) = \mathrm{var}_{\neg 1s}(F)$, that is, all singular variables are non-1-singular.*
   (b) *The main clauses of the singular variables coincide (that is, there is $C \in F$ such that for all singular literals x for F we have $x \in C$).*
   (c) *If $F \in \mathcal{ESMU}$ then $|\mathrm{var}_s(F)| = 2$.*
2. *If $F \in \mathcal{ESMU}$ then $F \in \mathcal{CFIMU}$.*

Proof. Part 1a follows by Part 2a of Corollary 14, and Part 1b follows by Lemma 24. Now Parts 1c, 2 follow from Lemma 40.   □

By Corollary 39 we obtain from Part 2 of Corollary 41:

Theorem 42. $\mathcal{ESMU} \subset \mathcal{CFIMU}$.

Applications to $\mathcal{MU}_{\delta=2}$ Consider $k \in \mathbb{N}$ and $F \in \mathcal{MU}_{\delta=k}$. Assume that we know the isomorphism types of $\mathcal{MU}'_{\delta=k}$. If $F \in \mathcal{CFIMU}$, then we can speak of *the type* of F as the (unique) isomorphism type of the elements of sDP(F). We now show that for $F \in \mathcal{MU}_{\delta=2}$ these assumptions are fulfilled. First we recall the fundamental classification:

Theorem 43. *[5] For $n \geq 2$ let addition of indices be modulo n, and define*
$$\mathcal{F}_n := \{\{v_1, \ldots, v_n\}, \{\overline{v_1}, \ldots, \overline{v_n}\}\} \cup \{\{\overline{v_i}, v_{i+1}\} : i \in \{1, \ldots, n\}\} \in \mathcal{MU}'_{\delta=2}. \text{ Now}$$
for $F \in \mathcal{MU}'_{\delta=2}$ we have $F \cong \mathcal{F}_{n(F)}$.

Theorem 44. $\mathcal{MU}_{\delta=2} \subseteq \mathcal{CFIMU}$.

Proof. The first proof is obtained by applying Corollary 36 and the observation that non-isomorphic elements of $\mathcal{MU}'_{\delta=2}$ have different numbers of variables. The second proof is obtained by applying Theorem 42 and the fact that $\mathcal{MU}'_{\delta=2} \subseteq \mathcal{SMU}$, whence $\mathcal{MU}'_{\delta=2} \subseteq \mathcal{ESMU}$.   □

7   Conclusion and Open Problems

We have discussed questions regarding confluence of singular DP-reduction on minimally unsatisfiable clause-sets. Besides various detailed characterisations, we obtained the invariance of the length of maximal sDP-reduction-sequences, confluence for saturated and confluence modulo isomorphism for eventually saturated clause-sets. The main open questions regarding these aspects are:

– Are there other interesting classes for which we can show confluence resp. confluence mod isomorphism of singular DP-reduction?
– Can we characterise $\mathcal{CFMU}$ and/or $\mathcal{CFIMU}$? Especially, what is the decision complexity of these classes?

As a first application of our results, around Theorem 44 we considered the types of (arbitrary) elements of $\mathcal{MU}_{\delta=2}$. This detailed knowledge is a stepping stone for the determination of the isomorphism types of the elements of $\mathcal{MU}'_{\delta=3}$, which we have obtained meanwhile (to be published; based on a mixture of general insights into the structure of $\mathcal{MU}$ and (very) detailed investigations into $\mathcal{MU}_{\delta\leq2}$). A typical application of Theorem 44 considers a class of $F \in \mathcal{MU}_{\delta=2}$ given by some extremal condition, and then characterisation of these F becomes possible once one knows they must reduce to some *unique* $\mathcal{F}_n$, and so by inverse sDP-reduction one can reconstruct F from $\mathcal{F}_n$. The major open problem of the field is the classification (of isomorphism types) of $\mathcal{MU}'_{\delta=k}$ for arbitrary k. Finally, regarding the potential applications as discussed in the introduction, applying singular DP-reductions in algorithms searching for MUS's is a natural next step.

References

1. Aharoni, R., Linial, N.: Minimal non-two-colorable hypergraphs and minimal unsatisfiable formulas. Journal of Combinatorial Theory A 43, 196–204 (1986)
2. Davydov, G., Davydova, I., Büning, H.K.: An efficient algorithm for the minimal unsatisfiability problem for a subclass of CNF. Annals of Mathematics and Artificial Intelligence 23, 229–245 (1998)
3. Fleischner, H., Kullmann, O., Szeider, S.: Polynomial–time recognition of minimal unsatisfiable formulas with fixed clause–variable difference. Theoretical Computer Science 289(1), 503–516 (2002)
4. Fliti, T., Reynaud, G.: Sizes of minimally unsatisfiable conjunctive normal forms. Faculté des Sciences de Luminy, Dpt. Mathematique-Informatique, 13288 Marseille, France (November 1994)
5. Büning, H.K.: On subclasses of minimal unsatisfiable formulas. Discrete Applied Mathematics 107, 83–98 (2000)
6. Büning, H.K., Kullmann, O.: Minimal unsatisfiability and autarkies. In: Biere, A., Heule, M.J.H., van Maaren, H., Walsh, T. (eds.) Handbook of Satisfiability. Frontiers in Artificial Intelligence and Applications, vol. 185, ch. 11, pp. 339–401. IOS Press (February 2009)
7. Kullmann, O.: Obere und untere Schranken für die Komplexität von aussagenlogischen Resolutionsbeweisen und Klassen von SAT-Algorithmen. Master's thesis, Johann Wolfgang Goethe-Universität Frankfurt am Main (Upper and lower bounds for the complexity of propositional resolution proofs and classes of SAT algorithm; Diplomarbeit am Fachbereich Mathematik) (April 1992) (in German)
8. Kullmann, O.: An application of matroid theory to the SAT problem. In: Fifteenth Annual IEEE Conference on Computational Complexity, pp. 116–124. IEEE Computer Society (July 2000)
9. Kullmann, O.: Lean clause-sets: Generalizations of minimally unsatisfiable clause-sets. Discrete Applied Mathematics 130, 209–249 (2003)
10. Kullmann, O.: Constraint satisfaction problems in clausal form I: Autarkies and deficiency. Fundamenta Informaticae 109(1), 27–81 (2011)
11. Kullmann, O.: Constraint satisfaction problems in clausal form II: Minimal unsatisfiability and conflict structure. Fundamenta Informaticae 109(1), 83–119 (2011)
12. Kullmann, O., Luckhardt, H.: Deciding propositional tautologies: Algorithms and their complexity. Preprint, 82 pages; the ps-file can be obtained (January 1997), http://cs.swan.ac.uk/~csoliver/Artikel/tg.ps
13. Kullmann, O., Luckhardt, H.: Algorithms for SAT/TAUT decision based on various measures. Preprint, 71 pages; the ps-file can be obtained (February 1999), http://cs.swan.ac.uk/~csoliver/Artikel/TAUT.ps
14. Kullmann, O., Zhao, X.: On Variables with Few Occurrences in Conjunctive Normal Forms. In: Sakallah, K.A., Simon, L. (eds.) SAT 2011. LNCS, vol. 6695, pp. 33–46. Springer, Heidelberg (2011)
15. Kullmann, O., Zhao, X.: On Davis-Putnam reductions for minimally unsatisfiable clause-sets. Technical Report arXiv:1202.2600v3 [cs.DM], arXiv (May 2012)
16. Marques-Silva, J.: Computing minimally unsatisfiable subformulas: State of the art and future directions. Journal of Multiple-Valued Logic and Soft Computing (to appear, 2012)
17. Szeider, S.: Minimal unsatisfiable formulas with bounded clause-variable difference are fixed-parameter tractable. Journal of Computer and System Sciences 69(4), 656–674 (2004)
18. Zhao, X., Decheng, D.: Two tractable subclasses of minimal unsatisfiable formulas. Science in China (Series A) 42(7), 720–731 (1999)

Improvements to Core-Guided
Binary Search for MaxSAT

Antonio Morgado[1], Federico Heras[1], and Joao Marques-Silva[1,2]

[1] CASL, University College Dublin, Ireland
[2] IST/INESC-ID, Lisbon, Portugal

Abstract. *Maximum Satisfiability* (MaxSAT) and its weighted variants are well-known optimization formulations of *Boolean Satisfiability* (*SAT*). Motivated by practical applications, recent years have seen the development of *core-guided algorithms* for MaxSAT. Among these, *core-guided binary search with disjoint cores* (BCD) represents a recent robust solution. This paper identifies a number of inefficiencies in the original BCD algorithm, related with the computation of *lower* and *upper bounds* during the execution of the algorithm, and develops solutions for them. In addition, the paper proposes two additional novel techniques, which can be implemented on top of core-guided MaxSAT algorithms that maintain both lower and upper bounds. Experimental results, obtained on representative problem instances, indicate that the proposed optimizations yield significant performance gains, and allow solving more problem instances.

1 Introduction

Maximum Satisfiability (MaxSAT) and its variants, namely (*Weighted*) (*Partial*) MaxSAT, find a growing number of practical applications. Concrete recent examples include hardware design debugging [19] and fault localization in C code [9]. In addition, reference applications that use *Pseudo-Boolean Optimization* (*PBO*) can be cast as MaxSAT [7,4]. Another major application of MaxSAT is in algorithms for *Minimal Unsatisfiable Subset* (*MUS*) enumeration [13]. Indeed, the most efficient MUS enumeration algorithms build on MaxSAT algorithms for computing all *Maximal Satisfiable Subsets* (*MSSes*) and, from these, MUSes can be enumerated using a standard *hitting set* approach [13,18]. The variety of relevant applications of MUS enumeration (e.g. [13,1]), further highlights the practical significance of efficient MaxSAT algorithms.

Motivated by the practical applications of MaxSAT, recent years have witnessed a large number of MaxSAT algorithms being proposed. MaxSAT approaches for solving practical problem instances differ significantly from early work on MaxSAT [12,7]. These approaches are characterized by guiding the search with *unsatisfiable subformulas* [20] and are referred to as *core-guided* MaxSAT algorithms [6,16,14,2,3]. Recent work has proposed two core-guided versions of binary search for MaxSAT [8]. These

---

* This work is partially supported by SFI grant BEACON (09/IN.1/I2618), and by FCT grants ATTEST (CMU-PT/ELE/0009/2009) and POLARIS (PTDC/EIA-CCO/123051/2010).

A. Cimatti and R. Sebastiani (Eds.): SAT 2012, LNCS 7317, pp. 284–297, 2012.
© Springer-Verlag Berlin Heidelberg 2012

include a basic version (BC) and a version that maintains a set of disjoint unsatisfiable cores (BCD). The BCD algorithm was shown to be one of the most efficient on a comprehensive set of problem instances from recent MaxSAT evaluations. Nevertheless, recent detailed analysis of BCD revealed a number of possible inefficiencies, that result from relaxed and conservative maintenance of lower and upper bounds.

This paper addresses the inefficiencies in the original BCD algorithm, and develops a number of key optimizations. These optimizations can be categorized as: (i) modifications to how the upper bound of each disjoint core is initialized, updated, and an associated maintenance of a *global upper bound*; (ii) modifications on how the lower bounds are updated when disjoint cores are *merged*; and (iii) techniques for refining the lower bound so that it reflects a feasible sum of weights. The previous optimizations are implemented in a new algorithm, BCD2, that often requires fewer SAT solver calls than BCD. The paper also proves the correctness of BCD2 and shows that BCD2 is significantly more efficient than BCD on a comprehensive set of benchmarks from recent MaxSAT Evaluations.

In addition, the paper proposes two novel techniques, that can be implemented on top of any core-guided MaxSAT algorithm that maintains both lower and upper bounds, namely the *hardening rule* and *biased search*. The *hardening rule*, which has been extensively used in *branch and bound* algorithms [5,12,11,7], is adapted for core-guided binary search algorithms. As a result, many soft clauses can be declared *hard*. Binary search algorithms always compute the middle value between a lower bound and an upper bound. The *biased search* technique allows biasing the search with the outcomes of the previous iterations and compute a value between the lower and upper bounds, though *not necessarily* the middle one.

The remainder of the paper is organized as follows. Section 2 introduces the MaxSAT problem and core-guided binary search MaxSAT algorithms. Section 3 details the inefficiencies of BCD, and develops a new improved algorithm for core-guided binary search with disjoint cores (BCD2). Section 4 presents the hardening rule and biased search techniques for core-guided MaxSAT. Section 5 evaluates the performance of the algorithms with the proposed techniques. Section 6 presents some concluding remarks.

2   Preliminaries

Let $X = \{x_1, x_2, \ldots, x_n\}$ be a set of Boolean variables. A *literal* l is either a variable x_i or its negation $\bar{x}_i$. A *clause* c is a disjunction of literals. A clause may also be regarded as a set of literals. An *assignment* $\mathcal{A}$ is a mapping $\mathcal{A} : X \to \{0,1\}$ which satisfies (unsatisfies) a Boolean variable x if $\mathcal{A}(x) = 1$ ($\mathcal{A}(x) = 0$). Assignments can be extended in a natural way for literals (l) and clauses (c):

$$\mathcal{A}(l) = \begin{cases} \mathcal{A}(x), & \text{if } l = x \\ 1 - \mathcal{A}(x), & \text{if } l = \neg x \end{cases} \qquad \mathcal{A}(c) = \max\{\mathcal{A}(l) \mid l \in c\}$$

Assignments can also be regarded as set of literals, in which case the assignment $\mathcal{A}$ satisfies (unsatisfies) a variable x if $x \in \mathcal{A}$ ($\bar{x} \in \mathcal{A}$). A *complete assignment* contains a literal for each variable, otherwise is a *partial assignment*. A CNF formula φ is a set of clauses. A *model* is a complete assignment that satisfies all the clauses in a CNF formula φ. The *Propositional Satisfiability Problem* (*SAT*) is the problem of deciding whether there exists a model for a given formula. Given an unsatisfiable formula φ, a subset of clauses φ_C (i.e. $\varphi_C \subseteq \varphi$) whose conjunction is still unsatisfiable is called an *unsatisfiable core* of the original formula. Modern SAT solvers can be instructed to generate an unsatisfiable core for unsatisfiable formulas [20]. A *weighted* clause is a pair (c, w), where c is a clause and w is the cost of its falsification, also called its *weight*. Many real problems contain clauses that *must* be satisfied. Such clauses are called *mandatory* (or *hard*) and are associated with a special weight $\top$. Non-mandatory clauses are also called *soft* clauses. A *weighted* formula in *conjunctive normal form* (WCNF) φ is a set of weighted clauses. For MaxSAT, a *model* is a complete assignment $\mathcal{A}$ that satisfies all mandatory clauses. The *cost of a model* is the sum of weights of the soft clauses that it falsifies. Given a WCNF formula, *Weighted Partial MaxSAT* is the problem of finding a model of minimum cost.

Core-Guided Binary Search Algorithms for MaxSAT. Several MaxSAT solvers in the literature are based on iteratively calling a SAT solver and refining a *lower bound*, an *upper bound* or both [6,2,3,16,14,8,10]. *Core-guided MaxSAT algorithms* are those that additionally take advantage of unsatisfiable cores computed at each unsatisfiable iteration to guide the search [6,2,3,16,14,8], (some of which use binary search [8]).

Auxiliary notation is introduced to describe core-guided binary search MaxSAT algorithms. The remainder of the paper assumes a WCNF formula φ with m soft clauses. Core-guided algorithms use *relaxation variables*, which are fresh Boolean variables. The algorithms described add *at most* one relaxation variable to each soft clause. The process of adding a relaxation variable to a clause, is referred to as *relaxing* the clause. Relaxation variables are maintained in a set R, and it is assumed that relaxation variable r_i is associated to the soft clause c_i with weight w_i, $1 \leq i \leq m$. In order to add relaxation variables to soft clauses, the algorithms use the function $Relax(R, \varphi, \psi)$ which receives a set of existing relaxation variables R, a WCNF formula φ and a set of soft clauses ψ and returns the pair (R_o, φ_o). φ_o corresponds to φ whose soft clauses included in ψ have been augmented with fresh relaxation variables. R_o corresponds to R augmented with the relaxation variables added in φ_o. Given the set of relaxation variables in R, the algorithms add *cardinality / pseudo-Boolean constraints* [4] and translate them to hard clauses. Such constraints usually state that the sum of the weights of the relaxed clauses is less than or equal to a specific value K (AtMostK with $\sum_{i=1}^{m} w_i r_i \leq K$). The algorithms use the following functions:

- *Soft*(φ) returns the set of all *soft* clauses in φ.
- *SATSolver*(φ) makes a call to the SAT solver and returns a triple $(st, \varphi_C, \mathcal{A})$, where st is the status of the formula φ, that is whether φ is satisfiable (SAT or UNSAT). If st =UNSAT, then φ_C contains an unsatisfiable core of φ, and if st =SAT, then $\mathcal{A}$ corresponds to a complete satisfying assignment of φ. Throughout the paper, by abuse of notation, st is referred to as the outcome of the SAT solver.
- *CNF*(c) returns a set of clauses that encode the constraint c into CNF.

Core-guided binary search (BC) and its extension *with disjoint cores* (BCD) [8] compute both a lower bound and an upper bound and have been shown to be very robust

Algorithm 1. BCD

Input: φ

1   $(\varphi_W, \varphi_S, C, lastA) \leftarrow (\varphi, \text{Soft}(\varphi), \emptyset, \emptyset)$ // C - set of disj. core's information

2   **repeat**

3      $\forall_{C_i \in C}, \nu_i \leftarrow (\lambda_i + 1 = \mu_i) ? \mu_i : \lfloor \frac{\mu_i + \lambda_i}{2} \rfloor$

4      $(st, \varphi_C, A) \leftarrow \text{SATSolver}(\varphi_W \cup \bigcup_{C_i \in C} \text{CNF}(\sum_{r_j \in R_i} w_j \cdot r_j \leq \nu_i))$

5   **if** $st = SAT$ **then**

6        $lastA \leftarrow A$

7        $\forall_{C_i \in C}, \mu_i \leftarrow \sum_{r_j \in R_i} w_j \cdot A(r_j)$

8   **else**

9        $subC \leftarrow \text{Intersect}(\varphi_C, C)$ // $subC$ - set of disj. cores that intersect φ_C

10   **if** $\varphi_C \cap \varphi_S = \emptyset$ **and** $|subC| = |\{< R_s, \lambda_s, \nu_s, \mu_s >\}| = 1$ **then**

11         $\lambda_s \leftarrow \nu_s$

12   **else**

13         $(R_s, \varphi_W) \leftarrow \text{Relax}(\emptyset, \varphi_W, \varphi_C \cap \varphi_S)$

14         $(\lambda_s, \mu_s) \leftarrow (0, \sum_{r_j \in R_s} w_j + 1)$

15         $\forall_{C_i \in subC}, (R_s, \lambda_s, \mu_s) \leftarrow (R_s \cup R_i, \lambda_s + \lambda_i, \mu_s + \mu_i)$

16         $C \leftarrow C \setminus subC \cup \{< R_s, \lambda_s, 0, \mu_s >\}$

17   **end**

18   **end**

19   **until** $\forall_{C_i \in C} \lambda_i + 1 \geq \mu_i$

20   **return** $lastA$

---

approaches for MaxSAT solving (in terms of number of solved instances). In what follows, the most sophisticated version (BCD) is briefly overviewed.

The pseudo-code of BCD is shown in Algorithm 1. BCD maintains information about disjoint cores in a set C (initially empty). Whenever a new core is found, a new entry C_s in C is created, that contains the set of relaxation variables R_s in the core (after relaxing required soft clauses), a lower bound λ_s, an upper bound μ_s, and the current middle value ν_s, i.e. $C_s =< R_s, \lambda_s, \nu_s, \mu_s >$. The algorithm iterates while there exists a C_i for which $\lambda_i + 1 < \mu_i$ (line 19). Before calling the SAT solver, for each $C_i \in C$, the middle value ν_i is computed with the current bounds and an AtMostK constraint is added to the working formula (lines 3-4). If the SAT solver returns SAT, the algorithm iterates over each core $C_i \in C$ and its upper bound μ_i is updated according to the satisfying assignment A (lines 6-7). If the SAT solver returns UNSAT, then the set $subC$ is computed which contains every C_i in C that intersects the current core (i.e. $subC \subseteq C$, line 9). If no soft clause needs to be relaxed and $|subC| = 1$, then $subC = \{< R_s, \lambda_s, \nu_s, \mu_s >\}$ and λ_s is updated to ν_s (line 11). Otherwise, all the required soft clauses are relaxed, an entry for the new core C_s is added to C, which aggregates the information of the previous cores in $subC$, and each $C_i \in subC$ is removed from C (lines 13-16).

A concept similar to disjoint cores (namely covers) is used by the core-guided (non binary search) algorithm WPM2 [3] coupled with the constraints to add in each iteration.

3 Improving BCD

Detailed analysis of BCD has revealed two key inefficiencies, both related with how the lower and upper bounds are computed and updated. The first observation is that BCD does *not* maintain a *global upper bound*. When the SAT solver outcome is satisfiable

---

Algorithm 2. BCD2

Input: φ
1 $(\varphi_W, \varphi_S) \leftarrow (\varphi, \text{Soft}(\varphi))$
2 $\forall_{c_j \in \varphi_S},\ \sigma_j \leftarrow w_j$
3 $(\mathcal{C}, \mathcal{A}_\mu, \mu) \leftarrow (\emptyset, \emptyset, 1 + \sum_{c_j \in \varphi_S} \sigma_j)$
4 **repeat**
5       $\forall_{C_i \in \mathcal{C}},\ \nu_i \leftarrow \lfloor \frac{\lambda_i + \epsilon_i}{2} \rfloor$
6       $(st, \varphi_C, \mathcal{A}) \leftarrow \text{SATSolver}(\varphi_W \cup \bigcup_{C_i \in \mathcal{C}} \text{CNF}(\sum_{r_j \in R_i} w_j \cdot r_j \leq \nu_i))$
7   **if** $st = SAT$ **then**
8            $\forall_{c_j \in \varphi_S},\ \sigma_j \leftarrow 0$
9            $\forall_{C_i \in \mathcal{C}} \forall_{r_j \in R_i},\ \sigma_j \leftarrow w_j \cdot (1 - \mathcal{A}(c_j \setminus \{r_j\}))$ // $c_j \in \varphi_W$ and $r_j \in c_j$
10           $\forall_{C_i \in \mathcal{C}},\ \epsilon_i \leftarrow \sum_{r_j \in R_i} \sigma_j$
11           $(\mu, \mathcal{A}_\mu) \leftarrow (\sum_{C_i \in \mathcal{C}} \sum_{r_j \in R_i} \sigma_j, \mathcal{A})$
12   **else**
13           $subC \leftarrow \text{Intersect}(\varphi_C, \mathcal{C})$
14   **if** $\varphi_C \cap \varphi_S = \emptyset$ **and** $|subC| = |\{< R_s, \lambda_s, \nu_s, \epsilon_s >\}| = 1$ **then**
15                $\lambda_s \leftarrow \text{Refine}(\{w_j\}_{r_j \in R_S},\ \nu_s)$
16   **else**
17                $(R_s, \varphi_W) \leftarrow \text{Relax}(\bigcup_{C_i \in subC} R_i,\ \varphi_W,\ \varphi_C \cap \varphi_S)$
18                $\Delta \leftarrow \min\{1 + \min\{\nu_i - \lambda_i \mid C_i \in subC\}, \min\{w_j \mid r_j \text{ is a new relax. var.}\}\}$
19                $\lambda_s \leftarrow \text{Refine}(\{w_j\}_{r_j \in R_s},\ \sum_{C_i \in subC} \lambda_i + \Delta - 1)$
20                $\epsilon_s \leftarrow ((\mathcal{A}_\mu = \emptyset) ? 1 : 0) + \sum_{r_j \in R_s} \sigma_j$
21                $\mathcal{C} \leftarrow \mathcal{C} \setminus subC \cup \{< R_s, \lambda_s, 0, \epsilon_s >\}$
22   **end**
23   **end**
24 **until** $\sum_{C_i \in \mathcal{C}} \lambda_i = \sum_{C_i \in \mathcal{C}} \epsilon_i = \mu$
25 **return** $\mathcal{A}_\mu$

(SAT), each μ_i value is updated for each disjoint core $C_i \in \mathcal{C}$, with an overall sum given by $K_1 = \sum_{C_i \in \mathcal{C}} \mu_i$. However, after merging disjoint cores, if the SAT solver outcome is again SAT, it can happen that $K_2 = \sum_{C_i \in \mathcal{C}} \mu_i > K_1$. Although this issue does not affect the correctness of the algorithm, it can result in a number of iterations higher than needed to compute the optimum. The second observation is that the lower bound updates for each disjoint core are conservative. A more careful analysis of how the algorithm works allows devising significantly more aggressive lower bound updates. Again, the main consequence of using conservative lower bounds is that this can result in a number of iterations higher than needed to compute the optimum.

This section presents the new algorithm BCD2. Although similar to BCD, BCD2 proposes key optimizations that address the inefficiencies described above. As the experimental results demonstrate, these optimizations lead to significant performance gains, that can be explained by a reduced number of iterations.

The pseudo-code of BCD2 is shown in Algorithm 2. The organization of BCD2 is similar to the organization of BCD but with important differences. The first difference between BCD and BCD2 is the way the algorithms use the information of the upper bounds. As stated before, BCD does not maintain a global upper bound, and as such, whenever an upper bound is needed, then the worst case scenario is used. Concretely in line 14 of BCD, the upper bound is updated with the weights of the new relaxed clauses.

On the other hand, BCD2 keeps a *global upper bound* μ and its corresponding assignment $\mathcal{A}_\mu$. More importantly it maintains the cost of each soft clause for the current global upper bound. In order to achieve this, BCD2 associates with each soft clause j a variable σ_j that represents the contribution of the clause to the overall cost of the global upper bound. σ_j can take as value either 0 or w_j (the weight of the soft clause j) depending on whether $\mathcal{A}_\mu$ unsatisfies the clause or not. In contrast to BCD, the contribution of soft clauses is with respect to the original variables. As such in line 9 of BCD2, the update of σ_j considers the satisfiability of the clause c_j without the relaxation variable $(w_j \cdot (1 - \mathcal{A}(c_j \setminus \{r_j\})))$, rather than the satisfiability of the relaxation variable $(w_j \cdot \mathcal{A}(r_j))$ as in BCD (line 7). Considering the satisfiability of the original soft clause instead of the associated relaxation variable, has the benefit of tightening the upper bound on assignments that satisfy the clause without the relaxation variable but still satisfy the relaxation variable.

Unlike BCD, BCD2 does not maintain upper bounds in the disjoint cores. Instead, each disjoint core C_i maintains an *estimate* ϵ_i that represents the contribution of the disjoint core to the cost of the global upper bound. Each ϵ_i takes the role of the upper bounds μ_i in BCD, with updates that respect the last satisfying assignment. The difference is that in BCD2, the updates of the estimates, done in lines 10 and 20, include the contribution of the soft clauses to the global upper bound (stored in the σ_j variables).

The use of σ_j variables in the computation of estimates ϵ_i, allow BCD2 to use the information of the current upper bound assignment for a tighter bound, specifically, when merging cores with soft clauses not previously relaxed. The contribution of the newly relaxed clauses in the update of ϵ_i in line 20, is dependent on a previous discovery of a satisfying assignment. Before the first satisfying assignment is found, the contribution is the same as in BCD, that is the weight of the soft clause ($\sigma_j = w_j$, initialization of σ_j in line 2 of BCD2), whereas after the first satisfying assignment, newly relaxed clauses are satisfied by $\mathcal{A}_\mu$ (thus $\sigma_j = 0$ from line 8) and its contribution to ϵ_i is 0.

The reason why the ϵ_i variables are called estimates is that, unlike the upper bound μ_i of BCD, the ϵ_i variables are allowed to have a value lower than the cost of the optimum model restricted to the clauses associated to the disjoint core. In such situations ϵ_i is said to be *optimistic* and represents a local optimum of a MaxSAT model. BCD2 can shift ϵ_i away from the local optimum by merging with different cores as needed.

Example 1. Consider an execution of the algorithm with the current working formula $\varphi_W = \varphi^S \cup \varphi^H$, where $\varphi^S = \{(x_1 \vee r_1, 5), (x_2 \vee r_2, 10), (x_3 \vee r_3, 30), (x_4 \vee r_4, 10)\}$ and $\varphi^H = \{(\neg x_1 \vee \neg x_2), (\neg x_2 \vee \neg x_3), (\neg x_3 \vee \neg x_4)\}$. Consider the upper bound assignment $\mathcal{A}_\mu = \{x_1 = x_3 = r_2 = r_4 = 0, x_2 = x_4 = r_1 = r_3 = 1\}$ with a cost of 35, and two disjoint cores $C_1 = < R_1 = \{r_1, r_2\}, \lambda_1 = 5, \nu_1 = 5, \epsilon_1 = 5 >, C_2 = < \{r_3, r_4\}, 10, 20, 30 >$.

The optimum cost of φ is 20. Considering the optimum model, the contribution of the clauses associated to C_1 is 10 which is lower than ϵ_1, thus ϵ_1 is *optimistic*. The next core returned by the SAT solver merges C_1 and C_2 into a new disjoint core C_3 with $\epsilon_3 = 35$.

Another improvement in BCD2 is the way the lower bound is computed when merging cores. In this case, BCD2 proposes a stronger update in lines 18 and 19, which corresponds to the expression in Equation 1.

$$\sum_{C_i \in subC} \lambda_i + \min\left\{1 + \min\{\nu_i - \lambda_i | C_i \in subC\}, \min\{w_j | r_j \text{ new relax. var.}\}\right\} \quad (1)$$

The update of the lower bound of the merged disjoint cores in Equation 1, is obtained by summing all the previous lower bounds, as is done by BCD in line 15, but also by adding an increment Δ (line 18 in BCD2). The rationale for the increment Δ comes as a justification for obtaining the current core. At this point of the algorithm, there are three possible reasons why the current core was obtained: (i) one or more of the newly relaxed soft clauses has a non-zero contribution to the cost of the final optimum model; (ii) one or more of the disjoint cores is unable to satisfy the corresponding constraint $\sum_{r_j \in R_i} w_j \cdot r_j \leq \nu_i$; (iii) a combination of the previous two.

Suppose that the reason for obtaining the current core is as stated in (i). Since the number of newly relaxed soft clauses with a non-zero contribution is unknown, then Δ corresponds to the weight of the relaxation variable with the lowest weight, that is, in this case $\Delta = \min\{w_j | r_j \text{ new relax. var.}\}$.

Consider now that the reason for obtaining the current core is as stated in (ii). Then at least one of the disjoint cores merged, requires its lower bound to be increased from λ_i to $\nu_i + 1$ (an increment of $1 + \nu_i - \lambda_i$). Since it is unknown which disjoint cores require to be increased, then in Δ is only considered the disjoint core with the lowest increment, that is $\Delta = 1 + \min\{\nu_i - \lambda_i | C_i \in sub\mathcal{C}\}$.

Finally, in the case of reason (iii), the increment Δ can be obtained by summing the increments corresponding to the previous reasons. Nevertheless, it is unknown exactly which of the three reasons explains the current core, then BCD2 uses as increment the minimum of the previous increments, thus obtaining the expression in Equation 1.

An additional difference between the algorithms is the use of the *Refine()* function to further improve the update of the lower bound in lines 15 and 19 of BCD2. The result of *Refine*($\{w_j\}, \lambda$) is the smallest integer greater than λ that can be obtained by summing a subset of the input weights $\{w_j\}$. In BCD2, *Refine*($\{w_j\}, \lambda$) starts by searching if all weights are equal, in which case the minimum sum of weights greater than λ is returned, otherwise, *subsetsum*($\{w_j\}, \lambda$) is computed as used by WPM2 [3].

Finally, the last difference between BCD and BCD2 is the stopping criteria. Given the new bounds, BCD2 stops when the sum the lower bounds of each disjoint core is the same as the global upper bound.

3.1 Proof of Correctness

This subsection proves the correctness of the BCD2 algorithm. First, the correctness of the updates of the lower bound are proven, followed by a proof of the invariant of BCD2. The section ends with a proof of the correctness of BCD2.

Proposition 1. *Consider a disjoint core C_s in the conditions of the update of λ_s in line 15. There is no MaxSAT model for which the clauses associated to C_s contribute to the cost with a value smaller than $Refine(\{w_j\}_{r_j \in R_s}, \nu_s)$.*

Proof. Consider an iteration where the SAT solver returned a core which only contains clauses previously relaxed, and that these clauses belong to the same disjoint core C_s.

For the purpose of contradiction, assume there is a model for which the clauses of C_s contribute with a cost lower than $\nu_s + 1$. Then the assignment of the model can be augmented with assignments to the relaxation variables, such that, each relaxation variable $r_i \in R_s$ is assigned true iff the assignment of the model does not satisfy the corresponding clause c_i. The augmented assignment is able to satisfy the constraint $\sum_{r_i \in R_s} w_i \cdot r_i \leq \nu_s$, all the hard clauses (because it is a MaxSAT model), and all the soft clauses (due to the assignments to the relaxation variables). Then the core returned by the SAT solver is not an unsatisfiable subformula, which is a contradiction, thus the update $\lambda_s \leftarrow \nu_s + 1$ is correct.

Since there is no model with $\lambda_s \leq \nu_s$, then the next value to consider for $\sum_{r_i \in R_s} w_i \cdot r_i$ is the minimum sum of subsets of $\{w_j\}_{r_j \in R_s}$ that is greater than ν_s. This corresponds to the value returned by $Refine(\{w_j\}_{r_j \in R_s}, \nu_s)$. Thus the update on line 15 is correct.

Proposition 2. *Consider the subset of disjoint cores $subC = \{C_1, \ldots, C_m\}$ and a new set of relaxation variables and Δ as in the conditions of line 19, then there is no MaxSAT model for which the clauses associated to the resulting disjoint core C_s contribute to the cost with a value smaller than $Refine(\{w_j\}_{r_j \in R_s}, \sum_{C_i \in subC} \lambda_i + \Delta - 1)$.*

Proof. There is no model with cost lower than $\sum_{C_i \in subC} \lambda_i$ because at this point of the algorithm, each disjoint core $C_i \in subC$ has been proved to have a lower bound of at least λ_i. Then the union of disjoint sets of the clauses of each C_i together with the clauses that just got relaxed have a cost of at least $\sum_{C_i} \lambda_i$ in any MaxSAT model.

Consider by contradiction, that there is a model, for which the clauses associated to the resulting disjoint core C_s, have a cost $costSol \in [\sum_{C_i \in subC} \lambda_i, \sum_{C_i \in subC} \lambda_i + \Delta[$. Two cases are considered.

1) In the first case, suppose that the model assigns to true at least one of the new relaxation variables (of the soft clauses that just got relaxed), and that the cost associated to that relaxation variable is w_{newRV}. Then,

$$w_{newRV} \geq \min\{w_j | r_j \text{ is a new relax. var.}\} \geq \Delta$$

Consider the contribution of all the clauses without the newly relaxed clause:

$$costSol - w_{newRV} \leq costSol - \Delta$$

but by contradiction $costSol < \sum_{C_i \in subC} \lambda_i + \Delta$ and then

$$costSol - w_{newRV} \leq costSol - \Delta < \sum_{C_i \in subC} \lambda_i$$

which means that the contribution of the remaining clauses is lower than $\sum_{C_i \in subC} \lambda_i$; but this is a contradiction (previously the cost of the union of clauses of $C_i \in subC$ was proven to be at least $\sum_{C_i \in subC} \lambda_i$).

2) In the second case suppose that the model assigns all newly relaxed clauses to false, then the contribution of the newly relaxed clauses is 0. Since by contradiction

$$costSol < \sum_{C_i \in subC} \lambda_i + \Delta \leq \sum_{C_i \in subC} \lambda_i + 1 + \min\{\nu_i - \lambda_i | C_i \in subC\}$$

then

$$costSol - \sum_{C_i \in subC \setminus \{C_1\}} \lambda_i \leq \lambda_1 + \min\{\nu_i - \lambda_i | C_i \in subC\} \leq \nu_1$$

Let $costSol\langle C_i \rangle$ be the contribution of the clauses of C_i to the cost of the model. Previously, was proven that $costSol\langle C_i \rangle \geq \lambda_i$, then

$$costSol\langle C_1 \rangle = costSol - \sum_{C_i \in subC \backslash \{C_1\}} costSol\langle C_i \rangle \leq costSol - \sum_{C_i \in subC \backslash \{C_1\}} \lambda_i \leq \nu_1$$

By analogy, for each of the disjoint cores merged $C_i \in subC$, $costSol\langle C_i \rangle \leq \nu_i$ Then the model is able to satisfy all the new soft clauses and the constraints $\sum_{r_j \in R_i} w_j \cdot r_j \leq \nu_i$. Since the model is a MaxSAT model, then it is also able satisfy all the hard clauses, meaning that the model is able to satisfy all the clauses in the core; but this is again a contradiction.

Since there is no model with $\lambda_s < \sum_{C_i \in subC} \lambda_i + \Delta$, then the next value to consider for $\sum_{r_i \in R_s} w_i \cdot r_i$ is the minimum sum of subsets of $\{w_j\}_{r_j \in R_s}$ that is greater than $\sum_{C_i \in subC} \lambda_i + \Delta - 1$. This corresponds to the value returned by $Refine(\{w_j\}_{r_j \in R_s}, \sum_{C_i \in subC} \lambda_i + \Delta - 1)$. Thus the update on line 19 is correct.

Proposition 3 (Invariant of BCD2). *Let opt be cost of the optimum model of a MaxSAT instance. During the execution of BCD2, the invariant $\sum_{C_i \in C} \lambda_i \leq opt \leq \mu$ holds.*

Proof. Initially C is empty, and $\sum_{C_i \in C} \lambda_i$ is 0. On the other hand, μ is initialized to $\sum_{(c_j, w_j) \in Soft(\varphi)} w_j + 1$. Since $0 \leq opt \leq \sum_{(c_j, w_j) \in Soft(\varphi)} w_j$, then initially the invariant holds.

Each λ_i is only updated on unsatisfiable iterations in lines 15 and 19 and each update was proved to be correct in Propositions 1 and 2, respectively. Then after the updates we are guaranteed that $\sum_{C_i \in C} \lambda_i \leq opt$.

Consider now a satisfiable iteration. Assume for the sake of contradiction that μ is updated such that $\mu < opt$. Then the assignment returned by the SAT solver can be extended with assignments to new relaxation variables (one for each clause not yet relaxed). In particular, these variables can be assigned value false. Then, the sum of the weights of the relaxation variables assigned value true is lower than opt which is a contradiction since, by definition, the sum of weights of relaxed clauses is an upper bound on the optimum MaxSAT model.

Proposition 4. *For any disjoint core C_s, the invariant $\lambda_s \leq \epsilon_s$ holds.*

Proof. The values of variables ϵ_i are only updated in lines 10 and 20 (see Algorithm 2). The updates are due to assignments that are models to the MaxSAT formula, and represent the cost of the model with respect to the clauses associated to the disjoint core C_i. Line 20 also considers the case where no model has been found yet, and updates ϵ_i to one plus the sum of all the weights of the soft clauses considered.

On the other hand, the values of variables λ_i are only updated in lines 15 and 19. In Propositions 1 and 2, was proven that there is no MaxSAT model with a cost smaller than the update of the lower bound in lines 15 and 19 (with respect to the clauses associated with the resulting core C_s). Hence, $\lambda_s \leq \epsilon_s$ for each disjoint core C_s.

Proposition 5. *BCD2 is correct and returns the optimum model for any WCNF formula.*

Proof. The algorithm performs binary search on the range of values $\{\sum_{C_i \in \mathcal{C}} \lambda_i, \ldots, \mu\}$. In each iteration the algorithm asks for a model with a cost at most $\sum_{C_i \in \mathcal{C}} \nu_i$. Due to the assignment of each ν_i in line 5 and Proposition 4, then $\sum_{C_i \in \mathcal{C}} \lambda_i \leq \sum_{C_i \in \mathcal{C}} \nu_i \leq \mu$. If the SAT solver returns with a satisfiable answer, then μ is updated to a lower value than the current upper bound (due to the added constraints). If the SAT solver returns with an unsatisfiable answer, then either $\sum_{C_i \in \mathcal{C}} \lambda_i$ increases or more than one of the disjoint cores are merged. Since the number of clauses to be relaxed is bounded by the number of soft clauses, then the maximum number of merges of disjoint cores is also bounded (disjoint cores only contain clauses that are relaxed). Thus the number of iterations where the algorithm does not increase the sum $\sum_{C_i \in \mathcal{C}} \lambda_i$, is bounded.

Finally, Proposition 3 proves that during the execution of the algorithm, there is always an optimum MaxSAT model between the bounds. Since the bounds are integer numbers, then the algorithm is guaranteed to stop with the optimum MaxSAT model.

4 Additional Techniques

This section introduces two additional techniques to improve the performance of core-guided binary search algorithms, namely, the hardening rule and biased search.

4.1 Hardening Rule

The *hardening rule* is widely used in *branch and bound* (*BB*) algorithms for MaxSAT [12,11,7] which are based on a *systematic enumeration* of all possible assignments, where large subsets of *useless* assignments are discarded by computing *upper* and *lower bounds* on the cost of the optimum model. Whenever the weight of a soft clause plus the lower bound reaches the upper bound, the clause can be *made* hard. Indeed, the hardening rule was introduced in the most primitive BB algorithm for MaxSAT in the literature [5], but nowadays is still not used in core-guided MaxSAT algorithms. In what follows, a first integration of the hardening rule is proposed for core-guided MaxSAT algorithms that maintain both a lower bound and upper bound. To explain the idea, each soft clause (c, w) is extended with two weights (c, w, w') where w is the original weight and w' represents the weight of the clause after its *contributions* to the lower bound have been deducted. w' will be referred as the *deducted* weight. Let φ_d be a set of soft clauses involved in an increment d of the global lower bound. Then, the deducted weight of all the soft clauses in φ_d needs to be decreased by d. As a result, the hardening rule is applied taking into account the deducted weight rather than the original one. Hence, the hardening rule is shown in Equation 2

$$\text{if } w' + \lambda \geq \mu \text{ then } (c, w, w') \text{ can be replaced by } (c, \top, \top) \tag{2}$$

Let (c, w, w') be a soft clause that is made hard due to the hardening rule. There are two situations. If the soft clause has no relaxation variable, the weight of the clause is just replaced by $\top$. If the soft clause has a relaxation variable, the weight is updated to $\top$ and additionally, the relaxation variable is removed.

Example 2. Consider the formula $\{(x, 3, 3), (\bar{x}, 4, 4), (y, 3, 3), \ldots\}$. An initial upper bound $\mu = 5$ is obtained using any heuristic [8]. An initial lower bound $\lambda = 3$ can be obtained due to an unsatisfiable core between the two first clauses. The minimum weight for the conflicting clauses $(x, 3, 3)$ and $(\bar{x}, 4, 4)$ is 3. The resulting formula is $\{(x, 3, 0), (\bar{x}, 4, 1), (y, 3, 3), \dots\}$ with $\lambda = 3$. Then, the hardening rule can be applied to the clause $(y, 3, 3)$ given that $3 + 3 \geq 5$. Hence, $(y, 3, 3)$ is replaced by $(y, \top, \top)$. The current formula is $\{(x, 3, 0), (\bar{x}, 4, 1), (y, \top, \top), \dots\}$ with $\lambda = 3$ and $\mu = 5$.

The integration of the hardening rule in BCD2 is as follows. Assume BCD2 maintains internally the deducted weight of each soft clause, then any of the initial lower bounds introduced in [8] can be used. Such lower bounds iteratively compute unsatisfiable cores until a satisfiable instance is reached. For each unsatisfiable core, the minimum weight is subtracted to the deducted weight of each soft clause in the core.

Assume any arbitrary iteration of the main loop of BCD2. Let $\lambda = \sum_{C_i \in \mathcal{C}} \lambda_i$ be the global lower bound and let $\mu = \sum_{C_i \in \mathcal{C}} \epsilon_i$ be the global upper bound, before the call to the SAT solver (line 6). After the call to the SAT solver, there are two possibilities:

– The SAT solver returns satisfiable (SAT). The global upper bound μ is updated and the hardening rule is checked with the new global upper bound.
– The SAT solver returns unsatisfiable (UNSAT) and the global lower bound is increased. Let $\lambda' = \sum_{C_i \in \mathcal{C}} \lambda_i$ be the new global lower bound in line 23. Let d be the difference between the previous and the current global lower bounds, i.e., $d = \lambda' - \lambda$. Such increment d is due to the disjoint core C_s in line 15 or in line 21. Hence, the deducted weight of each soft clause in the proper disjoint core C_s is decreased by d. Afterwards, the hardening rule is checked.

4.2   Biased Search

At each iteration, binary search algorithms compute a middle value ν between an upper bound μ and a lower bound λ (i.e. $\nu \leftarrow \lfloor \frac{\mu + \lambda}{2} \rfloor$). However, when the cost of the optimum model is close to one of the bounds, binary search can make several iterations before realizing that. In fact, QMAXSAT (0.4 version) solver [10] alternates iterations which compute the middle value between the bounds, and iterations which use the value of the upper bound. As such, QMAXSAT favors the discovery of models with a cost closer to the upper bound. Note that QMAXSAT was the best performing solver on recent MaxSAT Evaluations in the *partial MaxSAT industrial* category.

This paper proposes to compute a value between the lower bound and upper bound (i.e. $\nu \in [\lambda, \mu]$) based on the previous iterations. Two counters are maintained. A counter of the iterations that returned satisfiable (SAT) $nsat$, and a counter of the iterations that returned unsatisfiable (UNSAT) $nunsat$. Both counters are initialized to 1. At each iteration of the binary search algorithm the following percentage is computed:

$$p = nunsat / (nunsat + nsat)$$

The expression compares the number of unsatisfiable iterations against the total number of iterations, and gives a value closer to the bound with fewer outcomes in terms of a percentage. The value ν to be considered at each iteration is $\nu = \lambda + p \times (\mu - \lambda)$.

Note that the QMAXSAT approach is similar to always alternating the percentage p between 50% (middle value) and 100% (upper bound). The integration in BCD2 is straightforward. For each disjoint core C_i with estimate of the upper bound ϵ_i and lower bound λ_i, BCD2 computes the value ν_i as $\nu_i = \lambda_i + p \times (\epsilon_i - \lambda_i)$.

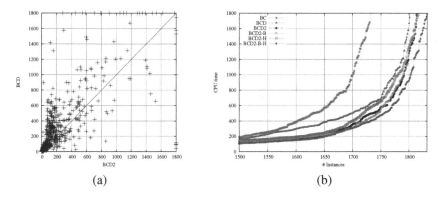

Fig. 1. (a) Scatter plot of BCD vs BCD2, (b) Cactus plot of BC, BCD, BCD2 and BCD2 with additional techniques

5 Experimental Evaluation

Experiments were conducted on a HPC cluster with 50 nodes, each node is a CPU Xeon E5450 3GHz, 32GB RAM and Linux. For each run, the time limit was set to 1800 seconds and the memory limit to 4GB. BCD2 and the additional techniques were implemented in the MSUNCORE [17] system, and compared against BC and BCD[1].

Figure 1 presents results on the performance of BCD2 (from Section 3) and the new techniques (from Section 4) in all of the *non-random* instances from 2009-2011 MaxSAT Evaluations (for a total of 2615 instances). The scatter plot (Figure 1.a) shows a comparison of the original BCD [8] with BCD2 (as described in Section 3). Note that BCD2 (1813) solves 12 more instances than BCD (1801). The scatter plot indicates that in general BCD requires larger run times than BCD2. A more detailed analysis indicates that, out of 1305 instances where the performance difference between BCD and BCD2 exceeds 20%, BCD2 outperforms BCD in 918, whereas BCD outperforms BCD2 in 387. Moreover, over the 1793 instances solved by both BCD and BCD2, the total number of SAT solver calls for BCD is 124907 and for BCD2 is 68690. This represents an average of 31.5 fewer SAT solver calls per instance for BCD2 (from 69.7 to 38.3), i.e. close to 50% fewer calls in BCD2 than in BCD on average. The difference is quite significant; it demonstrates the effectiveness of the new algorithm, but also indirectly suggests that some of the SAT solver calls, being closer to the optimum, may be harder for BCD2 than for BCD. Nevertheless, BCD2 consistently outperforms BCD overall.

The cactus plot (Figure 1.b) shows the run times for BCD, BCD2, BCD2 with hardening rule (BCD2-H), BCD2 with biased search (BCD2-B) and BCD2 with both techniques (BCD2-B-H). The original core-guided binary search algorithm [8] (BC) is also included. The performance difference between BCD and BCD2 is conclusive, and confirmed by the area below each plot. For the vast majority of instances, BCD2 outperforms BCD. The hardening rule (BCD2-H) allows solving 3 additional instances than BCD2, whereas biased search (BCD2-B) allows solving one more instance. However,

---

[1] Observe that in [8], BCD was shown to solve more instances than a representative sample of MaxSAT solvers.

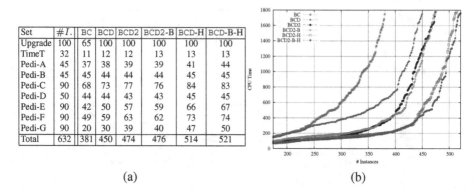

Set	#I.	BC	BCD	BCD2	BCD2-B	BCD-H	BCD-B-H
Upgrade	100	65	100	100	100	100	100
TimeT	32	11	12	12	13	13	13
Pedi-A	45	37	38	39	39	41	44
Pedi-B	45	45	44	44	44	45	45
Pedi-C	90	68	73	77	76	84	83
Pedi-D	50	44	44	43	43	45	45
Pedi-E	90	42	50	57	59	66	67
Pedi-F	90	49	59	63	62	73	74
Pedi-G	90	20	30	39	40	47	50
Total	632	381	450	474	476	514	521

(a)   (b)

Fig. 2. (a) Table number of solved instances per algorithm (b) Cactus plot with the different algorithms

the integration of both techniques (BCD2-B-H) allows solving 1832 instances, i.e. 19 more instances than BCD2 and 31 more than the original BCD. As expected, BC is the worst performing algorithm (solves 1730 instances), and indirectly demonstrates that maintaining disjoint cores is essential to obtain a more robust algorithm.

The effect of the more accurate bounds maintained by BCD2 and the additional techniques is even more significant on weighted partial MaxSAT industrial instances. A second experiment, see Figure 2, shows the results for 100 *upgradeability* instances, 32 *timetabling* instances [3] and 500 *haplotyping with pedigrees* instances [15]. Observe that the haplotyping with pedigrees instances are divided in 7 sets (A, B, C, D, E, F, G). The results are summarized in the table of Figure 2.a. The first column shows the name of benchmark set. The second column shows the total number of instances in the set. The remaining columns show the total number of solved instances within the time and memory limits by BC, BCD and the different versions of BCD2. The same results are presented with a cactus plot in Figure 2.b to highlight the runtimes.

BC is again the worst performing algorithm, and is the only approach unable to solve the 100 upgradeability problems. BCD outperforms BC and solves 69 more instances. BCD2 is clearly better than BCD, being able to solve 26 more instances. Biased search (BCD2-B) has small effect and solves 2 more instances than BCD2. The hardening rule (BCD2-H) is quite helpful on these instances and solves 40 more instances than BCD2. Finally, the integration of the two new techniques (BCD2-B-H) allows solving 521 instances, i.e. 47 more instances than BCD2 and 71 more than the original BCD.

6   Conclusions

This paper proposes a number of improvements to a recently proposed MaxSAT algorithm [8] that implements core-guided binary search. The first improvement addresses the organization of the original algorithm, and modifies the algorithm to (i) maintain a global upper bound, that results in tighter local upper bounds for each disjoint core; and (ii) use of more aggressive lower bounding techniques. The improvements to the upper and lower bound result in significant reduction in the number of SAT solver calls made by the algorithm. The second improvement consists of two techniques that can be implemented on top of any core-guided algorithm that uses lower and upper bounds. One of the techniques is referred to as the hardening rule and has been extensively used in branch-and-bound algorithms [5,12,11,7], but not in core-guided algorithms. The second technique is referred to as *biased search*, and is shown to work effectively with the hardening rule. Experimental results, obtained on a comprehensive set of instances from past MaxSAT Evaluations, demonstrates that the new algorithm BCD2 significantly outperforms (an already quite robust) BCD.

References

1. Andraus, Z.S., Liffiton, M.H., Sakallah, K.A.: Reveal: A Formal Verification Tool for Verilog Designs. In: Cervesato, I., Veith, H., Voronkov, A. (eds.) LPAR 2008. LNCS (LNAI), vol. 5330, pp. 343–352. Springer, Heidelberg (2008)
2. Ansótegui, C., Bonet, M.L., Levy, J.: Solving (Weighted) Partial MaxSAT through Satisfiability Testing. In: Kullmann, O. (ed.) SAT 2009. LNCS, vol. 5584, pp. 427–440. Springer, Heidelberg (2009)
3. Ansótegui, C., Bonet, M.L., Levy, J.: A new algorithm for weighted partial maxsat. In: AAAI Conference on Artificial Intelligence. AAAI (2010)
4. Biere, A., Heule, M., van Maaren, H., Walsh, T. (eds.): Handbook of Satisfiability (2009)
5. Borchers, B., Furman, J.: A two-phase exact algorithm for max-sat and weighted max-sat problems. J. Comb. Optim. 2(4), 299–306 (1998)
6. Fu, Z., Malik, S.: On Solving the Partial MAX-SAT Problem. In: Biere, A., Gomes, C.P. (eds.) SAT 2006. LNCS, vol. 4121, pp. 252–265. Springer, Heidelberg (2006)
7. Heras, F., Larrosa, J., Oliveras, A.: MiniMaxSat: An efficient weighted Max-SAT solver. JAIR 31, 1–32 (2008)
8. Heras, F., Morgado, A., Marques-Silva, J.: Core-guided binary search algorithms for maximum satisfiability. In: AAAI (2011)
9. Jose, M., Majumdar, R.: Cause clue clauses: error localization using maximum satisfiability. In: PLDI, pp. 437–446 (2011)
10. Koshimura, M., Zhang, T., Fujita, H., Hasegawa, R.: QMaxSAT: A partial Max-SAT solver. JSAT, 95–100 (2012)
11. Larrosa, J., Heras, F., de Givry, S.: A logical approach to efficient Max-SAT solving. Artificial Intelligence 172(2-3), 204–233 (2008)
12. Li, C.M., Manyà, F., Planes, J.: New inference rules for Max-SAT. Journal of Artificial Intelligence Research 30, 321–359 (2007)
13. Liffiton, M.H., Sakallah, K.A.: Algorithms for computing minimal unsatisfiable subsets of constraints. J. Autom. Reasoning 40(1), 1–33 (2008)
14. Manquinho, V., Marques-Silva, J., Planes, J.: Algorithms for Weighted Boolean Optimization. In: Kullmann, O. (ed.) SAT 2009. LNCS, vol. 5584, pp. 495–508. Springer, Heidelberg (2009)
15. Marques-Silva, J., Argelich, J., Graca, A., Lynce, I.: Boolean lexicographic optimization: Algorithms and applications. Annals of Mathematics and A. I., 1–27 (2011)
16. Marques-Silva, J., Planes, J.: Algorithms for maximum satisfiability using unsatisfiable cores. In: DATE, pp. 408–413 (2008)
17. Morgado, A., Heras, F., Marques-Silva, J.: The MSUnCore MaxSAT solver. In: POS (2011)
18. Reiter, R.: A theory of diagnosis from first principles. Artif. Intell. 32(1), 57–95 (1987)
19. Safarpour, S., Mangassarian, H., Veneris, A., Liffiton, M.H., Sakallah, K.A.: Improved design debugging using maximum satisfiability. In: FMCAD (2007)
20. Zhang, L., Malik, S.: Validating sat solvers using an independent resolution-based checker: Practical implementations and other applications. In: DATE, pp. 10880–10885 (2003)

On Efficient Computation of Variable MUSes

Anton Belov[1], Alexander Ivrii[3], Arie Matsliah[3], and Joao Marques-Silva[1,2]

[1] CASL, University College Dublin, Ireland
[2] IST/INESC-ID, Lisbon, Portugal
[3] IBM Research – Haifa, Israel

Abstract. In this paper we address the following problem: given an unsatisfiable CNF formula $\mathcal{F}$, find a minimal subset of variables of $\mathcal{F}$ that constitutes the set of variables in *some* unsatisfiable core of $\mathcal{F}$. This problem, known as *variable MUS (VMUS)* computation problem, captures the need to reduce the number of *variables* that appear in unsatisfiable cores. Previous work on computation of VMUSes proposed a number of algorithms for solving the problem. However, the proposed algorithms lack all of the important optimization techniques that have been recently developed in the context of (clausal) MUS computation. We show that these optimization techniques can be adopted for VMUS computation problem and result in multiple orders magnitude speed-ups on industrial application benchmarks. In addition, we demonstrate that in practice VMUSes can often be computed faster than MUSes, even when state-of-the-art optimizations are used in both contexts.

1 Introduction

Concise descriptions of the sources of inconsistency in unsatisfiable CNF formulas have traditionally been associated with Minimally Unsatisfiable Subformulas (MUSes). An MUS of a CNF formula is an unsatisfiable subset of its clauses that is minimal in the sense that any of its proper subsets is satisfiable. Development of efficient algorithms for computation of MUSes is an active area of research motivated by many applications originating from industry [10,6,11,17,12]. The most recent generation of MUS extraction algorithms is capable of handling large industrial formulas efficiently.

Additional ways of capturing sources of inconsistency in CNF formulas have been proposed. For example, in [9,12] the inconsistency is analysed in terms of sets of clauses (the so called *groups* of clauses); efficient algorithms for the computation of *group-MUSes* have been developed in [12,16]. Sources of inconsistency can also be described in terms of the sets of the *variables* of the formula. One such description, the *variable-MUS (VMUS)*, has been proposed in [4] — a variable-MUS of an unsatisfiable CNF formula $\mathcal{F}$ is a subset V of its variables that constitutes the set of variables of some unsatisfiable subformula of $\mathcal{F}$ and is minimal in the sense that no proper subset of V has this property. While [4] does not develop any VMUS extraction algorithms, in [6] several such algorithms have been proposed, and their applications have been pointed out. However, the proposed algorithms lack all the optimization techniques parallel to

* This work is partially supported by SFI grant BEACON (09/IN.1/I2618), and by FCT grants ATTEST (CMU-PT/ELE/0009/2009) and POLARIS (PTDC/EIA-CCO/123051/2010).

A. Cimatti and R. Sebastiani (Eds.): SAT 2012, LNCS 7317, pp. 298–311, 2012.

those that have been recently developed in the context of (clausal) MUS computation (e.g. [10,2]) and are known to be essential for handling large industrial instances. This observation motivates the development of novel optimization techniques for VMUS computation algorithms.

Beside a pure scientific interest in the development of efficient algorithms for VMUS computation, this line of research is motivated by a number of possible industrially-relevant applications of VMUSes (e.g. [5]) and other related variable-based descriptions of inconsistency in CNF formulas. To this end, in this paper we make the following contributions. We formalize the VMUS computation problem and its extensions, and establish basic theoretical properties. We describe a number of optimization techniques to the basic VMUS computation algorithm presented in [6] and demonstrate empirically the multiple-order of magnitude improvements in the performance of the algorithm on the set of industrially-relevant benchmarks used for the evaluation of MUS extractors in SAT Competition 2011. We develop a relaxation-variable based constructive algorithm for VMUS extraction, based on the ideas proposed in [10]. We also describe a number of indirect approaches whereby the VMUS computation problem is translated to group-MUS computation problem, and evaluate these approaches empirically. Finally, we describe a number of potential industrial applications of VMUSes and its extensions.

2   Preliminaries

We focus on formulas in CNF (*formulas*, from hence on), which we treat as (finite) (multi-)sets of clauses. We assume that clauses do not contain duplicate variables.

Given a formula $\mathcal{F}$ we denote the set of variables that occur in $\mathcal{F}$ by $Var(\mathcal{F})$, and the set of variables that occur in a clause $C \in \mathcal{F}$ by $Var(C)$. An *assignment* τ for $\mathcal{F}$ is a map $\tau : Var(\mathcal{F}) \to \{0, 1\}$. By $\tau|_{\neg x}$ we denote the assignment $(\tau \setminus \{\langle x, \tau(x) \rangle\}) \cup \{\langle x, 1 - \tau(x) \rangle\}$. Assignments are extended to clauses and formulas according to the semantics of classical propositional logic. By $Unsat(\mathcal{F}, \tau)$ we denote the set of clauses of $\mathcal{F}$ falsified by τ. If $\tau(\mathcal{F}) = 1$, then τ is a *model* of $\mathcal{F}$. If a formula $\mathcal{F}$ has (resp. does not have) a model, then $\mathcal{F}$ is *satisfiable* (resp. *unsatisfiable*). By SAT (resp. UNSAT) we denote the set of all satisfiable (resp. unsatisfiable) CNF formulas.

A CNF formula $\mathcal{F}$ is *minimally unsatisfiable* if (i) $\mathcal{F} \in$ UNSAT, and (ii) for any clause $C \in \mathcal{F}$, $\mathcal{F} \setminus \{C\} \in$ SAT. We denote the set of minimally unsatisfiable CNF formulas by MU. A CNF formula $\mathcal{F}'$ is a *minimally unsatisfiable subformula (MUS)* of a formula $\mathcal{F}$ if $\mathcal{F}' \subseteq \mathcal{F}$ and $\mathcal{F}' \in$ MU. The set of MUSes of a CNF formula $\mathcal{F}$ is denoted by MUS($\mathcal{F}$). (In general, a given unsatisfiable formula $\mathcal{F}$ may have more than one MUS.)

A clause $C \in \mathcal{F}$ is *necessary* for $\mathcal{F}$ (cf. [8]) if $\mathcal{F} \in$ UNSAT and $\mathcal{F} \setminus \{C\} \in$ SAT. Necessary clauses are often referred to as *transition* clauses. The set of all necessary clauses of $\mathcal{F}$ is precisely $\bigcap$MUS($\mathcal{F}$). Thus $\mathcal{F} \in$ MU if and only if every clause of $\mathcal{F}$ is necessary. The problem of deciding whether a given CNF formula is in MU is DP-complete [13].

Motivated by several applications, minimal unsatisfiability and related concepts have been extended to CNF formulas where clauses are partitioned into disjoint sets called *groups* [9,12].

Definition 1 (Group-Oriented MUS). *Given an explicitly partitioned unsatisfiable CNF formula* $\mathcal{F} = \mathcal{G}_0 \cup \cdots \cup \mathcal{G}_n$, *a group oriented MUS (or, group-MUS) of* $\mathcal{F}$ *is a subset* $\mathcal{F}' = \mathcal{G}_0 \cup \mathcal{G}_{i_1} \cup \cdots \cup \mathcal{G}_{i_k}$ *of* $\mathcal{F}$ *such that* $\mathcal{F}'$ *is unsatisfiable and, for every* $1 \le j \le k$, $\mathcal{F}' \setminus \mathcal{G}_{i_j}$ *is satisfiable.*

3   Variable-MUS Computation Problem, and Generalizations

In this section we review the formal definition of VMUS computation problem and some of its basic properties, and generalize it in two ways (motivated by applications): the *interesting variables MUS computation problem (IVMUS)* and the *group-VMUS computation problem (GVMUS)*, the later being related to (clausal) group-MUS.

3.1   VMUS

Variable-MUSes of CNF formula $\mathcal{F}$ are defined in terms of subformulas *induced* by subsets of $Var(\mathcal{F})$[1].

Definition 2 (Induced subformula [4,6]). *Let* $\mathcal{F}$ *be a CNF formula, and let* V *be a subset of* $Var(\mathcal{F})$. *The subformula of* $\mathcal{F}$ *induced by* V *is the formula* $\mathcal{F}|_V = \{C \mid C \in \mathcal{F} \text{ and } Var(C) \subseteq V\}$.

Thus, $\mathcal{F}|_V$ is the set of *all* clauses of $\mathcal{F}$ that are defined *only* on variables from V. Alternatively, if we consider the variables in $Var(\mathcal{F}) \setminus V$ as *removed*, then $\mathcal{F}|_V$ is obtained from $\mathcal{F}$ by removing all clauses that contain at least one of the removed variables. Note that in general $Var(\mathcal{F}|_V)$ may be a strict subset of V — consider for example $\mathcal{F}|_{\{p,q\}} = \{(p)\}$ for $\mathcal{F} = \{(p \vee q \vee r), (p)\}$.

   Clearly $V_1 \subset V_2 \subseteq Var(\mathcal{F})$ implies $\mathcal{F}|_{V_1} \subseteq \mathcal{F}|_{V_2}$, and so variable minimal unsatisfiability can be well defined as follows.

Definition 3 (Variable minimally unsatisfiable formula, VMU [4]). *A CNF formula* $\mathcal{F}$ *is called* variable minimally unsatisfiable *if* $\mathcal{F} \in$ UNSAT *and for any* $V \subset Var(\mathcal{F})$, $\mathcal{F}|_V \in$ SAT. *The set of all such CNF formulas* $\mathcal{F}$ *is denoted by* VMU.

It is not difficult to see that MU $\subset$ VMU: clearly, every minimally unsatisfiable formula is variable minimally unsatisfiable, while, for example, the formula $\{(p), (\neg p \vee q), (\neg q), (p \vee \neg q)\}$ is in VMU, but not in MU. Nevertheless, as shown in [4], just like MU, the language VMU is DP-complete. The complexity of decision problems associated with various subclasses of VMU is also given in [4].

   A *variable-MUS*, or VMUS, of an unsatisfiable CNF formula $\mathcal{F}$, is defined by analogy with (clausal) MUS as follows.

Definition 4 (Variable-MUS, VMUS [6]). *Let* $\mathcal{F}$ *be unsatisfiable CNF formula. Then, a set* $V \subseteq Var(\mathcal{F})$ *is a* variable-MUS (VMUS) *of* $\mathcal{F}$ *if* $\mathcal{F}|_V \in$ VMU, *or, explicitly,* $\mathcal{F}|_V \in$ UNSAT, *and for any* $V' \subset V$, $\mathcal{F}|_{V'} \in$ SAT[2]. *The set of all VMUSes of* $\mathcal{F}$ *is denoted by* VMUS($\mathcal{F}$).

---

[1] This is the terminology used in [6]; in [4] these subformulas are called *projections* of $\mathcal{F}$.
[2] Note that, in general, for $\mathcal{F}' = \mathcal{F}|_V$ and $V' \subset V$, we have $\mathcal{F}'|_{V'} = \mathcal{F}|_{V'}$.

Thus, a VMUS V of an unsatisfiable formula $\mathcal{F}$ is a subset of $Var(\mathcal{F})$ that has exactly the property we are interested in: it is the set of variables of some unsatisfiable core $\mathcal{F}'$ of $\mathcal{F}$, and for no other unsatisfiable core $\mathcal{F}''$ of $\mathcal{F}$, $Var(\mathcal{F}'')$ is a strict subset of V.

Example 1. Let $\mathcal{F} = \{C_1, \ldots, C_5\}$, where $C_1 = (p)$, $C_2 = (q)$, $C_3 = (\neg p \vee \neg q)$, $C_4 = (\neg p \vee r)$, and $C_5 = (\neg q \vee \neg r)$. $\mathcal{F}$ is unsatisfiable, however is not variable minimally unsatisfiable, because for $V = \{p, q\}$, $\mathcal{F}|_V = \{C_1, C_2, C_3\} \in$ UNSAT. On the other hand, V is a VMUS of $\mathcal{F}$. Note that $\mathcal{F}' = \{C_1, C_2, C_4, C_5\}$ is a (clausal) MUS of F, however $Var(F') = \{p, q, r\}$ is *not* a VMUS of $\mathcal{F}$. (This example contradicts the claim from [6] that variables of any clausal MUS constitute a VMUS.)

We conclude this subsection with a definition of a *necessary variable*, analogous to that of a necessary clause.

Definition 5 (Necessary variable). *Let $\mathcal{F}$ be a CNF formula. A variable $v \in Var(\mathcal{F})$ is necessary for $\mathcal{F}$ if $\mathcal{F} \in$ UNSAT, and $\mathcal{F}|_{Var(\mathcal{F})\setminus\{v\}} \in$ SAT.*

Thus, $\mathcal{F} \in$ VMU if and only if every variable in $Var(\mathcal{F})$ is necessary for $\mathcal{F}$, and so V is a VMUS of $\mathcal{F}$, if every variable in V is necessary for $\mathcal{F}|_V$. Furthermore, the set of all necessary variables of $\mathcal{F}$ is precisely $\bigcap$VMUS$(\mathcal{F})$. The following proposition establishes the property of necessary variables required for ensuring the correctness of VMUS computation algorithms presented in this paper.

Proposition 1 (Monotonicity). *Let $\mathcal{F}$ be an unsatisfiable CNF formula. If $v \in Var(\mathcal{F})$ is necessary for $\mathcal{F}$, then it is also necessary for any unsatisfiable subset $\mathcal{F}'$ of $\mathcal{F}$.*

Finally, from the perspective of clausal MUSes of an unsatisfiable CNF formula $\mathcal{F}$ we have that $v \in Var(\mathcal{F})$ is necessary for $\mathcal{F}$ if and only if every MUS of $\mathcal{F}$ includes some clause containing v.

3.2 Generalizations

The generalizations of VMUS computation problem developed in this section are motivated by some of the applications, which we describe in Section 5. In the first generalization, the set $Var(\mathcal{F})$ is partitioned into the set of *interesting* variables I and the set of *uninteresting* variables U, that is $Var(\mathcal{F}) = I \uplus U$, and the VMUSes are computed in terms of interesting variables only.

Definition 6 (Interesting-VMUS, IVMUS). *Let $\mathcal{F}$ be an unsatisfiable CNF formula, and let I and U be a partition of $Var(\mathcal{F})$ into the sets of interesting and uninteresting variables, respectively. Then, the interesting-VMUS (IVMUS) of $\mathcal{F}$ is a set of variables $V \subseteq I$ such that $\mathcal{F}|_{V \cup U} \in$ UNSAT, and for any $V' \subset V$, $\mathcal{F}|_{V' \cup U} \in SAT$.*

Thus, the uninteresting variables play the role analogous to that of the clauses in group $\mathcal{G}_0$ in group-MUS computation problem, which represent the clauses outside of the interesting constraints [12].

Clearly, for a formula $\mathcal{F}$, VMUS computation problem is a special case of IVMUS computation problem when all variables are interesting, i.e. $I = Var(\mathcal{F})$. Note that, as opposed to VMUSes, IVMUSes can be empty even if $I \neq \emptyset$. Consider, for example, the formula $\mathcal{F} = \{(p), (\neg p), (p \vee q), (\neg p \vee q), (\neg q)\}$. If $I = \{q\}$ and $U = \{p\}$, then IVMUS of $\mathcal{F}$ is empty set because we can remove all clauses with q while preserving unsatisfiability. If, on the other hand, $I = \{p\}$ and $U = \{q\}$, then IVMUS of $\mathcal{F}$ is $\{p\}$. This is also the case when $I = \{p, q\}$ and $U = \emptyset$.

Further generalization of IVMUS (and of VMUS) computation problem can be obtained by partitioning the set of interesting variables into disjoint *groups* of variables, and then analyzing the minimal unsatisfiability of the formula in terms of these groups. This is the parallel of group-MUS computation problem in the clausal context.

Definition 7 (Group-VMUS, GVMUS). *Let $\mathcal{F}$ be an unsatisfiable CNF formula, and let $U, I_1, \ldots, I_n$ be a partition of $Var(\mathcal{F})$ into the set of uninteresting variables U and the sets I_i of interesting variables called* groups. *Then, the* group-VMUS (GVMUS) *of $\mathcal{F}$ is a set of groups $G \subseteq \{I_1, \ldots, I_n\}$, such that for $V = U \cup \bigcup_{I \in G} I$, $\mathcal{F}|_V \in$ UNSAT, and for any $G' \subset G$ and $V' = U \cup \bigcup_{I \in G'} I$, the formula $\mathcal{F}|_{V'}$ is in SAT.*

Thus, IVMUS computation problem is a special case of GVMUS computation problem where each group contains exactly one interesting variable.

4   Algorithms for VMUS Computation

The algorithms described in this section are based on iterative invocations of a SAT solver. Specifically, the function call $\text{SAT}(\mathcal{F})$ accepts a CNF formula $\mathcal{F}$ and returns a tuple $\langle \text{st}, \tau, \mathcal{U} \rangle$ with the following semantics: if $\mathcal{F} \in$ SAT, then st is set to *true* and τ is a model of $\mathcal{F}$; otherwise $\text{st} = false$ and $\mathcal{U} \subseteq \mathcal{F}$ is an unsatisfiable core of $\mathcal{F}$[3].

We introduce an additional notation. Given a CNF formula $\mathcal{F}$ and a variable $v \in Var(\mathcal{F})$, by $\mathcal{F}^v = \{C \mid C \in \mathcal{F} \text{ and } v \in Var(C)\}$ we denote the set of clauses of $\mathcal{F}$ that contain v. Note that $\mathcal{F}|_{Var(\mathcal{F})\setminus\{v\}} = \mathcal{F} \setminus \mathcal{F}^v$.

4.1   Hybrid VMUS Computation

Without the optimizations on lines RR, REF and VMR, Algorithm 1 (VHYB) represents a basic destructive algorithm for VMUS computation, similarly to the algorithm Removal proposed in [6]. Note that it also closely mimics the organization of a *hybrid* algorithm for MUS computation (cf. [11]).

The algorithm accepts an unsatisfiable CNF formula $\mathcal{F}$ as input, and maintains three datastructures: the set of necessary variables V (initially empty), the *working set of variables* V_w (initialized to $Var(\mathcal{F})$), and the *working formula* $\mathcal{F}_w$ (initialized to $\mathcal{F}$).

On each iteration of the while-loop (line 4) the algorithm removes a variable v from V_w and tests whether v is necessary for $\mathcal{F}_w$ (see Definition 5). This test is performed by invoking a SAT solver on the formula $\mathcal{F}_w \setminus \mathcal{F}_w^v$ (line 7). If the formula is unsatisfiable, v is not necessary, and the clauses of $\mathcal{F}_w^v$ are removed from $\mathcal{F}_w$ (line 9). Otherwise, v is necessary, and it is added to V.

---

[3] Some of the modern SAT solvers have the capability of producing unsatisfiable cores (although not necessarily minimal); for SAT solvers without this capability we can set $\mathcal{U} = \mathcal{F}$.

Algorithm 1. VHYB($\mathcal{F}$) — Hybrid VMUS Computation

Input : Unsatisfiable CNF Formula $\mathcal{F}$
Output: VMUS V of $\mathcal{F}$
begin

1	$V \leftarrow \emptyset$	// VMUS under-approximation
2	$V_w \leftarrow Var(\mathcal{F})$	// Working set of variables
3	$\mathcal{F}_w \leftarrow \mathcal{F}$	// Working formula
4	**while** $V_w \neq \emptyset$ **do** // Inv: $\mathcal{F}_w = \mathcal{F}|_{V \cup V_w}$ and $\forall v \in V$ is nec. for $\mathcal{F}_w$	
5	$\quad v \leftarrow$ PickVariable(V_w)	
6	$\quad V_w \leftarrow V_w \setminus \{v\}$	
7	$\quad (\mathsf{st}, \tau, \mathcal{U}) = \text{SAT}(\mathcal{F}_w \setminus \mathcal{F}_w^v)$	
RR	$\quad \mathcal{R} \leftarrow \text{CNF}(\neg \mathcal{F}_w^v)$	// Redundancy removal
RR	$\quad (\mathsf{st}, \tau, \mathcal{U}) = \text{SAT}((\mathcal{F}_w \setminus \mathcal{F}_w^v) \cup \mathcal{R})$	
8	$\quad$ **if** st = false **then**	// v is not necessary for $\mathcal{F}_w$
REF	$\quad\quad$ **if** $\mathcal{U} \cap \mathcal{R} = \emptyset$ **then** $V_w \leftarrow V_w \cap Var(\mathcal{U})$;	// Refinement
9	$\quad\quad \mathcal{F}_w \leftarrow \mathcal{F}_w|_{V \cup V_w}$	
10	$\quad$ **else**	// v is necessary for $\mathcal{F}_w$
11	$\quad\quad V \leftarrow V \cup \{v\}$	
VMR	$\quad\quad$ VModelRotation($\mathcal{F}_w, V, \tau$)	// $v \in V$ after this call
VMR	$\quad\quad V_w \leftarrow V_w \setminus V$	
12	**return** V	// $V \in$ VMUS($\mathcal{F}$) and $\mathcal{F}_w = \mathcal{F}|_V \in$ VMU

end

The correctness of the algorithm follows from the the loop invariant presented on line 4. This invariant is trivially satisfied prior to any iteration of the loop, and its inductiveness can be easily established from the pseudocode. Since on every iteration one variable is removed from V_w, the loop eventually terminates, and on termination it holds that $\mathcal{F}_w = \mathcal{F}|_V$, and that every variable in V is necessary for $\mathcal{F}_w$. Hence, $\mathcal{F}|_V \in$ VMU and $V \in$ VMUS($\mathcal{F}$).

It comes as no surprise that the basic algorithm described above is not efficient for large CNF formulas — on every iteration exactly one variable is removed from V_w, and so the algorithm makes exactly $Var(\mathcal{F})$ SAT solver calls. This lack of scalability is demonstrated clearly in our experimental evaluation, presented in Section 6. In the context of MUS extraction a number of crucial optimization techniques have been proposed — these include clause-set refinement [10] to remove multiple unnecessary clauses in a single SAT solver call, recursive model rotation [1] to detect multiple necessary clauses in a single SAT solver call, and redundancy removal [17,10] to make SAT instances easier to solve. We now describe the way these techniques can be adopted in the setting of VMUS computation.

Redundancy Removal. The idea behind the *redundancy removal* technique is to add certain constraints to the formula $\mathcal{F}_w \setminus \mathcal{F}_w^v$ prior to the invocation of a SAT solver on line 7. These additional constraints are taken to be the clauses of the Plaisted-Greenbaum CNF transformation [14] of the propositional formula $\neg \mathcal{F}_w^v$, denoted by $CNF(\neg \mathcal{F}_w^v)$. These clauses are constructed as follows: for each $C \in \mathcal{F}_w^v$, create an auxiliary variable a_C, and the binary clauses $(\neg a_C \vee \neg l)$, for each literal $l \in C$; then add a clause $(\bigvee_{C \in \mathcal{F}_w^v} a_C)$. The correctness of the redundancy removal technique is guaranteed by the following proposition.

Proposition 2. *Let $\mathcal{F}$ be an unsatisfiable CNF formula. Then for any $v \in Var(\mathcal{F})$, $\mathcal{F} \setminus \mathcal{F}^v \in$ SAT if and only if $(\mathcal{F} \setminus \mathcal{F}^v) \cup CNF(\neg \mathcal{F}^v) \in$ SAT.*

Proof. Let τ be any model of $\mathcal{F} \setminus \mathcal{F}^v$. Since $\mathcal{F} \in$ UNSAT, for any $\tau' \supset \tau$ that extends τ to $Var(\mathcal{F}^v) \setminus Var(\mathcal{F})$, we have $\tau'(\mathcal{F}^v) = 0$. Thus, for some extension τ'' of τ', $\tau''(CNF(\neg \mathcal{F}^v)) = 1$ [4]. Therefore τ'' is a model of $(\mathcal{F} \setminus \mathcal{F}^v) \cup CNF(\neg \mathcal{F}^v)$. The opposite direction holds trivially. □

The technique is integrated into VHYB by replacing line 7 with the two lines labeled RR. Even though the additional constraints imposed by $CNF(\neg \mathcal{F}^v)$ are redundant, they can help the SAT solver to prune the search space more efficiently, and in fact our experiments show that in practice this method leads to an improved performance.

Variable-Set Refinement. *Variable-set refinement* is a technique for detection of unnecessary variables that takes advantage of the capability of modern SAT solvers to produce unsatisfiable cores. The technique is based on the following observation.

Proposition 3 (Variable-Set Refinement). *Let $\mathcal{U}$ be an unsatisfiable core of $\mathcal{F} \in$ UNSAT. Then, there exists $V \subseteq Var(\mathcal{U})$ such that $V \in$ VMUS$(\mathcal{F})$.*

Proof. Take V to be any VMUS of the formula $\mathcal{F}|_{Var(\mathcal{U})}$. □

When the outcome of SAT solver call is UNSAT (line 8), the unsatisfiable core $\mathcal{U}$ of the formula $(\mathcal{F}_w \setminus \mathcal{F}_w^v) \cup \mathcal{R}$, where $\mathcal{R} = CNF(\neg \mathcal{F}_w^v)$, can be used in the following way: if $\mathcal{U}$ does not include any of the clauses of $\mathcal{R}$, then any variable of the working set V_w outside of the set $Var(\mathcal{U})$ is not necessary for $\mathcal{F}_w$, and thus can be removed from V_w (line REF). This is because the condition $\mathcal{U} \cap \mathcal{R} = \emptyset$ guarantees that $\mathcal{U}$ is an unsatisfiable core of the formula $(\mathcal{F}_w \setminus \mathcal{F}_w^v)$, and so Proposition 3 applies. Note that since the set V contains variables that are necessary for $\mathcal{F}_w$, it must be that $V \subseteq Var(\mathcal{U})$. If the unsatisfiable core $\mathcal{U}$ does include some of the clauses of $\mathcal{R}$, the formula $\mathcal{U} \setminus \mathcal{R}$ could be satisfiable, and so some of the variables necessary for $\mathcal{F}_w$ might be outside $Var(\mathcal{U})$. As an example, consider $\mathcal{F} = \{(p \vee q), (p \vee \neg q), (\neg p \vee r), (\neg p \vee \neg r), (\neg p \vee s), (\neg p \vee \neg s)\}$. Suppose we try to remove the variable r first. $\mathcal{F}^r = \{(\neg p \vee r), (\neg p \vee \neg r)\}$, and so $\mathcal{F} \setminus \mathcal{F}^r \in$ UNSAT. As the clauses of $\mathcal{R} = CNF(\neg \mathcal{F}^r)$ imply the unit clause (p), the returned unsatisfiable core might consist only of the clauses $\{(\neg p \vee s), (\neg p \vee \neg s)\}$ together with the clauses of $\mathcal{R}$. Note that the variable q is necessary for $\mathcal{F}$ but not included in this core.

It should be noted that, just as in the case of MUS extraction, variable-set refinement is a *crucial* optimization technique that leads to dramatic (multiple orders of magnitude) speed-ups in VMUS computation — see Section 6 for the empirical data.

---

[4] τ'' extends τ' by assigning values to the auxiliary variables introduced by Plaisted-Greenbaum transform.

Algorithm 2. VModelRotation($\mathcal{F}, V, \tau$) — Variable-based Model Rotation

---

Input: $\mathcal{F}$ — an unsatisfiable CNF formula
: V — a set of necessary variables $\mathcal{F}$
: τ — a model of $\mathcal{F} \setminus \mathcal{F}^v$ for some $v \in Var(\mathcal{F})$
Effect: V contains v and possibly additional necessary variables of $\mathcal{F}$

```
1  begin
2  |   V' ← ∩_{C∈Unsat(F,τ)} Var(C)        // all common variables;  v ∈ V'
3  |   foreach v ∈ V' do
4  |   |   if v ∉ V then
5  |   |   |   V ← V ∪ {v}                 // v is a new necessary variable
6  |   |   |   τ' ← τ|_{¬v}
7  |   |   └   VModelRotation(F, V, τ')
8  end
```

Variable-Based Model Rotation (VMR). *Variable-based model rotation (VMR)* is a technique for detection of multiple necessary variables in a single SAT call. The technique makes use of the satisfying assignment τ returned by the SAT solver on line RR of Algorithm 1 (or line 7 if redundancy removal is not used). The technique uses the following property.

Proposition 4. *Let $\mathcal{F}$ be an unsatisfiable CNF formula, let τ be an assignment to $Var(\mathcal{F})$, and let $V = \bigcap_{C\in Unsat(\mathcal{F},\tau)} Var(C)$. Then any variable in V is necessary for $\mathcal{F}$.*

Proof. Take any $v \in V$. Clearly $Unsat(\mathcal{F}, \tau) \subseteq \mathcal{F}^v$, and so $\mathcal{F}|_{Var(\mathcal{F})\setminus\{v\}} = \mathcal{F}\setminus\mathcal{F}^v \subseteq \mathcal{F} \setminus Unsat(\mathcal{F}, \tau)$. Since $\mathcal{F} \setminus Unsat(\mathcal{F}, \tau) \in$ SAT, we have $\mathcal{F}|_{Var(\mathcal{F})\setminus\{v\}} \in$ SAT □

An assignment τ is called a *witness for necessity of a variable v in $\mathcal{F}$* if satisfies the condition $v \in \bigcap_{C\in Unsat(\mathcal{F},\tau)} Var(C)$. Note that in the context of Algorithm 1 the assignment τ returned by the SAT solver in the case the formula $(\mathcal{F}_w \setminus \mathcal{F}_w^v) \cup \mathcal{R}$ is satisfiable is exactly a witness of necessity of v in $\mathcal{F}_w$ — it must be that $Unsat(\mathcal{F}_w, \tau) \subseteq \mathcal{F}_w^v$ and so v appears in all clauses of $Unsat(\mathcal{F}_w, \tau)$ (and, in fact, in the same polarity). Furthermore, it is possible that there are other variables shared among the clauses of $Unsat(\mathcal{F}_w, \tau)$. These variables can be immediately declared as necessary for $\mathcal{F}_w$, *without additional SAT calls*. A special case of particular practical importance is when $Unsat(\mathcal{F}_w, \tau)$ consists of a single clause — in this case all of the other variables of this clause are necessary as well.

Next, suppose that τ is a witness for (the necessity of) v. Note that the assignment $\tau' = \tau|_{\neg v}$, obtained by flipping the value of v in τ, is *also* a witness for v — indeed all the clauses in $Unsat(\mathcal{F}_w, \tau)$ are satisfied by τ', and all the clauses falsified by τ' must share the variable v (in the opposite polarity). Thus τ' can be analyzed in the same manner as τ — if there are any variables beside v that are shared among the clauses of $Unsat(\mathcal{F}_w, \tau')$, then these variables are necessary for $\mathcal{F}_w$ and τ' is the witness for each of these variables. This leads to the recursive process of detection of necessary variables and construction of witnesses, which we refer to as *variable-based model rotation (VMR)*.

The algorithm for VMR is presented in Algorithm 2. VMR is integrated into Algorithm 1 by replacing line 11 with the lines labelled VMR. Note that the necessary variable v detected by the SAT call in Algorithm 1 is always added to the set V as a result of VMR. The purpose of the if-statement on line 4 of Algorithm 2 is to prevent the algorithm from re-detecting variables that are already known to be necessary. In turn, this bounds the number of recursive calls to VMR during the execution of VMUS computation algorithm – this bound is twice the number of variables in the computed VMUS. Thus VMR is a light-weight technique for detection of necessary variables. In practice, however, the technique is extremely effective — as demonstrated in Section 6, VMR results in very significant performance improvements (up to the factor of 20).

The basic VMR algorithm described above allows for many various modifications. For example, one can allow the algorithm to re-visit variables already known to be necessary — it is very likely that a different initial model will lead to discovering a different set of new necessary variables. However, it is important to bound the total running time of the algorithm (and in particular to avoid loops). One specific modification of this type, which we refer to as *extended* VMR (EVMR), takes advantage of the diversity of the initial models passed to Algorithm 2 over the life-time of VMUS computation. In this variant the set V in Algorithm 2 should be thought of as the set of variables found necessary in the current invocation of VMR from Algorithm 1 (in other words, Algorithm 1 always calls Algorithm 2 with $V = \emptyset$, and at the end updates its own set V of all necessary variables to include the newly discovered variables). Although in the worst case the number of recursive calls to VMR can grow quadratically, in our experiments such growth was not observed. On the other hand, the number of necessary variables detected by EVMR was on average 5% higher than that of VMR alone (recall that every additional variable saves a potentially expensive SAT call). Overall, the EVMR modification has a positive impact on the performance of Algorithm 1.

Extensions. Algorithm 1 can be extended to handle the generalizations to IVMUS or to GVMUS introduced in section 3. These extensions, with the exception of VMR in GVMUS setting, are rather straightforward, and we do not describe them explicitly.

4.2   Relaxation-Variables Approach

A constructive MUS extraction algorithm based on relaxing clauses and AtMost1 constraint has been proposed in [10]. The idea is to augment each clause C_i of the unsatisfiable input formula $\mathcal{F}$ with a fresh *relaxation* variable a_i, thus replacing C_i with $(a_i \vee C_i)$ in $\mathcal{F}$. Furthermore, the CNF representation of the constraint $\sum a_i \leq 1$ is added to the modified formula. The resulting formula $\mathcal{F}'$ is then checked for satisfiability. If $\mathcal{F}'$ is satisfiable, then since the original formula $\mathcal{F}$ is unsatisfiable, exactly one of the variables a_i must be set to true — the associated clause C_i is then necessary for $\mathcal{F}$. The algorithm is quite special in that it essentially offloads the task of *searching* for a necessary clause to the SAT solver.

This algorithm can be extended to VMUS in the following way. The first step consists of relaxing variables instead of clauses. Let v be a variable of $\mathcal{F}$. The positive literals of v are replaced with a new variable v_p, and the negative literals of v are replaced with a new variable v_n. If both v_n and v_p are assigned value 1, then variable v

is said to be *relaxed*. To control the relaxation of v, another variable v_r is used, and the constraint $(\neg v_p \vee \neg v_n \vee v_r)$ is added to the formula. Note that when v_r is set to 1, v_p and v_n can be assigned by the SAT solver to 1 freely, effectively removing the clauses of $\mathcal{F}^v$. Otherwise, the constraint represents the relationship between v_p and v_n. The new variables are represented by sets V_P, V_N and V_R, where V denotes the initial set of variables. Now, a variable is necessary for $\mathcal{F}$ if by relaxing (or removing) it the formula becomes satisfiable — similar to the clausal version of the algorithm, this condition achieved with the constraint $\sum_{v_r \in V_R} v_r = 1$.

4.3   Reduction to Group-MUS Computation

An alternative approach to handle the VMUS computation problem is to provide a reduction to a MUS (or to a group-MUS) computation — especially in the light of the extensive amount of tools capable solving the latter problem efficiently. We sketch here two possible reductions, one suited for dense formulas (with #vars $\ll$ #clauses), and the other for sparse formulas (with #vars $\approx$ #clauses).

Reduction from VMUS to Group-MUS (for Dense Formulas). Let $\mathcal{F}$ be a CNF formula. We translate it to $\mathcal{F}'$ as follows. For each variable $v_i \in Vars(\mathcal{F})$ introduce two new variables p_i and n_i in $\mathcal{F}'$, and translate all clauses $C \in \mathcal{F}$ to clauses $C' \in \mathcal{F}'$ over variables $\{p_i, n_i\}$ by replacing every positive occurrence of v_i with p_i, and every negative occurrence v_i with n_i. As an example, the clause $(v_1 \vee \neg v_3 \vee v_4)$ gets translated to the clause $(p_1 \vee n_3 \vee p_4)$. Additionally, add $|Vars(\mathcal{F})|$ pairs of clauses $\{(p_i, n_i), (\neg p_i, \neg n_i)\}$ to $\mathcal{F}'$ (forcing the positive and negative representatives to have opposite values).

Define $\mathcal{G}_0 = \{C'\}$ to be the group consisting of all of the translated clauses, and for each i define the group $\mathcal{G}_i = \{(p_i, n_i), (\neg p_i, \neg n_i)\}$.

Proposition 5. *Any group-MUS S of $\mathcal{F}'$ (w.r.t. groups $\mathcal{G}_i$) corresponds to some VMUS V of the original formula $\mathcal{F}$, where $v_i \in V$ iff $\mathcal{G}_i \in S$.*

(Proof follows by construction.) It is also straightforward to extend this reduction to be applicable for IVMUS and GVMUS problems.

Reduction from VMUS to Group-MUS (for Sparse Formulas). Let $\mathcal{F}$ be a CNF formula. We translate it to $\mathcal{F}'$ as follows. For each variable $v_i \in Vars(\mathcal{F})$ introduce an activation variable a_i, and translate all clauses $C \in \mathcal{F}$ to clauses $C' \in \mathcal{F}'$ that contain the same set of literals plus their corresponding activation literals (namely C' is double in size). As an example, the clause $(v_1 \vee \neg v_3 \vee v_4)$ gets translated to the clause $(v_1 \vee a_1 \vee \neg v3 \vee a_3 \vee v_4 \vee a_4)$. Additionally, add $|Vars(\mathcal{F})|$ unit clauses $(\neg a_i)$ to $\mathcal{F}'$.

Define $\mathcal{G}_0 = \{C'\}$ to be the group consisting of all of the translated clauses, and for each i define the group $\mathcal{G}_i = \{(\neg a_i)\}$.

Proposition 6. *Any group-MUS S of $\mathcal{F}'$ (w.r.t. groups $\mathcal{G}_i$) corresponds to some VMUS V of the original formula $\mathcal{F}$, where $v_i \in V$ iff $\mathcal{G}_i \in S$.*

Here too, proof follows by construction, and the reduction is easily extended to IVMUS and GVMUS problems.

5   Applications of VMUSes

A number of applications of VMUSes have been proposed in the previous work. In [6] the authors use VMUSes to search for vertex critical subgraphs in the context of graph coloring problem. There the variables of the CNF representation of a graph correspond to its vertices, and thus a removal of a variable represents the operation of the removal of a vertex from the graph. The authors of [5] propose to use VMUSes for computing abstractions in the context of the abstraction-refinement approach to model checking. We describe a possible application of GVMUSes in the similar context below. We also describe a possible application of IVMUSes to minimization of satisfying assignments.

Minimizing Satisfying Assignments. Many formal verification techniques require the ability to efficiently generalize bad or interesting assignments to inputs of a circuit — the assignments which by propagation induce a value of 1 on one of the circuit's outputs. Similarly to [15], the problem is formalized as follows. Let F be a CNF with variables $Var(F) = J \uplus W \uplus \{o\}$ separated into a set of input variables J, a set of auxiliary variables W, and the property output o. We require that F satisfies the following property: for any satisfying assignment A to $F \wedge o$, the formula $A_J \wedge F \wedge \neg o$ is unsatisfiable, where A_J denotes the restriction of A to J. In this setting the set of "important" variables I is a subset of J and the set of of "unimportant" variables $U = (J \setminus I) \cup W \cup \{o\}$ consists of the remaining variables. Given such a formula F and an assignment A to $F \wedge o$, the goal is to find a minimal subset V of I which is still sufficient to force the value of 1 on the output, i.e. that the formula $A_V \wedge A_{J \setminus I} \wedge F \wedge \neg o$ remains unsatisfiable. This is precisely the IVMUS problem defined in Section 3.

Proof-Based Abstraction. A popular technique to reduce the size of models in model checking is to create a quality over-approximation of the design under verification. In the abstraction refinement approach one usually unrolls the design for a certain number k of time-frames and runs a SAT solver with the property that there is no erroneous execution within this bound. If the SAT solver returns UNSAT, the unsatisfiable core of the problem is analyzed. In the latch-based abstraction (cf [7]) any latch that does not appear in the core is abstracted away, and it is usually beneficial to remove as many latches as computationally feasible. Though MUS algorithms have been developed for this problem (e.g. [12]), a translation to GVMUS problem is also possible: for every latch L in the design, create a group $\mathcal{G}_L$ of CNF variables corresponding to L in every time-frame of the unrolled design; all the remaining CNF variables are "noninteresting" and put into $\mathcal{G}_0$. The minimization of the set of "interesting" variable groups while preserving the unsatisfiability is precisely the GVMUS problem (cf. Section 3).

Remark 1. An extra care should be taken when reducing problems to VMUS (or generalizations) in applications where CNF formulas arise from encoding of circuits. Certain variables should be split into two copies — a variable x is split into x_i and x_o if it satisfies the following conditions: (i) x represents an output of a gate/latch in the circuit, and (ii) the fan-out of the corresponding gate/latch is greater than 1. For every such x, the clauses that encode $x_i = x_o$ are added to the formula encoding the circuit, and the representative variable for the gate/latch is taken to be x_i.

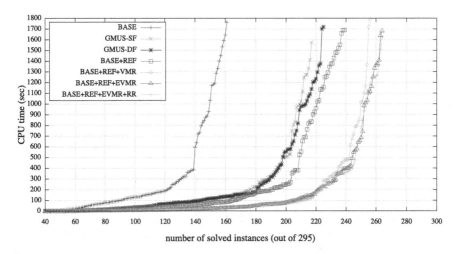

Fig. 1. Cactus plot: BASE with various optimizations; translations to GMUS

6   Experimental Study

We implemented the VMUS extraction algorithm VHYB (Algorithm 1) and all of the optimization techniques described in Section 4 on top of the MUS extraction framework of our state-of-the-art (group-)MUS extractor MUSer2[5]. In addition, we implemented the translations to GMUS described in Section 4.3. We did not implement the relaxation-variable based algorithm from Section 4.2, since, in the context of (clausal) MUS extraction, this approach is notably less effective than the hybrid approach (cf. [10]).

To evaluate the effectiveness of the proposed techniques we performed a comprehensive experimental study on 295 benchmarks used in the MUS track of SAT Competition 2011 [6]. These benchmark instances originate from various industrial applications of SAT, including hardware bounded model checking, FPGA routing, hardware and software verification, equivalence checking, abstraction refinement, design debugging, functional decomposition and bioinformatics. The experiments were performed on an HPC cluster, where each node is dual quad-core Intel Xeon E5450 3 GHz with 32 GB of memory. The timeout was set to 1800 seconds, and memory was limited to 4 GB. In our experiments, MUSer2, and its VMUS-oriented version, was configured to use picosat-935 [3] in incremental mode as a SAT solver.

The cactus plot in Fig. 1 provides an overview of the results of our experimental study[7]. The legend in this, and subsequent, plots is as follows: BASE represents the implementation of VHYB without any of the optimizations (i.e. this is the equivalent of the Removal algorithm from [6]); +REF represents the addition of variable-set refinement; +VMR (resp. +EVMR) represents the addition of VMR (resp. EVMR); +RR represents the addition of redundancy removal; GMUS-SF (resp. GMUS-DF) represents the results

---

[5] http://logos.ucd.ie/wiki/doku.php?id=muser

[6] http://www.satcompetition.org/. Note that the website lists 300 benchmarks in MUS category — this discrepancy is due to duplicate instances.

[7] Data is available at http://logos.ucd.ie/paper-results/sat12-vmus

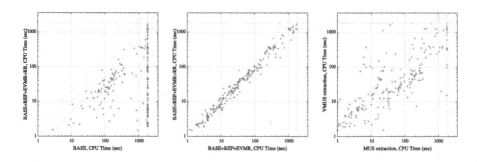

Fig. 2. Selected scatter plots (timeout 1800 sec.)

of the group-MUS version of MUSer2 on the GMUS translation for sparse (resp. dense) formulas. A number of observations can be made. We note that the addition of variable-set refinement has a dramatic effect on the performance of VHYB allowing to solve 80 more instances than BASE within the timeout. The addition of VMR to the variable-set refinement results in another huge leap in the performance of VHYB, demonstrating the positive impact of the extended version of VMR (EVMR). We also conclude that the proposed translations of VMUS computation problem to GMUS computation does not allow to extract VMUSes efficiently, despite the fact that the group-MUS extractor MUSer2 used in these experiments implements the group-MUS analog of the all crucial optimizations described in this paper — the exceptions are the *extended* model rotation and redundancy removal.

The scatter plots in Fig. 2 present additional views on the results of our experimental evaluation. The plot on the left-hand side demonstrates the combined effect of all the optimization techniques described in this paper to the VMUS extraction algorithm of [6]. Note the positive impact on the effectiveness of the algorithm, resulting in up to 3 orders of magnitude improvement in runtime of VMUS computation. The plot in the center of Fig. 2 allows to take a closer look at the effects of redundancy removal. While the results are mixed, a careful examination of the plot reveals an overall positive impact (close to 15%) of the technique on the runtime of HYB. The plot on the right-hand side of Fig. 2 deserves special attention. The plot compares the runtimes of VMUS extraction with the runtime of MUS extraction on the same set of instances. MUS extraction was performed with MUSer2, and we employed all the relevant optimizations, including the extended version of recursive model rotation [1]. We observe that in many cases it is cheaper to compute a VMUS of an instance, rather than the (clausal) MUS. Note that in the presence of the current MUS optimization techniques, such as clause set refinement and recursive model rotation, this observation is not trivial, and suggests that some of the current applications of MUS extraction algorithms could be reconsidered with VMUSes in mind.

7   Conclusion

In this paper we re-visit the VMUS computation problem and propose a number of its extensions. We develop a state-of-the-art VMUS extraction algorithm VHYB by intro-ducing a number of optimization techniques to the basic VMUS extraction algorithm proposed in [6]. In addition, we present a relaxation-variable based constructive algorithm for VMUS extraction, and a number of translations of VMUS computation problem to group-MUS computation problem. We demonstrate empirically that the optimization techniques employed in VHYB lead to significant improvements in the runtime of VMUS computation on a set of industrially-relevant benchmarks. We also show that the indirect approach via group-MUS computation can be significantly less efficient than VHYB. Finally, we demonstrate that in many cases computation of VMUSes is cheaper than the computation of MUSes. This observation motivates re-evaluation of the current applications of MUSes, further investigation of the applications of VMUSes and its extensions, and the development of relevant algorithms.

Acknowledgements. We thank the anonymous referees for insightful comments.

References

1. Belov, A., Marques-Silva, J.: Accelerating MUS extraction with recursive model rotation. In: Formal Methods in Computer-Aided Design (2011)
2. Belov, A., Marques-Silva, J.: Minimally Unsatisfiable Boolean Circuits. In: Sakallah, K.A., Simon, L. (eds.) SAT 2011. LNCS, vol. 6695, pp. 145–158. Springer, Heidelberg (2011)
3. Biere, A.: Picosat essentials. Journal on Satisfiability, Boolean Modeling and Computation 4, 75–97 (2008)
4. Chen, Z., Ding, D.: Variable Minimal Unsatisfiability. In: Cai, J.-Y., Cooper, S.B., Li, A. (eds.) TAMC 2006. LNCS, vol. 3959, pp. 262–273. Springer, Heidelberg (2006)
5. Chen, Z.-Y., Tao, Z.-H., Büning, H.K., Wang, L.-F.: Applying variable minimal unsatisfiability in model checking. Journal of Software 19(1), 39–47 (2008)
6. Desrosiers, C., Galinier, P., Hertz, A., Paroz, S.: Using heuristics to find minimal unsatisfiable subformulas in satisfiability problems. J. Comb. Optim. 18(2), 124–150 (2009)
7. Gupta, A., Ganai, M.K., Yang, Z., Ashar, P.: Iterative abstraction using sat-based bmc with proof analysis. In: ICCAD, pp. 416–423 (2003)
8. Kullmann, O., Lynce, I., Marques-Silva, J.: Categorisation of Clauses in Conjunctive Normal Forms: Minimally Unsatisfiable Sub-clause-sets and the Lean Kernel. In: Biere, A., Gomes, C.P. (eds.) SAT 2006. LNCS, vol. 4121, pp. 22–35. Springer, Heidelberg (2006)
9. Liffiton, M.H., Sakallah, K.A.: Algorithms for computing minimal unsatisfiable subsets of constraints. J. Autom. Reasoning 40(1), 1–33 (2008)
10. Marques-Silva, J., Lynce, I.: On Improving MUS Extraction Algorithms. In: Sakallah, K.A., Simon, L. (eds.) SAT 2011. LNCS, vol. 6695, pp. 159–173. Springer, Heidelberg (2011)
11. Marques-Silva, J.: Minimal unsatisfiability: Models, algorithms and applications. In: International Symposium on Multiple-Valued Logic, pp. 9–14 (2010)
12. Nadel, A.: Boosting minimal unsatisfiable core extraction. In: Formal Methods in Computer-Aided Design (October 2010)
13. Papadimitriou, C.H., Wolfe, D.: The complexity of facets resolved. J. Comput. Syst. Sci. 37(1), 2–13 (1988)
14. Plaisted, D.A., Greenbaum, S.: A structure-preserving clause form translation. Journal of Symbolic Computation 2(3), 293–304 (1986)
15. Ravi, K., Somenzi, F.: Minimal Assignments for Bounded Model Checking. In: Jensen, K., Podelski, A. (eds.) TACAS 2004. LNCS, vol. 2988, pp. 31–45. Springer, Heidelberg (2004)
16. Ryvchin, V., Strichman, O.: Faster Extraction of High-Level Minimal Unsatisfiable Cores. In: Sakallah, K.A., Simon, L. (eds.) SAT 2011. LNCS, vol. 6695, pp. 174–187. Springer, Heidelberg (2011)
17. van Maaren, H., Wieringa, S.: Finding Guaranteed MUSes Fast. In: Kleine Büning, H., Zhao, X. (eds.) SAT 2008. LNCS, vol. 4996, pp. 291–304. Springer, Heidelberg (2008)

Interpolant Strength Revisited

Georg Weissenbacher[1,2]

[1] Princeton University
[2] Vienna University of Technology, Austria

Craig's interpolation theorem has numerous applications in model checking, automated reasoning, and synthesis. There is a variety of interpolation systems which derive interpolants from refutation proofs; these systems are ad-hoc and rigid in the sense that they provide exactly one interpolant for a given proof. In previous work, we introduced a parametrised interpolation system which subsumes existing interpolation methods for propositional resolution proofs and enables the systematic variation of the logical strength and the elimination of non-essential variables in interpolants. In this paper, we generalise this system to propositional hyper-resolution proofs and discuss its application to proofs generated by contemporary SAT solvers. Finally, we show that, when applied to local (or split) proofs, our extension generalises two existing interpolation systems for first-order logic and relates them in logical strength.

1 Introduction

Craig interpolation [5] has proven to be an effective heuristic in applications such as model checking, where it is used as an approximate method for computing invariants of transition systems [18], and synthesis, where interpolants represent deterministic implementations of specifications given as relations [14]. The intrinsic properties of interpolants enable concise abstractions in verification and smaller circuits in synthesis. Intuitively, stronger interpolants provide more precision, and interpolants with fewer variables lead to smaller designs. However, interpolation is mostly treated as a black box, leaving no room for a systematic exploration of the solution space. In addition, the use of different interpolation systems complicates a comparison of their interpolants. We present a novel framework which generalises a number of existing interpolation techniques and supports a systematic variation and comparison of the generated interpolants.

Contributions. We present a novel *parametrised* interpolation system which extends our previous work on propositional interpolation [7].
- The extended system supports hyper-resolution (see § 3) and allows for systematic variation of the logical strength (with an additional degree of freedom over [7]) and the elimination of non-essential literals [6] in interpolants.
- We discuss (in § 4) the application of our interpolation system to hyper-resolution steps (introduced by pre-processing [10], for instance) and refutations generated by contemporary SAT solvers such as MiniSAT [8].
- When applied to local (or split) proofs [13], the extended interpolation system generalises the existing interpolation systems for first-order logic presented in [15] and [25] and relates them in logical strength (§ 5).

A. Cimatti and R. Sebastiani (Eds.): SAT 2012, LNCS 7317, pp. 312–326, 2012.
© Springer-Verlag Berlin Heidelberg 2012

2   Background

This section introduces our notation (§ 2.1) and restates the main results of our previous paper on labelled interpolation systems [7] in § 2.2.

2.1   Formulae and Proofs

In our setting, the term *formula* refers to either a propositional logic formula or a formula in standard first-order logic.

Propositional Formulae. We work in the standard setting of propositional logic over a set X of propositional variables, the logical constants T and F (denoting true and false, respectively), and the standard logical connectives $\wedge$, $\vee$, $\Rightarrow$, and $\neg$ (denoting conjunction, disjunction, implication, and negation, respectively).

Moreover, let $\mathtt{Lit}_X = \{x, \overline{x} \mid x \in X\}$ be the set of literals over X, where $\overline{x}$ is short for $\neg x$. We write $var(t)$ for the variable occurring in the literal $t \in \mathtt{Lit}_X$. A clause C is a set of literals. The empty clause $\square$ contains no literals and is used interchangeably with F. The disjunction of two clauses C and D is their union, denoted $C \vee D$, which is further simplified to $C \vee t$ if D is the singleton $\{t\}$. A propositional formula in Conjunctive Normal Form (CNF) is a conjunction of clauses, also represented as a set of clauses.

First-Order Logic. The logical connectives from propositional logic carry over into first-order logic. We fix an enumerable set of variables, function and predicate symbols over which formulae are built in the usual manner. The *vocabulary* of a formula A is the set of its function and predicate symbols. $\mathcal{L}(A)$ refers to the set of well-formed formulae which can be built over the vocabulary of A.

Variables may be universally ($\forall$) or existentially ($\exists$) quantified. A formula is *closed* if all its variables are quantified and *ground* if it contains no variables. As previously, conjunctions of formulae are also represented as sets.

Given a formula A in either first-order or propositional logic, we use $\mathrm{Var}(A)$ to denote the set of free (unquantified) variables in A.

Inference Rules and Proofs. We write $A_1, \cdots, A_n \models A$ to denote that the formula A holds in all models of $A_1, \ldots, A_n$ (where $n \geq 0$). An inference rule

$$\frac{A_1 \quad \cdots \quad A_n}{A} \tag{1}$$

associates zero or more *premises* (or *antecedents*) $A_1, \ldots, A_n$ with a *conclusion* A. The inference rule (1) is *sound* if $A_1, \ldots, A_n \models A$ holds. A (sound) inference system $\mathcal{I}$ is a set of (sound) inference rules.

Propositional *resolution*, for example, is a sound inference rule stating that an assignment satisfying the clauses $C \vee x$ and $D \vee \overline{x}$ also satisfies $C \vee D$:

$$\frac{C \vee x \quad D \vee \overline{x}}{C \vee D} \quad [\text{Res}]$$

The clauses $C \vee x$ and $D \vee \overline{x}$ are the *antecedents*, x is the *pivot*, and $C \vee D$ is the *resolvent*. $\mathrm{Res}(C, D, x)$ denotes the resolvent of C and D with the pivot x.

Definition 1 (Proof). *A proof (or derivation) P in an inference system $\mathcal{I}_P$ is a directed acyclic graph $(V_P, E_P, \ell_P, \mathsf{s}_P)$, where V_P is a set of vertices, E_P is a set of edges, ℓ_P is a function mapping vertices to formulae, and $\mathsf{s}_P \in V_P$ is the sink vertex. An initial vertex has in-degree 0. All other vertices are internal and have in-degree ≥ 1. The sink has out-degree 0. Each internal vertex v with edges $(v_1, v), \ldots, (v_m, v) \in E_P$ is associated with an inference rule $\mathsf{Inf} \in \mathcal{I}_P$, i.e.,*

$$\frac{\ell_P(v_1) \quad \cdots \quad \ell_P(v_m)}{\ell_P(v)} \quad [\mathsf{Inf}].$$

The subscripts above are dropped if clear. A vertex v_i in P is a *parent* of v_j if $(v_i, v_j) \in E_P$. A proof P is a *refutation* if $\ell_P(\mathsf{s}_P) = \mathsf{F}$. Let A and B conjunctive formulae. A refutation P is an (A, B)-*refutation* of an unsatisfiable formula $A \wedge B$ if $\ell_P(v)$ is a conjunct of A or a conjunct of B for each initial vertex $v \in V_P$. A proof is *closed* (*ground*, respectively) if $\ell_P(v)$ is closed (ground) for all $v \in V_P$.

In the following, we use the propositional resolution calculus to instantiate Definition 1.

Definition 2 (Resolution Proof). *A resolution proof R is a proof in the inference system comprising only the resolution rule Res. Consequently, ℓ_R maps each vertex $v \in V_R$ to a clause, and all internal vertices have in-degree 2. Let piv_R be the function mapping internal vertices to pivot variables. For an internal vertex v and $(v_1, v), (v_2, v) \in E_R$, $\ell_R(v) = \mathrm{Res}(\ell_R(v_1), \ell_R(v_2), piv_R(v))$.*

Note that the value of ℓ_R at internal vertices is determined by that of ℓ_R at initial vertices and the pivot function piv_R. We write v^+ for the parent of v with $piv(v)$ in $\ell(v^+)$ and v^- for the parent with $\neg piv(v)$ in $\ell(v^-)$. A resolution proof R is a *resolution refutation* if $\ell_R(\mathsf{s}_R) = \square$.

2.2   Interpolation Systems and Labelling Functions

There are numerous variants and definitions of Craig's interpolation theorem [5]. We use the definition of a *Craig-Robinson interpolant* given by Harrison [11]:

Definition 3 (Interpolant). *A Craig-Robinson interpolant for a pair of formulae (A, B), where $A \wedge B$ is unsatisfiable, is a formula I whose free variables, function and predicate symbols occur in both A and B, such that $A \Rightarrow I$, and $B \Rightarrow \neg I$ holds.*

Craig's interpolation theorem guarantees the existence of such an interpolant for unsatisfiable pairs of formulae (A, B) in first order logic. Consequently, it also holds in the propositional setting, where the conditions of Definition 3 reduce to $A \Rightarrow I$, $B \Rightarrow \neg I$, and $\mathrm{Var}(I) \subseteq \mathrm{Var}(A) \cap \mathrm{Var}(B)$.

Numerous techniques to construct interpolants have been proposed (c.f. § 6). In particular, there is a class of algorithms that derive interpolants from proofs;

the first such algorithm for the sequent calculus is present in Maehara's constructive proof [17] of Craig's theorem. In this paper, we focus on *interpolation systems* that construct an interpolant from an (A, B)-refutation by mapping the vertices of a resolution proof to a formula called the *partial interpolant*.

Formally, an interpolation system ltp is a function that given an (A, B)-refutation R yields a function, denoted $\mathsf{ltp}(R, A, B)$, from vertices in R to formulae over $\mathrm{Var}(A) \cap \mathrm{Var}(B)$. An interpolation system is *correct* if for every (A, B)-refutation R with sink $\mathbf{s}$, it holds that $\mathsf{ltp}(R, A, B)(\mathbf{s})$ is an interpolant for (A, B). We write $\mathsf{ltp}(R)$ for $\mathsf{ltp}(R, A, B)(\mathbf{s})$ when A and B are clear. Let v be a vertex in an (A, B)-refutation R. The pair $(\ell(v), \mathsf{ltp}(R, A, B)(v))$ is an *annotated clause* and is written $\ell(v) [\mathsf{ltp}(R, A, B)(v)]$ in accordance with [19].

In the following, we review the labelled interpolation systems we introduced in [7]. This approach generalises several existing propositional interpolation systems presented by Huang [12], Krajíček [16] and Pudlák [21], and McMillan [18]. A distinguishing feature of a labelled interpolation system is that it assigns an individual label $\mathsf{c} \in \{\bot, \mathsf{a}, \mathsf{b}, \mathsf{ab}\}$ to *each literal* in the resolution refutation.

Definition 4 (Labelling Function). *Let $(\mathcal{S}, \sqsubseteq, \sqcap, \sqcup)$ be the lattice below, where $\mathcal{S} = \{\bot, \mathsf{a}, \mathsf{b}, \mathsf{ab}\}$ is a set of symbols and $\sqsubseteq$, $\sqcap$ and $\sqcup$ are defined by the Hasse diagram to the right. A labelling function $L_R : V_R \times \mathtt{Lit} \to \mathcal{S}$ for a refutation R over a set of literals $\mathtt{Lit}$ satisfies that for all $v \in V_R$ and $t \in \mathtt{Lit}$:*

1. *$L_R(v, t) = \bot$ iff $t \notin \ell_R(v)$*
2. *$L_R(v, t) = L_R(v_1, t) \sqcup \cdots \sqcup L_R(v_m, t)$ for an internal vertex v, its parents $\{v_1, \cdots, v_m\}$, and literal $t \in \ell_R(v)$.*

Due to condition (2) above, the labels of literals at initial vertices completely determine the labelling function for literals at internal vertices. The following condition ensures that a labelling function respects the *locality* of a literal t with respect to (A, B). A literal t is *A-local* and therefore labelled a if $\mathrm{var}(t) \in \mathrm{Var}(A) \setminus \mathrm{Var}(B)$. Conversely, t is *B-local* and therefore labelled b if $\mathrm{var}(t) \in \mathrm{Var}(B) \setminus \mathrm{Var}(A)$. Literals t for which $\mathrm{var}(t) \in \mathrm{Var}(A) \cap \mathrm{Var}(B)$ are *shared* and can be labelled a, b, or ab (which generalises existing interpolation systems).

Definition 5 (Locality). *A labelling function for an (A, B)-refutation R preserves locality if for any initial vertex v and literal t in R*
1. *$\mathsf{a} \sqsubseteq L(v, t)$ implies that $\mathrm{var}(t) \in \mathrm{Var}(A)$, and*
2. *$\mathsf{b} \sqsubseteq L(v, t)$ implies that $\mathrm{var}(t) \in \mathrm{Var}(B)$.*

For a given labelling function L, we define the downward *projection* of a clause at a vertex v with respect to $\mathsf{c} \in \mathcal{S}$ as $\ell(v)\!\downarrow_{\mathsf{c},L} \stackrel{\text{def}}{=} \{t \in \ell(v) \mid L(v, t) \sqsubseteq \mathsf{c}\}$. and the upward projection $\ell(v)\!\uparrow_{\mathsf{c},L}$ as $\ell(v)\!\uparrow_{\mathsf{c},L} \stackrel{\text{def}}{=} \{t \in \ell(v) \mid \mathsf{c} \sqsubseteq L(v, t)\}$. The subscript L is omitted if clear from the context.

Definition 6 (Labelled Interpolation System for Resolution). *Let L be a locality preserving labelling function for an (A, B)-refutation R. The labelled interpolation system $\mathsf{ltp}(L)$ maps vertices in R to partial interpolants as follows:*

For an initial vertex v with $\ell(v) = C$

(A-clause) $\dfrac{}{C \quad [C\lfloor_\mathsf{b}]}$ if $C \in A$ (B-clause) $\dfrac{}{C \quad [\neg(C\lfloor_\mathsf{a})]}$ if $C \in B$

For an internal vertex v with $piv(v) = x$, $\ell(v^+) = C_1 \vee x$ and $\ell(v^-) = C_2 \vee \overline{x}$

$$\frac{C_1 \vee x \quad [I_1] \qquad C_2 \vee \overline{x} \quad [I_2]}{C_1 \vee C_2 \quad [I_3]}$$

(A-Res)   if $L(v^+, x) \sqcup L(v^-, \overline{x}) = \mathsf{a}$, $I_3 \overset{def}{=} \quad I_1 \vee I_2$

(AB-Res) if $L(v^+, x) \sqcup L(v^-, \overline{x}) = \mathsf{ab}$, $I_3 \overset{def}{=} (x \vee I_1) \wedge (\overline{x} \vee I_2)$

(B-Res)   if $L(v^+, x) \sqcup L(v^-, \overline{x}) = \mathsf{b}$, $I_3 \overset{def}{=} \quad I_1 \wedge I_2$

Labelling functions provide control over the interpolants constructed from a resolution proof. Firstly, labelled interpolation systems support the elimination of *non-essential* (*peripheral* [24], respectively) variables from interpolants [6]. Secondly, labelled interpolation systems – and their respective interpolants – are ordered by logical strength. A labelled interpolation system $\mathsf{Itp}(L)$ is *stronger than* $\mathsf{Itp}(L')$ if for all refutations R, $\mathsf{Itp}(L, R) \Rightarrow \mathsf{Itp}(L', R)$. The partial order $\preceq$ on labelling functions (first introduced in [7]) guarantees an ordering in strength:

Definition 7 (Strength Order). *We define the total order $\preceq$ on the lattice $S = \{\bot, \mathsf{a}, \mathsf{b}, \mathsf{ab}\}$ as $\mathsf{b} \preceq \mathsf{ab} \preceq \mathsf{a} \preceq \bot$ (c.f. the Hasse diagram to the right). Let L and L' be labelling functions for an (A, B)-refutation R. The function L is stronger than L', denoted $L \preceq L'$, if for all $v \in V_R$ and $t \in \ell(v)$, $L(v, t) \preceq L'(v, t)$.*

Theorem 2 in [7] shows that if L is a stronger labelling function than L', the interpolant obtained from $\mathsf{Itp}(L)$ logically implies the one obtained from $\mathsf{Itp}(L')$.

3 Interpolation for Hyper-resolution

In this section, we extend labelled interpolation systems to a richer inference system, in particular, the inference system comprising (propositional) *hyper-resolution* [22]. Hyper-resolution is a condensation of a derivation consisting of several resolutions and avoids the construction of intermediate clauses. Hyper-resolution has several applications in propositional satisfiability checking, such as pre-processing [10] of formulae or as an integral part of the solver (e.g., [2]).

Positive hyper-resolution combines a single clause (called the *nucleus*) containing n negative literals $\overline{x}_1, \ldots, \overline{x}_n$ and n *satellite* clauses each of which contains one of the corresponding non-negated literals x_i (where $1 \leq i \leq n$):

$$\frac{\overbrace{(C_1 \vee x_1) \quad \cdots \quad (C_n \vee x_n)}^{\text{satellites}} \quad \overbrace{(\overline{x}_1 \vee \cdots \vee \overline{x}_n \vee D)}^{\text{nucleus}}}{\bigvee_{i=1}^{n} C_i \vee D} \quad \text{[HyRes]}$$

In *negative* hyper-resolution the rôles of x_i and $\overline{x}_i$ are exchanged.

Definition 8 (Hyper-resolution Proof). *A* hyper-resolution proof R *is a proof using only the inference rule* HyRes. *Accordingly,* ℓ_R *maps each vertex* $v \in V_R$ *to a clause, and all internal vertices have in-degree* ≥ 2. *Each internal vertex v has* $n \geq 1$ *parents* $v_1^+, \ldots, v_n^+$ *such that* $\ell_R(v_i^+) = C_i \vee x_i$ *and one parent* v^- *with* $\ell_R(v^-) = \overline{x}_1 \vee \cdots \vee \overline{x}_n \vee D$, *and consequently,* $\ell_R(v) = \bigvee_{i=1}^{n} C_i \vee D$.

The definition of labelling functions (Definition 4) readily applies to hyper-resolution proofs. Note that $\preceq$ is not a total order on labelling functions. Lemma 1 (a generalisation of Lemma 3 in [7] to hyper-resolution proofs) enables a comparison of labelling functions based solely on the values at the *initial* vertices.

Lemma 1. *Let L and L' be labelling functions for an (A, B)-refutation R. If $L(v, t) \preceq L'(v, t)$ for all initial vertices v and literals $t \in \ell(v)$, then $L \preceq L'$.*

The following definition provides a labelled interpolation system for hyper-resolution proofs.

Definition 9 (Labelled Interpolation System for Hyper-resolution). *Let L be a locality preserving labelling function for an (A, B)-refutation R, where R is a hyper-resolution proof. The labelled interpolation system* ltp(L) *maps vertices in R to partial interpolants as defined below.[1]*

For an initial vertex v with $\ell(v) = C$

$(A$-clause$) \quad \dfrac{}{C \quad [C|_{\mathsf{b}}]}$ *if* $C \in A$ $(B$-clause$) \quad \dfrac{}{C \quad [\neg(C|_{\mathsf{a}})]}$ *if* $C \in B$

For an internal vertex v with predecessors $\{v_1^+, \ldots, v_n^+, v^-\}$ (where $n \geq 1$) with $\ell(v_i^+) = (C_i \vee x_i)$, for $1 \leq i \leq n$, and $\ell(v^-) = (D \vee \overline{x}_1 \vee \cdots \vee \overline{x}_n)$

$$\frac{C_1 \vee x_1 \quad [I_1] \qquad \cdots \qquad C_n \vee x_n \quad [I_n] \qquad \overline{x}_1 \vee \cdots \vee \overline{x}_n \vee D \quad [I_{n+1}]}{\bigvee_{i=1}^{n} C_i \vee D \quad [I]}$$

$(A$-HyRes$) \quad$ *if* $\forall i \in \{1..n\} . L(v_i^+, x_i) \sqcup L(v^-, \overline{x}_i) = \mathsf{a}, \quad I \overset{\text{def}}{=} \bigvee_{i=1}^{n+1} I_i$

$(AB$-HyRes$)$ *if* $\forall i \in \{1..n\} . L(v_i^+, x_i) \sqcup L(v^-, \overline{x}_i) = \mathsf{ab},$

$\qquad\qquad$ *1.)* $\quad I \overset{\text{def}}{=} \bigwedge_{i=1}^{n}(x_i \vee I_i) \wedge (I_{n+1} \vee \bigvee_{i=1}^{n} \overline{x}_i), \quad$ *or*

$\qquad\qquad$ *2.)* $\quad I \overset{\text{def}}{=} \bigvee_{i=1}^{n}(\overline{x}_i \wedge I_i) \vee (I_{n+1} \wedge \bigwedge_{i=1}^{n} x_i)$

$(B$-HyRes$) \quad$ *if* $\forall i \in \{1..n\} . L(v_i^+, x_i) \sqcup L(v^-, \overline{x}_i) = \mathsf{b}, \quad I \overset{\text{def}}{=} \bigwedge_{i=1}^{n+1} I_i$

The system can be easily extended to *negative* hyper-resolution. In fact, ltp can be generalised by replacing the variables $x_1, \ldots, x_n$ in the definition with

---

[1] Note that unlike the interpolation system for ordinary resolution proofs presented in Definition 6, ltp is not total for hyper-resolution proofs (see discussion in § 4).

literals $t_1, \ldots, t_n$, since the proofs of our theorems below are not phase-sensitive. We avoid this generalisation to simplify the presentation.

Note that the interpolation system leaves us a choice for internal nodes $AB-HyRes$. We will use ltp_1 (ltp_2, respectively) to refer to the interpolation system that always chooses case 1 (case 2, respectively). Note furthermore that Definition 6 and Definition 9 are equivalent in the special case where $n = 1$.

Before we turn to the correctness of our novel interpolation system, we point out the limitation stated in Footnote 1. There are labelling functions L and proofs R for which the function $\mathsf{ltp}(L, R)$ is not total. This restriction is imposed by the case split in Definition 9 which requires the pivots of the hyper-resolution step to be uniformly labelled. We address this issue in § 4 and present a provisional conditional correctness result.

Theorem 1 (Correctness). *For any (A, B)-refutation R (where R is a hyper-resolution proof) and locality preserving labelling function L, $\mathsf{ltp}(L, R)$ (if defined) is an interpolant for (A, B).*

The proof[2] of Theorem 1 establishes that for each vertex $v \in V_R$ with $\ell_R(v) = C$ and $I = \mathsf{ltp}(L, R)(v)$, the following conditions hold:

- $A \wedge \neg(C\!\restriction_{\mathsf{a}, L}) \Rightarrow I$,
- $B \wedge \neg(C\!\restriction_{\mathsf{b}, L}) \Rightarrow \neg I$, and
- $\mathrm{Var}(I) \subseteq \mathrm{Var}(A) \cap \mathrm{Var}(B)$.

For $\ell_R(\mathsf{s}) = \square$, this establishes the correctness of the system.

We emphasise that Theorem 1 does not constrain the choice for the case $AB-HyRes$. Since both $\mathsf{ltp}_1(L, R)$ and $\mathsf{ltp}_2(R, L)$ satisfy the conditions above, this choice does not affect the correctness of the interpolation system. In fact, it is valid to *mix* both systems by defining a choice function $\chi : V_R \to \{1, 2\}$ which determines which interpolation system is chosen at each internal node. We use $\mathsf{ltp}_\chi(L, R)$ to denote the resulting interpolation system. This modification, however, may have an impact on the logical strength of the resulting interpolant.

Theorem 2. *Let the hyper-resolution proof R be an (A, B)-refutation and L be a locality preserving labelling function. Moreover, let $\mathsf{ltp}_\chi(L, R)$ and $\mathsf{ltp}_{\chi'}(L, R)$ be labelled interpolation systems (defined for L, R) with the choice functions χ and χ', respectively. Then $\mathsf{ltp}_\chi(L, R) \Rightarrow \mathsf{ltp}_{\chi'}(L, R)$ if $\chi(v) \leq \chi'(v)$ for all internal vertices $v \in V_R$.*

Proof sketch: This follows (by structural induction over R) from

$$\left(\bigwedge_{i=1}^{n}(x_i \vee I_i) \wedge (I_{n+1} \vee \bigvee_{i=1}^{n} \overline{x}_i)\right) \Rightarrow \left(\bigvee_{i=1}^{n}(\overline{x}_i \wedge I_i) \vee (I_{n+1} \wedge \bigwedge_{i=1}^{n} x_i)\right). \quad \blacksquare$$

Note that the converse implication does not hold; a simple counterexample for an internal vertex with $n = 2$ is the assignment $x_1 = x_2 = \mathsf{F}$, $I_1 = \mathsf{T}$, and $I_2 = I_3 = \mathsf{F}$.

The final theorem in this section extends the result of Theorem 2 in [7] to hyper-resolution proofs:

---

[2] All proofs can be found in an extended version of the paper available from the author's website (http://www.georg.weissenbacher.name).

Theorem 3. *If L and L' are labelling functions for an (A, B)-refutation R (R being a hyper-resolution proof) and $L \preceq L'$ such that $\mathsf{ltp}_i(L, R)$ as well as $\mathsf{ltp}_i(L', R)$ are defined, then $\mathsf{ltp}_i(L, R) \Rightarrow \mathsf{ltp}_i(L', R)$ (for a fixed $i \in \{1, 2\}$).*

The proof of Theorem 3, is led by structural induction over R. For any vertex v in R, let I_v and I'_v be the partial interpolants due to $\mathsf{ltp}_i(L, R)$ and $\mathsf{ltp}_i(L', R)$, respectively. We show that $I_v \Rightarrow I'_v \vee \{t \in \ell_R(v) \mid L(v, t) \sqcup L'(v, t) = \mathsf{ab}\}$ for all vertices v, establishing $I_v \Rightarrow I'_v$ for the sink to show that $\mathsf{ltp}_i(L, R) \Rightarrow \mathsf{ltp}_i(L', R)$.

Theorems 2 and 3 enable us to fine-tune the strength of interpolants, since the sets of all labelling and choice functions ordered by $\preceq$ and $\leq$, respectively, form complete lattices (c.f. [7, Theorem 3]). Finally, we remark that the Theorems 2 and 3 are orthogonal. The former fixes the labelling function L, whereas the latter fixes the choice function χ.

4 Hyper-resolution and Resolution Chains

Contemporary proof-logging SAT solvers typically generate compacted proofs. MINISAT [8], for example, discards all intermediate resolvents generated during the construction of a conflict clause and retains only resolution chains.

Definition 10 (Chain). *A (resolution) chain of length n is a tuple consisting of an input clause D_0 and an ordered sequence of clause-pivot pairs $\langle C_i, x_i \rangle$ (where $1 \leq i \leq n$). The final resolvent D_n of a resolution chain is defined inductively as $D_i = \mathrm{Res}(D_{i-1}, C_i, x_i)$.*

If D_0 is a nucleus and $C_1, \ldots, C_n$ are suitable satellites, the chain can be replaced by a hyper-resolution step if its conclusion D_n satisfies the HyRes rule. In general, this may not be the case: in the presence of merge literals [1], the final resolvent of a chain may depend on the order of the ordinary resolution steps. For example, the chain $(\{\overline{x}_1, x_2\}, [\langle \{\overline{x}_2\}, x_2 \rangle, \langle \{x_1, x_2\}, x_1 \rangle])$ yields the resolvent $\{x_2\}$, whereas swapping the clause-pivot pairs leads to the resolvent $\square$. This is because the literal x_2 is re-introduced after being eliminated in the original chain, while it is *merged* and eliminated once and for all in the modified chain.

In the absence of merge literals, this issue does not arise. The following definition is a generalisation of *merge-free* edges (c.f. [7, § 5.1]) to chains.

Definition 11 (Strongly Merge-Free). *A chain*

$$(D_0, [\langle t_1 \vee C_1, \mathrm{var}(t_1) \rangle, \ldots, \langle t_n \vee C_n, \mathrm{var}(t_n) \rangle])$$

is strongly merge-free if $\{\overline{t}_1, \cdots, \overline{t}_n\} \cap C_i = \emptyset$ for all $1 \leq i \leq n$.

Strongly merge-free chains are insensitive to changes in the order of the resolution steps in the sense that any permutation of the clause-pivot sequence still represents a valid resolution proof (an immediate consequence of [7, Lemma 4]) with the final resolvent $(D_0 \setminus \{\overline{t}_1, \ldots, \overline{t}_n\}) \vee \bigvee_{i=1}^{n} C_i$. This property is stronger than just requiring that the sequence of resolution steps defined by a chain contains no merge literals; it demands that $\{\bar{t}_0, \ldots, \bar{t}_n\} \subseteq D_0$.[3]

Corollary 1. *Any strongly merge-free chain*

$$(\bar{x}_1 \vee \cdots \vee \bar{x}_n \vee D_0, [\langle x_1 \vee C_1, x_1 \rangle, \ldots, \langle x_n \vee C_n, x_n \rangle])$$

corresponds to a hyper-resolution step

$$\frac{(C_1 \vee x_1) \quad \cdots \quad (C_n \vee x_n) \quad (\bar{x}_1 \vee \cdots \vee \bar{x}_n \vee D)}{\bigvee_{i=1}^{n} C_i \vee D}.$$

Consequently, Definition 11 provides a sufficient (but not necessary) condition for replacing chains with hyper-resolution steps. We emphasise that Corollary 1 can be generalised by replacing the variables $x_1, \ldots, x_n$ in the respective definitions with literals $t_1, \ldots, t_n$ (c.f. § 3).

By definition, a single chain can be split into two consecutive chains, with the final resolvent of the first acting as the input clause of the second, without affecting the final result. Therefore, chains that are not merge-free can be split repeatedly until the resulting sub-sequences become strongly merge-free.

A further incentive for splitting is to enable interpolation. By splitting hyper-resolution steps whose literals are not uniformly labelled (recall the remark in § 3) we can *always* generate a labelled refutation for which Itp is a total function. The following example illustrates this transformation for a single resolution step:

$$\frac{(\overset{a}{\bar{x}_1} \vee C_1)\,(\overset{ab}{\bar{x}_2} \vee C_2)\,(\overset{a}{\bar{x}_3} \vee C_3)\,(\overset{a}{\bar{x}_4} \vee C_4) \quad (\overset{a}{\bar{x}_1} \vee \overset{a}{x_2} \vee \overset{a}{x_3} \vee \overset{b}{x_4} \vee D)}{C_1 \vee C_2 \vee C_3 \vee C_4 \vee D}$$

$$\updownarrow$$

$$\frac{(\overset{ab}{\bar{x}_2} \vee C_2)\,(\overset{a}{\bar{x}_4} \vee C_4) \quad \dfrac{(\overset{a}{\bar{x}_1} \vee C_1)\,(\overset{a}{\bar{x}_3} \vee C_3) \quad (\overset{a}{\bar{x}_1} \vee \overset{a}{x_2} \vee \overset{a}{x_3} \vee \overset{b}{x_4} \vee D)}{(\overset{a}{x_2} \vee \overset{b}{x_4} \vee C_1 \vee C_3 \vee D)}\ [A\text{-HyRes}]}{C_1 \vee C_2 \vee C_3 \vee C_4 \vee D}\ [AB\text{-HyRes}]$$

Each hyper-resolution step may need to be rewritten into at most three uniformly labelled steps (a, b, ab), thus changing the proof structure. Note that the results on the relative strength of interpolants in § 3 naturally only apply if both proofs have the same structure. The effect of the order of resolution steps on interpolants is discussed in [7, § 5.2] and exceeds the scope of this paper.

5   Local Refutations and Hyper-resolution

Jhala and McMillan demonstrate in [13, Theorem 3] that the applicability of propositional interpolation systems is not restricted to propositional logic. If a

---

[3] This condition, however, can be met by extending the chain with an additional resolution step $\mathrm{Res}(D_0, t \vee \bar{t} \vee T, \mathrm{var}(t))$ for any $t \in D_0$, which introduces the missing literals $T \subseteq \{\bar{t}_0, \ldots, \bar{t}_n\}$. This transformation is valid since $t \vee \bar{t} \vee T$ is a tautology.

first-order refutation R has a certain structure, namely if for each inference step in R the antecedents as well as the conclusion are either entirely in $\mathcal{L}(A)$ or in $\mathcal{L}(B)$, then one can use a propositional interpolation system (such as the ones in § 2.2 and § 3) to construct an interpolant that is a Boolean combination of the formulae in R. Kovács and Voronkov subsequently arrived at a similar result [15].

We recapitulate the results from [13,15] before we proceed to show that our interpolation system from Definition 9 generalises the system of [15] as well as a variation of [15] presented in [25].

Definition 12 (Local Refutation). *An (A, B)-refutation R in a given inference system for first-order logic is local if there exists a total partitioning function $\pi_R : V_R \to \{A, B\}$ such that for all edges $(v_1, v_2) \in E_R$ we have $\ell_R(v_1), \ell_R(v_2) \in \mathcal{L}(\pi_R(v_2))$.*

While proofs in general do *not* have this property, there is a variety of decision procedures that yield local (ground) refutations. The construction of local proofs is addressed in [13,20,9,15], to name only a few.

The following operation, which resembles the constructions in [15, Lemma 8], [13, Theorem 3], and [9, Section 5.5]), extracts a premise in $\mathcal{L}(A)$ ($\mathcal{L}(B)$, respectively) for a vertex $v \in V_R$ with $\pi(v) = A$ ($\pi(v) = B$, respectively) from a local refutation R.

Definition 13 (A-Premise, B-Premise). *Let R be a local (A, B)-refutation with partitioning function π, and let $v \in V_R$ such that $\pi(v) = A$. Then*

$$A\text{-}premise\,(v) \stackrel{\text{def}}{=}$$
$$\{u \mid (u, v) \in E_R \text{ and } \pi(u) = B \text{ or } u \text{ is initial}\} \,\cup$$
$$\bigcup \{A\text{-}premise\,(u) \mid (u, v) \in E_R \text{ and } \pi(u) = A\}.$$

B-premise(v) is defined analogously.

Intuitively, A-premise(v) comprises the leaves of the largest sub-derivation S rooted at v such that $\pi(u) = A$ for all internal vertices $u \in V_S$.[4] If the underlying inference system is sound, we have $\{\ell(u) \mid u \in A\text{-}premise(v)\} \models \ell(v)$. If, moreover, $\ell(v)$ as well as all formulae of A-premise(v) are *closed*, we make the following observation (c.f. related results in [15, Lemma 1] and [9, Lemma 3]):

Corollary 2. *Let R be a local closed refutation in a sound inference system, and let $v \in V_R$ an internal vertex such that $\pi_R(v) = A$. Then, the following Horn clause is a tautology:*

$$\bigvee_{u \in A\text{-}premise(v)} \neg \ell_R(u) \vee \ell_R(v) \tag{2}$$

A similar claim holds for the case in which $\pi(v) = B$.

---

[4] In particular, it is possible to choose π_R in such a manner that S is the largest sub-derivation rooted at v in R such that $\ell_R(u) \in \mathcal{L}(A)$ for all $u \in V_S$. This corresponds to the setting in [15, Lemma 8].

Corollary 2 is a pivotal element in our proof of the following theorem:

Theorem 4. *(c.f. [13, Theorem 3]) Let R be a closed local (A, B)-refutation in a sound inference system. Then one can extract a Craig-Robinson interpolant from R using a propositional interpolation system.*

Proof: Let $v \in V_R$ be such that $\pi(v) = A$. If v is initial, then either A or B contains the unit clause $C_v = \ell(v)$. Otherwise, according to Corollary 2, the clause $C_v = (\{\neg\ell(u) \mid u \in A\text{-premise}(v)\} \vee \ell(v))$ is tautological (and therefore implied by A). Moreover, it follows from Definition 12 that if $u \in A\text{-premise}(v)$ is not an initial vertex of R then $\ell_R(u) \in \mathcal{L}(A) \cap \mathcal{L}(B)$ holds. Accordingly, $C_v \in \mathcal{L}(A)$, and we add C_v to A. A similar argument holds for $v \in V_R$ with $\pi(v) = B$.

By construction, the resulting set of clauses C_v, $v \in V_R$, is propositionally unsatisfiable [13,15]; also, each clause is implied by either A or B. Moreover, all literals with $t \in \mathcal{L}(A) \setminus \mathcal{L}(B)$ ($t \in \mathcal{L}(B) \setminus \mathcal{L}(A)$, respectively) are local to A (B, respectively). Accordingly, it is possible to construct an interpolant for (A, B) using the interpolation systems presented in § 2.2 and § 3. ∎

Kovács and Voronkov avoid the explicit construction of a resolution proof by defining their interpolation system directly on the local proof [15, Theorem 11]:

Definition 14. *Let R be a local and closed (A, B)-refutation. The interpolation system ltp_{KV} maps vertices $v \in V_R$ for which $\ell_R(v) \in \mathcal{L}(A) \cap \mathcal{L}(B)$ holds to partial interpolants as defined below.*

For an initial vertex v

$(A\text{-clause}) \; \dfrac{}{\ell(v) \quad [\ell(v)]} \; \textit{if } \ell(v) \in A$ $\quad (B\text{-clause}) \; \dfrac{}{\ell(v) \quad [\neg\ell(v)]} \; \textit{if } \ell(v) \in B$

For an internal vertex v with $\{v_1, \ldots, v_n\} = \pi(v)\text{-premise}(v)$ such that
$$\ell(v_i) \in \mathcal{L}(A) \cap \mathcal{L}(B) \textit{ for } 1 \leq i \leq m \leq n \textit{ and}$$
$$\ell(v_j) \notin \mathcal{L}(A) \cap \mathcal{L}(B) \textit{ for } m < j \leq n.$$

$$\frac{\ell(v_1) \quad [I_1] \quad \cdots \quad \ell(v_m) \quad [I_m] \quad \ell(v_{m+1}) \quad \cdots \quad \ell(v_n)}{\ell(v) \quad [I]}$$

$(A\text{-justified}) \; \textit{if } \pi(v) = A, \; I \overset{\text{def}}{=} \bigwedge_{i=1}^{m}(\ell(v_i) \vee I_i) \wedge \bigvee_{i=1}^{m} \neg\ell(v_i)$

$(B\text{-justified}) \; \textit{if } \pi(v) = B, \; I \overset{\text{def}}{=} \bigwedge_{i=1}^{m}(\ell(v_i) \vee I_i)$

Remark. In addition to the condition in Definition 12, Kovács and Voronkov require that for each $v \in V_R$ with predecessors $v_1, \ldots, v_n$, $\ell(v) \in \mathcal{L}(A) \cap \mathcal{L}(B)$ if $\ell(v_i) \in \mathcal{L}(A) \cap \mathcal{L}(B)$ for all $i \in \{1..n\}$. A local derivation satisfying this condition is *symbol-eliminating*, i.e., it does not introduce "irrelevant" symbols. This technical detail allows the leaves of R to be merely implied by A (or B, respectively), while preserving the correctness of the interpolation system. This effectively enables interpolation for *non-closed* formulae (A, B).

We proceed to show one of the main results of this paper, namely that our interpolation system Itp from Definition 9 is able to simulate the interpolation system Itp_{KV}.

Theorem 5. *Let R be a local and closed (A, B)-refutation. Then we can construct a hyper-resolution refutation H of (A, B) and a locality preserving labelling function L such that for each $v \in V_R$ with $\ell_R(v) \in \mathcal{L}(A) \cap \mathcal{L}(B)$ there exists a corresponding vertex $u \in V_H$ such that $\mathsf{Itp}_{KV}(R)(v) \Leftrightarrow \mathsf{Itp}_1(L, H)(u)$.*

Proof sketch: We demonstrate that it is possible to construct a hyper-resolution refutation H of (A, B) in which each internal step of Itp_{KV} is simulated using *two* hyper-resolution steps. The induction hypothesis is that for each internal vertex $v \in V_R$ with $\{v_1, \ldots, v_n\} = \pi(v)$-premise$(v)$ and m as in Definition 14, we have vertices $\{u_1, \ldots, u_n\} \subseteq V_H$ such that

1. $\ell_H(u_i) = \ell_R(v_i)$ for $1 \leq i \leq n$, and
2. $\mathsf{Itp}_1(L, H)(u_i) \Leftrightarrow \mathsf{Itp}_{KV}(R)(v_i)$ for $1 \leq i \leq m$, and
3. $\mathsf{Itp}_1(L, H)(u_j) = \begin{cases} \mathsf{F} & \text{if } \ell(v_j) \in A \\ \mathsf{T} & \text{if } \ell(v_j) \in B \end{cases}$ for $m < j \leq n$.

We add an auxiliary vertex labelled with the clause $\neg \ell_H(u_1) \vee \cdots \vee \neg \ell_H(u_n) \vee \ell_R(v)$, which, by Corollary 2 and by Definition 12, can be regarded as element of formula $\pi(v)$ (see proof of Theorem 4). The first hyper-resolution step eliminates the literals local to $\pi(v)$; the interpolants and labels are indicated for $\pi(v) = A$:

$$\frac{\overset{a}{\ell_H(u_{m+1})} \ [\mathsf{F}] \ \cdots \ \overset{a}{\ell_H(u_n)} \ [\mathsf{F}] \qquad \overset{a}{(\neg \ell_H(u_{m+1})} \vee \cdots \vee \neg \ell_H(u_n) \vee \ell_R(v)) \ [\mathsf{F}]}{\overset{ab}{(\neg \ell_H(u_1)} \vee \cdots \vee \overset{ab}{\neg \ell_H(u_m)} \vee \overset{a}{\ell_R(v)}) \quad [\mathsf{F}]}$$

The second hyper-resolution step eliminates the shared literals $\ell_H(u_i)$ (for $1 \leq i \leq m$). Again, the labels and interpolants are for the case that $\pi(v) = A$:

$$\frac{\overset{ab}{\ell_H(u_1)} \ [I_1] \ \cdots \ \overset{ab}{\ell_H(u_m)} \ [I_m] \qquad \overset{ab}{(\neg \ell_H(u_1)} \vee \cdots \vee \overset{ab}{\neg \ell_H(u_m)} \vee \overset{a}{\ell_R(v)}) \ [\mathsf{F}]}{\overset{a}{\ell_R(v)} \quad [\bigwedge_{i=1}^m (\ell_H(u_i) \vee I_i)] \wedge (\mathsf{F} \vee \bigvee_{i=1}^m \neg \ell_H(u_i))]}$$

The sink of this resolution step is the vertex $u \in V_H$ such that $\ell_H(u) = \ell_R(v)$ and $\mathsf{Itp}_1(L, H)(u) = \mathsf{Itp}_{KV}(v)$. ∎

We proceed to show that our system for hyper-resolution also generalises another existing interpolation system for local refutations. In [25], we introduced the following variation of the interpolation system in Definition 14:

Definition 15. *Let Itp_W be the interpolation system as described in Definition 14, except for the following modification:*

(*A*-justified) if $\pi(v) = A$, $I \overset{\text{def}}{=} \bigvee_{i=1}^m (\neg \ell(v_i) \wedge I_i)$
(*B*-justified) if $\pi(v) = B$, $I \overset{\text{def}}{=} \bigvee_{i=1}^m (\neg \ell(v_i) \wedge I_i) \vee \bigwedge_{i=1}^m \ell(v_i)$

The following theorem states that the interpolation system in Definition 9 is powerful enough to simulate ltp_W.

Theorem 6. *Let R be a local and closed (A, B)-refutation. Then we can construct a hyper-resolution refutation H of (A, B) and a locality preserving labelling function L such that for each $v \in V_R$ with $\ell_R(v) \in \mathcal{L}(A) \cap \mathcal{L}(B)$ there exists a corresponding vertex $u \in V_H$ such that $\mathsf{ltp}_W(R)(v) \Leftrightarrow \mathsf{ltp}_2(L, H)(u)$.*

The proof is essentially equivalent to the proof of Theorem 5. Moreover, as a consequence of Theorem 2, ltp_{KV} is *stronger* than ltp_W.

Corollary 3. *Let R be a closed local (A, B)-refutation in a sound inference system. Then $\mathsf{ltp}_{KV}(R) \Rightarrow \mathsf{ltp}_W(R)$.*

6   Related Work

There is a vastly growing number of different interpolation techniques; a recent survey of interpolation in decision procedures is provided by [3]. An exposition of interpolation techniques for SMT solvers can be found in [4]. The work of Yorsh and Musuvathi [26] enables the combination of theory-specific and propositional interpolation techniques [12,16,21,18,7].

The novel interpolation system presented in Section 3 extends our prior work on propositional interpolation systems [7]. The idea of using labelling functions (initially introduced in [24] in the context of LTL vacuity detection to determine the *peripherality* of variables in resolution proofs) is common to both approaches.

A number of interpolation techniques provide local proofs (e.g., [13,20,9,15]). Not all interpolation techniques are based on local proofs, though: McMillan's interpolating inference system for equality logic with uninterpreted functions and linear arithmetic [19], for instance, performs an implicit conversion of the proof, and the approach presented in [23] avoids the construction of proofs altogether.

7   Consequences and Conclusion

We present a novel interpolation system for hyper-resolution proofs which generalises our previous work [7]. By applying our technique to local proofs, we combine a number of first-order [15,25] and propositional interpolation techniques [12,16,21,18] into one *uniform* interpolation approach. As in [13], our approach avoids an explicit theory combination step [26]. Therefore, it enables the variation of interpolant strength and the elimination of non-essential literals across the theory boundary. Finally, by defining a rule that addresses hyper-resolution steps (introduced by pre-processing or extracted from resolution chains), we avoid the construction of intermediate partial interpolants. An experimental evaluation of the benefit on overhead and interpolant size is future work.

References

1. Andrews, P.B.: Resolution with merging. J. ACM 15, 367–381 (1968)
2. Bacchus, F.: Enhancing Davis Putnam with extended binary clause reasoning. In: IAAI, pp. 613–619. AAAI Press / MIT Press (2002)
3. Bonacina, M.P., Johansson, M.: On Interpolation in Decision Procedures. In: Brünnler, K., Metcalfe, G. (eds.) TABLEAUX 2011. LNCS, vol. 6793, pp. 1–16. Springer, Heidelberg (2011)
4. Cimatti, A., Griggio, A., Sebastiani, R.: Efficient generation of Craig interpolants in satisfiability modulo theories. In: TOCL (2010)
5. Craig, W.: Linear reasoning. A new form of the Herbrand-Gentzen theorem. J. Symbolic Logic 22, 250–268 (1957)
6. D'Silva, V.: Propositional Interpolation and Abstract Interpretation. In: Gordon, A.D. (ed.) ESOP 2010. LNCS, vol. 6012, pp. 185–204. Springer, Heidelberg (2010)
7. D'Silva, V., Kroening, D., Purandare, M., Weissenbacher, G.: Interpolant Strength. In: Barthe, G., Hermenegildo, M. (eds.) VMCAI 2010. LNCS, vol. 5944, pp. 129–145. Springer, Heidelberg (2010)
8. Eén, N., Sörensson, N.: An Extensible SAT-solver. In: Giunchiglia, E., Tacchella, A. (eds.) SAT 2003. LNCS, vol. 2919, pp. 502–518. Springer, Heidelberg (2004)
9. Fuchs, A., Goel, A., Grundy, J., Krstić, S., Tinelli, C.: Ground Interpolation for the Theory of Equality. In: Kowalewski, S., Philippou, A. (eds.) TACAS 2009. LNCS, vol. 5505, pp. 413–427. Springer, Heidelberg (2009)
10. Gershman, R., Strichman, O.: Cost-Effective Hyper-Resolution for Preprocessing CNF Formulas. In: Bacchus, F., Walsh, T. (eds.) SAT 2005. LNCS, vol. 3569, pp. 423–429. Springer, Heidelberg (2005)
11. Harrison, J.: Handbook of Practical Logic and Automated Reasoning. Cambridge University Press (2009)
12. Huang, G.: Constructing Craig Interpolation Formulas. In: Li, M., Du, D.-Z. (eds.) COCOON 1995. LNCS, vol. 959, pp. 181–190. Springer, Heidelberg (1995)
13. Jhala, R., McMillan, K.L.: A Practical and Complete Approach to Predicate Refinement. In: Hermanns, H. (ed.) TACAS 2006. LNCS, vol. 3920, pp. 459–473. Springer, Heidelberg (2006)
14. Jiang, J.-H.R., Lin, H.-P., Hung, W.-L.: Interpolating functions from large Boolean relations. In: ICCAD, pp. 779–784. ACM (2009)
15. Kovács, L., Voronkov, A.: Interpolation and Symbol Elimination. In: Schmidt, R.A. (ed.) CADE 2009. LNCS, vol. 5663, pp. 199–213. Springer, Heidelberg (2009)
16. Krajíček, J.: Interpolation theorems, lower bounds for proof systems, and independence results for bounded arithmetic. J. Symbolic Logic 62, 457–486 (1997)
17. Maehara, S.: On the interpolation theorem of Craig. Sûgaku 12, 235–237 (1961)
18. McMillan, K.L.: Interpolation and SAT-Based Model Checking. In: Hunt Jr., W.A., Somenzi, F. (eds.) CAV 2003. LNCS, vol. 2725, pp. 1–13. Springer, Heidelberg (2003)
19. McMillan, K.L.: An interpolating theorem prover. TCS 345(1), 101–121 (2005)
20. McMillan, K.L.: Quantified Invariant Generation Using an Interpolating Saturation Prover. In: Ramakrishnan, C.R., Rehof, J. (eds.) TACAS 2008. LNCS, vol. 4963, pp. 413–427. Springer, Heidelberg (2008)
21. Pudlák, P.: Lower bounds for resolution and cutting plane proofs and monotone computations. J. Symbolic Logic 62, 981–998 (1997)
22. Robinson, J.: Automatic deduction with hyper-resolution. J. Comp. Math. 1 (1965)

23. Rybalchenko, A., Sofronie-Stokkermans, V.: Constraint Solving for Interpolation. In: Cook, B., Podelski, A. (eds.) VMCAI 2007. LNCS, vol. 4349, pp. 346–362. Springer, Heidelberg (2007)
24. Simmonds, J., Davies, J., Gurfinkel, A., Chechik, M.: Exploiting resolution proofs to speed up LTL vacuity detection for BMC. STTT 12, 319–335 (2010)
25. Weissenbacher, G.: Program Analysis with Interpolants. PhD thesis, Oxford (2010)
26. Yorsh, G., Musuvathi, M.: A Combination Method for Generating Interpolants. In: Nieuwenhuis, R. (ed.) CADE 2005. LNCS (LNAI), vol. 3632, pp. 353–368. Springer, Heidelberg (2005)

Exponential Lower Bounds for DPLL Algorithms on Satisfiable Random 3-CNF Formulas

Dimitris Achlioptas[1,2,3] and Ricardo Menchaca-Mendez[3]

[1] University of Athens, Greece
[2] CTI, Greece
[3] University of California, Santa Cruz, USA

Abstract. We consider the performance of a number of DPLL algorithms on random 3-CNF formulas with n variables and $m = rn$ clauses. A long series of papers analyzing so-called "myopic" DPLL algorithms has provided a sequence of lower bounds for their satisfiability threshold. Indeed, for each myopic algorithm $\mathcal{A}$ it is known that there exists an algorithm-specific clause-density, $r_{\mathcal{A}}$, such that if $r < r_{\mathcal{A}}$, the algorithm finds a satisfying assignment in linear time. For example, $r_{\mathcal{A}}$ equals $8/3 = 2.66..$ for ORDERRED-DLL and $3.003...$ for GENERALIZED UNIT CLAUSE. We prove that for densities well within the provably satisfiable regime, every backtracking extension of either of these algorithms takes *exponential* time. Specifically, all extensions of ORDERRED-DLL take exponential time for $r > 2.78$ and the same is true for GENERALIZED UNIT CLAUSE for all $r > 3.1$. Our results imply exponential lower bounds for many other myopic algorithms for densities similarly close to the corresponding $r_{\mathcal{A}}$.

1 Introduction

The problem of determining the satisfiability of Boolean formulas is central to computational complexity. Moreover, it is of tremendous practical interest as it arises naturally in numerous settings. Random CNF formulas have emerged as a mathematically tractable vehicle for studying the performance of satisfiability algorithms and proof systems. For a given set of n Boolean variables, let B_k denote the set of all possible disjunctions of k non-complementary literals on the variables (k-clauses). A random k-SAT formula $F_k(n, m)$ is formed by selecting uniformly and independently m clauses from B_k and taking their conjunction.

We will be interested in random formulas from an asymptotic point of view, i.e., as the number of variables grows. In particular, we will say that a sequence of random events $\mathcal{E}_n$ occurs *with high probability (w.h.p.)* if $\lim_{n \to \infty} \Pr[\mathcal{E}_n] = 1$. In this context, the ratio of constraints-to-variables, $r = m/n$, known as density, plays a fundamental role as most interesting monotone properties are believed to exhibit 0-1 laws with respect to density. Perhaps the best known example is the satisfiability property.

Conjecture 1. For each $k \geq 3$, there exists a constant r_k such that for any $\epsilon > 0$,

$$\lim_{n \to \infty} \Pr[F_k(n, (r_k - \epsilon)n)] = 1, \quad \text{and} \quad \lim_{n \to \infty} \Pr[F_k(n, (r_k + \epsilon)n)] = 0 .$$

A. Cimatti and R. Sebastiani (Eds.): SAT 2012, LNCS 7317, pp. 327–340, 2012.
© Springer-Verlag Berlin Heidelberg 2012

The satisfiability threshold conjecture above has attracted a lot of attention in computer science, mathematics and statistical physics. At this point, neither the value, nor even the existence of r_k has been established. In a breakthrough result, Friedgut [15] gave a very general condition for a monotone property to have a *non-uniform* sharp threshold. In particular, his result yields the statement of the conjecture if one replaces r_k with a function $r_k(n)$. For $k = 3$, the best known bounds are $3.52 < r_3 < 4.49$, due to results in [13] and [19], respectively.

A key feature of random k-CNF formulas is that their underlying hypergraph is locally tree-like for every finite density, i.e., for both satisfiable and unsatisfiable formulas. One implication of this fact is that the formula induced by any finite-depth neighborhood of any variable is highly under-constrained. As a result, unsatisfiability comes about due to long-range interactions between variables something that appears hard to capture by efficient algorithms. In particular, random formulas have been shown to be hard both for proof systems, e.g., in the seminal work of Chvátal and Szemerédi on resolution [10], and, more recently, for some of the most sophisticated satisfiability algorithms known [11]. More generally, for the connections of random formulas to proof-complexity and computational-hardness see the surveys by Beame and Pitassi [7] and Cook and Mitchell [12], respectively.

The last decade has seen a great deal of rigorous results on random CNF formulas, including a proliferation of upper and lower bounds for the satisfiability threshold. Equally importantly, random CNF formulas have been the domain of an extensive exchange of ideas between computer science and statistical physics, including the discovery of the clustering phenomenon [21,20], establishing it rigorously [3], and relating it to algorithmic performance [11]. In this work we take another step in this direction by taking a technique from mathematical physics, the *interpolation method* [18,14,24,6], and using it to derive rigorous upper bounds for the satisfiability threshold of random CNF formulas that are mixtures or 2- and 3-clauses. As we discuss below, such formulas arise naturally as residual formulas in the analysis of satisfiability algorithms and their unsatisfiability implies exponential lower bounds for the running time of a large class of algorithms. Our main result is the following.

Theorem 1. *Let F be a random CNF formula on n variables with $(1 - \epsilon)n$ random 2-clauses, and $(1 + \epsilon)n$ random 3-clauses. W.h.p. F is unsatisfiable for $\epsilon = 10^{-4}$.*

Our method for proving Theorem 1 involves estimating an infinite sum with no close form, any truncation of which yields a rigorous bound. The choice of 10^{-4} is rather arbitrary as our methods can deliver arbitrarily small $\epsilon > 0$, given enough computational resources. We have chosen $\epsilon = 10^{-4}$ as it can be checked readily with very modest computation.

2   Background and Motivation

Many algorithms for finding satisfying assignments for CNF formulas operate by building a partial assignment step by step. These algorithms commit to the assignments made at each step and operate on a *residual formula*, in which clauses already satisfied have been removed, while the remaining clauses have been shortened by the removal of their falsified literals. We call such algorithms *forward search* algorithms. During the execution of any such algorithm a partial assignment may produce clauses of size 1 (unit clauses) in the residual formula which in turn create additional *forced* choices in the partial assignment, since the variables appearing in unit clauses have only one possible assignment if the formula is to be satisfied. The choices made by a forward search algorithm when no unit clauses are present are called *free*.

A large class of natural DPLL algorithms are "myopic" in that their free-step choices are based on local considerations in terms of the underlying hypergraph. Perhaps the simplest such algorithm is ORDERRED-DLL which performs unit-clause propagation but, otherwise, sets variables in some a priori fixed random order/sign. Another example is GENERALIZED UNIT CLAUSE (GUC) [22,16], where in each step a random literal in a random shortest clause is assigned true. The key property of myopic algorithms that makes their analysis mathematically tractable is the following (indeed, this can be seen as a definition of myopic algorithms): as long as the algorithm has never backtracked, the residual formula is uniformly random conditional on its number of 2- and 3-clauses (unit-clauses are satisfied as soon as they occur).

To analyze the performance of myopic algorithms on random formulas one employs the standard technique of approximating the mean path of Markov chains by differential equations in order to keep track of the 2- and 3-clause density of the residual formula. As is well understood, in the large n limit, both of these densities behave as deterministic functions, for every myopic algorithm. In the absence of backtracking, i.e., if the algorithm continues blithely on after a 0-clause is generated, this means that for any given initial 3-clause density r, we can model the algorithm's behavior as a continuous 2-dimensional curve $(d_2^r(x), d_3^r(x))$ of the 2- and 3-clause density, where $x \in [0, 1]$ denotes the fraction of assigned variables. Since the 2-SAT satisfiability threshold [10,17] is $r_2 = 1$, it follows that for any initial 3-clause density $r > 0$ and every $\gamma > 0$ such that $d_2^r(x) < 1$ for all $x \in [0, \gamma)$, the probability that no 0-clause is ever generated is bounded away from 0. Indeed, to determine the threshold r_A for each myopic algorithm it suffices to determine the largest r such that $d_2^r(x) < 1$ for all $x \in [0, 1)$. This is because as long as $d_2^r(x) < 1$, w.h.p. 0-clauses are generated for trivial local reasons. In particular, as was shown by Frieze and Suen [16], there exists a very simple form of backtracking which never flips the value of any variable more than once, such that endowing any myopic algorithm with this backtracking boosts its probability of finding a satisfying assignment to $1 - o(1)$ for all $r < r_A$.

To understand what happens for $r > r_A$, let us consider what happens if one gives as input to a myopic algorithm $\mathcal{A}$ a random 3-CNF formula of density $r > r_A$, but only runs the algorithm for $x_0 \cdot n$ steps where x_0 is such that $d_2^r(x) < 1$ for all $x \in [0, x_0)$. Up to that point, the algorithm will have either not backtracked at all, or backtracked for trivial local reasons, so that the residual formula will be a mixture of random 2- and 3-clauses in which the 2-clauses alone are satisfiable. Naturally, if the residual formula is satisfiable the algorithm still has a chance of finding a satisfying assignment in polynomial time. But what happens if this mixture, as a whole, is unsatisfiable? How fast will it discover this and backtrack? In [2] it was shown that the resolution complexity of unsatisfiable random mixtures of 2- and 3-clauses in which the 2-clause are satisfiable is exponential. Since every DPLL algorithm produces a resolution proof of unsatisfiability, it follows that if the residual mixture is unsatisfiable, the algorithm will take exponential time to establish its unsatisfiability.

To delineate satisfiable from unsatisfiable mixtures, define Δ_c to be the largest Δ such that for every $\epsilon > 0$, a mixture of $(1 - \epsilon)n$ 2-clauses and Δn 3-clauses is w.h.p. satisfiable. In [4] it was proven that $2/3 \leq \Delta_c < 2.28...$ The upper bound, combined with the differential equations analysis mentioned above was used in [2] to prove that if ORDERED-DLL is started with $3.81n$ random 3-clauses it will reach a stage where the residual formula has exponential resolution complexity (and, therefore, take exponential time on such formulas). Similarly, for GUC started with $3.98n$ random 3-clauses.

By establishing $\Delta_c < 1.001$, the exact same analysis as in [2] allows us to prove that each of these algorithms fails for much lower densities, well within the proven satisfiable regime. Specifically, while ORDERED-DLL succeeds in finding a satisfying assignment in linear time up to $8/3 = 2.66...$ we prove that it already requires exponential time at $r > 2.71$. Similarly, while GUC succeeds in linear time for $r < 3.003$, we prove that it requires exponential time at $r > 3.1$. We state both of this results more precisely in the next section, after discussing the different types of backtracking that one can consider.

We note that these two explicit results for ORDERED-DLL and GUC are simply indicative and Theorem 1 can be applied to prove similar bounds for all myopic algorithms. This includes all algorithms in [1] and many others. In fact, our Theorem 1 can be generalized to random mixtures of 2- and 3-clauses with a given degree sequence, thus also covering algorithms such as the one in [19].

2.1 Backtracking

When a path in the search tree leads to a contradiction, the algorithm must begin backtracking by undoing all the (forced) choices up to the last free choice and flipping the assignment to that variable. From there, perhaps the simplest option would be for the algorithm to act as if it had reached this point without backtracking and apply the original heuristic to decide which variable(s) to set next.

As long as the 2-clause density stays below 1 it is not hard to show that any such backtracking w.h.p. is due to trivial "local" reasons and can be fixed by changing the value of $O(\log n)$ variables (typically $O(1)$ variables suffice). From a technical point of view, though, such backtracking (minimally) disturbs the uniform randomness property of the residual formula, enough to make the statement of crisp mathematical statements cumbersome.

An alternative heuristic, due to Frieze and Suen [16], which we call FS-backtracking is the following: when a contradiction is reached, record the portion of the assignment between the last free choice and the contradiction; these literals become *hot*. After flipping the value of the last free choice, instead of making the choice that the original heuristic would suggest, give priority to the complements of the hot literals in the order that they appeared; once the hot literals are exhausted continue as with the original heuristic. FS-backtracking is quite natural in that this last part of the partial assignment got us into trouble in the first place.

A key property of FS-backtracking that is useful in analysis is that as long as the value of each variable in a partial assignment has been flipped at most once, the residual formula is perfectly uniformly random conditional on the number of clauses of each size. We emphasize that while the original motivation for introducing FS-backtracking is technical, such backtracking is, in fact, a genuinely good algorithmic idea. Specifically, on random 3-CNF formulas with densities between 3.8 and 4.0, large experiments show that the histogram of run-times of FS-backtracking is significantly better than simple backtracking. We will denote a forward search algorithm $\mathcal{A}$ extended with FS-backtracking by $\mathcal{A}$-FS.

Let us say that a DPLL algorithm is at a *t-stage* if precisely t variables have been set.

Definition 1. *Let* $\epsilon = 10^{-4}$. *A t-stage of a DPLL algorithm is* bad *if the residual formula at that stage is the union of a random 2-CNF formula with* $(1-\epsilon)(n-t)$ *2-clauses and a random 3-CNF formula with* $(1+\epsilon)(n-t)$ *clauses, where* $t \le n/2$.

Definition 1 is identical to that of bad stages in [2], except that they have $2.28+\epsilon$ instead of our $1 + \epsilon$, since $\Delta_c \le 2.28$ was the best known bound prior to our work. Proceeding exactly as in [2], i.e., by using the differential equations method to determine the smallest initial 3-clause density such that the algorithm eventually reaches a bad stage, we get the following.

Lemma 1. *Let* $\Delta_{\text{ORDERED-DLL}} = 2.71$ *and let* $\Delta_{\text{GUC}} = 3.1$.

1. *For each* $\mathcal{A} \in \{\text{ORDERED-DLL,GUC}\}$, *an execution of any backtracking extension of* $\mathcal{A}$ *on a random 3-CNF formula with* $\Delta_{\mathcal{A}} \cdot n$ *clauses reaches a* bad *t-stage with constant probability.*
2. *For each* $\mathcal{A} \in \{\text{ORDERED-DLL,GUC}\}$, *an execution of algorithm* $\mathcal{A}$-FS *on a random 3-CNF formula with* $\Delta_{\mathcal{A}} \cdot n$ *clauses reaches a* bad *t-stage w.h.p.*

Theorem 2. *Let* $\Delta_{\text{UC}} = \Delta_{\text{ORDERED-DLL}} = 2.71$ *and let* $\Delta_{\text{GUC}} = 3.1$.

1. *For each* $A \in \{\text{ORDERED-DLL,GUC}\}$, *an execution of any backtracking extension of* A *on a random 3-CNF formula with* $\Delta_A n$ *clauses takes time* $2^{\Omega(n)}$ *with constant probability.*
2. *For each* $A \in \{\text{ORDERED-DLL,GUC}\}$, *an execution of algorithm* A-FS *on a random 3-CNF formula with* $\Delta_A n$ *clauses takes time* $2^{\Omega(n)}$ *w.h.p.*

2.2 Proving Upper Bounds for Satisfiability Thresholds

The simplest upper bound on the satisfiability threshold of random k-CNF formulas comes from taking the union bound over all assignments $\sigma \in \{0,1\}^n$ of the probability that each one is satisfying. That is,

$$\Pr[F_k(n, rn) \text{ is satisfiable}] \le \sum_\sigma \Pr[\sigma \text{ satisfies } F_k(n, rn)] = \left[2(1 - 2^{-k})^r\right]^n \to 0 \ ,$$

for all $r > r_k^*$, where $2(1 - 2^{-k})^{r_k^*} = 1$. It is easy to see that $r_k^*/(2^k \ln 2) \to 1$.

Note that the above argument holds even for $k = n$, in which case satisfiability reduces to the coupon collector's problem over $\{0,1\}^n$. By standard results, in this case the number of clauses to cover the cube is very close to $2^n \ln(2^n) = n2^n \ln 2$.

Aggregating these assignments, for the sake of comparison, into groups of size 2^{n-k} so that they are comparable to k-clauses, recovers the union bound above. In other words, the simplicity (and the weakness) of the union bound is that it treats the 2^{n-k} *distinct* assignments forbidden by each k-clause as a random multiset of the same size. As one can imagine, this phenomenon is stronger as the cubes are larger, i.e., for smaller values of k. For example, $r_3^* = 5.19..$, but a long series of increasingly sophisticated results has culminated with the bound $r_3 < 4.49$ by Díaz et al. [13]. In the extreme case $k = 1$, the birthday paradox readily implies that a collection of $\Theta(n^{1/2})$ random 1-clauses is *w.h.p.* unsatisfiable, yet the union bound only gives $r_1 \leq 1$, i.e., requires $\Omega(n)$ 1-clauses.

For the special case $k = 2$, it has long been shown, independently, by Chvátal and Reed [9] and Goerdt [17] that $r_2 = 1$. In all these proofs, the fact $r_2 \leq 1$ is established by exploiting the existence of succinct certificates of unsatisfiability for 2-SAT enabling proofs that proceed by identifying "most likely" unsatisfiable subformulas in the evolution of $F_2(n, rn)$. Intriguingly, early non-rigorous arguments of statistical physics [23] recover the fact $r_2 \leq 1$ *without* relying on the fact that 2-SAT is in P. It is precisely this feature that we exploit in the present work.

2.3   Applying the Interpolation Method to Random CNFs Formulas

To prove Theorem 1 we abandon standard combinatorial proofs of unsatisfiability and turn to a remarkable tool developed by Francesco Guerra [18], called the *interpolation method*, to deal with the Sherrignton Kirkpatrick model (SK) of statistical physics. Following Guerra's breakthrough, Franz and Leone [14], in a very important paper, applied the interpolation method to random k-SAT and random XOR-SAT to prove that certain expressions derived via the non-rigorous replica method of statistical physics for these problems, correspond to rigorous lower bounds for the free energy of each problem. As such, these expressions can, in principle, be used to derive upper bounds for the satisfiability threshold of each problem, yet this involves the solution of certain functional equations that appear beyond analytical penetration. In [24], Panchenko and Talagrand showed that the results of [14] can be derived in a simpler and uniform way, unifying the treatment of different levels of Parisi's Replica Symmetry Breaking.

In a recent paper [6], Bayati, Gamarnik and Tetali, showed that a combinatorial analogue of the interpolation method can be used to elegantly derive an approximate subadditivity property for a number of CSPs on Erdős-Renyi and regular random graphs. This allowed them to prove the existence of a number of limits in these problems, including the existence of a limit for the size of the largest independent set in a random regular graph. At the same time, the simplicity of their combinatorial approach comes at the cost of losing the capacity to yield quantitative bounds for the associated limiting quantities.

To overcome these problems we will apply a recently developed [5] *energetic* interpolation method to random mixtures of 2- and 3-clauses. This, implicitly, exploits the second order nature of random 2-SAT phase transitions to gain in computational tractability.

3 The Energetic Interpolation Method

In the more standard models of random k-SAT, the number of clauses m is fixed (not a random variable). The interpolation method requires that m is a *Poisson random variable* with mean $\mathbb{E}[m] = rn$. Since the standard deviation of the Poisson distribution is the square root of its mean we have $m = (1 + o(1))rn$ *w.h.p.*, thus not affecting any asymptotic results regarding densities.

We shall work with the random variable $H_{n,r}(\sigma)$, know as the Hamiltonian, counting the number of unsatisfied clauses in the instance for each $\sigma \in \{0,1\}^n$. We will sometimes refer to $H_{n,r}(\sigma)$ as the energy function. The goal of the method is to compute lower bounds on the following quantity

$$\xi_r = n^{-1}\mathbb{E}\left[\min_{\sigma \in \{0,1\}^n} H_{n,r}(\sigma)\right] . \tag{1}$$

Note that proving $\liminf_{n \to \infty} \xi_r > 0$ implies that the satisfiability threshold is upper bounded by r, since the random variable $n^{-1}\min_\sigma H_{n,r}(\sigma)$ is known to be concentrated in a $o(n)$ window [8].

Given $\sigma = (x_1, x_2, \ldots, x_n)$ we will write $H_{n,r}(\sigma)$ as the sum of m functions $E_a(x_{a_1}, \ldots, x_{a_k})$, one for each clause a. That is, $E_a(x_{a_1}, \ldots, x_{a_k}) = 1$ if the associated clause is not satisfied and 0 otherwise. The basic object of the energetic interpolation method is a modified energy function that interpolates between $H_{n,r}(\sigma)$ and the energy function of a dramatically simpler (and fully tractable) model. Specifically, for $t \in [0,1]$, let

$$H_{n,r,t}(x_1, \ldots, x_n) = \sum_{m=1}^{m_t} E_{a_m}(x_{a_{m,1}}, \ldots, x_{a_{m,k}}) + \sum_{i=1}^{n}\sum_{j=1}^{k_{i,t}} \hat{h}_{i,j}(x_i) , \tag{2}$$

where m_t is a Poisson random variable with mean $\mathbb{E}[m_t] = trn$, the $k_{i,t}$'s are i.i.d. Poisson random variables with mean $\mathbb{E}[k_{i,t}] = (1 - t)kr$, and the functions $\hat{h}_{i,j}(\cdot)$ are i.i.d. random functions distributed as the function of (4) below. Before delving into the meaning of the random functions $\hat{h}_{i,j}(\cdot)$, which are the heart of the method, let us first make a few observations about (2). To begin with, note that for $t = 1$, equation (2) is simply the energy function of the original model, i.e., a sum of m functions counting whether each clause has been violated or not. On the other hand, for $t < 1$, we see that, in expectation, $(1 - t)m$ of these clause-functions have been replaced by k times as many $\hat{h}$-functions, each of which takes as input the value of a single variable in the assignment. A good way to think about this replacement is as a decombinatorialization of the energy function wherein (combinatorial) k-ary functions are replaced by univariate functions. As one can imagine, for $t = 0$ the model is fully tractable. In particular, if

$$\xi_r(t) = \frac{1}{n}\mathbb{E}\left[\min_{\sigma \in \{0,1\}^n} H_{n,r,t}(\sigma)\right] , \tag{3}$$

one can readily compute $\xi_r(0)$.

The main idea of the interpolation method is to select the univariate functions $\hat{h}_i(\cdot)$ independently, from a probability distribution that reflects aspects of the geometry of the underlying solution space. A particularly appealing aspect of the energetic interpolation method is that it projects all information about the geometry of the solution space into a single probability p, which can be interpreted as the probability that a variable picked at random will be frozen, i.e., have the same value in all optimal assignments. The method then delivers a valid bound for *any* choice of $p \in [0, 1]$ and the bound is then optimized by choosing the best value of p, i.e., performing a single-parameter search.

Let "1", "0", and "$*$" denote the binary functions $\{h(0) = 1, h(1) = 0\}$, $\{h(0) = 0, h(1) = 1\}$, and $\{h(0) = 0, h(1) = 0\}$ respectively. One can think of function 1 as being 1 (unhappy) when the input is not 1, of function 0 as being 1 when the input is not 0, and of function $*$ as never being 1. Let $h(x)$ be a random function in $\{$"0", "1", "$*$"$\}$ with $\Pr(h(\cdot) = $ "1"$) = \Pr(h(\cdot) = $ "0"$) = p/2$ and let the random function $\hat{h}(x)$ be defined as follows

$$\hat{h}(x) = \min_{y_1,\ldots,y_{k-1}} \left\{ E(y_1, .., y_{k-1}, x) + \sum_{i=1}^{k-1} h_i(y_i) \right\} , \qquad (4)$$

where $E(\cdot)$ is a random clause-function and the functions $h_i(\cdot)$ are i.i.d. random functions distributed as $h(x)$ i.e. $\hat{h}(\cdot) = $ "1" with probability $2^{-k}p^{k-1}$.

The main point of the interpolation method is that as t goes from 1 to 0, we can control in the change of $\xi_r(t)$, hence the name. Specifically, one has the following.

Theorem 3 ([5]).

$$\xi_r \geq \xi_r(0) - r(k-1)2^{-k}p^k . \qquad (5)$$

Determining $\xi_r(0)$ is a tractable task and establishing $\xi_r > 0$ implies that $F_k(n, rn)$ is w.h.p. unsatisfiable.

For completeness, we present the proof of Theorem 3 in Appendix A.

4   Energy Density Bounds for $(2 + p)$-SAT

Let $F_{2,3}(n, \epsilon, \Delta)$ denote a random CNF formula over n variables consisting of m_2 random 2-CNF clauses, where m_2 is a Poisson random variable with mean $\mathbb{E}[m_2] = (1 - \epsilon)n$ and m_3 random 3-CNF clauses, where m_3 is a Poisson random variable with mean $\mathbb{E}[m_3] = \Delta n$. Thus, the energy function is now

$$H_{n,\epsilon,\Delta}(\sigma) = H_{n,1-\epsilon}^{(2)}(\sigma) + H_{n,\Delta}^{(3)}(\sigma) ,$$

where $H_{n,1-\epsilon}^{(2)}(\sigma)$ is the k-SAT energy function for $k = 2$ and $r = 1 - \epsilon$, while $H_{n,\Delta}^{(3)}(\sigma)$ is the energy function for $k = 3$ and $r = \Delta$. Similarly the interpolation function is the sum of the two independent interpolation functions corresponding to $k = 2$ and $k = 3$, i.e.,

$$H_{n,\epsilon,\Delta,t}(x_1,\ldots,x_n) = H_{n,1-\epsilon,t}^{(2)}(x_1,\ldots,x_n) + H_{n,\Delta,t}^{(3)}(x_1,\ldots,x_n) . \qquad (6)$$

Letting

$$\xi_{\epsilon,\Delta}(t) = n^{-1}\mathbb{E}\left[\min_{\sigma \in \{0,1\}^n} H_{n,\epsilon,\Delta,t}(\sigma) \right] , \qquad (7)$$

the analogue of Theorem 3 for random mixtures of 2- and 3-clauses is the following.

Theorem 4. *For every value of $p \in [0, 1]$,*

$$\xi_{\epsilon,\Delta} \geq \xi_{\epsilon,\Delta}(0) - \frac{1}{4}(1 - \epsilon)p^2 - \frac{1}{4}\Delta p^3 \ . \tag{8}$$

Proof. As with Theorem 3, the probability joint distribution implicit in the expectation of $\xi_{\epsilon,\Delta}(t)$ can be written as the product of Poisson random functions, due to the independence among the random variables appearing in $H_{n,\epsilon,\Delta,t}(x_1, \ldots, x_n)$. Now, the derivative with respect to t gives rise to two independent set of equations similar to the ones in (14) and (15) for $k = 2$ and $k = 3$, where the base energy function is $H_{n,\epsilon,\Delta}(\sigma)$. Since all the relevant properties of the mixture are captured by its set of frozen variables, the theorem follows simply by applying the proof of (12) in Theorem 3 twice.

5   Application to $(2 + p)$-SAT

In this section we give first an analytical expression for $\xi_{\epsilon,\Delta}(0)$ and then compute a lower bound for it. We have

$$\xi_{\epsilon,\Delta}(0) = n^{-1}\mathbb{E}\left[\min_{\sigma \in \{0,1\}^n} H_{n,\epsilon,\Delta,0}(\sigma)\right]$$

$$= n^{-1}\mathbb{E}\left[\min_{\sigma \in \{0,1\}^n} \left(\sum_{i=1}^n \left(\sum_{j=1}^{k_{2,i}} \hat{h}_{2,i,j}(x_i) + \sum_{j=1}^{k_{3,i}} \hat{h}_{3,i,j}(x_i)\right)\right)\right]$$

$$= n^{-1}\mathbb{E}\left[\sum_{i=1}^n \min_{x_i \in \{0,1\}} \left(\sum_{j=1}^{k_{2,i}} \hat{h}_{2,i,j}(x_i) + \sum_{j=1}^{k_{3,i}} \hat{h}_{3,i,j}(x_i)\right)\right]$$

$$= n^{-1}\sum_{i=1}^n \mathbb{E}\left[\min_{x_i \in \{0,1\}} \left(\sum_{j=1}^{k_{2,i}} \hat{h}_{2,i,j}(x_i) + \sum_{j=1}^{k_{3,i}} \hat{h}_{3,i,j}(x_i)\right)\right] \ ,$$

where the $k_{2,i}$'s and the $k_{3,i}$'s are Poisson random variables with means $2(1-\epsilon)$ and 3Δ respectively, as defined in (6). Note now that the n expectations in the above summation are identical, thus

$$\xi_{\epsilon,\Delta}(0) = \mathbb{E}\left[\min_{x \in \{0,1\}} \left(\sum_{j=1}^{s_2} \hat{h}_{2,j}(x) + \sum_{j=1}^{s_3} \hat{h}_{3,j}(x)\right)\right] \ , \tag{9}$$

where s_2 and s_3 are Poisson random variables with means $2(1-\epsilon)$ and 3Δ respectively, and the functions $\hat{h}_{2,j}(\cdot)$ and $\hat{h}_{3,j}(\cdot)$ are i.i.d. copies of the function $\hat{h}(\cdot)$ in (4) for $k = 2$ and $k = 3$, respectively, i.e., random functions in {"0", "1", " * "} with $\Pr(\hat{h}_{k,j}(\cdot) = $ "1") = $\Pr(\hat{h}_{k,j}(\cdot) = $ "0") = $2^{-k}p^{k-1}$.

Let $l_{k,0}$, $l_{k,1}$, and $l_{k,*}$ denote the number "0", "1", and "*" functions, respectively among the $\hat{h}_{k,j}(\cdot)$ functions inside the summation in (9). Conditional on the value of s_k, the random vector $(l_{k,0}, l_{k,1}, l_{k,*})$ is distributed as a multinomial random vector with s_k trials and probability vector $(2^{-k}p^{k-1}, 2^{-k}p^{k-1}, 1 - 2^{-k+1}p^{k-1})$, therefore,

$$
\xi_{\epsilon,\Delta}(0) = \sum_{x=0}^{\infty} \sum_{y=0}^{\infty} \sum_{l_{2,0}=0}^{x} \sum_{l_{2,1}=0}^{x-l_{2,0}} \sum_{l_{3,0}=0}^{y} \sum_{l_{3,1}=0}^{y-l_{2,0}} \min\{l_{2,0} + l_{3,0}, l_{2,1} + l_{3,1}\} \times
$$
$$
\text{Poi}(2(1 - \epsilon), x)\text{Multi}(l_{2,0}, l_{2,1}, x - l_{2,0} - l_{2,1}) \times
$$
$$
\text{Poi}(3\Delta, y)\text{Multi}(l_{3,0}, l_{3,1}, y - l_{3,0} - l_{3,1}) ,
$$

where $\text{Multi}(\cdot, \cdot, \cdot)$ denotes the multinomial density function.

Changing the limits of all summations to infinity, does not change the value of $\xi_{\epsilon,\Delta}(0)$, since $\text{Multi}(\cdot, \cdot, \cdot)$ evaluates to zero for negative numbers, hence, we can interchange the order of the summations to get

$$
\xi_{\epsilon,\Delta}(0) = \sum_{l_{2,0}=0}^{\infty} \sum_{l_{2,1}=0}^{\infty} \sum_{l_{3,0}=0}^{\infty} \sum_{l_{3,1}=0}^{\infty} \min\{l_{2,0} + l_{3,0}, l_{2,1} + l_{3,1}\} \times
$$
$$
\sum_{x=0}^{\infty} \text{Poi}(2(1 - \epsilon), x)\text{Multi}(l_{2,0}, l_{2,1}, x - l_{2,0} - l_{2,1}) \times
$$
$$
\sum_{y=0}^{\infty} \text{Poi}(3\Delta, y)\text{Multi}(l_{3,0}, l_{3,1}, y - l_{3,0} - l_{3,1}) .
$$

The last equation can be simplified by summing out the randomness in the Poisson random variables. The result is that $l_{2,0}$ and $l_{2,1}$ become two independent Poisson random variables with mean $\frac{1}{2}(1 - \epsilon)p$, that is,

$$
\sum_{x=0}^{\infty} \text{Poi}(2(1 - \epsilon), x)\text{Multi}(l_{2,0}, l_{2,1}, x - l_{2,0} - l_{2,1}) =
$$
$$
\text{Poi}((1 - \epsilon)p/2, l_{2,0}) \times \text{Poi}((1 - \epsilon)p/2, l_{2,1}) .
$$

Similarly, $l_{3,0}$ and $l_{3,1}$ become two independent Poisson random variables with mean $\frac{3}{8}\Delta p^2$. Moreover, letting

$$
\lambda = \frac{1}{2}(1 - \epsilon)p + \frac{3}{8}\Delta p^2 ,
$$

we see that $l_0 = l_{2,0} + l_{3,0}$ is itself a Poisson random variable with mean λ, since the sum of two independent Poisson random variables with means λ_1 and λ_2 is a Poisson random variable with mean $\lambda = \lambda_1 + \lambda_2$. Thus,

$$
\xi_{\epsilon,\Delta}(0) = \sum_{l_0=0}^{\infty} \sum_{l_1=0}^{\infty} \min\{l_0, l_1\} \times \text{Poi}(\lambda, l_0) \times \text{Poi}(\lambda, l_1) ,
$$

i.e., $\xi_{\epsilon,\Delta}(0)$ is the expected value of the minimum of two independent Poisson random variables l_0, l_1 with mean λ. Consequently, the bound of Theorem 4 becomes

$$
\xi_{\epsilon,\Delta} \geq \mathbb{E}\left[\min\{l_0, l_1\}\right] - \frac{1}{4}(1 - \epsilon)p^2 - \frac{1}{4}\Delta p^3 . \tag{10}
$$

Finally, we note that

$$\mathbb{E}\left[\min\{l_0, l_1\}\right] = \sum_{i=0}^{\infty} i \left(2\mathrm{Poi}(\lambda, i) \left(1 - \sum_{j=0}^{i-1} \mathrm{Poi}(\lambda, j) \right) - (\mathrm{Poi}(\lambda, i))^2 \right) . \quad (11)$$

Thus, to compute lower bounds for (10) is enough to truncate (11) at any value of i. In particular, by letting $\epsilon = 0.0001$, $\Delta = 1.0001$ and $i = 50$, we get that for $p = 1.2 \cdot 10^{-3}$ the truncated version of (10) is greater than 0, implying that a random CNF formula with $0.9999n$ 2-clauses and $1.0001n$ 3-clauses is w.h.p. unsatisfiable.

References

1. Achlioptas, D.: Lower bounds for random 3-sat via differential equations. Theoretical Computer Science 265(1-2), 159–185 (2001)
2. Achlioptas, D., Beame, P., Molloy, M.: A sharp threshold in proof complexity yields lower bounds for satisfiability search. Journal of Computer and System Sciences 68(2), 238–268 (2004)
3. Achlioptas, D., Coja-Oghlan, A.: Algorithmic barriers from phase transitions. In: IEEE 49th Annual IEEE Symposium on Foundations of Computer Science, FOCS 2008, pp. 793–802. IEEE (2008)
4. Achlioptas, D., Kirousis, L.M., Kranakis, E., Krizanc, D.: Rigorous results for random (2+ p)-SAT. Theoretical Computer Science 265(1), 109–129 (2001)
5. Achlioptas, D., Menchaca-Mendez, R.: Unsatisfiability bounds for random csps from an energetic interpolation method (2012) (to appear in ICALP 2012)
6. Bayati, M., Gamarnik, D., Tetali, P.: Combinatorial approach to the interpolation method and scaling limits in sparse random graphs. In: STOC 2010, pp. 105–114 (2010)
7. Beame, P., Pitassi, T.: Propositional proof complexity: Past, present and future. Current Trends in Theoretical Computer Science, 42–70 (2001)
8. Broder, A.Z., Frieze, A.M., Upfal, E.: On the satisfiability and maximum satisfiability of random 3-cnf formulas. In: Proceedings of the Fourth Annual ACM-SIAM Symposium on Discrete Algorithms, SODA 1993, pp. 322–330. Society for Industrial and Applied Mathematics, Philadelphia (1993),
   http://dl.acm.org/citation.cfm?id=313559.313794
9. Chvátal, V., Reed, B.: Mick gets some (the odds are on his side) [satisfiability]. In: Proceedings 33rd Annual Symposium on Foundations of Computer Science, pp. 620–627. IEEE (1992)
10. Chvatal, V., Szemeredi, E.: Many hard examples for resolution. Journal of the Association for Computing Machinery 35(4), 759–768 (1988)
11. Coja-Oghlan, A.: On belief propagation guided decimation for random k-sat. In: Proceedings of the Twenty-Second Annual ACM-SIAM Symposium on Discrete Algorithms, pp. 957–966. SIAM (2011)
12. Cook, S., Mitchell, D.: Finding hard instances of the satisfiability problem. In: Satisfiability Problem: Theory and Applications: DIMACS Workshop, March 11-13, vol. 35, p. 1. Amer. Mathematical Society (1997)
13. Díaz, J., Kirousis, L., Mitsche, D., Pérez-Giménez, X.: On the satisfiability threshold of formulas with three literals per clause. Theoretical Computer Science 410(30-32), 2920–2934 (2009)

14. Franz, S., Leone, M.: Replica bounds for optimization problems and diluted spin systems. Journal of Statistical Physics 111(3), 535–564 (2003)
15. Friedgut, E.: Sharp thresholds of graph properties, and the k-sat problem. J. Amer. Math. Soc. 12, 1017–1054 (1998)
16. Frieze, A., Suen, S.: Analysis of Two Simple Heuristics on a Random Instance ofk-sat. Journal of Algorithms 20(2), 312–355 (1996)
17. Goerdt, A.: A threshold for unsatisfiability. Journal of Computer and System Sciences 53(3), 469–486 (1996)
18. Guerra, F., Toninelli, F.L.: The thermodynamic limit in mean field spin glass models. Communications in Mathematical Physics 230(1), 71–79 (2002)
19. Kaporis, A.C., Kirousis, L.M., Lalas, E.G.: The probabilistic analysis of a greedy satisfiability algorithm. Random Structures & Algorithms 28(4), 444–480 (2006)
20. Mézard, M., Mora, T., Zecchina, R.: Clustering of solutions in the random satisfiability problem. Physical Review Letters 94(19), 197205 (2005)
21. Mézard, M., Parisi, G., Zecchina, R.: Analytic and algorithmic solution of random satisfiability problems. Science 297(5582), 812–815 (2002)
22. Ming-Te, C., Franco, J.: Probabilistic analysis of a generalization of the unit-clause literal selection heuristics for the <i> k</i> satisfiability problem. Information Sciences 51(3), 289–314 (1990)
23. Monasson, R., Zecchina, R.: Statistical mechanics of the random k-satisfiability model. Phys. Rev. E 56, 1357–1370 (1997),
http://link.aps.org/doi/10.1103/PhysRevE.56.1357
24. Panchenko, D., Talagrand, M.: Bounds for diluted mean-fields spin glass models. Probability Theory and Related Fields 130(3), 319–336 (2004)

A   Proof of Theorem 3

Proof. Since $\xi_r(1) = \xi_r(0) + \int_0^1 \xi_r'(t)dt$ and since $r(k-1)2^{-k}p^k$ does not depend on t, it suffices to show that $-r(k-1)2^{-k}p^k$ is an lower bound for $\xi_r'(t)$, i.e., we have to show that for all $t \in [0,1]$,

$$\xi_r'(t) \geq -r(k-1)2^{-k}p^k \ . \qquad (12)$$

We begin by computing $\xi_r'(t)$. Let $\min_\sigma H_m(\sigma)$ and $\min_\sigma H_{k_i}(\sigma)$ denote the random variable $\min_\sigma H_{n,t}(\sigma)$ conditioned on the values of the random variables m_t and $k_{i,t}$ respectively, that is

$$\min_\sigma H_m(\sigma) = \min_\sigma H_{n,r,t}(\sigma)\Big|_{m_t=m} \quad \text{and} \quad \min_\sigma H_{k_i}(\sigma) = \min_\sigma H_{n,r,t}(\sigma)\Big|_{k_{i,t}=k_i}$$

and more generally

$$\min_\sigma H_{m,k_1,\ldots,k_n}(\sigma) = \min_\sigma H_{n,r,t}(\sigma)\Big|_{m_t=m,k_{1,t}=k_1,\ldots,k_{n,t}=k_n} \ .$$

Denote the Poisson density function with mean μ as $\text{Poi}(\mu, z) = e^{-\mu}(\mu^z/z!)$. Since the random variable m_t and the random variables $k_{i,t}$ are independent, we can write the expectation in (3) as

$$\xi_r(t) = \sum_{m,k_1,\ldots,k_n} \text{Poi}(trn,m) \prod_{i=1}^{n} \text{Poi}((1-t)rk,k_i) \frac{1}{n}\mathbb{E}[\min_{\sigma} H_{m,k_1,\ldots,k_n}(\sigma)] \ .$$

By differentiating $\xi_r(t)$ with respect to t we get

$$\xi_r'(t) = \sum_{m=0}^{\infty} \frac{\partial}{\partial t}\text{Poi}(trn,m)\frac{1}{n}\mathbb{E}[\min H_m(\sigma)] + \tag{13}$$

$$\sum_{i=1}^{n}\sum_{k_i=0}^{\infty} \frac{\partial}{\partial t}\text{Poi}((1-t)rk,k_i)\frac{1}{n}\mathbb{E}[\min H_{k_i}(\sigma)] \ .$$

Recall now that $(\partial/\partial t)\text{Poi}(trn,m) = -rn\text{Poi}(trn,m) + rn\text{Poi}(trn,m-1)$. Thus, the derivative with respect to t in the first summation in (13) can be written as

$$-r\sum_{m=0}^{\infty} \text{Poi}(trn,m)\mathbb{E}[\min H_m(\sigma)] + r\sum_{m=1}^{\infty} \text{Poi}(trn,m-1)\mathbb{E}[\min H_m(\sigma)] =$$

$$r\sum_{m=0}^{\infty} \text{Poi}(trn,m)\left[\mathbb{E}[\min H_{m+1}] - \mathbb{E}[\min H_m]\right] \ . \tag{14}$$

Similarly, the derivatives in the double sum in (13) with respect to t can be written as

$$-rk\frac{1}{n}\sum_{i=1}^{n}\sum_{k_i=0}^{\infty} \text{Poi}((1-t)rk,k_i)\left[\mathbb{E}[\min H_{k_i+1}] - \mathbb{E}[\min H_{k_i}]\right] \ . \tag{15}$$

Now, a crucial observation is that (14) is r times the expected value of the change in $\min H$ after adding a random clause, while (15) is $-rk$ times the expected value of the change in $\min H$ after adding a single $\hat{h}$ function whose argument is a variable selected uniformly at random. Thus, to establish (12) we need to show that the expected change in $\min H$ caused by adding a random clause minus k times the expected change caused by adding a random function $\hat{h}$ is at most $-r(k-1)2^{-k}p^k$. Equivalently, we need to:

1. Consider the experiment:
   - Select: (i) a random formula H from the distribution $H_{n,r,t}$, (ii) a random clause c, (iii) a random variable $x \in \{x_1,\ldots,x_n\}$, and (iv) a random $\hat{h}$-function.
   - Let $H' = H(\sigma) + E_c$, $H'' = H(\sigma) + \hat{h}(x)$.
   - Let $Y = (\min H' - \min H) - k(\min H'' - \min H)$.
2. Prove that $\mathbb{E}Y$, over the choice of $H, c, \hat{h}$, is at most $-r(k-1)2^{-k}p^k$.

The averaging task in Step 2 above appears quite daunting, as we need to average over H. The reason we can establish the desired conclusion is that, in fact, something far stronger holds. Namely, we will prove that for *every* realization of H, the conditional expectation of Y, i.e., the expectation over only c, h and $\hat{h}$, satisfies the desired inequality.

Specifically, Let $H_0(\cdot)$ denote any realization of $H_{n,r,t}(\cdot)$. Let $C^* \subseteq \{0,1\}^n$ be the set of optimal assignments in H_0. A variable x_i is frozen if its value is the same in all optimal assignments. Let O^* be the set of frozen variables corresponding to H_0. We are going to compute the expected value in the change of $\min_\sigma\{H_0(\sigma)\}$ after adding a new factor node $E_{a_{new}}(x_{a_{new},1}, \ldots, x_{a_{new},k})$ and after adding an individual factor $\hat{h}_{new}(\cdot)$ to a variable selected u.a.r.

Adding a new factor $E_{new}(x_{1,new}, \ldots, x_{k,new})$ will change the minimum value by 1 iff all the variables appearing in E_{new} are frozen, i.e., $\{x_{a_{new},1}, \ldots, x_{a_{new},k}\} \subseteq O^*$, and the sign of all the frozen variables in the clause associated with E_{new} is not equal to its frozen value. For, otherwise, any non frozen variable could be adjusted to make the new factor zero. The probability that $\{x_{a_{new},1}, \ldots, x_{a_{new},k}\} \subseteq O^*$ is $(|O^*|/n)^k$, since each of the variables in a random clause are selected uniformly at random with replacement. Thus,

$$\mathbb{E}\left[\min_\sigma\{H_0(\sigma) + E(x_{a_{new},1}, \ldots, x_{a_{new},k})\}\right] - \min_\sigma\{H_0(\sigma)\} = 2^{-k}\left(\frac{|O^*|}{n}\right)^k .$$

The change after adding a new individual factor $\hat{h}_{new}(\cdot)$ to variable selected uniformly at random can be computed in a similar way. In this case the minimum value will change by 1 only if the selected variable x is frozen and if the new factor forces the variable x to take its non-frozen value. This requires that both of the following occur:

- The variable x is frozen to the opposite sign from the one it has in the random clause $E(y_1, \ldots, y_{k-1}, x)$ in the added factor. This event occurs with probability $|O^*|/(2n)$.
- The $k-1$ random functions, distributed as $h(\cdot)$, are all "0" or "1" functions and the $(k-1)$-tuple $(y_1^*, \ldots, y_{k-1}^*)$ that minimizes $\sum h_i(y_i)$ does not satisfy the random clause in the factor. This event occurs with probability $(p/2)^{k-1}$.

Therefore,

$$\mathbb{E}\left[\min_\sigma\{H_0(\sigma) + \hat{h}_{new}(x)\}\right] - \min_\sigma\{H_0(\sigma)\} = 2^{-k}p^{k-1}\frac{|O^*|}{n} .$$

Thus the value $\xi'(t)$ conditional on H_0 is

$$\xi'(t)|H_0 = r2^{-k}\left(\frac{|O^*|}{n}\right)^k - rk2^{-k}p^{k-1}\frac{|O^*|}{n} .$$

We finish the proof by noting that

$$-\xi'(t)|H_0 - r(k-1)2^{-k}p^k = 2^{-k}\left(-r\left(\frac{|O^*|}{n}\right)^k + rkp^{k-1}\frac{|O^*|}{n} - r(k-1)p^k\right)$$

is always non-positive since the polynomial $F(x,p) = x^k - kp^{k-1}x + (k-1)p^k \geq 0$ for all $0 \leq x, p \leq 1$. To see this last statement note that:

- $F(0,p), F(1,p), F(x,0), F(x,1) \geq 0$.
- The derivative of F with respect to p is 0 only when $p = x$, wherein $F(x,x) = 0$.

Parameterized Complexity of Weighted Satisfiability Problems

Nadia Creignou[1,*] and Heribert Vollmer[2,**]

[1] Laboratoire d'Informatique Fondamentale de Marseille, CNRS UMR 7279,
Aix-Marseille Université, 163 avenue de Luminy, F-13288 Marseille Cedex 9, France
`creignou@lif.univ-mrs.fr`
[2] Institut für Theoretische Informatik, Leibniz Universität Hannover, Appelstr. 4,
30167 Hannover, Germany
`vollmer@thi.uni-hannover.de`

Abstract. We consider the weighted satisfiability problem for Boolean circuits and propositional formulæ, where the weight of an assignment is the number of variables set to true. We study the parameterized complexity of these problems and initiate a systematic study of the complexity of its fragments. Only the monotone fragment has been considered so far and proven to be of same complexity as the unrestricted problems. Here, we consider all fragments obtained by semantically restricting circuits or formulæ to contain only gates (connectives) from a fixed set B of Boolean functions. We obtain a dichotomy result by showing that for each such B, the weighted satisfiability problems are either W[P]-complete (for circuits) or W[SAT]-complete (for formulæ) or efficiently solvable. We also consider the related counting problems.

1 Introduction

Satisfiability of circuits and formulæ are fundamental problems, which are the core of many complexity classes. This is true not only in the "classical" complexity setting but also in parameterized complexity theory. Here, with each problem instance we associate a parameter. Instances with the same parameter are thought to share a common structure. A parameterized problem is fixed-parameter tractable (in FPT) if it can be solved in polynomial time for each fixed value of the parameter, where the degree of the polynomial does not depend on the parameter. Much like in the classical setting, to give evidence that certain algorithmic problems are not in FPT one shows that they are complete for superclasses of FPT, like the classes in the so-called W-hierarchy.

Weighted satisfiability (where the weight of a solution is given by the number of variables assigned true) gives rise to a parameterized version of the problems of satisfiability of circuits or formulæ. The goal is then to decide the existence of satisfying assignments of weight exactly k, where k is the parameter. From a complexity theoretic viewpoint, these parameterized problems are very hard since they are W[P]-complete for circuits and W[SAT]-complete for formulæ (see, e.g., [8]).

---

* Supported by Agence Nationale de la Recherche under grant ANR-09-BLAN-0011-01.
** Work done while on leave at the University of Oxford. Supported by DFG VO 630/6-2 and EPSRC EP/G055114/1.

A. Cimatti and R. Sebastiani (Eds.): SAT 2012, LNCS 7317, pp. 341–354, 2012.
© Springer-Verlag Berlin Heidelberg 2012

This intractability result raises the question for restrictions leading to fragments of lower complexity. Concerning formulæ such restrictions have been considered in previous work. Indeed Marx [10] studied the parameterized complexity of satisfiability problems in the famous Schaefer's framework where formulæ are restricted to generalized conjunctive normal form with clauses from a fixed set of relations (the constraint language). He obtained a dichotomy classification by showing that for every possible constraint language the weighted satisfiability problem for generalized CNF formulæ is either in FPT or W[1]-complete (thus, in any case, much lower than the W[SAT]-completeness for general weighted SAT). A similar yet different approach is not to restrict the *syntactic shape* of the formulæ by stipulating a certain normal form but rather to require formulæ to be constructed from a restricted set of Boolean functions B (in contrast to the Schaefer framework, one might say that these are *semantic* restrictions). Such formulæ are called B-*formulæ*. This approach has first been taken by Lewis, who showed that deciding satisfiability of B-formulæ is NP-complete if and only if the set of Boolean functions B has the ability to express the negation of implication $\not\rightarrow$ [9]. Since then this approach has been applied to a wide range of algorithmic problems from the area of circuits [14,3] or propositional formulæ in, e.g., temporal logics [2] or non-monotonic logics [5].

The goal of this paper is to follow this approach and to show that Post's lattice allows to completely classify the complexity of weighted satisfiability for all possible sets of allowed Boolean functions. We consider both circuits and formulæ, and the complexity of deciding whether they admit a satisfying assignment of weight exactly k. We show that depending on the set B of allowed connectives the parameterized weighted satisfiability problem is either W[P]-complete (for circuits) and W[SAT]-complete for formulæ, or in P. More precisely, we prove that the complexity of these problems is W[P]-complete or W[SAT]-complete (depending on whether they concern circuits or formulæ) as soon as B can express either the function $x \wedge (y \vee z)$, or any 2-threshold function as for example the ternary majority function. The problem becomes solvable in polynomial time in all remaining cases. Thus, in a sense, we exactly pinpoint the reason for intractability of weighted satisfiability by exhibiting which Boolean functions make the problem hard.

Besides the decision problem, we study the complexity of the corresponding counting problems. We prove here also a dichotomy theorem in showing that the problems are either #W[P]-complete (or #W[SAT]-complete), or in FP. The frontier of this dichotomy is not the same as in the decision case, since some tractable decision problems, as, e.g., the weighted satisfiability problem in which only the connective $\rightarrow$ is allowed, become hard counting problems.

Our results are summarized in Fig. 1. White sets B of Boolean functions lead to easy problems, black sets lead to hard problems. The gray colored nodes correspond to those sets B for which the decision problems are easy, but the counting problems are hard.

The rest of the paper is structured as follows. We first give the necessary preliminaries. Afterwards, we define the weighted satisfiability considered herein. We then classify the parameterized complexity of these problems. Next, we consider the counting problems and finally conclude with a discussion of the results.

2 Preliminaries

Parameterized complexity. We assume familiarity with the basic classes and reducibility notions from parameterized complexity theory, see, e.g., [8,12], such as FPT, W[P], W[SAT], fpt-reductions and Turing fpt-reductions.

Boolean circuits and propositional formulae. We assume familiarity with propositional logic. A *Boolean function* is an n-ary function $f \colon \{0,1\}^n \to \{0,1\}$. We define *Boolean circuits* (see also [17]) in the standard way as directed acyclic graphs with each node of in-degree $k > 0$ labeled by a Boolean function of arity k. For non-commutative functions, there is in addition an ordering on the incoming edges. Nodes of in-degree 0 are either labeled as Boolean constants 0 or 1, or as input nodes. In addition, one node of out-degree 0 is labeled as the output node. We think of the input nodes as being numbered $1, \ldots, n$. This definition of a Boolean circuit corresponds to the intuitive idea that a circuit consists of a set of gates which are either input gates, or compute some Boolean function with arguments taken from the predecessor gates. The value computed by the circuit is the result computed in the distinguished output-gate. So, a circuit $\mathcal{C}$ with n input nodes naturally computes an n-ary Boolean function, we denote it by $f_{\mathcal{C}}$.

We denote the value computed by $\mathcal{C}$ on input $a \in \{0,1\}^n$ by $\mathcal{C}(a)$. If $\mathcal{C}(a) = 1$, we say that a *satisfies* $\mathcal{C}$. We call $\mathcal{C}$ *satisfiable* if there is some tuple $a \in \{0,1\}^n$ that satisfies $\mathcal{C}$. We define the *weight of a tuple* $a = (a_1, \ldots, a_n) \in \{0,1\}^n$ to be $\sum_{i=1}^{n} a_i$, the number of 1-entries of a. We say $\mathcal{C}$ is k-*satisfiable* if it is satisfied by a tuple of weight k. A circuit $\mathcal{C}$ is *monotone* if for all $a = (a_1, \ldots, a_n) \in \{0,1\}^n$ such that $\mathcal{C}(a) = 1, a_1 \leq b_1, \ldots a_n \leq b_n$ implies $\mathcal{C}(b_1, \ldots b_n) = 1$.

A formula φ is a circuit where the underlying graph forms a tree. Hence, such circuits can always be written as a formula in the usual string representation without growing significantly in size. For a general circuit, the length of its "formula representation" can be exponential in the size of the original circuit. Further we denote by $\varphi[\alpha/\beta]$ the formula obtained from φ by replacing all occurrences of α with β. The set $\mathrm{Var}(\varphi)$ denotes the set of variables occurring in the formula

Deciding the k-satisfiability of a Boolean circuit, p-WCIRCUIT-SAT, where k is taken to be the parameter, is of fundamental importance for parameterized complexity theory. Indeed, p-WCIRCUIT-SAT is W[P]-complete under fpt-reductions (see [8, Theorem 3.9]). Deciding the k-satisfiability of a Boolean formula, p-WSAT, is W[SAT]-complete by definition.

Given B a finite set of Boolean functions, a B-circuit (resp. a B-formula) is a Boolean circuit (a formula) using only functions (connectives) from B.

Clones of Boolean functions. A *clone* is a set of Boolean functions that is closed under superposition, *i.e.*, it contains all projections (that is, the functions $f(a_1, \ldots, a_n) = a_k$ for $1 \leq k \leq n$ and $n \in \mathbb{N}$) and is closed under arbitrary composition. Let B be a finite set of Boolean functions. We denote by $[B]$ the smallest clone containing B and call B a *base* for $[B]$. The set $[B]$ corresponds to the set of all Boolean functions that can be computed by B-circuits. All closed classes of Boolean functions were identified by Post ([13]). Post also found a finite base for each of them and detected their inclusion structure, hence the name of *Post's lattice* (see Figure 1).

In order to define the clones, we require the following notions, where f is an n-ary Boolean function:

- f is *c-reproducing* if $f(c, \ldots, c) = c$, $c \in \{0, 1\}$.
- f is *monotonic* (or, *monotone*) if $a_1 \le b_1, a_2 \le b_2, \ldots, a_n \le b_n$ implies $f(a_1, \ldots, a_n) \le f(b_1, \ldots, b_n)$.
- f is *c-separating of degree* k if for all $A \subseteq f^{-1}(c)$ of size $|A| = k$ there exists an $i \in \{1, \ldots, n\}$ such that $(a_1, \ldots, a_n) \in A$ implies $a_i = c$, $c \in \{0, 1\}$.
- f is *c-separating* if f is c-separating of degree $|f^{-1}(c)|$.
- f is *self-dual* if $f(x_1, \ldots, x_n) \equiv \neg f(\neg x_1, \ldots, \neg x_n)$.
- f is *affine* if it is of the form $f(x_1, \ldots, x_n) = x_1 \oplus \cdots \oplus x_n \oplus c$ with $c \in \{0, 1\}$.

In the following we will often use well-known Boolean functions, as $\wedge, \vee, \neg, \oplus, \rightarrow$ the implication function and the ternary majority operation maj (defined by $\mathrm{maj}(x_1, x_1, x_3) = 1$ if and only if $x_1 + x_2 + x_3 \ge 2$). We will also refer to *q-threshold functions* as functions f verifying $f(x_1, \ldots, x_n) = 1$ if and only if $\sum_{i=1}^{n} x_i \ge q$. Observe that maj is thus a ternary 2-threshold function. More generally T_k^n will denote the k-threshold function of arity n.

A list of all clones with definitions and finite bases is given in Table 1 on page 345, see also, *e.g.*, [4]. Clones of particular importance in this paper, either because they are of technical importance or because they mark points in Post's lattice where the complexity of our problems changes, are the following:

- The clone of all Boolean functions $\mathsf{BF} = [\wedge, \neg] = [\wedge, \vee, \neg, 0, 1]$.
- The monotonic clones M_*, e.g., $\mathsf{M}_2 = [\wedge, \vee]$ and $\mathsf{M} = [\wedge, \vee, 0, 1]$.
- The dual clones D_*, e.g., $\mathsf{D}_2 = [\mathrm{maj}]$.
- The disjunctive clones V_*, e.g., $\mathsf{V} = [\vee, 0, 1]$.
- The conjunctive clones E_*, e.g., $\mathsf{E} = [\wedge, 0, 1]$.
- The affine clones L_*, e.g., $\mathsf{L} = [\oplus, 0, 1]$.
- The implication clone S_0^*, e.g., $\mathsf{S}_0 = [\rightarrow]$.

We will often add some function $f \notin C$ to a clone C and consider the clone $C' = [C \cup \{f\}]$ generated out of C and f. With Post's lattice one can determine this C' quite easily: it is the lowest clone above C that contains f. The following list contains identities we will frequently use.

- $[\mathsf{S}_{02} \cup \{0, 1\}] = [\mathsf{S}_{12} \cup \{0, 1\}] = \mathsf{BF}$
- $[\mathsf{S}_{00} \cup \{0, 1\}] = [\mathsf{D}_2 \cup \{0, 1\}] = [\mathsf{M}_2 \cup \{0, 1\}] = [\mathsf{S}_{10} \cup \{0, 1\}] = \mathsf{M}$
- $[\mathsf{S}_{10} \cup \{1\}] = [\mathsf{M}_2 \cup \{1\}] = \mathsf{M}_1$
- $[\mathsf{S}_{00} \cup \{0\}] = [\mathsf{M}_2 \cup \{0\}] = \mathsf{M}_0$
- $[\mathsf{D}_2 \cup \{1\}] = \mathsf{S}_{01}^2$, $[\mathsf{D}_2 \cup \{0\}] = \mathsf{S}_{11}^2$

Let f be an n-ary Boolean function. A B-formula φ such that $\mathrm{Var}(\varphi) \supseteq \{x_1, \ldots, x_n\}$ is a *B-representation* of $f(x_1, \ldots, x_n)$ if it holds that $\sigma: \mathrm{Var}(\varphi) \rightarrow \{0, 1\}$ satisfies φ if and only $f(\sigma(x_1), \ldots, \sigma(x_n)) = 1$. Such a B-representation exists for every $f \in [B]$. Yet, it may happen that the B-representation of some function uses some input variable more than once.

Example 1. Let $h(x, y) = x \wedge \neg y$. An $\{h\}$-representation of the function $x \wedge y$ is $h(x, h(x, y))$.

Table 1. The list of all Boolean clones with definitions and bases, where T_k^n denotes the k-threshold function of arity n,

Clone	Definition	Base
BF	All Boolean functions	$\{x \wedge y, \neg x\}$
R_0	$\{f \in BF \mid f$ is 0-reproducing$\}$	$\{x \wedge y, x \oplus y\}$
R_1	$\{f \in BF \mid f$ is 1-reproducing$\}$	$\{x \vee y, x \leftrightarrow y\}$
R_2	$R_0 \cap R_1$	$\{x \vee y, x \wedge (y \leftrightarrow z)\}$
M	$\{f \in BF \mid f$ is monotone$\}$	$\{x \wedge y, x \vee y, 0, 1\}$
M_0	$M \cap R_0$	$\{x \wedge y, x \vee y, 0\}$
M_1	$M \cap R_1$	$\{x \wedge y, x \vee y, 1\}$
M_2	$M \cap R_2$	$\{x \wedge y, x \vee y\}$
S_0	$\{f \in BF \mid f$ is 0-separating$\}$	$\{x \to y\}$
S_0^n	$\{f \in BF \mid f$ is 0-separating of degree $n\}$	$\{x \to y, T_2^{n+1}\}$
S_1	$\{f \in BF \mid f$ is 1-separating$\}$	$\{x \nrightarrow y\}$
S_1^n	$\{f \in BF \mid f$ is 1-separating of degree $n\}$	$\{x \nrightarrow y, T_n^{n+1}\}$
S_{02}^n	$S_0^n \cap R_2$	$\{x \vee (y \wedge \neg z), T_2^{n+1}\}$
S_{02}	$S_0 \cap R_2$	$\{x \vee (y \wedge \neg z)\}$
S_{01}^n	$S_0^n \cap M$	$\{T_2^{n+1}, 1\}$
S_{01}	$S_0 \cap M$	$\{x \vee (y \wedge z), 1\}$
S_{00}^n	$S_0^n \cap R_2 \cap M$	$\{x \vee (y \wedge z), T_2^3\}$ if $n = 2$, $\{T_2^{n+1}\}$ if $n \geq 3$
S_{00}	$S_0 \cap R_2 \cap M$	$\{x \vee (y \wedge z)\}$
S_{12}^n	$S_1^n \cap R_2$	$\{x \wedge (y \vee \neg z), T_n^{n+1}\}$
S_{12}	$S_1 \cap R_2$	$\{x \wedge (y \vee \neg z)\}$
S_{11}^n	$S_1^n \cap M$	$\{T_n^{n+1}, 0\}$
S_{11}	$S_1 \cap M$	$\{x \wedge (y \vee z), 0\}$
S_{10}^n	$S_1^n \cap R_2 \cap M$	$\{x \wedge (y \vee z), T_2^3\}$ if $n = 2$, $\{T_n^{n+1}\}$ if $n \geq 3$
S_{10}	$S_1 \cap R_2 \cap M$	$\{x \wedge (y \vee z)\}$
D	$\{f \in BF \mid f$ is self-dual$\}$	$\{\mathrm{maj}(x, \neg y, \neg z)\}$
D_1	$D \cap R_2$	$\{\mathrm{maj}(x, y, \neg z)\}$
D_2	$D \cap M$	$\{\mathrm{maj}(x, y, z)\}$
L	$\{f \in BF \mid f$ is affine$\}$	$\{x \oplus y, 1\}$
L_0	$L \cap R_0$	$\{x \oplus y\}$
L_1	$L \cap R_1$	$\{x \leftrightarrow y\}$
L_2	$L \cap R_2$	$\{x \oplus y \oplus z\}$
L_3	$L \cap D$	$\{x \oplus y \oplus z \oplus 1\}$
E	$\{f \in BF \mid f$ is constant or a conjunction$\}$	$\{x \wedge y, 0, 1\}$
E_0	$E \cap R_0$	$\{x \wedge y, 0\}$
E_1	$E \cap R_1$	$\{x \wedge y, 1\}$
E_2	$E \cap R_2$	$\{x \wedge y\}$
V	$\{f \in BF \mid f$ is constant or a disjunction$\}$	$\{x \vee y, 0, 1\}$
V_0	$V \cap R_0$	$\{x \vee y, 0\}$
V_1	$V \cap R_1$	$\{x \vee y, 1\}$
V_2	$V \cap R_2$	$\{x \vee y\}$
N	$\{f \in BF \mid f$ is essentially unary$\}$	$\{\neg x, 0, 1\}$
N_2	$N \cap D$	$\{\neg x\}$
I	$\{f \in BF \mid f$ is constant or a projection$\}$	$\{\mathrm{id}, 0, 1\}$
I_0	$I \cap R_0$	$\{\mathrm{id}, 0\}$
I_1	$I \cap R_1$	$\{\mathrm{id}, 1\}$
I_2	$I \cap R_2$	$\{\mathrm{id}\}$

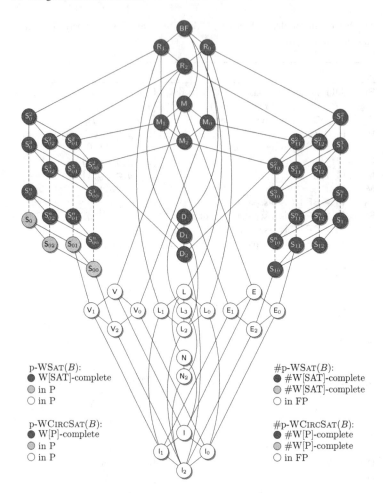

Fig. 1. Graph of all Boolean clones

3   Weighted Satisfiability Problems

Let B be a finite set of Boolean functions. We are interested in the complexity of weighted satisfiability problems as a function of the set B of allowed connectives. We define *weighted satisfiability problems for B-circuits* and for *B-formulæ* as follows:

Problem:	p-WSAT(B)
Input:	a B-formula φ and $k \in \mathbb{N}$
Parameter:	k
Question:	Does φ have a satisfying assignment of weight exactly k?

The corresponding problem for circuits is denoted by p-WCIRCSAT(B).

Our goal is to obtain a complexity classification of these problems according to B. Observe that if B_1 and B_2 are two finite sets of Boolean functions such that $B_1 \subseteq [B_2]$, then every function of B_1 can be expressed by a B_2-formula, its so-called B_2-*representation*. This provides a canonical fpt-reduction from p-WCIRCSAT(B_1) to p-WCIRCSAT(B_2): Given $(\mathcal{C}_1, k)$ an input of the first problem, construct in logarithmic space the pair $(\mathcal{C}_2, k)$ in which $\mathcal{C}_2$ is obtained from $\mathcal{C}_1$ in replacing all B_1-functions by their B_2-representation (observe that since B_1 and B_2 are not part of the input the cost of computing these representations is not taken into account in the complexity of the reduction). However, since the B_2-representation of some function may use some input variable more than once (see Example 1) this reduction is not necessarily polynomial when we turn to formulæ. In order to avoid an exponential blow-up when dealing with formulæ we will seek *short representations*, i.e., representations in which every variable appears exactly once. We say that a set B *efficiently implements* an n-ary function f if there is a B-formula φ that is equivalent to $f(x_1, \ldots, x_n)$ and in which each x_i appears exactly once. In that case the above reduction applies to formulæ as well, and we get a canonical fpt-reduction from p-WSAT($\{f\}$) to p-WCIRCSAT(B).

4   The Complexity of Weighted Satisfiability Problems

In the following B denotes a finite set of Boolean functions,

Lemma 2. p-WSAT(B) *is in* W[SAT] *and* p-WCIRCSAT(B) *is in* W[P].

Proof. For circuits it follows from the fact that $\{\wedge, \vee, \neg\}$ can represent all Boolean functions. Therefore, p-WCIRCSAT(B) $\leq^{\mathrm{fpt}}$ p-WCIRCSAT($\{\wedge, \vee, \neg\}$), the latter being in W[P] by definition, thus proving membership in W[P]. For formulæ the fact that p-WSAT(B) $\leq^{\mathrm{fpt}}$ p-WSAT($\{\wedge, \vee, \neg\}$) follows from [16, Theorem 4.7].   □

Lemma 3. *If B contains the constants 0 and 1, and if* M $\subseteq [B]$*, then* p-WSAT(B) *is* W[SAT]*-complete, and* p-WCIRCSAT(B) *is* W[P]*-complete.*

Proof. It is proved in [1] that p-WSAT($\{\wedge, \vee\}$) is W[SAT]-complete and p-WCIRCSAT($\{\wedge, \vee\}$) is W[P]-complete. Since M $\subseteq [B]$, either $[B] =$ M or $[B] =$ BF. In both cases, since by assumption B contains the two constants and according to [15, Lemma 4], B efficiently implements the functions $\wedge$ and $\vee$. This shows that p-WSAT($\{\wedge, \vee\}$) $\leq^{\mathrm{fpt}}$ p-WSAT(B), thus concluding the proof for formulæ. The same reduction actually shows the hardness result for the circuit problem.   □

The following lemma shows that we can freely use the constant 1 as soon as the function $\wedge$ can be computed by a B-circuit.

Lemma 4. *If $[B]$ contains the conjunction function then* p-WSAT($B \cup \{1\}$) *fpt-reduces to* p-WSAT(B) *and* p-WCIRCSAT($B \cup \{1\}$) *fpt-reduces to* p-WSAT(B).

Proof. Let φ be a $B \cup \{1\}$-formula. Let $\varphi' := \varphi[1/t] \wedge t$, where t is a fresh variable. Since $\wedge \in [B]$ the formula φ' can be represented by a B-formula. Moreover, it is clear that φ has a satisfying assignment of weight k if and only if φ' has a satisfying assignment of weight $k + 1$, thus showing that p-WSAT($B \cup \{1\}$) $\leq^{\mathrm{fpt}}$ p-WSAT(B). The same proof shows a similar result for circuits.   □

Dealing with the constant 0 requires additional tricks. First let us introduce variants of our problems as technical tools: p-WSAT$^+(B)$ and p-WCIRCSAT$^+(B)$ denote our original problems restricted to monotone instances. Let us observe that Lemmas 3 and 4 still hold for these variants. Obviously proving hardness for these variants is enough for proving hardness for the original problems.

Definition 5. *Let l be a positive integer. A formula ψ is l-costly if every satisfying assignment of ψ has weight at least l.*

Lemma 6. *If for every non-negative integer k there exists a $(k+1)$-costly B-formula, then we obtain the reductions* p-WSAT$^+(B \cup \{0\})$ $\leq^{\text{fpt}}$ p-WSAT$^+(B)$ *and* p-WCIRCSAT$^+(B \cup \{0\})$ $\leq^{\text{fpt}}$ p-WCIRCSAT$^+(B)$.

Proof. Let φ be a monotone $B \cup \{0\}$-formula. Let ψ be a $(k+1)$-costly B-formula over m variables. Consider the B-formula φ' obtained from φ in replacing every occurrence of 0 by $\psi(y_1, \ldots, y_m)$, where the y_i's are fresh variables. If there is a satisfying assignment for φ of weight k, then it can be extended to a satisfying assignment of φ' of same weight by setting all the y_i's to 0. Conversely, any truth assignment to the variables of φ' of weight k makes the formula ψ false (since it is $(k+1)$-costly). Therefore, the restriction of such an assignment to the variables of φ provides a satisfying assignment of φ of weight at most k. Since by assumption φ is monotone this implies that φ is k-satisfiable. To sum up, φ has a satisfying assignment of weight k if and only if φ' has a satisfying assignment of weight k, thus concluding the proof. The same proof holds for circuits. Observe that in the reduction the size of φ' is in the worse case the size of φ times the size of ψ, which is an arbitrary function depending on k. Therefore, the reduction here is an fpt-reduction, but not a many-one-log-space reduction as in Lemma 4. □

Lemma 7. *If $[B]$ contains some q-threshold function (of arbitrary arity $n \geq q$) where $q \geq 2$, then there exists an l-costly B-formula for any $l \geq 1$.*

Proof. Build a balanced tree of depth d whose gates are q-threshold functions, with $q \geq 2$. Every satisfying assignment of this tree must have weight at least q^d. Thus, in choosing d large enough, say such that $q^d > l$, we can build this way an l-costly B-formula. □

Lemma 8. *If $D_2 \subseteq [B]$, then* p-WSAT(B) *is* W[SAT]*-complete and* p-WCIRCSAT(B) *is* W[P]*-complete.*

Proof. Suppose that $D_2 \subseteq [B]$. Note that in this case $M \subseteq B \cup \{0, 1\}$. Therefore, p-WSAT$^+(B \cup \{0, 1\})$ is W[SAT]-complete and p-WCIRCSAT$^+(B \cup \{0, 1\})$ is W[P]-complete according to Lemma 3. The hardness result for the formula problem is proved by the following sequence of reductions: p-WSAT$^+(B \cup \{0, 1\})$ $\leq^{\text{fpt}}$ p-WSAT$^+(B \cup \{0\})$ $\leq^{\text{fpt}}$ p-WSAT$^+(B)$ $\leq^{\text{fpt}}$ p-WSAT(B). The first reduction holds according to Lemma 4 since $[B \cup \{0\}]$ is a superset of S_{11}^2, and thus $\wedge \in [B]$. Observe that B can express the ternary majority function (see Table 1), which is a 2-threshold function. Thus, the second reduction follows from Lemma 7 and Lemma 6. As mentioned above the last reduction is trivial. The same sequence of reductions provides the desired result for circuits. □

Lemma 9. *If* $S_{10} \subseteq [B]$, *then* p-WSAT(B) *is* W[SAT]-*complete, and* p-WCIRCSAT(B) *is* W[P]-*complete.*

Proof. Suppose that $S_{10} \subseteq [B]$. We still have $M \subseteq [B \cup \{0, 1\}]$, therefore p-WSAT$^+(B \cup \{0, 1\})$ is W[SAT]-complete according to Lemma 3. Hardness is proved by the following sequence of reductions: p-WSAT$^+(B \cup \{0, 1\}) \leq^{\text{fpt}}$ p-WSAT$^+(B \cup \{1\}) \leq^{\text{fpt}}$ p-WSAT$^+(B)$. The first reduction holds according to Lemma 7 and Lemma 6. Indeed $[B \cup \{1\}]$ is a superset of M_1, and thus a superset of D_2. For this reason $[B \cup \{1\}]$ contains the ternary majority function, which is a 2-threshold function. The second reduction follows from Lemma 4 since $\wedge \in S_{10}$ and thus $\wedge \in [B]$. The same sequence of reductions provides the desired result for circuits. □

Lemma 10. *If* $S_{00}^n \subseteq [B]$ *for some* $n \geq 2$, *then* p-WSAT(B) *is* W[SAT]-*complete and* p-WCIRCSAT$_\leq(B)$ *is* W[P]-*complete.*

Proof. Observe that $M = [S_{00}^n \cup \{0, 1\}]$, and hence that $M \subseteq [B \cup \{0, 1\}]$. Thus, according to Lemma 3, p-WSAT$^+(B \cup \{0, 1\})$ is W[SAT]-complete. The lemma follows then from the following sequence of reductions: p-WSAT$^+(B \cup \{0, 1\}) \leq^{\text{fpt}}$ p-WSAT$^+(B \cup \{0\}) \leq^{\text{fpt}}$ p-WSAT$^+(B)$. The first reduction holds according to Lemma 4 since $M_0 = [S_{00}^n \cup \{0\}]$, and hence $\wedge \in [B \cup \{0\}]$. Observe that S_{00}^n contains the 2-threshold function T_{n+1}^2. Hence, the second reduction follows from Lemma 6 and Lemma 7. The same sequence of reductions provides the desired result for circuits. □

In the following lemmas we prove tractability results for circuit problems (tractability for formulæ follows trivially as a special case).

Lemma 11. *If* $[B] \subseteq V$, *or* $[B] \subseteq E$, *or* $[B] \subseteq L$, *then* p-WSAT(B) *and* p-WCIRCSAT(B) *are in* P.

Proof. The basic idea is to compute a normal form of the functions computed by such B-circuits, from which it is easy to decide whether the circuits are k-satisfiable.

First, let $V_2 \subseteq [B] \subseteq V$. Let $C(x_1, \ldots, x_n)$ be a B-circuit. The Boolean function described by C can be expressed as $f_C(x_1, \ldots, x_n) = a_0 \vee (a_1 \wedge x_1) \vee \ldots \vee (a_n \wedge x_n)$, where the a_i's are in $\{0, 1\}$. The values a_i, where $0 \leq i \leq n$, can be determined easily by using the following simple facts: $a_0 = 0$ if and only if $f_C(0, \ldots, 0) = 0$ and $a_i = 0$ for $1 \leq i \leq n$ if and only if $a_0 = 0$ and $f_C(0^{i-1}, 1, 0^{n-i}) = 0$. This can be checked in polynomial time since the value problem for B-circuits is known to be in P (see [14]).We conclude that the normal form, from which deciding k-satisfiability is easy, can be computed efficiently.

Tractability for $E_2 \subseteq [B] \subseteq E$ follows as above in computing the dual normal form, i.e., $f_C(x_1, \ldots, x_n) = a_0 \wedge (a_1 \vee x_1) \wedge \ldots \wedge (a_n \vee x_n)$.

Finally let $L_2 \subseteq [B] \subseteq L$. The proof follows by computing the normal form as $f_C(x_1, \ldots, x_n) = a_0 \oplus (a_1 \wedge x_1) \oplus \ldots \oplus (a_n \wedge x_n)$. Similar to the above the values a_i's can be easily determined by n well-chosen oracles to the circuit value problem. Again deciding k-satisfiability is easy from this normal form. □

Lemma 12. *If* $S_{00} \subseteq [B] \subseteq S_0$ *then* p-WSAT(B) *and* p-WCIRCSAT(B) *are in* P.

Proof. Observe that $\rightarrow$ is the basis of S_0. Since $B \subseteq S_0$ any B-circuit can be transformed in logarithmic space into an $\{\rightarrow\}$-circuit in locally replacing each gate $f \in B$ by its $\{\rightarrow\}$-representation. Note that such a circuit has satisfying assignments of all possible weights (except may be the all-0 one). To see this, start from the output gate and go backwards. At every gate take backwards the edge corresponding to the right argument of the implication. Thus we get a path from the output gate to a 'target-gate' which is either a variable or the constant 1 (the constant 0 does not appear by assumption). In case of a variable, setting this variable to 1 is sufficient to satisfy the circuit. Therefore from this we can build satisfying assignments of any weight ≥ 1. If by the described procedure we reach the constant 1, then the circuit represents a tautology. The special case of the all-0 assignment has to be dealt with separately.   □

We are now in a position to state our main result for weighted satisfiability decision problems. Indeed a careful examination of Post's lattice shows that the above lemmas cover all the lattice and thus provide a complete classification.

Theorem 13. *Let B be a finite set of Boolean functions.*

1. *If* $D_2 \subseteq [B]$ *or* $S_{10} \subseteq [B]$ *or* $S_{00}^n \subseteq [B]$ *for some* $n \geq 2$, *then* p-WSAT(B) *is* W[SAT]*-complete, and* p-WCIRCSAT(B) *is* W[P]*-complete under* fpt*-reductions.*
2. *In all other cases* p-WSAT(B) *and* p-WCIRCSAT(B) *are in* P.

5 Complexity of the Counting Problems

There are natural counting problems associated with the decision problems studied above.

> *Problem:*   p-#WSAT(B)
> *Input:*   a B-formula φ and $k \in \mathbb{N}$
> *Parameter:* k
> *Output:*   Number of satisfying assignments for φ of weight exactly k

The corresponding problem for circuits is denoted by p-#WCIRCSAT(B).

In the following, as proposed in [8], we use $\leq^{\mathrm{fpt}}$ to designate a parsimonious fpt-reduction, while $\leq^{\mathrm{fpt\text{-}T}}$ will refer to a Turing fpt-reduction. Let us now introduce two complexity classes, which are the counting analogues of W[SAT] and W[P]. To the best of our knowledge, the class #W[SAT] has not been considered in the literature so far.

Definition 14. *The class* #W[SAT] *is the closure of* p-#WSAT$(\{\wedge, \vee, \neg\})$ *under* fpt-*parsimonious reductions, that is*

$$\#W[SAT] := [\text{p-}\#\text{WSAT}(\{\wedge, \vee, \neg\})]^{\mathrm{fpt}}.$$

Whereas it was originally defined in terms of counting accepting runs of a κ-restricted nondeterministic Turing machines, the class #W[P] can be defined in a similar way (see, e.g., [8, page 366]):

Definition 15. *The class* #W[P] *is the closure of* p-#WCIRCSAT($\{\wedge, \vee, \neg\}$) *under* fpt-*parsimonious reductions, that is*

$$\#W[P] := [p\text{-}\#WCIRCSAT(\{\wedge, \vee, \neg\})]^{\text{fpt}}.$$

Proposition 16. p-#WSAT(B) *is in* #W[SAT] *and* p-#WCIRCSAT(B) *is in* #W[P].

Proof. This follows from the proof of Lemma 2 in observing that all reductions are parsimonious.   □

We first state two lemmas which allow to take care of the constants. The first one is simply the observation that the reduction in Lemma 4 is parsimonious.

Lemma 17. *If* $[B]$ *contains the conjunction function then* p-#WSAT($B \cup \{1\}$) *parsimoniously fpt-reduces to* p-#WSAT(B) *and* p-#WCIRCSAT($B \cup \{1\}$) *parsimoniously fpt-reduces to* p-#WCIRCSAT(B).

Given a formula φ, let $\#\text{Sat}_k(\varphi)$ denote its number of satisfying assignments of weight exactly k.

Lemma 18. *If* $[B]$ *contains the disjunction function then* p-#WSAT($B \cup \{0\}$) *Turing fpt-reduces to* p-#WSAT(B) *and* p-#WCIRCSAT($B \cup \{0\}$) *Turing fpt-reduces to* p-#WCIRCSAT(B).

Proof. Let φ be a $B \cup \{0\}$-formula. Let $\varphi' := \varphi[0/f] \vee f$, where f is a fresh variable. Since $\vee \in [B]$ the formula φ' can be represented by a B-formula. Moreover, $\#\text{Sat}_k(\varphi') = \#\text{Sat}_k(\varphi) + \binom{n}{k-1}$, thus showing that p-#WSAT($B \cup \{0\}$) $\leq^{\text{fpt}}$ p-WSAT(B). The reduction consists of a precomputation phase, one oracle call, and then some postcomputation, namely the summation of the result from the oracle and the binomial coefficient; hence it is actually a 1-Turing fpt-reduction. The same reduction holds for circuits.   □

Lemma 19. *If* $\text{M}_2 \subseteq [B]$, *then* p-#WSAT(B) *is complete for the class* #W[SAT] *and* p-#WCIRCSAT(B) *is complete for* #W[P], *both under Turing* fpt-*reductions.*

Proof. First we prove that p-#WSAT($\{\wedge, \vee, \neg\}$) $\leq^{\text{fpt-T}}$ p-#WSAT($\{\wedge, \vee\}$). Let φ be a $\{\wedge, \vee, \neg\}$-formula. Without loss of generality one can suppose that φ is in negation normal form, NNF. Indeed, if it is not the case one can transform it in NNF in polynomial time in pushing the negation symbols in front of variables in applying de Morgan's laws and the double negation elimination. Now we use a well-known reduction to express a general formula as conjunction of a monotone and a negated monotone formula. The formula $\varphi(\bar{x}) = \varphi(x_1, \ldots, x_n)$ is mapped to $\psi(x_1, \ldots, x_n, y_1, \ldots, y_n)$ where $\neg x_i$ is replaced by a fresh variable y_i. This gives

$$\#\text{Sat}_k(\varphi(\bar{x})) = \#\text{Sat}_k(\psi(\bar{x}, \bar{y}) \wedge \bigwedge_{i=1}^{n}(x_i \vee y_i) \wedge \neg \bigvee_{i=1}^{n}(x_i \wedge y_i))$$
$$= \#\text{Sat}_k(\alpha(\bar{x}, \bar{y}) \wedge \neg\beta(\bar{x}, \bar{y}))$$

where α and β are $\{\wedge, \vee\}$-formulæ defined by

$$\alpha(\bar{x}, \bar{y}) = \psi(\bar{x}, \bar{y}) \wedge \bigwedge_{i=1}^{n}(x_i \vee y_i) \text{ and } \beta(\bar{x}, \bar{y}) = \bigvee_{i=1}^{n}(x_i \wedge y_i).$$

Thus we have

$$\#\mathrm{Sat}_k(\varphi) = \#\mathrm{Sat}_k(\alpha) - \#\mathrm{Sat}_k(\alpha \wedge \beta).$$

Indeed, if a k-assignment satisfies α but not $\alpha \wedge \beta$, then it satisfies $\alpha \wedge \neg\beta$. Conversely a k-assignment that satisfies $\alpha \wedge \neg\beta$, satisfies α and does not satisfy $\alpha \wedge \beta$. This proves that p-$\#\mathrm{WSAT}(\{\wedge, \vee, \neg\})$ 2-Turing fpt-reduces to p-$\#\mathrm{WSAT}(\{\wedge, \vee\})$.

Now, $M \subseteq [B \cup \{0,1\}]$, hence there are short $B \cup \{0,1\}$-representations of $\wedge$ and $\vee$ [15], therefore p-$\#\mathrm{WSAT}(\{\wedge, \vee\}) \leq^{\mathrm{fpt}}$ p-$\#\mathrm{WSAT}(B \cup \{0,1\})$. Since both $\vee$ and $\wedge$ are in M_2, and thus in B we can get rid of the constants by applying successively Lemmas 18 and 17.

For circuits p-$\#\mathrm{WCIRCSAT}(\{\wedge, \vee, \neg\})$ is $\#\mathrm{W}[\mathrm{P}]$-complete by definition (see Def. 15). The completeness of p-$\#\mathrm{WCIRCSAT}(\{\wedge, \vee\})$ results from the fact that the reduction from the weighted satisfiability of Boolean circuits to the weighted satisfiability of monotone circuits given in [8, Thm 3.14] is parsimonious . Then the same sequence of reductions as in the case of formulæ allows to conclude.   □

Lemma 20. *If* $S_{10} \subseteq [B]$, *then* p-$\#\mathrm{WSAT}(B)$ *is complete for the class* $\#\mathrm{W}[\mathrm{SAT}]$ *and* p-$\#\mathrm{WCIRCSAT}(B)$ *is complete for* $\#\mathrm{W}[\mathrm{P}]$, *both under Turing fpt-reductions.*

Proof. Observe that $M_1 \subseteq [B \cup \{1\}]$, therefore, by Lemma 19, p-$\#\mathrm{WSAT}(B \cup \{1\})$ is $\#\mathrm{W}[\mathrm{SAT}]$-complete. The result is then obtained by the reduction p-$\#\mathrm{WSAT}(B \cup \{1\}) \leq^{\mathrm{fpt}}$ p-$\#\mathrm{WSAT}(B)$, which follows from Lemma 17 since $\wedge \in S_{10} \subseteq [B]$. A similar proof provides the result for circuits.   □

Lemma 21. *If* $S_{00} \subseteq [B]$, *then* p-$\#\mathrm{WSAT}(B)$ *is complete for the class* $\#\mathrm{W}[\mathrm{SAT}]$ *and* p-$\#\mathrm{WCIRCSAT}(B)$ *is complete for* $\#\mathrm{W}[\mathrm{P}]$-complete, *both under Turing fpt-reductions.*

Proof. Similar to the proof above in using the fact that p-$\#\mathrm{WSAT}(B \cup \{0\})$ is $\#\mathrm{W}[\mathrm{SAT}]$-complete and the reduction p-$\#\mathrm{WSAT}(B \cup \{0\}) \leq^{\mathrm{fpt\text{-}T}}$ p-$\#\mathrm{WSAT}(B)$, which is obtained through Lemma 18.   □

Lemma 22. *If* $D_2 \subseteq [B]$, *then* p-$\#\mathrm{WSAT}(B)$ *is* $\#\mathrm{W}[\mathrm{SAT}]$-complete *and* p-$\#\mathrm{WCIRCSAT}(B)$ *is* $\#\mathrm{W}[\mathrm{P}]$-complete, *both under Turing fpt-reductions.*

Proof. Observe that $[D_2 \cup \{1\}] = S_{01}^2$, therefore according to Lemma 21 we get hardness for p-$\#\mathrm{WSAT}(B \cup \{1\})$. It remains to show that p-$\#\mathrm{WSAT}(B \cup \{1\}) \leq^{\mathrm{fpt}}$ p-$\#\mathrm{WSAT}(B)$. For this we will use some specific functions g_l which belong to D_2 (for they are self-dual) and which are defined as follows:

$$g_l(x_1, \ldots, x_l, 0) = x_1 \wedge \ldots \wedge x_l, \text{ and } g_l(x_1, \ldots, x_l, 1) = x_1 \vee \ldots \vee x_l.$$

Let maj denote the ternary majority function, which is also a function from D_2. Let φ be a $B \cup \{1\}$-formula. Consider φ' defined by:

$$\varphi' := \mathrm{maj}(\varphi[1/t], t, g_{k+2}(y_1, \ldots, y_{k+2}, t)),$$

where t and the y_i's are fresh variables. Then we map (φ, k) to $(\varphi', k+1)$. Observe that every assignment which sets t to 0 and that satisfies φ' has at least weight $k+2$

(and thus is too costly). Now consider assignments that set t to 1 and that satisfies φ'. Either they satisfy $g_{k+2}(y_1, \ldots, y_{k+2}, 1) = y_1 \vee \ldots \vee y_{k+2}$ or they don't. In the latter case they have to satisfy φ. To sum up we have the following equality:

$$\#\mathrm{Sat}_k(\varphi') = \sum_{j=1}^{k-1} \binom{k+2}{j} \binom{n}{k-j-1} + \#\mathrm{Sat}_{(k-1)}(\varphi).$$

As in Lemma 18 above, we thus obtain a 1-Turing fpt-reduction. Note that we do not obtain a Turing polynomial-time-reduction, since the time required to compute the B-representation of g_{k+2} maybe too high; we only have parameterized polynomial-time. $\qquad\square$

Lemma 23. *If $B \subseteq \mathsf{V}$, or $B \subseteq \mathsf{E}$, or $B \subseteq \mathsf{L}$, then p-$\#\mathrm{WSAT}(B)$ is in FP.*

Proof. Easy after having computed the normal form as in Lemma 11.

We are now in a position to state the full classification for the counting problems.

Theorem 24. *Let B be a finite set of Boolean functions.*

1. *If $\mathsf{D}_2 \subseteq [B]$ or $\mathsf{S}_{10} \subseteq [B]$ or $\mathsf{S}_{00} \subseteq [B]$, then p-$\#\mathrm{WSAT}(B)$ is $\#\mathrm{W}[\mathrm{SAT}]$-complete and p-$\#\mathrm{WCIRCSAT}(B)$ is $\#\mathrm{W}[\mathrm{P}]$-complete, both under Turing fpt-reductions.*
2. *In all other cases p-$\#\mathrm{WSAT}(B)$ and p-$\#\mathrm{WCIRCSAT}(B)$ are in FP.*

6   Conclusion

In this paper we obtained a complete classification of the parameterized complexity of the weighted satisfiability problem, depending on the Boolean functions allowed to appear, both for formulas and for Boolean circuits, and both in the decision and in the counting context. It may seem a little disappointing not to see any involved FPT algorithm in our classification, contrary to the classification of Marx [10] in Schaefer's framework that revealed some nontrivial FPT algorithms. However let us advocate that Post's framework does not seem well adapted to such nontrivial algorithms. Indeed in the classifications that appeared in the literature in the past, the tractable cases usually turned out to be trivially algorithmically solvable (with the possible exception of auto-epistemic logic [5] in which a nontrivial algorithm was developed for the affine fragment).

Parameterized counting complexity was introduced in [7,11], but surprisingly is not much developed so far. We see our paper also as a contribution to this study. While the class $\#\mathrm{W}[\mathrm{P}]$ was introduced in [7] in analogy to $\mathrm{W}[\mathrm{P}]$, we here introduced $\#\mathrm{W}[\mathrm{SAT}]$ in analogy to $\mathrm{W}[\mathrm{SAT}]$, and we present natural complete satisfiability problems for both classes. So we believe our study makes a step towards a better understanding of counting problems within parameterized complexity.

One might also consider the variants of the weighted satisfiability problem in which the task is to find a satisfying assignment of weight at most k or at least k. Preliminary results exist for the monotone fragment, see, e.g., [8,6]. We leave further results for these variants as future work.

Acknowledgement. We are grateful to Arne Meier (Hannover) and Steffen Reith (Wiesbaden) for helpful discussions.

References

1. Abrahamson, K.R., Downey, R.G., Fellows, M.R.: Fixed-parameter tractability and completeness IV: On completeness for W[P] and PSPACE analogues. Annals of Pure and Applied Logic 73(3), 235–276 (1995)
2. Bauland, M., Schneider, T., Schnoor, H., Schnoor, I., Vollmer, H.: The complexity of generalized satisfiability for linear temporal logic. Logical Methods in Computer Science 5(1) (2008)
3. Böhler, E., Creignou, N., Galota, M., Reith, S., Schnoor, H., Vollmer, H.: Boolean circuits as a representation for Boolean functions: Efficient algorithms and hard problems. Logical Methods in Computer Science (to appear, 2012)
4. Böhler, E., Creignou, N., Reith, S., Vollmer, H.: Playing with Boolean blocks I: Post's lattice with applications to complexity theory. SIGACT News 34(4), 38–52 (2003)
5. Creignou, N., Meier, A., Thomas, M., Vollmer, H.: The complexity of reasoning for fragments of autoepistemic logic. ACM Transactions on Computational Logic (to appear, 2012)
6. Dantchev, S., Martin, B., Szeider, S.: Parameterized proof complexity. Computational Complexity 20(1), 51–85 (2011)
7. Flum, J., Grohe, M.: The parameterized complexity of counting problems. SIAM J. Comput. 33(4), 892–922 (2004)
8. Flum, J., Grohe, M.: Parameterized Complexity Theory. Springer (2006)
9. Lewis, H.: Satisfiability problems for propositional calculi. Mathematical Systems Theory 13, 45–53 (1979)
10. Marx, D.: Parameterized complexity of constraint satisfaction problems. Computational Complexity 14(2), 153–183 (2005)
11. McCartin, C.: Parameterized counting problems. Annals of Pure and Applied Logic 138(1-3), 147–182 (2006)
12. Niedermeier, R.: Invitation to Fixed-Parameter Algorithms. Oxford University Press (2006)
13. Post, E.: The two-valued iterative systems of mathematical logic. Annals of Mathematical Studies 5, 1–122 (1941)
14. Reith, S., Wagner, K.: The complexity of problems defined by Boolean circuits. In: Proceedings Mathematical Foundation of Informatics (MFI 1999), pp. 141–156. World Science Publishing (2005)
15. Schnoor, H.: The complexity of model checking for boolean formulas. Int. J. Found. Comput. Sci. 21(3), 289–309 (2010)
16. Thomas, M.: On the applicability of Post's lattice. CoRR, abs/1007.2924 (2010)
17. Vollmer, H.: Introduction to Circuit Complexity. Springer, Heidelberg (1999)

Fixed-Parameter Tractability of Satisfying beyond the Number of Variables

Robert Crowston[1], Gregory Gutin[1], Mark Jones[1], Venkatesh Raman[2], Saket Saurabh[2], and Anders Yeo[3]

[1] Royal Holloway, University of London, Egham, Surrey TW20 0EX, UK
[2] The Institute of Mathematical Sciences, Chennai 600 113, India
[3] University of Johannesburg, Auckland Park, 2006 South Africa

Abstract. We consider a CNF formula F as a multiset of clauses: $F = \{c_1, \ldots, c_m\}$. The set of variables of F will be denoted by $V(F)$. Let B_F denote the bipartite graph with partite sets $V(F)$ and F and an edge between $v \in V(F)$ and $c \in F$ if $v \in c$ or $\bar{v} \in c$. The matching number $\nu(F)$ of F is the size of a maximum matching in B_F. In our main result, we prove that the following parameterization of MAXSAT is fixed-parameter tractable: Given a formula F, decide whether we can satisfy at least $\nu(F) + k$ clauses in F, where k is the parameter.

A formula F is called variable-matched if $\nu(F) = |V(F)|$. Let $\delta(F) = |F| - |V(F)|$ and $\delta^*(F) = \max_{F' \subseteq F} \delta(F')$. Our main result implies fixed-parameter tractability of MAXSAT parameterized by $\delta(F)$ for variable-matched formulas F; this complements related results of Kullmann (2000) and Szeider (2004) for MAXSAT parameterized by $\delta^*(F)$.

To prove our main result, we obtain an $O((2e)^{2k}k^{O(\log k)}(m+n)^{O(1)})$-time algorithm for the following parameterization of the HITTING SET problem: given a collection $\mathcal{C}$ of m subsets of a ground set U of n elements, decide whether there is $X \subseteq U$ such that $C \cap X \neq \emptyset$ for each $C \in \mathcal{C}$ and $|X| \leq m - k$, where k is the parameter. This improves an algorithm that follows from a kernelization result of Gutin, Jones and Yeo (2011).

1 Introduction

In this paper we study a parameterization of MAXSAT. We consider a CNF formula F as a multiset of clauses: $F = \{c_1, \ldots, c_m\}$. (We allow repetition of clauses.) We assume that no clause contains both a variable and its negation, and no clause is empty. The set of variables of F will be denoted by $V(F)$, and for a clause c, $V(c) = V(\{c\})$. A *truth assignment* is a function $\tau : V(F) \to \{\text{TRUE, FALSE}\}$. A truth assignment τ *satisfies* a clause C if there exists $x \in V(F)$ such that $x \in C$ and $\tau(x) = \text{TRUE}$, or $\bar{x} \in C$ and $\tau(x) = \text{FALSE}$. We will denote the number of clauses in F satisfied by τ as $\text{sat}_\tau(F)$ and the maximum value of $\text{sat}_\tau(F)$, over all τ, as $\text{sat}(F)$.

Let B_F denote the bipartite graph with partite sets $V(F)$ and F with an edge between $v \in V(F)$ and $c \in F$ if $v \in V(c)$. The *matching number* $\nu(F)$ of F is the size of a maximum matching in B_F. Clearly $\text{sat}(F) \geq \nu(F)$.

The problem we study in this paper is as follows.

A. Cimatti and R. Sebastiani (Eds.): SAT 2012, LNCS 7317, pp. 355–368, 2012.
© Springer-Verlag Berlin Heidelberg 2012

$(\nu(F) + k)$-SAT
Instance: A CNF formula F and a positive integer α.
Parameter: $k = \alpha - \nu(F)$.
Question: Is sat$(F) \geq \alpha$?

In our main result, we show that $(\nu(F)+k)$-SAT is fixed-parameter tractable by obtaining an algorithm with running time $O((2e)^{2k}k^{O(\log k)}(n + m)^{O(1)})$, where e is the base of the natural logarithm. (We provide basic definitions on parameterized algorithms and complexity, including kernelization, in the next section.)

The *deficiency* $\delta(F)$ of a formula F is $|F| - |V(F)|$; the *maximum deficiency* $\delta^*(F) = \max_{F' \subseteq F} \delta(F')$. A formula F is called *variable-matched* if $\nu(F) = |V(F)|$. Our main result implies fixed-parameter tractability of MAXSAT parameterized by $\delta(F)$ for variable-matched formulas F.

There are two related results: Kullmann [13] obtained an $O(n^{O(\delta^*(F))})$-time algorithm for solving MAXSAT for formulas F with n variables and Szeider [19] gave an $O(f(\delta^*(F))n^4)$- algorithm for the problem, where f is a function depending on $\delta^*(F)$ only. Note that we cannot just drop the condition of being variable-matched and expect a a similar algorithm: it is not hard to see that the satisfiability problem remains NP-complete for formulas F with $\delta(F) = 0$.

A formula F is *minimal unsatisfiable* if it is unsatisfiable but $F \setminus c$ is satisfiable for every clause $c \in F$. Papadimitriou and Wolfe [17] showed that recognition of minimal unsatisfiable CNF formulas is complete for the complexity class D^P. Kleine Büning [11] conjectured that for a fixed integer k, it can be decided in polynomial time whether a formula F with $\delta(F) \leq k$ is minimal unsatisfiable. Independently, Kullmann [13] and Fleischner and Szeider [8] (see also [7]) resolved this conjecture by showing that minimal unsatisfiable formulas with n variables and $n + k$ clauses can be recognized in $n^{O(k)}$ time. Later, Szeider [19] showed that the problem is fixed-parameter tractable by obtaining an algorithm of running time $O(2^k n^4)$. Note that Szeider's results follow from his results mentioned in the previous paragraph and the well-known fact that $\delta^*(F) = \delta(F)$ holds for every minimal unsatisfiable formula F. Since every minimal unsatisfiable formula is variable-matched, our main result also implies fixed-parameter tractability of recognizing minimal unsatisfiable formula with n variables and $n + k$ clauses, parameterized by k.

To obtain our main result, we introduce some reduction rules and branching steps and reduce the problem to a parameterized version of HITTING SET, namely, $(m - k)$-HITTING SET defined below. Let H be a hypergraph. A set $S \subseteq V(H)$ is called a *hitting set* if $e \cap S \neq \emptyset$ for all $e \in E(H)$.

$(m - k)$-HITTING SET
Instance: A hypergraph H ($n = |V(H)|$, $m = |E(H)|$) and a positive integer k.
Parameter: k.
Question: Does there exist a hitting set $S \subseteq V(H)$ of size $m - k$?

Gutin et al. [10] showed that $(m-k)$-HITTING SET is fixed-parameter tractable by obtaining a kernel for the problem. The kernel result immediately implies a

$2^{O(k^2)}(m+n)^{O(1)}$-time algorithm for the problem. Here we obtain a faster algorithm for this problem that runs in $O((2e)^{2k}k^{O(\log k)}(m+n)^{O(1)})$ time using the color-coding technique. This happens to be the dominating step for solving the $(\nu(F)+k)$-SAT problem.

It has also been shown in [10] that the $(m-k)$-HITTING SET problem cannot have a kernel whose size is polynomial in k unless NP $\subseteq$ coNP/poly. In this paper, we give a parameter preserving reduction from this problem to the $(\nu(F)+k)$-SAT problem, thereby showing that $(\nu(F)+k)$-SAT problem has no polynomial sized kernel unless NP $\subseteq$ coNP/poly.

Organization of the Rest of the Paper. In Section 2, we provide additional terminology and notation and some preliminary results. In Section 3, we give a sequence of polynomial time preprocessing rules on the given input and justify their correctness. In Section 4, we give two simple branching rules and reduce the resulting input to a $(m-k)$-HITTING SET problem instance. Section 5 gives an improved fixed-parameter algorithm for the $(m-k)$-HITTING SET problem using color coding. Section 6 summarizes the entire algorithm, and show its correctness and analyzes its running time. Section 7 shows the hardness of kernelization result. Section 8 concludes with some remarks.

2   Additional Terminology, Notation and Preliminaries

Graphs and Hypergraphs. For a subset X of vertices of a graph G, $N_G(X)$ denotes the set of all neighbors of vertices in X. When G is clear from the context, we write $N(X)$ instead of $N_G(X)$. A matching *saturates* all end-vertices of its edges. For a bipartite graph $G = (V_1, V_2; E)$, the classical Hall's matching theorem states that G has a matching that saturates every vertex of V_1 if and only if $|N(X)| \geq |X|$ for every subset X of V_1. The next lemma follows from Hall's matching theorem: add d vertices to V_2, each adjacent to every vertex in V_1.

Lemma 1. *Let* $G = (V_1, V_2; E)$ *be a bipartite graph, and suppose that for all subsets* $X \subseteq V_1$, $|N(X)| \geq |X| - d$ *for some* $d \geq 0$. *Then* $\nu(G) \geq |V_1| - d$.

A *hypergraph* $H = (V(H), \mathcal{F})$ consists of a nonempty set $V(H)$ of *vertices* and a family $\mathcal{F}$ of nonempty subsets of V called *edges* of H ($\mathcal{F}$ is often denoted $E(H)$). Note that $\mathcal{F}$ may have *parallel* edges, i.e., copies of the same subset of $V(H)$. For any vertex $v \in V(H)$, and any $\mathcal{E} \subseteq \mathcal{F}$, $\mathcal{E}[v]$ is the set of edges in $\mathcal{E}$ containing v, $N[v]$ is the set of all vertices contained in edges of $\mathcal{F}[v]$, and the *degree* of v is $d(v) = |\mathcal{F}[v]|$. For a subset T of vertices, $\mathcal{F}[T] = \bigcup_{v \in T} \mathcal{F}[v]$.

CNF Formulas. For a subset X of the variables of CNF formula F, F_X denotes the subset of F consisting of all clauses c such that $V(c) \cap X \neq \emptyset$. A formula F is called q-*expanding* if $|X| + q \leq |F_X|$ for each $X \subseteq V(F)$. Note that, by Hall's matching theorem, a formula is variable-matched if and only if it is 0-expanding. For $x \in V(F)$, $n(x)$ and $n(\bar{x})$ denote the number of clauses containing x and the number of clauses containing $\bar{x}$, respectively. Given a matching M, an *alternating path* is a path in which the edges belong alternatively to M and not to M.

A function $\pi : U \to \{\text{TRUE, FALSE}\}$, where U is a subset of $V(F)$, is called a *partial truth assignment*. A partial truth assignment $\pi : U \to \{\text{TRUE, FALSE}\}$ is an *autarky* if π satisfies all clauses of F_U. We have the following:

Lemma 2. *[4] Let $\pi : U \to \{\text{TRUE, FALSE}\}$ be an autarky for a CNF formula F and let γ be any truth assignment on $V(F) \backslash U$. Then for the combined assignment $\tau := \pi\gamma$, it holds that $\mathrm{sat}_\tau(F) = |F_U| + \mathrm{sat}_\gamma(F \backslash F_U)$. Clearly, τ can be constructed in polynomial time given π and γ.*

Autarkies were first introduced in [16]; they are the subject of much study, see, e.g., [7,14,19], and see [12] for an overview.

Parameterized Complexity. A *parameterized problem* is a subset $L \subseteq \Sigma^* \times \mathbb{N}$ over a finite alphabet Σ. L is *fixed-parameter tractable* if the membership of an instance (x, k) in $\Sigma^* \times \mathbb{N}$ can be decided in time $f(k)|x|^{O(1)}$, where f is a function of the *parameter* k only [6,9]. Given a parameterized problem L, a *kernelization of L* is a polynomial-time algorithm that maps an instance (x, k) to an instance (x', k') (the *kernel*) such that (i) $(x, k) \in L$ if and only if $(x', k') \in L$, (ii) $k' \le h(k)$, and (iii) $|x'| \le g(k)$ for some functions h and g. It is well-known [6,9] that a decidable parameterized problem L is fixed-parameter tractable if and only if it has a kernel. Polynomial-size kernels are of main interest, due to applications [6,9], but unfortunately not all fixed-parameter problems have such kernels unless coNP$\subseteq$NP/poly, see, e.g., [2,3,5].

For a positive integer q, let $[q] = \{1, \ldots, q\}$.

3   Preprocessing Rules

In this section we give preprocessing rules and their correctness.

Let F be the given CNF formula on n variables and m clauses with a maximum matching M on B_F, the variable-clause bipartite graph corresponding to F. Let α be a given integer and recall that our goal is to check whether $\mathrm{sat}(F) \ge \alpha$. For each preprocessing rule below, we let (F', α') be the instance resulting by the application of the rule on (F, α). We say that a rule is *valid* if (F, α) is a YES instance if and only if (F', α') a YES instance.

Reduction Rule 1. *Let x be a variable such that $n(x) = 0$ (respectively $n(\bar{x}) = 0$). Set $x = \text{FALSE}$ ($x = \text{TRUE}$) and remove all the clauses that contain $\bar{x}$ (x). Reduce α by $n(\bar{x})$ (respectively $n(x)$).*

The proof of the following lemma is immediate.

Lemma 3. *If $n(x) = 0$ (respectively $n(\bar{x}) = 0$) then $\mathrm{sat}(F) = \mathrm{sat}(F') + n(\bar{x})$ (respectively $\mathrm{sat}(F) = \mathrm{sat}(F') + n(x)$) and so Rule 1 is valid.*

Reduction Rule 2. *Let $n(x) = n(\bar{x}) = 1$ and let c' and c'' be the two clauses containing x and $\bar{x}$, respectively. Let $c^* = (c' - x) \cup (c'' - \bar{x})$ and let F' be obtained from F be deleting c' and c'' and adding the clause c^*. Reduce α by 1.*

Lemma 4. $\mathrm{sat}(F) = \mathrm{sat}(F') + 1$ *and so Rule 2 is valid.*

Proof. Consider any assignment for F. If it satisfies both c' and c'', then the same assignment will satisfy c^*. So when restricted to variables of F', it will satisfy at least $sat(F) - 1$ clauses of F'. Thus $sat(F') \geq sat(F) - 1$ which is equivalent to $sat(F) \leq sat(F') + 1$. Similarly if an assignment γ to F' satisfies c^* then at least one of c', c'' is satisfied by γ. Therefore by setting x true if γ satisfies c'' and false otherwise, we can extend γ to an assignment on F that satisfies both of c', c''. On the other hand, if c^* is not satisfied by γ then neither c' nor c'' are satisfied by γ, and any extension of γ will satisfy exactly one of c', c''. Therefore in either case $sat(F) \geq sat(F') + 1$. We conclude that $\mathrm{sat}(F) = \mathrm{sat}(F') + 1$, as required.   □

Our next reduction rule is based on the following lemma proved in Fleischner et al. [7, Lemma 10], Kullmann [14, Lemma 7.7] and Szeider [19, Lemma 9].

Lemma 5. *Let F be a CNF formula. Given a maximum matching in B_F, in time $O(|F|)$ we can find an autarky $\pi : U \to \{\text{TRUE}, \text{FALSE}\}$ such that $F \setminus F_U$ is 1-expanding.*

Reduction Rule 3. *Find an autarky $\pi : U \to \{\text{TRUE}, \text{FALSE}\}$ such that $F \setminus F_U$ is 1-expanding. Set $F' = F \setminus F_U$ and reduce α by $|F_U|$.*

The next lemma follows from Lemma 2.

Lemma 6. $\mathrm{sat}(F) = \mathrm{sat}(F') + |F_U|$ *and so Rule 3 is valid.*

After exhaustive application of Rule 3, we may assume that the resulting formula is 1-expanding. For the next reduction rule, we need the following results.

Theorem 1 ([19]). *Given a variable-matched formula F, with $|F| = |V(F)| + 1$, we can decide whether F is satisfiable in time $O(|V(F)|^3)$.*

Consider a bipartite graph $G = (A, B; E)$. We say that G is *q-expanding* if for all $A' \subseteq A$, $|N_G(A')| \geq |A'| + q$. Clearly, a formula F is q-expanding if and only if B_F is q-expanding. From a bipartite graph $G = (A, B; E)$, $x \in A$ and $q \geq 1$, we obtain a bipartite graph G_{qx}, by adding new vertices $x_1, \ldots, x_q$ to A and adding edges such that new vertices have exactly the same neighborhood as x, that is, $G_{qx} = (A \cup \{x_1, \ldots, x_q\}, B; E \cup \{(x_i, y) : (x, y) \in E\})$. The following result is well known.

Lemma 7. [15, Theorem 1.3.6] *Let $G = (A, B; E)$ be a 0-expanding bipartite graph. Then G is q-expanding if and only if G_{qx} is 0-expanding for all $x \in A$.*

Lemma 8. *Let $G = (A, B; E)$ be a 1-expanding bipartite graph. In polynomial time, we can check whether G is 2-expanding, and if it is not, find a set $S \subseteq A$ such that $|N_G(S)| = |S| + 1$.*

Proof. Let $x \in A$. By Hall's Matching Theorem, G_{2x} is 0-expanding if and only if $\nu(G_{2x}) = |A| + 2$. Since we can check the last condition in polynomial time, by Lemma 7 we can decide whether G is 2-expanding in polynomial time. So, assume that G is not 2-expanding and we know this because G_{2y} is not 0-expanding for some $y \in A$. By Lemma 3(4) in Szeider [19], in polynomial time, we can find a set $T \subseteq A \cup \{y_1, y_2\}$ such that $|N_{G_{2y}}(T)| < |T|$. Since G is 1-expanding, $y_1, y_2 \in T$ and $|N_{G_{2y}}(T)| = |T| - 1$. Hence, $|S| + 1 = |N_G(S)|$, where $S = T \setminus \{y_1, y_2\}$. $\qquad \square$

For a formula F and a set $S \subseteq V(F)$, $F[S]$ denotes the formula obtained from F_S by deleting all variables not in S.

Reduction Rule 4. *Let F be a 1-expanding formula and let $B = B_F$. Using Lemma 8, check if F is 2-expanding and otherwise find a set $S \subseteq V(F)$ with $|N_B(S)| = |S| + 1$. Let M be a matching that saturates S in $B[S \cup N_B(S)]$ (that exists as $B[S \cup N_B(S)]$ is 1-expanding). Use Theorem 1 to decide whether $F[S]$ is satisfiable, and proceed as follows.*

$F[S]$ **is satisfiable:** *Obtain a new formula F' by removing all clauses in $N_B(S)$ from F. Reduce α by $|N_B(S)|$.*

$F[S]$ **is not satisfiable:** *Let y denote the clause in $N_B(S)$ that is not matched to a variable in S by M. Let Y be the set of all clauses in $N_B(S)$ that can be reached from y with an M-alternating path in $B[S \cup N_B(S)]$. (We will argue later that $Y = N_B(S)$.) Let c' be the clause obtained by deleting all variables in S from $\cup_{c'' \in Y} c''$. That is, a literal l belongs to c' if and only if it belongs to some clause in Y and is not a variable from S. Obtain a new formula F' by removing all clauses in $N_B(S)$ from F and adding c'. Reduce α by $|S|$.*

Lemma 9. *If $F[S]$ is satisfiable then $\mathrm{sat}(F) = \mathrm{sat}(F') + |N_B(S)|$ else $\mathrm{sat}(F) = \mathrm{sat}(F') + |S|$ and thus Rule 4 is valid.*

Proof. We consider two cases.
Case 1: $F[S]$ **is satisfiable.** Since $F[S]$ is satisfiable, any truth assignment for F' can be extended to an assignment for F that satisfies every clause in $N_B(S)$. Therefore $\mathrm{sat}(F) \geq \mathrm{sat}(F') + |N_B(S)|$. The inequality $\mathrm{sat}(F) \leq \mathrm{sat}(F') + |N_B(S)|$ follows from the fact, any optimal truth assignment to F when restricted to F' will satisfy at least $\mathrm{sat}(F) - |N_B(S)|$ clauses.
Case 2: $F[S]$ **is not satisfiable.** Let $F'' = F' \setminus c'$. As any optimum truth assignment to F will satisfy at least $\mathrm{sat}(F) - |N_B(S)|$ clauses of F'', it follows that $sat(F) \leq \mathrm{sat}(F'') + |N_B(S)| \leq \mathrm{sat}(F') + |N_B(S)|$.

Let S' be the set of variables in S appearing in clauses in Y. Since Y is made up of clauses that are reachable in $B_{F[S]}$ by an M-alternating path from the single unmatched clause y, $|Y| = |S'| + 1$. It follows that $|N_B(S) \setminus Y| = |S \setminus S'|$, and M matches every clause in $N_B(S) \setminus Y$ with a variable in $S \setminus S'$. Furthermore, $N_B(S \setminus S') \cap Y = \emptyset$ as otherwise the matching partners of some elements of $S \setminus S'$ would have been reachable by an M-alternating path from y, contradicting the definition of Y and S'. Thus $S \setminus S'$ has an autarky such that $F \setminus F_{S \setminus S'}$ is 1-expanding which would have been detected by Rule 3, hence $S \setminus S' = \emptyset$ or

$S = S'$. That is, all clauses in $N_B(S)$ are reachable from the unmatched clause y by an M-alternating path.

Suppose that there exists an assignment γ to F', that satisfies $\text{sat}(F')$ clauses of F' that also satisfies c'. Then there exists a clause $c'' \in Y$ that is satisfied by γ. As c'' is reachable from y by an M-alternating path, we can modify M to include y and exclude c'', by taking the symmetric difference of the matching and the M-alternating path from y to c''. This will give a matching saturating S and $Y \setminus c''$, and we use this matching to extend the assignment γ to one which satisfies all of $N_B(S) \backslash c''$. We therefore have satisfied all the clauses of $N_B(S)$. Therefore since c'' is satisfied in F'' but does not appear in F, we have satisfied extra $|N_B(S)| - 1 = |S|$ clauses. Suppose on the other hand that every assignment γ for F'' that satisfies $\text{sat}(F'')$ clauses does not satisfy c''. We can use the matching on $B[S \cup N_B(S)]$ to satisfy $|N_B(S)| - 1$ clauses in $N_B(S)$, which would give us an additional $|S|$ clauses in $N_B(S)$. Thus $\text{sat}(F) \geq \text{sat}(F'') + |S|$.

As $|N_B(S)| = |S| + 1$, it suffices to show that $\text{sat}(F) < \text{sat}(F'') + |N_B(S)|$. Suppose that there exists an assignment γ to F that satisfies $\text{sat}(F'') + |N_B(S)|$ clauses, then it must satisfy all the clauses of $N_B(S)$ and $\text{sat}(F'')$ clauses of F''''. As $F[S]$ is not satisfiable, variables in S alone can not satisfy all of $N_B(S)$. Hence there exists a clause $c'' \in N_B(S)$ such that there is a variable $y \in c'' \setminus S$ that satisfies c''. But then $c'' \in Y$ and hence $y \in c'$ and hence c' would be satisfiable by γ, a contradiction as γ satisfies $\text{sat}(F')$ clauses of F'. □

4   Branching Rules and Reduction to $(m - k)$-Hitting Set

Our algorithm first applies Reduction Rules 1, 2, 3 and 4 exhaustively on (F, α). Then it applies two branching rules we describe below, in the following order.

Branching on a variable x means that the algorithm constructs two instances of the problem, one by substituting $x = \text{TRUE}$ and simplifying the instance and the other by substituting $x = \text{FALSE}$ and simplifying the instance. Branching on x or y being false means that the algorithm constructs two instances of the problem, one by substituting $x = \text{FALSE}$ and simplifying the instance and the other by substituting $y = \text{FALSE}$ and simplifying the instance. Simplifying an instance is done as follows. For any clause c, if c contains a literal x with $x = \text{TRUE}$, or a literal $\bar{x}$ with $x = \text{FALSE}$, we remove c and reduce α by 1. If c contains a literal x with $x = \text{FALSE}$, or a literal $\bar{x}$ with $x = \text{TRUE}$, and c contains other literals, remove x from c. If c consists of the single literal $x = \text{FALSE}$ or $\bar{x}$ with $x = \text{TRUE}$, remove c.

A branching rule is correct if the instance on which it is applied is a YES-instance if and only if the simplified instance of (at least) one of the branches is a YES-instance.

Branching Rule 1. *If $n(x) \geq 2$ and $n(\bar{x}) \geq 2$ then we branch on x.*

Before attempting to apply Branching Rule 2, we apply the following rearranging step: For all variables x such that $n(\bar{x}) = 1$, swap literals x and $\bar{x}$ in all clauses. Observe now that for every variable $n(x) = 1$ and $n(\bar{x}) \geq 2$.

Branching Rule 2. *If there is a clause c such that positive literals* $x, y \in c$ *then we branch on x being false or y being false.*

Branching Rule 1 is exhaustive and thus its correctness also follows. When we reach Branching Rule 2 for every variable $n(x) = 1$ and $n(\bar{x}) \geq 2$. As $n(x) = 1$ and $n(y) = 1$ we note that c is the only clause containing these literals. Therefore there exists an optimal solution with x or y being false (if they are both true just change one of them to false). Thus, we have the following:

Lemma 10. *Branching Rules 1 and 2 are correct.*

Let (F, α) be the given instance on which Reduction Rules 1, 2, 3 and 4, and Branching Rules 1 and 2 do not apply. Observe that for such an instance F the following holds:

1. For every variable x, $n(x) = 1$ and $n(\bar{x}) \geq 2$.
2. Every clause contains at most one positive literal.

We call a formula F satisfying the above properties *special*. In what follows we describe an algorithm for our problem on special instances. Let $c(x)$ denote the *unique* clause containing positive literal x. We can obtain a matching saturating $V(F)$ in B_F by taking the edge connecting the variable x and the clause $c(x)$. We denote the resulting matching by M_u.

We first describe a transformation that will be helpful in reducing our problem to $(m - k)$-HITTING SET. Given a formula F we obtain a new formula F' by changing the clauses of F as follows. If there exists some $c(x)$ such that $|c(x)| \geq 2$, do the following. Let $c' = c(x) - x$ (that is, c' contains the same literals as $c(x)$ except for x) and add c' to all clauses containing the literal $\bar{x}$. Furthermore remove c' from $c(x)$ (which results in $c(x) = (x)$ and therefore $|c(x)| = 1$).

Next we prove the validity of the above transformation.

Lemma 11. *Let F' be the formula obtained by applying the transformation described above on F. Then $\mathrm{sat}(F') = \mathrm{sat}(F)$ and $\nu(B_F) = \nu(B_{F'})$.*

Proof. We note that the matching M_u remains a matching in $B_{F'}$ and thus $\nu(B_F) = \nu(B_{F'})$. Let γ be any truth assignment to the variables in F (and F') and note that if c' is false under γ then F and F' satisfy exactly the same clauses under γ (as we add and subtract something false to the clauses). So assume that c' is true under γ.

If γ maximizes the number of satisfied clauses in F then clearly we may assume that x is false (as $c(x)$ is true due to c'). Now let γ' be equal to γ except the value of x has been flipped to true. Note that exactly the same clauses are satisfied in F and F' by γ and γ', respectively. Analogously, if an assignment maximizes the number of satisfied clauses in F' we may assume that x is true and by changing it to false we satisfy equally many clauses in F. Hence, $\mathrm{sat}(F') = \mathrm{sat}(F)$. $\square$

Given an instance (F, α) we apply the above transformation repeatedly until no longer possible and obtain an instance (F', α) such that $\nu(B_F) = \nu(B_{F'})$. Thus, by the above transformation we may assume that $|c(x)| = 1$ for all $x \in V(F)$.

For simplicity of presentation we denote the transformed instance by (F, α). Let C^* denote all clauses that are not matched by M_u (and therefore only contain negated literals). We associate a hypergraph H^* with the transformed instance. Let H^* be the hypergraph with vertex set $V(F)$ and edge set $E^* = \{V(c) \mid c \in C^*\}$.

We now show the following equivalence between $(\nu(F) + k)$-SAT on special instances and $(m - k)$-HITTING SET.

Lemma 12. *Let (F, α) be the transformed instance and H^* be the hypergraph associated with it. Then $\mathrm{sat}(F) \geq \alpha$ if and only if there is a hitting set in H^* of size at most $|E(H^*)| - k$, where $k = \alpha - \nu(F)$.*

Proof. We start with a simple observation about an assignment satisfying the maximum number of clauses of F. There exists an optimal truth assignment to F, such that all clauses in C^* are true. Assume that this is not the case and let γ be an optimal truth assignment satisfying as many clauses from C^* as possible and assume that $c \in C^*$ is not satisfied. Let $\bar{x} \in c$ be an arbitrary literal and note that $\gamma(x) = \mathrm{TRUE}$. However, changing x to false does not decrease the number of satisfied clauses in F and increases the number of satisfied clauses in C^*.

Now we show that $\mathrm{sat}(F) \geq \alpha$ if and only if there is a hitting set in H^* of size at most $|E(H^*)| - k$. Assume that γ is an optimal truth assignment to F, such that all clauses in C^* are true. Let $U \subseteq V(F)$ be all variables that are false in γ and note that U is a hitting set in H^*. Analogously if U' is a hitting set in H^* then by letting all variables in U' be false and all other variables in $V(F)$ be true we get a truth assignment that satisfies $|F| - |U'|$ constraints in F. Therefore if $\tau(H^*)$ is the size of a minimum hitting set in H^* we have $\mathrm{sat}(F) = |F| - \tau(H^*)$. Hence, $\mathrm{sat}(F) = |F| - \tau(H^*) = |V(F)| + |C^*| - \tau(H^*)$ and thus $\mathrm{sat}(F) \geq \alpha$ if and only if $|C^*| - \tau(H^*) \geq k$, which is equivalent to $\tau(H^*) \leq |E(H^*)| - k$. $\square$

Therefore our problem is fixed-parameter tractable on special instances, by the following known result.

Theorem 2 ([10]). *There exists an algorithm for $(m - k)$-HITTING SET running in time $2^{O(k^2)} + O((n + m)^{O(1)})$.*

In the next section we give a faster algorithm for $(\nu(F) + k)$-SAT on special instances by giving a faster algorithm for $(m - k)$-HITTING SET.

5   Improved FPT Algorithm for $(m - k)$-Hitting Set

To obtain a faster algorithm for $(m - k)$-HITTING SET, we utilize the following concept of k-*mini-hitting set* introduced in [10].

Definition 1. *Let $H = (V, \mathcal{F})$ be a hypergraph and k be a nonnegative integer. A k-mini-hitting set is a set $S_{\mathrm{MINI}} \subseteq V$ such that $|S_{\mathrm{MINI}}| \leq k$ and $|\mathcal{F}[S_{\mathrm{MINI}}]| \geq |S_{\mathrm{MINI}}| + k$.*

Lemma 13 ([10]). *A hypergraph H has a hitting set of size at most $m - k$ if and only if it has a k-mini-hitting set. Moreover, given a k-mini-hitting set S_{MINI}, we can construct a hitting set S with $|S| \leq m - k$ such that $S_{\text{MINI}} \subseteq S$ in polynomial time.*

Next we give an algorithm that finds a k-mini-hitting set S_{MINI} if it exists, in time $c^k(m + n)^{O(1)}$, where c is a constant. We first describe a randomized algorithm based on color-coding [1] and then derandomize it using hash functions. Let $\chi : E(H) \to [q]$ be a function. For a subset $S \subseteq V(H)$, $\chi(S)$ denotes the maximum subset $X \subseteq [q]$ such that for all $i \in X$ there exists an edge $e \in E(H)$ with $\chi(e) = i$ and $e \cap S \neq \emptyset$. A subset $S \subseteq V(H)$ is called a *colorful hitting set* if $\chi(S) = [q]$. We now give a procedure that given a coloring function χ finds a minimum colorful hitting set, if it exists. This algorithm will be useful in obtaining a k-mini-hitting set S_{MINI}.

Lemma 14. *Given a hypergraph H and a coloring function $\chi : E(H) \to [q]$, we can find a minimum colorful hitting set if there exists one in time $O(2^q(m+n))$.*

Proof. We first check whether for every $i \in [q]$, $\chi^{-1}(i) \neq \emptyset$. If for any i we have that $\chi^{-1}(i) = \emptyset$, then we return that there is no colorful hitting set. So we may assume that for all $i \in [q]$, $\chi^{-1}(i) \neq \emptyset$. We will give an algorithm using dynamic programming over subsets of $[q]$. Let γ be an array of size 2^q indexed by the subsets of $[q]$. For a subset $X \subseteq [q]$ by $\gamma[X]$ we denote the size of a smallest set $W \subseteq V(H)$ such that $\chi(W) = X$. We obtain a recurrence for $\gamma[X]$ as follows:

$$\gamma[X] = \begin{cases} \min_{(v \in V(H), \chi(\{v\}) \cap X \neq \emptyset)}\{1 + \gamma[X \setminus \chi(\{v\})]\} & \text{if } |X| \geq 1, \\ 0 & \text{if } X = \emptyset. \end{cases}$$

The correctness of the above recurrence is clear. The algorithm computes $\gamma[[q]]$ by filling the γ in the order of increasing set sizes. Clearly, each cell can be filled in time $O(n + m)$ and thus the whole array can be filled in time $O(2^q(n + m))$. The size of a minimum colorful hitting set is given by $\gamma[[q]]$. We can obtain a minimum colorful hitting set by the routine back-tracking. $\square$

Now we describe a randomized procedure to obtain a k-mini-hitting set S_{MINI} in a hypergraph H, if there exists one. We do the following for each possible value p of $|S_{\text{MINI}}|$ (that is, for $1 \leq p \leq k$). Color $E(H)$ uniformly at random with colors from $[p + k]$; we denote this random coloring by χ. Assume that there is a k-mini-hitting set S_{MINI} of size p and some $p + k$ edges $e_1, \ldots, e_{p+k}$ such that for all $i \in [p + k]$, $e_i \cap S_{\text{MINI}} \neq \emptyset$. The probability that for all $1 \leq i < j \leq p + k$ we have that $\chi(e_i) \neq \chi(e_j)$ is $\frac{(p+k)!}{(p+k)^{p+k}} \geq e^{-(p+k)} \geq e^{-2k}$. Now, using Lemma 14 we can test in time $O(2^{p+k}(m + n))$ whether there is a colorful hitting set of size at most p. Thus with probability at least e^{-2k} we can find a S_{MINI}, if there exits one. To boost the probability we repeat the procedure e^{2k} times and thus in time $O((2e)^{2k}(m+n)^{O(1)})$ we find a S_{MINI}, if there exists one, with probability at least $1 - (1 - \frac{1}{e^{2k}})^{e^{2k}} \geq \frac{1}{2}$. If we obtained S_{MINI} then using Lemma 13 we can construct a hitting set of H of size at most $m - k$.

To derandomize the procedure, we need to replace the first step of the procedure where we color the edges of $E(H)$ uniformly at random from the set $[p+k]$ to a deterministic one. This is done by making use of an $(m, p+k, p+k)$-*perfect hash family*. An $(m, p+k, p+k)$-perfect hash family, $\mathcal{H}$, is a set of functions from $[m]$ to $[p+k]$ such that for every subset $S \subseteq [m]$ of size $p+k$ there exists a function $f \in \mathcal{H}$ such that f is injective on S. That is, for all $i, j \in S$, $f(i) \neq f(j)$. There exists a construction of an $(m, p+k, p+k)$-perfect hash family of size $O(e^{p+k} \cdot k^{O(\log k)} \cdot \log m)$ and one can produce this family in time linear in the output size [18]. Using an $(m, p+k, p+k)$-perfect hash family $\mathcal{H}$ of size at most $O(e^{2k} \cdot k^{O(\log k)} \cdot \log m)$ rather than a random coloring we get the desired deterministic algorithm. To see this, it is enough to observe that if there is a subset $S_{\text{mini}} \subseteq V(H)$ such that $|\mathcal{F}[S_{\text{MINI}}]| \geq |S_{\text{MINI}}| + k$ then there exists a coloring $f \in \mathcal{H}$ such that the $p+k$ edges $e_1, \ldots, e_{p+k}$ that intersect S_{mini} are distinctly colored. So if we generate all colorings from $\mathcal{H}$ we will encounter the desired f. Hence for the given f, when we apply Lemma 14 we get the desired result. This concludes the description. The total time of the derandomized algorithm is $O(k 2^{2k}(m+n) e^{2k} \cdot k^{O(\log k)} \cdot \log m)$.

Theorem 3. *There exists an algorithm solving* $(m-k)$-HITTING SET *in time* $O((2e)^{2k} k^{O(\log k)} (m+n)^{O(1)})$.

By Theorem 3 and the transformation discussed in Section 4 we have the following Theorem.

Theorem 4. *There exists an algorithm solving a special instance of* $(\nu(F)+k)$-SAT *in time* $O((2e)^{2k} k^{O(\log k)} (m+n)^{O(1)})$.

6   Complete Algorithm, Correctness and Analysis

The complete algorithm for an instance (F, α) is as follows.

Find a maximum matching M on B_F and let $k = \alpha - |M|$. If $k \leq 0$, return YES. Otherwise, apply Reduction Rules 1 to 4, whichever is applicable, in that order and then run the algorithm on the reduced instance and return the answer. If none of the Reduction Rules apply, then apply Branching Rule 1 if possible, to get two instances (F', α') and (F'', α''). Run the algorithm on both instances; if one of them returns YES, return YES, otherwise return NO. If Branching Rule 1 does not apply then we rearrange the formula and attempt to apply Branching Rule 2 in the same way. Finally if $k > 0$ and none of the reduction or branching rules apply, then we have for all variables x, $n(x) = 1$ and every clause contains at most one positive literal, i.e. (F, α) is a special instance. Then solve the problem by reducing to an instance of $(m-k)$-HITTING SET and solving in time $O((2e)^{2k} k^{O(\log k)} (m+n)^{O(1)})$ as described in Sections 4 and 5.

Correctness of all the preprocessing rules and the branching rules follows from Lemmata 3, 4, 6, 9 and 10.

Analysis of the algorithm. Let (F, α) be the input instance. Let $\mu(F) = \mu = \alpha - \nu(F)$ be the measure. We will first show that our preprocessing rules do not increase this measure. Following this, we will prove a lower bound on the decrease in the measure occurring as a result of the branching, thus allowing us to bound the running time of the algorithm in terms of the measure μ. For each case, we let (F', α') be the instance resulting by the application of the rule or branch. Also let M' be a maximum matching of $B_{F'}$.

Reduction Rule 1. We consider the case when $n(x) = 0$; the other case when $n(\bar{x}) = 0$ is analogous. We know that $\alpha' = \alpha - n(\bar{x})$ and $\nu(F') \geq \nu(F) - n(\bar{x})$ as removing $n(\bar{x})$ clause can only decrease the matching size by $n(\bar{x})$. This implies that $\mu(F) - \mu(F') = \alpha - \nu(F) - \alpha' + \nu(F') = (\alpha - \alpha') + (\nu(F') - \nu(F)) \geq n(\bar{x}) - n(\bar{x})$. Thus, $\mu(F') \leq \mu(F)$.

Reduction Rule 2. We know that $\alpha' = \alpha - 1$. We show that $\nu(F') \geq \nu(F) - 1$. In this case we remove the clauses c' and c'' and add $c^* = (c' - x) \cup (c'' - \bar{x})$. We can obtain a matching of size $\nu(F) - 1$ in $B_{F'}$ as follows. If at most one of the c' and c'' is the end-point of some matching edge in M then removing that edge gives a matching of size $\nu(F) - 1$ for $B_{F'}$. So let us assume that some edges (a, c') and (b, c'') are in M. Clearly, either $a \neq x$ or $b \neq x$. Assume $a \neq x$. Then $M \setminus \{(a, c'), (b, c'')\} \cup \{(a, c^*)\}$ is a matching of size $\nu(F) - 1$ in $B_{F'}$. Thus, we conclude that $\mu(F') \leq \mu(F)$.

Reduction Rule 3. The proof is the same as in the case of Reduction Rule 1.

Reduction Rule 4: The proof that $\mu(F') \leq \mu(F)$ in the case when $F[S]$ is satisfiable is the same as in the case of Reduction Rule 1 and in the case when $F[S]$ is not satisfiable is the same as in the case of Reduction Rule 2.

Branching Rule 1. Consider the case when we set $x = \text{TRUE}$. In this case, $\alpha' = \alpha - n(x)$. Also, since no reduction rules are applicable we have that F is 2-expanding. Hence, $\nu(F) = |V(F)|$. We will show that after the removal of $n(x)$ clauses, the matching size will remain at least $\nu(F) - n(x) + 1$ $(= |V(F)| - n(x) + 1 = |V(F')| - n(x) + 2.)$ This will imply that $\mu(F') \leq \mu(F) - 1$. By Lemma 1, it suffices to show that in $B' = B_{F'}$, every subset $S \subseteq V(F')$, $|N_{B'}(S)| \geq |S| - (n(x) - 2)$. The only clauses that have been removed by the simplification process after setting $x = \text{TRUE}$ are those where x appears positively and the singleton clauses $(\bar{x})$. hence, the only edges of $G[S \cup N_B[S]]$ that are missing in $N_{B'}(S)$ from $N_B(S)$ are those corresponding to clauses that contain x as a pure literal and some variable in S. Thus, $|N_{B'}(S)| \geq |S| + 2 - n(x) = |S| - (n(x) - 2)$ (as F is 2-expanding).

The case when we set $x = \text{FALSE}$ is similar to the case when we set $x = \text{TRUE}$. Here, also we can show that $\mu(F') \leq \mu(F) - 1$. Thus, we get two instances, with each instance (F', α') having $\mu(F') \leq \mu(F) - 1$.

Branching Rule 2. The analysis here is the same as for Branching Rule 1 and again we get two instances with $\mu(F') \leq \mu(F) - 1$.

We therefore have a depth-bounded search tree of depth at most $\mu = \alpha - \nu(F) = k$, in which any branching splits an instance into two instances. Thus the search tree has at most 2^k instances. As each reduction and branching rule takes polynomial time, and an instance to which none of the rules apply can be solved in time $O((2e)^{2\mu}\mu^{O(\log \mu)}(m + n)^{O(1)})$, we have that the total running time of the algorithm is at most $O((2e)^{2k}k^{O(\log k)}(n + m)^{O(1)})$.

Theorem 5. *There is an algorithm solving $(\nu(F) + k)$-SAT in time* $O((2e)^{2k}k^{O(\log k)}(n + m)^{O(1)})$.

7   Hardness of Kernelization

In this section, we show that $(\nu(F)+k)$-SAT does not have a polynomial kernel, unless coNP $\subseteq$ NP/poly. To do this, we use the concept of a *polynomial time and parameter transformation* [3,5]: Let L and Q be parameterized problems. We say a polynomial time computable function $f : \Sigma^* \times \mathbb{N} \to \Sigma^* \times \mathbb{N}$ is a *polynomial time and parameter transformation* from L to Q if there exists a polynomial $p : \mathbb{N} \to \mathbb{N}$ such that for any $(x, k) \in \Sigma^* \times \mathbb{N}, (x, k) \in L$ if and only if $f(x, k) = (x', k') \in Q$, and $k' \leq p(k)$.

Lemma 15. [3, Theorem 3] *Let L and Q be parameterized problems, and suppose that L^c and Q^c are the derived classical problems. Suppose that L^c is NP-complete, and $Q^c \in NP$. Suppose that f is a polynomial time and parameter transformation from L to Q. Then, if Q has a polynomial kernel, then L has a polynomial kernel.*

Theorem 6. *$(\nu(F)+k)$-SAT has no polynomial kernel, unless coNP $\subseteq$ NP/poly.*

Proof. By [10, Theorem 3], there is no polynomial sized kernel for the problem of deciding whether a hypergraph has a hitting set of size $|E(H)|-k$, where k is the parameter unless coNP $\subseteq$ NP/poly. We prove the theorem by a polynomial time parameter preserving reduction from this problem. Then the theorem follows from Lemma 15, as $(\nu(F) + k)$-SAT is NP-complete.

Given a hypergraph H, construct a CNF formula F as follows. Let the variables of F be the vertices of H. For each variable x, let the unit clause (x) be a clause in F. For every edge E in H, let c_E be the clause containing the literal $\bar{x}$ for every $x \in E$. Observe that F is matched, and that H has a hitting set of size $|E(H)| - k$ if and only if $sat(F) \geq n + k$.   □

8   Conclusion

We have shown that for any CNF formula F, it is fixed-parameter tractable to decide if F has a satisfiable subformula containing α clauses, where $\alpha - \nu(F)$ is the parameter. Our result implies fixed-parameter tractability for the problem of deciding satisfiability of F when F is variable-matched and $\delta(F) \leq k$, where k is the parameter. In addition, we show that the problem does not have a polynomial kernel unless coNP $\subseteq$ NP/poly.

If every clause contains exactly two literals then it is well known that we can satisfy at least $3m/4$ clauses. From this, and by applying Reduction Rules 1 and 2, we can get a linear kernel for this version of the $(\nu(F) + k)$-SAT problem. It would be nice to see whether a linear or a polynomial sized kernel exists for the $(\nu(F) + k)$-SAT problem if every clause has exactly r literals.

References

1. Alon, N., Yuster, R., Zwick, U.: Color-coding. J. ACM 42(4), 844–856 (1995)
2. Bodlaender, H.L., Downey, R.G., Fellows, M.R., Hermelin, D.: On problems without polynomial kernels. J. Comput. System Sci. 75(8), 423–434 (2009)
3. Bodlaender, H.L., Thomassé, S., Yeo, A.: Kernel Bounds for Disjoint Cycles and Disjoint Paths. In: Fiat, A., Sanders, P. (eds.) ESA 2009. LNCS, vol. 5757, pp. 635–646. Springer, Heidelberg (2009)
4. Crowston, R., Gutin, G., Jones, M., Yeo, A.: A New Lower Bound on the Maximum Number of Satisfied clauses in Max-SAT and its algorithmic applications. Algorithmica, doi:10.1007/s00453-011-9550-1
5. Dom, M., Lokshtanov, D., Saurabh, S.: Incompressibility through Colors and IDs. In: Albers, S., Marchetti-Spaccamela, A., Matias, Y., Nikoletseas, S., Thomas, W. (eds.) ICALP 2009, Part I. LNCS, vol. 5555, pp. 378–389. Springer, Heidelberg (2009)
6. Downey, R.G., Fellows, M.R.: Parameterized Complexity. Springer (1999)
7. Fleischner, H., Kullmann, O., Szeider, S.: Polynomial-time recognition of minimal unsatisfiable formulas with fixed clause-variable difference. Theor. Comput. Sci. 289(1), 503–516 (2002)
8. Fleischner, H., Szeider, S.: Polynomial-time recognition of minimal unsatisfiable formulas with fixed clause-variable difference. Electronic Colloquium on Computational Complexity (ECCC) 7(49) (2000)
9. Flum, J., Grohe, M.: Parameterized Complexity Theory. Springer (2006)
10. Gutin, G., Jones, M., Yeo, A.: Kernels for below-upper-bound parameterizations of the hitting set and directed dominating set problems. Theor. Comput. Sci. 412(41), 5744–5751 (2011)
11. Kleine Büning, H.: On subclasses of minimal unsatisfiable formulas. Discrete Applied Mathematics 107(1-3), 83–98 (2000)
12. Kleine Büning, H., Kullmann, O.: Minimal Unsatisfiability and Autarkies. In: Handbook of Satisfiability, ch. 11, pp. 339–401
13. Kullmann, O.: An application of matroid theory to the sat problem. In: IEEE Conference on Computational Complexity, pp. 116–124 (2000)
14. Kullmann, O.: Lean clause-sets: Generalizations of minimally unsatisfiable clause-sets. Discr. Appl. Math. 130, 209–249 (2003)
15. Lovász, L., Plummer, M.D.: Matching theory. AMS Chelsea Publ. (2009)
16. Monien, B., Speckenmeyer, E.: Solving satisfiability in less than 2^n steps. Discr. Appl. Math. 10, 287–295 (1985)
17. Papadimitriou, C.H., Wolfe, D.: The complexity of facets resolved. J. Comput. Syst. Sci. 37(1), 2–13 (1988)
18. Srinivasan, A.: Improved approximations of packing and covering problems. In: STOC 1995, pp. 268–276 (1995)
19. Szeider, S.: Minimal unsatisfiable formulas with bounded clause-variable difference are fixed-parameter tractable. J. Comput. Syst. Sci. 69(4), 656–674 (2004)

Finding Efficient Circuits for Ensemble Computation*

Matti Järvisalo[1], Petteri Kaski[2], Mikko Koivisto[1], and Janne H. Korhonen[1]

[1] HIIT & Department of Computer Science, University of Helsinki, Finland
[2] HIIT & Department of Information and Computer Science, Aalto University, Finland

Abstract. Given a Boolean function as input, a fundamental problem is to find a Boolean circuit with the least number of elementary gates (AND, OR, NOT) that computes the function. The problem generalises naturally to the setting of *multiple* Boolean functions: find the smallest Boolean circuit that computes *all* the functions simultaneously. We study an NP-complete variant of this problem titled *Ensemble Computation* and, especially, its relationship to the Boolean satisfiability (SAT) problem from both the theoretical and practical perspectives, under the two monotone circuit classes: OR-circuits and SUM-circuits. Our main result relates the existence of nontrivial algorithms for CNF-SAT with the problem of rewriting in subquadratic time a given OR-circuit to a SUM-circuit. Furthermore, by developing a SAT encoding for the ensemble computation problem and by employing state-of-the-art SAT solvers, we search for concrete instances that would witness a substantial separation between the size of optimal OR-circuits and optimal SUM-circuits. Our encoding allows for exhaustively checking all small witness candidates. Searching over larger witness candidates presents an interesting challenge for current SAT solver technology.

1 Introduction

A fundamental problem in computer science both from the theoretical and practical perspectives is program optimisation, i.e., the task of finding the most efficient sequence of elementary operations that carries out a specified computation. As a concrete example, suppose we have eight variables $x_1, x_2, \ldots, x_8$ and our task is to compute each of the eight sums depicted in Fig. 1. What is the minimum number of SUM gates that implement this computation?

This is an instance of a problem that plays a key role in Valiant's study [18] of circuit complexity over a monotone versus a universal basis; Fig. 1 displays Valiant's solution. More generally, the problem is an instantiation of the NP-complete *Ensemble Computation* problem [8]:

> *(SUM-)Ensemble Computation.* Given as input a collection Q of nonempty subsets of a finite set P and a nonnegative integer b, decide (yes/no) whether there is a sequence
>
> $$Z_1 \leftarrow L_1 \cup R_1, \quad Z_2 \leftarrow L_2 \cup R_2, \quad \ldots \quad, \quad Z_b \leftarrow L_b \cup R_b$$

* This research is supported in part by Academy of Finland (grants 132812 and 251170 (MJ), 252083 and 256287 (PK), and 125637 (MK)), and by Helsinki Doctoral Programme in Computer Science - Advanced Computing and Intelligent Systems (JK).

A. Cimatti and R. Sebastiani (Eds.): SAT 2012, LNCS 7317, pp. 369–382, 2012.

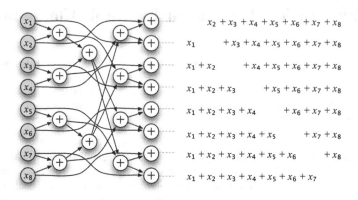

Fig. 1. An instance of ensemble computation (right) and a circuit that solves it (left)

of union operations, where

(a) for all $1 \leq j \leq b$ the sets L_j and R_j belong to $\{\{x\} : x \in P\} \cup \{Z_1, Z_2, \ldots, Z_{j-1}\}$,
(b) for all $1 \leq j \leq b$ the sets L_j and R_j are disjoint, and
(c) the collection $\{Z_1, Z_2, \ldots, Z_b\}$ contains $\mathcal{Q}$.

It is also known that *SUM-Ensemble Computation* remains NP-complete even if the requirement (b) is removed, that is, the unions need not be disjoint [8]; we call this variant *OR-Ensemble Computation*. Stated in different but equivalent terms, each set A in $\mathcal{Q}$ in an instance of *SUM-Ensemble Computation* specifies a subset of the variables in P whose sum must be computed. The question is to decide whether b arithmetic gates suffice to evaluate all the sums in the ensemble. An instance of *OR-Ensemble Computation* asks the same question but with sums replaced by ORs of Boolean variables, and with SUM-gates replaced by OR-gates. We will refer to the corresponding circuits as *SUM-circuits* and *OR-circuits*.

Despite the fundamental nature of these two variants of monotone computation, little seems to be known about their relative power. In particular, here we focus the following open questions:

(Q1) Given an OR-circuit for a collection $\mathcal{Q}$, how efficiently can it be rewritten as a SUM-circuit?
(Q2) Are there collections $\mathcal{Q}$ that require a significantly larger SUM-circuit than an OR-circuit?

Answering these questions would advance our understanding of the computational advantage of, in algebraic terms, idempotent computation (e.g. the maximum of variables) over non-idempotent computation (e.g. the sum of variables); the ability to express the former succinctly in terms of the latter underlies recent advances in algebraic and combinatorial algorithms [2]. Interestingly, it turns out that the questions have strong connections to Boolean satisfiability (SAT) both from the theoretical and practical perspectives, as will be shown in this paper.

As the main theoretical contribution, we establish a connection between (Q1) and the existence of non-trivial algorithms for CNF-SAT. In particular, we show (Theorem 2) that the existence of a subquadratic-time rewriting algorithm implies a nontrivial algorithm for general CNF-SAT (without restrictions on clause length), i.e., an algorithm for CNF-SAT that runs in time $O(2^{cn}m^2n)$ for a constant $0 < c < 1$ that is independent of the number of variables n and the number of clauses m. It should be noted that the existence of such an algorithm for CNF-SAT is a question that has attracted substantial theoretical interest recently [3,14,16,21]. In particular, such an algorithm would contradict the *Strong Exponential Time Hypothesis* [11], and would have significant implications also for the exponential-time complexity of other hard problems beyond SAT. Intuitively, our result suggests that the relationship of the two circuit classes may be complicated and that the difference in the circuit sizes could be large for some collections $\mathcal{Q}$. Furthermore, we show (Proposition 2) that our main result is tight in the sense that (Q1) admits an quadratic-time algorithm.

Complementing our main theoretical result, we address (Q2) from the practical perspective. While it is easy to present concrete instances for which the difference in size between optimal SUM-circuits and OR-circuits is small, finding instances that witness even a factor-2 separation between the number of arithmetic gates is a non-trivial challenge. In fact, our best construction (Theorem 1) achieves this factor only asymptotically, leaving open the question whether there are small witnesses achieving factor 2. As the main practical contribution, we employ state-of-the-art SAT solvers for studying this witness finding task by developing a SAT encoding for finding the optimal circuits for a given ensemble. We show experimentally that our encoding allows for exhaustively checking all small witness candidates. On the other hand, searching over larger witness candidates presents an interesting challenge for current SAT solvers.

As for related earlier work, SAT solvers have been suggested for designing small circuits [4,6,7,12,13], albeit of different types than the ones studied in this work. However, our focus here is especially in circuits implementing an *ensemble* of Boolean functions. A further key motivation that sets this work apart from earlier work is that our interest is not only to find efficient circuits, but also to discover witnesses (ensembles) that separate SUM-circuits and OR-circuits.

2   OR-Circuits, SUM-Circuits, and Rewriting

We begin with some key definitions and basic results related to OR- and SUM-circuits and the task of rewriting an OR-circuit into a SUM-circuit: We show that a SUM-circuit may require asymptotically at least twice as many arithmetic gates as an OR-circuit, and present two rewriting algorithms, one of which rewrites a given OR-circuit with g gates in $O(g^2)$ time into a SUM-circuit. In particular, a SUM-circuit requires at most g times as many arithmetic gates as an OR-circuit.

2.1   Definitions

For basic graph-theoretic terminology we refer to West's introduction [19]. A *circuit* is a directed acyclic graph C whose every node has in-degree either 0 or 2. Each node of

C is a *gate*. The gates of C are partitioned into two sets: each gate with in-degree 0 is an *input gate*, and each gate with in-degree 2 is an *arithmetic gate*. The *size* of C is the number $g = g(C)$ of gates in C. We write $p = p(C)$ for the number of input gates in C. For example, the directed acyclic graph depicted on the left in Fig. 1 is a circuit with 26 gates that partition into 8 input gates and 18 arithmetic gates.

The *support* of a gate z in C is the set of all input gates x such that there is a directed path in C from x to z. The *weight* of a gate z is the size of its support. All gates have weight at least one, with equality if and only if a gate is an input gate. For example, in Fig. 1 the five columns of gates consist of gates that have weight 1, 2, 4, 6, and 7, respectively.

In what follows we study two classes of circuits, where the second class is properly contained within the first class. First, every circuit is an *OR-circuit*. Second, a circuit C is a *SUM-circuit* if for every gate z and for every input gate x it holds that there is at most one directed path in C from x to z.

We adopt the convention of using the operator symbols "$\vee$" and "$+$" on the arithmetic gates to indicate the type of a circuit. Fig. 2 below displays an example of both types of circuits. We observe that the circuit on the left in Fig. 2 is not a SUM-circuit because the bottom right gate can be reached from the input x_1 along two distinct directed paths.

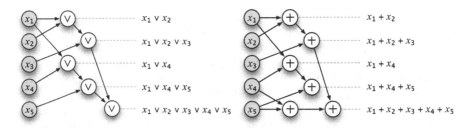

Fig. 2. An OR-circuit (left) and a SUM-circuit (right)

Let $(P, \mathcal{Q})$ be an instance of ensemble computation, that is, let P be a finite set and let $\mathcal{Q}$ be a set of nonempty subsets of P. We adopt the convention that for a SUM-ensemble all circuits considered are SUM-circuits, and for an OR-ensemble all circuits considered are OR-circuits. We say that a circuit C *solves* the instance $(P, \mathcal{Q})$ if (a) the set of input gates of C is P; and (b) for each $A \in \mathcal{Q}$, there exists a gate in C whose support is A. The *size* of the solution is the size of C. A solution to $(P, \mathcal{Q})$ is *optimal* if it has the minimum size over all possible solutions. A circuit C' *implements* a circuit C if for every gate z of C there is a gate z' of C' such that z and z' have the same support. A *circuit rewriting algorithm* takes as input a circuit C and outputs (i) a circuit C' that implements C; and (ii) a mapping $z \mapsto z'$ that identifies each gate z in C with a corresponding gate z' in C'.

2.2 Bounds for Separation

The size of an optimal solution to an instance $(P, \mathcal{Q})$ is dependent on whether we are considering an OR-ensemble or a SUM-ensemble. To see this, let us consider Fig. 2.

Observe that both circuits solve the same instance (P, Q), but only the circuit on the right is a SUM-circuit. We claim that both circuits are optimal. Indeed, observe that the instance has five distinct sets of size at least 2. At least one arithmetic gate is required for each distinct set of size at least 2. Thus, the circuit on the left in Fig. 2 is optimal. Analogously, on the right in Fig. 2 at least four arithmetic gates are required to compute the first four sets in the instance, after which at least two further SUM-gates are required to produce the fifth set because the first four sets intersect pairwise.

The following construction shows that asymptotically (that is, by taking a large enough h and w) at least twice the number of arithmetic gates may be required in an optimal SUM-circuit compared with an optimal OR-circuit.

Theorem 1. *For all $h, w = 1, 2, \ldots$ there exists an ensemble whose optimal OR-circuit has $(h + 1)w - 1$ arithmetic gates and whose optimal SUM-circuit has $(2w - 1)h$ arithmetic gates.*

Proof. Take $P = \{x_0\} \cup \{x_{i,j} : i = 1, 2, \ldots, h; j = 1, 2, \ldots, w\}$ and let Q consist of the following sets. For each $j = 1, 2, \ldots, w$ and for each $i = 1, 2, \ldots, h$, insert the set $\{x_0, x_{1,j}, x_{2,j}, \ldots, x_{i,j}\}$ to Q. Let us say that this set belongs to *chain j*. Finally, insert the set P into Q. Let us call this set the *top*. In total Q thus has $hw + 1$ sets, and the largest set (that is, the top) has size $hw + 1$.

Every OR-circuit that solves (P, Q) must use one OR-gate for each element in each chain for a total of hw gates. Excluding the element x_0 which occurs in all sets in Q, the top has size hw, and the largest sets in each chain have size h. Thus, at least $w - 1$ OR-gates are required to construct the top. In particular, an optimum OR-circuit that solves (P, Q) has $hw + w - 1 = (h + 1)w - 1$ arithmetic gates.

Next consider an arbitrary SUM-circuit that solves (P, Q). Observe that each chain requires h distinct SUM-gates, each of which has x_0 in its support. There are hw such SUM-gates in total, at most one of which may be shared in the subcircuit that computes the top. Such a shared SUM-gate has weight at most $h + 1$, whereas the top has weight $hw + 1$. Thus the subcircuit that computes the top can share weight at most $h + 1$ and must use non-shared SUM-gates to accumulate the remaining weight (if any), which requires $h(w - 1)$ gates. Thus, the SUM-circuit requires at least $hw + h(w - 1) = (2w - 1)h$ arithmetic gates. □

Remark 1. Traditional nonconstructive tools for deriving lower bounds to circuit size appear difficult to employ for this type of separation between two monotone circuit classes. Indeed, it is easy to show using standard counting arguments that most ensembles (P, Q) with $|P| = |Q| = r$ require $\Omega(r^2/\log r)$ gates for both OR- and SUM-circuits, but showing that there exist ensembles where the required SUM-circuit is significantly larger than a sufficient OR-circuit appears inaccessible to such tools.

2.3   Upper Bounds for Rewriting

Let us now proceed to study the algorithmic task of rewriting a given OR-circuit into a SUM-circuit. In particular, our interest is to quantify the number of extra gates required. We start with the observation that no extra gates are required if all gates in the given OR-circuit have weight at most 4.

Proposition 1. *Every OR-circuit with g gates of weight at most 4 can be rewritten into a SUM-circuit with g gates. Moreover, there is an algorithm with running time $O(g)$ that rewrites the circuit.*

Proof. Let C be an OR-circuit with g gates given as input. First, topologically sort the nodes of C in time $O(g)$. Then, compute the support of each gate by assigning unique singleton sets at the input gates and evaluating the gates in topological order. Finally, proceed in topological order and rewrite the gates of the circuit using the following rules. Input gates do not require rewriting. Furthermore, every OR-gate of weight 2 can be trivially replaced with a SUM-gate. Each OR-gate z with weight 3 either has the property that the in-neighbours z_1, z_2 of z have disjoint supports (in which case we may trivially replace z with a SUM-gate) or z_1, z_2 have weight at least 2. In the latter case, if at least one of z_1, z_2 has weight 3 (say, z_1), we may delete z and replace it with z_1; otherwise rewrite z so that one of its in-neighbours is z_1 and the other in-neighbour is the appropriate input gate. Each OR-gate z with weight 4 either has in-neighbours z_1, z_2 with disjoint supports or z_1, z_2 have weight at least 3 and at least 2, respectively. Again we may either delete z or rewrite z so that one of its in-neighbours is z_1 and the other in-neighbour is the appropriate input gate. It is immediate that this rewriting can be carried out in time $O(g)$.   □

Next we observe that an OR-circuit can always be rewritten into a SUM-circuit with at most g times the number of gates in the OR-circuit.

Proposition 2. *There exists an algorithm that in time $O(g^2)$ rewrites a given OR-circuit with g gates into a SUM-circuit.*

Proof. The algorithm operates as follows. Let C be an OR-circuit with g gates and p input gates given as input. Topologically sort the nodes of C in time $O(g)$. Suppose the input gates of C are $x_1, x_2, \ldots, x_p$. Associate with each of the g gates an array of p bits. Then, iterate through the gates of C in topological order. For each input gate x_j, initialise the bit array associated with x_j so that the jth bit is set to 1 and the other bits are set to 0. For each OR-gate z with in-neighbours z_1, z_2, assign the bit array associated with z to be the union of the bit arrays associated with z_1 and z_2. This step takes time $O(gp)$. Finally, iterate through the gates of C. For each arithmetic gate z, output a SUM-circuit that computes the sum of the at most p inputs specified by the bit array associated with z. This requires at most $p - 1$ SUM-gates for each z. The algorithm takes $O(gp)$ time and outputs a circuit with $O(gp)$ gates. The claim follows because $p \leq g$.   □

3   Subquadratic Rewriting Implies Faster CNF-SAT

Complementing the quadratic-time algorithm in Proposition 2, this section studies the possibility of developing fast (subquadratic-time) algorithms for rewriting OR-circuits as SUM-circuits. In particular, we show that the existence of such a subquadratic-time rewriting algorithm would, surprisingly, yield a non-trivial algorithm for general CNF-SAT (cf. Refs. [16,21] and [20, Theorem 5]).

Theorem 2. *Let $0 < \epsilon \leq 1$. If there is an algorithm that in time $O(g^{2-\epsilon})$ rewrites a given OR-circuit with g gates into a SUM-circuit, then there is an algorithm that solves CNF-SAT in time $O\big(2^{(1-\epsilon/2)n}m^{2-\epsilon}n\big)$, where n is the number of variables and m is the number of clauses.*

Proof. Let $0 < \epsilon \leq 1$ be fixed and let A be a circuit rewriting algorithm with the stated properties. We present an algorithm for CNF-SAT. Let an instance of CNF-SAT given as input consist of the variables $x_1, x_2, \ldots, x_n$ and the clauses $C_1, C_2, \ldots, C_m$. Without loss of generality (by inserting one variable as necessary), we may assume that n is even. Call the variables $x_1, x_2, \ldots, x_{n/2}$ *low* variables and the variables $x_{n/2+1}, x_{n/2+2}, \ldots, x_n$ *high* variables. The algorithm operates in three steps.

In the first step, the algorithm constructs the following OR-circuit. First let us observe that there are $2^{n/2}$ distinct ways to assign truth values (0 or 1) to the low variables. Each of these assignments indexes an input gate to the circuit. Next, for each clause C_i, we construct a subcircuit that takes the OR of all input gates that *do not satisfy* the clause C_i, that is, the input gate indexed by an assignment a to the low variables is in the OR if and only if no literal in C_i is satisfied by a. For each C_i, this subcircuit requires at most $2^{n/2} - 1$ OR-gates. Let us refer to the output gate of this subcircuit as gate C_i. Finally, for each assignment b to the high variables, construct a subcircuit that takes the OR of all gates C_i such that the clause C_i is *not satisfied* by b. Let us refer to the output gate of this subcircuit as gate b. The constructed circuit has $p = 2^{n/2}$ inputs and $g \leq m(2^{n/2} - 1) + 2^{n/2}(m - 1) = O(2^{n/2}m)$ gates. The construction time for the circuit is $O(2^{n/2}mn)$.

In the second step, the algorithm rewrites the constructed OR-circuit using algorithm A as a subroutine into a SUM-circuit in time $O(g^{2-\epsilon})$, that is, in time $O(2^{(1-\epsilon/2)n}m^{2-\epsilon})$. In particular, the number of gates in the SUM-circuit is $G = O(2^{(1-\epsilon/2)n}m^{2-\epsilon})$. For a gate z in the OR-circuit, let us write z' for the corresponding gate in the SUM-circuit.

In the third step, the algorithm assigns the value 1 to each input a' in the SUM-circuit (any other inputs are assigned to 0), and evaluates the SUM-circuit over the integers using $O(2^{(1-\epsilon/2)n}m^{2-\epsilon})$ additions of $O(n)$-bit integers. If there exists a gate b' that evaluates to a value less than $2^{n/2}$, the algorithm outputs "satisfiable"; otherwise the algorithm outputs "unsatisfiable". The running time of the algorithm is $O(2^{(1-\epsilon/2)n}m^{2-\epsilon}n)$.

To see that the algorithm is correct, observe that in the OR-circuit, the input a occurs in the support of b if and only if there is a clause C_i such that neither a nor b satisfies C_i. Equivalently, the assignment (a, b) into the n variables is not satisfying (because it does not satisfy the clause C_i). The rewrite into a SUM-circuit enables us to infer the presence of an a' that does not occur in the support of b' by counting the number of a' that do occur in the support of b'. SUM-gates ensure that each input in the support of b' is counted exactly once.   □

Theorem 2 thus demonstrates that unless the strong exponential time hypothesis [11] fails, there is no subquadratic-time algorithm for rewriting arbitrary OR-circuits into SUM-circuits.

4   Finding Small Circuits Using SAT Solvers

We next develop a SAT encoding for deciding whether a given ensemble has a circuit of a given size.

4.1   SAT Encoding

We start by giving a representation of an OR- or SUM-circuit as a binary matrix. This representation then gives us a straightforward way to encode the circuit existence problem as a propositional formula.

Let $(P, \mathcal{Q})$ be an OR- or SUM-ensemble and let C be a circuit of size g that solves $(P, \mathcal{Q})$. For convenience, let us assume that $|P| = p$, $|\mathcal{Q}| = q$ and $P = \{1, 2, \ldots, p\}$. Furthermore, we note that outputs corresponding to sets of size 1 are directly provided by the input gates, and we may thus assume that $\mathcal{Q}$ does not contain sets of size 1. The circuit C can be represented as a $g \times p$ binary matrix M as follows. Fix a topological ordering $z_1, z_2, \ldots, z_g$ of the gates of C such that $z_i = i$ for all i with $1 \leq i \leq p$ (recall that we identify the input gates with elements of P). Each row i of the matrix M now corresponds to the support of the gate z_i so that for all $1 \leq j \leq p$ we have $M_{i,j} = 1$ if j is in the support of z_i and $M_{i,j} = 0$ otherwise. In particular, for all $1 \leq i \leq p$ we have $M_{i,i} = 1$ and $M_{i,j} = 0$ for all $j \neq i$. Figure 3 displays an example.

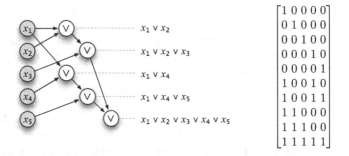

Fig. 3. An OR-circuit (left) and a matrix describing the circuit (right)

Now, C (viewed as an OR-circuit) solves $(P, \mathcal{Q})$ if and only if the matrix M satisfies

(a) for all i with $1 \leq i \leq p$ it holds that $M_{i,i} = 1$ and $M_{i,j} = 0$ for all $j \neq i$,
(b) for all i with $p + 1 \leq i \leq g$ there exist k and ℓ such that $1 \leq k < \ell < i$ and for all j with $1 \leq j \leq p$ it holds that $M_{i,j} = 1$ if and only if $M_{k,j} = 1$ or $M_{\ell,j} = 1$, and
(c) for every set A in $\mathcal{Q}$ there exists an i with $1 \leq i \leq g$ such that for all j with $1 \leq j \leq p$ it holds that $M_{i,j} = 1$ if $j \in A$ and $M_{i,j} = 0$ otherwise.

Similarly, C (viewed as a SUM-circuit) solves $(P, \mathcal{Q})$ if and only if the matrix M satisfies conditions (a), (c), and

(b') for all i with $p + 1 \leq i \leq g$ there exist k and ℓ such that $1 \leq k < \ell < i$ and for all j with $1 \leq j \leq p$ it holds that $M_{i,j} = 1$ if and only if $M_{k,j} = 1$ or $M_{\ell,j} = 1$ and that $M_{k,j} = 0$ or $M_{\ell,j} = 0$.

Based on the above observations, we encode an ensemble computation instance as SAT instance as follows. Given an OR-ensemble (P, Q) and integer g as input, we construct a propositional logic formula φ over variables $M_{i,j}$, where $1 \leq i \leq g$ and $1 \leq j \leq p$, so that any assignment into variables $M_{i,j}$ satisfying φ gives us a matrix that satisfies conditions (a)–(c). We encode condition (a) as

$$\alpha = \bigwedge_{i=1}^{p} \left(M_{i,i} \wedge \bigwedge_{j \neq i} \neg M_{i,j} \right).$$

Similarly, we encode the conditions (b) and (c), respectively, as

$$\beta = \bigwedge_{i=p+1}^{g} \bigvee_{k=1}^{i-2} \bigvee_{\ell=k+1}^{i-1} \bigwedge_{j=1}^{p} \left((M_{k,j} \vee M_{\ell,j}) \leftrightarrow M_{i,j} \right), \quad \text{and}$$

$$\gamma = \bigwedge_{A \in Q} \bigvee_{i=p+1}^{g} \left[\left(\bigwedge_{j \in A} M_{i,j} \right) \wedge \left(\bigwedge_{j \notin A} \neg M_{i,j} \right) \right].$$

The desired formula φ is then $\varphi = \alpha \wedge \beta \wedge \gamma$. For a SUM-ensemble, we replace β with

$$\beta' = \bigwedge_{i=p+1}^{g} \bigvee_{k=1}^{i-2} \bigvee_{\ell=k+1}^{i-1} \bigwedge_{j=1}^{p} \left(((M_{k,j} \vee M_{\ell,j}) \leftrightarrow M_{i,j}) \wedge (\neg M_{k,j} \vee \neg M_{\ell,j}) \right).$$

4.2 Practical Considerations

There are several optimisations that can be used to tune this encoding to speed up SAT solving. The resulting SAT instances have a high number of symmetries, as any circuit can be represented as a matrix using any topological ordering of the gates. This makes especially the unsatisfiable instances difficult to tackle with SAT solver. To alleviate this problem, we constrain the rows i for $p + 1 \leq i \leq g$ appear in lexicographic order, so that any circuit that solves (P, Q) has a unique valid matrix representation. Indeed, we note that the lexicographic ordering of the gate supports (viewed as binary strings) is a topological ordering. We insert this constraint to the SAT encoding as the formula

$$\bigwedge_{i=p+2}^{g} \bigwedge_{k=p+1}^{i-1} \left[(M_{i,1} \vee \neg M_{k,1}) \wedge \bigwedge_{j_1=2}^{p} \left(\left(\bigwedge_{j_2=1}^{j_1-1} (M_{i,j_2} \leftrightarrow M_{k,j_2}) \right) \rightarrow (M_{i,j_1} \vee \neg M_{k,j_1}) \right) \right].$$

We obtain further speedup by constraining the first t arithmetic gates to have small supports. Indeed, the ith arithmetic gate in any topological order has weight at most $i + 1$. Thus, we fix $t = 6$ in the experiments and insert the formula

$$\bigwedge_{i=1}^{t} \bigwedge_{\substack{S \subseteq P \\ |S| = i+2}} \neg \left(\bigwedge_{j \in S} M_{p+i,j} \right).$$

Further tuning is possible if Q is an *antichain*, that is, if there are no distinct $A, B \in Q$ with $A \subseteq B$. In this case an optimal circuit C has the property that every gate whose support is in $\mathcal{Q}$ has out-degree 0. Thus, provided that we do not use the lexicographical ordering of gates as above, we may assume that the gates corresponding to sets in $\mathcal{Q}$ are the last gates in the circuit, and moreover, their respective order is any fixed order. Thus, if $\mathcal{Q} = \{A_1, A_2, \ldots, A_q\}$ is an antichain, we can replace γ with

$$\bigwedge_{i=1}^{q} \left[\left(\bigwedge_{j \in A_j} M_{g-q+i,j} \right) \wedge \left(\bigwedge_{j \notin A_j} \neg M_{g-q+i,j} \right) \right]$$

to obtain a smaller formula. Finally, we note that we can be combine this with the lexicographic ordering by requiring that only rows i for $p + 1 \leq i \leq g - q$ are in lexicographic order.

5   Experiments

We report on two series of experiments with the developed encoding and state-of-the-art SAT solvers: (a) an exhaustive study of small ensembles aimed at understanding the separation between OR-circuits and SUM-circuits, and (b) a study of the scalability of our encoding by benchmarking different solvers on specific structured ensembles.

5.1   Instance Generation and Experimental Setup

For both series of experiments, the problem instances given to SAT solvers were generated by translating the encoding in Sect. 4 into CNF. We used the symmetry breaking constraints and antichain optimisations described in Sect. 4.2; without these, most instances could not solved by any of the solvers.

The formula encoding an input ensemble $(P, \mathcal{Q})$ and a target number of gates g was first translated into a Boolean circuit and then into CNF using the bc2cnf encoder (http://users.ics.tkk.fi/tjunttil/circuits/), which implements the standard Tseitin encoding [17]. The instance generator and a set of interesting handpicked CNF-level benchmark instances are available at

http://cs.helsinki.fi/u/jazkorho/sat2012/.

When working with an ensemble, the size of the optimal OR-circuit or optimal SUM-circuit is not generally known. Thus, we structured the experiments for a given ensemble $(P, \mathcal{Q})$ with $|P| = p$ and $|\mathcal{Q}| = q$ as a sequence of jobs that keeps the ensemble $(P, \mathcal{Q})$ fixed and varies the target number of gates g. We start from a value of g for which a circuit is known to exist $(p(1 + q))$ and then decrease the value in steps of 1 until we hit an unsatisfiable instance at $g = u$; an optimal circuit then has $g = u + 1$ gates.

The experiments were run on Dell PowerEdge M610 blade servers with two quad-core 2.53-GHz Intel Xeon processors and 32 GB of memory. We report the user times recorded via time under Linux (kernel version 2.6.38). In the timed benchmarking runs we ran one simultaneous job on a single server, but in the explorative experiments we ran multiple jobs per server in parallel. SAT solvers used were Minisat 2.2.0 [5] and Lingeling 587f [1] (two CDCL solvers among the best for application instances), Clasp 2.0.4 [9] (CDCL solver, one of the best for crafted instances), and March_rw [10] (a DPLL-lookahead solver, one of the best for unsatisfiable random instances).

5.2 Optimal Circuits for All Small Ensembles

We say that two ensembles $(P, \mathcal{Q}_1)$ and $(P, \mathcal{Q}_2)$ are *isomorphic* if there is a permutation of P that takes $\mathcal{Q}_1$ to $\mathcal{Q}_2$. The optimal circuit size is clearly an isomorphism invariant of an ensemble, implying that in an exhaustive study it suffices to consider one ensemble from each isomorphism class.

We carried out an exhaustive study of all nonisomorphic ensembles $(P, \mathcal{Q})$ across the three parameter ranges (i) $p = 5$ and $2 \leq q \leq 7$, (ii) $p = 6$ and $2 \leq q \leq 7$, and (iii) $p = 7$ and $2 \leq q \leq 6$ subject to the following additional constraints: (a) every set in $\mathcal{Q}$ has size at least 2, (b) every set in $\mathcal{Q}$ contains at least two points in P that each occur in at least two sets in $\mathcal{Q}$, and (c) the ensemble is connected (when viewed as a hypergraph with vertex set P and edge set $\mathcal{Q}$). We generated the ensembles using the genbg tool that is part of the canonical labelling package nauty [15].

For all of the generated 1,434,897 nonisomorphic ensembles, we successfully determined the optimum OR-circuit size and the optimum SUM-circuit size in approximately 4 months of total CPU time using Minisat. Among the instances considered, we found no instance where the gap between the two optima is more than one gate. The smallest ensembles in terms of the parameters p and q where we observed a gap of one gate occurred for $p = 5$ and $q = 5$, for exactly 3 nonisomorphic ensembles; one of the ensembles with accompanying optimal circuits is displayed in Fig. 2. A further analysis of the results led to Theorem 1 and Proposition 1.

After this work the next open parameters for exhaustive study are $p = 7$ and $q = 7$ with 13,180,128 nonisomorphic ensembles.

In general, the large number of isomorphism classes for larger p and q makes an exhaustive search prohibitively time-consuming. A natural idea would be to randomly sample ensembles with given parameters to find an ensemble witnessing a large separation between optimal OR- and SUM-circuits. However, as highlighted in Remark 1, most ensembles require both a large OR-circuit and a large SUM-circuit, suggesting that random sampling would mostly give instances with small difference between optimal OR- and SUM-circuits. This intuition was experimentally supported as follows. We generated random ensembles $(P, \mathcal{Q})$ by setting $P = \{1, 2, \ldots, p\}$ and drawing uniformly at random a $\mathcal{Q}$ consisting of q subsets of P of size at least 2. We generated 1,000 instances for $p = q = 9$ and for $p = q = 10$. Among these instances, we found only one instance (with $p = q = 10$) where the gap between the optimal OR-circuit and and the optimal SUM-circuit was 2, while we know that instances with larger separation do exist for these parameters. However, there were 49 instances with $p = q = 10$ where the optimal circuit sizes were not found within a 6-hour time limit.

5.3 Scaling on Structured Ensembles

To test the scalability of our encoding and to benchmark different solvers, we also studied two parameterised families of structured ensembles for varying family parameters and target number of gates g. The first family is illustrated by the Valiant's construction in Fig. 1 for $p = 8$. This family is parameterised by the number of inputs p, with $P = \{1, 2, \ldots, p\}$ and $\mathcal{Q} = \{P \setminus \{i\} : i \in P\}$. As benchmarks we generated CNF instances for $p = 8, 9, 10, 11$ and $g = 2p, 2p + 1, \ldots, 2p + 20$ using the SUM-encoding and the antichain optimisation. The second family is given in Theorem 1 and is parameterised by two parameters h and w. As benchmarks we generated CNF instances for $h = 3$ and $w = 5$ and $g = 32, 33, \dots, 52$ using both the OR-encoding and the SUM-encoding.

The results for the two benchmark families are reported in Figs. 4 and 5. The solver March_rw was omitted from the second benchmark due to its poor performance on the first benchmark family. In an attempt to facilitate finding upper bounds for even larger instances, we also tested the local search solver SATTIME2011, which performed notably well on satisfiable crafted instances in the 2011 SAT Competition. However, in our experiments on instances from the satisfiable regime, SATTIME2011 was unable to find the solution within the 3600-second time limit already for the ensembles in Fig. 4 with $p = 8$ and $g = 26, 27, 28$.

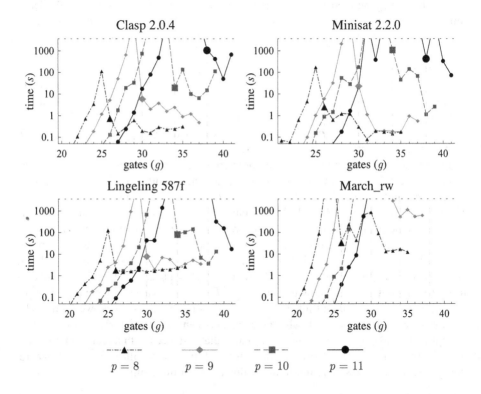

Fig. 4. Solution times for different SAT solvers as a function of the number of gates on SUM-ensembles corresponding to Valiant's construction (Fig. 1). The data points highlighted with larger markers and a vertical dashed line indicate the smallest circuits found. The horizontal dashed line at 3600 seconds is the timeout limit for each run. As the instance size p grows, the unsatisfiable instances with g just below the size of the optimal circuit rapidly become very difficult to solve.

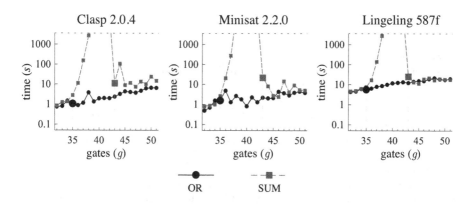

Fig. 5. Solution times for different SAT solvers as a function of the number of gates on OR- and SUM-ensembles from Theorem 1 with parameters $w = 5$ and $h = 3$. The data points highlighted with larger markers and a vertical dashed line indicate the smallest circuits found. The horizontal dashed line at 3600 seconds is the timeout limit for each run. The optimal OR circuit is small, and SAT solvers have no difficulty in finding it.

6 Conclusions

We studied the relative power of OR-circuits and SUM-circuits for ensemble computation, and developed tight connections to Boolean satisfiability from both the theoretical and practical perspectives. As the main theoretical contribution, we showed that, while OR-circuits can be rewritten in quadratic-time into SUM-circuits, a subquadratic-time rewriting algorithm would imply that general CNF-SAT has non-trivial algorithms, which would contradict the strong exponential time hypothesis. From the practical perspective, we developed a SAT encoding for finding smallest SUM- and OR-circuits for a given ensemble. State-of-the-art SAT solvers proved to be a highly useful tool for studying the separation of these two circuit classes. Using the developed encoding, we were able to exhaustively establish the optimum OR-circuit and SUM-circuit sizes for all small instances, which contributed to our analytical understanding of the problem and led to the theoretical results presented in this paper. Our publicly available instance generator may also be of independent interest as a means of generating interesting benchmarks.

Larger, structured instances provide interesting challenges for current state-of-the-art SAT solver technology. Further developments either on the encoding or the solver level—including tuning SAT solvers especially for this problem—would allow for providing further understanding to the problem of separating different circuit classes.

Acknowledgment. We thank Teppo Niinimäki for insight concerning the construction in Theorem 1.

References

1. Biere, A.: Lingeling, Plingeling, PicoSAT and PrecoSAT at SAT Race 2010. FMV Technical Report 10/1, Johannes Kepler University, Linz, Austria (2010)
2. Björklund, A., Husfeldt, T., Kaski, P., Koivisto, M., Nederlof, J., Parviainen, P.: Fast zeta transforms for lattices with few irreducibles. In: Proc. SODA, pp. 1436–1444. SIAM (2012)
3. Dantsin, E., Wolpert, A.: On Moderately Exponential Time for SAT. In: Strichman, O., Szeider, S. (eds.) SAT 2010. LNCS, vol. 6175, pp. 313–325. Springer, Heidelberg (2010)
4. Demenkov, E., Kojevnikov, A., Kulikov, A.S., Yaroslavtsev, G.: New upper bounds on the Boolean circuit complexity of symmetric functions. Inf. Process. Lett. 110(7), 264–267 (2010)
5. Eén, N., Sörensson, N.: An Extensible SAT-solver. In: Giunchiglia, E., Tacchella, A. (eds.) SAT 2003. LNCS, vol. 2919, pp. 502–518. Springer, Heidelberg (2004)
6. Estrada, G.G.: A Note on Designing Logical Circuits Using SAT. In: Tyrrell, A.M., Haddow, P.C., Torresen, J. (eds.) ICES 2003. LNCS, vol. 2606, pp. 410–421. Springer, Heidelberg (2003)
7. Fuhs, C., Schneider-Kamp, P.: Synthesizing Shortest Linear Straight-Line Programs over GF(2) Using SAT. In: Strichman, O., Szeider, S. (eds.) SAT 2010. LNCS, vol. 6175, pp. 71–84. Springer, Heidelberg (2010)
8. Garey, M.R., Johnson, D.S.: Computers and Intractability: A Guide to the Theory of NP-Completeness. W.H. Freeman and Company (1979)
9. Gebser, M., Kaufmann, B., Schaub, T.: The Conflict-Driven Answer Set Solver *clasp*: Progress Report. In: Erdem, E., Lin, F., Schaub, T. (eds.) LPNMR 2009. LNCS, vol. 5753, pp. 509–514. Springer, Heidelberg (2009)
10. Heule, M., Dufour, M., van Zwieten, J.E., van Maaren, H.: March_eq: Implementing Additional Reasoning into an Efficient Look-Ahead SAT Solver. In: Hoos, H.H., Mitchell, D.G. (eds.) SAT 2004. LNCS, vol. 3542, pp. 345–359. Springer, Heidelberg (2005)
11. Impagliazzo, R., Paturi, R.: On the complexity of k-SAT. J. Comput. System Sci. 62(2), 367–375 (2001)
12. Kamath, A.P., Karmarkar, N.K., Ramakrishnan, K.G., Resende, M.G.C.: An interior point approach to Boolean vector function synthesis. In: Proc. MWSCAS, pp. 185–189. IEEE (1993)
13. Kojevnikov, A., Kulikov, A.S., Yaroslavtsev, G.: Finding Efficient Circuits Using SAT-Solvers. In: Kullmann, O. (ed.) SAT 2009. LNCS, vol. 5584, pp. 32–44. Springer, Heidelberg (2009)
14. Lokshtanov, D., Marx, D., Saurabh, S.: Known algorithms on graphs of bounded treewidth are probably optimal. In: Proc. SODA, pp. 777–789. SIAM (2010)
15. McKay, B.: nauty user's guide. Tech. Rep. TR-CS-90-02, Australian National University, Department of Computer Science (1990)
16. Pătraşcu, M., Williams, R.: On the possibility of faster SAT algorithms. In: Proc. SODA, pp. 1065–1075. SIAM (2010)
17. Tseitin, G.S.: On the complexity of derivation in propositional calculus. In: Automation of Reasoning 2: Classical Papers on Computational Logic 1967-1970, pp. 466–483. Springer, Heidelberg (1983)
18. Valiant, L.G.: Negation is powerless for Boolean slice functions. SIAM J. Comput. 15(2), 531–535 (1986)
19. West, D.B.: Introduction to graph theory. Prentice Hall Inc., Upper Saddle River (1996)
20. Williams, R.: A new algorithm for optimal 2-constraint satisfaction and its implications. Theoret. Comput. Sci. 348(2-3), 357–365 (2005)
21. Williams, R.: Improving exhaustive search implies superpolynomial lower bounds. In: Proc. STOC, pp. 231–240. ACM (2010)

Conflict-Driven XOR-Clause Learning*

Tero Laitinen, Tommi Junttila, and Ilkka Niemelä

Aalto University
Department of Information and Computer Science
PO Box 15400, FI-00076 Aalto, Finland
{Tero.Laitinen,Tommi.Junttila,Ilkka.Niemela}@aalto.fi

Abstract. Modern conflict-driven clause learning (CDCL) SAT solvers are very good in solving conjunctive normal form (CNF) formulas. However, some application problems involve lots of parity (xor) constraints which are not necessarily efficiently handled if translated into CNF. This paper studies solving CNF formulas augmented with xor-clauses in the DPLL(XOR) framework where a CDCL SAT solver is coupled with a separate xor-reasoning module. New techniques for analyzing xor-reasoning derivations are developed, allowing one to obtain smaller CNF clausal explanations for xor-implied literals and also to derive and learn new xor-clauses. It is proven that these new techniques allow very short unsatisfiability proofs for some formulas whose CNF translations do not have polynomial size resolution proofs, even when a very simple xor-reasoning module capable only of unit propagation is applied. The efficiency of the proposed techniques is evaluated on a set of challenging logical cryptanalysis instances.

1 Introduction

Modern propositional satisfiability (SAT) solvers (see e.g. [1]) have been successfully applied in a number of industrial application domains. Propositional satisfiability instances are typically encoded in conjunctive normal form (CNF) which allows very efficient Boolean constraint propagation and conflict-driven clause learning (CDCL) techniques. However, such CNF encodings may not allow optimal exploitation of the problem structure in the presence of parity (xor) constraints; such constraints are abundant especially in the logical cryptanalysis domain and also present in circuit verification and bounded model checking. An instance consisting only of parity constraints can be solved in polynomial time using Gaussian elimination, but even state-of-the-art SAT solvers relying only on basic Boolean constraint propagation and CDCL can scale poorly on the corresponding CNF encoding.

In this paper we develop new techniques for exploiting structural properties of xor constraints (i.e. linear equations modulo 2) in the recently introduced DPLL(XOR) framework [2,3] where a problem instance is given as a combination of CNF and xor-clauses. In the framework a CDCL SAT solver takes care of the CNF part while a separate xor-reasoning module performs propagation on the xor-clauses. In this paper we introduce new techniques for explaining why a literal was implied or why a conflict

---

* This work has been financially supported by the Academy of Finland under the Finnish Centre of Excellence in Computational Inference (COIN).

A. Cimatti and R. Sebastiani (Eds.): SAT 2012, LNCS 7317, pp. 383–396, 2012.

occurred in the xor-clauses part; such explanations are needed by the CDCL part. The new core idea is to not see xor-level propagations as implications but as linear arithmetic equations. As a result, the new proposed *parity explanation* techniques can (i) provide smaller clausal explanations for the CDCL part, and also (ii) derive new xor-clauses that can then be learned in the xor-clauses part. The goal of learning new xor-clauses is, similarly to clause learning in CDCL solvers, to enhance the deduction capabilities of the reasoning engine. We introduce the new techniques on a very simple xor-reasoning module allowing only unit propagation on xor-clauses and prove that, even when new xor-clauses are not learned, the resulting system with parity explanations can efficiently solve parity problems whose CNF translations are very hard for resolution. We then show that the new parity explanation techniques also extend to more general xor-reasoning modules, for instance to the one in [2] capable of equivalence reasoning in addition to unit propagation. Finally, we experimentally evaluate the effect of the proposed techniques on a challenging benchmark set modelling cryptographic attacks.

Related Work. In [4] a calculus combining basic DPLL without clause learning and Gauss elimination is proposed; their Gauss rules are similar to the general rule $\oplus$-Gen we use in Sect. 8. The solvers EqSatz [5], lsat [6], and March_eq [7] incorporate parity reasoning into DPLL without clause learning, extracting parity constraint information from CNF and during look-ahead, and exploiting it during the preprocessing phase and search. MoRsat [8] extracts parity constraints from a CNF formula, uses them for simplification during preprocessing, and proposes a watched literal scheme for unit propagation on parity constraints. Cryptominisat [9,10], like our approach, accepts a combination of CNF and xor-clauses as input. It uses the computationally relatively expensive Gaussian elimination as the xor-reasoning method and by default only applies it at the first levels of the search; we apply lighter weight xor-reasoning at all search levels. Standard clause learning is supported in MoRsat and Cryptominisat; our deduction system characterization of xor-reasoning allows us to exploit the linear properties of xor-clauses to obtain smaller CNF explanations of xor-implied literals and xor-conflicts as well as to derive and learn new xor-clauses.

2   Preliminaries

An *atom* is either a propositional variable or the special symbol $\top$ which denotes the constant "true". A *literal* is an atom A or its negation $\neg A$; we identify $\neg \top$ with $\bot$ and $\neg \neg A$ with A. A traditional, non-exclusive *or-clause* is a disjunction $l_1 \vee \cdots \vee l_n$ of literals. An *xor-clause* is an expression of form $l_1 \oplus \cdots \oplus l_n$, where $l_1, \ldots, l_n$ are literals and the symbol $\oplus$ stands for the exclusive logical or. In the rest of the paper, we implicitly assume that each xor-clause is in a *normal form* such that (i) each atom occurs at most once in it, and (ii) all the literals in it are positive. The unique (up to reordering of the atoms) normal form for an xor-clause can be obtained by applying the following rewrite rules in any order until saturation: (i) $\neg A \oplus C \rightsquigarrow A \oplus \top \oplus C$, and (ii) $A \oplus A \oplus C \rightsquigarrow C$, where C is a possibly empty xor-clause and A is an atom. For instance, the normal form of $\neg x_1 \oplus x_2 \oplus x_3 \oplus x_3$ is $x_1 \oplus x_2 \oplus \top$, while the normal form of $x_1 \oplus x_1$ is the empty xor-clause (). We say that an xor-clause is *unary* if it is either of form x or $x \oplus \top$ for some variable x; we will identify $x \oplus \top$ with the literal $\neg x$. An xor-clause is *binary* (*ternary*) if its normal form has two (three) variables. A *clause* is either an or-clause or an xor-clause.

A *truth assignment* π is a set of literals such that $\top \in \pi$ and $\forall l \in \pi : \neg l \notin \pi$. We define the "satisfies" relation $\models$ between a truth assignment π and logical constructs as follows: (i) if l is a literal, then $\pi \models l$ iff $l \in \pi$, (ii) if $C = (l_1 \vee \cdots \vee l_n)$ is an or-clause, then $\pi \models C$ iff $\pi \models l_i$ for some $l_i \in \{l_1, \ldots, l_n\}$, and (iii) if $C = (l_1 \oplus \cdots \oplus l_n)$ is an xor-clause, then $\pi \models C$ iff π is total for C (i.e. $\forall 1 \leq i \leq n : l_i \in \pi \vee \neg l_i \in \pi$) and $\pi \models l_i$ for an odd number of literals of C. Observe that no truth assignment satisfies the empty or-clause () or the empty xor-clause (), i.e. these clauses are synonyms for $\bot$.

A *cnf-xor formula* ϕ is a conjunction of clauses, expressible as a conjunction

$$\phi = \phi_{\mathrm{or}} \wedge \phi_{\mathrm{xor}}, \tag{1}$$

where ϕ_{or} is a conjunction of or-clauses and ϕ_{xor} is a conjunction of xor-clauses. A truth assignment π *satisfies* ϕ, denoted by $\pi \models \phi$, if it satisfies each clause in it; ϕ is called *satisfiable* if there exists such a truth assignment satisfying it, and *unsatisfiable* otherwise. The *cnf-xor satisfiability* problem studied in this paper is to decide whether a given cnf-xor formula has a satisfying truth assignment. A formula ϕ' is a *logical consequence* of a formula ϕ, denoted by $\phi \models \phi'$, if $\pi \models \phi$ implies $\pi \models \phi'$ for all truth assignments π. The set of variables occurring in a formula ϕ is denoted by $\mathrm{vars}(\phi)$, and $\mathrm{lits}(\phi) = \{x, \neg x \mid x \in \mathrm{vars}(\phi)\}$ is the set of literals over $\mathrm{vars}(\phi)$. We use $C[A/D]$ to denote the (normal form) xor-clause that is identical to C except that all occurrences of the atom A in C are substituted with D once. For instance, $(x_1 \oplus x_2 \oplus x_3)[x_1/(x_1 \oplus x_3)] = x_1 \oplus x_3 \oplus x_2 \oplus x_3 = x_1 \oplus x_2$.

3   The DPLL(XOR) Framework

The idea in the DPLL(XOR) framework [2] for satisfiability solving of cnf-xor formulas $\phi = \phi_{\mathrm{or}} \wedge \phi_{\mathrm{xor}}$ is similar to that in the DPLL(T) framework for solving satisfiability of quantifier-free first-order formulas modulo a background theory T (SMT, see e.g. [11,12]). In DPLL(XOR), see Fig. 1 for a high-level pseudo-code, one employs a conflict-driven clause learning (CDCL) SAT solver (see e.g. [1]) to search for a satisfying truth assignment π over all the variables in $\phi = \phi_{\mathrm{or}} \wedge \phi_{\mathrm{xor}}$.[1] The CDCL-part takes care of the usual unit clause propagation on the cnf-part ϕ_{or} of the formula (line 4 in Fig. 1), conflict analysis and non-chronological backtracking (line 15–17), and heuristic selection of decision literals (lines 19–20) which extend the current partial truth assignment π towards a total one.

To handle the parity constraints in the xor-part ϕ_{xor}, an *xor-reasoning module* M is coupled with the CDCL solver. The values assigned in π to the variables in $\mathrm{vars}(\phi_{\mathrm{xor}})$ by the CDCL solver are communicated as *xor-assumption literals* to the module (with the ASSIGN method on line 6 of the pseudo-code). If $l_1, \ldots, l_m$ are the xor-assumptions communicated to the module so far, then the DEDUCE method (invoked on line 7) of the module is used to deduce a (possibly empty) list of *xor-implied literals* $\hat{l}$ that are logical consequences of the xor-part ϕ_{xor} and xor-assumptions, i.e. literals for which

---

[1] See [2] for a discussion on handling "xor-internal" variables occurring in ϕ_{xor} but not in ϕ_{or}.

solve($\phi = \phi_{or} \wedge \phi_{xor}$):
1. initialize xor-reasoning module M with ϕ_{xor}
2. $\pi = \langle \rangle$ /*the truth assignment*/
3. while true:
4. $(\pi', confl) = \text{UNITPROP}(\phi_{or}, \pi)$ /*unit propagation*/
5.   if not $confl$:   /*apply xor-reasoning*/
6.   for each literal l in π' but not in π: $M.\text{ASSIGN}(l)$
7. $(\hat{l}_1, ..., \hat{l}_k) = M.\text{DEDUCE}()$
8.   for $i = 1$ to k:
9. $C = M.\text{EXPLAIN}(\hat{l}_i)$
10.   if $\hat{l}_i = \bot$ or $\neg \hat{l}_i \in \pi'$: $confl = C$, break
11.   else if $\hat{l}_i \notin \pi'$: add $\hat{l}_i$ to π' with the implying or-clause C
12.   if $k > 0$ and not $confl$:
13. $\pi = \pi'$; continue   /*unit propagate further*/
14.   let $\pi = \pi'$
15.   if $confl$:   /*standard Boolean conflict analysis*/
16.   analyze conflict, learn a conflict clause
17.   backjump or return "unsatisfiable" if not possible
18.   else:
19.   add a heuristically selected unassigned literal in ϕ to π
20.   or return "satisfiable" if no such variable exists

Fig. 1. The essential skeleton of the DPLL(XOR) framework

$\phi_{xor} \wedge l_1 \wedge ... \wedge l_m \models \hat{l}$ holds. These xor-implied literals can then be added to the current truth assignment π (line 11) and the CDCL part invoked again to perform unit clause propagation on these. The conflict analysis engine of CDCL solvers requires that each implied (i.e. non-decision) literal has an *implying clause*, i.e. an or-clause that forces the value of the literal by unit propagation on the values of literals appearing earlier in the truth assignment (which at the implementation level is a sequence of literals instead of a set). For this purpose the xor-reasoning module has a method EXPLAIN that, for each xor-implied literal $\hat{l}$, gives an or-clause C of form $l'_1 \wedge ... \wedge l'_k \Rightarrow \hat{l}$, i.e. $\neg l'_1 \vee ... \vee \neg l'_k \vee \hat{l}$, such that (i) C is a logical consequence of ϕ_{xor}, and (ii) $l'_1, ..., l'_k$ are xor-assumptions made or xor-implied literals returned before $\hat{l}$. An important special case occurs when the "false" literal $\bot$ is returned as an xor-implied literal (line 10), i.e. when an *xor-conflict* occurs; this implies that $\phi_{xor} \wedge l_1 \wedge ... \wedge l_m$ is unsatisfiable. In such a case, the clause returned by the EXPLAIN method is used as the unsatisfied clause $confl$ initiating the conflict analysis engine of the CDCL part (lines 10 and 15–17).

In addition to the ASSIGN, DEDUCE, and EXPLAIN methods, an xor-reasoning module must also implement methods that allow xor-assumptions to be retracted from the solver in order to allow backtracking in synchronization with the CDCL part (line 17).

Naturally, there are many *xor-module integration strategies* that can be considered in addition to the one described in the above pseudo-code. For instance, the xor-explanations for the xor-implied literals can be computed always (as in the pseudo-code for the sake of simplicity) or only when needed in the CDCL-part conflict analysis (as in a real implementation for efficiency reasons).

$$\oplus\text{-Unit}^{+}: \frac{x \quad C}{C\,[x/\top]} \qquad \oplus\text{-Unit}^{-}: \frac{x \oplus \top \quad C}{C\,[x/\bot]}$$

Fig. 2. Inference rules of UP; the symbol x is variable and C is an xor-clause

4 The Xor-Reasoning Module "UP"

To illustrate our new parity-based techniques, we first introduce a very simple xor-reasoning module "UP" which only performs unit propagation on xor-clauses. As such it can only perform the same deduction as CNF-level unit propagation would on the CNF translation of the xor-clauses. However, with our new parity-based xor-implied literal explanation techniques (Sect. 5) we can deduce much stronger clauses (Sect. 6) and also new xor-clauses that can be learned (Sect. 7). In Sect. 8 we then generalize the results to other xor-reasoning modules such as the the one in [2] incorporating also equivalence reasoning.

As explained above, given a conjunction of xor-clauses ϕ_{xor} and a sequence $l_1, \ldots, l_k$ of xor-assumption literals, the goal of an xor-reasoning module is to deduce xor-implied literals and xor-conflicts over $\psi = \phi_{\text{xor}} \wedge l_1 \wedge \cdots \wedge l_k$. To do this, the UP-module implements a deduction system with the inference rules shown in Fig. 2. An UP-*derivation* on ψ is a finite, vertex-labeled directed acyclic graph $G = \langle V, E, L \rangle$, where each vertex $v \in V$ is labeled with an xor-clause $L(v)$ and the following holds for each vertex v:

1. v has no incoming edges (i.e. is an *input vertex*) and $L(v)$ is an xor-clause in ψ, or
2. v has two incoming edges originating from vertices v_1 and v_2, and $L(v)$ is derived from $L(v_1)$ and $L(v_2)$ by using one of the inference rules.

As an example, Fig. 3 shows a UP-derivation for $\phi_{\text{xor}} \wedge (\neg a) \wedge (d) \wedge (\neg b)$, where $\phi_{\text{xor}} = (a \oplus b \oplus c) \wedge (c \oplus d \oplus e) \wedge (c \oplus e \oplus f)$ (please ignore the "cut" lines for now). An xor-clause C is UP-*derivable* on ψ, denoted by $\psi \vdash_{\text{UP}} C$, if there exists a UP-derivation on ψ that contains a vertex labeled with C; the UP-derivable unary xor-clauses are the xor-implied literals that the UP-module returns when its DEDUCE method is called. In Fig. 3, the literal f is UP-derivable and the UP-module returns f as an xor-implied literal after $\neg a$, d, and $\neg b$ are given as xor-assumptions. As a direct consequence of the definition of xor-derivations and the soundness of the inference rules, it holds that if an xor-derivation on ψ contains a vertex labeled with the xor-clause C, then C is a logical consequence of ψ, i.e. $\psi \vdash_{\text{UP}} C$ implies $\psi \models C$. A UP-derivation on ψ is a UP-*refutation of* ψ if it contains a vertex labeled with the false literal $\bot$; in this case, ψ is unsatisfiable. A UP-derivation G on ψ is *saturated* if for each unary xor-clause C such that $\psi \vdash_{\text{UP}} C$ it holds that there is a vertex v in G with the label $L(v) = C$. Note that UP is not refutationally complete,

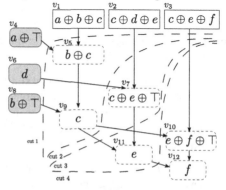

Fig. 3. A UP-derivation

e.g. there is no UP-refutation of the unsatisfiable conjunction $(a \oplus b) \wedge (a \oplus b \oplus \top)$. However, it is "eventually refutationally complete" in the DPLL(XOR) setting: if each variable in ψ occurs in a unary clause in ψ, then the empty clause is UP-derivable iff ψ is unsatisfiable; thus when the CDCL SAT solver has assigned a value to all variables in ϕ_{xor}, the UP-module can check whether all the xor-clauses are satisfied.

As explained in the previous section, the CDCL part of the DPLL(XOR) framework requires an implying or-clause for each xor-implied literal. These can be computed by interpreting the $\oplus$-Unit$^+$ and $\oplus$-Unit$^-$ rules as implications

$$(x) \wedge C \Rightarrow C\,[x/\top] \tag{2}$$
$$(x \oplus \top) \wedge C \Rightarrow C\,[x/\bot] \tag{3}$$

respectively, and recursively expanding the xor-implied literal with the left-hand side conjunctions of these until a certain cut of the UP-derivation is reached. Formally, a *cut* of a UP-derivation $G = \langle V, E, L \rangle$ is a partitioning (V_a, V_b) of V. A *cut for a non-input vertex* $v \in V$ is a cut (V_a, V_b) such that (i) $v \in V_b$, and (ii) if $v' \in V$ is an input vertex and there is a path from v' to v, then $v' \in V_a$. Now assume a UP-derivation $G = \langle V, E, L \rangle$ for $\phi_{\text{xor}} \wedge l_1 \wedge ... \wedge l_k$. For each non-input node v in G, and each cut $W = \langle V_a, V_b \rangle$ of G for v, the *implicative explanation* of v under W is the conjunction $Expl(v, W) = f_W(v)$, there f_W is recursively defined as follows:

E1 If u is an input node with $L(u) \in \phi_{\text{xor}}$, then $f_W(u) = \top$.
E2 If u is an input node with $L(u) \in \{l_1, ..., l_k\}$, then $f_W(u) = L(u)$.
E3 If u is a non-input node in V_a, then $f_W(u) = L(u)$.
E4 If u is a non-input node in V_b, then $f_W(u) = f_W(u_1) \wedge f_W(u_2)$, where u_1 and u_2 are the source nodes of the two edges incoming to u.

Based on Eqs. (2) and (3), it is easy to see that $\phi_{\text{xor}} \models Expl(v, W) \Rightarrow L(v)$ holds. The implicative explanation $Expl(v, W)$ can in fact be read directly from the cut W as in [2]: $Expl(v, W) = \bigwedge_{u \in \text{reasons}(W)} L(u)$, where $\text{reasons}(W) = \{u \in V_a \mid L(u) \notin \phi_{\text{xor}} \wedge \exists u' \in V_b : \langle u, u' \rangle \in E\}$ is the *reason set* for W. A cut W is *cnf-compatible* if $L(u)$ is a unary xor-clause for each $u \in \text{reasons}(W)$. Thus if the cut W is cnf-compatible, then $Expl(v, W) \Rightarrow L(v)$ is the required or-clause implying the xor-implied literal $L(v)$.

Example 1. Consider again the UP-derivation on $\phi_{\text{xor}} \wedge (\neg a) \wedge (d) \wedge (\neg b)$ in Fig. 3. It has four cuts, 1–4, for the vertex v_{12}, corresponding to the explanations $\neg a \wedge d \wedge \neg b$, $c \wedge d$, $c \wedge (c \oplus e \oplus \top)$, and $e \wedge c$, respectively. The non-cnf-compatible cut 3 cannot be used to give an implying or-clause for the xor-implied literal f but the others can; the one corresponding to the cut 2 is $(\neg c \vee \neg d \vee f)$. ♣

The UP-derivation bears an important similarity with "traditional" implication graph of a SAT solver where each vertex represents a variable assignment: graph partitions are used to derive clausal explanations for implied literals. Different partitioning schemes for such implication graphs have been studied in [13], and we can directly adopt some of them for our analysis. A cut $W = (V_a, V_b)$ for a non-input vertex v is:

1. *closest cut* if W is the cnf-compatible cut with the smallest possible V_b part. Observe that each implying or-clause derived from these cuts is a clausification of a single xor-clause; e.g., $(\neg c \vee \neg e \vee f)$ obtained from the cut 4 in Fig. 3.

2. *first UIP cut* if W is the cut with the largest possible V_a part such that reasons(W) contains either the latest xor-assumption vertex or exactly one of its successors.
3. *furthest cut* if V_b is maximal. Note that furthest cuts are also cnf-compatible as their reason sets consist only of xor-assumptions.

In the implementation of the UP-module, we use a modified version of the 2-watched literals scheme first presented in [14] for or-clauses; all but one of the *variables* in an xor-clause need to be assigned before the xor-clause implies the last one. Thus it suffices to have two *watched variables*. MoRsat [8] uses the same data structure for all clauses and has 2×2 watched literals for xor-clauses. Cryptominisat [9] uses a scheme similar to ours except that it manipulates the polarities of literals in an xor-clause while we take the polarities into account in the explanation phase. Because of this implementation technique, the implementation does not consider the non-unary non-input vertices in UP-derivations; despite this, Thm. 3 does hold also for the implemented inference system.

5 Parity Explanations

So far in this paper, as well as in our previous works [2,3], we have used the inference rules in an "implicative way". For instance, we have implicitly read the $\oplus$-Unit$^+$ rule as

if the xor-clauses (x) *and* C hold, *then* $C[x/\top]$ also holds.

Similarly, the implicative explanation for an xor-implied literal $\hat{l}$ labelling a non-input node v under a cnf-compatible cut W has been defined to be a conjunction $Expl(v, W)$ of literals with $\phi_{xor} \models Expl(v, W) \Rightarrow \hat{l}$ holding. We now propose an alternative method allowing us to compute a *parity explanation* $Expl_\oplus(v, W)$ that is an xor-clause such that

$$\phi_{xor} \models Expl_\oplus(v, W) \Leftrightarrow \hat{l}$$

holds. The variables in $Expl_\oplus(v, W)$ will always be a subset of the variables in the implicative explanation $Expl(v, W)$ computed on the same cut.

The key observation for computing parity explanations is that the inference rules can in fact also be read as *equations* over xor-clauses under some provisos. As an example, the $\oplus$-Unit$^+$ rule can be seen as the equation $(x) \oplus C \oplus \top \Leftrightarrow C[x/\top]$ *provided that* (i) $x \in C$, and (ii) C is in normal form. That is, taking the exclusive-or of the two premises and the constant true gives us the consequence clause of the rule. The provisos are easy to fulfill: (i) we have already assumed all xor-clauses to be in normal form, and (ii) applying the rule when $x \notin C$ is redundant and can thus be disallowed. The reasoning is analogous for the $\oplus$-Unit$^-$ rule and thus for UP rules we have the equations:

$$(x) \oplus C \oplus \top \Leftrightarrow C[x/\top] \tag{4}$$
$$(x \oplus \top) \oplus C \oplus \top \Leftrightarrow C[x/\bot] \tag{5}$$

As all the UP-rules can be interpreted as equations of form "left-premise xor right-premise xor true equals consequence", we can expand any xor-clause C in a node of a UP-derivation by iteratively replacing it with the left hand side of the corresponding equation. As a result, we will get an xor-clause that is logically equivalent to C; from this, we can eliminate the xor-clauses in ϕ_{xor} and get an xor-clause D such that $\phi_{xor} \models D \Leftrightarrow C$. Formally, assume a UP-derivation $G = \langle V, E, L \rangle$ for $\phi_{xor} \wedge l_1 \wedge ... \wedge l_k$. For each non-input node v in G, and each cut $W = \langle V_a, V_b \rangle$ of G for v, the *parity explanation* of v under W is $Expl_\oplus(v, W) = f_W(v)$, there f_W is recursively defined as earlier for $Expl(v, W)$ except that the case "E4" is replaced by

E4 If u is a non-input node in V_b, then $f_W(u) = f_W(u_1) \oplus f_W(u_2) \oplus \top$, where u_1 and u_2 are the source nodes of the two edges incoming to u.

We now illustrate parity explanations and show that they can be smaller (in the sense of containing fewer variables) than implicative explanations:

Example 2. Consider again the UP-derivation given in Fig. 3. Take the cut 4 first; we get $Expl_\oplus(v_{12}, W) = c \oplus e \oplus \top$. Now $\phi_{xor} \models Expl_\oplus(v_{12}, W) \Leftrightarrow L(v_{12})$ holds as $(c \oplus e \oplus \top) \Leftrightarrow f$, i.e. $c \oplus e \oplus f$, is an xor-clause in ϕ_{xor}. Observe that the implicative explanation $c \wedge e$ of v_{12} under the cut is just one conjunct in the disjunctive normal form $(c \wedge e) \vee (\neg c \wedge \neg e)$ of $c \oplus e \oplus \top$.

On the other hand, under the cut 2 we get $Expl_\oplus(v_{12}, W) = d$. Now $\phi_{xor} \models Expl_\oplus(v_{12}, W) \Leftrightarrow L(v_{12})$ as $d \Leftrightarrow f$, i.e. $d \oplus f \oplus \top$, is a linear combination of the xor-clauses in ϕ_{xor}. Note that the implicative explanation for v_{12} under the cut is $(c \wedge d)$, and no cnf-compatible cut for v_{12} gives the implicative explanation (d) for v_{12}.   ♣

We observe that $vars(Expl_\oplus(v, W)) \subseteq vars(Expl(v, W))$ by comparing the definitions of $Expl(v, W)$ and $Expl_\oplus(v, W)$. The correctness of $Expl_\oplus(v, W)$, formalized in the following theorem, can be established by induction and using Eqs. (4) and (5).

Theorem 1. *Let* $G = \langle V, E, L \rangle$ *be a* UP-*derivation on* $\phi_{xor} \wedge l_1 \wedge \cdots \wedge l_k$, v *a node in it, and* $W = \langle V_a, V_b \rangle$ *a cut for* v. *It holds that* $\phi_{xor} \models Expl_\oplus(v, W) \Leftrightarrow L(v)$.

Recall that the CNF-part solver requires an implying or-clause C for each xor-implied literal, forcing the value of the literal by unit propagation. A parity explanation can be used to get such implying or-clause by taking the implicative explanation as a basis and omitting the literals on variables not occurring in the parity explanation:

Theorem 2. *Let* $G = \langle V, E, L \rangle$ *be a* UP-*derivation on* $\phi_{xor} \wedge l_1 \wedge \cdots \wedge l_k$, v *a node with* $L(v) = \hat{l}$ *in it, and* $W = \langle V_a, V_b \rangle$ *a cnf-compatible cut for* v. *Then* $\phi_{xor} \models (\bigwedge_{u \in S} L(u)) \Rightarrow \hat{l}$, *where* $S = \{u \in reasons(W) \mid vars(L(u)) \subseteq vars(Expl_\oplus(v, W))\}$.

Observing that only expressions of the type $f_W(u)$ occurring an odd number of times in the expression $f_W(v)$ remain in $Expl_\oplus(v, W)$, we can derive a more efficient graph traversal method for computing parity explanations. That is, when computing a parity explanation for a node, we traverse the derivation backwards from it in a breadth-first order. If we come to a node u and note that its traversal is requested because an even number of its successors have been traversed, then we don't need to traverse u further or include $L(u)$ in the explanation if u was on the "reason side" V_a of the cut.

Example 3. Consider again the UP-derivation in Fig. 3 and the cnf-compatible cut 1 for v_{12}. When we traverse the derivation backwards, we observe that the node v_9 has an even number of traversed successors; we thus don't traverse it (and consequently neither

v_8, v_5, v_4 or v_1). On the other hand, v_6 has an odd number of traversed successors and it is included when computing $Expl_\oplus(v_{12}, W)$. Thus we get $Expl_\oplus(v_{12}, W) = L(v_6) = (d)$ and the implying or-clause for f is $d \Rightarrow f$, i.e. $(\neg d \lor f)$. ♣

Although parity explanations can be computed quite fast using graph traversal as explained above, this can still be computationally prohibitive on "xor-intensive" instances because a single CNF-level conflict analysis may require that implying or-clauses for hundreds of xor-implied literals are computed. In our current implementation, we compute the closest cnf-compatible cut (for which parity explanations are very fast to compute but equal to implicative explanations and produce clausifications of single xor-clauses as implying or-clauses) for an xor-implied literal $\hat{l}$ when an explanation is needed in the regular conflict analysis. The computationally more expensive furthest cut is used if an explanation is needed again in the conflict-clause minimization phase of minisat.

6   Resolution Cannot Polynomially Simulate Parity Explanations

Intuitively, as parity explanations can contain fewer variables than implicative explanations, the implying or-clauses derived from them should help pruning the remaining search space of the CDCL solver better. We now show that, in theory, parity explanations can indeed be very effective as they can allow small refutations for some formula classes whose CNF translations do not have polynomial size resolution proofs. To do this, we use the hard formulas defined in [15]; these are derived from a class of graphs which we will refer to as "parity graphs". A *parity graph* is an undirected, connected, edge-labeled graph $G = \langle V, E \rangle$ where each node $v \in V$ is labeled with a *charge* $c(v) \in \{\bot, \top\}$ and each edge $\langle v, u \rangle \in E$ is labeled with a distinct variable. The *total charge* $c(G) = \bigoplus_{v \in V} c(v)$ of an parity graph G is the parity of all node charges. Given a node v, define the xor-clause $\alpha(v) = q_1 \oplus \ldots \oplus q_n \oplus c(v) \oplus \top$, where $q_1, \ldots, q_n$ are the variables used as labels in the edges connected to v, and $xorclauses(G) = \bigwedge_{v \in V} \alpha(v)$. For an xor-clause C over n variables, let $cnf(C)$ denote the equivalent CNF formula, i.e. the conjunction of 2^{n-1} clauses with n literals in each. Define $clauses(G) = \bigwedge_{v \in V} cnf(\alpha(v))$.

As proven in Lemma 4.1 in [15], $xorclauses(G)$ and $clauses(G)$ are unsatisfiable if and only if $c(G) = \top$. The unsatisfiable formulas derived from parity graphs can be very hard for resolution: there is an infinite sequence $G_1, G_2, \ldots$ of degree-bounded parity graphs such that $c(G_i) = \top$ for each i and the following holds:

Lemma 1 (Thm. 5.7 of [15]). *There is a constant $c > 1$ such that for sufficiently large m, any resolution refutation of* $clauses(G_m)$ *contains c^n distinct clauses, where* $clauses(G_m)$ *is of length $\mathcal{O}(n)$, $n = m^2$.*

We now present our key result on parity explanations: for *any* parity graph G with $c(G_i) = \top$, the formula $xorclauses(G)$ can be refuted with a *single* parity explanation after a number of xor-assumptions have been made:

Theorem 3. *Let $G = \langle V, E \rangle$ be a parity graph such that $c(G) = \top$. There is a UP-refutation for* $xorclauses(G) \land q_1 \cdots \land q_k$ *for some xor-assumptions $q_1, \ldots, q_k$, a node*

v with $L(v) = \perp$ in it, and a cut $W = \langle V_a, V_b \rangle$ for v such that $Expl_\oplus(v, W) = \top$. Thus xorclauses$(G) \models (\top \Leftrightarrow \perp)$, showing xorclauses$(G)$ unsatisfiable.

By recalling that CDCL SAT solvers are equally powerful to resolution [16], and that unit propagation on xor-clauses can be efficiently simulated by unit propagation their CNF translation, we get the following:

Corollary 1. *There are families of unsatisfiable cnf-xor formulas for which DPLL(XOR) using UP-module (i) has polynomial sized proofs if parity explanations are allowed, but (ii) does not have such if the "classic" implicative explanations are used.*

In practice, the CDCL part does not usually make the correct xor-assumptions needed to compute the empty implying or-clause, but if parity explanations are used in learning as explained in the next section, instances generated from parity graphs can be solved very fast.

7   Learning Parity Explanations

As explained in Sect. 5, parity explanations can be used to derive implying or-clauses, required by the conflict analysis engine of the CDCL solver, that are shorter than those derived by the classic implicative explanations. In addition to this, parity explanations can be used to derive *new xor-clauses* that are logical consequences of ϕ_{xor}; these xor-clauses D can then be *learned*, meaning that ϕ_{xor} is extended to $\phi_{xor} \wedge D$, the goal being to increase the deduction power of the xor-reasoning module. As an example, consider again Ex. 2 and recall that the parity explanation for v_{12} under the cut 2 is d. Now $\phi_{xor} \models (d \Leftrightarrow f)$, i.e. $\phi_{xor} \models (d \oplus f \oplus \top)$, holds, and we can extend ϕ_{xor} to $\phi'_{xor} = \phi_{xor} \wedge (d \oplus f \oplus \top)$ while preserving all the satisfying truth assignments. In fact, it is not possible to deduce f from $\phi_{xor} \wedge (d)$ by using UP, but f can be deduced from $\phi'_{xor} \wedge (d)$. Thus learning new xor-clauses derived from parity explanations can increase the deduction power of the UP inference system in a way similar to conflict-driven clause learning increasing the power of unit propagation in CDCL SAT solvers.

However, if all such derived xor-clauses are learned, it is possible to learn the same xor-clause many times, as illustrated in the following example and Fig. 4.

Example 4. Let $\phi_{xor} = (a \oplus b \oplus c \oplus \top) \wedge (b \oplus c \oplus d \oplus e) \wedge \ldots$ and assume that CNF part solver gives its first decision level literals a and $\neg c$ as xor-assumptions to the UP-module; the module deduces b and returns it to the CNF solver. At the next decision level the CNF part guesses d, gives it to UP-module, which deduces e, returns it to the CNF part, and the CNF part propagates it so that a conflict occurs. Now the xor-implied literal e is explained and a new xor-clause $D = (a \oplus d \oplus e \oplus \top)$ is learned in ϕ_{xor}. After this the CNF part backtracks, implies $\neg d$ at the decision level 1, and gives it to the UP-module; the module can then deduce $\neg e$ *without* using D. If $\neg e$ is now explained, the same "new" xor-clause $(a \oplus d \oplus e \oplus \top)$ can be derived.   ♣

The example illustrates a commonly occurring case in which a derived xor-clause contains two or more literals on the latest decision level (e and d in the example); in such a case, the xor-clause may already exist in ϕ_{xor}. A conservative approach to avoid learning the same xor-clause twice, under the reasonable assumption that the CNF and xor-reasoning module parts saturate their propagations before new heuristic decisions are

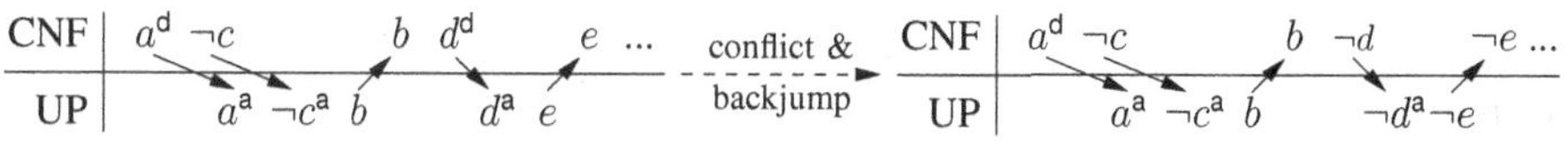

Fig. 4. Communication between CNF part and UP-module in a case when duplicate xor-clauses are learned; the d and a superscripts denote decision literals and xor-assumptions, respectively

made, is to disregard derived xor-clauses that have two or more variables assigned on the latest decision level. If a learned xor-clause for xor-implied literal $\hat{l}$ does not have other literals on the latest decision level, it can be used to infer $\hat{l}$ with fewer decision literals. Note that it may also happen that an implying or-clause for an xor-implied literal $\hat{l}$ does not contain any literals besides $\hat{l}$ on the latest decision level; the CNF part may then compute a conflict clause that does not have any literals on the current decision level, which needs to be treated appropriately.

In order to avoid slowing down propagation in our implementation, we store and remove learned xor-clauses using a strategy adopted from minisat: the maximum number of learned xor-clauses is increased at each restart and the "least active" learned xor-clauses are removed when necessary. However, using the conservative approach to learning xor-clauses, the total number of learned xor-clauses rarely exceeds the number of original xor-clauses.

8   General Xor-Derivations

So far in this paper we have considered a very simple xor-reasoning module capable only of unit propagation. We can in fact extend the introduced concepts to more general inference systems and derivations. Define an *xor-derivation* similarly to UP-derivation except that there is only one inference rule, $\oplus$-Gen : $\dfrac{C_1 \quad C_2}{C_1 \oplus C_2 \oplus \top}$, where C_1 and C_2 are xor-clauses. The inference rule $\oplus$-Gen is a generalization of the rules Gauss$^-$ and Gauss$^+$ in [4]. Now Thms. 1 and 2 can be shown to hold for such derivations as well.

As another concrete example of xor-reasoning module implementing a sub-class of $\oplus$-Gen, consider the Subst module presented in [2]. In addition to the unit propagation rules of UP in Fig. 2, it has inference rules allowing equivalence reasoning:

$$\oplus\text{-Eqv}^+ : \frac{x \oplus y \oplus \top \quad C}{C\,[x/y]} \qquad\qquad \oplus\text{-Eqv}^- : \frac{x \oplus y \quad C}{C\,[x/(y \oplus \top)]}$$

where the symbols x and y are variables while C is an xor-clause in the normal form with an occurrence of x. Note that these Subst rules are indeed instances of the more general inference rule $\oplus$-Gen. For instance, given two xor-clauses $C_1 = (c \oplus d \oplus \top)$ and $C_2 = (b \oplus d \oplus e)$, the Subst-system can produce the xor-clause $C_2\,[d/c] = (b \oplus c \oplus e)$ which is also inferred by $\oplus$-Gen: $(C_1 \oplus C_2 \oplus \top) = ((c \oplus d \oplus \top) \oplus (b \oplus d \oplus e) \oplus \top) = (b \oplus c \oplus e)$.

Subst-derivations are defined similarly to UP-derivations. As an example, Fig. 5

shows a Subst-derivation on $\phi_{\text{xor}} \wedge (a)$, where
$\phi_{\text{xor}} = (a \oplus b \oplus c) \wedge (a \oplus c \oplus d) \wedge (b \oplus d \oplus e)$.
The literal e is Subst-derivable on $\phi_{\text{xor}} \wedge (a)$;
the xor-reasoning module returns e as an xor-
implied literal on ϕ_{xor} after a is given as
an xor-assumption. The cnf-compatible cut
1 for the literal e gives the implicative explanation (a) and thus the implying or-clause
$(\neg a \vee e)$ for e. Parity explanations are defined
for Subst in the same way as for UP; the parity explanation for the literal e in the figure

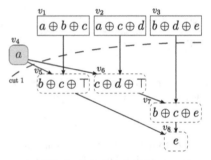

Fig. 5. A Subst-derivation

is $\top$ and thus the implying or-clause for e is (e). Observe that e is *not* UP-derivable
from $\phi_{\text{xor}} \wedge (a)$, i.e. Subst is a stronger deduction system than UP in this sense.

Parity explanations can also be computed in another xor-reasoning module, EC presented in [3], that is based on equivalence class manipulation. We omit this construction
due to space constraints.

9   Experimental Results

We have implemented a prototype solver integrating both xor-reasoning modules (UP
and Subst) to minisat [17] (version 2.0 core) solver. In the experiments we focus on the
domain of logical cryptanalysis by modeling a "known cipher stream" attack on stream
ciphers Bivium, Crypto-1, Grain, Hitag2, and Trivium. To evaluate the performance
of the proposed techniques, we include both unsatisfiable and satisfiable instances. In
the unsatisfiable instances, generated with grain-of-salt [18], the task is to recover the
internal cipher state when 256 output stream bits are given. This is infeasible in practice,
so the instances are made easier and also unsatisfiable by assigning a large enough
number of internal state bits randomly. Thus, the task becomes to prove that it is not
possible to assign the remaining bits of the internal cipher state so that the output would
match the given bits. To include also satisfiable instances we modeled a different kind
of attack on the ciphers Grain, Hitag2 and Trivium where the task is to recover the
full key when a small number of cipher stream bits are given. In the attack, the IV and
a number of key stream bits are given. There are far fewer generated cipher stream
bits than key bits, so a number of keys probably produce the same prefix of the cipher
stream. All instances were converted into (i) the standard DIMACS CNF format, and
(ii) a DIMACS-like format allowing xor-clauses as well. Structurally these instances
are interesting for benchmarking xor-reasoning as they have a large number of tightly
connected xor-clauses combined with a significant CNF part.

We first compare the following solver configurations: (i) unmodified minisat, (ii)
up: minisat with watched variable based unit propagation on xor-clauses, (iii) up-pexp:
up extended with parity explanations, (iv) up-pexp-learn: up-pexp extended with xor-
clause learning, and (v) up-subst-learn: up using Subst-module to compute parity xor-
explanations and xor-clause learning. The reference configuration up computes closest
cnf-compatible cut parity explanations, and the other configurations use also furthest
cuts selectively as described in Sect. 5. We also tested first UIP cuts, but the performance
did not improve.

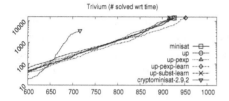 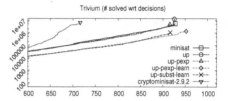

Fig. 6. Number of solved instances with regard to time and decisions on satisfiable Trivium benchmark (1020 instances, 51 instances per generated cipher stream length ranging from 1 to 20 bits)

Solver	Bivium			Crypto-1			Grain			Hitag2			Trivium		
	#	Dec.	Time	#	Dec.	Time	#	Dec.	Time	#	Dec.	Time	#	Dec.	Time
minisat	51	834	80.9	51	781	691.1	1	-	-	35	428	440.0	51	55	5.7
up	51	985	127.7	51	1488	1751.8	51	40	13.8	39	291	403.9	51	59	8.0
up-pexp	51	1040	147.8	51	1487	1748.2	51	35	10.9	37	124	148.0	51	62	9.8
up-pexp-learn	47	651	114.0	51	1215	1563.0	36	122	87.7	37	222	255.4	51	24	3.7
up-subst-learn	47	616	336.4	51	1037	2329.5	37	70	90.3	36	215	374.8	51	29	12.9
cryptominisat-2.9.2	51	588	89.8	51	0	0.06	51	89	10.4	51	0	0.07	51	71	6.04

Fig. 7. Results of the unsatisfiable benchmarks showing the number of solved instances (#) within the 4h time limit, median of decisions ($\times 10^3$), and median of solving time

The results for the satisfiable Trivium benchmarks are shown in Fig. 6. Learning xor-clauses reduces the number of decisions needed substantially, and in the case of the computationally less expensive UP reasoning module this is also reflected in the solving time and in the number of solved instances. On the other satisfiable benchmark sets learning new xor-clauses also reduced the number of required decisions significantly but the number of propagations per decision is also greatly increased due to increased deduction power and the reduction is not really reflected in the solving time.

The results for the unsatisfiable benchmarks are shown in Fig. 7. Parity explanations reduce decisions on Grain and Hitag2, leading to fastest solving time. Learning parity explanations reduces explanations on all benchmarks except Grain and gives the best solving time on Trivium. Equivalence reasoning seems to reduce decisions slightly with the cost of increased solving time. Obviously more work has to be done to improve data structures and adjust heuristics so that the theoretical power of parity explanations and xor-clause learning can be fully seen also in practice.

We also ran cryptominisat version 2.9.2 [10] on the benchmarks. As shown in Figs. 6 and 7, it performs (i) extremely well on the unsatisfiable Crypto-1 and Hitag2 instances due to "failed literal detection" and other techniques, but (ii) not so well on our satisfiable Trivium instances, probably due to differences in restart policies or other heuristics.

10 Conclusions

We have shown how to compute linearity exploiting parity explanations for literals deduced in an xor-reasoning module. Such explanations can be used (i) to produce more compact clausal explanations for the conflict analysis engine of a CDCL solver incorporating the xor-reasoning module, and (ii) to derive new parity constraints that can be learned in order to boost the deduction power of the xor-reasoning module. It has been proven that parity explanations allow very short refutations of some formulas whose CNF translations do not have polynomial size resolution proofs, even when using a simple xor-reasoning module capable only of unit-propagation. The experimental evaluation suggests that parity explanations and xor-clause learning can be efficiently implemented and demonstrates promising performance improvements also in practice.

References

1. Marques-Silva, J., Lynce, I., Malik, S.: Conflict-driven clause learning SAT solvers. In: Handbook of Satisfiability. IOS Press (2009)
2. Laitinen, T., Junttila, T., Niemelä, I.: Extending clause learning DPLL with parity reasoning. In: Proc. ECAI 2010, pp. 21–26. IOS Press (2010)
3. Laitinen, T., Junttila, T., Niemelä, I.: Equivalence class based parity reasoning with DPLL(XOR). In: Proc. ICTAI 2011. IEEE (2011)
4. Baumgartner, P., Massacci, F.: The Taming of the (X)OR. In: Palamidessi, C., Moniz Pereira, L., Lloyd, J.W., Dahl, V., Furbach, U., Kerber, M., Lau, K.-K., Sagiv, Y., Stuckey, P.J. (eds.) CL 2000. LNCS (LNAI), vol. 1861, pp. 508–522. Springer, Heidelberg (2000)
5. Li, C.M.: Integrating equivalency reasoning into Davis-Putnam procedure. In: Proc. AAAI/IAAI 2000, pp. 291–296. AAAI Press (2000)
6. Ostrowski, R., Grégoire, É., Mazure, B., Saïs, L.: Recovering and Exploiting Structural Knowledge from CNF Formulas. In: Van Hentenryck, P. (ed.) CP 2002. LNCS, vol. 2470, pp. 185–199. Springer, Heidelberg (2002)
7. Heule, M., Dufour, M., van Zwieten, J.E., van Maaren, H.: March_eq: Implementing Additional Reasoning into an Efficient Look-Ahead SAT Solver. In: Hoos, H.H., Mitchell, D.G. (eds.) SAT 2004. LNCS, vol. 3542, pp. 345–359. Springer, Heidelberg (2005)
8. Chen, J.: Building a Hybrid SAT Solver via Conflict-Driven, Look-Ahead and XOR Reasoning Techniques. In: Kullmann, O. (ed.) SAT 2009. LNCS, vol. 5584, pp. 298–311. Springer, Heidelberg (2009)
9. Soos, M., Nohl, K., Castelluccia, C.: Extending SAT Solvers to Cryptographic Problems. In: Kullmann, O. (ed.) SAT 2009. LNCS, vol. 5584, pp. 244–257. Springer, Heidelberg (2009)
10. Soos, M.: Enhanced gaussian elimination in DPLL-based SAT solvers. In: Pragmatics of SAT, Edinburgh, Scotland, GB, pp. 1–1 (2010)
11. Nieuwenhuis, R., Oliveras, A., Tinelli, C.: Solving SAT and SAT modulo theories: From an abstract Davis-Putnam-Logemann-Loveland procedure to DPLL(T). Journal of the ACM 53(6), 937–977 (2006)
12. Barrett, C., Sebastiani, R., Seshia, S.A., Tinelli, C.: Satisfiability modulo theories. In: Handbook of Satisfiability. IOS Press (2009)
13. Zhang, L., Madigan, C.F., Moskewicz, M.W., Malik, S.: Efficient conflict driven learning in boolean satisfiability solver. In: Proc. ICCAD 2001, pp. 279–285. IEEE Press (2001)
14. Moskewicz, M.W., Madigan, C.F., Zhao, Y., Zhang, L., Malik, S.: Chaff: Engineering an efficient SAT solver. In: Proc. DAC 2001, pp. 530–535. ACM (2001)
15. Urquhart, A.: Hard examples for resolution. J. ACM 34(1), 209–219 (1987)
16. Pipatsrisawat, K., Darwiche, A.: On the power of clause-learning SAT solvers as resolution engines. Artificial Intelligence 175(2), 512–525 (2011)
17. Eén, N., Sörensson, N.: An Extensible SAT-solver. In: Giunchiglia, E., Tacchella, A. (eds.) SAT 2003. LNCS, vol. 2919, pp. 502–518. Springer, Heidelberg (2004)
18. Soos, M.: Grain of salt — an automated way to test stream ciphers through SAT solvers. In: Tools 2010: Proceedings of the Workshop on Tools for Cryptanalysis 2010, pp. 1–2. RHUL (2010)

Perfect Hashing and CNF Encodings of Cardinality Constraints

Yael Ben-Haim, Alexander Ivrii, Oded Margalit, and Arie Matsliah

IBM R&D Labs in Israel
Haifa University Campus, Mount Carmel, Haifa, 31905, Israel
{yaelbh,alexi,odedm,ariem}@il.ibm.com

Abstract. We study the problem of encoding cardinality constraints (threshold functions) on Boolean variables into CNF. Specifically, we propose new encodings based on (perfect) hashing that are efficient in terms of the number of clauses, auxiliary variables, and propagation strength. We compare the properties of our encodings to known ones, and provide experimental results evaluating their practical effectiveness.

1 Introduction

Modern Boolean satisfiability (SAT) solvers are powerful tools, capable of solving many practical problems with millions of variables within minutes. They come well-tuned off-the-shelf, allowing non-expert users to solve a wide range of complex problems quickly. However, using a SAT solver to solve general *constraint satisfaction problems* (CSPs) requires encoding the problem into strict *conjunctive normal form* (CNF). The method of encoding can dramatically affect the runtime of a SAT solver and its memory consumption. Hence, the problem of encoding CSPs into CNF is well studied within both the CSP and SAT communities.

In this paper, we consider the special (and probably the most common) case of encoding cardinality constraints (on Boolean variables) of the form $\leq_k(X_1, \ldots, X_n)$ into CNF. The $\leq_k(X_1, \ldots, X_n)$ constraint, on variables $X_1, \ldots, X_n$, is satisfied if at most k of them are assigned TRUE. A CNF encoding of $\leq_k(X_1, \ldots, X_n)$ is a formula F on variables $X_1, \ldots, X_n$ and (possibly) additional auxiliary variables $Y_1, \ldots, Y_\ell$, satisfying the following conditions:

- for any assignment x to $X_1, \ldots, X_n$ with at most k TRUEs (in short $|x| \leq k$), there is an assignment $y = y_1 y_2 \cdots y_\ell$ to $Y_1, \ldots, Y_\ell$ such that $(x \cup y) \models F$;
- if x is an assignment to $X_1, \ldots, X_n$ with $|x| > k$, then for all assignments y to $Y_1, \ldots, Y_\ell$, $(x \cup y) \not\models F$.

The constraints $\circ_k(X_1, \ldots, X_n)$, $\circ \in \{<, >, \geq, =\}$, and their CNF encodings are defined similarly, and can be all translated to at most two[1] constraints of the form $\leq_k(X_1, \ldots, X_n)$. Therefore, throughout this paper (except in Section 4.2), we will focus on the $\leq_k(X_1, \ldots, X_n)$ type, with a further restriction of $0 < k < n$ (the cases $k = 0$ and $k = n$ can be handled trivially).

---

[1] One in all cases other than '='.

A. Cimatti and R. Sebastiani (Eds.): SAT 2012, LNCS 7317, pp. 397–409, 2012.

Cardinality constraints arise naturally in many problems with optimization flavor (e.g., "Is there a solution in which at most/at least k of the variables (from a certain set) are set to TRUE/FALSE?"), and in problems where multi-valued variables are expressed with Boolean variables (e.g., "Is there a solution in which exactly one of the Boolean variables representing each multi-valued variable is set to TRUE?"). Furthermore, any symmetric function on n Boolean variables (i.e., a function whose value depends only on the number of input variables assigned TRUE) can be expressed as a disjunction of at most n cardinality constraints.

1.1 Efficient Encodings and Related Work

An efficient CNF encoding F of the cardinality constraint $\leq_k(X_1, \ldots, X_n)$ has the following characteristics:

- Few clauses.
- Few auxiliary variables.
- One (or better both) of the following properties preserving arc-consistency under unit propagation:

  $\mathbf{P}_{\text{confl}}$: for any partial assignment $\hat{x}$ to X with $|\hat{x}| > k$, the formula F under $\hat{x}$ implies the empty clause (contradiction) with unit propagation (UP);

  $\mathbf{P}_{\text{extend}}$: for any partial assignment $\hat{x}$ to X with $|\hat{x}| = k$, the formula F under $\hat{x}$ assigns FALSE to all unassigned variables in X with UP (note that this property implies the first one, but not vice versa).

In short, a (c, a, p) *encoding* is an encoding with $c(n, k)$ clauses, $a(n, k)$ auxiliary variables and propagation strength $p \in \{\mathbf{P}_{\text{confl}}, \mathbf{P}_{\text{extend}}, \emptyset\}$.

The naive (often called *binomial*) encoding of $\leq_k(X_1, \ldots, X_n)$ has parameters $(\binom{n}{k+1}, 0, \mathbf{P}_{\text{extend}})$; it is a conjunction of all[2] $\binom{n}{k+1}$ clauses of the form $(\neg X_{i_1} \vee \cdots \vee \neg X_{i_{k+1}})$. The exponential dependance of its size on k makes the naive encoding impractical except for small n and k. The following table summarizes several interesting methods for encoding the $\leq_k(X_1, \ldots, X_n)$ constraint into CNF efficiently:

Name	parameters	origin
Sequential counter	$(O(kn), O(kn), \mathbf{P}_{\text{extend}})$	[Sin05]
Parallel counter	$(O(n), O(n), \emptyset)$	[Sin05]
Binary*	$(O(kn \log n), O(kn), \emptyset)$	[FPDN05, FG10]
Product	$(O(kn + k^2 n^{k/(k+1)}), O(kn^{k/(k+1)}), \mathbf{P}_{\text{extend}})$	[FG10]
Commander	$(O(2^{2k} n), O(kn), \mathbf{P}_{\text{extend}})$	[FG10]
Sorting networks	$(O(n \log^2 n), O(n \log^2 n), \mathbf{P}_{\text{extend}})$	[ES06]
Cardinality networks	$(O(n \log^2 k), O(n \log^2 k), \mathbf{P}_{\text{extend}})$	[ANOR09]
Totalizer	$(O(n^2), O(n \log^n), \mathbf{P}_{\text{extend}})$	[BB03]
Linear	$(O(n), O(n), \emptyset)$	[War98]

(* The binary encoding for $k = 1$ has parameters $(O(n \log n), \log n, \mathbf{P}_{\text{extend}})$ [FPDN05].) While the parallel-counter based encoding is smallest in size, it lacks the propagation strength that is crucial for SAT solver performance. On the other hand, encodings with

---

[2] Recall that we assume $k < n$.

strong propagation properties have a size that is super-linear in n, making them impractical for very large n and k. In their recent work, Frisch and Giannaros [FG10] discuss these tradeoffs, and survey and compare several known and new encodings. As [FG10] conclude, the sequential-counter based encoding from [Sin05] seems to perform better than other known encodings in practice (for problems that are not considered too small), and it is the method we use here as a benchmark against our encodings[3].

1.2 Our Contribution

PHF-Based Encoding. In Section 4, we propose a method for encoding cardinality constraints into CNF with perfect hash functions (PHFs) (see the definition in Section 2). The high-level idea behind this method is to reduce (using PHFs) the problem of encoding $\leq_k(X_1, \ldots, X_n)$ to several disjoint problems of encoding $\leq_k(Y_1, \ldots, Y_r)$ for $r \ll n$. From the special structure of PHFs, we get the following: 1) any (partial) assignment that satisfies the original constraint can be extended to an assignment that satisfies all the smaller constraints; 2) any assignment falsifying the original constraint must also falsify one of the smaller constraints. Hence, it suffices to enforce the smaller constraints only, using any of the known encodings at hand. Furthermore, this reduction inherits propagation strength and incrementality[4] of the used encodings. When constructed using the sequential-counter based method (on the reduced constraints), we obtain an encoding that uses the fewest number of auxiliary variables among all encodings listed in the table above. The parameters of our encodings are of the form $(nk^c \log n, k^c \log n, \mathbf{P}_{\text{extend}})$, where c depends on the PHF used (and can be as small as 4).

Encoding At-Least-k. In Section 4.2, we describe how to encode the $\geq_k(X_1, \ldots, X_n)$ constraint using PHFs. Although $\geq_k(X_1, \ldots, X_n)$ can be trivially reformulated in terms of an at-most type constraint (namely, $\leq_{n-k}(\neg X_1, \ldots, \neg X_n)$), the problem with this transformation is that for $1 < k \ll n/2$ it blows up the size of the encoding in all but the non-propagating methods. We show how PHF and sequential-counter based encodings can handle $\geq_k(X_1, \ldots, X_n)$ constraints natively, without any size penalties. In fact, the encodings we get have parameters $(k^c \log n, k^c \log n, \mathbf{P}_{\text{confl}})$, where c again is a small constant depending on the PHF used. Note that here both the number of clauses and the number of auxiliary variables are sublinear in n (however, the clauses themselves are larger).

Hybrid Encoding. While interesting from a theoretical point of view, our experience shows that the PHF based methods become practical only when k is significantly smaller than n (e.g., a small constant versus hundreds of thousands). This is because the

---

[3] The performance of the parallel-counter based encodings from [Sin05] was not measured in [FG10]. However, as our results in Section 6 indicate, despite being smallest in size, their lack in propagation strength makes them much worse in practice than the sequential-counter based encodings.

[4] Incrementality is the property of an encoding that allows tightening the cardinality bound by setting values to some variable(s). This property is useful when applying a sequence of decreasing/increasing cardinality constraints in the process of search for the optimal value.

number of copies of the reduced problem (corresponding to the number of hash functions in a perfect hash family) grows quadratically in k. To overcome the impracticality of the PHF based encodings, we propose a hybrid encoding, combining simple (non-perfect) hashing, parallel-counter, and sequential-counter based encodings. The idea here is similar: reduce one big problem to several small ones (but fewer than before), and enforce the smaller constraints with the sequential-counter method. In contrast to the previous construction, here we do not require perfect reduction; that is, we may have a situation where, under some assignment, the original constraint is falsified, but all the small ones to which we reduced are satisfied. Therefore, to make the encoding still correct, we add the parallel-counter based encoding on the original variables. Due to the exponential coverage vs the number-of-copies nature of hashing, we can guarantee that the strong propagation properties of the sequential-counter method are preserved for most of the assignments, even when reducing to only a single small problem. Namely, this hybrid encoding enjoys partial $\mathbf{P}_{\text{extend}}$ propagation strength, in the sense that most (but not all) of the partial assignments with $\geq k$ TRUEs are forcing propagation as required. However, it is much smaller in size (e.g., for $k \ll n$, an encoding that propagates most ($> 99\%$) of the partial assignments has only $O(n)$ clauses and auxiliary variables, and is nearly the same in size as the asymptotically-optimal parallel-counter based encoding). In addition, the hybrid encoding is simpler to implement. It can be viewed as a simple way to augment the parallel-counter encoding with propagation strength using a small number of small sequential counters.

Experimental Evaluation. The experimental evaluation in Section 6 compares various versions of the hybrid encoding to the sequential and parallel-counter based methods on a benchmark set containing encodings of different optimization and scheduling problems, kindly provided to us by the authors of [ANOR09].

It is clear that there is a sharp time-memory tradeoff between the counter-based encodings. Namely, whenever it is possible, with respect to memory limitations, to encode the constraints with the sequential-counter based method, the solver performance improves dramatically, compared to the parallel-counter based encoding. On the other hand, the parallel-counter based encoding is very efficient in terms of memory, but slower due to lack of propagation strength.

Our bottom-line conclusion is that with the hybrid encoding, we can enjoy both worlds: it is as fast as the sequential-counter based encoding (even faster), and its memory consumption is very close to that of the parallel-counter based encoding.

2   Perfect Hashing

An (n, ℓ, r, m) perfect hash family (PHF) is a collection $\mathcal{H} = h_1, \ldots, h_m$ of functions mapping $[n] = \{1, \ldots, n\}$ to $[r]$ that satisfies the following property: for every subset $S \subseteq [n]$ of size $|S| \leq \ell$, there is $i \in [m]$ such that $|h_i(S)| = |S|$. Namely, at least one of the functions hashes S perfectly, with no collisions. Fixing ℓ and r, we usually look for the smallest $m = m(n)$ for which an (n, ℓ, r, m) perfect hash family exists.

Naturally, the task becomes easier when r is larger than ℓ (and is impossible when $r < \ell$). For the case $r = \ell$, m can be bounded by $O(\ell e^\ell \log n)$. Allowing $r = \ell^2$,

m can be bounded by $O(\ell \log n)$. These upper bounds can be obtained using standard probabilistic arguments (for example, see [CLRS09]). However, constructing such families explicitly and efficiently imposes additional penalty on their size. See [KM03] and [FK84] for more details on explicit constructions of PHFs.

Before we describe how PHFs are useful for cardinality-constraint encodings, we sketch three simple constructions of PHFs for small ℓ and r. The first construction is straightforward, the second is (to our knowledge) new, and the third one was proposed to us by Noga Alon.

3 Perfect Hashing – Constructions

3.1 $(n, 2, 2, \lceil \log n \rceil)$ PHF

Set $m \triangleq \lceil \log n \rceil$. For $i \in [m]$ and $j \in [n]$, let $h_i(j)$ map to the ith bit of j, when j is written in binary base. Since every two numbers differ in at least one bit, the functions h_i form a $(n, 2, 2, \lceil \log n \rceil)$ PHF.

3.2 $(n, 3, 3, \lceil \log_3^2 n \rceil)$ PHF

For $i_1, i_2 \in [\log_3 n]$, and $j \in [n]$, write j in ternary base and define

$$h_{i_1, i_2}(j) = \begin{cases} (i_1\text{th digit of } j + i_2\text{th digit of } j) \mod 3, & i_1 < i_2 \\ i_1\text{th digit of } j, & i_1 = i_2 \\ (i_1\text{th digit of } j - i_2\text{th digit of } j) \mod 3, & i_1 > i_2 \end{cases}$$

We now prove the property of perfect hashing. Let j_1, j_2, and j_3 be three different indices $\in [n]$. We need to show the existence of i_1, i_2 such that

$$h_{i_1, i_2}(j_1) \neq h_{i_1, i_2}(j_2), \; h_{i_1, i_2}(j_1) \neq h_{i_1, i_2}(j_3),$$

and $h_{i_1, i_2}(j_2) \neq h_{i_1, i_2}(j_3)$. Since $j_1 \neq j_2$, there exists i_1 such that the i_1th digit of j_1 differs from the i_1th digit of j_2. If the i_1th digits of j_3 has the (remaining) third value, then h_{i_1, i_1} separates the three j's, and we are done. Otherwise, j_3 has the same i_1th digit as, w.l.o.g, j_1. So let i_2 be a digit in which j_1 differs from j_3. W.l.o.g $i_1 < i_2$. If all three j's have different i_2th digits, h_{i_2, i_2} separates the three j's, and we are done. Otherwise, j_2 has the same i_2th digit as either j_1 or j_3. W.l.o.g, the i_2th digit of j_2 is the same as the i_2th digit of j_1; this is depicted in the following table:

j	i_1th digit	i_2th digit
j_1	x	z
j_2	y	z
j_3	x	w

Here $x \neq y$ and $z \neq w$ are four ternary digits. Since $x + z \neq y + z$ and $x + z \neq x + w$, the only way h_{i_1, i_2} will not separate the three j's is if $y + z = x + w$. Similarly, the only way h_{i_2, i_1} will not separate j_1, j_2, and j_3 is if $y - z = x - w$. But adding these two equalities yields the contradiction $x = y$, so either h_{i_1, i_2} or h_{i_2, i_1} must separate j_1, j_2, and j_3.

3.3 $(n, \ell, \tilde{O}(\ell^2 \log n), \ell^2 \log n)$ PHF

This construction is useful in practice for $\ell \ll n$.

Let $p_1, \ldots, p_m$ be the first m prime numbers. For every $i \in [m]$, let the function h_i map $[n]$ to $\{0, \ldots, p_i - 1\}$ as follows:

$$h_i(j) = j \mod p_i.$$

Now let $S \subseteq [n]$ be a set that is not hashed perfectly by any of the functions h_i. This means that for every $i \in [m]$ there are $j \neq j' \in S$ so that

$$j = j' \mod p_i.$$

Equivalently, for all $i \in [m]$

$$p_i \mid \prod_{j < j' \in S} (j' - j),$$

and since p_i are primes,

$$p_1 \cdot p_2 \cdots p_m \mid \prod_{j < j' \in S} (j' - j).$$

Since $p_1 \cdot p_2 \cdots p_m > 2^m$ and $\prod_{j < j' \in S}(j' - j) < n^{|S|^2}$, we must have $m < |S|^2 \log n$. In other words, setting $m = \ell^2 \log n$ assures that every set of size $\leq \ell$ is perfectly hashed by some function h_i.

The largest prime used, p_m, is the upper bound on r in this construction, which by the Prime Number Theorem is bounded by $O(m \ln m)$. Combined with the inequality above, we get

$$r < O(|S|^2 \log n (\log |S| + \log \log n)) = \tilde{O}(\ell^2 \log n).$$

4 New Encodings Based on PHFs

In this section, we describe how to encode the $\leq_k (X_1, \ldots, X_n)$ and $\geq_k (X_1, \ldots, X_n)$ constraints into CNF using PHFs. We start with the general parameterized constructions, analyze them, and then instantiate them with several different parameters for comparison against known encodings.

4.1 Encoding the $\leq_k (X_1, \ldots, X_n)$ Constraint

Fix $r > k$ and an $(n, k + 1, r, m)$ perfect hash family $\mathcal{H} = h_1, \ldots, h_m : [n] \rightarrow [r]$. Perform the following steps for all $i \in [m]$:

1. Introduce r auxiliary variables $Y_1^i, \ldots, Y_r^i$.
2. Encode implications $X_j \rightarrow Y_{h_i(j)}^i$ (using binary clauses $(\neg X_j \vee Y_{h_i(j)}^i)$) for all $j \in [n]$.
3. Encode an $\leq_k (Y_1^i, \ldots, Y_r^i)$ constraint using the sequential-counter based encoding from [Sin05].

The final CNF encoding of $\leq_k (X_1, \ldots, X_n)$ is the conjunction of all the clauses generated in Step 2 and Step 3 for $i = 1, \ldots, m$.

Correctness. To verify that the encoding is correct, assume first that we start with an assignment x to $X = X_1, \ldots, X_n$ with weight $> k$. Let $b_1, \ldots, b_{k+1}$ be the first $k + 1$ indices corresponding to variables assigned TRUE in x. Since $\mathcal{H}$ is an $(n, k + 1, r, m)$ perfect hash family, there must be $i \in [m]$ so that $h_i(b_j) \neq h_i(b_{j'})$ for all $1 \leq j < j' \leq k + 1$. In addition, by the implication clauses introduced in Step 2, the $k + 1$ distinct Y_j^i variables to which $h_i(b_j)$ are mapped (for $j = 1, \ldots, k + 1$) are forced to be TRUE. Consequently, the $\leq_k(Y_1^i, \ldots, Y_r^i)$ constraint encoding introduced in Step 3 for that particular index i is falsified.

Now assume that we start with an assignment x to X with weight $\leq k$. Then, by construction, each one of the i copies of the auxiliary variables can be assigned values of overall weight $\leq k$, and furthermore, the corresponding $\leq_k(Y_1^i, \ldots, Y_r^i)$ constraints can all be simultaneously satisfied.

Size. The encoding requires $m \cdot n + m \cdot O(r \cdot k)$ clauses (the first term comes from Step 2, and the second term comes from Step 3) and it uses $m \cdot r + m \cdot O(r \cdot k)$ auxiliary variables (the first term is the Y variables, and the second term comes from the sequential counter encodings on the Y's).

Propagation strength. The encoding is $\mathbf{P}_{\text{extend}}$: Suppose that a partial assignment $\hat{x}$ to X has exactly k variables set to TRUE. We show that unit propagation sets all other variables to FALSE. Let $b_1, \ldots, b_k$ be the k indices corresponding to variables assigned TRUE, and let b_{k+1} be the index of any other variable. Using the perfect hashing property, take h_i so that $h_i(b_j) \neq h_i(b_{j'})$ for all $1 \leq j < j' \leq k + 1$. The implications clauses of Step 2 imply by unit propagation a partial assignment $\hat{y}$ on $Y_1^i, \ldots, Y_r^i$ with $Y_{h_i(b_j)}^i$ set to TRUE for $1 \leq j \leq k$. Since the sequential counter (Step 3) is $\mathbf{P}_{\text{extend}}$, unit propagation extends $\hat{y}$ by assigning FALSE to all other variables in that copy, including the variable $Y_{h_i(b_{k+1})}^i$. This triggers again the binary implications from Step 2, and unit propagation assigns $X_{b_{k+1}}$ to FALSE.

Instantiation. Using the probabilistic[5] PHFs with $\ell = k+1, r = \ell^2$ and $m = O(\ell \log n)$ (see Section 2), we get $(O(nk \log n + k^4 \log n), O(k^4 \log n), \mathbf{P}_{\text{extend}})$ encoding for $\leq_k(X_1, \ldots, X_n)$. Compared, e.g., to the sequential-counter based encoding, this construction uses slightly more clauses (a factor of $\log n$), but for large enough n it consumes a significantly smaller number of variables ($k^4 \log n$ vs kn). Furthermore, apart from the naive encoding (whose size is exponential in k) this encoding uses the fewest number of auxiliary variables among all encoding methods listed above.

Instantiating this construction with $k = 1$ and $\ell = r = 2$ based on the corresponding PHF from Section 3.1, we get $(n \log n, \log n, \mathbf{P}_{\text{extend}})$ encoding for $\leq_1(X_1, \ldots, X_n)$, with properties similar to the binary encoding [FPDN05]. The PHFs from Sections 3.2 (for $k = 2$) and 3.3 (for any k) yield to $(n \log^2 n, \log^2 n, \mathbf{P}_{\text{extend}})$ and $(O(k^2 n \log n + k^5 \log^2 n(\log k + \log \log n)), O(k^5 \log n), \mathbf{P}_{\text{extend}})$ encodings, respectively.

---

[5] We stress that probabilistic arguments are only required to prove the existence of such families; in practice, one can find such PHFs by brute-force search, and then hard-code them in the software once and for all. In the cost of size penalty, there are also derandomized versions of the probabilistic constructions [AN96].

4.2 Encoding the $\geq_k(X_1, \ldots, X_n)$ Constraint

The straightforward way to implement the at-least constraints is to negate the at-most constraints: having at least k TRUEs out of n variables is equivalent to having at most $n - k$ TRUEs out of the n negated literals. The problem with this simple transformation is that moving from $k \ll n$ to $k \approx n$ imposes a big size penalty for all but the non-propagating encodings (this is due to the fact that encoding size grows with k). In this section we show how the PHF-based encodings can handle the $\geq_k(X_1, \ldots, X_n)$ constraint natively. To this end, we need to invert the sequential-counter based encoding.

A Sequential-Counter Based Encoding for the $\geq_k(X_1, \ldots, X_n)$ Constraint

Construction. The sequential-counter circuit described in [Sin05] is an example of a monotone circuit with n inputs $X_1, \ldots, X_n$ and a single output Z computing the $\geq_k(X_1, \ldots, X_n)$ constraint. In other words, given a full assignment to all of the X_is, Z evaluates to TRUE if and only if the number of X_is assigned to TRUE is at least k. One can convert this circuit into CNF via the Tseitsin encoding (see also Theorem 2 from [BKNW09] and Theorems 7 and 8 from [Bai11] for the connection between monotone circuits and the corresponding CNF encodings), and to add the unit clause (Z) enforcing the output of the circuit to be TRUE. The resulting CNF has $O(kn)$ clauses, $O(kn)$ auxiliary variables[6], correctly encodes the $\geq_k(X_1, \ldots, X_n)$ constraint, and is $\mathbf{P}_{\text{extend}}$.

Optimization. One can optimize the above construction due to polarity considerations, encoding the $\geq_k(X_1, \ldots, X_n)$ constraint as follows:

- Introduce the auxiliary variables $Y_{i,j}$ for all $i \in [n]$ and $j \in [k]$.
- Introduce the clauses:
    1. $(\neg Y_{1,1} \vee X_1)$,
    2. $(\neg Y_{1,j})$ for for $1 < j \leq k$,
    3. $(\neg Y_{i,j} \vee Y_{i-1,j-1})$ for $1 < i \leq n$ and $1 < j \leq k$,
    4. $(\neg Y_{i,j} \vee Y_{i-1,j} \vee X_i)$ for $1 < i \leq n$ and $1 \leq j \leq k$,
    5. $(Y_{n,k})$.

The Boolean variables $Y_{i,j}$ represent the statements "at least j out of $X_1, \ldots, X_i$ are TRUE". The clauses from (3) enforce the implications $Y_{i,j} \to Y_{i-1,j-1}$, and the clauses from (4) enforce the implications $Y_{i,j} \to (Y_{i-1,j} \vee X_i)$. The unit clause (5) corresponds to the "output of the circuit". (The unit clause is not added when using this encoding as a building block.) This version of the $\geq_k(X_1, \ldots, X_n)$ encoding has roughly $2kn$ clauses and kn auxiliary variables, and is $\mathbf{P}_{\text{extend}}$.

A PHF-Based Encoding for the $\geq_k(X_1, \ldots, X_n)$ Constraint

Fix $r > k$ and an (n, k, r, m) perfect hash family $\mathcal{H} = h_1, \ldots, h_m : [n] \to [r]$. Perform the following steps for each $i \in [m]$:

1. Introduce an auxiliary variable Z_i and r auxiliary variables $Y_1^i, \ldots, Y_r^i$.
2. Introduce implications $Y_d^i \to (\bigcup_{h_i(j)=d} X_j)$ for all $d \in [r]$.

---

[6] Here we can see the advantage of having $k \ll n$.

3. Encode the $\geq_k(Y_1^i, \ldots, Y_r^i)$ using the sequential-counter based construction described above. Let Z_i denote the variable corresponding to the circuit's output.

The final CNF encoding is the conjunction of the clauses in Step 2, the clauses in Step 3 for $i = 1, \ldots, m$, and the clause $(\bigcup_{i=1}^m Z_i)$.

Correctness. Let x be an assignment to $X_1, \ldots, X_n$ with weight $\geq k$, and let $b_1, \ldots, b_s$ ($s \geq k$), be the indices corresponding to the variables assigned TRUE in x. We can extend this assignment to Y_j^is by assigning Y_j^i to TRUE whenever $Y_j^i = h_i(b_j)$ for some j; this assignment satisfies all of the clauses in Step 2. This assignment (uniquely) extends to satisfy all of the clauses in Step 3, and for every i, Z_i is true if and only if the weight of the assignment to $Y_1^i, \ldots, Y_m^i$ is at least k. Since H is an (n, k, r, m) PHF, there exists at least one $i \in [m]$ with $h_i(b_j) \neq h_i(b_{j'})$ for all $1 \leq j < j' \leq k$. Consequently, for that index i, Z_i is TRUE and the remaining clause $(\bigcup_{i=1}^m Z_i)$ is also satisfied.

For the other direction, assume that x has weight $< k$. Then by Step 2 for all $i \in [m]$, there are at most $k - 1$ values of d for which Y_d^i can be TRUE. It follows by Step 3 that each Z_i is always FALSE, in conflict with the clause $(\bigcup_{i=1}^m Z_i)$.

Properties. The encoding has $O(m \cdot r \cdot k)$ clauses[7] and $O(m \cdot r \cdot k)$ variables, and is $\mathbf{P}_{\text{confl}}$. The $\mathbf{P}_{\text{confl}}$ property can be shown directly, but we do not present it here since it also follows by Theorem 2 from [BKNW09] and Theorems 7 and 8 from [Bai11]. We leave as an open question if it is possible to modify the construction to be able to enforce the stronger $\mathbf{P}_{\text{extend}}$ property as well.

Instantiation. Using the probabilistic PHFs with $\ell = k+1$, $r = \ell^2$ and $m = O(\ell \log n)$ (see Section 2), we get $(O(k^4 \log n), O(k^4 \log n), \mathbf{P}_{\text{confl}})$ encoding for $\geq_k(X_1, \ldots, X_n)$. The PHFs from Sections 3.2 (for $k = 2$) and 3.3 (for any k) yield $(\log^2 n, \log^2 n, \mathbf{P}_{\text{confl}})$ and $(O(k^5 \log n), O(k^5 \log n), \mathbf{P}_{\text{confl}})$ encodings, respectively.

5   Hybrid Encodings

In this section we show how to augment the parallel-counter based encoding (which is nearly optimal in size) with partial propagation strength, using (non-perfect) hash functions.

Given c hash functions $\mathcal{H} = \{h_i : [n] \to [r]\}_{i=1}^c$, a hybrid encoding of the $\leq_k(X_1, \ldots, X_n)$ constraint is formed as follows:

- Enforce the $\leq_k(X_1, \ldots, X_n)$ constraint on $X_1, \ldots, X_n$ using the parallel-counter based method.
- For each $1 \leq i \leq c$:
  - Add implications $X_j \implies Y_{h_i[j]}^i$ for all $j \in [n]$.
  - Enforce the $\leq_k(Y_1^i, \ldots, Y_r^i)$ constraint on $Y_1^i, \ldots, Y_r^i$ using one of the known methods.

---

[7] Note that the number of clauses is significantly smaller than for the at-most constraint, however the clauses themselves are larger.

The correctness of this encoding follows by construction. It has $O\left(c \cdot (n + cl(k, r))\right)$ clauses and uses $O\left(n + c \cdot aux(k, r)\right)$ auxiliary variables, where $cl(k, r)$ and $aux(k, r)$ are the number of clauses and auxiliary variables used by the method of encoding applied to enforce each of the $\leq_k(Y_1^i, \ldots, Y_r^i)$ constraints. When using one of the encodings with $\mathbf{P}_{\text{extend}}$ we also inherit a "partial $\mathbf{P}_{\text{extend}}$", where ρ-fraction of all partial assignments with weight $\leq k$ are propagating as required. The parameter ρ depends on c, r, k and n, and a crude lower bound on it can be obtained with elementary calculation. In practice, however, we observed that setting $r = 2k$ has already a positive effect on propagation strength, even when using only one hash function (i.e. with $c = 1$). In Section 6 we evaluate hybrid encodings that use the sequential-counter based method to enforce the constraints $\leq_k(Y_1^i, \ldots, Y_r^i)$ with $c = 1, 2, 3, 5$ and $r = 2k$.

6   Experiments

In this section we compare various versions of the hybrid encoding to the sequential and parallel-counter based encodings[8] on a benchmark set produced from the Partial Max-SAT division of the Third Max-SAT evaluation[9]. The benchmarks were produced (and kindly provided to us) by the authors of [ANOR09]. Out of roughly 14,000 benchmarks we extracted all those (1344) instances containing at least one cardinality constraint with $1 < k < n$ and $n > 10k$.

6.1   Setting

We ran the experiments on a Linux based workstation with Intel Xeon E5430 processor and 8GB of RAM. The encoding time was not counted, and it was negligible for all but the sequential-counter based method (where it was still small compared to solving time). As a SAT solver we used *Minisat 2.2.0* (with preprocessing enabled).

6.2   Results – Conclusion

Our bottom-line conclusion is that with the simplest version of the hybrid encoding, where $r = 2k$ and $c = 1$ (see Section 5), we can enjoy both of the advantages of the counter based encodings: the hybrid encoding is even faster than the sequential-counter based encoding, and its memory consumption is very close to that of the parallel-counter based encoding.

Another interesting observation is that the performance of hybrid encodings deteriorates (quite consistently) when the number of hash functions increase. This may be explained by the fact that the number of new tuples perfectly hashed by each additional hash function decreases exponentially faster than the increase in the encoding size.

---

[8] As a byproduct, our experiments compare the two counter-based methods from [Sin05] among themselves.

[9] See http://www.maxsat.udl.cat/08/
index.php?disp=submitted-benchmarks.

6.3 Results – Details

The following notation is used in all tables/plots below.

- The sequential and parallel-counter based encodings are denoted by Seq and Par respectively.
- Hybrid encodings with $r = 2k$ and 1,2,3, and 5 hash functions (copies), all using the sequential-counter based encoding for the small constraints, are denoted by HybSeq1, HybSeq2, HybSeq3, and HybSeq5 respectively.
- A hybrid encoding with $r = 2k$ and 1 hash function, using the parallel-counter based encoding for the small constraints, is denoted by HybPar1.

We analyze several aspects of encodings' performance:

Table 1. Cell i, j indicates the number of instances (out of 1344) for which the encoding in the ith row is strictly faster than the encoding in the jth column

	Seq	HybSeq1	HybSeq2	HybSeq3	HybSeq5	Par	HybPar1
Seq	0	362	479	626	909	803	491
HybSeq1	963	0	919	1028	1158	918	723
HybSeq2	847	398	0	938	1138	803	553
HybSeq3	702	301	376	0	1060	743	464
HybSeq5	428	175	191	270	0	645	284
Par	520	403	518	589	687	0	355
HybPar	830	590	776	866	1044	965	0

Table 2. Counting for each encoding the number of instances on which it was faster than all other encodings

	Seq	HybSeq1	HybSeq2	HybSeq3	HybSeq5	Par	HybPar1
#fastest	173	422	179	83	19	142	326

Table 3. Means and medians of various measures for each encoding

	Seq	HybSeq1	HybSeq2	HybSeq3	HybSeq5	Par	HybPar1
Medians							
Run time (seconds)	1.54	1.09	1.31	1.48	1.72	1.68	1.13
Number of clauses	186886	107678	127632	147978	181534	88876	95360
Number of variables	116624	72544	80354	88898	103414	65042	65828
Number of decisions	1835	15128	13280	14636	16536	103340	43192
Number of conflicts	208	2470	2214	2128	2126	9992	5429
Memory consumption (MB)	39	24	26	29	34	22	23
Means							
Run time (seconds)	16.57	15.08	15.68	17.34	21.76	31.76	18.9
Number of clauses	560204	218825	296379	371470	521646	144145	158424
Number of variables	391897	192718	228558	264372	336001	156930	158953
Number of decisions	42825	438311	185851	179692	181976	1991480	799956
Number of conflicts	23596	41508	37271	40083	42076	113130	58261
Memory consumption (MB)	123	52	66	77	101	42	43

- Size of the resulting CNF formula, as reflected by the number of clauses, the number of variables, and memory consumption of the SAT solver.
- Solver run time (counting parsing and preprocessing as well).
- Propagation strength, as reflected by the number of decisions and the number of conflicts during the run of the SAT solver.

7   Concluding Remarks and Future Work

We presented new methods to encode cardinality constraints into CNF based on perfect hashing. This approach, coupled with existing methods, leads to encodings with sublinear number of clauses and auxiliary variables, and strong propagation properties.

From a practical perspective, we proposed to use the hybrid approach (with non-perfect hashing) to boost the performance of existing counter based encodings.

We list the following directions for further research.

Optimizing the Ingredients: Several components of our encodings can be tuned or replaced for performance optimization: the underlying encoding applied on the reduced problems, the values of r and c (see Sections 2 and 5), and the underlying perfect hash family constructions. The amount of freedom is large, and we are convinced that the experimental results can be further improved.

Application to More General Constraints: It is possible that similar approach, based on perfect hash families and other related combinatorial structures like block designs, can be used for efficient encoding of other Pseudo-Boolean constraints.

Acknowledgments. We thank the authors of [ANOR09] for sharing with us their benchmark set, and Noga Alon for proposing the construction in Section 3.3. We also thank the anonymous reviewer for pointing out [BKNW09] and [Bai11], whose results were partially reproved in the initial version of this manuscript.

References

[AN96]   Alon, N., Naor, M.: Derandomization, witnesses for Boolean matrix multiplication and construction of perfect hash functions. Algorithmica 16(4/5), 434–449 (1996)

[ANOR09]   Asín, R., Nieuwenhuis, R., Oliveras, A., Rodríguez-Carbonell, E.: Cardinality Networks and Their Applications. In: Kullmann, O. (ed.) SAT 2009. LNCS, vol. 5584, pp. 167–180. Springer, Heidelberg (2009)

[Bai11]   Bailleux, O.: On the expressive power of unit resolution. Technical Report arXiv:1106.3498 (June 2011)

[BB03]   Bailleux, O., Boufkhad, Y.: Efficient CNF Encoding of Boolean Cardinality Constraints. In: Rossi, F. (ed.) CP 2003. LNCS, vol. 2833, pp. 108–122. Springer, Heidelberg (2003)

[BKNW09] Bessiere, C., Katsirelos, G., Narodytska, N., Walsh, T.: Circuit complexity and decompositions of global constraints. In: Proceedings of the 21st International Joint Conference on Artifical Intelligence, IJCAI 2009, pp. 412–418. Morgan Kaufmann Publishers Inc., San Francisco (2009)

[CLRS09]   Cormen, T.H., Leiserson, C.E., Rivest, R.L., Stein, C.: Introduction to Algorithms, 3rd edn. MIT Press (2009)

[ES06]   Eén, N., Sörensson, N.: Translating Pseudo-Boolean constraints into SAT. JSAT 2(1-4), 1–26 (2006)

[FG10]   Frisch, A.M., Giannaros, P.A.: SAT encodings of the at-most-k constraint. In: ModRef (2010)

[FK84]   Fredman, M.L., Komlós, J.: On the size of separating systems and families of perfect hash functions. SIAM Journal on Algebraic and Discrete Methods 5(1), 61–68 (1984)

[FPDN05]   Frisch, A.M., Peugniez, T.J., Doggett, A.J., Nightingale, P.: Solving non-Boolean satisfiability problems with stochastic local search: A comparison of encodings. J. Autom. Reasoning 35(1-3), 143–179 (2005)

[KM03]   Kim, K.-M.: Perfect hash families: Constructions and applications. Master Thesis (2003)

[Sin05]   Sinz, C.: Towards an Optimal CNF Encoding of Boolean Cardinality Constraints. In: van Beek, P. (ed.) CP 2005. LNCS, vol. 3709, pp. 827–831. Springer, Heidelberg (2005)

[War98]   Warners, J.P.: A linear-time transformation of linear inequalities into conjunctive normal form. Inf. Process. Lett. 68(2), 63–69 (1998)

The Community Structure of SAT Formulas[*]

Carlos Ansótegui[1], Jesús Giráldez-Cru[2], and Jordi Levy[2]

[1] DIEI, Univ. de Lleida
carlos@diei.udl.cat
[2] IIIA-CSIC
{jgiraldez,levy}@iiia.csic.es

Abstract. The research community on complex networks has developed techniques of analysis and algorithms that can be used by the SAT community to improve our knowledge about the structure of industrial SAT instances. It is often argued that modern SAT solvers are able to exploit this *hidden structure*, without a precise definition of this notion.

In this paper, we show that most industrial SAT instances have a high modularity that is not present in random instances. We also show that successful techniques, like learning, (indirectly) take into account this community structure. Our experimental study reveal that most learnt clauses are local on one of those modules or communities.

1 Introduction

In recent years, SAT solvers efficiency solving industrial instances has undergone a great advance, mainly motivated by the introduction of lazy data-structures, learning mechanisms and activity-based heuristics [11,18]. This improvement is not shown when dealing with randomly generated SAT instances. The reason for this difference seems to be the existence of a structure in industrial instances [25].

In parallel, there have been significant advances in our understanding of complex networks, a subject that has focused the attention of statistical physicists. The introduction of these network analysis techniques could help us to understand the nature of SAT instances, and could contribute to further improve the efficiency of SAT solvers. Watts and Strogatz [24] introduce the notion of *small world*, the first model of complex networks, as an alternative to the classical random graph models. Walsh [23] analyzes the small world topology of many graphs associated with search problems in AI. He also shows that the cost of solving these search problems can have a *heavy-tailed distribution*. Gomes et al. [14,15] propose the use of *randomization* and *rapid restart* techniques to prevent solvers from falling on the long tail of such kinds of distributions.

The notion of structure has been addressed in previous work [14,16,13,17,3]. In [22] it is proposed a method to generate more realistic random SAT problems based on the notions of *characteristic path length* and *clustering coefficient*. Here

---

[*] This research has been partially founded by the CICYT research projects TASSAT (TIN2010-20967-C04-01/03) and ARINF (TIN2009-14704-C03-01).

A. Cimatti and R. Sebastiani (Eds.): SAT 2012, LNCS 7317, pp. 410–423, 2012.

we use a distinct notion of modularity. In [6], it is shown that many SAT instances can be decomposed into *connected components*, and how to handle them within a SAT solver. They discuss how this component structure can be used to improve the performance of SAT solvers. However, their experimental investigation shows that this is not enough to solve more efficiently SAT instances. The notion of *community* is more general than the notion of *connected components*. In particular, it allows the existence of (a few) connections between communities. As we discus later, industrial SAT instances use to have a connected component containing more than the 99% of the variables. Also, in [1] some techniques are proposed to reason with multiple knowledge bases that overlap in content. In particular, they discuss strategies to induce a partitioning of the axioms, that will help to improve the efficiency of reasoning.

In this paper we propose the use of techniques for detecting the *community structure* of SAT instances. In particular, we apply the notion of *modularity* [19] to detect these communities. We also discuss how existing conflict directed clause learning algorithms and activity-based heuristics already take advantage, *indirectly*, of this community structure. Activity-based heuristics [18] rely on the idea of giving higher priority to the variables that are involved in (recent) conflicts. By focusing on a sub-space, the covered spaces tend to coalesce, and there are more opportunities for resolution since most of the variables are common.

2   Preliminaries

Given a set of Boolean variables $X = \{x_1, \ldots, x_n\}$, a *literal* is an expression of the form x_i or $\neg x_i$. A *clause* c of length s is a disjunction of s literals, $l_1 \vee \ldots \vee l_s$. We say that s is the size of c, noted $|c|$, and that $x \in c$, if c contains the literal x or $\neg x$. A *CNF formula* or *SAT instance* of length t is a conjunction of t clauses, $c_1 \wedge \ldots \wedge c_t$.

An (undirected) graph is a pair (V, w) where V is a set of vertices and $w : V \times V \to \mathbb{R}^+$ satisfies $w(x, y) = w(y, x)$. This definition generalizes the classical notion of graph (V, E), where $E \subseteq V \times V$, by taking $w(x, y) = 1$ if $(x, y) \in E$ and $w(x, y) = 0$ otherwise. The degree of a vertex x is defined as $\deg(x) = \sum_{y \in V} w(x, y)$. A bipartite graph is a tuple (V_1, V_2, w) where $w : V_1 \times V_2 \to \mathbb{R}^+$.

Given a SAT instance, we construct two graphs, following two models. In the Variable Incidence Graph model (VIG, for short), vertices represent variables, and edges represent the existence of a clause relating two variables. A clause $x_1 \vee \ldots \vee x_n$ results into $\binom{n}{2}$ edges, one for every pair of variables. Notice also that there can be more than one clause relating two given variables. To preserve this information we put a higher weight on edges connecting variables related by more clauses. Moreover, to give the same relevance to all clauses, we ponderate the contribution of a clause to an edge by $1/\binom{n}{2}$. This way, the sum of the weights of the edges generated by a clause is always one. In the Clause-Variable Incidence Graph model (CVIG, for short), vertices represent either variables or clauses, and edges represent the occurrence of a variable in a clause. Like in the VIG model, we try to assign the same relevance to all clauses, thus every edge connecting a variable x with a clause C containing it has weight $1/|C|$. This way, the sum of the weights of the edges generated by a clause is also one in this model.

Definition 1 (Variable Incidence Graph (VIG)). *Given a SAT instance Γ over the set of variables X, its variable incidence graph is a graph (X, w) with set of vertices the set of Boolean variables, and weight function:*

$$w(x,y) = \sum_{\substack{c \in \Gamma \\ x,y \in c}} \frac{1}{\binom{|c|}{2}}$$

Definition 2 (Clause-Variable Incidence Graph (CVIG)). *Given a SAT instance Γ over the set of variables X, its clause-variable incidence graph is a bipartite graph $(X, \{c \mid c \in \Gamma\}, w)$, with vertices the set of variables and the set of clauses, and weight function:*

$$w(x,c) = \begin{cases} 1/|C| & \text{if } x \in c \\ 0 & \text{otherwise} \end{cases}$$

3   Modularity in Large-Scale Graphs

To analyze the structure of a SAT instance we will use the notion of *modularity* introduced by [20]. This property is defined for a graph and a specific *partition* of its vertices into *communities*, and measures the adequacy of the partition in the sense that most of the edges are within a community and few of them connect vertices of distinct communities. The modularity of a graph is then the maximal modularity for all possible partitions of its vertices. Obviously, measured this way, the maximal modularity would be obtained putting all vertices in the same community. To avoid this problem, Newman and Girvan define modularity as the fraction of edges connecting vertices of the same community minus the expected fraction of edges for a random graph with the same number of vertices and same degree.

Definition 3 (Modularity of a Graph). *Given a graph $G = (V, w)$ and a partition $P = \{P_1, \ldots, P_n\}$ of its vertices, we define their modularity as*

$$Q(G, P) = \sum_{P_i \in P} \frac{\sum\limits_{x,y \in P_i} w(x,y)}{\sum\limits_{x,y \in V} w(x,y)} - \left(\frac{\sum\limits_{x \in P_i} \deg(x)}{\sum\limits_{x \in V} \deg(x)} \right)^2$$

We call the first term of this formula the inner edges fraction, *IEF for short, and the second term the* expected inner edges fraction, *IEF^e for short. Then, $Q = IEF - IEF^e$.*

The (optimal) modularity of a graph is the maximal modularity, for any possible partition of its vertices: $Q(G) = \min\{Q(G, P) \mid P\}$.

Since the IEF and the IEF^e of a graph are both in the range $[0, 1]$, and, for the partition given by a single community, both have value 1, the optimal modularity of graph will be in the range $[0, 1]$. In practice, Q values for networks showing a strong community structure range from 0.3 to 0.7, higher values are rare [20].

There has not been an agreement on the definition of modularity for bipartite graphs. Here we will use the notion proposed by [4] that extends Newman and Girvan's definition by restricting the random graphs used in the computation of the IEF^e to be bipartite. In this definition, communities may contain vertices of V_1 and of V_2.

Definition 4 (Modularity of a Bipartite Graph). *Given a graph* $G = (V_1, V_2, w)$ *and a partition* $P = \{P_1, \ldots, P_n\}$ *of its vertices, we define their modularity as*

$$Q(G, P) = \sum_{P_i \in P} \frac{\sum_{\substack{x \in P_i \cap V_1 \\ y \in P_i \cap V_2}} w(x, y)}{\sum_{\substack{x \in V_1 \\ y \in V_2}} w(x, y)} - \frac{\sum_{x \in P_i \cap V_1} deg(x)}{\sum_{x \in V_1} deg(x)} \cdot \frac{\sum_{y \in P_i \cap V_2} deg(y)}{\sum_{y \in V_2} deg(y)}$$

There exist a wide variety of algorithms for computing the modularity of a graph. Moreover, there exist alternative notions and definitions of modularity for analyzing the community structure of a network. See [12] for a survey in the field. The decision version of modularity maximization is NP-complete [8]. All the modularity-based algorithms proposed in the literature return an approximated lower bound for the modularity. They include greedy methods, methods based on simulated annealing, on spectral analysis of graphs, etc. Most of them have a complexity that make them inadequate to study the structure of an industrial SAT instance. There are algorithms specially designed to deal with large-scale networks, like the greedy algorithms for modularity optimization [19,9], the label propagation-based algorithm [21] and the method based on graph folding [7].

The first described algorithm for modularity maximization is a *greedy method* of Newman [19]. This algorithm starts by assigning every vertex to a distinct community. Then, it proceeds by joining the pair of communities that result in a bigger increase of the modularity value. The algorithm finishes when no community joining results in an increase of the modularity. In other words, it is a greedy gradient-guided optimization algorithm. The algorithm may also return a dendogram of the successive partitions found. Obviously, the obtained partition may be a local maximum. In [9] the data structures used in this basic algorithm are optimized, using among other data structures for sparse matrices. The complexity of this refined algorithm is $\mathcal{O}(m\, d\, \log n)$, where d is the depth of the dendogram (i.e. the number of joining steps), m the number of edges and n the number of vertices. They argue that d may be approximated by $\log n$, assuming that the dendogram is a balanced tree, and the sizes of the communities are similar. However, this is not true for the graphs we have analyzed, where the sizes of the communities are not homogeneous. This algorithm has not been able to finish, for none of our SAT instances, with a run-time limit of one hour.

An alternative algorithm is the *Label Propagation Algorithm (LPA)* proposed by [21] (see Algorithm 1). Initially, all vertices are assigned to a distinct label, e.g., its identifier. Then, the algorithm proceeds by re-assigning to every vertex the label that is more frequent among its neighbors. The procedure ends when every vertex is assigned a label that is maximal among its neighbors. The order in which the vertices update their labels in every iteration is chosen randomly. In case of a tie between maximal labels, the winning label is also chosen randomly. The algorithm returns the partition defined by the vertices sharing the same label. The label propagation algorithm has a near linear complexity. However, it has been shown experimentally that the partitions it computes have a worse modularity than the partitions computed by the Newman's greedy algorithm.

The *Graph Folding Algorithm (GFA)* proposed in [7] (see Algorithm 2) improves the Label Propagation Algorithm in two directions. The idea of moving one node from one community to another following a greedy strategy is the same, but, instead of selecting the community where the node has more neighbors, it selects the community where the movement would most increase the modularity. Second, once no movement of node from community to community can increase the modularity (we have reached a (local) modularity maximum), we allow to merge communities. For this purpose we construct a new graph where nodes are the communities of the old graph, and where edges are weighted with the sum of the weights of the edges connecting both communities. Then, we apply again the greedy algorithm to the new graph. This folding process is repeated till no modularity increase is possible.

4   Modularity of SAT Instances

We have computed the modularity of the SAT instances used in the 2010 SAT Race Finals (see http://baldur.iti.uka.de/sat-race-2010/). They are 100 instances grouped into 16 families. These families are also classified as cryptography, hardware verification, software verification and mixed, according to their application area. All instances are *industrial*, in the sense that their solubility has an industrial or practical application. However, they are expected to show a distinct nature.

We have observed that all instances of the same family have a similar modularity. Therefore, in Table 1, we only show the median of these values. We present the modularities obtained by LPA on the graphs VIG and CVIG, and by GFA on the graphs VIG. We have re-implemented both algorithms, and in the case of LPA, we have developed a new algorithm adapted for bi-partite graphs. The GFA algorithm is not yet adapted for bipartite graphs. We also study the *connected components* as in [6].

We have to remark that both algorithms give a lower bound on the modularity, hence we can take the maximum of both measures as a lower bound. Having this in mind, we can conclude that, except for the grieu family, all families show a clear community structure with values of Q around 0.8. In other kind of networks values greater than 0.7 are rare, therefore the values obtained for SAT instances can be considered as exceptionally high.

Algorithm 1: Label Propagation Algorithm (LPA). The function most_freq_label returns the label that is most frequent among a set of vertices. In case of tie, it randomly chooses one of the maximal labels.

---

1 **function** most_freq_label(v, N)
2     $SL := \{L[v] \mid v \in N\}$;
3   **for** $l \in SL$ **do**
4   freq$[l] := \sum_{\substack{v' \in N \\ l = L[v']}} w(v, v')$
5   Max $:= \{l \in SL \mid \text{freq}[l] = \max\{\text{freq}[l] \mid l \in SL\}\}$;
6   **return** random_choose(Max)

Input: Graph $G = (X, w)$
Output: Label L

7 **for** $x \in X$ **do** $L[x] := x$; $freq[x] := 0$;;
8 **repeat**
9   ord $:=$ shuffle(X);
10   changes $:=$ false;
11   **for** $i \in X$ **do**
12       $(l, f) :=$ most_freq_label(i,neighbors(i));
13   changes $:=$ changes $\lor f > \text{freq}[i]$;
14   L$[i] := l$;
15   freq$[i] := f$
16 **until** $\neg changes$;

As one could expect, we obtain better values with GFA than with the LPA algorithm. The reason for this better performance is that, whereas in the LPA we use the most frequent label among neighbors (in order to assign a new community to a node), in the GFA we select the label leading to a bigger increase in the modularity. The latter is clearly a better strategy for obtaining a bigger resulting modularity. Moreover, in the GFA a further step is added where communities can be merged, when no movement of a single node from one community to another leads to a modularity increase.

If we compare the modularity values for the VIG model (obtained with the LPA) with the same values for the CVIG model, we can conclude that, in general, these values are higher for the CVIG model. It could be concluded that the loss of information, during the *projection* of the bipartite CVIG graph into the VIG graph, may destroy part of the modular structure. However, this is not completely true. Suppose that the instance has no modular structure at all, but all clauses are binary. We can construct a partition as follows: put every variable into a distinct community, and every clause into the same community of one of its variables. Using this partition, half of the edges will be internal, i.e. IEF $= 0.5$, IEF^e will be nearly zero, and $Q \approx 0.5$. Therefore, we have to take into account that using Barber's modularity definition for bipartite graphs, as we do, if vertex degrees are small, modularity can be quite big compared with Newman's modularity.

Algorithm 2: Graph Folding algorithm (GFA)

---

1 **function** $OneLevel(GraphG = (X, w)) : Label\ L$
2   **foreach** $i \in X$ **do** $L[i] := i$ **repeat**
3     $changes := false;$
4   **foreach** $i \in X$ **do**
5       $bestinc := 0;$
6   **foreach** $c \in \{c \mid \exists j. w(i, j) \neq 0 \wedge L[j] = c\}$ **do**
7         $inc :=$
          $\sum_{L(j)=c} w(i, j) - arity(i) \cdot \sum_{L[j]=c} arity(j) / \sum_{j \in X} arity(j);$
8   **if** $inc > bestinc$ **then**
9           $L[i] := c; bestinc := inc; changes := true;$

10   **until** $\neg changes;$
11   **return** $L;$

12 **function** $Fold(Graph\ G_1, Label\ L) : Graph\ G_2$
13   $X_2 = \{c \mid \forall i, j \in c. L[i] = L[j];$
14   $w_2(c_1, c_2) = \sum_{i \in c_1, j \in c_2} w_2(i, j);$
15   **return** $G_2 = (X_2, w_2);$

Input: Graph $G = (X, w)$
Output: Label L_1
16 **foreach** $i \in X$ **do** $L_1[i] := i;$
17 $L_2 := OneLevel(G);$
18 **while** $Modularity(G, L_1) < Modularity(G, L_2)$ **do**
19   $L_1 := L1 \circ L2;$
20   $G = Fold(G, L_2);$
21   $L_2 := OneLevel(G);$

We also report results on the number of communities ($|P|$) and the fraction of vertices belonging to the largest community ($larg$) expressed as a percentage. If all communities have a similar size, then $larg \approx 1/|P|$. In some cases, like palacios and mizh, we have $|P| \gg 1/larg$. This means that the community structure corresponds to a big (or few) big central communities surrounded by a multitude of small communities. In some cases, the sizes of communities seem to follow a power-law distribution (this is something we would have to check). The existence of a big community implies an expected inner fraction close to one, hence a modularity close to zero.

In both algorithms, in every iteration we have to visit all neighbors of every node. Therefore, the cost of an iteration is linear in the number of edges of the graph. We observe although than GFA usually needs more iterations than LPA. This is because, after folding the graph, we can do further iterations, and even several graph foldings.

We have also studied the *connected components* of these instances as in [6]. As we can see, almost all instances have a single connected component, or almost all variables are included in the same one. Hence the rest of connected components contain just a few variables. Therefore, the modularity gives us much

Table 1. Modularity of 2010 SAT Race instances, using LPA and GFA. Q stands for modularity, $|P|$ for number of communities, *larg.* for fraction of vertices in the largest community, and *iter.* for number of iterations of the algorithm.

Family (#instanc.)	Variable IG								Clause-Variable IG				Connect. Comp.	
	LPA				GFA				LPA					
	Q	$|P|$	larg.	iter.	Q	$|P|$	larg.	iter.	Q	$|P|$	larg.	iter.	$|P|$	larg.
cripto desgen(4)	0.88	532	0.8	36	0.95	97	2	37	0.75	3639	1	25	1	100
md5gen(3)	0.61	7176	0.1	16	0.88	38	7	40	0.78	7904	0.1	42	1	100
mizh(8)	0.00	33	99	6	0.74	30	9	33	0.67	5189	31	43	1	100
hard. ver. ibm(4)	0.81	3017	0.6	9	0.95	723	4	32	0.77	19743	0.2	140	70	99
manolios(16)	0.30	66	81	9	0.89	37	9	81	0.76	6080	1	26	1	100
velev (10)	0.47	8	68	9	0.69	12	30	24	0.30	1476	77	31	1	100
mixed anbulagan(8)	0.55	30902	0.1	11	0.91	90	2	43	0.72	46689	0.6	26	1	100
bioinf(6)	0.61	87	44	3	0.67	60	17	22	0.64	94027	15	10	1	100
diagnosis(4)	0.61	20279	0.7	15	0.95	68	3	43	0.65	85928	0.1	42	1	100
grieu(3)	0	1	100	2	0.23	9	14	11	0	1	100	14	1	100
jarvisalo(1)	0.57	260	5	8	0.76	19	9	26	0.71	336	1	11	1	100
palacios(3)	0.14	1864	96	58	0.93	1802	6	13	0.76	2899	0.4	35	1	100
soft. ver. babic(2)	0.68	34033	8	54	0.90	6944	10	141	0.73	59743	4	53	41	99
bitverif(5)	0.48	3	57	4	0.87	24	6	29	0.76	33276	0.4	8	1	100
fuhs(4)	0.02	18	99	43	0.81	43	7	30	0.67	12617	0.8	28	1	100
nec(17)	0.07	107	96	31	0.93	65	14	124	0.79	23826	0.8	114	1	100
post(2)	0.36	$3 \cdot 10^6$	53	54	0.81	$3 \cdot 10^6$	9	262	0.72	$3 \cdot 10^6$	6	49	224	99

more information about the structure of the formula. Notice that a connected component can be structured into several communities.

In Table 2 we show the results for the modularity computed after preprocessing the formula with the Satelite preprocessor [10]. The modularities are computed using the GFA algorithm. Satelite is an algorithm that applies variable elimination techniques. We can see that these transformations almost does not affect to the modularity of the formula. However, it eliminates almost all the small unconnected components of the formula.

5   Modularity of the Learnt Clauses

Most modern SAT solvers, based on variants of the DPLL schema, transform the formula during the proof or the satisfying assignment search. Therefore, the natural question is: even if the original formula shows a community structure, could it be the case that this structure is quickly destroyed during the search process? Moreover, most SAT solvers also incorporate learning techniques that introduce new learnt clauses to the original formula. Therefore, a second question is: how these new clauses affect to the community structure of the formula? Finally, even if the value of the modularity is not altered, it can be the case that the communities are changed.

Table 2. Modularity (computed with GFA) after and before preprocessing the formula with the Satelite preprocessor [10]

Family		Orig. Form.	Preprocessed Formula				
			Modularity			Connect. Comp.	
		Q	Q	$|P|$	larg.	$|P|$	larg.
cripto.	desgen (4)	0.951	0.929	81	3.1	1	100
	md5gen (3)	0.884	0.884	18	8.5	1	100
	mizh (8)	0.741	0.741	18	9.5	1	100
hard. ver.	ibm (4)	0.950	0.905	26	6.1	1	100
	manolios (16)	0.890	0.800	16	14.9	1	100
	velev (10)	0.689	0.687	6	30.3	1	100
mixed	anbulagan (8)	0.909	0.913	47	5.1	1	100
	bioinf (6)	0.673	0.657	25	11.1	2	99.9
	diagnosis (4)	0.952	0.950	65	3.6	1	100
	grieu (3)	0.235	0.235	9	14.3	1	100
	jarvisalo (1)	0.758	0.722	11	13.1	1	100
	palacios (3)	0.928	0.848	17	10.76	1	100
soft. ver.	babic (2)	0.901	0.875	23	9.7	1	100
	bitverif (5)	0.875	0.833	19	7.3	1	100
	fuhs (4)	0.805	0.743	32	9.6	1	100
	nec (17)	0.929	0.879	37	10.3	1	100

We have conducted a series of experiments to answer to the previous questions. We use the picosat SAT solver [5] (version 846), since it incorporates a conflict directed clause learning algorithm, activity-based heuristics, and restarting strategies.

In Table 3 we show the values of the original modularity compared with the modularity obtained after adding the learnt clauses to the original formula. We can observe that the modularity weakly decreases with the learnt clauses, but it is still meaningful. Therefore, learning does not completely destroy the organization of the formula into weakly connected communities.

The question now is, even if the modularity does not decreases very much, could it be the case that the communities have changed? In other words, there are still communities, but are they distinct communities?

If a considerable part of learning is performed locally inside one or a few communities, then the communities will not change. We have conducted another experiment to see if this is true. For the VIG model, we use the original formula to get a partition of the vertices, i.e. of the variables, into communities. Then, we use modularity as a *quality measure* to see how good is the same partition, applied to the graph obtained from the set of learnt clauses. Notice that modularity is a function of two parameters, in this case: the graph is the graph *containing* the learnt clauses, and the partition is computed for the formula *without* the learnt clauses. Since both graphs (the original formula and the learnt clauses) have the same set of vertices (the set of variables), this can be done directly.

Table 3. Modularity (computed with GFA) of the original formula, and of the original formula with learnt clauses included

Family		Orig. Form. Q	Orig. + Learnt Formula Modularity		
			Q	$|P|$	larg.
cripto.	desgen (2)	**0.951**	**0.561**	53	13.0
	md5gen (3)	**0.884**	**0.838**	19	8.0
	mizh (1)	**0.741**	**0.705**	28	11.4
hard. ver.	ibm (3)	**0.950**	**0.912**	752	6.7
	manolios (14)	**0.890**	**0.776**	31	11.2
	velev (1)	**0.689**	**0.558**	6	30.1
mixed	anbulagan (4)	**0.909**	**0.876**	84	2.6
	bioinf (5)	**0.673**	**0.287**	32	58.7
	grieu (3)	0.235	0.085	6	35.0
	palacios (2)	**0.928**	**0.851**	2289	7.4
soft. ver.	babic (2)	**0.901**	**0.904**	6942	10.7
	fuhs (1)	**0.805**	**0.670**	24	7.6
	nec (15)	**0.929**	**0.936**	62	7.3

Table 4. Modularity (computed by LPA) of the formula containing the first 100 learnt clauses, and all learnt clauses. In the first column we show the modularity of the original formula.

Family	VIG			CVIG		
	orig.	first 100	all	orig	first 100	all
desgen (1)	**0.89**	**0.74**	0.08	**0.77**	0.28	0.09
md5gen (1)	**0.61**	**0.74**	0.02	**0.78**	**0.96**	0.02
ibm (2)	**0.84**	**0.60**	0.47	**0.81**	0.58	0.29
manolios (10)	0.21	0.04	0.10	**0.76**	0.11	0.09
anbulagan (2)	**0.56**	0.16	0.01	**0.87**	0.10	0.04
bioinf (4)	**0.62**	**0.46**	0.06	**0.68**	**0.69**	0.15
grieu (1)	0.00	0.00	0.00	0.00	0.00	0.00
babic (2)	**0.68**	**0.36**	0.36	**0.71**	**0.33**	**0.33**
fuhs (1)	**0.66**	**0.59**	0.14	**0.71**	**0.78**	0.07
nec (10)	0.12	0.01	0.12	**0.78**	**0.49**	0.24

For the CVIG model we must take into account that the graph contains variables and clauses as vertices. Therefore, the procedure is more complicated. We use the original formula to get a partition. We remove from this partition all clauses, leaving the variables. Then, we construct the CVIG graph for the set of learnt clauses. The partition classifies the variables of this second graph into communities, but not the clauses. To do this, we assign to each clause the community of variables where it has more of its variables included. In other words, given the labels of the variables we apply a single iteration of the label propagation algorithm to find the labels of the clauses.

Table 5. Modularity (computed by LPA) of the learnt clauses that have been contributed to prove the unsatisfiability of the original formula. Like in Table 4 we show results for the first 100 learnt clauses, and for all clauses. We only show results for unsatisfiable formulas.

Family	VIG			CVIG		
	orig.	first 100	all	orig.	first 100	all
md5gen (1)	**0.61**	**0.74**	0.02	**0.78**	**0.96**	0.02
ibm (2)	**0.84**	**0.50**	**0.43**	**0.81**	**0.58**	**0.42**
anbulagan (1)	**0.55**	0.03	0.00	**0.87**	0.13	0.05
manolios (9)	0.20	0.07	0.10	**0.76**	0.27	0.15
nec (10)	0.12	0.05	0.05	**0.78**	**0.30**	0.22

We want to see how fast is the community structure degraded along the execution process of a SAT solver. Therefore, we have repeated the experiment for just the first 100 learnt clauses and for all the learnt clauses. We also want to know the influence of the quality of the learnt clauses. Therefore, we also repeat the experiment for all the learnt clauses (Table 4), and only using the clauses that participate in the proof of unsatisfiability (Table 5). Notice that Table 5 contains fewer entries than Table 4 because we can only consider unsatisfiable instances. Notice also that picosat is not able to solve all 2010 SAT Race instances, therefore Tables 4 and 5 contain fewer instances than Table 1. The analysis of the tables shows us that the CVIG model gives better results for the original formula and the first 100 learnt clauses, but equivalent results if we consider all learnt clauses. There are not significant differences if we use all learnt clauses, or just the clauses that participate in the refutation. Finally, there is a drop-off in the modularity (in the quality of the original partition) as we incorporate more learnt clauses. This means that, if we use explicitly the community structure to improve the efficiency of a SAT solver, to overcome this problem, we would have to recompute the partition (after some number of variable assignments or after a unit clause is learnt) to adjust it to the modified formula, like in [6].

It is worth to remark that, for the experiments in Table 4, the modularity for the VIG and the CVIG models, and the first 100 learnt clauses, is respectively 0.72 and 0.59. This means that, in the VIG model, around 72% of the first 100 learnt clauses could also be learnt working locally in each one of the communities. However, the percentage of learnt clauses that connect distinct communities is very significant.

6   Modularity of Random Formulas

We have also conducted a study of the modularity of 100 random 3-CNF SAT instances of 10^4 variables for different clause variable ratios (α). For this experiment we used the GFA on the VIG model. Table 6 shows the results. As we can see, the modularity of random instances is only significant for very low variable ratios, i.e., on the leftist SAT easy side. This is due to the presence of a large

Table 6. Modularity (computed with GFA) of random formulas varying the clause variable ratio (α), and for $n = 10^4$ variables

n	α	Q	$|P|$	larg.	iter
10^4	1	0.486	545	3.8	54
10^4	1.5	0.353	146	5.1	52
10^4	2	0.280	53	6.8	51
10^4	3	0.217	14	15.5	64
10^4	4	0.178	11	14.8	54
10^4	4.25	0.170	11	14.6	53
10^4	4.5	0.163	11	14.7	53
10^4	5	0.152	11	14.3	51
10^4	6	0.133	12	13.9	53
10^4	7	0.120	10	15.0	56
10^4	8	0.138	6	25.0	50
10^4	9	0.130	6	24.3	49
10^4	10	0.123	6	24.4	47

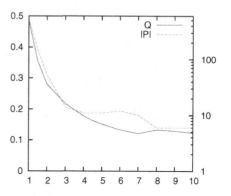

Table 7. Modularity (computed with GFA) of random formulas at the peak transition region (clause variable ratio α=4.25), varying the number of variables (n)

n	α	Q	$|P|$	larg.	iter
10^2	4.25	0.177	6	14.5	11
10^3	4.25	0.187	10.5	11.4	35
10^4	4.25	0.170	11	12.2	53
10^5	4.25	0.151	14	6.8	102
10^6	4.25	0.151	14	5.7	167

quantity of very small communities. Notice, that as α increases, the variables get more connected but without following any particular structure, and the number of communities highly decreases. Even for low values of α, the modularity is not as high as for industrial instances, confirming their distinct nature. We do not observe any abrupt change in the phase transition point.

As a second experiment with random formulas, we wanted to investigate the modularity at the peak transition region for an increasing number of variables. Table 7 shows the results. As we can see, the modularity is very low and it tends to slightly decrease as the number of variables increases, and seems to tend to a particular value (0.15 for the phase transition point).

Finally, as with industrial instances, we wanted to evaluate the impact on modularity of the preprocessing with Satelite [10], and the effect of adding all the learnt clauses needed to solve the formula by Picosat [5]. Table 8 shows the results. The preprocessing has almost no impact on the modularity of the formula, except for α=1, because the preprocessing already solves the formula. With respect to the addition of learnt clauses, it is interesting to observe that in the peak transition region $\alpha = 4.25$, we get the lowest modularity. A possible explanation is that at the peak region we find the hardest instances, and in

Table 8. Modularity (computed with GFA) of random formulas with 300 variables varying the clause variable ratio after and before preprocessing the formula with the Satelite preprocessor [10], and with all learnt clauses included

n	α	Orig.	Preproc.	Learnt	Connect. Comp.
300	1	0.459	0	0.453	1
300	2	0.291	0.235	0.291	1
300	4	0.190	0.188	0.073	1
300	4.25	0.183	0.182	0.041	1
300	4.5	0.177	0.177	0.045	1
300	6	0.150	0.150	0.120	1
300	10	0.112	0.112	0.171	1

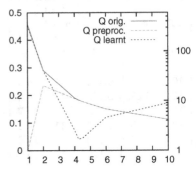

order to solve them the learnt clauses added by the solver tend to connect more communities. In [2] we observed that random SAT instances have not a scale-free structure[1], but that the addition of learnt clauses makes the formula clearly scale-free. On the contrary, we observe here that modularity tends to decrease with learning.

7   Conclusions

The research community on complex networks has developed techniques of analysis and algorithms that can be used by the SAT community to improve our knowledge about the structure of industrial SAT instances, and, as result, to improve the efficiency of SAT solvers.

In this paper we address the first systematic study of the community structure of SAT instances, finding a clear evidence of such structure in most analyzed instances. In fact, some features, like Moskewicz's activity-based heuristics, were already designed thinking on the existence of this kind of structure. Here we go a step further, and we propose the use of an algorithm that is able to compute the communities of a SAT instance. It verifies the assumption about the existence of this community structure. The algorithm could also be used directly by SAT solvers to focus their search.

References

1. Amir, E., McIlraith, S.A.: Partition-based logical reasoning for first-order and propositional theories. Artif. Intell. 162(1-2), 49–88 (2005)
2. Ansótegui, C., Bonet, M.L., Levy, J.: On the Structure of Industrial SAT Instances. In: Gent, I.P. (ed.) CP 2009. LNCS, vol. 5732, pp. 127–141. Springer, Heidelberg (2009)

---

[1] This is another structural quality of formulas that is being used by modern SAT solvers.

3. Audemard, G., Simon, L.: Predicting learnt clauses quality in modern SAT solvers. In: IJCAI, pp. 399–404 (2009)
4. Barber, M.J.: Modularity and community detection in bipartite networks. Phys. Rev. E 76(6), 066102 (2007)
5. Biere, A.: Picosat essentials. JSAT 4(2-4), 75–97 (2008)
6. Biere, A., Sinz, C.: Decomposing SAT problems into connected components. JSAT 2(1-4), 201–208 (2006)
7. Blondel, V.D., Guillaume, J.L., Lambiotte, R., Lefebvre, E.: Fast unfolding of communities in large networks. Journal of Statistical Mechanics: Theory and Experiment 2008 (10), P10008 (2008)
8. Brandes, U., Delling, D., Gaertler, M., Görke, R., Hoefer, M., Nikoloski, Z., Wagner, D.: On modularity – NP-completeness and beyond. Tech. rep., Faculty of Informatics, Universität Karlsruhe (TH), Tech. Rep. a 2006-19 (2006)
9. Clauset, A., Newman, M.E.J., Moore, C.: Finding community structure in very large networks. Phys. Rev. E 70(6), 066111 (2004)
10. Eén, N., Biere, A.: Effective Preprocessing in SAT Through Variable and Clause Elimination. In: Bacchus, F., Walsh, T. (eds.) SAT 2005. LNCS, vol. 3569, pp. 61–75. Springer, Heidelberg (2005)
11. Eén, N., Sörensson, N.: An Extensible SAT-solver. In: Giunchiglia, E., Tacchella, A. (eds.) SAT 2003. LNCS, vol. 2919, pp. 502–518. Springer, Heidelberg (2004)
12. Fortunato, S.: Community detection in graphs. Physics Reports 486(3-5), 75–174 (2010)
13. Gent, I.P., Hoos, H.H., Prosser, P., Walsh, T.: Morphing: Combining structure and randomness. In: AAAI/IAAI, pp. 654–660 (1999)
14. Gomes, C.P., Selman, B.: Problem structure in the presence of perturbations. In: AAAI/IAAI, pp. 221–226 (1997)
15. Gomes, C.P., Selman, B., Kautz, H.A.: Boosting combinatorial search through randomization. In: Proc. of the 15th Nat. Conf. on Artificial Intelligence (AAAI 1998), pp. 431–437 (1998)
16. Hogg, T.: Refining the phase transition in combinatorial search. Artif. Intell. 81(1-2), 127–154 (1996)
17. Järvisalo, M., Niemelä, I.: The effect of structural branching on the efficiency of clause learning SAT solving: An experimental study. J. Algorithms 63, 90–113 (2008)
18. Moskewicz, M.W., Madigan, C.F., Zhao, Y., Zhang, L., Malik, S.: Chaff: Engineering an efficient SAT solver. In: DAC, pp. 530–535 (2001)
19. Newman, M.E.J.: Fast algorithm for detecting community structure in networks. Phys. Rev. E 69(6), 066133 (2004)
20. Newman, M.E.J., Girvan, M.: Finding and evaluating community structure in networks. Phys. Rev. E 69(2), 026113 (2004)
21. Raghavan, U.N., Albert, R., Kumara, S.: Near linear time algorithm to detect community structures in large-scale networks. Phys. Rev. E 76(3), 036106 (2007)
22. Slater, A.: Modelling more realistic SAT problems. In: Australian Joint Conference on Artificial Intelligence, pp. 591–602 (2002)
23. Walsh, T.: Search in a small world. In: IJCAI, pp. 1172–1177 (1999)
24. Watts, D.J., Strogatz, S.H.: Collective dynamics of 'small-world' networks. Nature 393(6684), 440–442 (1998)
25. Williams, R., Gomes, C.P., Selman, B.: Backdoors to typical case complexity. In: IJCAI, pp. 1173–1178 (2003)

SATLab: X-Raying Random k-SAT
(Tool Presentation)

Thomas Hugel

I3S - UMR 7271 - Université de Nice-Sophia & CNRS
BP 121 - 06903 Sophia Antipolis Cedex - France
hugel@i3s.unice.fr

Abstract. In the random k-SAT model, probabilistic calculations are often limited to the first and second moments, thus giving an idea of the average behavior, whereas what happens with high probability can significantly differ from this average behavior. In these conditions, we believe that the handiest way to understand what really happens in random k-SAT is experimenting. Experimental evidence may then give some hints hopefully leading to fruitful calculations.

Also, when you design a solver, you may want to test it on real instances before you possibly prove some of its nice properties.

However doing experiments can also be tedious, because you must generate random instances, then measure the properties you want to test and eventually you would even like to make your results accessible through a suitable graph. All this implies lots of repetitive tasks, and in order to automate them we developed a GUI-software called SATLab.

1 Introduction

Random k-SAT has been widely studied for the following reasons:

1. it exhibits a phase transition of satisfiability at a given ratio C between the number of clauses and the number of variables; it was experimentally observed that this transition would occur at a ratio $C \simeq 4.25$ for the standard 3-SAT model, in which all clauses are drawn uniformly and independently (see [16]);
2. it is assumed to be difficult even on average, especially in the neighborhood of the above phase transition, which makes this transition particularly interesting (see figure 1).

There have been numerous attempts to locate precisely the observed phase transition. Statistical physicists gave very tight bounds on the location of the threshold [15,13], however rigorous mathematical bounds are still far from it [1,9,10,7]. Meanwhile some light has been shed onto the structure of the solutions space and a clustering phenomenon has been investigated [17]. The existence of an exponential number of solutions clusters has been established under some parameters [2,14]. Not unconnected with that phenomenon is the question whether

A. Cimatti and R. Sebastiani (Eds.): SAT 2012, LNCS 7317, pp. 424–429, 2012.

non-trivial cores of solutions exist [12]. A core is a set of variables constraining each other; all solutions in a given cluster have the same core.

Handling all these complex properties is a mathematical challenge; therefore nurtering intuition through experiments seems to be essential in their understanding. Another area where experiments on random instances are needed is the design of new solvers. To these ends we introduce SATLab.

2   SATLab's Features

2.1   Installing SATLab

SATLab was written in Perl/Tk and plots the graphs through gnuplot. SATLab has a dedicated website where you can download it and browse through its documentation: www.pratum.org/satlab.

2.2   What SATLab Does

SATLab draws some random formulas in a given *random model*, finds some solutions with a SAT-*solver*, analyzes all of these raw data to infer *properties* you want to investigate and finally makes them available on a *graph*. For example you can see on figure 1 the well-known complexity peak in random 3-SAT for various complete solvers.

On this example, what we call *first variable* is C (the ratio between clauses and variables), whereas Solver is the *second variable*. The first and second variables may be chosen at the user's discretion, enabling numerous possibilities of graphs. You can also notice that the y-axis was chosen to be *logarithmic*, which is an option for both axes. You can also tune the number of solvings you want to be

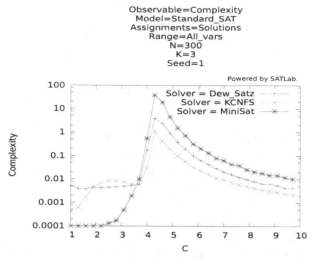

Fig. 1. Algorithmic time complexity of random 3-SAT with respect to C

averaged over at every point of the graph. The main numeric parameters are N (the number of variables), C (the ratio number of clauses / number of variables), K (the clause width) and **Seed** (a random seed in order to generate different formulas).

Moreover the GUI is designed in order to eliminate as far as possible the irrelevant parameters (depending on the requested observation).

2.3   SATLab's Universe

Besides the aforementioned numeric parameters, SATLab can handle various kinds of formulas, solvers and properties:

Models of Formulas: standard SAT(where clauses are drawn uniformly and independently), planted SAT(with a hidden solution), standard NAE-SAT(not all equal sat, i.e. where each clause is drawn together with its opposite clause), planted NAE-SAT, and many other planted models described in detail on the website.

Observables: measures on single assignments (such as frozen variables, cores, the surface of true literals occurrences, the fraction of uniquely satisfied clauses, the number of clauses a variable flip would break/make etc.), measures on pairs of assignments (such as Hamming distances, independence of some of the former quantities), time complexity, a formula generator etc.

Assignments: solutions or random assignments (the latters can be used as benchmarks for the formers).

Range: the range of variables to consider: either all variables, or variables with a given number of occurrences, or fixed variables etc.

Solvers: various kinds of solvers are embedded in SATLab: random walk solvers (UBCSAT [19]), the statistical physics solver SP [5] and some resolution based solvers (Dew_Satz [3], KCNFS [6] and MiniSat [8]).

SATLab was written in a modular way in object-oriented Perl, so this list may be extended by adding corresponding Perl modules to it.

3   Case Study: The Structure of the Solutions Space

Figure 2 is the outcome of an experiment designed to check the clustering phenomenon of solutions. We asked WalkSAT [18] to find 100 solutions of a random instance of 3-SAT at the ratio $C = 4.2$ with $N = 1000$ variables, and to plot an histogram of the Hamming similarities between every couple of solutions. What we mean by *Hamming similarity* between two assignments is just the proportion of variables assigned the same value in both assignments.

You can see two typical distances, which corroborates the presence of clusters. For high values of N there is only one typical distance, maybe because the number of clusters grows faster than their size, so that 2 random solutions will almost surely belong to different clusters. What is interesting though, is the fact

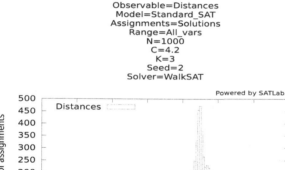

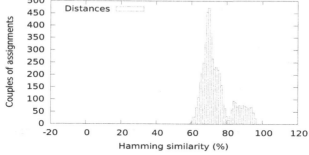

Fig. 2. Such Hamming distances between solutions corroborate the clustering phenomenon

that these solutions of 3-SAT are correlated with respect to their Hamming distances. Namely their Hamming distances are not centered around 50%, contrary to solutions of random 3-NAE-SAT. This non-independence between solutions in 3-SAT may explain why the second moment method fails on it (see also [1]).

An explanation of the correlation between solutions might be the presence of frozen variables (a variable is called *frozen* when it takes the same value in every solution). So we asked SATLab to find 100 solutions and to report the number of variables assigned 1 in a given proportion of solutions: this is figure 3. Frozen variables are variables assigned 1 in 100% or in 0% of the solutions. You can see that just below the satisfiability threshold, some variables tend to be frozen or quasi-frozen.

[2] showed that for large values of k, in a region just below the satisfiability threshold, almost all solutions of a random k-SAT instance have a non-trivial core (a *core* is a set of variables constraining each other; the core is *trivial* when this set is empty). On the contrary, experiments conducted by [11] tend to show that in random 3-SAT almost all solutions have a trivial core. With SATLab we could observe (cf. figure 4) that in 3-SAT at $C = 4.2$:

1. most solutions seem indeed to have a trivial core;
2. non-trivial cores seem to exist nevertheless, insofar as formulas typically seem to have a small number of solutions whose cores are made up of around 80% of the variables.

The question of the existence of the non-trivial cores is important, because if they were proved to exist just below the satisfiability threshold, then the bounds on their existence contained in [12,4] would become upper bounds on the satisfiability threshold.

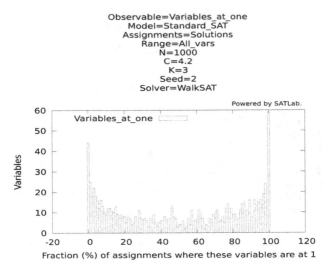

Fig. 3. Number of variables assigned 1 in a given proportion of solutions. Frozen variables appear on the left and right sides.

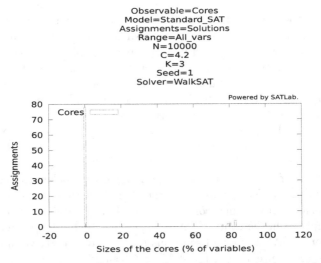

Fig. 4. Most solutions have a trivial core, but non-trivial cores do exist

Acknowledgments. SATLab was developed mainly when the author was a PhD student at Université Denis Diderot Paris 7. The author is very grateful to his supervisor Yacine Boufkhad for lots of interesting suggestions, his valuable feedback and his encouragement.

The author is also grateful to the solvers' authors who kindly permitted him to embed them into SATLab.

References

1. Achlioptas, D., Peres, Y.: The Threshold for Random k-SAT is 2^k ln2 - O(k). JAMS: Journal of the American Mathematical Society 17, 947–973 (2004)
2. Achlioptas, D., Ricci-Tersenghi, F.: On the solution-space geometry of random constraint satisfaction problems. In: STOC, pp. 130–139. ACM Press (2006)
3. Anbulagan: Dew Satz: Integration of Lookahead Saturation with Restrictions into Satz. In: SAT Competition, pp. 1–2 (2005)
4. Boufkhad, Y., Hugel, T.: Estimating satisfiability. Discrete Applied Mathematics 160(1-2), 61–80 (2012)
5. Braunstein, A., Mézard, M., Zecchina, R.: Survey propagation: An algorithm for satisfiability. Random Structures and Algorithms 27(2), 201–226 (2005)
6. Dequen, G., Dubois, O.: kcnfs: An Efficient Solver for Random k-SAT Formulae. In: Giunchiglia, E., Tacchella, A. (eds.) SAT 2003. LNCS, vol. 2919, pp. 486–501. Springer, Heidelberg (2004)
7. Díaz, J., Kirousis, L.M., Mitsche, D., Pérez-Giménez, X.: On the satisfiability threshold of formulas with three literals per clause. Theoretical Computer Science 410(30-32), 2920–2934 (2009)
8. Een, N., Sörensson, N.: MiniSat — A SAT Solver with Conflict-Clause Minimization. In: SAT (2005)
9. Hajiaghayi, M.T., Sorkin, G.B.: The satisfiability threshold of random 3-SAT is at least 3.52. IBM Research Report RC22942 (2003)
10. Kaporis, A.C., Lalas, E.G., Kirousis, L.M.: The probabilistic analysis of a greedy satisfiability algorithm. Random Structures and Algorithms 28(4), 444–480 (2006)
11. Maneva, E.N., Mossel, E., Wainwright, M.J.: A new look at survey propagation and its generalizations. Journal of the ACM 54(4), 2–41 (2007)
12. Maneva, E.N., Sinclair, A.: On the satisfiability threshold and clustering of solutions of random 3-SAT formulas. Theor. Comput. Sci. 407(1-3), 359–369 (2008)
13. Mertens, S., Mézard, M., Zecchina, R.: Threshold values of Random K-SAT from the cavity method. Random Structures and Algorithms 28(3), 340–373 (2006)
14. Mézard, M., Mora, T., Zecchina, R.: Clustering of Solutions in the Random Satisfiability Problem. Physical Review Letters 94(19), 1–4 (2005)
15. Mézard, M., Zecchina, R.: Random K-satisfiability problem: From an analytic solution to an efficient algorithm. Physical Review E 66(5), 1–27 (2002)
16. Mitchell, D.G., Selman, B., Levesque, H.: Hard and easy distributions of SAT problems. In: Proceedings of the 10th Nat. Conf. on A.I, pp. 459–465 (1992)
17. Parkes, A.J.: Clustering at the phase transition. In: Proc. of the 14th Nat. Conf. on AI, pp. 340–345 (1997)
18. Selman, B., Kautz, H., Cohen, B.: Local Search Strategies for Satisfiability Testing. In: Trick, M., Johnson, D.S. (eds.) Proceedings of the Second DIMACS Challange on Cliques, Coloring, and Satisfiability (1995)
19. Tompkins, D.A.D., Hoos, H.H.: UBCSAT: An Implementation and Experimentation Environment for SLS Algorithms for SAT and MAX-SAT. In: Hoos, H.H., Mitchell, D.G. (eds.) SAT 2004. LNCS, vol. 3542, pp. 306–320. Springer, Heidelberg (2005)

Resolution-Based Certificate Extraction for QBF (Tool Presentation)*

Aina Niemetz, Mathias Preiner,
Florian Lonsing, Martina Seidl, and Armin Biere

Institute for Formal Models and Verification
Johannes Kepler University, Linz, Austria
http://fmv.jku.at/

Abstract. A certificate of (un)satisfiability for a quantified Boolean formula (QBF) represents sets of assignments to the variables, which act as witnesses for its truth value. Certificates are highly requested for practical applications of QBF like formal verification and model checking. We present an integrated set of tools realizing resolution-based certificate extraction for QBF in prenex conjunctive normal form. Starting from resolution proofs produced by the solver DepQBF, we describe the workflow consisting of proof checking, certificate extraction, and certificate checking. We implemented the steps of that workflow in stand-alone tools and carried out comprehensive experiments. Our results demonstrate the practical applicability of resolution-based certificate extraction.

1 Introduction

Over the last 10 years, several approaches and tools supporting the generation of certificates for *quantified Boolean formulae* (QBF) have been presented. An overview of the status quo of 2009 is given in [7]. Initially, the main goal was to use certificates for validating the results of QBF solvers instead of relying on majority votes (only). Therefore, independent verifiers to check the output of a QBF solver have been developed. Such solver outputs are either clause/cube resolution proofs, or functions representing variable assignments. In case of so-called *Skolem/Herbrand* functions as output, we obtain QBF (counter-)models [3] and we gain further information on the solution of a solved problem, e.g., the path to a bad state in case of model checking. Their extraction, however, is not directly applicable if the successful variant of DPLL style procedures for QBF is employed. Most of these tools and solvers are not maintained anymore.

More recently, the circuit solver CirQit [5] has been extended to produce Q-refutations for true and false QBFs. Due to the circuit representation, solving the negated formula does not involve any expensive transformations. Furthermore, (partial) solution strategies can be extracted. However, CirQit cannot exploit its full strength on formulae in prenex conjunctive normal form (PCNF).

* This work was partially funded by the Vienna Science and Technology Fund (WWTF) under grant ICT10-018 and by the Austrian Science Fund (FWF) under NFN Grant S11408-N23 (RiSE).

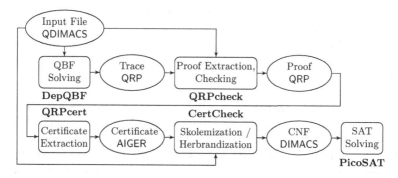

Fig. 1. Certification workflow

Based on the resolution generation tools discussed above, recently the prototype ResQu [2], which implements an approach to extract Skolem/Herbrand functions from resolution proofs, was presented. Given a resolution proof for true or false QBF, such functions can be extracted in linear time.

In this paper, we present a framework for generating QBF certificates from resolution proofs obtained by the state-of-the-art solver DepQBF [6]. We further performed an empirical study on the applicability of resolution-based extraction of QBF certificates and discuss strengths and limitations of this approach.

2 The Certification Framework at a Glance

We provide a complete and solver independent framework to certify and validate the results of the state-of-the-art QBF solver DepQBF [6]. The workflow is shown in Fig. 1. The framework consists of loosely coupled stand-alone tools, which support both proof extraction and checking (QRPcheck) as well as certificate extraction and validation (QRPcert, CertCheck and PicoSAT).

Trace Extraction. We instrumented DepQBF to output traces in our novel, text-based QRP format. A trace represents the set of all resolution sequences involved in generating learnt constraints in a search-based QBF solver. Our approach of tracing is similar to [8], except that each single resolution step has *exactly* two antecedents. This way, exponential worst-case behaviour during reconstruction of resolvents from unordered lists of antecedents is avoided [4]. Alternatively, the QIR format [1] used in [5] allows multiple antecedents but predefines the ordering in which resolvents should be reconstructed. As far as resolution is concerned, QRP proofs can be checked in deterministic log space, a desirable property of proof formats suggested in [4].

Proof Extraction and Checking. QRPcheck is a proof checker for resolution-based traces and proofs of (un)satisfiability in QRP format. Starting with the empty constraint, QRPcheck extracts the proof from a given trace on-the-fly while checking each proof step incrementally. Parts irrelevant for deriving the empty constraint are omitted. In case of a proof of satisfiability, QRPcheck allows to check that initial (input) cubes, generated during constraint learning, can be extended to satisfy the matrix. For this propositional check we are using the SAT solver PicoSAT.[1] In order to handle very large traces and proofs we map the input file to memory using virtual memory mechanisms ("mmap").

Certificate Extraction. QRPcert is a tool for extracting Skolem/Herbrand function-based QBF certificates from resolution proofs in QRP format. It implements the algorithm of [2] in reverse topological order. Hence, irrelevant parts are omitted as in proof extraction using QRPcheck. Extracted certificates are represented as AIGs[2], which are simplified by common basic simplification techniques, including structural hashing and constant propagation.

SAT-Based Certificate Checking. The tool CertCheck transforms the given input formula into an AIG and merges the result with the certificate by substituting each existentially (universally) quantified input variable with its Skolem (Herbrand) function. The resulting AIG is first translated into CNF via Tseitin transformation and then checked for being tautological (unsatisfiable). We validate the correctness of the certificate by checking the resulting CNF with the SAT solver PicoSAT.

3   Experiments

We applied our framework on the benchmark sets of the QBF competitions 2008 and 2010 consisting of 3326 and 568 formulas, respectively[3]. We considered only those 1229 and 362 formulas solved by DepQBF within 900 seconds. Instead of advanced dependency schemes, we used the orderings of quantifier prefixes in DepQBF. All experiments[4] were performed on 2.83 GHz Intel Core 2 Quad machines with 8 GB of memory running Ubuntu 9.04. Time and memory limits for the whole certification workflow were set to 1800 seconds and 7 GB, respectively.

Out of 362 solved instances of the QBFEVAL'10 benchmark set, our framework was able to check 348 proofs and extract 337 certificates, of which 275 were validated successfully. DepQBF required almost 5000 seconds for solving and tracing the 275 instances that were validated by PicoSAT, whereas the certification of those instances needed about 5600 seconds. On 14 instances, QRPcheck ran out of memory, as the file size of the traces produced by DepQBF were 16 GB on average with a maximum of 27 GB. QRPcert ran out of memory on 11 proofs with an average file size of 3.6 GB and a maximum of 5.9 GB.

The largest number of instances (62) were lost during the validation process. PicoSAT timed out on 17 instances and ran out of memory on 45 instances. From 62 instances that were not validated by PicoSAT, 51 instances are part of the

---

[1] http://www.fmv.jku.at/picosat

[2] ASCII AIGER format: http://fmv.jku.at/aiger/FORMAT.aiger

[3] Available at http://www.qbflib.org/index_eval.php

[4] Log files and binaries are available from http://fmv.jku.at/cdepqbf/.

'mqm' family, which consists of a total of 128 formulae with 70 instances being unsatisfiable and 58 satisfiable. PicoSAT was able to validate all 70 unsatisfiable instances, but did not succeed in validating even one satisfiable instance. This is due to the fact that proofs of satisfiability tend to grow much larger than proofs of unsatisfiability mostly because of the size of the initial cubes. For example, the proofs of the 51 instances that were not validated by PicoSAT have 40000 intial cubes on average, where each cube has an average size of 970 literals.

We evaluated the runtime of each component of the framework w.r.t. the 275 certified instances. First, we compared the time required by DepQBF for solving and tracing to the aggregated time needed by QRPcheck, QRPcert, CertCheck, and PicoSAT for certification. Figure 2a shows that the whole certification process requires marginally more time than DepQBF on average. It also shows that only a few instances are responsible for the certification process being slower in total runtime than DepQBF. In fact, three instances require more than 59% of the total certification runtime, in contrast to 34% of total solving time.

We further compared the runtime of each component of the framework, which is depicted in Fig. 2b. More than 77% of the certification time is required for validating the certificates with PicoSAT, where three instances require over 58% of the total certification time. Extracting and checking proofs with QRPcheck requires about 20% of the total certification time, which typically involves heavy I/O operations in case proof extraction is enabled. Considering all checked instances, disabling proof extraction saves about 54% of the runtime of QRPcheck. The extraction of certificates and the CNF conversion takes a small fraction of the total certification time, which is approximately 2% and 1%, respectively.

Table 1 summarizes the results. Certification heavily depends on whether an instance is satisfiable or unsatisfiable, especially for certificate validation. On average over 91% of the solved unsatisfiable instances were certified in about 61% of the solving time. For 74% of the solved satisfiable instances, the certification took over four times the solving time.

Certificate validation requires most of the time. Particularly vaildating satisfiable instances is time-consuming. Given the QBFEVAL'10 set, traces of satisfiable instances are on average 2-3 times larger than traces of unsatisfiable instances and further contain in the worst case 13 million steps with 1.4 billion literals and 18 million steps with 1.3 billion literals, respectively. The difference between proofs of satisfiability and unsatisfiability is even larger by a factor of eight on average.

An interesting property of the generated AIG certificates is the number of and-gates involved, where certificates of satisfiability are on average (and in the median) over 100 times larger than certificates of unsatisfiability. The maximum number of and-gates generated for AIG certificates of satisfiability and unsatisfiability are 147 million resp. 10 million and-gates. Compared to certificates of unsatisfiability, CNFs generated for validating certificates of satisfiability are on average up to 70 times larger and contain in the worst-case over 10 times more clauses with a maximum of 440 million clauses.On certain instances, the file size

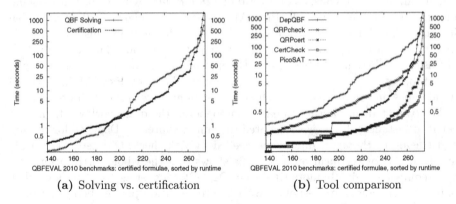

(a) Solving vs. certification   **(b)** Tool comparison

Fig. 2. Runtime comparison, all instances with solving time ≥ 0.2s considered

of traces were enourmous with almost 52 GB and 27 GB in the worst-case in the QBFEVAL'08 and QBFEVAL'10 benchmark sets, respectively.

Finally, we investigated the 14 instances (4 sat., 10 unsat.) of the QBFE-VAL'10 benchmark set that were not checked by QRPcheck due to given memory constraints. The corresponding traces had an average file size of 16 GB (with 27 GB as a maximum) and 17 million steps with 3.3 billion literals on average. For these 14 instances, we lifted the previous memory limit of 7 GB and rerun the experiments on a machine with 96 GB and a time limit of 3600 seconds. As a consequence, we were able to certify 12 out of 14 instances. On two instances, PicoSAT timed out while validating CNFs with 3 and 30 million clauses, respectively. Certification of the other 12 instances took less than 4600 seconds in total, whereas DepQBF required over 7700 seconds for solving and tracing altogether.

Average (median) time for the whole workflow on all 14 instances was 1412 (1014) seconds. The size of extracted proofs ranges from 85% to 0.0001% relative to trace size, with a maximum of 14.6 GB and a minimum of 13 kB. Average (median) number of steps was 18 (10) million in traces, and 4 (0.1) million in extracted proofs. The ratio of proof size over trace size in the number of steps for each of the 14 instances was 0.23 on average. As an extreme case, the certified satisfiable instance blocks_enc_2_b3_ser--opt-9_shuffled.qdimacs resulted in a trace of more than 50 million steps and 18 GB file size for which a proof of only 38 (!) steps and 47 kB file size was extracted. Average (median) memory usage for the whole workflow was 19 (18) GB with a maximum of 28 GB.

4   Discussion

In this paper, we presented the first framework for complete and robust certification of QBF using the state-of-the-art QBF solver DepQBF. We presented solver independent tools for proof extraction, proof checking, certificate extraction and certificate validation. We further performed an extensive evaluation on recent benchmark sets, which shows that our framework is able to extract certificates

Table 1. Solved (sv), checked (ch), extracted (ex), validated (va) inst., runtimes

		Instances				Total Time [s]			
		sv	ch	ex	va	DepQBF	QRPcheck	QRPcert	PicoSAT
2008	sat	494	476	464	397	3502.9	911.6	95.3	13874.1
	unsat	735	690	685	673	9863.7	2938.1	831.8	2639.8
	total	1229	1166	1149	1070	13366.6	3849.7	927.1	16513.9
2010	sat	157	153	143	86	701.8	80.1	30.9	3247.0
	unsat	205	195	194	189	4241.9	1011.5	86.8	1090.0
	total	362	348	337	275	4943.7	1091.7	117.6	4337.0

for over 90% of solved instances. Further, we were able to validate over 80% of extracted certificates, which all were proved correct.

In future work, we consider to extend DepQBF to maintain proofs internally in order to extract certificates directly from the solver. We also plan to extend QRPcert to support advanced dependency schemes as applied in DepQBF. Further, we want to improve the process of certificate validation as it is considered to be a bottleneck in the current framework. We believe that this technology will finally actually enable the application of QBF solving in practice, both in already proposed as well as new applications.

References

1. QIR Proof Format (Version 1.0),
   http://users.soe.ucsc.edu/~avg/ProofChecker/qir-proof-grammar.txt
2. Balabanov, V., Jiang, J.-H.R.: Resolution Proofs and Skolem Functions in QBF Evaluation and Applications. In: Gopalakrishnan, G., Qadeer, S. (eds.) CAV 2011. LNCS, vol. 6806, pp. 149–164. Springer, Heidelberg (2011)
3. Kleine Büning, H., Zhao, X.: On Models for Quantified Boolean Formulas. In: Lenski, W. (ed.) Logic versus Approximation. LNCS, vol. 3075, pp. 18–32. Springer, Heidelberg (2004)
4. Van Gelder, A.: Verifying Propositional Unsatisfiability: Pitfalls to Avoid. In: Marques-Silva, J., Sakallah, K.A. (eds.) SAT 2007. LNCS, vol. 4501, pp. 328–333. Springer, Heidelberg (2007)
5. Goultiaeva, A., Van Gelder, A., Bacchus, F.: A Uniform Approach for Generating Proofs and Strategies for Both True and False QBF Formulas. In: IJCAI 2011, pp. 546–553. IJCAI/AAAI (2011)
6. Lonsing, F., Biere, A.: Integrating Dependency Schemes in Search-Based QBF Solvers. In: Strichman, O., Szeider, S. (eds.) SAT 2010. LNCS, vol. 6175, pp. 158–171. Springer, Heidelberg (2010)
7. Narizzano, M., Peschiera, C., Pulina, L., Tacchella, A.: Evaluating and Certifying QBFs: A Comparison of State-of-the-Art Tools. AI Commun. 22(4), 191–210 (2009)
8. Yu, Y., Malik, S.: Validating the Result of a Quantified Boolean Formula (QBF) Solver: Theory and Practice. In: ASP-DAC, pp. 1047–1051. ACM Press (2005)

Coprocessor 2.0 – A Flexible CNF Simplifier (Tool Presentation)

Norbert Manthey

Knowledge Representation and Reasoning Group
Technische Universität Dresden, 01062 Dresden, Germany
norbert@janeway.inf.tu-dresden.de

Abstract. This paper presents the CNF simplifier COPROCESSOR 2.0, an extension of COPROCESSOR [1]. It implements almost all currently known simplification techniques in a modular way and provides access to each single technique to execute them independently. Disabling pre-processing for a set of variables is also possible and enables to apply simplifications also for incremental SAT solving. Experiments show that COPROCESSOR 2.0 performs better than its predecessor or SATELITE [2].

1 Introduction

Simplifying CNF formulae before passing them to SAT solvers increases the power of the overall tool chain. SAT solvers that do not implement an integrated preprocessor most of the time utilize the CNF simplifier SATELITE [2], which has been published in 2005. Since that time, new simplification methods like *blocked clause elimination* [3] or *hidden tautology elimination* [4] have been developed, that reduce the formula further. Most of these new techniques are implemented in the CNF preprocessor COPROCESSOR 2.0 in a modular way[1]. By granting access to each single technique and the possibility to rearrange the execution order, this tool can be widely used to remove the redundancy of encoded formulae. Further features include the ability to rewrite *At-Most-One* constraints in an *Exactly-One* constraint and to exclude *white-listed* variables, i.e. variables whose set of models should not change, from preprocessing. This technique is especially valuable, if the preprocessed formula is used for solving MaxSat by re-encoding the instance [5], or for incremental SAT solving [6], where clauses are added to the formula after solving. By keeping the models of the variables of these clauses, adding them can be done safely whereas the rest of the formula can still be reduced.

In contrast to its predecessor COPROCESSOR [1], COPROCESSOR 2.0 can also process huge formulae, by limiting expensive techniques. Consequentially, the performance of the next tool in the execution chain, for example SAT solvers, is also improved on those instances. An empirical analysis on the performance of the SAT solver GLUCOSE 2 combined with either COPROCESSOR 2.0 or SATELITE [2]

---

[1] The tool as well as a README file and a description of first steps is available at http://tools.computational-logic.org.

A. Cimatti and R. Sebastiani (Eds.): SAT 2012, LNCS 7317, pp. 436–441, 2012.
© Springer-Verlag Berlin Heidelberg 2012

reveals that by using the new tool application instances of recent SAT competitions can be solved faster[2].

After an overview on preprocessing and tool details in Section 2, a short tool comparison is given in Section 3.

2   Preprocessing

Simplification techniques can be divided into two categories: (i) model preserving and (ii) satisfiability preserving techniques. Implemented simplifications of the first category are *Boolean constraint propagation*(BCP), *self-subsuming resolution*(SUS), *pure literal*(PL), *hidden tautology elimination*(HTE) [4], *failed literal elimination*(FLE) [7], *probing*(PRB) [7,8] and *clause vivification*(VIV) [9]. The *unhiding* techniques(UHD) presented in [10] also belong to this category. COPROCESSOR 2.0 also implements all other techniques from [10]. All above techniques can be applied if further clauses should be added to the formula instead of only showing satisfiability. For FLE and PRB, a double lookahead [8] is used to find more implications. The 1st-UIP conflict analysis is used to retrieve even more powerful failed literals than in [7]. Additionally, the implementation of COPROCESSOR 2.0 already contains another two unpublished simplification methods that preserve equivalence[3]. Both techniques try to reduce the size of the formula by utilizing extended resolution.

The second category, satisfiability preserving techniques, alters the set of models of the formula and also eliminates variables from the formula. COPROCESSOR 2.0 implements *blocked clause elimination*(BCE) [3], *variable elimination*(BVE) [2] and *equivalence elimination*(EE) [11]. The required undo-information for postprocessing the model of the simplified formula is stored as described in [1].

COPROCESSOR 2.0 is also able to rewrite parts of the formula as described in [12]. An extension to this algorithm enables the tool to rewrite *At-Most-One* constraints also by the *2-product* encoding [13], if this encoding produces less clauses. Currently, only at-most-one constraints are found that are part of an exactly-one constraint. The algorithms iterates over all clauses C of the formula and checks whether the current clause is part of an exactly-one constraint. This check is done by looking whether for all literals $l, l' \in C$ there is a binary clause $\{\bar{l}, \bar{l'}\}$. Another feature that is kept from the predecessor is *variable renaming*: After all simplifications have been applied, the variables are renamed so that they are consecutive again.

Finally, a *white-list* and *black-list* for variables can be specified. Variables on the white-list are only used in equivalence preserving techniques to guarantee that their set of models is not altered. Blacklisted variables are eliminated in any case, even if the number of clauses increases. These two techniques can be used for example to calculate the projection of a formula with respect to a certain set of variables [14]. The white-list is also valuable if the input formula encodes an optimization problem or is part of an incremental SAT instance. If the white-list

---

[2] Instances and solvers are taken from www.satcompetition.org.
[3] Submitted for publication.

contains all variables for which clauses might be added to the formula, e.g. by decreasing the upper bound of the optimization function, the formula is still valid, even if variables of the remaining formula have been eliminated.

2.1 Limited Preprocessing

Obviously, the highest simplification can be achieved by executing all technique until completion. However, for big formulae too much time is consumed whereas solving the original problem might be much faster. Thus, for the most time consuming techniques, namely SUS, BVE, BCE, PRB, VIV and HTE, limits are introduced. Whenever the number of execution steps in the algorithm reaches its limit, the algorithm is not executed any longer. Additionally, a timeout for the whole preprocessor can be specified.

2.2 Simplifying During Search

A trade-off between run time and simplification power can be to apply the simplification also during search, as for example used in lingeling [15] and Crypto-MiniSat [16]. In the SAT solver RISS [17] that natively uses COPROCESSOR 2.0 all simplification techniques could be used for preprocessing. Each technique is randomized during simplification, so that not the same simplification is tried multiple times without noticing that it cannot be applied. The implementation of each algorithm is aware of learned clauses, because not all algorithms can treat both types of clauses equally. The following example shows this effect for SUS: Let the formula be $\{\{a, b, c, \overline{d}\}, \{a, d\}\}$. A learned clause could be the clause $\{a, b, c\}$. By subsumption, $\{a, b, c, \overline{d}\}$ is removed, and afterwards the removal strategy of the solver could decide to remove $\{a, b, c\}$ again. The invalid resulting formula is $\{\{a, d\}\}$. The correct behavior would torn $\{a, b, c\}$ into a clause of the original formula that cannot be removed by the removal strategy.

2.3 Implementation Details

COPROCESSOR 2.0 schedules its simplification algorithms in a predefined way, if the user does not define an order. This order is: BCP, PL, SUS, EE, UHD, HTE, BCE, BVE, PRB and VIV. The cheap techniques are executed first, more expensive simplifications like BVE are scheduled thereafter and finally deduction methods as PRB and VIV are applied. Whenever a technique can achieve a reduction, the procedure starts from top with UP again. Furthermore, for each simplification algorithm a queue of elements to process is used to only try simplifications on promising literals.

To provide access to single techniques, the execution order of algorithms can be given to the preprocessor. A simple grammar enables to also run a combination of techniques until a fix point is reached, before the next technique is executed. This feature is especially valuable if new techniques should be tested, or to normalize formulae before using another tool. For the normalization the formula can also be sorted before it is printed.

The preprocessor is part of the SAT solver RISS 2.0 and is implemented in C++. Priority queues are implemented as binary heap. Sets are implemented as a combination of a look-up table and a vector, because the number of elements is known before the algorithm execution. Finally, whenever it should be checked whether elements of a clause C are part of another clause, a vector is used that stores a value i_l per element l and a reference value $r \geq 1$. Whenever $i_l = r$, the element is in the set, otherwise it is not. Removing the element l is done by setting $i_l = 0$. If all elements in the set should be removed, r is incremented. When r reaches the maximum value of its type, all values i have to be set to $i = 0$ and r is set to $r = 1$. This way, a single value per literal is sufficient for set checks with very small overhead. Literals or clause references are not used as index, because the vector is shared among all techniques and collisions should be avoided. Besides the *occurrence lists* for clauses, that store a list of all clauses C per literal l where $l \in C$, PRB also maintains the data structures for the two-watched-literal unit propagation. In total, most of these structures trade space for time so that the memory consumption of the tool can be very high for large formula.

Since COPROCESSOR 2.0 is a complex tool that needs to take care about its current state, the interaction of all implemented techniques and still should provide a fast implementation, the preprocessor has been tested with CNF debugging tools [18]. Within 48 hours no single error was found for more than 20 million generated test instances. *Lingeling* [15] has been used as SAT solver for these experiments.

3   Empirical Results

In this section the default configuration of COPROCESSOR 2.0 is evaluated. Due to the fact that both unpublished techniques are disabled during the experiments, the performance of the preprocessor combined with a SAT solver is very similar to the performance with the SatElite preprocessor, because the performance of the SAT solvers does not benefit as much of the other techniques as from BVE, and COPROCESSOR 2.0 even limits this technique.

The reduction of application and crafted instances with respect to the number of clauses is presented for the three preprocessors SATELITE, COPROCESSOR and COPROCESSOR 2.0. In total, 658 application instances and 560 crafted instances from the SAT competitions and SAT races since 2008 have been used as the underlying benchmark[2]. Table 1 shows three kinds of data: (i) the number of clauses in the formula relative to the input formula after the input formula has been processed by the given tool, (ii) the time it took to handle the instance and (iii) the total number of instances that have been altered by the tool. For both the clause reduction and the used time, the median value is presented for the whole benchmark set. Whereas the reduction is only measured on instances that have been finished to be processed, the time is cumulated over all benchmark instances. Preprocessing a formula is aborted after a timeout of 10 minutes.

In total, COPROCESSOR times out on 20 instances, COPROCESSOR 2.0 on 15 and SATELITE times out on no instance. On the other hand, COPROCESSOR 2.0

Table 1. Relative Reduction of the Number of Clauses

Preprocessors	SATELITE			COPROCESSOR			COPROCESSOR 2.0		
Category	clauses	time	touched	clauses	time	touched	clauses	time	touched
Application	62.1 %	4 s	639	55.9 %	17 s	642	61.8 %	7 s	647
Crafted	98.8 %	2 s	399	101.1 %	3 s	513	90.4 %	3 s	435

can handle more instances than its competitors. Since a run time limit can be specified, timeouts are no problem in praxis. Comparing the reduction of the tools, COPROCESSOR results in the best reduction of the number of clauses and the other two tools behave almost similar, only the run time of COPROCESSOR 2.0 is a little higher. For crafted instances the picture is different: COPROCESSOR 2.0 can reduce the size of the formula much more than the other two tools by keeping a comparable run time. COPROCESSOR even increases the number of clauses in the formula. This effect has already been seen in [1] and is caused by a technique based on extended resolution that is not yet used in COPROCESSOR 2.0.

With the default configuration, the performance of state-of-the-art SAT solvers, such as GLUCOSE 2, can be improved slightly for the benchmark instances we used for the comparison. Improvements can be achieved by tuning the execution order of the simplification techniques and their execution limits. Since each SAT solver behaves different, the tool is not tuned for a specific SAT solver and benchmark.

4   Conclusion

We present a CNF simplifier that can replace the most widely used preprocessor SATELITE, because it provides similar run time, better reduction, more flexibility and other valuable features.

The flexibility comes on the one hand from the huge number of simplification techniques and on the other hand from the ability to choose the execution order and execution limits per simplification technique. Since adding new techniques is comparably simple, COPROCESSOR 2.0 will be updated with more methods, for example to use *hidden literal addition* [4] also during *blocked clause elimination*. By turning on and off single techniques and the ability to exclude variables from the simplification, CNF simplifications now also reach neighboring fields such as incremental SAT solving of MaxSat. Therefore, COPROCESSOR 2.0 is a nice tool for both researchers and applicants that encode instances from a high level description into SAT or that want to apply simplifications before they process a CNF formula.

References

1. Manthey, N.: Coprocessor – a Standalone SAT Preprocessor. In: Proceedings of the 25th Workshop on Logic Programming, WLP 2011, Infsys Research Report 1843-11-06, Technische Universität Wien, 99–104 (2011)
2. Eén, N., Biere, A.: Effective Preprocessing in SAT Through Variable and Clause Elimination. In: Bacchus, F., Walsh, T. (eds.) SAT 2005. LNCS, vol. 3569, pp. 61–75. Springer, Heidelberg (2005)
3. Järvisalo, M., Biere, A., Heule, M.: Blocked Clause Elimination. In: Esparza, J., Majumdar, R. (eds.) TACAS 2010. LNCS, vol. 6015, pp. 129–144. Springer, Heidelberg (2010)
4. Heule, M., Järvisalo, M., Biere, A.: Clause Elimination Procedures for CNF Formulas. In: Fermüller, C.G., Voronkov, A. (eds.) LPAR-17. LNCS, vol. 6397, pp. 357–371. Springer, Heidelberg (2010)
5. Berre, D.L., Parrain, A.: The Sat4j library, release 2.2. JSAT 7(2-3), 56–59 (2010)
6. Eén, N., Sörensson, N.: Temporal induction by incremental SAT solving. Electr. Notes Theor. Comput. Sci. 89(4), 543–560 (2003)
7. Lynce, I., Marques-Silva, J.: Probing-Based Preprocessing Techniques for Propositional Satisfiability. In: Proceedings of the 15th IEEE International Conference on Tools with Artificial Intelligence, ICTAI 2003. IEEE Computer Society (2003)
8. Li, C.M.: A constraint-based approach to narrow search trees for satisfiability. Inf. Process. Lett. 71, 75–80 (1999)
9. Piette, C., Hamadi, Y., Saïs, L.: Vivifying propositional clausal formulae. In: 18th European Conference on Artificial Intelligence (ECAI 2008), Patras, Greece, pp. 525–529 (2008)
10. Heule, M.J.H., Järvisalo, M., Biere, A.: Efficient CNF Simplification Based on Binary Implication Graphs. In: Sakallah, K.A., Simon, L. (eds.) SAT 2011. LNCS, vol. 6695, pp. 201–215. Springer, Heidelberg (2011)
11. Van Gelder, A.: Toward leaner binary-clause reasoning in a satisfiability solver. Ann. Math. Artif. Intell. 43(1), 239–253 (2005)
12. Manthey, N., Steinke, P.: Quadratic Direct Encoding vs. Linear Order Encoding. Technical Report 3, Knowledge Representation and Reasoning Group, Technische Universität Dresden, 01062 Dresden, Germany (2011)
13. Chen, J.: A New SAT Encoding of the At-Most-One Constraint. In: Proceedings of ModRef 2011 (2011)
14. Wernhard, C.: Computing with Logic as Operator Elimination: The ToyElim System. In: Proceedings of the 25th Workshop on Logic Programming, WLP 2011, Infsys Research Report 1843-11-06, Technische Universität Wien, pp. 94–98 (2011)
15. Biere, A.: Lingeling, Plingeling, PicoSAT and PrecoSAT at SAT Race 2010. Technical Report 10/1, Institute for Formal Models and Verification, Johannes Kepler University (2010)
16. Soos, M.: CryptoMiniSat 2.5.0. In: SAT Race Competitive Event Booklet (2010)
17. Manthey, N.: Solver Description of riss and priss. Technical Report 2, Knowledge Representation and Reasoning Group, Technische Universität Dresden, 01062 Dresden, Germany (2012)
18. Brummayer, R., Biere, A.: Fuzzing and delta-debugging smt solvers. In: Proceedings of the 7th International Workshop on Satisfiability Modulo Theories, SMT 2009, pp. 1–5. ACM, New York (2009)

SMT-RAT: An SMT-Compliant Nonlinear Real Arithmetic Toolbox (Tool Presentation)

Florian Corzilius, Ulrich Loup, Sebastian Junges, and Erika Ábrahám

RWTH Aachen University, Germany

Abstract. We present SMT-RAT, a C++ toolbox offering theory solver modules for the development of SMT solvers for *nonlinear real arithmetic (NRA)*. NRA is an important but hard-to-solve theory and only fragments of it can be handled by some of the currently available SMT solvers. Our toolbox contains modules implementing the *virtual substitution* method, the *cylindrical algebraic decomposition* method, a *Gröbner bases simplifier* and a *general simplifier*. These modules can be combined according to a user-defined strategy in order to exploit their advantages.

1 Introduction

The *Satisfiability-Modulo-Theories* (SMT) problem is the problem of checking *SMT formulas*, i.e. Boolean combinations of constraints of one or more theories, for satisfiability. SMT solvers use a SAT solver to find satisfying solutions for the Boolean skeleton of an input SMT formula, which are in turn checked for consistency with other decision procedures for the underlying theories.

The last decade brought great achievements in the field of SMT solving. For instance, the SMT-LIB standard defines a common input format for SMT solvers and provides the community with benchmarks for different theories. In addition, SMT competitions motivate the development and improvement of SMT solvers. Nowadays, different efficient SMT solvers are available for several theories, e.g., for linear real arithmetic. However, only a few solvers support *nonlinear real arithmetic (NRA)*, the theory of the reals with addition and multiplication.

Nonlinear real arithmetic was shown to be decidable by Tarski [16]. Though the worst-case time complexity of solving real-arithmetic formulas is doubly exponential in the number of variables [18], its existential fragment, which is addressed by SMT solving, can be solved in exponential time [13]. One of the most widely used decision procedures for NRA is the *cylindrical algebraic decomposition (CAD)* method [6]. Other well-known methods use, e.g., *Gröbner bases* [17] or the *realization of sign conditions* [2]. Some incomplete methods based on, e.g., *interval constraint propagation (ICP)* [12] and *virtual substitution (VS)* [19], can handle significant fragments and, even though they have the same worst-case complexity as the complete methods, they are more efficient in practice. Moreover, they are well-suited for a combination with complete methods, to which they pass reduced sub-problems.

A. Cimatti and R. Sebastiani (Eds.): SAT 2012, LNCS 7317, pp. 442–448, 2012.

The methods mentioned above are implemented in different tools. For example, QEPCAD [4] implements the CAD method, the Redlog package [11] of the computer algebra system Reduce offers an optimized combination of the VS, the CAD, and Gröbner bases methods, and RAHD [15] combines different methods by a user-defined strategy. The strength of these tools lies in solving *conjunctions* of real-arithmetic constraints, but they are not designed for formulas with arbitrary Boolean structures. A natural way to join the advantages of these tools with those of SAT solving suggests their embedding in an SMT solver.

There are some SMT solvers available which support fragments of NRA. Z3 [20] applies an optimized combination of linear arithmetic decision procedures, ICP and the VS method. MiniSmt tries to reduce NRA problems to linear real arithmetic and HySAT/iSAT uses ICP. All these SMT solvers are incomplete for NRA, i.e., they can not check satisfiability for all real-arithmetic SMT formulas.

The development of a complete SMT solver for NRA is problematic because the aforementioned algebraic decision procedures are not *SMT-compliant*, i.e., they do not fulfill the requirements for the embedding into an efficient SMT solver. Firstly, in less lazy SMT solving, theory solvers should be able to work *incrementally*, i.e., if they determine the satisfiability of a set of constraints, they should be able to check an extended set of constraints on the basis of the previous result. Secondly, in case a constraint set is unsatisfiable, theory solvers should be able to compute an *infeasible subset* as explanation. Thirdly, they must be able to *backtrack* according to the search of the SAT solver.

In this paper, we present the open-source C++ toolbox SMT-RAT, which implements real-arithmetic constraint simplifier and theory solver modules suited for the embedding into an SMT solver. Besides standard libraries, SMT-RAT invokes only the libraries GiNaC and GiNaCRA [14]. The source code with all modules and a manual with examples can be found at http://smtrat.sourceforge.net/. Our toolbox SMT-RAT offers an incremental implementation of the VS method [1,7], which can generate infeasible subsets and supports backtracking. It also provides two incremental implementations of the CAD method. One can handle the multivariate case, whereas the other one is specialized on univariate instances only and can generate infeasible subsets. Furthermore, two simplifier modules are available based on smart simplifications [10] and Gröbner bases, respectively.

This is the first release of our toolbox. At this stage, we do not aim at competing with state-of-the-art solvers in all categories. For example, we do not yet offer extensive simplifiers, ICP, or theory solver modules for linear arithmetic. The main advantages of our toolbox lie in offering (1) *complete* SMT-compliant decision procedures, (2) the possibility to *combine* theory solvers according to a user-defined strategy, and (3) a *modular* and *extendable open-source* implementation. Syntactically, our strategies are more simple than those proposed in [9]. However, we choose a procedure depending on not only the formula but also on the history of solving, offering a very flexible approach.

We use SMT-RAT to extend the open-source SMT solver OpenSMT [5] by the theory NRA. First experimental results comparing this tool with Z3 and CVC3

[8] indicate that for some highly nonlinear benchmark sets we are able to solve a much larger number of instances, but for some other benchmark sets we still need further improvements.

In the following, we first give some preliminaries in Section 2 and a short introduction to the toolbox design in Section 3. We give some experimental results in Section 4 and conclude the paper in Section 5.

2   Satisfiability Modulo Real Arithmetic

SMT solving denotes an algorithmic framework for solving Boolean combinations of constraints from some theories. SMT solvers combine a SAT solver computing satisfying assignments for the Boolean structure of the SMT formula with procedures to check the consistency of theory constraints. For more details on SMT related topics we refer to [3, Ch. 26].

We consider *NRA formulas* φ, which are Boolean combinations of *constraints* c comparing *polynomials* p to 0. A polynomial p can be a constant, a variable x, or a composition of polynomials by addition, subtraction or multiplication:

$$
\begin{aligned}
p &::= &0 &\mid 1 &\mid x &\mid (p+p) \mid (p-p) \mid (p \cdot p) \\
c &::= &p = 0 &\mid p < 0 &\mid p > 0 \\
\varphi &::= &c &\mid (\neg\varphi) &\mid (\varphi \wedge \varphi) \mid (\exists x \varphi)
\end{aligned}
$$

The semantics of NRA formulas is defined as usual.

Given a polynomial $p = a_1 x_1^{e_{1,1}} \cdots x_n^{e_{n,1}} + \cdots + a_k x_1^{e_{1,k}} \cdots x_n^{e_{n,k}}$ in monomial normal form, by $\deg(p) := \max_{1 \leq j \leq k}(\sum_{i=1}^{n} e_{i,j})$ we denote the *degree of* p. We call an NRA formula φ *linear* if $\deg(p) \leq 1$ for all polynomials p in φ, and *nonlinear* otherwise. *Linear real arithmetic (LRA)* formulas are linear NRA formulas.

3   Toolbox Design

Our toolbox has a modular C++ class design which can be used to compose NRA theory solvers for an SMT embedding in a *dynamic* and *hierarchic* fashion. Our NRA theory solvers are instances of **Manager**, which offers an interface to communicate with the environment and which coordinates the satisfiability check according to a user-defined **Strategy**. Such a strategy combines basic NRA theory solver modules, derived from **Module**. Figure 1 shows an example configuration. Moreover, a **Java**-based graphical user interface can be used for an intuitive and user-friendly specification of strategies (and the automatic generation of a corresponding **Strategy** class). Next, we briefly describe these concepts. For more details we refer to the manual of SMT-RAT at http://smtrat.sourceforge.net/manual/manual.pdf.

The Formula Class. Formula instances contain, besides a sequence of NRA constraints, a bitvector storing some information about the problem and the history of its check. E.g., there is a bit which is 1 if some of the constraints are equations. Also for each module there is a bit which is 1 if the module was

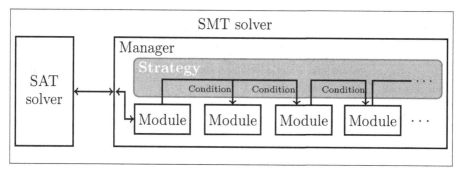

Fig. 1. A snapshot of an SMT-RAT composition embedded in an SMT solver

already invoked on the given problem. Such information can be used to specify conditions under which a procedure should be invoked for a certain problem.

The Module Class. A *module* is an SMT-compliant implementation of a procedure (e.g., constraint simplifier, an incomplete procedure or a complete decision procedure) which can be used for the satisfiability check of NRA formulas. A module's interface allows to add constraints, to push and pop backtrack points, to check the so far added constraints for consistency and to obtain an infeasible subset of these constraints if they are detected to be inconsistent.

Modules have the opportunity to call other modules (*backends*) on sub-problems. A novel achievement of our toolbox is that this call hierarchy is *dynamic* and guided by a *user-defined strategy*. Currently, we only support sequential execution, parallel solving is planned for later releases.

Inheritance can be used to extend existing modules. Besides the basic type Module, our toolbox offers five sub-classes. SimplifierModule (SI_M) implements smart simplifications [10], while GroebnerModule (GS_M) simplifies equation systems using Gröbner bases and probably detects inconsistency. The CAD method is implemented in UnivariateCADModule (UC_M) for the univariate case and in CADModule (MC_M) for the general multivariate case. The last module class VSModule (VS_M) implements a version of the VS method as published in [7].

The Strategy Class. SMT-RAT offers a framework to integrate single modules to powerful and efficient *composed theory solvers*. The composition relies on a user-defined Strategy that specifies for each Formula instance which module should be used for its check. A strategy is basically a sequence of condition-module pairs. For each Formula instance, it determines the first module whose condition evaluates to true on the bitvector of the formula. E.g., the strategy "c_1 ? (m_1) : (c_2 ? (m_2) : (m_3))" determines m_1 as module type for φ if the bitvector of φ fulfills the condition c_1. If φ does not fulfill c_1 but c_2, then an instance of m_2 is called, otherwise of m_3.

The Manager Class. The Manager contains references to the available module instances and to the user-defined strategy. It manages, on the one hand, the creation and linking of the modules, and, on the other hand, the communication between them and the environment, e.g., the frontend of an SMT solver.

Table 1. Running times [sec] of Rat_1, Rat_2, Z3 and CVC3 on four benchmarks

	Rat_1		Rat_2		Z3		CVC3	
	solved	acc. time	solved	acc. time	solved	acc. time	solved	acc. time
bouncing ball	43/52	4226.24	43/52	424.63	0/52	0.00	0/52	0.00
etcs	2/5	136.15	2/5	135.05	1/5	42.00	1/5	0.11
rectangular pos.	16/22	305.54	16/22	299.54	22/22	27.29	0/22	0.00
zankl	22/166	26.30	22/166	25.81	62/166	1138.96	9/166	2.86

4   Experimental Results

All experiments were performed on an Intel® Core™ i7 CPU at 2.80 GHz with 4 GB RAM with Gentoo Linux. We defined two strategies

$$c_1 \; ? \; (MC_M) \; : \; (c_2 \; ? \; (UC_M) \; : \; (c_4 \; ? \; (VS_M) \; : \; (SI_M)))$$
$$\text{and} \quad c_1 \; ? \; (MC_M) \; : \; (c_2 \; ? \; (UC_M) \; : \; (c_3 \; ? \; (VS_M) \; : \; (c_4 \; ? \; (GS_M) \; : \; (SI_M))))$$

where c_1, c_2, c_3 and c_4 hold if the UC_M, the VS_M, the GS_M and SI_M was invoked and could not solve the given formula, respectively. We embedded two theory solver components using these strategies into OpenSMT, yielding the SMT solvers Rat_1 and Rat_2, respectively, which we compared to CVC3 2.4 and Z3 3.1, the latter being the winner of last year's SMT competition for NRA.

Table 1 shows the running times in seconds on four benchmark sets with the timeout of 150 seconds. The first one models the nonlinear movement of a bouncing ball which may drop into a hole. The second one is a nonlinear version of the European Train Control System benchmark set. The third one contains problems to checks whether a given set of rectangles fits in a given area. The last benchmark set stems from the SMT competition in 2011.

The results show, that we can solve many examples which Z3 and CVC3 cannot solve. However, Z3 does a better job in the last two benchmark sets, where the major part of the formula is linear. Here it can benefit from its ICP and Simplex solver checking the linear fragment. Nevertheless, the results point out that we can build efficient SMT solvers for NRA using OpenSMT and SMT-RAT. Furthermore, it indicates that extending SMT-RAT by modules, e.g. implementing Simplex or ICP, would lead to significant improvements.

5   Conclusion and Future Work

SMT-RAT is a toolbox contributing several SMT-compliant simplifier and theory solver modules and a framework to combine them according to a user-defined strategy. Experimental results show that an SMT solver enriched by SMT-RAT for solving NRA can compete with state-of-the-art SMT solvers and even solve instances, which they cannot solve.

The design of SMT-RAT aims at modularity, extensibility, and the easy adding of new modules. Moreover, we plan to improve the performance of SMT-RAT compositions by modules implementing, e.g., Simplex and ICP. Furthermore, we want to extend the framework to allow parallel calls of modules, theory propagation, and shared heuristics.

References

1. Ábrahám, E., et al.: A lazy SMT-solver for a non-linear subset of real algebra. In: Proc. of SMT 2010 (2010)
2. Basu, S., Pollack, R., Roy, M.: Algorithms in Real Algebraic Geometry. Springer (2010)
3. Biere, A., Heule, M., van Maaren, H., Walsh, T. (eds.): Handbook of Satisfiability. Frontiers in Artificial Intelligence and Applications, vol. 185. IOS Press (2009)
4. Brown, C.W.: QEPCAD B: A program for computing with semi-algebraic sets using CADs. SIGSAM Bulletin 37(4), 97–108 (2003)
5. Bruttomesso, R., Pek, E., Sharygina, N., Tsitovich, A.: The OpenSMT Solver. In: Esparza, J., Majumdar, R. (eds.) TACAS 2010. LNCS, vol. 6015, pp. 150–153. Springer, Heidelberg (2010)
6. Collins, G.E.: Quantifier Elimination for Real Closed Fields by Cylindrical Algebraic Decomposition. In: Brakhage, H. (ed.) GI-Fachtagung 1975. LNCS, vol. 33, pp. 134–183. Springer, Heidelberg (1975)
7. Corzilius, F., Ábrahám, E.: Virtual Substitution for SMT-Solving. In: Owe, O., Steffen, M., Telle, J.A. (eds.) FCT 2011. LNCS, vol. 6914, pp. 360–371. Springer, Heidelberg (2011)
8. http://cs.nyu.edu/acsys/cvc3/
9. de Moura, L., Passmore, G.O.: The strategy challenge in SMT solving, http://research.microsoft.com/en-us/um/people/leonardo/mp-smt-strategy.pdf
10. Dolzmann, A., Sturm, T.: Simplification of quantifier-free formulas over ordered fields. Journal of Symbolic Computation 24, 209–231 (1995)
11. Dolzmann, A., Sturm, T.: REDLOG: Computer algebra meets computer logic. SIGSAM Bulletin 31(2), 2–9 (1997)
12. Fränzle, M., et al.: Efficient solving of large non-linear arithmetic constraint systems with complex Boolean structure. Journal on Satisfiability, Boolean Modeling and Computation 1(3-4), 209–236 (2007)
13. Heintz, J., Roy, M.F., Solernó, P.: On the theoretical and practical complexity of the existential theory of the reals. The Computer Journal 36(5), 427–431 (1993)
14. Loup, U., Ábrahám, E.: GiNaCRA: A C++ Library for Real Algebraic Computations. In: Bobaru, M., Havelund, K., Holzmann, G.J., Joshi, R. (eds.) NFM 2011. LNCS, vol. 6617, pp. 512–517. Springer, Heidelberg (2011)
15. Passmore, G.O., Jackson, P.B.: Combined Decision Techniques for the Existential Theory of the Reals. In: Carette, J., Dixon, L., Coen, C.S., Watt, S.M. (eds.) MKM 2009, Held as Part of CICM 2009. LNCS, vol. 5625, pp. 122–137. Springer, Heidelberg (2009)
16. Tarski, A.: A Decision Method for Elementary Algebra and Geometry. University of California Press (1948)

17. Weispfenning, V.: A new approach to quantifier elimination for real algebra. In: Quantifier Elimination and Cylindrical Algebraic Decomposition. Texts and Monographs in Symbolic Computation, pp. 376–392. Springer (1998)
18. Weispfenning, V.: The complexity of linear problems in fields. Journal of Symbolic Computation 5(1-2), 3–27 (1988)
19. Weispfenning, V.: Quantifier elimination for real algebra – The quadratic case and beyond. Applicable Algebra in Engineering, Communication and Computing 8(2), 85–101 (1997)
20. http://research.microsoft.com/en-us/um/redmond/projects/z3/

CoPAn: Exploring Recurring Patterns in Conflict Analysis of CDCL SAT Solvers (Tool Presentation)*

Stephan Kottler, Christian Zielke, Paul Seitz, and Michael Kaufmann

University of Tübingen, Germany

Abstract. Even though the CDCL algorithm and current SAT solvers perform tremendously well for many industrial instances, the performance is highly sensitive to specific parameter settings. Slight modifications may cause completely different solving behaviors for the same benchmark. A fast run is often related to learning of 'good' clauses.

Our tool CoPAn allows the user for an in-depth analysis of conflicts and the process of creating learnt clauses. Particularly we focus on isomorphic patterns within the resolution operation for different conflicts. Common proof logging output of any CDCL solver can be adapted to configure the analysis of CoPAn in multiple ways.

1 Introduction and Motivation

Though the vast success of the CDCL approach to SAT solving [12,16,6] is well-documented, it is not fully understood, why small changes in the choice of parameters may cause significantly different behavior of the solver. With the tool presented in this paper, we provide a perspective to find an answer for the question about subtle differences between successful and rather bad solver runs.

Our tool CoPAn, an abbreviation for **Co**nflict **P**attern **An**alysis, can be used to analyse the complete learning and conflict analysis of a CDCL solver using the common proof logging output of the systems. Due to the use of efficient external data structures, CoPAn manages to cope with a big amount of logged data.

The influence of learning to the SAT solvers' efficiency is undeniable [10]. On the one hand new measures for the quality of learnt clauses based on the observation of CDCL solvers on industrial instances [4] were proposed. On the other hand it is very promising to turn away from these static measures that cause a definitive elimination of clauses and focus on a dynamical handling of learnt clauses [3]. Therefore changes in learning schemes can lead to a considerable speed-up of the solving process (see [15,4,8]).

We are convinced that CoPAn can help to obtain a better understanding about when and how good clauses are learnt by the SAT solver. We provide a tool for in-depth analysis of conflicts and the associated process of producing learnt clauses.

---

* This work was supported by DFG-SPP 1307, project "Structure-based Algorithm Engineering for SAT-Solving"

A. Cimatti and R. Sebastiani (Eds.): SAT 2012, LNCS 7317, pp. 449–455, 2012.

In Section 2 we introduce CoPAn and its usability, we present its most important features, and delineate the option of enhancing the data linked to each clause such that the focus of analysis can be adapted to meet the requirements of different users. Section 3 covers the theoretical and algorithmic background of the tool. We summarize the functionality in Section 4. The tool can be downloaded from `http://algo.inf.uni-tuebingen.de/?site=forschung/sat/CoPAn`.

2   The Application

CoPAn has been designed for analyzing CDCL like solvers in terms of their learning and conflict characteristics. This chapter describes the main functionality of the tool and the most important and new ideas of the analysis. The vast functionality of CoPAn can all be used by the GUI version. Moreover, all procedures where visualization is not required can also be run as batch jobs to allow for extensive unattended computations.

2.1   Patterns in Clause Learning

Practical research in SAT solving has shown that the performance of a CDCL solver to solve one benchmark often depends (amongst others) on the quality of learning and on rules about when and how to reduce the set of learnt clauses [6,5,4,3]. It has been shown long time ago that learning the first unit implication point (FUIP) clearly outperforms other learning schemes [16]. By FUIP those clauses are learnt that are very close to the conflict itself. However, a learning scheme does not imply anything about the way how resolution operation are applied nor about the number of resolutions to terminate learning when the first unit implication point is reached.

In contrast to any generic learning scheme we focus on different resolution trees that are applied to learn new clauses. At first CoPAn allows for visualizing the resolution trees that were logged by a CDCL solver. The main focus, however, is put on the patterns that can be observed in the resolution trees of different conflicts rather than pure visualization. In difference to the implication graph as it is used in [12,15] we consider a resolution graph (tree) that contains one node for each clause contributing to the conflict. For each resolution operation an edge is drawn. In CoPAn we consider the resolution graph with additional edges, a so-called clashing graph: An edge between two nodes (clauses) is drawn iff they share exactly one clashing literal (see [11]).

Definition 1. *Two conflicts c_1, c_2 exhibit the same resolution pattern iff their unlabeled resolution graphs g_1, g_2 are isomorphic. If a resolution graph g_1 is isomorphic to a subgraph of another resolution graph g_2 we say that the conflicts c_1 and c_2 have common subpatterns.*

The decision to focus on clashing graphs rather than on common implication graphs is due to its direct relation to resolution. Isomorphic subpatterns in different clashing graphs are likely to allow for similar resolution operation.

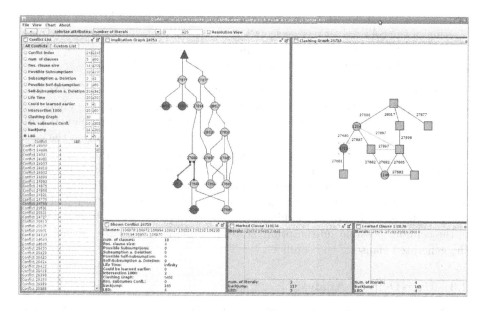

Fig. 1. Main view of the CoPAn GUI

2.2   The CoPAn GUI

Figure 1 shows the main view of the CoPAn GUI. The GUI variant always expects one solver run for a particular benchmark to be loaded. Via the panel on the left side any conflict can be selected for visualization and deeper interactive analysis. The left graph shows the common implication graph of the selected conflict. In the right panel the clashing graph is depicted. The subgraph with black edges specifies the resolution operation for the selected conflict. Green edges indicate alternative resolution operations. In contrast to the clashing rule red edges can be shown to connect clauses with equal literals. Elliptic nodes indicate learnt clauses. These nodes may be selected in the graph view to jump to the according conflict to trace reasons of a conflict.

The three panels at the bottom state some additional information about the selected conflict, the selected node (clause) and the generated clause.

One of the most advanced features in CoPAn is the search and filter functionality that goes along with resolution patterns. Any resolution pattern that is shown in the main window can be used to filter all conflicts of the solver's run. This allows a user to search for conflicts that exhibit similar resolution patterns even if the pattern under consideration only constitutes a subgraph of another learning operation. In Section 3 we describe how these computationally expensive operations are realized for suitable interactive work.

To allow for in-depth analysis of conflicts CoPAn enables filtering and search for several different properties and their combinations. Basically, there are two kinds of properties of conflicts:

– User-defined properties that can be specified by additional logging information of the SAT solver (see section 2.3). A typical user-defined property may
be the LBD value [4] or the backjumping level of a conflict.
– Pre-defined properties that require extra computational work.

The demand of pre-defined properties arose when an early version of CoPAn had
been implemented and new, more complex queries became desirable. There are
several papers that point out or implicitly use the fact that two learnt clauses may
contain a high percentage of equal literals [9,2,14]. Moreover, it is often the case
that a learnt clause subsumes another clause. This motivated the implementation
of subsumption checks. CoPAn can now indicate correlations between similar
resolution patterns and subsumption of conflicts. In Figure 2 a particular conflict
of the solving process of instance *li-exam-63* is selected. In the left panel only
those conflicts are listed that exhibit an isomorphic resolution pattern. A value
in the column 'possible subsumption' states how many clauses are subsumed by
the generated clause of the conflict.

2.3   User-Defined Analysis

To make CoPAn usable for different or changing groups of interest the tool
offers a configurable and extendable interface for analysis. For common use a
SAT solver has to log each clause that was created during conflict analysis.
Optionally, arbitrary properties can be added to the clause as simple key-value
pairs (e.g. a pair 'b 3' could be added to indicate that this conflict caused a
backjump to level 3). CoPAn allows for filtering and searching for user-defined
properties. A user may be interested in the question if conflicts with the same
resolution pattern have similar LBD values or any new measure of quality. For
a detailed description we refer the reader to the user guide which is available
online.

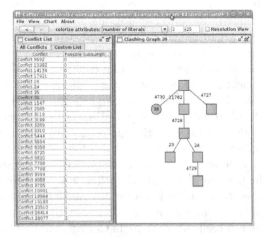

Fig. 2. Relating subsumption and resolution patterns

3   Theoretical and Algorithmic Background

Learning from conflicting assignments causes a CDCL based SAT solver to generate large amounts of clauses during the solving process. Printing these clauses and the resolution operations produces millions of bytes of data in short time. To cope with as many conflicts as possible any tool operating with these data volumes has to use highly efficient data structures that support fast lookups.

To reduce additional software dependencies and for user convenience we do not use any kind of database system for these lookups, but rather use an efficient implementation of the well-known data structure *B+-tree*. With the use of indices for every implemented filter we ensure the required efficient lookup costs and therefore a smooth behavior of the GUI tool, as well as fast computation of query results. These index structures have to be built in advance of any following investigation and analysis and require the majority of a typical work flow. But once these *B+-trees* and corresponding index structures are built further operations can be executed vastly efficient.

A main feature of CoPAn is the ability to compare the structure of clashing graphs and different isomorphic patterns within these graphs. For each conflict that arose during SAT solving a clashing graph is created containing one node for each clause that contributed to this conflict. An edge is drawn iff two clauses have exactly one clashing literal [11]. In general, it is unknown whether the graph isomorphism problem is in P or NP-complete [7], whereas the subgraph isomorphism problem is known to be NP-complete.

Therefore we only consider those edges of a clashing graph that were actually used for resolution to create the learnt clause. This is a tree structure with the clause causing the conflict as the root node. Given the fact that the tree isomorphism problem is known to be in P, we are able to efficiently detect isomorphic clashing graph structures of conflicts in linear time [1].

The use of clashing graphs to compare the structures of conflicts results in a loss of information regarding the resolution sequence within the solving process, but is essential to check for isomorphisms efficiently. CoPAn is able to detect and depict alternative resolutions within the clashing graph structures.

4   Summary

We present our tool CoPAn that was designed to analyse the learning behaviour of SAT solvers. Schemes and rules for learning have been proven to be significant for the performance of CDCL solvers. In contrast to previous analysis of learning we focus on patterns in resolution graphs. To the knowledge of the authors CoPAn is the first application that analyses resolution patterns of SAT solvers. It facilitates the study of correlations between several properties of learnt clauses. A typical task could be to search for significantly common properties among conflicts with isomorphic resolution pattern. Properties of clauses may be well-known attributes such as backjumping distance [12], activity [13,6], LBD value [4] but also user-defined properties. The most important characteristics of CoPAn can be summarized:

- Search for isomorphic resolution patterns (within one instance or within a group of several instances)
- Subsumption checks of (subsets of) clauses
- Specification of user-defined properties that can be linked to clauses
- GUI to visualize conflicts, to explore the effect of various filters, and to trace learning and proofs interactively
- Preprocessing of logged data to build efficient data structures and indices for further examinations
- Creation of diagrams for applied filters to plot the distribution of attributes
- Batch processing to analyse sets of instances

We are convinced that CoPAn can support developers of SAT solvers to analyse the behavior of their solvers. Moreover, CoPAn may be used to evaluate new quality measures for learnt clauses.

For future work we plan to realize more advanced isomorphism tests, not only restricted to tree-like patterns. Moreover, the complete clashing graph of a SAT instance shall be analyzable such that a user may search for isomorphic patterns within the complete instance. Furthermore, we plan to incorporate the feedback and suggestions of users and solver engineers.

References

1. Aho, A.V., Hopcroft, J.E.: The Design and Analysis of Computer Algorithms. Addison-Wesley, Boston (1974)
2. Audemard, G., Katsirelos, G., Simon, L.: A restriction of extended resolution for clause learning sat solvers. In: AAAI 2010, pp. 10–15 (2010)
3. Audemard, G., Lagniez, J.-M., Mazure, B., Saïs, L.: On Freezing and Reactivating Learnt Clauses. In: Sakallah, K.A., Simon, L. (eds.) SAT 2011. LNCS, vol. 6695, pp. 188–200. Springer, Heidelberg (2011)
4. Audemard, G., Simon, L.: Predicting learnt clauses quality in modern sat solvers. In: IJCAI, pp. 399–404 (2009)
5. Biere, A.: Picosat essentials. JSAT 4, 75–97 (2008)
6. Eén, N., Sörensson, N.: An Extensible SAT-solver. In: Giunchiglia, E., Tacchella, A. (eds.) SAT 2003. LNCS, vol. 2919, pp. 502–518. Springer, Heidelberg (2004)
7. Garey, M.R., Johnson, D.S.: Computers and Intractability; A Guide to the Theory of NP-Completeness. W. H. Freeman & Co. (1990)
8. Hamadi, Y., Jabbour, S., Sais, L.: Learning for Dynamic Subsumption. International Journal on Artificial Intelligence Tools 19(4) (2010)
9. Huang, J.: Extended clause learning. Artif. Intell. 174(15), 1277–1284 (2010)
10. Katebi, H., Sakallah, K.A., Marques-Silva, J.P.: Empirical Study of the Anatomy of Modern Sat Solvers. In: Sakallah, K.A., Simon, L. (eds.) SAT 2011. LNCS, vol. 6695, pp. 343–356. Springer, Heidelberg (2011)
11. Kullmann, O.: The Combinatorics of Conflicts between Clauses. In: Giunchiglia, E., Tacchella, A. (eds.) SAT 2003. LNCS, vol. 2919, pp. 426–440. Springer, Heidelberg (2004)
12. Marques-Silva, J.P., Sakallah, K.A.: Grasp: A search algorithm for propositional satisfiability. IEEE Trans. Comput. 48(5), 506–521 (1999)

13. Moskewicz, M.W., Madigan, C.F., Zhao, Y., Zhang, L., Malik, S.: Chaff: engineering an efficient SAT solver. In: DAC (2001)
14. Sinz, C.: Compressing Propositional Proofs by Common Subproof Extraction. In: Moreno Díaz, R., Pichler, F., Quesada Arencibia, A. (eds.) EUROCAST 2007. LNCS, vol. 4739, pp. 547–555. Springer, Heidelberg (2007)
15. Sörensson, N., Biere, A.: Minimizing Learned Clauses. In: Kullmann, O. (ed.) SAT 2009. LNCS, vol. 5584, pp. 237–243. Springer, Heidelberg (2009)
16. Zhang, L., Madigan, C.F., Moskewicz, M.H., Malik, S.: Efficient conflict driven learning in a Boolean satisfiability solver. In: ICCAD 2001 (2001)

Azucar: A SAT-Based CSP Solver Using Compact Order Encoding (Tool Presentation)

Tomoya Tanjo[1], Naoyuki Tamura[2], and Mutsunori Banbara[2]

[1] Transdisciplinary Research Integration Center, Japan
[2] Information Science and Technology Center, Kobe University, Japan
tanjo@nii.ac.jp, {tamura,banbara}@kobe-u.ac.jp

Abstract. This paper describes a SAT-based CSP solver Azucar. Azucar solves a finite CSP by encoding it into a SAT instance using the compact order encoding and then solving the encoded SAT instance with an external SAT solver. In the compact order encoding, each integer variable is represented by using a numeral system of base $B \geq 2$ and each digit is encoded by using the order encoding. Azucar is developed as a new version of an award-winning SAT-based CSP solver Sugar. Through some experiments, we confirmed Azucar can encode and solve very large domain sized CSP instances which Sugar can not encode, and shows better performance for Open-shop scheduling problems and the Cabinet problems of the CSP Solver Competition benchmark.

1 Introduction

A (finite) Constraint Satisfaction Problem (CSP) is a combinatorial problem to find an assignment which satisfies all given constraints on finite domain variables [1]. A SAT-based CSP solver is a program which solves a CSP by encoding it to SAT [2] and searching solutions by SAT solvers.

There have been several SAT-based CSP solvers developed, such as Sugar [1] [3], FznTini [4], SAT4J CSP [5], and others. Especially, Sugar became a winner of several categories at the recent International CSP Solver Competitions in two consecutive years. It uses order encoding [6] which shows good performance on various applications [6–8] and is known as the only SAT encoding reducing tractable CSP to tractable SAT [9]. In the order encoding, the Unit Propagation in SAT solvers corresponds to the Bounds Propagation in CSP solvers.

In this paper, we describe a SAT-based CSP solver Azucar [2]. It uses a new SAT encoding method named compact order encoding [10–12], in which each integer variable is divided into digits by using a numeral system of base $B \geq 2$ and each digit is encoded by using the order encoding. Therefore, it is equivalent to the order encoding [6] when $B \geq d$ and it is equivalent to the log encoding [13]

---

[1] http://bach.istc.kobe-u.ac.jp/sugar/
[2] http://code.google.com/p/azucar-solver/

A. Cimatti and R. Sebastiani (Eds.): SAT 2012, LNCS 7317, pp. 456–462, 2012.

when $B = 2$ where d is the maximum domain size. In that sense, the compact order encoding is the generalization and integration of both encodings.

Size of generated SAT instances by the compact order encoding using two or more digits are much smaller than those by the direct [14], support [15], and order encodings. Therefore, it is another encoding applicable to large domain sized CSPs besides the log and log-support [16] encodings. In the compact order encoding, the Unit Propagation in SAT solvers corresponds to the Bounds Propagation in the most significant digit in CSP solvers. Therefore, the conflicts are likely to be detected with fewer decisions than the log and log-support encodings. These observations are confirmed through the experimental results on Open-Shop Scheduling problems with large domain sizes in which the compact order encoding is about 5 times faster than the log encoding on average.

Azucar is a first implementation of the compact order encoding and it is developed as an enhancement version of Sugar. User can specify either the number of digits m or the base B as the command line option of Azucar. When $m = 1$ is specified, Azucar uses the order encoding to encode the given CSP. The log encoding is used when $B = 2$ is specified. If user specifies neither m nor B, Azucar uses $m = 2$ by default. In various problems, Azucar with $m \in \{2, 3\}$ shows the better performance than Sugar especially for large domain sized CSP.

2   Compact Order Encoding

The basic idea of the compact order encoding is the use of a numeral system of base $B \geq 2$ [10–12]. That is, each integer variable x is represented by a summation $\sum_{i=0}^{m-1} B^i x^{(i)}$ where $m = \lceil \log_B d \rceil$ and $0 \leq x^{(i)} < B$ for all integer variables $x^{(i)}$, and each $x^{(i)}$ is encoded by using the order encoding. As described in Section 1, the compact order encoding is equivalent to the order encoding [6] when $B \geq d$ and it is equivalent to the log encoding [13] when $B = 2$.

In this paper, we will show some examples to encode an integer variable and a constraint by using the compact order encoding. More details about the compact order encoding are described in [10–12].

For example, when we choose $B = 3$, the integer variable $x \in \{0..8\}$ is divided into two digits as follows. Each $x^{(i)}$ represents the i-th digit of x and $x^{(1)}$ represents the most significant digit.

$$x^{(1)}, x^{(0)} \in \{0..2\}$$

By using the order encoding, these propositional variables are introduced where $p(x^{(i)} \leq a)$ is defined as true if and only if the comparison $x^{(i)} \leq a$ holds. $p(x^{(1)} \leq 2)$ is not necessary since $x^{(1)} \leq 2$ is always true.

$$p(x^{(1)} \leq 0) \quad p(x^{(1)} \leq 1)$$
$$p(x^{(0)} \leq 0) \quad p(x^{(0)} \leq 1)$$

To represents the order of propositional variables, these two clauses are required. For instance, $\neg p(x^{(1)} \leq 0) \vee p(x^{(1)} \leq 1)$ represents $x^{(1)} \leq 0 \Rightarrow x^{(1)} \leq 1$.

$$\neg p(x^{(1)} \leq 0) \vee p(x^{(1)} \leq 1) \qquad \neg p(x^{(0)} \leq 0) \vee p(x^{(0)} \leq 1)$$

Constraint	Order Encoding	Compact Order Encoding	Log Encoding
$x \leq a$	$O(1)$	$O(m)$	$O(\log_2 d)$
$x \leq y$	$O(d)$	$O(mB)$	$O(\log_2 d)$
$z = x + a$	$O(d)$	$O(mB)$	$O(\log_2 d)$
$z = x + y$	$O(d^2)$	$O(mB^2)$	$O(\log_2 d)$
$z = xy$	$O(d^2)$	$O(mB^3 + m^2 B^2)$	$O(\log_2^2 d)$

Fig. 1. Comparison of different encodings on the number of SAT-encoded clauses

Each constraint is divided into digit-wise constraints and then encoded by using the order encoding. For example, a constraint $x \leq y$ ($x, y \in \{0..8\}$) is encoded into the following clauses where p is a new propositional variable which represents $\neg(x^{(0)} \leq y^{(0)})$. C_0 and C_1 represent $x^{(1)} \leq y^{(1)}$, C_2, C_3 and C_4 represent $p \to x^{(1)} \leq y^{(1)} - 1$, and C_5 and C_6 represent $\neg(x^{(0)} \leq y^{(0)}) \to p$.

$$C_0: \quad p(x^{(1)} \leq 0) \vee \neg p(y^{(1)} \leq 0)$$
$$C_1: \quad p(x^{(1)} \leq 1) \vee \neg p(y^{(1)} \leq 1)$$
$$C_2: \quad \neg p \vee \neg p(y^{(1)} \leq 0)$$
$$C_3: \neg p \vee p(x^{(1)} \leq 0) \vee \neg p(y^{(1)} \leq 1)$$
$$C_4: \quad \neg p \vee p(x^{(1)} \leq 1)$$
$$C_5: \ p \vee p(x^{(0)} \leq 0) \vee \neg p(y^{(0)} \leq 0)$$
$$C_6: \ p \vee p(x^{(0)} \leq 1) \vee \neg p(y^{(0)} \leq 1)$$

Let d be the domain size of integer variables, B be the base and $m = \lceil \log_B d \rceil$ be the number of digits. Fig. 1 shows the number of clauses required to encode each constraint. In the compact order encoding, each addition $z = x + y$ and multiplication $z = xy$ are encoded into $O(mB^2)$ and $O(mB^3 + m^2 B^2)$ clauses respectively. It is much less than $O(d^2)$ clauses of the order encoding and thus it can be applicable to large domain CSP.

We also show the relations between the Unit Propagation in SAT solvers in each encodings and the constraint propagation in CSP solvers. In the order encoding, the Unit Propagation in SAT solvers corresponds to the Bounds Propagation in CSP solvers. The compact order encoding can achieve the Bounds Propagation in the most significant digit while the log encoding achieves the Bounds Propagation in the most significant bit. Therefore the compact order encoding can detect the conflicts earlier and thus it can solve CSP faster than the log encoding.

3   Azucar Implementation

Azucar is an open-source SAT-based CSP solver distributed under the BSD 3-clause license. Azucar encodes a CSP into SAT by using the compact order encoding, and then the SAT-encoded instance are solved by an external SAT solver such as MiniSat [17], SAT4J [18] or GlueMiniSat [3]. Azucar can handle finite CSP over integers written in Lisp-like input format or XCSP 2.1 format [4] which is used in the 2009 International CSP Solver Competition. Azucar can receive one of these options where d is the maximum domain size of integer variables.

- -b B: Azucar uses the numeral system of base B (i.e. $m = \lceil \log_B d \rceil$).
- -m m: Azucar divides each integer variable into at most m digits (i.e. $B = \lceil \sqrt[m]{d} \rceil$).

The encoder and decoder are written in Java, and the frontend of Azucar is written in Perl.

4   Performance Evaluation

To evaluate the scalability and efficiency of our encoding used in Azucar, we used 85 Open-Shop Scheduling problems with very large domain sizes, which are generated from "j7" and "j8" by Brucker et al. by multiplying the process times by some constant factor c. The factor c is varied within 10^i ($i \in \{0, 1, 2, 3, 4\}$). For example, when $c = 10^4$, the maximum domain size d becomes about 10^7.

We compare four different encodings: the order encoding which is used in Sugar, the compact order encoding with $m \in \{2, 3\}$, and the log encoding. For each instance, we set its makespan to the optimum value minus one and then encode it into SAT. Such SAT-encoded instances are unsatisfiable. We use the MiniSat solver [17] as a backend SAT solver.

Factor c	Domain Size d	#Instances	Order Encoding	Compact Order Encoding $m = 2$	$m = 3$	Log Encoding
1	10^3	17	13	**14**	**14**	**14**
10	10^4	17	12	**13**	**13**	**13**
10^2	10^5	17	8	**13**	**13**	12
10^3	10^6	17	0	**14**	13	12
10^4	10^7	17	0	12	**13**	**13**
	Total	85	34	**66**	**66**	63

Fig. 2. Benchmark results of different encodings on the number of solved instances for OSS benchmark set by Brucker et al. with multiplication factor c

Fig. 2 shows the number of solved instances within 3600 seconds by four solvers. All times were collected on a Linux machine with Intel Xeon 3.0 GHz, 16GB Memory. "Domain size d" indicates the approximate average of domain size of integer variables. We highlight the best number of solved instances.

The compact order encoding solved the most instances for any factor c and totally 66 out of 85 instances rather than 63 by the log encoding and 34 by the

---

[3] http://glueminisat.nabelab.org/
[4] http://www.cril.univ-artois.fr/CPAI08/XCSP2_1.pdf

order encoding. The compact order encoding with $m = 3$ can be highly scalable with the growth of c compared with the order encoding. For example, when $c = 1000$, it solved 13 out of 17 instances (76%), while none (0%) by the order encoding due to the memory limitation. Moreover, it is fastest on average when $c \geq 10$. For example, it solved about 5 times faster than the log encoding when $d \approx 10^7$.

Fig. 3 shows the cactus plot of benchmark results in which the number of solved instances is on the x-axis and the CPU time is on the y-axis. The compact order encoding solved the most instances for almost any CPU time limit. For example, the compact order encoding with $m = 3$ solved 61 instances within 600 seconds while the order encoding was 25, the compact order encoding with $m = 2$ was 56, and the log encoding was 53.

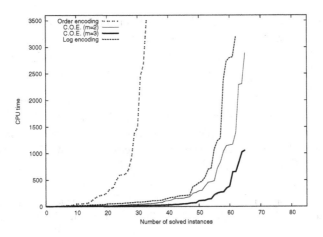

Fig. 3. Cactus plot of various encodings for 85 OSS instances

To evaluate the efficiency of our encoding for smaller domain CSP, we also used graph coloring problems published in Computational Symposium on Graph Coloring and its Generalizations [5] and we confirmed that the compact order encoding can solve the almost same number of instances compared with the order encoding even when the domain size is less than 10^2.

Finally, to evaluate the efficiency of Azucar for large domain CSP, we also used the Cabinet problems in GLOBAL category in the CSP Solver Competitions and we confirmed that Azucar is over 1.7 times faster than Sugar on average.

5   Conclusion

In this paper, we described a SAT-based CSP solver Azucar. Through some experiments, Azucar with $m \in \{2, 3\}$ shows the better performance than Sugar

---

[5] http://mat.gsia.cmu.edu/COLOR04/

especially for large domain sized CSP. Finally, although the compact order encoding used in Azucar is developed to encode CSP, it can be applicable to other problems dealing with the arithmetic constraints.

Acknowledgment. The authors would like to thank Prof. K. Inoue from the National Institute of Informatics for his valuable comments. This work is supported in part by Grant-in-Aid for Scientific Research (B) 2430000.

References

1. Rossi, F., van Beek, P., Walsh, T.: Handbook of Constraint Programming. Elsevier Science Inc. (2006)
2. Prestwich, S.D.: CNF encodings. In: Handbook of Satisfiability, pp. 75–97. IOS Press (2009)
3. Tamura, N., Tanjo, T., Banbara, M.: System description of a SAT-based CSP solver Sugar. In: Proceedings of the 3rd International CSP Solver Competition, pp. 71–75 (2008)
4. Huang, J.: Universal Booleanization of Constraint Models. In: Stuckey, P.J. (ed.) CP 2008. LNCS, vol. 5202, pp. 144–158. Springer, Heidelberg (2008)
5. Le Berre, D., Lynce, I.: CSP2SAT4J: A simple CSP to SAT translator. In: Proceedings of the 2nd International CSP Solver Competition, pp. 43–54 (2008)
6. Tamura, N., Taga, A., Kitagawa, S., Banbara, M.: Compiling finite linear CSP into SAT. Constraints 14(2), 254–272 (2009)
7. Soh, T., Inoue, K., Tamura, N., Banbara, M., Nabeshima, H.: A SAT-based method for solving the two-dimensional strip packing problem. Fundamenta Informaticae 102(3-4), 467–487 (2010)
8. Banbara, M., Matsunaka, H., Tamura, N., Inoue, K.: Generating Combinatorial Test Cases by Efficient SAT Encodings Suitable for CDCL SAT Solvers. In: Fermüller, C.G., Voronkov, A. (eds.) LPAR-17. LNCS, vol. 6397, pp. 112–126. Springer, Heidelberg (2010)
9. Petke, J., Jeavons, P.: The Order Encoding: From Tractable CSP to Tractable SAT. In: Sakallah, K.A., Simon, L. (eds.) SAT 2011. LNCS, vol. 6695, pp. 371–372. Springer, Heidelberg (2011)
10. Tanjo, T., Tamura, N., Banbara, M.: A Compact and Efficient SAT-Encoding of Finite Domain CSP. In: Sakallah, K.A., Simon, L. (eds.) SAT 2011. LNCS, vol. 6695, pp. 375–376. Springer, Heidelberg (2011)
11. Tanjo, T., Tamura, N., Banbara, M.: Proposal of a compact and efficient SAT encoding using a numeral system of any base. In: Proceedings of the 1st International Workshop on the Cross-Fertilization Between CSP and SAT, CSPSAT 2011 (2011)
12. Tanjo, T., Tamura, N., Banbara, M.: Towards a compact and efficient SAT-encoding of finite linear CSP. In: Proceedings of the 9th International Workshop on Constraint Modelling and Reformulation, ModRef 2010 (2010)
13. Iwama, K., Miyazaki, S.: SAT-variable complexity of hard combinatorial problems. In: Proceedings of the IFIP 13th World Computer Congress, pp. 253–258 (1994)
14. de Kleer, J.: A comparison of ATMS and CSP techniques. In: Proceedings of the 11th International Joint Conference on Artificial Intelligence (IJCAI 1989), pp. 290–296 (1989)

15. Kasif, S.: On the parallel complexity of discrete relaxation in constraint satisfaction networks. Artificial Intelligence 45(3), 275–286 (1990)
16. Gavanelli, M.: The Log-Support Encoding of CSP into SAT. In: Bessière, C. (ed.) CP 2007. LNCS, vol. 4741, pp. 815–822. Springer, Heidelberg (2007)
17. Eén, N., Sörensson, N.: An Extensible SAT-solver. In: Giunchiglia, E., Tacchella, A. (eds.) SAT 2003. LNCS, vol. 2919, pp. 502–518. Springer, Heidelberg (2004)
18. Berre, D.L., Parrain, A.: The Sat4j library, release 2.2. Journal on Satisfiability, Boolean Modeling and Computation 7(2–3), 56–64 (2010)

SatX10: A Scalable Plug&Play Parallel SAT Framework

(Tool Presentation)

Bard Bloom, David Grove, Benjamin Herta, Ashish Sabharwal,
Horst Samulowitz, and Vijay Saraswat

IBM Watson Research Center, New York, USA
{bardb,bherta,groved,ashish.sabharwal,samulowitz,vsaraswa}@us.ibm.com

Abstract. We propose a framework for SAT researchers to conveniently try out new ideas in the context of parallel SAT solving without the burden of dealing with all the underlying system issues that arise when implementing a massively parallel algorithm. The framework is based on the parallel execution language X10, and allows the parallel solver to easily run on both a single machine with multiple cores and across multiple machines, sharing information such as learned clauses.

1   Introduction

With tremendous progress made in the design of Boolean Satisfiability (SAT) solvers over the past two decades, a wide range of application areas have begun to exploit SAT as a powerful back end for declarative modeling of combinatorial (sub-)problems which are then solved by off-the-shelf or customized SAT solvers. Many interesting problems from areas such as software and hardware verification and design automation often translate into SAT instances with millions of variables and several million constraints. Surprisingly, such large instances are not out of reach of modern SAT solvers. This has led practitioners to push the boundary even further, resulting in plenty of harder instances that current SAT solvers cannot easily tackle.

In order to address this challenge, SAT researchers have looked to exploiting parallelism, especially with the advent of commodity hardware supporting many cores on a single machine, and of clusters with hundreds or even thousands of cores. However, the algorithmic and software engineering expertise required to design a highly efficient SAT solver is very different from that needed to most effectively optimize aspects such as communication between concurrently running solvers or search threads. Most SAT researchers do not possess deep knowledge of concurrency and parallelism issues (message passing, shared memory, locking, deadlocking etc). It is thus no surprise that state-of-the-art parallel SAT solvers often rely on a fairly straightforward combination of diversification and limited knowledge sharing (e.g., very short learned clauses), mainly on a single machine.

The goal of this work is to bridge this gap between SAT and systems expertise. We present a tool called **SatX10**, which provides a convenient plug&play framework for SAT researchers to try out new ideas in the context of parallel SAT

A. Cimatti and R. Sebastiani (Eds.): SAT 2012, LNCS 7317, pp. 463–468, 2012.

solving, without the burden of dealing with numerous systems issues. SatX10 is built using the X10 parallel programming language [2]. It allows one to incorporate and run any number of diverse solvers while sharing information using one of various communication methods. The choice of which solvers to run with what parameters is supplied at run-time through a configuration file. The same source code can be compiled to run on one node, across multiple nodes, and on a variety of computer architectures, networks, and operating systems. Thus, the SatX10 framework allows SAT researchers to focus on the solver design aspect, leaving an optimized parallel implementation and execution to X10.

We demonstrate the capabilities of SatX10 by incorporating into it four distinct DPLL-based SAT solvers that continuously exchange learned clauses of a specified maximum size while running on single or multiple nodes of a cluster.

The goal of this paper is not to present a state-of-the-art parallel SAT solver. Rather, we discuss the design of SatX10 and the API that must be implemented by any SAT solver to be included in SatX10. The SatX10 harness is available at http://x10-lang.org/satx10 under an appropriate open source license.

2   Background

We assume familiarity with the SAT problem and DPLL-based sequential systematic SAT solvers, which essentially are carefully designed enhancements of tree search, in particular *learning clauses* when they infer that a partial truth assignment cannot be extended to a full solution. These solvers typically make thousands of branching decisions per second and infer hundreds to thousands of new clauses per second. Most of the successful parallel SAT solvers are designed to run on a single machine. They exploit *diversification*, by simply launching multiple parameterizations of the same solver or of different solvers, and a very limited amount of *knowledge sharing*, typically through learned clauses. Three prominent examples are: ManySat [3], which won the parallel track in the SAT 2009 Competition and is based on different parameterizations of MiniSat and clause sharing; Plingeling [1], which was a winner in the 2011 SAT Competition (wall-clock category, Application instances) and runs multiple variations of lingeling while sharing only unit clauses; and ppfolio [4], which was placed first and second in the 2011 Competition and simply runs certain five solvers.

2.1   X10: A Parallelization Framework

X10 [2, 5, 7] is a modern programming language designed specifically for programming multi-core and clustered systems easily. Unlike C++ or Java, where threads and network communications are API calls, in X10, parallel operations are integral to the language itself, making it easy to write a single program that makes full use of the resources available in a cloud, GPUs, or other hardware.

X10 is a high-level language that gets compiled down into C++ or Java. Specifically, it runs on Java 1.6 VMs and Ethernet. When compiled to C++, X10 runs on x86, x86_64, PowerPC, Sun CPUs, and on the BlueGene/P, and

Ethernet and Infiniband interconnects (through an MPI implementation). It runs on the Linux, AIX, MacOS, Cygwin operating systems. Note that the *same* source program can be compiled for all these environments.

Importantly for SatX10, an X10 program can use existing libraries (e.g., sequential SAT solvers) written in C++ or Java. This permits us to build a parallel SAT solver by using many sequential SAT solvers (changed in modest ways) and using a small X10 program to launch them and permit communication between them. The X10 runtime handles all the underlying communications, thread scheduling, and other low-level system functions.

X10 follows the APGAS (Aschronous Partitioned Global Address Space) programming model. This model says that there are independent memory spaces available to a program, and the program can move data asynchronously between these memory spaces. In X10, these memory spaces are called *places*. Functions and other executable code in the form of closures can also move across places, to process the data where it resides.

X10's basic concurrency and distribution features most relevant for SAT solver designers include a `finish` block, an `async` block, and an `at` block:

- All activities, including parallel activities, inside of a `finish` block must complete before moving on past the finish block.
- The contents of an `async` block can execute in parallel with anything outside of the `async` block. This lets the programmer take advantage of multiple cores within the same place.
- `at` is a place-shifting operation. Code inside the at block is executed at place p, not locally. Any referenced data is automatically copied to p.

The X10 language has many other features, such as constrained types, GPU acceleration, atomic blocks, clocked computations, collectives, and others. These are not critical to the creation of SatX10 and will not be discussed here.

3   Building Parallel SAT Solvers with X10

The architecture of `SatX10` is shown in Fig. 1 as a high level schematic diagram. Its main components, discussed below, are designed with the goal of providing a generic way to incorporate a diverse set of SAT solvers and support information sharing across solvers. The basic integration of a SAT solver in `SatX10` requires only minimal changes to the SAT solver code and one additional header file to be created. Each constituent SAT solver is enhanced with an X10 part, which is then used for all parallelization and communication, done transparently by the X10 back end. We assume below that the solvers are written in C++.

A. `SatX10.x10` is the main X10 file that, at runtime, reads in the desired solver configuration from a ".ppconfig" file (indicating which solvers to launch with which parameters), sets up solvers at various "places", executes them, and sends a "kill" signal to all other solvers when one solver finishes. This file attaches to the solver at each place user-defined X10 "callbacks" that the

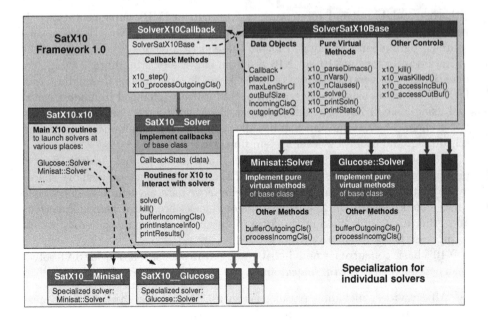

Fig. 1. A schematic diagram depicting the architecture of SatX10

solver can use to send messages to (or execute methods at) other places, such as sending a newly learned clause. While our current implementation uses a one-to-all, asynchronous communication strategy, X10 provides extensive support for other parallelization schemes such as clocked synchronization and periodic all-to-all information reduction and communication.

B. SolverSatX10Base.h provides the base C++ class that every SAT solver's main "solver" class must inherit from and implement some virtual methods for. These simple virtual methods include x10_solve(), which solves the instance and returns -1/0/1 based on whether the instance was found to be unsatisfiable/unknown/satisfiable; x10_printSoln(), which prints a satisfying assignment; x10_printStats(), which prints solver statistics; etc.

C. SatX10_Solver.h provides the generic C++ base class that, appropriately extended, acts as the interface between each solver and X10. It provides the implementation of the callback methods solvers use for communication. For each solver, a new header file is built with a solver-specific derived class that has a reference to the main "solver" object. It provides solver-specific routines, such as converting knowledge to be shared from solver-specific data types to the generic data types used in SatX10_Solver.h (e.g., converting a learnt clause in the solver's internal format to std::vector<int>). The main X10 routine creates at each place an object of such a derived class.

D. *Modifications to solver class*: As mentioned above, the main "solver" class in each SAT solver must inherit from SolverSatX10Base and implement its pure virtual methods. The entire code for each solver must also be put inside a unique `namespace` so as to avoid conflicting uses of the same object name across different solvers. Further, the main search routine should be modified to (i) periodically check an indicator variable, `x10_killed`, and abort search if it is set to `true`, (ii) call the appropriate "callback" method whenever it wants to send information to (or execute a method at) another place, (iii) periodically call `x10_step()` to probe the X10 back end for any incoming information (e.g., learned clauses or kill signal) sent by other places, and (iv) implement `x10_processIncomingClauses()`, which incorporates into the solver a list of clauses received from other concurrently running solvers. Note that depending on when in the search process the incoming clause is incorporated, one may need to properly define "watch" literals, possibly backtrack to a safe level, and satisfy other solver specific requirements.

To share other kinds of information, one can define methods similar to `x10_processIncomingClauses` and `x10_bufferOutgoingClauses`. The X10 compiler `x10c++` is used to compile everything into a single executable. The communication back end (shared memory, sockets, etc.) is specified as a compilation options, while solver configurations, number of solvers to run, hostnames across a cluster, etc., are specified conveniently at runtime.

4 Empirical Demonstration

The main objective of this section is to show that `SatX10` provides communication capabilities at a reasonable cost and that it allows effective deployment of a parallel SAT solver on a single machine as well as across multiple machines. As stated earlier, designing a parallel solver that outperforms all existing ones is not the goal of this work.

We used the `SatX10` framework to build a parallel SAT solver `MiMiGlCi`, composed of `Glucose 2.0`, `Cir_Minisat`, `Minisat 2.0`, and `Minisat 2.2.0`, with additional parameterizations (e.g., different restart strategies). As a test bed, we chose 30 instances of medium difficulty from various benchmark families from the application track of SAT Competition 2011 [6]. All experiments were conducted on 2.3 GHz AMD Opteron 2356 machines with two 4-core CPUs and 16 GB memory, InfiniBand network, running Red Hat Linux release 6.2.

The results of the evaluation are summarized in Fig. 2 in the form of the standard "cactus" plot, showing the maximum time (y-axis) need to solve a given number of instances (x-axis).

The two curves in the left-hand-side plot show the comparison on a single machine when `MiMiGlCi` is run on 8 places (i.e., with 8 sequential solvers running in parallel) while sharing clauses of maximum lengths 1 and 8, respectively. Here we see a clear gain in performance when sharing clauses of size up to 8, despite the overhead both on the communication side and for each solver to incorporate additional clauses. In general, what and how much to share must, of course, be carefully balanced out to achieve good performance.

The right-hand-side plot in Fig. 2 shows the results on multiple machines. Specifically, we report numbers for 8 places running on 8 different machines, and

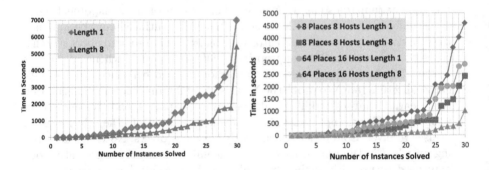

Fig. 2. Performance on a single machine (left) and on multiple machines (right)

64 places running on a total of 16 machines. The benefit of sharing clauses of length up to 8 is clear in this setting as well. In fact, 8 places sharing clauses of length at most 8 performed better than 64 places sharing only unit clauses, indicating that the ability to conveniently share more can be much more powerful than simply running more solvers. Not surprisingly, 64 places sharing clauses of length up to 8 performed the best, solving all instances in 1,000 seconds.

In summary, SatX10 provides a framework to easily build parallel SAT solvers composed of a diverse set of constituent solvers, with the capability of sharing information while executing various parameterizations of the constituent solvers on a single machine or multiple machines. The rich language of X10 underlying SatX10 handles all parallelization aspects, including parallel execution, and provides a uniform interface to all constituent SAT solvers. We hope this will serve as a useful tool in pushing the state of the art in parallel SAT solving.

Acknowledgement. SatX10 originated from an X10-based SAT solver named PolySat, which was developed in collaboration with David Cunningham.

References

[1] Biere, A.: Lingeling and friends at the sat competition 2011. Technical report, Johannes Kepler University (2011)
[2] Charles, P., Grothoff, C., Saraswat, V., Donawa, C., Kielstra, A., Ebcioglu, K., von Praun, C., Sarkar, V.: X10: an object-oriented approach to non-uniform cluster computing. In: OOPSLA 2005, San Diego, CA, USA, pp. 519–538 (2005)
[3] Hamadi, Y., Sais, L.: ManySAT: a parallel SAT solver. JSAT 6 (2009)
[4] Roussel, O.: Description of ppfolio. Technical report, Artois University (2011)
[5] Saraswat, V., Bloom, B., Peshansky, I., Tardieu, O., Grove, D.: Report on the experimental language, X10. Technical report, IBM Research (2011)
[6] SAT Competition (2011), www.satcompetition.org
[7] X10. X10 programming language web site (January 2010), http://x10-lang.org/

Improved Single Pass Algorithms
for Resolution Proof Reduction
(Poster Presentation)

Ashutosh Gupta

IST, Austria

An unsatisfiability proof is a series of applications of proof rules on an input formula to deduce *false*. Unsatisfiability proofs for a Boolean formula can find many applications in verification. For instance, one application is automatic learning of abstractions for unbounded model checking by analyzing proofs of program safety for bounded steps [6,5,4]. We can also learn unsatisfiable cores from unsatisfiability proofs, which are useful in locating errors in inconsistent specifications [10]. These proofs can be used by higher order theorem provers as sub-proofs of another proof [2].

One of the most widely used proof rules for Boolean formulas is the resolution rule, i.e., if $a \vee b$ and $\neg a \vee c$ holds then we can deduce $b \vee c$. In the application of the rule, a is known as *pivot*. A *resolution proof* is generated by applying resolution rule on the clauses of an unsatisfiable Boolean formula to deduce *false*. Modern SAT solvers (Boolean satisfiability checkers) implement some variation of DPLL that is enhanced with conflict driven clause learning [9,8]. Without incurring large additional cost on the solvers, we can generate a resolution proof from a run of the solvers on an unsatisfiable formula [11].

Due to the nature of the algorithms employed by SAT solvers, a generated resolution proof may contain redundant parts and a strictly smaller resolution proof can be obtained. Applications of the resolution proofs are sensitive to the proof size. Since minimizing resolution proofs is a hard problem [7], there has been significant interest in finding algorithms that partially minimize the resolution proofs generated by SAT solvers.

In [1], two low complexity algorithms for optimizing the proofs are presented. Our work is focused on one of the two, namely RECYCLE-PIVOTS. Lets consider a resolution step that produces a clause using some pivot p. The resolution step is called *redundant* if each deduction sequence from the clause to *false* contains a resolution step with the pivot p. A redundant resolution can easily be removed by local modifications in the proof structure. After removing a redundant resolution step, a strictly smaller proof is obtained. RECYCLE-PIVOTS traverses the proofs single time to remove the redundant resolutions partially. From each clause, the algorithm starts from the clause and follows the deduction sequences to find equal pivots. The algorithm stops looking for equal pivots if it reaches to a clause that is used to deduce more than one clause.

In this work, we developed *three algorithms* that are improved version of RECYCLE-PIVOTS. For the first algorithm, we observe that each literal from a clause must appear as a pivot somewhere in all the deduction sequences from

A. Cimatti and R. Sebastiani (Eds.): SAT 2012, LNCS 7317, pp. 469–470, 2012.
© Springer-Verlag Berlin Heidelberg 2012

the clause to *false*. Therefore, we can extend search of equal pivots among the literals from the stopping clause without incurring additional cost. For the second algorithm, we observe that the condition for the redundant resolutions can be defined recursively over the resolution proof structure. This observation leads to a single pass algorithm that covers even more redundancies but it requires an expensive operation at each clause in a proof. Note that the second algorithm does not remove all such redundancies because the removal of a redundancy may lead to exposure of more. Our third algorithm is parametrized. This algorithm applies the expensive second algorithm only for the clauses that are used to derive a number of clauses smaller than the parameter. The other clauses are handled as in the first algorithm. The parametrization reduces run time for the third algorithm but also reduces the coverage of the redundancy detection.

We have implemented our algorithms in OPENSMT [3] and applied them on unsatisfiable proofs of 198 examples from plain MUS track of SAT11 competition. The original algorithm removes 11.97% of clauses in the proofs of the examples. The first and the second algorithm additionally remove 0.89% and 10.57% of the clauses respectively. The third algorithm removes almost as many clauses as the second algorithm in lesser time for the parameter value as low as 10. We also observe similar pattern in reduction of the unsatisfiable cores of the examples.

References

1. Bar-Ilan, O., Fuhrmann, O., Hoory, S., Shacham, O., Strichman, O.: Linear-time reductions of resolution proofs. In: Haifa Verification Conference (2008)
2. Böhme, S., Nipkow, T.: Sledgehammer: Judgement Day. In: Giesl, J., Hähnle, R. (eds.) IJCAR 2010. LNCS, vol. 6173, pp. 107–121. Springer, Heidelberg (2010)
3. Bruttomesso, R., Pek, E., Sharygina, N., Tsitovich, A.: The OpenSMT Solver. In: Esparza, J., Majumdar, R. (eds.) TACAS 2010. LNCS, vol. 6015, pp. 150–153. Springer, Heidelberg (2010)
4. Henzinger, T.A., Jhala, R., Majumdar, R., McMillan, K.L.: Abstractions from proofs. In: POPL (2004)
5. McMillan, K.L.: An interpolating theorem prover. Theor. Comput. Sci. 345(1), 101–121 (2005)
6. McMillan, K.L., Amla, N.: Automatic Abstraction without Counterexamples. In: Garavel, H., Hatcliff, J. (eds.) TACAS 2003. LNCS, vol. 2619, pp. 2–17. Springer, Heidelberg (2003)
7. Papadimitriou, C.H., Wolfe, D.: The complexity of facets resolved. J. Comput. Syst. Sci. 37(1), 2–13 (1988)
8. Silva, J.P.M., Lynce, I., Malik, S.: Conflict-driven clause learning sat solvers. In: Handbook of Satisfiability, pp. 131–153 (2009)
9. Silva, J.P.M., Sakallah, K.A.: GRASP: A search algorithm for propositional satisfiability. IEEE Trans. Computers 48(5), 506–521 (1999)
10. Sinz, C., Kaiser, A., Küchlin, W.: Formal methods for the validation of automotive product configuration data. AI EDAM 17(1), 75–97 (2003)
11. Zhang, L., Malik, S.: Extracting small unsatisfiable cores from unsatisfiable boolean formulas. In: SAT (2003)

Creating Industrial-Like SAT Instances by Clustering and Reconstruction (Poster Presentation)*

Sebastian Burg[1,2], Stephan Kottler[2], and Michael Kaufmann[2]

[1] FZI, Karlsruhe, Germany
[2] University of Tübingen, Germany

1 Introduction

For the optimization of SAT solvers, it is crucial that a solver can be trained on a preferably large number of instances for general or domain specific problems. Especially for domain specific problems the set of available instances can be insufficiently small. In our approach we built large sets of instances by recombining several small snippets of different instances of a particular domain.

Also the fuzzer utility [3] builds industrial-like SAT instances by combining smaller pieces. However, these pieces are a combination of randomly created circuits and are not derived from an existing pool of instances. In Ansotegui [1] random pseudo-industrial instances are created in a more formal way.

2 Used Methodology

The presented approach requires small building blocks for the generation of new SAT instances. Thus, at first stage, blocks have to be identificied and extracted from existing instances. A block is a subset of clauses $S \subset C$ of an instance and induces to distinguish two different types of variables V: internal variables $\sigma(V)$ only appear within the subset of clauses and connector variables $\gamma(V)$ also appear in clauses outside the subset. In difference to [3], blocks are neither directly given nor can be generated from scratch.

The extraction of blocks is based on the connections between variables and clauses. The computation uses a weighted variable graph (VG) where an edge between two variables (vertices) exists iff the variables appear in at least one clause together. The weight of an edge states the number of common clauses for its incident variables. Using a VG, we apply different clustering algorithms [2] to split each instance in logically connected blocks. Edges whose two vertices are placed in different clusters are called connectors.

The clustering of a VG allows the use of different properties and constraints: e.g. variables may be clustered considering their weighted connectivity or the degree of coverage among the clauses they occur in. Connector variables within

* This work was supported by the DFG-SPP 1307, project StrAlEnSAT, and by the BMBF, projects SANITAS (grant 01M3088C) and RESCAR2.0 (grant 01M3195E).

A. Cimatti and R. Sebastiani (Eds.): SAT 2012, LNCS 7317, pp. 471–472, 2012.

one block S_i are partitioned regarding the blocks (e.g. $S_j, \ldots, S_k$) of their adjacent vertices: $\{\gamma(S_i)_j \cup \ldots \cup \gamma(S_i)_k\} = \gamma(S_i)$. The cardinality of partitions $|\gamma(S_i)_j| \ (= |\gamma(S_j)_i|)$ constitutes a general connectivity between any two blocks.

With the use of building blocks it is possible to rearrange existing information to generate new instances. Different blocks are joined via their connector variables. Joined variables will then get the same name in the newly created instance. For the joining of blocks we analyzed different approaches:

(i) External connector variables of different blocks are joined at random.
(ii) Select two blocks $S_i \neq S_j$ with a pair of equally sized partitions i.e. $\exists \, m, k$: $|\gamma(S_i)_m| = |\gamma(S_j)_k|$. Randomly join variables from $\gamma(S_i)_m$ and $\gamma(S_j)_k$.
(iii) In difference to (ii) allow the use of connectors of one block more than once. There may be S_i, S_j, S_t where both S_i and S_j are joined with S_t via $\gamma(S_t)$.

The more constraints are given, the more complex and time consuming the process of finding matching blocks is. Different strategies to tackle this issue may be applied: Building blocks in demand can be duplicated with renamed variables. Secondly, connectors that are too hard to be joined may remain unassigned, i.e. the according variable is treated as internal variable.

3   Results

This paper sketches a method to construct and rearrange several instances of different sizes based on a given set of SAT benchmarks. The size of a newly created instance depends on the constraints and number of blocks used.

To meassure the quality of our approach we compared the run-times for created instances compared to the runtimes of the orginal instances that were used for the assembling. Moreover, we used the set of features presented in [4] and compared some of the properties in the original and the generated instances. However, with this approach it is very hard to generate instances of a certain domain that sustain the relation of several properties of [4].

In future work we plan to extend the shown approach to combine extracted building blocks with generated blocks (such as in [3]). This could especially help to generate matching blocks on demand.

References

1. Ansótegui, C., Bonet, M.L., Levy, J.: Towards Industrial-Like Random SAT Instances. In: IJCAI, pp. 387–392 (2009)
2. Brandes, U., Delling, D., Gaertler, M., Görke, R., Hoefer, M., Nikoloski, Z., Wagner, D.: On Modularity Clustering. IEEE Trans. Knowl. Data Eng. 20(2), 172–188 (2008)
3. Brummayer, R., Lonsing, F., Biere, A.: Automated Testing and Debugging of SAT and QBF Solvers. In: Strichman, O., Szeider, S. (eds.) SAT 2010. LNCS, vol. 6175, pp. 44–57. Springer, Heidelberg (2010)
4. Nudelman, E., Leyton-Brown, K., Hoos, H.H., Devkar, A., Shoham, Y.: Understanding Random SAT: Beyond the Clauses-to-Variables Ratio. In: Wallace, M. (ed.) CP 2004. LNCS, vol. 3258, pp. 438–452. Springer, Heidelberg (2004)

Incremental QBF Preprocessing
for Partial Design Verification
(Poster Presentation)

Paolo Marin, Christian Miller, and Bernd Becker

University of Freiburg, Germany
{paolo,millerc,becker}@informatik.uni-freiburg.de
http://ira.informatik.uni-freiburg.de/

Bounded Model Checking (BMC) is a major verification method for finding errors in sequential circuits. BMC accomplishes this by iteratively unfolding a circuit k times, adding the negated property, and finally converting the BMC instance into a sequence of satisfiability (SAT) problems. When considering incomplete designs (i.e. those containing so-called blackboxes), we rather need the logic of Quantified Boolean Formulas (QBF) to obtain a more precise modeling of the unknown behavior of the blackbox. Here, we answer the question of *unrealizability* of a property, where finding a path of length k proves that the property is violated regardless of the implementation of the blackbox. To boost this task, solving blackbox BMC problems incrementally has been shown to be feasible [3], although the restrictions required in the preprocessing phase reduce its effectiveness. In this paper we enhance the verification procedure when using an off-the-shelf QBF solver, through a stronger preprocessing of the QBF formulas applied in an incremental fashion.

We started from the idea of preprocessing only the transition relation [2,3], we obtain $\Phi_{\text{TR-pp}}$ by avoiding the elimination of the current and next state variables s and s' tagging them as *don't touch*, as they will appear again in next unfoldings. We now improve the preprocessing of the QBF problem by keeping a more compact representation of each unfolding k in the QBF preprocessor, and adding only the new information of unfolding $k + 1$. What follow is what we name forward-incremental unfolding. Φ_{BMC} is a preprocessed QBF representation of $I_0 \wedge T_{0,1} \wedge \ldots \wedge T_{k-1,k}$ for any unfolding step k (where I_0 is the initial state, and $T_{i-1,i}$ is the transition relation from the time-frame $i - 1$ to i) which is permanently stored in the preprocessor. The state variables s^k in Φ_{BMC} are not touched, so that either the negated property $\Phi_{\neg P}^k$ or the next transition relation $\Phi_{\text{TR-pp}}^{k,k+1}$ can be connected to Φ_{BMC}. Φ_{BMC}^k adds $\Phi_{\neg P}^k$ to a copy Φ_{BMC}, and thus represents the k-th unfolding of the BMC problem. If Φ_{BMC}^k is satisfiable (this can be checked by any QBF solver) the algorithm terminates; otherwise the clauses of the next transition relation $\Phi_{\text{TR-pp}}^{k,k+1}$ are added incrementally to the preprocessor, the prefix of Φ_{BMC} being extended to the right. The new interface variables s_{k+1} are declared *don't touch* for this incremental run. Similarly, along the lines of [3], and in order to loosen some restrictions on preprocessing gate outputs, we perform preprocessing in a backward-incremental way. Here, the don't touch interface variables are the inputs of the circuit, and can be used for equivalence reasoning.

A. Cimatti and R. Sebastiani (Eds.): SAT 2012, LNCS 7317, pp. 473–474, 2012.

To try our incremental preprocessing methods, we modified sQueezeBF, the preprocessor built into QuBE7.2 [1], to take the QBF formulas incrementally, and added the possibility to tag variables as *don't touch* and treat them accordingly. As testbed, we used a range of circuit-based benchmarks built upon some VHDL designs from www.opencores.org, in which parts were blackboxed, and we show how the performance of the back-end solvers relies on different preprocessing methods. Table 1 gives the cumulative times needed to solve the whole benchmark set by a pool of modern QBF solvers, being the first group search-based (QuBE-nopp is QuBE with its preprocessor disabled), the second based on resolution/rewriting techniques, and the third based on portfolio approaches. For each setting, we provide both the cumulative time and the number of benchmarks solved. The search-based solvers without preprocessing are positively affected by incremental preprocessing. Rather, there is almost no difference for Quantor, while AIGSolve, which cannot find the Tseitin encoded logical gates destroyed by sQueezeBF, does not deal with preprocessed formulas at all. The same holds for Quaig. AQME can rather take advantage of our incremental preprocessing techniques, as it cannot properly handle very large formulas. Among all solvers, incremental preprocessing delivers the most robust performance.

Table 1. QBF results on incomplete circuit designs using various solvers. Times are given in seconds, adding a penalty of 7,200s for timeouts and memouts.

Solver	standard BMC procedure time (s)	#ps	transition relation preprocessing time (s)	#ps	backward incremental time (s)	#ps	forward incremental time (s)	#ps
QuBE	**15086.95**	21	59729.37	14	28231.85	21	38480.09	21
QuBE-nopp	66170.92	14	**5251.25**	22	8609.86	21	8392.63	21
DepQBF	78225.33	12	20223.02	20	17211.06	21	**16572.38**	21
Quantor	97591.99	9	98801.89	9	96924.74	9	**96772.89**	9
AIGSolve	**47739.95**	16	146406.03	3	151269.91	1	104319.68	8
AQME	96390.80	10	92883.30	10	**43972.26**	18	76689.74	13
Quaig	**83807.77**	11	98564.55	9	97791.62	9	98396.94	9

Acknowledgments. The authors would like to thank STAR-lab of the University of Genova, Italy, for fruitfully cooperating on QuBE. This work was partly supported by the German Research Council (DFG) as part of the SFB/TR 14 "Automatic Verification and Analysis of Complex Systems" (www.avacs.org).

References

1. Enrico: Enrico and Marin Paolo and Narizzano, Massimo. Journal of Satisfiability 7(8), 83–88 (2010)
2. Kupferschmid, S., Lewis, M., Schubert, T., Becker, B.: Incremental preprocessing methods for use in bmc. Formal Methods in System Design 39, 185–204 (2011)
3. Marin, P., Miller, C., Lewis, M., Becker, B.: Verification of Partial Designs Using Incremental QBF Solving. In: DATE(2012)

Concurrent Cube-and-Conquer
(Poster Presentation)*

Peter van der Tak[1], Marijn J.H. Heule[1,2], and Armin Biere[3]

[1] Delft University of Technology, The Netherlands
[2] University of Texas, Austin, United States
[3] Johannes Kepler University Linz, Austria

Satisfiability solvers targeting industrial instances are currently almost always based on conflict-driven clause learning (CDCL) [5]. This technique can success-fully solve very large instances. Yet on small, hard problems lookahead solvers [3] often perform better by applying much more reasoning in each search node and then recursively splitting the search space until a solution is found.

The *cube-and-conquer* (CC) approach [4] has shown that the two techniques can be combined, resulting in better performance particularly for very hard instances. The key insight is that lookahead solvers can be used to partition the search space into subproblems (called cubes) that are easy for a CDCL solver to solve. By first partitioning (*cube* phase) and then solving each cube (*conquer* phase), some instances can be solved within hours rather than days. This cube-and-conquer approach, particularly the *conquer* phase, is also easy to parallelize.

The challenge to make this technique work in practice lies in developing ef-fective heuristics to determine when to stop partitioning and start solving. The current heuristics already give strong results for very hard instances, but are far from optimal and require some fine tuning to work well with instances of differ-ent difficulty. For example, applying too much partitioning might actually result in a considerable increase of run time for easy instances. On the other hand, applying not enough partitioning reduces the benefits of cube-and-conquer.

The most important problem in developing an improved heuristic is that in the partitioning phase no information is available about how well the CDCL solver will perform on a cube. In CC's heuristics, performance of CDCL is assumed to be similar to that of lookahead: if lookahead refutes a cube, CDCL is expected to be able to refute similar cubes fast, and if CDCL would solve a cube fast, lookahead is expected to be able to refute it fast too. However, due to the different nature of lookahead and CDCL, this is not always true.

To improve cutoff heuristics, we propose *concurrent cube-and-conquer* (CCC): an online approach that runs the cube and conquer phases concurrently. When-ever the lookahead solver makes a new decision, this decision is sent to the CDCL solver, which adds it as an assumption [2]. If CDCL refutes a cube fast, it will refute it before lookahead makes another decision. This naturally cuts off easy branches, so that the cutoff heuristic is no longer necessary.

Although this basic version of CCC already achieves speedups, it can be im-proved further by applying a (slightly different) cutoff heuristic. This heuristic

* The 2nd and 3rd author are supported by FWF, NFN Grant S11408-N23 (RiSE).
The 2nd author is supported by DARPA contract number N66001-10-2-4087.

A. Cimatti and R. Sebastiani (Eds.): SAT 2012, LNCS 7317, pp. 475–476, 2012.

attempts to identify cubes similar to those that CDCL already solved, rather than estimating CDCL performance based on lookahead performance (as originally in CC). Cutting off has two advantages: often CDCL can already solve a cube efficiently without the last few decision variables; further partitioning these already easy cubes only hampers performance. Additionally, cutting off allows multiple cubes to be solved in parallel.

Other than improving performance of cube-and-conquer by replacing the cut-off heuristic, CCC also aims at solving another problem: on some instances (C)CC performs worse than CDCL regardless of the configuration of the solvers and heuristics. It seems that lookahead sometimes selects a decision l_{dec} which results in two subformulas $F \wedge l_{dec}$ and $F \wedge \neg l_{dec}$ that are not easier to solve separately by CDCL. If the decision is not relevant to CDCL search, (C)CC forces the CDCL solver to essentially solve the same problem twice. We propose two metrics that can detect this behavior, in which case CCC is aborted within 5 seconds and the problem is solved by CDCL alone.

Our experiments show that CCC works particularly well on crafted instances. Without selection of suitable instances, cube-and-conquer and CCC cannot compete with other solvers. However the proposed predictor based on CCC accurately selects instances for which cube-and-conquer techniques are not suitable and for which a CDCL search is preferred. It is thereby able to solve several more application and crafted instances than the CDCL and lookahead solvers it was based on. CCC solves 24 more crafted instances within one hour over all the SAT 2009 and 2011 competition instances than Plingeling [1], where both solvers use four threads. For application instances, Plingeling solves one more instance for a one hour timeout but CCC is slightly better for lower timeouts (anything below 2500 seconds).

We believe that CCC is particularly interesting as part of a portfolio solver, where our predictor can be used to predict whether to apply cube-and-conquer techniques. The authors of SATzilla specifically mention in their conclusion that identifying solvers that are only competitive for certain kinds of instances still has the potential to further improve SATzilla's performance substantially [6].

References

1. Biere, A.: Lingeling, Plingeling, PicoSAT and PrecoSAT at SAT Race 2010. FMV Technical Report 10/1, Johannes Kepler University, Linz, Austria (2010)
2. Eén, N., Sörensson, N.: Temporal induction by incremental SAT solving. ENTCS 89(4), 543–560 (2003)
3. Heule, M.J.H.: SmArT Solving: Tools and techniques for satisfiability solvers. PhD thesis, Delft University of Technology (2008)
4. Heule, M.J.H., Kullmann, O., Wieringa, S., Biere, A.: Cube and conquer: Guiding CDCL SAT solvers by lookaheads, Accepted for HVC (2011)
5. Marques-Silva, J.P., Lynce, I., Malik, S.: Conflict-Driven Clause Learning SAT Solvers. FAIA, vol. 185, ch. 4, pp. 131–153. IOS Press (February 2009)
6. Xu, L., Hutter, F., Hoos, H.H., Leyton-Brown, K.: Satzilla: Portfolio-based algorithm selection for sat. J. Artif. Intell. Res. (JAIR) 32, 565–606 (2008)

Satisfying versus Falsifying in Local Search for Satisfiability (Poster Presentation)

Chu Min Li and Yu Li

MIS, Université de Picardie Jules Verne, France
{chu-min.li,yu.li}@u-picardie.fr

During local search, clauses may frequently be satisfied or falsified. Modern SLS algorithms often exploit the falsifying history of clauses to select a variable to flip, together with variable properties such as score and age. The score of a variable x refers to the decrease in the number of unsatisfied clauses if x is flipped. The age of x refers to the number of steps done since the last time when x was flipped.

Novelty [5] and Novelty based SLS algorithms consider the youngest variable in a randomly chosen unsatisfied clause c, which is necessarily the last falsifying variable of c whose flipping made c from satisfied to unsatisfied. If the best variable according to scores in c is not the last falsifying variable of c, it is flipped, otherwise the second best variable is flipped with probability p, and the best variable is flipped with probability 1-p. TNM [4] extends Novelty by also considering the second last falsification of c, the third last falsification of c, and so on... If the best variable in c most recently and consecutively falsified c several times, TNM considerably increases the probability to flip the second best variable of c.

Another way to exploit the falsifying history of clauses is to define the weight of a clause to be the number of local minima in which the clause is unsatisfied, so that the objective function is to reduce the total weight of unsatisfied clauses.

In this paper, we propose a new heuristic by considering the satisfying history of clauses instead of their falsifying history, and by modifying Novelty as follows: If the best variable in c is not *the most recent satisfying variable* of c, flip it. Otherwise, flip the second best variable with probability p, and flip the best variable with probability 1-p. Here, the most recent **satisfying** variable in c is the variable whose flipping most recently made c from unsatisfied to satisfied. The intuition of the new heuristic is to avoid repeatedly satisfying c using the same variable.

Note that in a clause c, the most recent falsifying variable and the most recent satisfying variable can be the same variable. In this case, the variable flipped to make c from unsatisfied to satisfied was re-flipped later to make c from satisfied to unsatisfied (there can be other flips between the two flips), and vice versa. In our experiments using instances from the 2011 SAT competition, this is the case in the randomly selected unsatisfied clause in more than 95% steps for random 3-SAT. The percentage is less than 90% for random 5-SAT and 7-SAT, and for crafted instances. So the new heuristic is expected to behave similarly as Novelty on random 3-SAT, but differently for other SAT problems.

We propose a new SLS algorithm called SatTime that implements the new heuristic. Given a SAT instance ϕ to solve, SatTime first generates a random assignment and while the assignment does not satisfy ϕ, it repeatedly modifies the assignment as follows:

1. If there are promising decreasing variables, flip the oldest one;
2. Otherwise, randomly pick an unsatisfied clause c;

A. Cimatti and R. Sebastiani (Eds.): SAT 2012, LNCS 7317, pp. 477–478, 2012.
© Springer-Verlag Berlin Heidelberg 2012

3. With probability dp, make a diversification step with c. With probability 1-dp, consider the best and second best variables in c according to their score (breaking tie in favor of the least recently flipped one). If the best variable is not the most recent **satisfying** variable of c, then flip it. Otherwise, with probability p, flip the second best variable, and with probability 1-p, flip the best variable.

The promising decreasing variable and diversification probability dp were defined in [3]. Probability p is adapted during search according to [2] and $dp=p/10$. We also implement UnSatTime, which is the same as SatTime, except that the word "satisfying" is replaced with "**falsifying**" in UnSatTime.

We run ten times SatTime and UnSatTime on random and crafted instances from the 2011 SAT competition to compare the two heuristics respectively based on satisfying and falsifying variables. Each solver is given 5000 seconds to solve each instance as in the competition, but on a Macpro with XEON 2.8 Ghz (early 2008) under Macosx, which is slower than the computer in the competition. For random 3-SAT, SatTime and UnSatTime are similar as expected. For random 5-SAT and 7-SAT, SatTime solves in the average 152.6 instances while UnSatTime solves 140.1 instances. For crafted, SatTimes solves in the average 108.5 instances, while UnSatTime solves 103.2 instances.

SatTime participated in the 2011 SAT competition and won a silver medal in the random category[1]. Especially, SatTime beat easily all the Conflict Driven Clause Learning (CDCL) solvers in the crafted sat category (SatTime solved 109 instances while the best CDCL only solved 93 instances), although SLS has been considered less effective than CDCL for structured SAT problems for a long time. This is the first time that a SLS solver enters the final phase of the SAT competition in the crafted category and beats there all the CDCL algorithms on structured SAT problems.

In the future, we plan to improve SatTime by considering more satisfying variables of a clause (the 2nd last satisfying variable, the 3rd, and so on...), and by designing heuristics exploiting both satisfying history and falsifying history of clauses.

References

1. Balint, A., Fröhlich, A.: Improving Stochastic Local Search for SAT with a New Probability Distribution. In: Strichman, O., Szeider, S. (eds.) SAT 2010. LNCS, vol. 6175, pp. 10–15. Springer, Heidelberg (2010)
2. Hoos, H.: An Adaptive Noise Mechanism for WalkSAT. In: Proceedings of AAAI 2002, pp. 655–660. AAAI Press / The MIT Press (2002)
3. Li, C.-M., Huang, W.Q.: Diversification and Determinism in Local Search for Satisfiability. In: Bacchus, F., Walsh, T. (eds.) SAT 2005. LNCS, vol. 3569, pp. 158–172. Springer, Heidelberg (2005)
4. Li, C.-M., Wei, W., Li, Y.: Exploiting Historical Relationships of Clauses and Variables in Local Search for Satisfiability (Poster Presentation). In: Cimatti, A., Sebastiani, R. (eds.) SAT 2012. LNCS, vol. 7317, pp. 479–480. Springer, Heidelberg (2012)
5. McAllester, D.A., Selman, B., Kautz, H.: Evidence for invariant in local search. In: Proceedings of AAAI 1997, pp. 321–326 (1997)

---

[1] Just for competition purpose and for random 3-SAT, SatTime in the competition used a heuristic combining TNM [4] and Sparrow [1] that performs better for random 3-SAT than the new heuristic presented in this paper. Note that SatTime and UnSatTime are similar for random 3-SAT. For all other problems, SatTime in the competition was as presented in this paper.

Exploiting Historical Relationships of Clauses and Variables in Local Search for Satisfiability (Poster Presentation)

Chu Min Li, Wanxia Wei, and Yu Li

MIS, Université de Picardie Jules Verne, France

Variable properties such as score and age are used to select a variable to flip. The score of a variable x refers to the decrease in the number of unsatisfied clauses if x is flipped. The age of x refers to the number of steps done since the last time when x was flipped. If the best variable according to scores in a randomly chosen unsatisfied clause c is not the youngest in c, *Novelty* [4] flips this variable. Otherwise, with probability p (noise p), *Novelty* flips the second best variable, and with probability 1-p, *Novelty* flips the best variable. *Novelty+* [1] randomly flips a variable in c with probability wp and does as *Novelty* with probability 1-wp. *Novelty++* [3] flips the least recently flipped variable (oldest) in c with probability dp, and does as *Novelty* with probability 1-dp.

The above approaches just use the current properties of variables to select a variable to flip. These approaches are effective for the problems that do not present uneven distribution of variable and/or clause weights during the search. In other words, problems can be solved using these approaches when there are no clauses or variables whose weight is several times larger than the average during the search [6]. However, when solving hard random or structured SAT problems using these approaches, variable or clause weight distribution is often uneven. In this case, the falsification history of clauses and/or the flipping history of variables during the search should be exploited to select a variable to flip for the problems to be solved.

In this paper, we present a noise mechanism that exploits the history information to determine noise p. For a falsified clause c, let $var\_fals[c]$ denote the variable that most recently falsifies c and let $num\_fals[c]$ denote the number of the most recent consecutive falsifications of c due to the flipping of this variable. If the best variable in c is not $var\_fals[c]$, this variable is flipped. Otherwise, the second best variable is flipped with probability p, where p is determined as a function of $k=num\_fals[c]$: $\{20, 50, 65, 72, 78, 86, 90, 95, 98, 100\}$, i.e., p=0.2 if k=1, p=0.5 if k=2, ..., p=1 if $k \geq 10$. This probability vector was empirically turned using a subset of instances.

Another adaptive noise mechanism was introduced in [2] to automatically adjust noise during the search. We refer to this mechanism as Hoos's noise mechanism.

SLS solvers *TNM* and *adaptG² WSAT2011*, described in Fig. 1, are both based on *Novelty*, but use noise p determined by our noise mechanism at each **uneven** step, and use noise $p1$ adjusted by Hoos noise mechanism at each **even** step, to flip the second best variable in the randomly selected unsatisfied clause c when the best variable in c is the youngest in c. In *TNM*, a step is even if the variable weight distribution currently is even (e.g., all variable weights are smaller than $10 \times$ the average variable weight). In *adaptG² WSAT2011*, a step is even if c currently has small weight (e.g., smaller than $10 \times$ the average clause weight). Otherwise the step is uneven.

A. Cimatti and R. Sebastiani (Eds.): SAT 2012, LNCS 7317, pp. 479–480, 2012.

1: $A \leftarrow$ randomly generated truth assignment;
2: **for** *flip* $\leftarrow 1$ **to** *Maxsteps* **do**
3:    **if** A satisfies SAT-formula $\mathcal{F}$ **then return** A;
4:    adjust noise $p1$ according to Hoos's noise mechanism;
5:    **if** there is any promising decreasing variable
6:    **then** $y \leftarrow$ the promising decreasing variable with the largest score;
7:    **else** randomly select an unsatisfied clause c;
8:       **if** the current search step is even
9:       **then**
10:          $y \leftarrow$ heuristic *Novelty+* applied to c using probability $p1$ (and $wp=p1/10$);
11:       **else**
12:          determine noise p according to $var\_fals[c]$ and $num\_fals[c]$;
13:          $y \leftarrow$ heuristic *Novelty++* applied to c using probability p (and $dp=p/10$);
14:    $A \leftarrow A$ with y flipped;
15: **return** Solution not found;

Fig. 1. Algorithms *TNM* and *adaptG²WSAT2011*

Our noise mechanism is different from Hoos's noise mechanism in two respects. First, our mechanism uses the history of most recent consecutive falsifications of a clause due to the flipping of one variable, while Hoos's noise mechanism observes the improvement in the objective function. Second, the noise determined by our mechanism is clause-specific, while the noise adjusted by Hoos's noise mechanism is not.

Novelty, *Novelty+*, or *Novelty++* considers the last falsification of c. *TNM* and *adaptG²WSAT2011* extend them by considering more historical falsifications of c.

TNM won a gold medal in the 2009 SAT competition in the random category and is competitive for crafted instances. In the 2011 SAT competition, *ppfolio* won two gold medals in the crafted sat category, because *TNM* solved 75 instances, while its other four constituent algorithms solved 24, 18, 7, and 3 instances, respectively [5]. Solver *adaptG²WSAT2011* performs better than *TNM*. In the 2011 SAT competition, it solved 7 more instances than *TNM* in the random category in phase 1. In addition, *adaptG²WSAT2011* solved 99 instances in the crafted category in phase 1, while the best CDCL solver solved 81 instances in phase 1 (and 93 instances in phase 2).

References

1. Hoos, H.: On the run-time behavior of stochastic local search algorithms for SAT. In: Proceedings of AAAI 1999, pp. 661–666 (1999)
2. Hoos, H.: An adaptive noise mechanism for WalkSAT. In: Proceedings of AAAI 2002, pp. 655–660. AAAI Press / The MIT Press (2002)
3. Li, C.-M., Huang, W.Q.: Diversification and Determinism in Local Search for Satisfiability. In: Bacchus, F., Walsh, T. (eds.) SAT 2005. LNCS, vol. 3569, pp. 158–172. Springer, Heidelberg (2005)
4. McAllester, D.A., Selman, B., Kautz, H.: Evidence for invariant in local search. In: Proceedings of AAAI 1997, pp. 321–326 (1997)
5. Roussel, O.: Description of ppfolio (2011),
   http://www.cril.univ-artois.fr/SAT11/phase2.pdf
6. Wei, W., Li, C.-M., Zhang, H.: Switching among Non-Weighting, Clause Weighting, and Variable Weighting in Local Search for SAT. In: Stuckey, P.J. (ed.) CP 2008. LNCS, vol. 5202, pp. 313–326. Springer, Heidelberg (2008)

Towards Massively Parallel Local Search for SAT
(Poster Presentation)

Alejandro Arbelaez* and Philippe Codognet

JFLI - CNRS / University of Tokyo
{arbelaez,codognet}@is.s.u-tokyo.ac.jp

Introduction. Parallel portfolio-based algorithms have become a standard methodology for building parallel algorithms for SAT. In this methodology, different algorithms (or the same one with different random seeds) compete to solve a given problem instance. Moreover, the portfolio is usually equipped with cooperation, this way algorithms exchange important knowledge acquired during the search to solve a given problem instance. Portfolio algorithms based on complete solvers exchange learned clauses which are incorporated within each search engine (e.g. ManySAT [1] and plingeling), while those based on incomplete solvers [2] exchange the best assignment for the variables found so far in order to properly craft a new assignment for the variables to restart from. These strategies range from a voting mechanism where each algorithm in the portfolio suggests a value for each variable to probabilistic constructions.

In this paper, we focus on incomplete algorithms based on local search and concretely on Sparrow [3], the winner of the latest SAT competition which outperformed other local search participants in the *random* category.

Experiments. In the experiments reported in this paper, we consider two sets of instances of the SAT'11 competition: *random* and *crafted*. In both cases we only consider known SAT instances. This way, we consider a collection of 369 *random* and 145 *crafted* instances. All the experiments were performed on the Grid'5000 platform, in particular we used a 44-node cluster with 24 cores (2 AMD Opteron 6164 HE processors at 1.7 Ghz) and 44 GB of RAM per node.

We used openMPI to build our parallel solver on top of Sparrow, which is implemented in UBCSAT. Additionally, we equipped this solver with the cooperative strategy *Prob-NormalizedW* proposed in [2]. Notice that unlike [2] where the best portfolio construction was selecting different and complementary algorithms, the Sparrow solver outperforms other local search algorithms, at least for *random* instances. Therefore, here we only use independent copies of Sparrow.

Each instance was executed 10 times with a timeout of 5 minutes. Table 1 shows the total number of solved instances (#Sol) calculated as the median across the 10 runs, the Penalized Average Runtime (PAR) computes the average runtime (in seconds) but where unsolved instances contribute 10 times the time cutoff, and the speedup is calculated using the following formula: Speedup=PAR_1/PAR_p, where the sub-index indicates the number of cores.

---

* The first author was financially supported by the JSPS under the JSPS Postdoctoral Program and the *kakenhi* Grant-in-aid for Scientific Research.

A. Cimatti and R. Sebastiani (Eds.): SAT 2012, LNCS 7317, pp. 481–482, 2012.
© Springer-Verlag Berlin Heidelberg 2012

As can be observed in this table the cooperative portfolio performs better than the non-cooperative one for *random* instances when using up to 16 cores, however, after this point the performance degrades as the number of cores increases. We attribute this to the fact that increasing the number of cores includes more diversification, so that algorithms in the portfolio might restart with quasi-random assignment for the variables. Moreover, the communication overhead induced by MPI becomes another factor to consider. On the other hand, for *crafted* instances both portfolios exhibit a close performance up to 16 cores, after this point the non-cooperative portfolio shows a small performance improvement in the range of 32 to 256 cores. Then it seems that the algorithm reach a performance plateau, where little improvement is observed. Finally, it is worth mentioning that overall, parallel portfolios with and without cooperation exhibit a better speedup for *random* instances. We attribute this to the fact that solutions might be uniformly distributed in the search space, while that this is not necessarily the case for *crafted* instances.

Table 1. Performance evaluation of portfolios with and without cooperation

Total cores	random						crafted					
	Prob NormalizedW			No Cooperation			Prob NormalizedW			No Cooperation		
	#Sol	PAR	Speedup	#Sol	PAR	Speedup	#Sol	PAR	Speedup	#Sol	PAR	Speedup
1	272	743.7	–	272	743.7	–	85	1150.8	–	85	1150.8	–
4	318	408.9	1.81	309	467.9	1.58	93	985.7	1.16	92	995.8	1.15
8	338	306.9	2.42	326	365.4	2.03	94	941.6	1.22	97	912.2	1.26
16	338	256.8	2.89	335	281.6	2.64	101	861.2	1.33	101	851.5	1.35
32	334	320.9	2.31	342	235.1	3.16	102	821.3	1.40	104	799.4	1.43
64	292	601.6	1.23	346	187.8	3.96	102	811.5	1.41	104	756.2	1.52
128	272	726.0	1.02	355	152.0	4.89	103	823.5	1.39	105	737.3	1.56
256	265	779.8	0.95	357	125.0	5.94	101	830.3	1.38	108	708.2	1.62

Conclusions and Ongoing Work. This paper has presented an experimental analysis of parallel portfolios of local search algorithms for SAT with and without cooperation. Overall, the experiments show that cooperation is a powerful technique that helps to improve performance up to a given number of cores for *random* instances (usually 16 cores), after this point performance degrades. On the other hand, the simple non-cooperative scheme seems to scale reasonably well for a large number of cores. Our current work involves limiting cooperation to groups of solvers (e.g. 16 per group), each group exploits cooperation as proposed in [2] and limited information is exchanged between groups of solvers.

References

1. Hamadi, Y., Jabbour, S., Sais, L.: ManySAT: A Parallel SAT Solver. Journal on Satisfiability, Boolean Modeling and Computation, JSAT 6(4), 245–262 (2009)
2. Arbelaez, A., Hamadi, Y.: Improving Parallel Local Search for SAT. In: Coello, C.A.C. (ed.) LION 2011. LNCS, vol. 6683, pp. 46–60. Springer, Heidelberg (2011)
3. Balint, A., Fröhlich, A.: Improving Stochastic Local Search for SAT with a New Probability Distribution. In: Strichman, O., Szeider, S. (eds.) SAT 2010. LNCS, vol. 6175, pp. 10–15. Springer, Heidelberg (2010)

Optimizing MiniSAT Variable Orderings for the Relational Model Finder Kodkod (Poster Presentation)

Markus Iser, Mana Taghdiri, and Carsten Sinz

Karlsruhe Institute of Technology (KIT), Germany
markus.iser@student.kit.edu,
{mana.taghdiri, carsten.sinz}@kit.edu

Introduction. It is well-known that the order in which variables are processed in a DPLL-style SAT algorithm can have a substantial effect on its run-time. Different heuristics, such as VSIDS [2], have been proposed in the past to obtain good variable orderings. However, most of these orderings are general-purpose and do not take into account the additional structural information that is available on a higher problem description level. Thus, structural, problem-dependent strategies have been proposed (see, e.g., the work of Marques-Silva and Lynce on special strategies for cardinality constraints [1]).

In this paper, we propose a new, structural variable ordering heuristic that is specific to the relational model finder Kodkod [3]. Kodkod transforms first-order, relational logic formulas into equisatisfiable propositional formulas and solves them using a SAT solver. Structural properties of a Kodkod problem that can efficiently be extracted in its first-order relational representation get lost in its propositional encoding. Our proposed heuristic computes the "constrainedness" of relations, and gives priority to the propositional variables that stem from the most constrained relations. The constrainedness is computed from Kodkod's abstract syntax tree, and thus takes the structure of the original relational formula into account.

Kodkod is a model finder for a widely-used first-order relational logic—a constraint language that combines first-order logic and relational algebra, and is augmented with transitive closure, integer arithmetic, and cardinality operators. It has been used as the backend engine of several software analysis tools in order to find bugs in the design and implementation of various software. A *problem* in Kodkod consists of a universe declaration, a number of relation declarations, and a formula over those relations. The universe defines a finite set of uninterpreted atoms that can occur in the models of the problem. A relation declaration specifies both an upper and a lower bound on the value of that relation. An upper bound of a relation denotes all the tuples that the relation *may* contain, whereas a lower bound denotes all the tuples that it *must* contain. Relation bounds are used to specify various information such as partitioning the universe into types, or defining a partial model for the problem.

Method. Based on the observation that shuffling variables in the SAT encoding produced by Kodkod can have a tremendous effect on the run-time of the SAT

A. Cimatti and R. Sebastiani (Eds.): SAT 2012, LNCS 7317, pp. 483–484, 2012.

solver, we developed strategies to obtain better, Kodkod-specific orderings. We experimented with two ways to modify MiniSAT's standard variable ordering:

1. **Initializing VSIDS:** Instead of using MiniSAT's default initialization that assigns variables in the same order in which they occur in the CNF file, we used a Kodkod-specific initial order.
2. **Overriding VSIDS:** Here, we partition variables into subsets $(S_1, \ldots, S_k)$, and assign variables in S_i before any variable occurring in a subset S_j, $j > i$. Within each subset, we use VSIDS scores to order variables.

Kodkod-specific variable orderings are computed based on the *constraining effect* that a subformula exerts on a relation's possible values. For example, a subset formula $R \subseteq S$, written as "R in S" in Kodkod's language, forces the entries of the relation R to be *false*, as soon as the corresponding entries in S are *false*. Similarly, for a cardinality restriction, written as "#R <= c" in Kodkod, where the constant c is small, many entries in R have to be set to *false*. Thus, the constraining effect of a cardinality restriction is usually high. We therefore try to assign variables corresponding to such relations first. The intuition is that the number of unit propagations on the SAT level can be maximized thereby. In our approach, we iteratively compute the effect of *highly constraining operators* (like subset or cardinality restrictions) on the relations that occur in the constrained relational expressions. The effect is summarized as a *weight* that we assign to each relation.

Conclusion. We have implemented several variants of the heuristic outlined above in a modified version of MiniSAT. Experiments show that the initialization-based approach is superior to score-overriding. The initial scores blur rapidly while search advances, but, since MiniSAT is a learning solver, the impact of score-initialization lasts longer than the induced priorities remain intact.

Experiments with score-overriding reveal no evident trend regarding runtime. However, using an initialization-based strategy the SAT solving run-time can be improved considerably in many cases. On a test set of 91 Kodkod problems, run-times could be improved by a factor of two or more for 26 instances (only for 6 problems, the runtime got worse by a factor of two or more); the maximal speed-up was over 800, while the worst deterioration was less than a factor of 4. Run-time was improved on 58 instances and deteriorated on 33.

References

1. Marques-Silva, J., Lynce, I.: Towards Robust CNF Encodings of Cardinality Constraints. In: Bessière, C. (ed.) CP 2007. LNCS, vol. 4741, pp. 483–497. Springer, Heidelberg (2007)
2. Moskewicz, M.W., Madigan, C.F., Zhao, Y., Zhang, L., Malik, S.: Chaff: Engineering an efficient SAT solver. In: DAC 2001, pp. 530–535 (2001)
3. Torlak, E., Jackson, D.: Kodkod: A Relational Model Finder. In: Grumberg, O., Huth, M. (eds.) TACAS 2007. LNCS, vol. 4424, pp. 632–647. Springer, Heidelberg (2007)

A Cardinality Solver: More Expressive Constraints for Free (Poster Presentation)

Mark H. Liffiton and Jordyn C. Maglalang

Illinois Wesleyan University, Bloomington IL 61701, USA
{mliffito,jmaglala}@iwu.edu
http://www.iwu.edu/~mliffito/

Despite the semantic simplicity of cardinality constraints, the CNF encodings typically used to solve them invariably turn one constraint into a large number of CNF clauses and/or auxiliary variables. This incurs a significant cost, both in space complexity and in runtime, that could be avoided by reasoning about cardinality constraints directly within a solver. Adding a single, *native* cardinality constraint instead of numerous clauses and/or auxiliary variables avoids any space overhead and simplifies the solver's procedures for reasoning about that constraint. Inspired by the simple observation that clauses are cardinality constraints themselves, and thus cardinality constraints generalize clauses, this work seeks to answer the question: How much of the research on developing efficient CNF SAT solvers can be applied to solving cardinality constraints?

Additional motivation came from our experience with a native implementation of cardinality constraints that was included in some early versions of the MiniSAT solver [3] as a simple, unoptimized example of the solver's ability to easily incorporate non-clausal Boolean constraints. That ability incurred unwanted overhead and was removed in later versions, and the native cardinality constraint implementation received little attention compared to the work done on CNF encodings. In addition to those early versions of MiniSAT, there have been other implementations of cardinality constraints that could be considered "native," but we are aware of none that integrate the constraints into a SAT solver by simply extending the existing clauses to incur little to no overhead.

For example, any Pseudo-Boolean (PB) solver or Satisfiability Modulo Theories (SMT) solver that handles linear integer arithmetic can solve cardinality constraints directly, as their constraints subsume both clauses and cardinality. Numerous PB solvers have been developed by extending a SAT solver, but little attention was paid to their performance on CNF. We are aware of only one experimental comparison between a PB solver and its corresponding SAT solver on CNF instances [2], comparing PBChaff with ZChaff, and the PB version was found to be consistently slower; the extension to more expressive constraints came at a noticeable cost. On the contrary, by restricting the solver to cardinality constraints and not general PB constraints, the implementation in this work retains those properties and efficiencies.

Asín, et al. [1] evaluated an "SMT-based approach" to cardinality constraints that solved them without encoding them to CNF. The "SMT" implementation

A. Cimatti and R. Sebastiani (Eds.): SAT 2012, LNCS 7317, pp. 485–486, 2012.
© Springer-Verlag Berlin Heidelberg 2012

was created by "coupling" two solving engines, which does not permit the tight integration of cardinality into the SAT solver done in this work, and it did not perform well compared to CNF encodings. Marques-Silva and Lynce [4] explored modifications to a SAT solver that improved its efficiency when using a particular CNF encoding, but it still faced the inherent space complexity of such encodings and was limited to AtMost constraints with a bound of 1.

The aim of this work is to generalize a state-of-the-art SAT solver at negligible cost, producing a "cardinality solver" we call *MiniCARD*, and to exhibit the performance of MiniCARD compared to some of the best-performing CNF encodings of cardinality constraints. MiniCARD outperforms CNF encodings of cardinality constraints on all pure-cardinality instances tested, and instances with a mix of clauses and cardinality constraints exhibit mixed results indicating some effect beyond the performance of the constraints themselves. The modifications to the solver are minimal, and it retains its performance on pure CNF instances. Given the feasibility of achieving increased expressive power over CNF with minor, performance-neutral changes to a state-of-the-art CNF solver, it is well worth pursuing further research on cardinality solvers.

Several direction of future research are immediately suggested by this work. The first is to more completely evaluate the performance of MiniCARD relative to other means of solving cardinality constraints, especially SAT solvers with preprocessing. Investigations of specific instances are suggested as well, such as determining how different cardinality implementations affect applications like CAMUS and MSU4. And finally, cardinality constraints may be a better target than CNF for many types of problems and constraints for which CNF encodings have been developed, such as PB constraints. The greater expressive power of cardinality solvers with equivalent performance on clauses could enable encodings that are both simpler and more efficient than pure CNF encodings.

Acknowledgments. Thanks to Albert Oliveras for providing the MSU4 benchmarks and to Niklas Sörensson for helpful discussions and advice regarding MiniSAT.

References

1. Asín, R., Nieuwenhuis, R., Oliveras, A., Rodríguez-Carbonell, E.: Cardinality networks: a theoretical and empirical study. Constraints 16, 195–221 (2011)
2. Dixon, H.E.: Automating Psuedo-Boolean Inference within a DPLL Framework. Ph.D. thesis, University of Oregon (2004)
3. Eén, N., Sörensson, N.: An Extensible SAT-solver. In: Giunchiglia, E., Tacchella, A. (eds.) SAT 2003. LNCS, vol. 2919, pp. 502–518. Springer, Heidelberg (2004)
4. Marques-Silva, J., Lynce, I.: Towards Robust CNF Encodings of Cardinality Constraints. In: Bessière, C. (ed.) CP 2007. LNCS, vol. 4741, pp. 483–497. Springer, Heidelberg (2007)

Single-Solver Algorithms for 2QBF
(Poster Presentation)

Sam Bayless and Alan J. Hu

University of British Columbia, Vancouver, Canada

2QBF is a restriction of QBF, in which at most one quantifier alternation is allowed. This simplifying assumption makes the problem easier to reason about, and allows for simpler unit propagation and clause/cube learning procedures. We introduce two new 2QBF algorithms that take advantage of 2QBF specifically. The first improves upon earlier work by Ranjan, Tang, and Malik (2004), while the second introduces a new 'free' decision heuristic that doesn't need to respect quantifier order. Implementations of both new algorithms perform better than two state-of-the-art general QBF solvers on formal verification and AI planning instances.

Ranjan, Tang, and Malik [4] introduced an algorithm for 2QBF in which two standard SAT solvers cooperate to solve the formula; in brief, 'Solver B' solves the (complements of) the learnt cubes, while 'Solver A' solves the input formula ϕ under Solver B's current assignment to the universally quantified variables. The solvers iterate back and forth until either fails to find a satisfying solution.

We improved upon this algorithm so that it can be implemented in just a single augmented SAT solver, rather than two. This solver stores two different types of learnt clauses: a set $\phi_\exists$ of existential clauses (corresponding to the clauses in Solver A) and a set $\phi_\forall$ containing the complements of learnt cubes (corresponding to those in Solver B). As in a standard DPLL-based QBF solver, this algorithm requires all universals to be assigned before any existentials can be chosen as decision variables. This algorithm resembles a special case of standard cube-learning QBF solvers, however, we introduce some new termination conditions that are specific to 2QBF.

These termination conditions are sufficient to ensure that the solver never has to handle the case where the implication graph of a conflict contains both universally and existentially quantified literals at the same decision level. This dramatically simplifies clause/cube learning: any valid cut in the implication graph at the current decision level is a learnt existential clause iff the decision variable was existential, and is (the complement of) a learnt cube iff the decision variable was universal. In contrast, Quaffle-based QBF solvers require several additional conditions to be met to ensure that conflict resolution does not resolve learnt cubes with clauses [1], which complicate both clause learning and unit propagation; these conditions are implicitly met in 2QBF (so long as the two shortcuts above are handled), and are met by the standard 1-UIP clause learning algorithm [2] without modification. We implement this simple 2QBF algorithm in MINI2QBF, based on MINISAT (version 1.14), and find that it is faster than state-of-the-art QBF solvers DEPQBF [5] and QUBE [6] on real-world formal verification and AI planning instances (see Table 1).

A. Cimatti and R. Sebastiani (Eds.): SAT 2012, LNCS 7317, pp. 487–488, 2012.
© Springer-Verlag Berlin Heidelberg 2012

Table 1. Results on the 2QBF track instances from QBF Eval '10, after pre-processing with sQueezeBF [7]. 86 of the 2QBF track instances were solvable by pre-processing alone, leaving 114. Average runtimes exclude pre-processing, and are computed over only the 91 instances that all four solvers solved within the 900 second cutoff.

	MINI2QBF	FREE2QBF	DEPQBF	QUBE 7.2
Avg. Runtime	2.4s	20.4s	10.6s	10.2s
# Solved	99/114	95/114	94/114	93/114

Like standard DPLL-based QBF solvers, this algorithm requires the decision heuristic to respect the quantifier order: given a formula of the form $\forall xy \exists z \phi$, where ϕ is a CNF, x and y must be assigned (either through a decision or unit propagation) before the decision heuristic can pick the variable z as a decision. It has been a longstanding goal of the QBF community to produce a QBF solver with a 'free' decision heuristic that does not need to make decisions in quantifier order [3], however, the only such solver previously introduced (a second 2QBF algorithm from [4]) proved to be too slow for practical applications.

We modify MINI2QBF so that it has a free decision heuristic (using a different method than the one introduced in [4]). This solver is able to learn valid cubes and clauses by implicitly re-arranging the implication graph to prevent clauses and cubes from being resolved with each other during conflict analysis (which would otherwise produce 'mixed terms' that are neither valid cubes nor valid clauses). As a consequence of this re-arrangement the solver is biased towards deciding universal variables at earlier decision levels. We find that although this new solver, FREE2QBF, is slower than MINI2QBF, it is competitive with both general QBF solvers, making this the first *fast* 2QBF solver with a free decision heuristic.

References

1. Zhang, L., Malik, S.: Towards a Symmetric Treatment of Satisfaction and Conflicts in Quantified Boolean Formula Evaluation. In: Van Hentenryck, P. (ed.) CP 2002. LNCS, vol. 2470, pp. 185–199. Springer, Heidelberg (2002)
2. Zhang, L., Madigan, C., Moskewicz, M., Malik, S.: Efficient conflict-driven learning in a boolean satisfiability solver. In: Proceedings of the 2001 IEEE/ACM International Conference on Computer-Aided Design, pp. 279–285. IEEE Press (2001)
3. Prasad, M., Biere, A., Gupta, A.: A survey of recent advances in SAT-based formal verification. International Journal on Software Tools for Technology Transfer (STTT) 7(2), 156–173 (2005)
4. Ranjan, D., Tang, D., Malik, S.: A comparative study of 2QBF algorithms. In: The Seventh International Conference on Theory and Applications of Satisfiability Testing, SAT 2004 (2004)
5. Lonsing, F., Biere, A.: DepQBF: A dependency-aware QBF solver (system description). JSAT 7, 71–76 (2010)
6. Giunchiglia, E., Marin, P., Narizzano, M.: QuBE7.0 system description. Journal on Satisfiability, Boolean Modeling and Computation 7, 83–88 (2010)
7. Giunchiglia, E., Marin, P., Narizzano, M.: sQueezeBF: An Effective Preprocessor for QBFs Based on Equivalence Reasoning. In: Strichman, O., Szeider, S. (eds.) SAT 2010. LNCS, vol. 6175, pp. 85–98. Springer, Heidelberg (2010)

An Efficient Method for Solving UNSAT 3-SAT and Similar Instances via Static Decomposition (Poster Presentation)

Emir Demirović[1] and Haris Gavranović[2]

[1] Faculty of Natural Sciences, Department of Mathematics,
Zmaja od Bosne 33-35, 71 000 Sarajevo, Bosnia and Herzegovina
emir.demirovic@gmail.com
[2] BAO lab, 71 000 Sarajevo, Bosnia and Herzegovina
haris.gavranovic@gmail.com

We present work that we have done so far towards devising a method for solving a specific case of SAT problems: UNSAT 3-SAT. Our goal is not to improve general SAT solving, but to focus on improving current techniques to solve 3-SAT and instances which have 2-SAT and 3-SAT structures within them. We have restricted ourselves to UNSAT instances. Even with these restrictions, these types of instances are still important, as such occur in practice (e.g. Velev's hardware verification instances [5]).

We decompose the solution space into simpler pieces, which are then solved with an existing solver: MiniSat. One can view the decomposition as a preprocessing step, in which the original problem is decomposed into a number of smaller problems, which are then solved and combined to obtain the solution of the original problem. After processing each smaller problem, we try to extract important information obtained from the solution process, via clause learning (a modification of Minisat's clause learning strategy is used). Any existing solver may be used as the underlying solver, as long as it outputs sufficient information about the solution process.

A considerable amount of research has already been done in this direction. Look-Ahead techniques [7] and backbone heurististics [8] have proven to be good in the case of random 3-SAT. However, these approaches are notably weaker than Conflict Driven Clause Learning methods (e.g. [4], [1], [2]) on industrial instances, while these solvers are weaker on random instances. A hybrid approach has been proposed in [10]. For a foundation of branching heuristics, see [6]. For decomposition strategies, see [3] i [9].

Since we are dealing with 3-SAT and similar instances, when choosing variables for the decomposition we pick those variables which appear as frequently as possible within the instance, so that they simplify as many clauses as possible, but at the same time we want our chosen variables to guide and narrow the assignment processes for as many other variables as possible (e.g. via implications).

We have compared our solver D-Sat (Decomposition Sat) with MiniSat on 360 different instances. These include random UNSAT 3-SAT and industrial instances, which are not purely 3-SAT instances. Most of the instances tested have been used in SAT competitions.

A. Cimatti and R. Sebastiani (Eds.): SAT 2012, LNCS 7317, pp. 489–490, 2012.
© Springer-Verlag Berlin Heidelberg 2012

The experimental results have shown that D-sat performs better than MiniSat on both the random 3-SAT and certain types industrial instances, even though our solver uses MiniSat as its underlying solver. It is known that MiniSat does poorly on random instances, so the comparison on those instances is not of interest and in fact our solver still cannot compete with state-of-the-art random SAT solvers. However, results obtained on the industrial instances, which have a 2-SAT and 3-SAT structure, show promise, since MiniSat is known to perform good on these instances and our solver performs even better.

There remain a number of issues that must be handled in order to fully exploit our method. In some cases, clause learning can noteable hurt the execution time. Defining a good order for the assumptions generated would allow us to solve satisfiable instances much faster. Modifying an existing solver, rather than using one as our underlying solver or even writing a new solver could potentially leading to better results. These are only a few problems out of many that need to be addressed in order to achieve full potential of our method. Taking into consideration the experimental results obtained, we trust that further research in this direction will prove to be worthwhile.

References

1. Eén, N., Sörensson, N.: An Extensible SAT-solver. In: Giunchiglia, E., Tacchella, A. (eds.) SAT 2003. LNCS, vol. 2919, pp. 502–518. Springer, Heidelberg (2004)
2. Marques-Silva, J.P., Sakallah, K.A.: GRASP: a search algorithm for propositional satisfiability. IEEE Transactions on Computers 48(5), 506–521 (1999)
3. Gil, L., Flores, P., Silveira, L.M.: PMSat: a parallel version of minisat. Journal on Satisfiability, Boolean Modeling and Computation 6, 71–98 (2008)
4. Moskewicz, W.M., Madigan, F.C., Zhao, Y., Zhang, L., Malik, S.: Chaff: engineering an efficient SAT solver. In: Proceedings of the 38th Annual Design Automation Conference, DAC 2001. ACM, New York (2001)
5. Miroslav Velev's SAT Benchmarks,
   http://www.miroslav-velev.com/sat_benchmarks.html
6. Kullmann, O.: Fundaments of Branching Heuristics. In: Biere, A., Heule, M., van Maaren, H., Walsh, T. (eds.) Handbook of Satisfiability. IOS Press (February 2009)
7. Heule, M., van Maaren, H.: Look-Ahead Based SAT Solvers. In: Biere, A., Heule, M., van Maaren, H., Walsh, T. (eds.) Handbook of Satisfiability. IOS Press (February 2009)
8. Dubois, O., Dequen, G.: A backbone-search heuristic for efficient solving of hard 3-SAT formulae. In: IJCAI 2001 Proceedings of the 17th International Joint Conference on Artificial Intelligence, vol. 1, pp. 971–978. Morgan Kaufmann Publishers Inc, San Francisco (2001) ISBN:1-55860-812-5 978-1-558-60812-2
9. Hyvärinen, A.E.J., Junttila, T.A., Niemelä, I.: A Distribution Method for Solving SAT in Grids. In: Biere, A., Gomes, C.P. (eds.) SAT 2006. LNCS, vol. 4121, pp. 430–435. Springer, Heidelberg (2006)
10. Heule, M., Kullmann, O., Wieringa, S., Biere, A.: Cube and Conquer: Guiding CDCL SAT Solvers by Lookaheads. In: 7th Intl. Haifa Verification Conference (HVC 2011), Haifa, Israel (December 2011)

Intensification Search in Modern SAT Solvers (Poster Presentation)

Saïd Jabbour, Jerry Lonlac, and Lakhdar Saïs

CRIL, CNRS - Université Lille Nord de France
Rue Jean Souvraz SP-18, F-62307 Lens Cedex France

Restarts and activity based search are two important and correlated components of modern SAT solvers. On the one hand, updating the activity of the variables involved in conflict analysis aims to circumscribe the most relevant part of the Boolean formula. While restarts allow the solver to reorder the variables by focussing the search on this relevant subformula. This combination allows the solver to intensify the search and at the same time helps to avoid trashing. This well known strong connexion between restarts and variable ordering have also a direct consequence on clause learning. The effect of restarts on clause learning have been widely investigated (e.g. [1,5]). Our intuition is that if SAT solvers are able to solve efficiently application instances with millions of variables and clauses, this means that the most relevant part of the formula (or subset of variables) is of reasonable size. This is related to the observation previously made on the size of backdoor sets observed on many applications domains. In our previous work, and in the parallel portfolio ManySAT solver, we have shown how the two well known principles of diversification and intensification principles can be combined in the context of Masters/Slaves architecture [2]. The Masters perform an original search strategy, ensuring diversification, while the remaining units, classified as Slaves are there to intensify their master's strategy. By intensification we mean that the slave would explore "differently" around the search space explored by the Master.

In this paper, we propose to push forward this intensification based search by collecting the variables encountered during the last conflict analysis. More precisely, at each conflict, a bottom up traversal of the implication graph is achieved until the last UIP and the set of variables corresponding to the different visited nodes are collected in a queue. At each restart, the solver branch in priority on the collected variables. When all the variables from the queue are assigned, the solver follow the usual VSIDS branching heuristic [3]. In this way, the variables closest to the conflict side are assigned first using the progress saving literals polarities [4]. This simple intensification principle achieves significant improvements when integrated to MiniSAT 2.2 SAT solver. Our tests were done on Intel Xeon quadcore machines with 32GB of RAM running at 2.66 Ghz. For each instance, we used a timeout of 1 hour of CPU time. We used the whole set of application instances taken from the 2009 and 2011 SAT competitions. The number of different instances corresponds to 559. MiniSAT with our intensification strategy ($MiniSat + Intensification$) solves 12 more instances than MiniSat without intensification. These results are depicted in figure 1. It presents the cumulated time results i.e. the number of instances (x-axis) solved under a given amount of time in seconds (y-axis). The curve on the left (respectively

A. Cimatti and R. Sebastiani (Eds.): SAT 2012, LNCS 7317, pp. 491–492, 2012.
© Springer-Verlag Berlin Heidelberg 2012

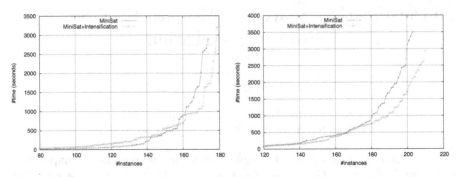

Fig. 1. Results on 2009-2011 SAT competitions (Applications category)

Table 1. Results on some families of application instances

instance	SAT?	MiniSat	MiniSat+I	instance	SAT?	MiniSat	MiniSat+I
goldb-heqc-alu4mul	N	140.94	128.64	9dlx_vliw_at_b_iq1	N	1964.90	28.86
goldb-heqc-term1mul	N	54.79	38.40	9dlx_vliw_at_b_iq2	N	–	70.07
goldb-heqc-i10mul	N	187.06	129.14	9dlx_vliw_at_b_iq7	N	2642.42	1282.10
goldb-heqc-dalumul	N	1663.95	181.31	9dlx_vliw_at_b_iq3	N	–	189.83
goldb-heqc-frg1mul	N	–	715.66	9dlx_vliw_at_b_iq4	N	1750.56	283.74
goldb-heqc-x1mul	N	–	2500.59	9dlx_vliw_at_b_iq8	N	2293.63	1633.04
velev-engi-uns-1.0-4nd	N	9.12	14.60	9dlx_vliw_at_b_iq9	N	2410.76	2572.97
velev-live-uns-2.0-ebuf	N	18.53	9.84	9dlx_vliw_at_b_iq5	N	1631.90	464.47
velev-pipe-sat-1.0-b7	Y	21.51	23.51	9dlx_vliw_at_b_iq6	N	1267.90	840.68
velev-vliw-uns-4.0-9C1	N	3378.39	112.87	velev-pipe-uns-1.0-8	N	–	414.21
velev-pipe-o-uns-1.1-6	N	619.57	28.26	velev-vliw-uns-4.0-9-i1	N	2095.14	592.52
velev-pipe-o-uns-1.0-7	N	446.30	441.81	velev-pipe-sat-1.0-b10	N	77.69	4.61

right) hand side, show the results obtained on satisfiable (respectively unsatis-fiable) instances. The improvements are even more important on unsatisfiable instances. Table 1 shows the results on some known families of SAT instances. Our motivation behind this paper, is to show that there remains many rooms for future improvements of the SAT variable ordering heuristics. To the better of our knowledge, this issue have not received much attention. The improvements obtained by our simple intensification strategy, suggests that further studies on the connection between restarts and variables ordering heuristics are needed.

References

1. Biere, A.: Adaptive Restart Strategies for Conflict Driven SAT Solvers. In: Kleine Büning, H., Zhao, X. (eds.) SAT 2008. LNCS, vol. 4996, pp. 28–33. Springer, Heidelberg (2008)
2. Guo, L., Hamadi, Y., Jabbour, S., Sais, L.: Diversification and Intensification in Parallel SAT Solving. In: Cohen, D. (ed.) CP 2010. LNCS, vol. 6308, pp. 252–265. Springer, Heidelberg (2010)
3. Moskewicz, M.W., Madigan, C.F., Zhao, Y., Zhang, L., Malik, S.: Chaff: Engineering an efficient SAT solver. In: DAC 2001, pp. 530–535 (2001)
4. Pipatsrisawat, K., Darwiche, A.: A Lightweight Component Caching Scheme for Satisfiability Solvers. In: Marques-Silva, J., Sakallah, K.A. (eds.) SAT 2007. LNCS, vol. 4501, pp. 294–299. Springer, Heidelberg (2007)
5. Pipatsrisawat, K., Darwiche, A.: Width-Based Restart Policies for Clause-Learning Satisfiability Solvers. In: Kullmann, O. (ed.) SAT 2009. LNCS, vol. 5584, pp. 341–355. Springer, Heidelberg (2009)

Using Term Rewriting to Solve Bit-Vector Arithmetic Problems
(Poster Presentation)

Iago Abal[1], Alcino Cunha[1], Joe Hurd[2], and Jorge Sousa Pinto[1]

[1] HASLab / INESC TEC & Universidade do Minho, Braga, Portugal
[2] Galois, Inc., Portland, OR, USA

Among many theories supported by SMT solvers, the theory of *finite-precision bit-vector arithmetic* is one of the most useful, for both hardware and software systems verification. This theory is also particularly useful for some specific domains such as cryptography, in which algorithms are naturally expressed in terms of bit-vectors. Cryptol is an example of a domain-specific language (DSL) and toolset for cryptography developed by Galois, Inc.; providing an SMT backend that relies on bit-vector decision procedures to certify the correctness of cryptographic specifications [3]. Most of these decision procedures use *bit-blasting* to reduce a bit-vector problem into pure propositional SAT. Unfortunately bit-blasting does not scale very well, especially in the presence of operators like multiplication or division. For example, the equality $x_{[n]}^2 - 1_{[n]} = (x_{[n]} + 1_{[n]}) \times (x_{[n]} - 1_{[n]})$ is a simple consequence of distributivity and associativity laws; but even for small values of n the bit-level representation of this formula is so huge that it is intractable by current SAT solvers. The main reason for this is the loss of high-level algebraic structure present in the original decision problem. The point here is that one can exploit algebraic properties concerning the domain of bit-vectors to rewrite this problem into an equisatisfiable, but computationally less hard, problem. For instance, the above equality can be proved *valid* as follows (subscripts are omitted for clarity): $x^2 - 1 = (x + 1) \times (x - 1)$ $\equiv$ {distributivity $\times$ 3; associativity} $x^2 - 1 = x^2 + x - x - 1$ $\equiv$ {inverse; right identity} $x^2 - 1 = x^2 - 1$ $\equiv$ {reflexivity} *true*. Modern SMT solvers already include a simplification phase that performs some rewriting on the input problem prior to bit-blasting [4]. Nevertheless, SMT solvers have to deal with a wide range of application domains, and hence the set of rewrite rules employed for simplification inevitably excludes many rules that are useful for some particular domains but may be inconvenient for others.

The present work was motivated by the difficulties reported by the Galois Cryptol team in achieving automatic equivalence checking for public-key cryptography (PKC). PKC is particularly hard because it involves multiplication and modular exponentiation on long bit-vectors. Hence, the bit-level representation of any PKC algorithm is usually so huge that such equivalence problems are too hard for current SAT solvers, unless a significant amount of rewriting is performed before bit-blasting. SMT solvers employing high-level rewriting-based techniques have been shown to be promising, but they are still insufficiently powerful to handle

A. Cimatti and R. Sebastiani (Eds.): SAT 2012, LNCS 7317, pp. 493–495, 2012.

hard problems, such as those resulting from PKC. This problem may be addressed by combining custom rewrite patterns, somehow encapsulating domain-specific proof strategies, with standard bit-vector decision procedures. Our first attempt consisted in extending SMT specifications with algebraic properties provided in the form of *quantified formulas*, expecting the SMT solver to use them as rewrite rules. Unfortunately, we have found that most of the times SMT solvers do not use these rules effectively, and even become quite unpredictable in the presence of universal quantifiers. After this failed attempt, we prototyped a rewriting system in Maude [1] that focuses on simplifying PKC equivalence problems. Employing a set of 200 handcrafted rewrite rules and a very simple rewriting strategy enabled us to achieve quite promising results. For instance, this system proved the correctness of a 16-bit peasant multiplier and SHA-1 implementations in a few seconds, while the 3.2 version of Z3 [2] times out (16 hours) for the peasant case and quickly runs out of memory (2 GB) solving the SHA-1 one. Using this rewriting system as a preprocessing step for Z3 we also achieved good speedups for some equivalence problems, such as a speedup of 2 for an 8-bit modular exponentiation algorithm.

Even though there is still considerable work to be done in order to reach a reasonable degree of automation for PKC equivalence checking, the above results show the potential of the term-rewriting approach. In the same way that proof assistants allow defining custom tactics to encapsulate specific proof techniques, our intention is to encode those proof tactics as rewrite patterns in the context of SMT solving. This allows simplifications that drastically reduce the size of the input problem before bit-blasting, leading to better overall performance. Ideally, SMT solvers should allow easy customization of their solving strategies with such rules —we are aware of some recent work in this direction. It is worth noting that we are not relying on complex combinations of rewriting strategies, which would make our approach more fragile and less scalable. Finally, Maude turned out to be a good platform for experimentation, but it significantly restricts the strategies that we could employ and presents some limitations with respect to achieving perfect subterm sharing. Thus we are presently working on a framework to specify custom rewriting-based simplifications for fixed-size bit-vector arithmetic, that should allow us to overtake the above limitations.

Acknowledgement. This work is funded by National Funds through the FCT - Fundação para a Ciência e a Tecnologia (Portuguese Foundation for Science and Technology) within project PTDC/EIA-CCO/105034/2008.

References

1. Clavel, M., Durán, F., Eker, S., Lincoln, P., Martí-Oliet, N., Meseguer, J., Talcott, C.: The Maude 2.0 System. In: Nieuwenhuis, R. (ed.) RTA 2003. LNCS, vol. 2706, pp. 76–87. Springer, Heidelberg (2003)

2. de Moura, L., Bjørner, N.: Z3: An Efficient SMT Solver. In: Ramakrishnan, C.R., Rehof, J. (eds.) TACAS 2008. LNCS, vol. 4963, pp. 337–340. Springer, Heidelberg (2008)
3. Erkök, L., Matthews, J.: Pragmatic equivalence and safety checking in Cryptol. In: Proceedings of the 3rd Workshop on Programming Languages Meets Program Verification, PLPV 2009, pp. 73–82. ACM, New York (2008)
4. Franzen, A.: Efficient Solving of the Satisfiability Modulo Bit-Vectors Problem and Some Extensions to SMT. Ph.D. thesis, University of Trento (March 2010)

Learning Polynomials over GF(2) in a SAT Solver
(Poster Presentation)

George Katsirelos[1] and Laurent Simon[2]

[1] INRA, Toulouse
george.katsirelos@toulouse.inra.fr
[2] LRI, Univ Paris 11
simon@lri.fr

1 Introduction

One potential direction for improving the performance of SAT solvers is by using a stronger underlying proof system, e.g., [1]. We propose a step in improving the learning architecture of SAT solvers and describe a learning scheme in the polynomial calculus with resolution (PCR), a proof system that generalizes both resolution and Gaussian elimination. The scheme fits the general structure of CDCL solvers, so many of the other techniques of CDCL solvers should be reusable.

The PCR proof system was introduced in [2]. In it, lines of a proof are polynomials, which are derived by summing two previous polynomials or multiplying a previous polynomial by a variable. The system also includes the axioms $x^2 - x = 0$, $\neg x^2 - \neg x = 0$ and $x + \neg x = 1$ for all variables x. In our approach, we use only polynomials over $GF(2)$. In this system, a clause $(a \vee b \vee \neg c)$ is expressed as the polynomial $\neg a \neg b c = 0$. A xor-clause $(a \oplus b \oplus \neg c)$ is also naturally expressed, as the polynomial $a + b + \neg c = 0$. However, neither a clause nor a xor clause can capture a general polynomial such as $xy + zw + pq + 1 = 0$. Note that the variables $\neg x$ are not necessary, as they can be replaced by $(1 + x)$ but using them can drastically reduce the number of monomials. When written as a sum of monomials, a global order on variables allows a canonical representation, unique for all equal polynomials.

There is significant previous work that addresses the efficient integration of XOR (or equivalence) reasoning techniques in SAT solvers, e.g. [3,4]. However, in these approaches, interaction between the CNF and XOR subproblems is limited to passing unit clauses from the CNF part to the XOR part and implied clauses from the XOR part to the CNF part.

2 Structure of the Solver

The structure and main loop of the proposed solver is identical to that of CDCL algorithms (not recalled here). However, the basic constraint stored in this scheme is a polynomial. Therefore, the operations we need to specify are propagating the implications of polynomials as we make decisions, and learning new polynomials when we encounter conflicts. The natural way to propagate polynomials is to decompose it into a set of clauses and one xor-clause. $c + \sum_{i=1}^{k} m_i = 0$ where $c \in \{0, 1\}$ and $m_i = \prod_{j=1}^{d} x_{ij}$ can be decomposed with one new variable y_{m_i} for each term m_i by the clauses that encode

A. Cimatti and R. Sebastiani (Eds.): SAT 2012, LNCS 7317, pp. 496–497, 2012.

$y_{m_i} \iff \bigwedge_{j=1}^{d} x_{ij}$ and the xor-clause $c + \sum_{i=1}^{k} y_{m_i} = 0$. Unfortunately, unit propagation on this decomposition is not complete. Consider the polynomial $ad + bd + 1 = 0$. Unit propagation on the corresponding CNF $x \iff a \wedge d, y \iff b \wedge d, x \oplus y$ does nothing, but all solutions have d set to true. We can improve this decomposition by factoring common subexpressions, but propagation remains incomplete. However, achieving complete propagation is too expensive and unnecessary.

In order to perform conflict analysis, we define a *polynomial resolution step*:

$$
\frac{\begin{array}{c} y p_1 + p_2 \\ y q_1 + q_2 \end{array}}{\begin{cases} p_1 q_2 + p_2 q_1 & \text{if } p_1 \neq q_1 \\ p_2 + q_2 & \text{if } p_1 = q_1 \end{cases}}
$$

This allows us to use polynomial resolution in much the same way as resolution: we keep track of the polynomial that forced each literal. On conflict, we iteratively resolve away the deepest variable until we get a polynomial that satisfies a stopping condition, such as having a single variable at the decision level. However, there are cases where polynomial resolution is less well behaved than resolution. First, during conflict analysis we may get a polynomial which contains no variable from the last decision level. Second, even if there exists a 1-UIP polynomial, it is not necessarily asserting. Third, a polynomial may contain variables which are assigned but which do not affect the satisfiability of that polynomial under the current assignment. Resolving on these variables may result in a tautology, so these variables have to be ignored. Additionally, the size of polynomials may grow quadratically with every polynomial resolution step. To keep their size in check, we propose several simplification procedures that can reduce their size, as well as a weakening procedure that, given a polynomial p gives a smaller and weaker polynomial p' such that $p' = 1 \implies p = 1$ and $p = 0 \implies p' = 0$.

In terms of its theoretical power, this solver is not any more powerful than a CDCL solver if the input is given in CNF and therefore also strictly less powerful than the unrestricted polynomial calculus. Thus, we apply a preprocessing step in which we detect XOR clauses and AND gates to extract implied polynomials, as in [5]. Given this preprocessing step, the solver is strictly more powerful than CDCL, as it clearly p-simulates resolution and additionally p-simulates Gaussian elimination but also more powerful than SMT with XOR reasoning [4].

References

1. Dixon, H.E., Ginsberg, M.L.: Inference methods for a pseudo-boolean satisfiability solver. In: AAAI 2002, pp. 635–640 (2002)
2. Alekhnovich, M., Ben-Sasson, E., Razborov, A.A., Wigderson, A.: Space complexity in propositional calculus. SIAM J. Comput. 31(4), 1184–1211 (2002)
3. Soos, M., Nohl, K., Castelluccia, C.: Extending SAT Solvers to Cryptographic Problems. In: Kullmann, O. (ed.) SAT 2009. LNCS, vol. 5584, pp. 244–257. Springer, Heidelberg (2009)
4. Laitinen, T., Junttila, T., Niemel, I.: Extending clause learning DPLL with parity reasoning. In: ECAI 2010, pp. 21–26 (2010)
5. Ostrowski, R., Grégoire, É., Mazure, B., Saïs, L.: Recovering and Exploiting Structural Knowledge from CNF Formulas. In: Van Hentenryck, P. (ed.) CP 2002. LNCS, vol. 2470, pp. 185–199. Springer, Heidelberg (2002)

Learning Back-Clauses in SAT
(Poster Presentation)

Ashish Sabharwal, Horst Samulowitz, and Meinolf Sellmann

IBM Watson Research Center, Yorktown Heights, NY 10598, USA
{ashish.sabharwal,samulowitz,meinolf}@us.ibm.com

In [3], SAT conflict analysis graphs were used to learn additional clauses, which we refer to as *back-clauses*. These clauses may be viewed as enabling the powerful notion of "probing": Back-clauses make inferences that would normally have to be deduced by setting a variable deliberately the other way and observing that unit propagation leads to a conflict. We show that short-cutting this process can in fact improve the performance of modern SAT solvers in theory and in practice. Based on out numerical results, it is suprising that back-clauses, proposed over a decade ago, are not yet part of standard clause-learning SAT solvers.

Back-Clauses. We assume familiarity with SAT conflict analysis [3,4]. Figure 1 shows an example formula and its conflict graph derived after branching on (-1), (+2), and (+3). A clause such as (1-2+5) in our notation may be thought of as $(x_1 \lor \neg x_2 \lor x_5)$. The corresponding first or rightmost UIP at the decision level is literal (-10) and the standard clause learnt from this conflict is (-4+10). In [3] it was found that, for any two consecutive UIPs at level L, we can infer that, under some *context* given by the literals on tree levels $< L$, the left UIP implies the right UIP. In our example, given 5, 7 implies -10. Written as a clause, this gives (-5-7-10). It makes sense to add this "back-clause" because unit propagation is incomplete and may in fact not be able to infer that, given 5, 10 implies -7. In our example, we can also infer that, given 2, 3 implies 7, or (-2-3+7). Note that these two clauses imply (-2-5-3-10), also under incomplete unit propagation. Since back-clauses in general have smaller "contexts" than traditional nogoods based on *all* UIPs, we conclude from the following proposition that adding all back-clauses between adjacent UIPs at level L is, in general, strictly stronger under unit propagation than adding all UIP nogoods at level L.

A. (1+4)	B. (1-2+5)
C. (-2-3-6)	D. (-3+6+7)
E. (-4+10+11)	F. (-5-7+8)
G. (-5-7+9)	H. (-8-9-10)
I. (10-11)	

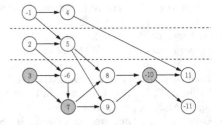

Fig. 1. An Example Formula and its Conflict Graph

A. Cimatti and R. Sebastiani (Eds.): SAT 2012, LNCS 7317, pp. 498–499, 2012.

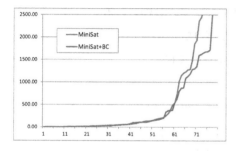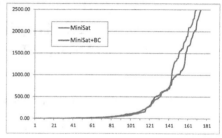

Fig. 2. Cactus plot showing the maximum time (y-axis, in seconds) needed by MiniSat with and without Back-Clauses to solve a given number of instances (x-axis). Left: SAT Race 2010 benchmark. Right: SAT Competition 2011 benchmark.

Proposition 1. *By adding the first UIP clause and back-clauses between every two consecutive UIPs at level L, we enable unit propagation to make all inferences that* all *traditional nogoods based on* all *UIPs at level L would.*

Empirical Evaluation. We added back-clause lerning as part of version 2.2.0 of MiniSat [5] and experimented with it on 2.3 GHz AMD Opteron 6134 machines with eight 4-core CPUs and 64 GB memory, running Scientific Linux release 6.1. As benchmarks we use all of the application instances from the 2010 SAT Race and the 2011 SAT Competition. Both MiniSat and MiniSat+BC (i.e., with back-clauses on the decision level) were configured to first simplify the formula using the SatELite preprocessor [2]. Figure 2 summarizes the results in terms of the commonly used cactus plot metric. We observe that learning back-clauses is clearly helpful for both benchmark sets, particularly for harder instances where the benefits of learning additional clauses become most noticeable.

In conclusion, we rediscovered the idea of learning back-clauses during search, first introduced in [3]. We showed that adding back-clauses is stronger than adding no-goods for all UIPs. We hope that our experiemntal findings will help this technique find its rightful place among modern SAT solving methods.

References

1. Bayardo, R.J.J., Schrag, R.C.: Using CSP look-back techniques to solve real-world SAT instances. In: Proceedings of AAAI 1997, pp. 203–208 (1997)
2. Eén, N., Biere, A.: Effective Preprocessing in SAT Through Variable and Clause Elimination. In: Bacchus, F., Walsh, T. (eds.) SAT 2005. LNCS, vol. 3569, pp. 61–75. Springer, Heidelberg (2005)
3. Marques-Silva, J.P., Sakallah, K.A.: GRASP: A Search Algorithm for Propositional Satisfiability. IEEE Trans. on Computers 48(5), 506–521
4. Moskewicz, M.W., Madigan, C.F., Zhao, Y., Zhang, L., Malik, S.: Chaff: Engineering an Efficient SAT Solver. In: Proc. DAC, pp. 530–535 (2001)
5. Sorensson, N., Eén, N.: MiniSAT 2.2.0 (2010), http://minisat.se
6. SAT Competition, http://www.satcomptition.org

Augmenting Clause Learning
with Implied Literals
(Poster Presentation)

Arie Matsliah[1], Ashish Sabharwal[2], and Horst Samulowitz[2]

[1] IBM Research, Haifa, Israel
ariem@il.ibm.com
[2] IBM Watson Research Center, Yorktown Heights, USA
{ashish.sabharwal,samulowitz}@us.ibm.com

There exist various approaches in SAT solvers that aim at extending inference based on unit propagation. For instance, *probing* [5] simply applies unit propagation of literals at the root node in order to detect failed literals [3] or to populate literal implication lists. The latter information can then, for instance, be used to shrink clauses by *hidden literal elimination* (e.g., if $a \mapsto b$ then $(a \vee b \vee c)$ can be reduced to $(b \vee c)$; cf. [4]).

Here we propose to strengthen clause learning by dynamically inferring literals that the newly learned clause entails. We say that a literal l is an *implied literal* for a clause C if all literals of C entail l. For instance, if $a \mapsto d$, $\neg b \mapsto d$, and $c \mapsto d$, then $(a \vee \neg b \vee c)$ entails d. While this insight has already been exploited in several methods (e.g., variations of *hyper binary resolution* and *hidden literal elimination*), we apply it to clause learning: when the SAT solver derives a new conflict clause c, we check if the literals in c imply a single or multiple literals which can then be propagated as new unit literals.

In order to employ this technique we first need to generate implication lists $L(l) = UnitPropagation(l)$ for each literal l. This is done at the root node of the search tree before the solving process starts and then periodically during search. During this computation, we also add not yet existing binary clauses corresponding to $\forall l \in L(p) : \neg l \mapsto \neg p$. As one might expect, we detect failed literals as well and add and propagate their negations as new unit literals; we do the same for all literals in the intersection of $L(p)$ and $L(\neg p)$ for all $p \in F$. Once the implied literal lists for each literal are computed, we iterate over the clauses in the original theory F and propagate all implied literals as new unit clauses.

Ideally we would like to augment these lists with new implications whenever new clauses have been learned by the solver. However, this operation can be computationally expensive and we must control how frequently it is performed. While learned clauses are usually 'forgotten' over time, we hold on to the implication lists and only extend them, when possible. Note that this can enable inference that might not be explicitly captured in the current clausal theory.

Whenever a new clause C is derived, we check whether $I = \bigcap_{l \in C} L(l)$ is non-empty. If so, we add all $p \in I$ as new unit clauses and backtrack the search to the root (i.e., restart). We reduce the computational effort of computing intersections by considering learned clauses of only a bounded length.

A. Cimatti and R. Sebastiani (Eds.): SAT 2012, LNCS 7317, pp. 500–501, 2012.
© Springer-Verlag Berlin Heidelberg 2012

Table 1. Performance of Glucose and Glucose+IL on the application category instances of SAT Competition 2011 (selected benchmark families and all instances)

Benchmarks	Glucose			Glucose+IL		
	% Solved	PAR10	#Dec. (G1k)	% Solved	PAR10	#Dec. (G1k)
Goldberg (8)	87.5	9,045	617,058	100.0	774	377,047
Grieu-VMPC (6)	83.3	12,715	8,109,876	100.0	1,286	4,809,073
Jarivsalo-AAAI10 (13)	100.0	485	522,951	100.0	224	218,455
Rintanen-Sokoban (6)	33.3	43,871	2,196,775	66.8	22,241	624,748
Competition 2011 (300)	71.3	19,280	472,059	72.0	18,778	377,712

We implemented this approach in Glucose 2.0 [1] and evaluated the performance on all 300 instances of the application track of SAT Competition 2011 [6]. All experiments were conducted on 2.3 GHz AMD Opteron 6134 machines with eight 4-core CPUs and 64 GB memory, running Scientific Linux release 6.1. We used a time limit of 6,500 sec (which roughly corresponds to the 5,000 sec timeout used in the competition), and limited activation of our method to instances with at most 2,000,000 clauses and 200,000 literals appearing in binary clauses. Implied literals are computed after 150 restarts and from then on in a geometrically increasing manner with a factor of 1.2. All runs used identical parameters (e.g., maximum learned clause length to check for implied literals).

Table 1 summarizes the results. For both versions of Glucose we show the percentage of instances solved, the PAR10 score (penalized average runtime where instances that time out are penalized with 10x the timeout), and the geometric mean shifted by $1,000$ of the number of decisions made on the instances solved by both approaches. The first four benchmark families highlight the potential of our approach. On all shown measures, the impact of implied literals is quite dramatic. For instance, on the Sokoban benchmark, adding inference based on implied literals doubles the number of solved instances. The final row shows that the new approach does not degrade performance of the baseline solver. In fact, it is able to solve more instances while making 20% fewer decisions.

We have extended this approach to also learn binary clauses when all but one literal in the learned clause imply a literal. Note that this corresponds to a generalized version of *hyper-binary resolution* [2].

References

1. Audemard, G., Simon, L.: Predicting learnt clauses quality in modern SAT solvers. In: IJCAI, pp. 399–404 (2009)
2. Bacchus, F.: Enhancing Davis Putnam with Extended Binary Clause Reasoning. In: AAAI (2002)
3. Freeman, J.: Improvements to Propositional Satisfiability Search Algorithms. PhD. Thesis, University of Pennsylvania (1995)
4. Heule, M.J.H., Järvisalo, M., Biere, A.: Efficient CNF Simplification Based on Binary Implication Graphs. In: Sakallah, K.A., Simon, L. (eds.) SAT 2011. LNCS, vol. 6695, pp. 201–215. Springer, Heidelberg (2011)
5. Lynce, I., Marques-Silva, J.: Probing-Based Preprocessing Techniques for Propositional Satisfiability. In: ICTAI, vol. 105 (2003)
6. SAT Competition (2011), http://www.satcompetition.org

Author Index